Günther Winkelmann (Editor)
Microbial Transport Systems

Günther Winkelmann (Ed.)

Microbial Transport Systems

WILEY-VCH

Weinheim – New York – Chichester – Brisbane – Singapore – Toronto

Editor

Günther Winkelmann
Institut für Mikrobiologie
Universität Tübingen
Auf der Morgenstelle 28
D-72076 Tübingen
Germany

■ This book was careful produced. Nevertheless, authors, editors and publisher do not warrant the information contained therein to be free of errors. Readers are advised to keep in mind that statements, data, illustrations, procedural details or other items may inadvertently be inaccurate.

Library of Congress Card No.: applied for

British Library Cataloguing-in-Publication Data:
A catalogue record for this book is available from the British Library

**Die Deutsche Bibliothek –
CIP-Cataloguing-in-Publication Data**
A catalogue record for this book is available from Die Deutsche Bibliothek

© Wiley-VCH Verlag GmbH,
D-69469 Weinheim, 2001

All rights reserved (including those of translation in other languages). No part of this book may be reproduced in any form – by photoprinting, microfilm, or any other means – nor transmitted or translated into machine language without written permission from the publishers. Registered names, trademarks, etc. used in this book, even when not specifically marked as such, are not to be considered unprotected by law.

printed in the Federal Republic of Germany

printed on acid-free paper

Composition Hagedorn Kommunikation
D-68519 Viernheim
Printing betz-druck gmbh
D-64291 Darmstadt
Bookbinding J. Schäffer GmbH & Co. KG.
D-67269 Grünstadt

ISBN 3-527-30304-9

Preface

"Please, pass the salt" is something that could be asked by microorganisms as well as gourmets. How do cells transport nutrients? An essential feature of all living organisms is the ability to accumulate nutrients against a concentration gradient and to excrete the various end products of metabolism. The topic of microbial transport systems involves a variety of other issues, such as generation of a membrane potential, homeostasis of ions, maintaining an osmotic balance, excretion of enzymes and toxins, the release of hormones and signals, drug resistance strategies, etc. The main cellular structure responsible for nutrient transport is the plasma membrane, which may be accompanied by an outer membrane in the case of gram-negative bacteria. Due to their long evolutionary development, microbial cells are the most diverse with respect to transport. The various mechanisms of solute transport across these membranes are so diverse that it is surprising that cells can manage the traffic of so many different compounds simultaneously. Cells obviously avoid traffic jams by two principal mechanisms, that is by up- or down-regulation and by energetic activation and inactivation of transporters and channels. Although a distinction between primary transporters (F-type ATPase, P-type ATPase, ABC-ATPase), secondary transporters (major facilitators, channels) and group translocation is generally made, many more strategies occur. While channel-type facilitated diffusion is common among pore-forming compounds, active transport against a concentration gradient occurs via ABC transporters, P-type ATPases, MFS transporters and group translocation. While some of these use direct ATP hydrolysis for transport, MFS transporters use indirect energy from a membrane potential, which in turn connects ion gradient to solute flow resulting in uniport, symport and antiport mechanisms.

This diversity of transport systems has necessitated the development of a transporter classification (TC) system (see Chapter 1 of Milton Saier).

It is the aim of the present book to demonstrate how some important nutrients are transported into the cells, how proteins are excreted and how the diverse transport mechanisms operate. Gene replacing techniques of transport genes, hydropathy plots, mutational analysis and structural and functional genomics are modern tools in transport biology which have led to unraveling the secrets of transport mechanisms. Although this book cannot be comprehensive it should inspire and encourage further studies. Including every topic on transport would generate a book three times this length and far too expensive – therefore, I hope to have selected the essentials.

My thanks go to all authors for their willingness to participate in this project and for producing their manuscripts so promptly. I am especially grateful to Carl J. Carrano, Volkmar Braun, Klaus Hantke, Dick van der Helm and Milton Saier for helpful suggestions and comments.

Tübingen
June 2001

Günther Winkelmann

Contents

Preface *V*

List of Authors *XIX*

Color Plates *XXIII*

1	**Families of Transporters: A Phylogenetic Overview** *1*	
1.1	Introduction *1*	
1.2	The TC System *1*	
1.3	The Value of Phylogenetic Classification *2*	
1.4	Phylogeny as Applied to Transporters *3*	
1.5	The Basis for Classification in the TC System *3*	
1.6	Classes of Transporters *4*	
1.7	Class 1: Channels/Pores *17*	
1.8	Class 2: Electrochemical Potential-driven Porters *17*	
1.9	Class 3: Primary Active Transporters *18*	
1.10	Class 4: Group Translocators *19*	
1.11	Class 8: Accessory Factors Involved in Transport *19*	
1.12	Class 9: Incompletely Characterized Transport Proteins *19*	
1.13	Transporters with Dual Modes of Energy Coupling *20*	
1.14	Transporters Exhibiting More than One Mode of Transport *20*	
1.15	Conclusions and Perspectives *21*	
	References *22*	
2	**Energy-transducing Ion Pumps in Bacteria: Structure and Function of ATP Synthases** *23*	
2.1	Introduction *23*	
2.2	Overview *23*	
2.3	Structure, Configuration, and Interaction of F_1 Subunits *25*	
2.4	Catalysis: Structural and Mechanistic Implications within the F_1 Complex *27*	
2.5	The F_1/F_O Interface: Contact Sites for Energy Transmission *31*	

2.6	Structure, Configuration, and Interaction of F_O Subunits *33*
2.7	Catalysis: Coupling Ion Translocation to ATP Synthesis *37*
	References *43*

3	**Sodium/Substrate Transport** *47*
3.1	Introduction *47*
3.2	Occurrence and Role of Na^+/Substrate Transport Systems *48*
3.2.1	General Considerations *48*
3.2.2	Elevated Temperatures *49*
3.2.3	Na^+-rich Environments *50*
3.2.4	High pH *50*
3.2.5	Citrate Fermentation *51*
3.2.6	Na^+/Substrate Transport in *Escherichia coli* *52*
3.2.7	Osmotic Stress *53*
3.3	Functional Properties of Na^+/Substrate Transport Systems *53*
3.3.1	General Considerations *53*
3.3.2	MelB *54*
3.3.3	PutP *55*
3.3.4	CitS *56*
3.4	Transporter Structure *57*
3.4.1	General Features *57*
3.4.2	MelB *58*
3.4.3	PutP and Other Members of the SSF *59*
3.4.4	CitS *61*
3.5	Structure–Function Relationships *62*
3.5.1	MelB *62*
3.5.1.1	Site of Ion Binding *62*
3.5.1.2	Sugar Binding and Functional Dynamics of MelB *63*
3.5.2	PutP *65*
3.5.2.1	Site of Na^+ Binding *65*
3.5.2.2	Regions Important for Proline Binding *67*
3.5.2.3	Functional Dynamics of PutP *68*
3.5.3	CitS *69*
3.6	Concluding Remarks and Perspective *69*
	References *70*

4	**Prokaryotic Binding Protein-dependent ABC Transporters** *77*
4.1	A Brief History of ABC Systems *77*
4.2	What is an ABC System? *79*
4.3	The Composition of the Prokaryotic ABC Transporters *80*
4.4	Associated Proteins and Signal Transduction Pathways *84*
4.5	The Components *85*
4.5.1	The Binding Proteins *85*
4.5.1.1	Substrate Recognition Sites are High-affinity Soluble Binding Proteins *85*

4.5.1.2	The Binding Test	86
4.5.1.3	Special Examples	86
4.5.1.4	Binding Proteins Undergo Conformational Changes upon Binding Substrate	87
4.5.1.5	The Crystal Structure	88
4.5.2	The Integral Transmembrane Domains (TMDs)	91
4.5.2.1	Organization	91
4.5.2.2	Composition and Structure	92
4.5.2.3	The Interaction of the TMDs with the Binding Protein	93
4.5.2.4	The Sequence	96
4.5.3	The ABC Subunit	97
4.5.3.1	The Sequence	97
4.5.3.2	The Localization	98
4.5.3.3	ATP Hydrolysis	98
4.5.3.4	The Crystal Structure of MalK from *Thermococcus litoralis*	101
4.5.3.5	The Asymmetry within the MalK Dimer	105
	References	108

5	**Glucose Transport by the Bacterial Phosphotransferase System (PTS): An Interface between Energy- and Signal Transduction**	**115**
5.1	Introduction	115
5.2	The Components of the PTS and Their Function	117
5.2.1	Distribution of the PTS	117
5.2.2	Modular Design and Classification	117
5.2.3	Active Sites	119
5.3	Structure and Function of the PTS Transporter for Glucose	119
5.3.1	The Genes *crr* (IIA^{Glc}) and *ptsG* (IICB^{Glc})	120
5.3.2	The IIA^{Glc} Subunit	120
5.3.3	The IICB^{Glc} Subunit	121
5.3.3.1	Structure and Function of the IIC Domain	122
5.3.3.2	Structure and Function of the IIB Domain	123
5.3.3.3	Structure and Function of the Linker Region	123
5.3.3.4	Mutants of IICB^{Glc}	124
5.4	Regulation by the PTS	129
5.4.1	Regulatory Role of IIA^{Glc}	131
5.4.2	Regulatory Role of IICB^{Glc}	132
5.5	Kinetic Properties of the Phosphorylation Cascade	133
	References	135

6	**Peptide Transport**	**139**
6.1	Introduction	139
6.2	Classification of Microbial Peptide Transport Systems	140
6.2.1	Classification Based upon Genome Sequencing	140
6.2.2	Classification Based upon Substrate Specificity	143

6.3	Peptide Transport in Prokaryotic Microorganisms *143*	
6.3.1	Gram-negative Bacteria *143*	
6.3.1.1	Enteric Bacteria *143*	
6.3.1.2	Rumen Bacteria *148*	
6.3.2	Gram-positive Bacteria *148*	
6.3.2.1	Lactic Acid Bacteria *148*	
6.3.2.2	Miscellaneous Organisms *150*	
6.4	Bacterial Peptide Transport Systems with Specific Functions and Substrates *151*	
6.4.1	Role of Peptides and Peptide Transporters in Microbial Communication *151*	
6.4.2	*Sap* Genes and Resistance to Antimicrobial Cationic Peptides *152*	
6.4.3	Uptake of Peptide Antibiotics *152*	
6.4.4	Polyamine Stimulation of OppA Synthesis and Sensitivity to Aminoglycoside Antibiotics *152*	
6.4.5	Role of MppA in Signaling Periplasmic Environmental Changes *153*	
6.4.6	Periplasmic Substrate Binding Proteins as Molecular Chaperones *153*	
6.4.7	Transport of δ-Aminolevulinic Acid *154*	
6.4.8	Transport of Glutathione *154*	
6.5	Peptide Transport in Eukaryotic Microorganisms *155*	
6.6	Structural Basis for Molecular Recognition of Substrates by Peptide Transporters *156*	
6.7	Exploitation of Peptide Transporters for Delivery of Therapeutic Compounds *160*	
	References *161*	
7	**Protein Export and Secretion in Gram-negative Bacteria** *165*	
7.1	Introduction *165*	
7.2	Protein Export *168*	
7.2.1	Sec Pathway *168*	
7.2.1.1	Introduction *168*	
7.2.1.2	Targeting to the Sec translocase: SRP and Trigger Factor SecA/B Routes *169*	
7.2.1.3	YidC, an Essential Component for Integration of Cytoplasmic Membrane Proteins *171*	
7.2.1.4	Oligomeric State of the Sec Translocase *173*	
7.2.2	Tat Pathway *173*	
7.2.2.1	Introduction *173*	
7.2.2.2	Genetic and Genomic Evidence for the *tat* Pathway in *Escherichia coli* *174*	
7.2.2.3	Functions and Interactions of the Tat Proteins *175*	
7.2.2.4	Role of the Tat Signal Peptide *176*	
7.2.2.5	Open Questions *177*	

7.3	Protein Secretion *178*	
7.3.1	*Sec*-Dependent Pathway:Type II Secretion Pathway *178*	
7.3.1.1	Type II Secretion Pathway with a Helper Domain Encoded by the Secreted Protein: The Autotransporter Mechanism *178*	
7.3.1.2	Type II Secretion Pathway with one Helper Protein *179*	
7.3.1.3	Type II Secretion Pathway with 11 to 12 Helper Proteins *180*	
7.3.2	*SEC*-independent Pathways *184*	
7.3.2.1	Type I Secretion Pathway – ABC Protein Secretion in Gram-negative Bacteria *184*	
7.3.2.2	Type III Secretion Pathway *192*	
7.3.2.3	Type IV Secretion System *198*	
7.4	Concluding Remarks *201*	
	References *202*	
8	**Bacterial Channel Forming Protein Toxins** *209*	
8.1	Toxins in Model Systems *210*	
8.2	Toxin Complexity *210*	
8.3	Classification of Channel Forming Proteins *211*	
8.4	Steps in Channel Formation *212*	
8.4.1	Binding to Target Cells *212*	
8.4.2	Activation *213*	
8.4.3	Oligomerization *213*	
8.4.4	Insertion *214*	
8.5	Consequences of Channel Formation *214*	
8.6	Toxins that Oligomerize to Produce Amphipathic β-Barrels *214*	
8.7	Toxins Forming Small β-Barrel Channels *215*	
8.7.1	Aerolysin *215*	
8.7.2	α-Toxin *217*	
8.7.3	Anthrax Protective Antigen *218*	
8.8	Toxins Forming Large β-Barrel Channels *219*	
8.8.1	The Cholesterol-dependent Toxins *219*	
8.9	The RTX Toxins *220*	
8.9.1	*Escherichia coli* HlyA *221*	
8.9.2	Pertussis CyaA *221*	
8.10	Ion Channel Forming Toxins *222*	
8.10.1	Channel Forming Colicins *222*	
8.10.2	*Bacillus thuringiensis* Cry Toxins *223*	
8.11	Other Channel Forming Toxins *224*	
	References *225*	
9	**Porins – Structure and Function** *227*	
9.1	Introduction *227*	
9.2	Structure of the Outer Membrane of Gram-negative Bacteria and Isolation of Porin Proteins *229*	

9.3	Model Membrane Studies with Porin Channels 230
9.4	Structure and Function of the General Diffusion Porins 234
9.5	Structure and Function of Specific Porins 237
9.6	The Inner and Outer Membrane Connector Channels 241
9.7	Conclusions 242
	References 243

10	**Aquaporins** 247
10.1	Introduction 247
10.2	Diversity of Species with Aquaporin Genes 248
10.3	Microbial Aquaporins 249
10.4	Structural Properties of Aquaporins 249
10.5	Functional Analysis of Aquaporins 250
10.6	Unspecific Aquaporins 251
10.7	Complexity of Microbial MIP-like Channel Genes 252
10.8	Gene Structures 253
10.9	Physiological Indications for Protein-mediated Membrane Water Transport 253
10.10	The Human Aquaporin 1 as a Model 254
10.11	The *Escherichia coli* Aquaporin Z 255
10.12	Physiological Relevance of Aquaporins 255
10.13	Glycerol Conducting Channels 256
10.13.1	Structure 256
10.13.2	Physiological Relevance of Glycerol Conducting Channels 257
	References 257

11	**Structures of Siderophore Receptors** 261
11.1	Introduction 261
11.1.1	Iron Transport 261
11.1.2	Siderophores 262
11.1.3	Siderophore Receptors 262
11.2	Biochemistry and Genetic Regulation of Siderophore Receptors 262
11.2.1	Chemistry 262
11.2.2	Genetic Regulation 263
11.3	Structures of FepA and FhuA 264
11.3.1	General 264
11.3.2	The β-Barrel and Periplasmic Loops 265
11.3.3	The N-terminal Domain 267
11.3.4	The Extracellular Loops 270
11.4	The FhuA Structures with Ligand 272
11.5	Is the FepA Structure the Liganded or Unliganded Form of the Protein? 275
11.6	Biochemical and Genetic Experiments 276
11.7	Binding and Mechanism 278
11.8	Proposed Mechanism 279

11.8.1	Overview	279
11.8.2	Binding of Ligand to Receptor	280
11.8.3	The TonB-dependent Transport	281
11.8.4	Homology	282
11.8.5	Experimental Evidence	283
11.9	Conclusions	285
	References	286

12	**Mechanisms of Bacterial Iron Transport**	**289**
12.1	Introduction	289
12.2	Transport of Fe^{3+}-Siderophores	291
12.2.1	Transport of Fe^{3+}-Siderophores Across the Outer Membrane of Gram-negative Bacteria	291
12.2.2	Transport of Fe^{3+}-Siderophores Across the Cytoplasmic Membrane by ABC Transporters	295
12.3	Bacterial Use of Fe^{3+} Contained in Transferrin and Lactoferrin	299
12.3.1	Bacterial Outer Membrane Proteins that Bind Transferrin and Lactoferrin and Transport Fe^{3+}	299
12.3.2	Transport of Fe^{3+} Across the Cytoplasmic Membrane	299
12.4	Bacterial Use of Heme	300
12.4.1	Bacterial Outer Membrane Transport Proteins for Heme	301
12.4.2	More than one Ton System for Certain Heme Transport Systems	303
12.5	Fe^{2+} Transport Systems	304
12.6	Regulation by Iron	304
12.6.1	Iron-dependent Repressors Regulate Iron Transport Systems	304
12.6.2	Regulation by Fe^{3+}	306
12.6.3	Regulation by Fe^{3+}-siderophores	306
12.6.4	Regulation of Outer Membrane Transport Protein Synthesis by Phase Variation	307
12.7	Outlook	307
	References	308

13	**Bacterial Zinc Transport**	**313**
13.1	Introduction	313
13.2	Exporters of Toxic Zn^{2+}	313
13.2.1	RND Family of Exporters	313
13.2.2	Cation Diffusion Facilitator	315
13.2.3	P-Type ATPases Export Cd^{2+} and Zn^{2+}	315
13.3	High-affinity Uptake Systems for Zn^{2+} are ABC Transporters	316
13.3.1	Binding Protein-dependent Zn^{2+} Uptake in Gram-positive Bacteria	316
13.3.2	Binding Protein-dependent Zn^{2+} Uptake in Gram-negative Bacteria	320
13.4	Low-affinity Zn^{2+} Uptake Systems	321
13.5	Concluding Remarks	322
	References	323

14 Bacterial Genes Controlling Manganese Accumulation 325
14.1 Introduction 325
14.1.1 Physicochemical Properties of Manganese 325
14.1.2 Physiological Role of Manganese in Bacteria 326
14.1.3 Effect of Manganese on Bacterial Growth 327
14.2 Manganese Transport in Bacteria 330
14.2.1 Overview of Biochemical Studies with Whole Cells and Membranes Vesicles 330
14.2.2 Genes Encoding Transport Systems for Manganese Acquisition 331
14.2.2.1 Primary Transport Systems 331
14.2.2.2 Secondary Transport Systems 335
14.2.3 Genes Encoding Transcription Factors Involved in Manganese Homeostasis 337
14.2.3.1 Fur and Fur-related Factors 337
14.2.3.2 DtxR and DtxR-related Factors 338
14.3 Importance of Manganese Transport in Bacterial Pathogenesis 339
14.4 Concluding Remarks 342
References 343

15 The Unusual Nature of Magnesium Transporters 347
15.1 Introduction 347
15.2 The Properties of Mg^{2+} 347
15.2.1 Chemistry 347
15.2.2 Association States of Magnesium 348
15.2.3 Technical Problems in Studying Magnesium 348
15.3 Prokaryotic Magnesium Transport 349
15.4 MgtE Magnesium Transporters 350
15.4.1 Genomics 350
15.4.2 Physiology 350
15.4.3 Structure and Mechanism 350
15.5 CorA Magnesium Transporter 351
15.5.1 Genomics 351
15.5.2 Physiology 352
15.5.3 Structure 354
15.6 MgtA/MgtB Mg^{2+} Transporters 355
15.6.1 Genomics 355
15.6.2 Structure 355
15.6.3 Physiology 356
15.6.4 The MgtC Protein 357
15.7 Conclusions and Perspective 357
References 359

16	**Bacterial Copper Transport** *361*	
16.1	Introduction *361*	
16.2	The New Subclass of Heavy Metal CPx-type ATPases *362*	
16.2.1	Membrane Topology of CPx-type ATPases *363*	
16.2.2	Role of the CPx Motif *364*	
16.2.3	N-Terminal Heavy Metal Binding Sites *365*	
16.2.4	The HP Locus *367*	
16.3	Copper Homeostasis in *Enterococcus hirae* *368*	
16.3.1	Function of CopA in Copper Uptake *369*	
16.3.2	Function of CopB in Copper Excretion *369*	
16.3.3	Regulation of Expression by Copper *370*	
16.4	Copper Resistance in *Escherichia coli* *371*	
16.4.1	Regulation of the *Escherichia coli* Copper ATPase *372*	
16.5	Synechococcal Copper ATPases *372*	
16.6	The *Helicobacter pylori* Copper ATPases *373*	
16.7	The Copper ATPase of *Listeria monocytogenes* *373*	
16.8	Other Copper Resistance Systems *374*	
16.9	Conclusion *375*	
	References *375*	
17	**Microbial Arsenite and Antimonite Transporters** *377*	
17.1	Introduction *377*	
17.1.1	Why Arsenic Transporters? *377*	
17.1.2	Efflux as a Mechanism for Resistance *377*	
17.2	Overall Architecture of the Plasmid-encoded Pump in *Escherichia coli* *378*	
17.2.1	ArsA *380*	
17.2.1.1	The Ligand (Arsenite/Antimonite) Binding Site *380*	
17.2.1.2	The Nucleotide Binding Sites *381*	
17.2.1.3	The DTAP Domain in ArsA *386*	
17.2.1.4	The Linker Region in ArsA *387*	
17.2.1.5	Variations on the ArsA Theme *387*	
17.2.1.6	Insights from the Crystal Structure of ArsA *389*	
17.2.2	ArsB *390*	
17.2.3	ArsC *391*	
17.3	Variations on the *Escherichia coli* Arsenic Transporter among Prokaryotes *391*	
17.4	Other Arsenic Transporters *392*	
17.5	Conclusion *393*	
	References *394*	
18	**Microbial Nickel Transport** *397*	
18.1	Introduction *397*	
18.2	Metabolic Roles of Nickel *398*	
18.2.1	Nickel as a Cofactor of Metalloenzymes *398*	

18.2.2	Nickel Toxicity *401*
18.2.3	Nickel Resistance *401*
18.3	Transport Systems Involved in Nickel Homeostasis *403*
18.4	High-affinity Nickel Uptake Systems *406*
18.4.1	ABC-type Nickel Transporters *407*
18.4.1.1	The Nik System of *Escherichia coli* *407*
18.4.1.2	Nik-related Transporters in Prokaryotes *408*
18.4.2	The Nickel/Cobalt Transporter Family *408*
18.4.2.1	Signature Motifs *408*
18.4.2.2	Significance in Microorganisms *409*
18.4.2.3	Substrate Specificity *412*
18.5	Perspective *413*
	References *414*

19	**Mitochondrial Copper Ion Transport** *419*
19.1	Introduction *419*
19.2	Mitochondrial Structure *419*
19.3	Mitochondrial Transport *420*
19.4	Assembly of Mitochondrial Cytochrome *c* Oxidase *422*
19.5	Copper Ion Delivery to Targets other than the Mitochondrion *426*
19.6	Copper Ion Transport to the Mitochondrion by Cox17 *429*
19.7	Co-metallochaperones in Cu Metallation of Cytochrome *c* Oxidase *431*
19.8	Terminal Oxidases in Prokaryotes *435*
19.9	Metallation of Prokaryotic Terminal Oxidases *437*
19.10	Postulated Model *440*
	References *442*

20	**Iron and Manganese Transporters in Yeast** *447*
20.1	Iron Transport in *Saccharomyces cerevisiae* *447*
20.1.1	Reduction of Iron at the Cell Surface *447*
20.1.2	Iron Translocation across the Plasma Membrane *448*
20.1.2.1	High-affinity Iron Uptake: The Requirement for a Multi-copper Oxidase *448*
20.1.2.2	The Iron–Copper Connection for High-affinity Iron Uptake *449*
20.1.2.3	Iron Transport by the Cell Surface Permease, FTR1 *449*
20.1.2.4	Low-affinity Iron Uptake at the Cell Surface *450*
20.1.3	Intracellular Iron Transport *450*
20.1.4	Regulation of Iron Transport *451*
20.2	Manganese Transport in *Saccharomyces cerevisiae* *452*
20.2.1	The Smf1p and Smf2p Members of the Nramp Family of Ion Transporters *452*
20.2.1.1	Transport of Heavy Metals by Smf1p and Smf2p *452*
20.2.1.2	Regulation of Smf1p and Smf2p by Bsd2p and Manganese Ions *453*

20.2.2	Manganese Transport in the Golgi Apparatus *455*	
20.2.2.1	Pmr1p: A Manganese Transporting ATPase *455*	
20.2.2.2	Ccc1p: A Manganese Homeostasis Protein Localized in the Golgi *456*	
20.2.2.3	Atx2p: An Antagonizer of Pmr1p? *456*	
20.2.3	Homeostasis of Cytosolic Manganese: A Possible Role for the *CDC1* Gene Product *456*	
20.2.4	The Yeast Vacuole and Manganese *457*	
20.3	Conclusions and Directions for the Future *457*	
	References *460*	
21	**Siderophore Transport in Fungi** *463*	
21.1	Introduction *463*	
21.2	Siderophore Classes and Properties *464*	
21.3	Siderophore Production and Biosynthesis *466*	
21.4	Evolutionary Aspects of Siderophores *467*	
21.5	Siderophore Transporters in *Saccharomyces cerevisiae* *468*	
21.5.1	SIT1 Transporter *468*	
21.5.2	TAF1 Transporter *469*	
21.5.3	ARN1 Transporter *469*	
21.5.4	Transporter for Ferrichromes *471*	
21.5.5	Transporter for Coprogens *472*	
21.5.6	ENB1 transporter *472*	
21.6	Energetics and Mechanisms *473*	
21.7	FRE Reductases in Siderophore Transport *474*	
21.8	Conclusions *477*	
	References *477*	

Index *481*

List of Authors

Karlheinz Altendorf
Fachbereich Biologie/Chemie
Universität Osnabrück
Barbarastr. 11
D-49069 Osnabrück
Germany
Phone: +49-541-969-2864
Fax: +49-541-969-2870
altendorf@biologie.uni-osnabrueck.de

Roland Benz
Lehrstuhl für Biotechnologie
Universität Würzburg
Am Hubland
D-97074 Würzburg
Germany
Phone: +49-931-8884501
roland.benz@mail.uni-wuerzburg.de

Winfried Boos
Fakultät für Biologie
Universität Konstanz
D-78457 Konstanz
Germany
Phone: +49-7531-88-2658
Winfried.Boos@uni-konstanz.de

Volkmar Braun
Universität Tübingen
Institut für Mikrobiologie
Auf der Morgenstelle 28
D-72076 Tübingen
Germany
Phone: +49-7071-29-72096
Fax: +49-7071-29-5843
volkmar.braun@uni-tuebingen.de

J. Thomas Buckley
Department of Biochemistry and
Microbiology
University of Victoria
Victoria, BC
Canada V8W 3P6
Canada
Phone: +1-250-721-7081
tbuckley@uvic.ca

Mathieu Cellier
INRS
Centre de Recherche en Santé Humaine
531, Bd. des Prairies
Laval, Quebec
Canada H7V 1B7
Canada
Phone: +1-450-687-5010
Fax: +1-450-686-5501
mathieu.cellier@iaf.uquebec.ca

List of Authors

Ranjan Chakraborty
Department of Chemistry and
Biochemistry
University of Oklahoma
620 Parrington Oval
Norman, OK 73019-0370
USA
Phone: +1-405-325-4811
Fax: +1-405-325-6111
dvdhelm@chemdept.chem.ou.edu

Valeria Culotta
Department of Biochemistry
John Hopkins University School of
Public Health
Baltimore, MD 21205
USA
Phone: +1-410-955-3029
Fax: +1-410-955-0116
vculatta@jhsph.eduo

Gabriele Deckers-Hebestreit
Fachbereich Biologie/Chemie
Universität Osnabrück
Barbarastr. 11
D-49069 Osnabrück
Germany
Phone: +49-541-969-2867/2809
deckers-hebestreit@biologie.
uni-osnabrueck.de

Philippe Delepelaire
Institut Pasteur
Unité des Membranes Bactériennes
25–28, Rue du Docteur Roux
75724 Paris Cedex 15
France
Phone: +33-1-4061-3666
Fax: +33-1-4568-8929
murield@pasteur.fr

Martin Eckert
Julius-von-Sachs-Institut für
Biowissenschaften
Julius-von-Sachs-Platz 2
D-97082 Würzburg
Germany
Phone: +49-931-8886133
eckert@botanik.uni-wuerzburg.de

Thomas Eitinger
Institut für Biologie
Humboldt-Universität zu Berlin
Chausseestr. 117
D-10115 Berlin
Germany
Phone: +49-30-2093-8103
Fax: +49-30-2093-8102
thomas.eitinger@rz.hu-berlin.de

Tanja Eppler
Fakultät für Biologie
Universität Konstanz
D-78457 Konstanz
Germany

Bernhard Erni
Departement für Chemie und
Biochemie
Universität Bern
Freiestraße
CH-3012 Bern
Switzerland
Phone: +41-31-6314343
Fax: +41-31-6314887
erni@ibc.unibe.ch

Jörg-Christian Greie
Fachbereich Biologie/Chemie
Universität Osnabrück
Barbarastr. 11
D-49069 Osnabrück
Germany
Phone: +49-541-969-2867/2809
greie@biologie.uni-osnabrueck.de

Klaus Hantke
Institut für Mikrobiologie
Universität Tübingen
Auf der Morgenstelle 28
D-72076 Tübingen
Germany
Phone: +49-7071-2974645
Fax: +49-7071-295843
hantke@uni-tuebingen.de

Heinrich Jung
Fachbereich Biologie/Chemie
Universität Osnabrück
Barbarastr. 11
D-49069 Osnabrück
Germany
Fax: +49-541-969-2870
jung_h@biologie.uni-osnabrueck.de

Ralf Kaldenhoff
Julius-von-Sachs-Institut für
Biowissenschaften
Julius-von-Sachs-Platz 2
D-97082 Würzburg
Germany
Phone: +49-931-8886107
Fax: +49-931-8886158
kaldenhoff@botanik.uni-wuerzburg.de

Parjit Kaur
Department of Biology
Georgia State University
Atlanta, GA 30303
USA
Phone: +1-404-651-3864
boppk@panther.gsu.edu

David G. Kehres
Department of Pharmacology
Case Western Reserve University
Cleveland, OH 44106-4965
USA
Phone: +1-216-368-6186
Fax: +1-216-368-3395
mem6@po.cwru.edu

Michael E. Maguire
Department of Pharmacology
Case Western Reserve University
Cleveland, OH 44106-4965
USA
Phone: +1-216-368-6186
Fax: +1-216-368-3395
mem6@po.cwru.edu

Neil J. Marshall
School of Biological Sciences
University of Wales Bangor
Bangor, Gwynedd LL57 2UW
UK
Phone: +44-1248-351151
Fax: +44-1248-371644
n.j.marshall@bangor.ac.uk

Keith McCall
Departments of Medicine and
Biochemistry
University of Utah Health Sciences
Center
Salt Lake City, UT 84132
USA

Thalia Nittis
Departments of Medicine and
Biochemistry
University of Utah Health Sciences
Center
Salt Lake City, UT 84132
USA

John W. Payne
School of Biological Sciences
University of Wales Bangor
Bangor, Gwynedd LL57 2UW
UK
Phone: +44-1248-382349
Fax: +44-1248-370731
j.w.payne@bangor.ac.uk

List of Authors

Matthew E. Portnoy
Department of Biochemistry
John Hopkins University School of
Public Health
Baltimore, MD 21205
USA
Phone: +1-410-955-9643
Fax: +1-410-955-0116
mportnoy@jhsph.edu

Milton H. Saier, Jr.
Department of Biology
University of California at San Diego
La Jolla, CA 92093-0116
USA
Phone: +1-858-534-4084
Fax: +1-858-534-7108
msaier@ucsd.edu

Marc Solioz
Department of Clinical Pharmacology
University of Berne
Murtenstraße 35
CH-3010 Bern
Switzerland
Phone: +41-31-632-3268
Fax: +41-31-632-4997
marc.solioz@ikp.unibe.ch

Dick van der Helm
Department of Chemistry and
Biochemistry
University of Oklahoma
620 Parrington Oval
Norman, OK 73019-0370
USA
dvdhelm@chemdept.chem.ou.edu

Cécile Wandersman
Institut Pasteur
Unité des Membranes Bactériennes
25–28, Rue du Docteur Roux
75724 Paris Cedex 15
France
Phone: +33-1-4061-3275
Fax: +33-1-4568-8790
cwander@pasteur.fr

Dennis R. Winge
Departments of Medicine and
Biochemistry
University of Utah Health Sciences
Center
Salt Lake City, UT 84132
USA
Dennis.winge@hsc.utah.edu

Günther Winkelmann
Institut für Mikrobiologie
Universität Tübingen
Auf der Morgenstelle 28
D-72076 Tübingen
Germany
Phone: +49-7071-2973094
Fax: +49-7071-295002
Winkelmann@uni-tuebingen.de

Color Plates

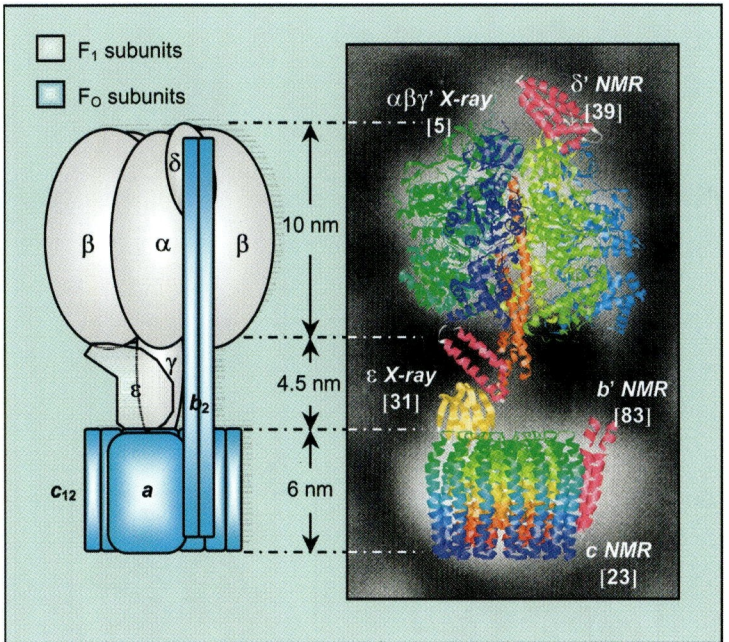

Chapter 2, Fig. 1. Schematic presentation of the F_1F_O ATP synthase. Overview of subunit assembly and modeling of available structural information from either NMR spectroscopy or X-ray crystallographic analysis into the electron density map of the *E. coli* F_1F_O complex (taken from [7] with kind permission from *Nature*). Corresponding references are quoted in brackets.

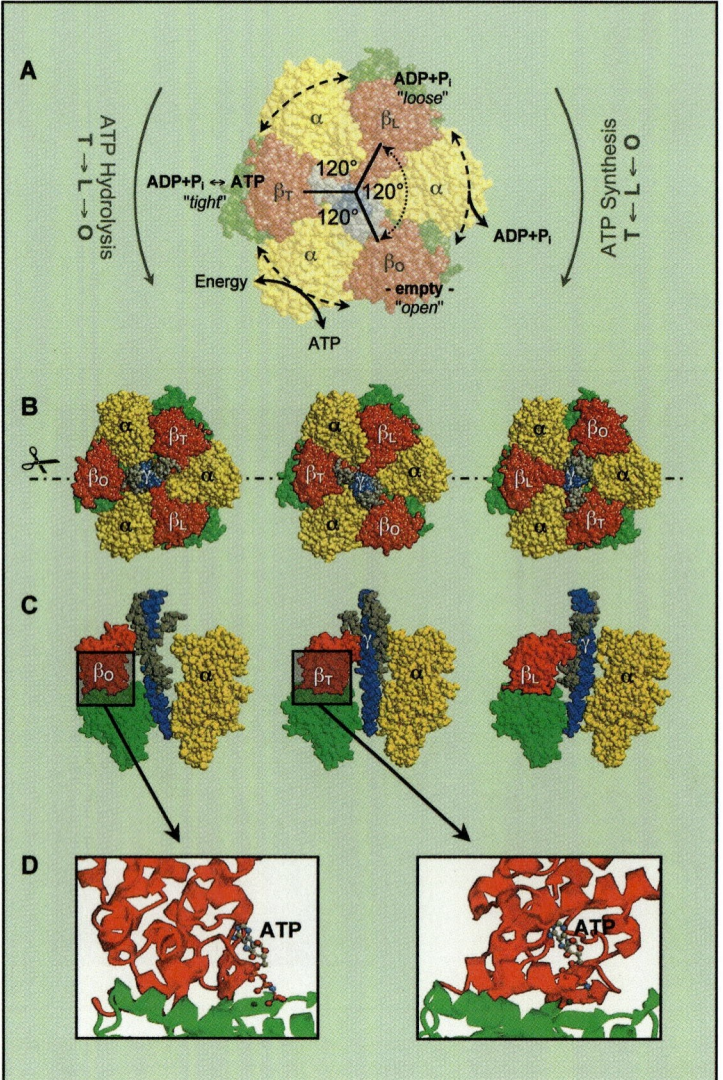

Chapter 2, Fig. 2. Catalysis within the F_1 complex – the binding change mechanism. **A** Different conformations assumed sequentially by each catalytic site during synthesis or hydrolysis of ATP as subunit γ rotates 120° within the $\alpha_3\beta_3$ hexamer. Sites are designated as "open" (β_O, no nucleotide bound), "loose" (β_L, ADP+P_i bound), and "tight" (β_T, interconversion of bound ADP + P_i and ATP). The sketch of the crystal structure from the bovine heart F_1 complex [5] is depicted as seen from the membrane. Clockwise rotation of subunit γ leads to ATP synthesis, whereas counter-clockwise rotation corresponds to ATP hydrolysis. Based on kinetic data it is likely that during steady state catalysis the "open" site is immediately occupied by another nucleotide. **B** Circulating conformational changes within the $\alpha_3\beta_3$ hexamer as subunit γ rotates stepwise at intervals of 120° each in counter-clockwise direction (i.e., ATP hydrolysis). **C** Cross-section through **B**. Nucleotide-dependent conformational changes within the C-terminal domain of the β-subunit during subunit γ rotation. Whereas the C-terminal domain undergoes spatio-temporal rearrangements during the catalytic cycle (red color), the N-terminal portion of subunit β (green) retains an approximately threefold symmetry around the rotational axis. The N- and C-terminal domain of subunit γ is depicted in gray and blue, respectively. **D** Clipping of the subunit β hinge region in either "open" (left) or "tight" (right) conformation. Refer to Sect. 4 for further details. Molecular sketches are kindly provided by Dr. G. Oster (Copyright © 2001, University of California, Berkeley).

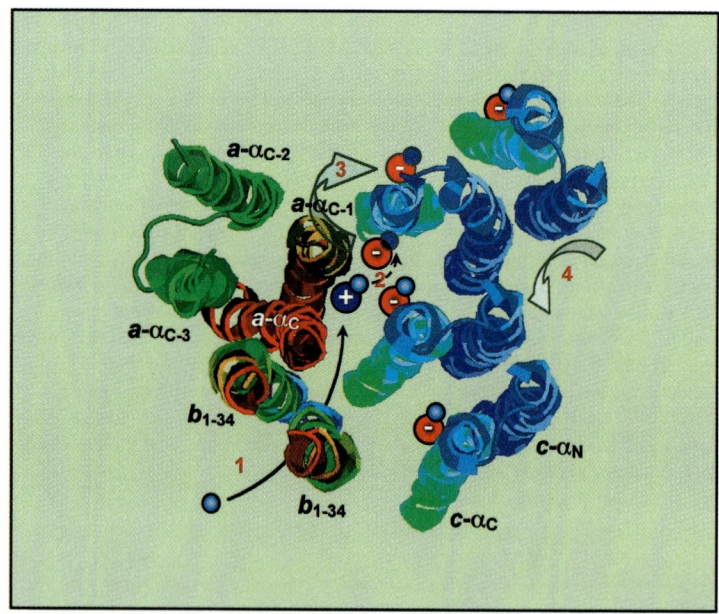

Chapter 2, Fig. 5. Hand-over-hand pattern of the proton translocation pathway within the assembled F_O complex. Structural sketches are shown from four of the c-subunits (both the N- and C-terminal helix, $c\text{-}\alpha_N$ and $c\text{-}\alpha_C$, respectively) as well as from the transmembrane domain of the subunit b dimer (b_{1-34}) and from the four C-terminal helices of subunit a ($a\text{-}\alpha_C - a\text{-}\alpha_{C\text{-}3}$) according to [92]. The assembly is presented as seen from the F_1 complex. The proposed functional cycle for the translocation of one proton is depicted according to the two-channel model established for the *E. coli* ATP synthase. The proton enters the complex via the inlet channel from the periplasmic side of the membrane, involving the positive stator charge aR210 (1). In the resting state, residue aR210 is sandwiched by both a protonated and a deprotonated cD61 side chain at the periphery of the subunit c oligomer. After proton transfer to cD61 (2), the C-terminal helix of the newly protonated monomer rotates 140° in order to adopt its protonated orientation (3), resulting in a fully protonated intermediate state of the oligomer. Simultaneously, by the interaction of cD61 and aR210 during helix rotation, the subunit c ring is pushed to rotate contrarily one step ahead (4), placing residue aR210 at the interface of the subsequent set of neighboring c-subunits. Concomitantly, residue cD61 of the next c-subunit loses its proton to the cytoplasmic side via the outlet channel (not shown), accompanied by rotation of the C-terminal helix in order to regenerate the deprotonated conformation of the resting state.

Color Plates | XXVII

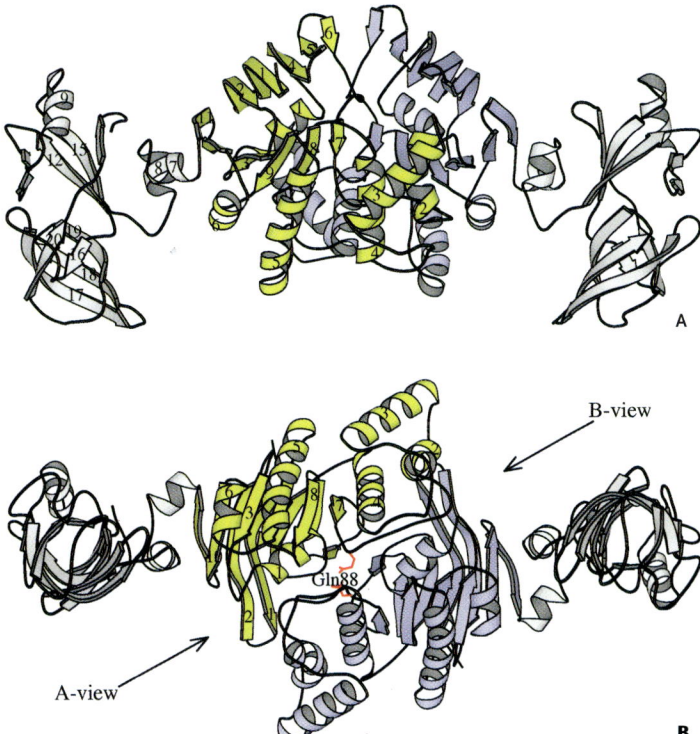

Chapter 4, Fig. 5. Ribbon representation of the *Thermococcus litoralis* MalK dimer. The A- and B-molecules are colored yellow and blue, respectively, except for both regulatory domains which are gray. Labels indicate numbers of strands and helices according to the secondary structure assignment given in Fig. 6. **(A)** The side view shows the extended dumbbell shape resulting from the two regulatory domains on either end and the central ATPase domain dimer. The pseudo-twofold symmetry axis is oriented vertically and runs through the center of the dimer. The strong involvement of helices 2 and 4 in dimerization is seen. The bottom part of the dimer is supposed to interact with the TMDs MalFG. **(B)** The bottom view along the pseudo-twofold axis shows the deviation from twofold symmetry. The helical layer of one monomer is seen in contact with the two upper layers containing the nucleotide binding site of the other monomer. The symmetry axis between strands 6 of both monomers seems to provide a mechanical hinge for the dimer. Residues Gln88 from both monomers are shown to demonstrate their close apposition. The A- and B-viewing directions are indicated. Taken from [31] with permission from the author and the publisher.

                                        Walker-A
E.c.MalK MASVQLQNVTKAWGEVVVSKDINLDIHEGEFVVFVGPSGCGKSTLLRMIAGLETIT   56
T.l.MalK MAGVRLVDVWKVFGEVTAVREMSLEVKDGEFMILLGPSGCGKTTTLRMIAGLEEPS   56

β1   β2   β3   α1

                              ----Lid----
E.c.MalK SGDLFIGEKRMNDTP------PAERGVGMVFQSYALYPHLSVAENMSFGLKPAGAK   106
T.l.MalK RGQIYIGDKLVADPEKGIFVPPKDRDIAMVFQSYALYPHMTVYDNIAFPLKLRKVP   112

β4 β5 β6 β7   α2

                              Signature motiv   Walker-B
E.c.MalK KEVINQRVNQVAEVLQLAHLLDRKPKALSGGQRQRVAIGRTLVAEPSVFLLDEPLS   162
T.l.MalK RQEIDQRVREVAELLGLTELLNRKPRELSGGQRQRVALGRAIVRKPQVFLMDEPLS   168

α3   α4   β8

D-loop   -Switch---
E.c.MalK NLDAALRVQMRIEISRLHKRLGRTMIYVTHDQVEAMTLADKIVVLDAGRVAQVGK   217
T.l.MalK NLDAKLRVRMRAELKKLQRQLGVTTIYVTHDQVEAMTMGDRIAVMNRGVLQQVGS   223

α5   β9   α6   β10   β11

E.c.MalK PLELYHPADRFVAGFIGSPKMNFLPVKVT-ATAIDQVQVELPMPNRQQVWLPVESR   273
T.l.MalK PDEVYDKPANTFVAGFIGSPPMNFLDAIVTEDGFVDFGEFRLKLLPDQFEVLGELGY 280

α7   α8   β12 β13 β14 α9

-------------------Regulatory region-------------------
E.c.MalK DVQVGANMSLGIRPEHLLPSDIA--DVI----LEGEVQVVEQLGNETQIHIQIPSIRQ 325
T.l.MalK VG----REVIFGIRPEDLYDAMFAQVRVPGENLVRAVVEIVENLGSERIVRLRVGGV-- 333

β15   β16   β17   β18 β19

E.c.MalK NLVYRQNDVVLVEEGATFAIGLPPERCHLFREDGTACRRLHKEPGV   371
T.l.MalK TFVGSFRSESRVREGVEVDVVFDMKKIHIFDKTTGKAIF------   372

β19   β20   β21

Chapter 4, Fig. 6. Alignment of *T. litoralis* MalK and *E. coli* MalK, conserved sequences. Secondary structure elements of *T. litoralis* MalK are indicated and numbered: helices are shown as cylinders, strands as arrows. Conserved regions are primed yellow. All marked mutations, found in *S. typhimurium* or in *E. coli*, respectively, are at conserved sites. The following mutations are shown: ★ (blue), cross-link with EAA loop in TMD [29]; # (green), suppressor to mutations in EAA [28]; ◇ (red), mutation causing superrepression [231]; ↓ (lilac), insertion causing loss of regulation [233]; ∇ (magenta), insertion causing no alteration in transport [233]; ● (turquoise), point mutation causing loss of regulation [230]. Positions with double mutations are primed gray. With modification taken from [31] with permission from the author and the publisher.

Chapter 4, Fig. 7. A view along the interface perpendicular to the pseudo-symmetry axis of *T. litoralis* MalK. From top to bottom, the three layers are seen: antiparallel sheet, mixed sheet with P-loop and helix 1, and helical layer. Coloring is as in Fig. 5 except that the conserved regions (Walker A, Walker B, signature motif, D-loop, switch from monomer A and the Lid region from monomer B) are marked in red with yellow outlines. Labels indicate numbers of strands and helices according to the numbering given in Fig. 6. Taken from [31] with permission from the author and the publisher.

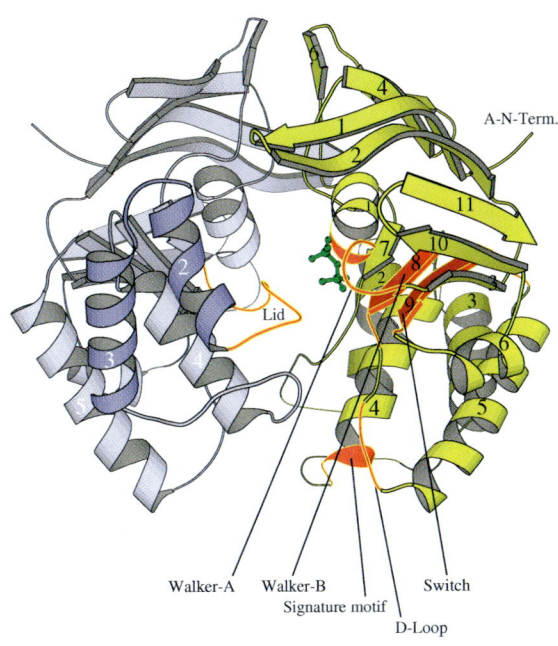

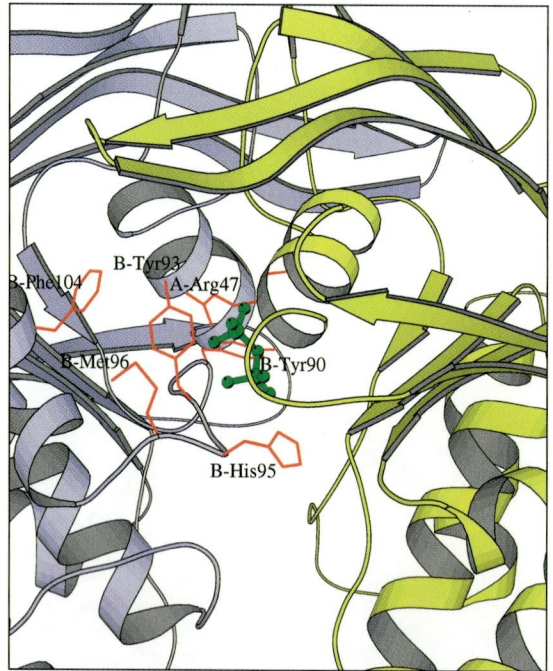

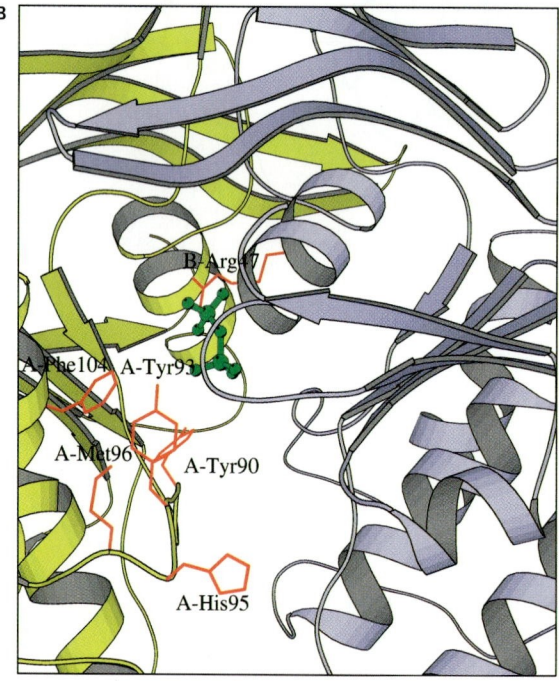

Chapter 4, Fig. 8. The asymmetry within the MalK dimer. Comparison of the A- and B-views. In the A-view (**A**) the interface appears to be narrower than in the corresponding B-view (**B**) due to the upward shift of the loop containing the Lid region. This has been emphasized by including the side chains of residues 90, 93, 96, and 104. Notably B-His95 approaches the pyrophosphate to hydrogen bonding distance in the A-view, while A-Tyr93 plays this role in the B-view. Arg47 shows a difference in side chain conformation in the two views. The two helices 2 and 3 are shifted outwards in the B-monomer by approximately 3 as compared to the A-monomer. Taken from [31] with permission from the author and the publisher.

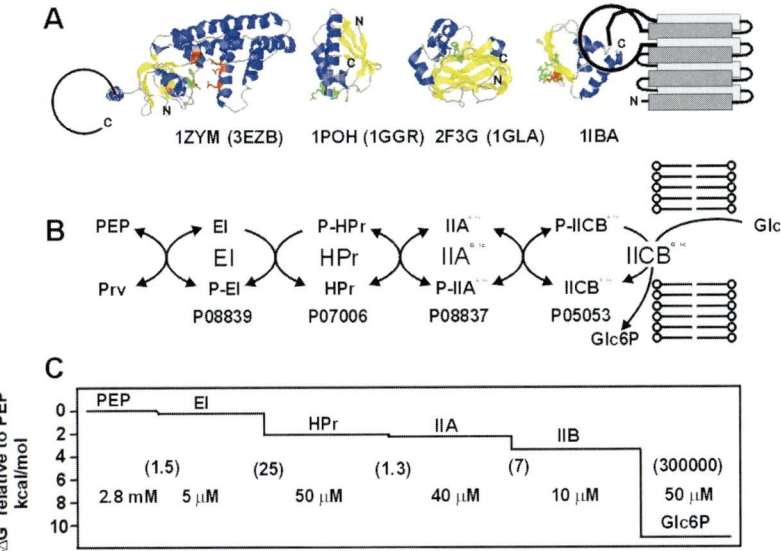

Chapter 5, Fig. 1. Structural, biochemical and thermodynamic aspects of the glucose PTS. (**A**) Cartoons of EI, HPr, IIA^{Glc}, and IICB^{Glc}. Active site residues are represented as sticks. The PDB access codes are indicated. Codes in parentheses refer to structures of protein–protein complexes. Of EI, only the structure of the N-terminal domain is known. (**B**) Phosphotransfer chain of the glucose PTS. Arrowheads emphasize the reversibility of the phosphotransfer reaction. The SWISS-PROT access codes are indicated. (**C**) Free energy differences, equilibrium constants (in parentheses) of the phosphotransfer reactions between PTS proteins and intracellular concentrations of the components. Notice the small difference between HPr and IIA^{Glc}, which allows equilibrium distribution of phosphoryl groups between different IIA units via HPr. (Fig. 2). The values are from [24].

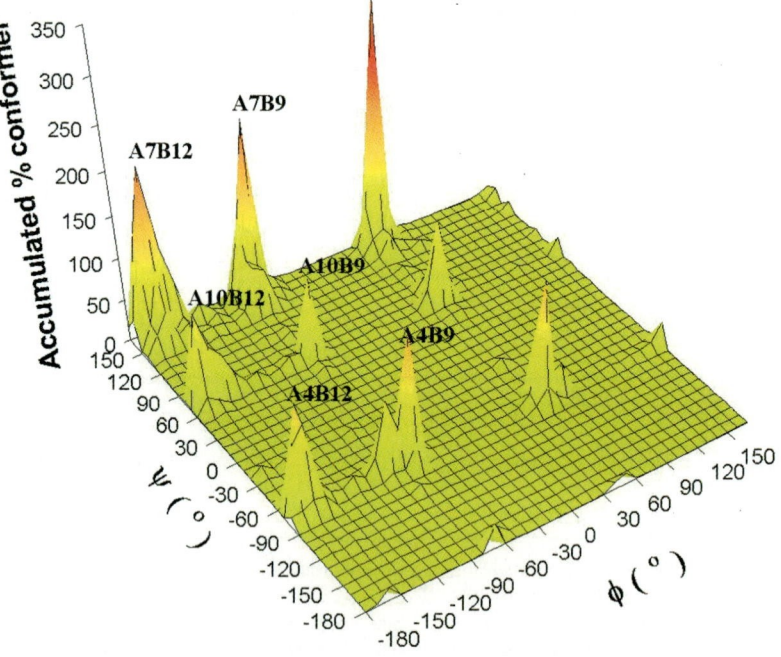

Chapter 6, Fig. 1. 3-D pseudo-Ramachandran plot for 50 dipeptides. The conformational forms recognized by Dpp-type (A7B9 and A7B12) and Tpp-type (A4B9, A4B12, A10B9, and A10B12) transporters are indicated above the relevant peaks.

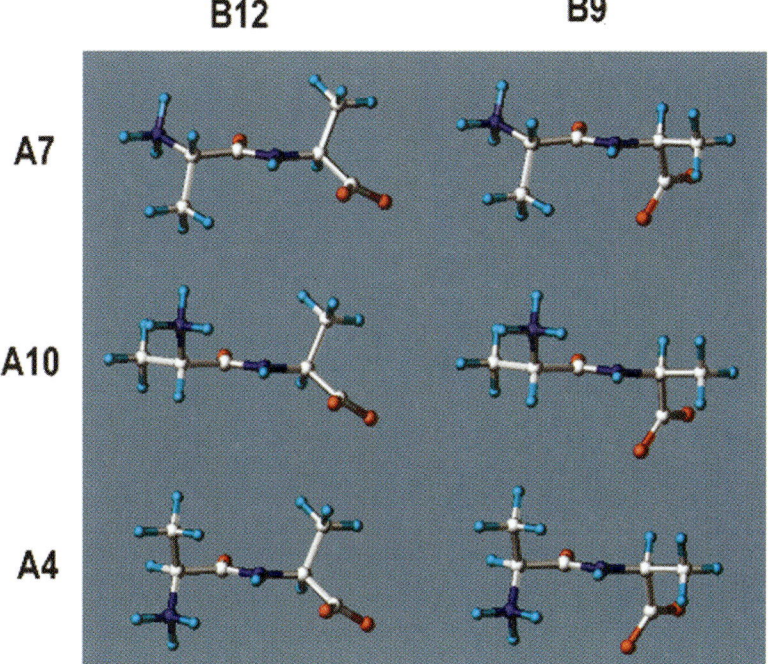

Chapter 6, Fig. 2. Ball-and-stick representations of AlaAla in the conformational forms recognized by Dpp-type and Tpp-type transporters. Dpp-type transporters recognize A7B9 and A7B12 conformers, whereas Tpp-type transporters recognize A4B9, A4B12, A10B9, and A10B12 conformers. The relative orientations of the central peptide bond of the conformers are constant throughout.

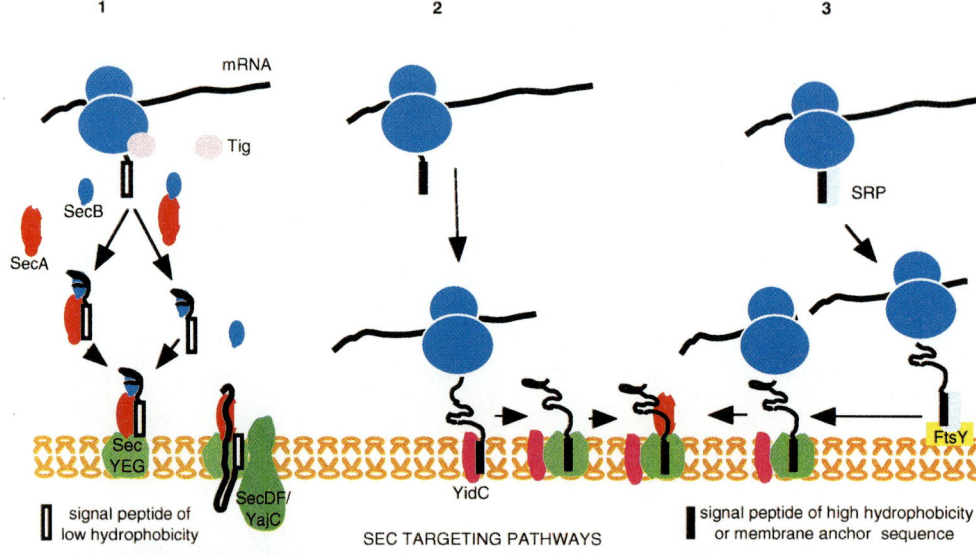

Chapter 7, Fig. 1. Sec targeting pathways. The different targeting pathways are shown (1, 2, 3) according to the nature of the signal sequence, whether of low hydrophobicity (Tig/SecB targeting pathway) or of high hydrophobicity, as a membrane anchor (SRP/FtsY pathway). Some proteins (e.g., M13 procoat) might be directly targeted to YidC. See text for additional details.

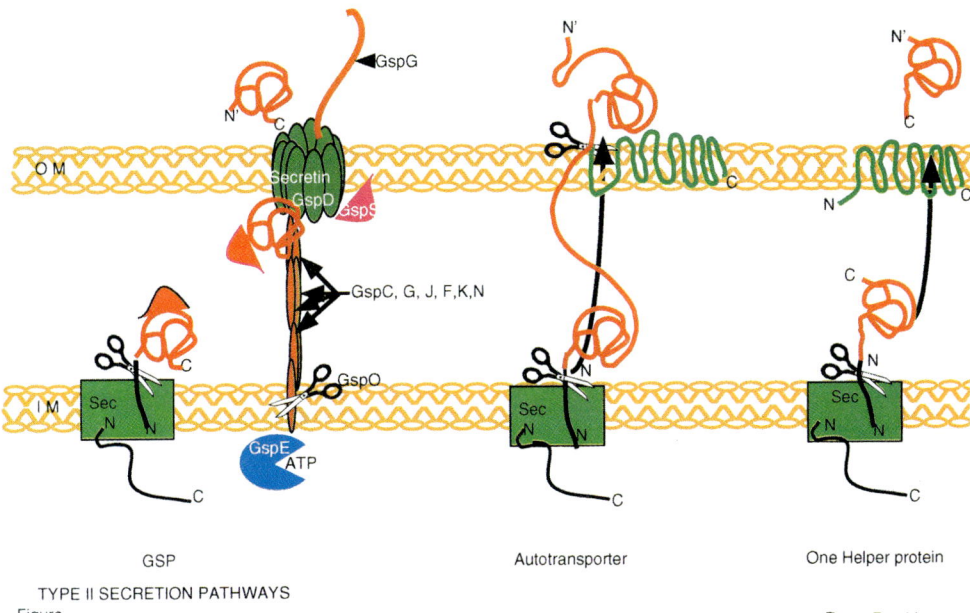

Chapter 7, Fig. 2. Type II secretion pathways. In all cases the precursor is recognized by the sec machinery and the signal sequence cleaved on the periplasmic side of the cytoplasmic membrane; subsequent transport across the outer membrane may involve no other protein (autotransporter mechanism) or one helper protein or a set of 11-12 proteins with a specific secretin in the outer membrane (GSP pathway). See text for additional details.

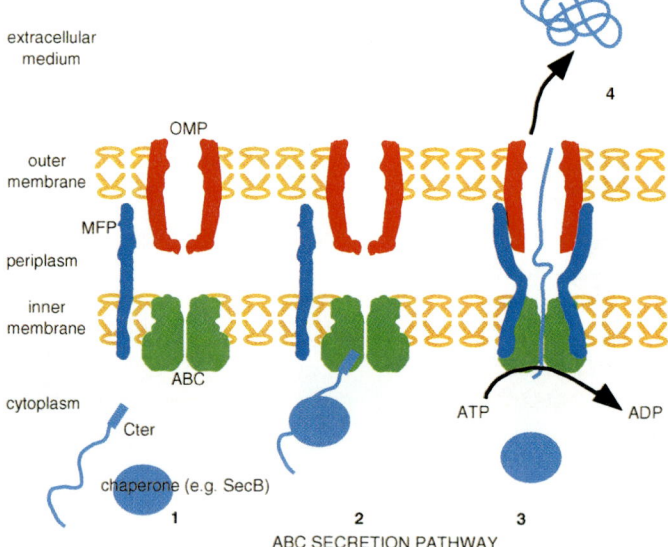

Chapter 7, Fig. 3. ABC or type I secretion pathway. The secreted protein is synthesized in the cytosol with a C-terminal secretion signal (1); it might associate with chaperones and interact with the ABC protein via the C-terminal secretion signal (2), triggering the functional association of the MFP and OMP components (3); the protein is then secreted (4) and the transporter is ready for a new round. See text for additional details.

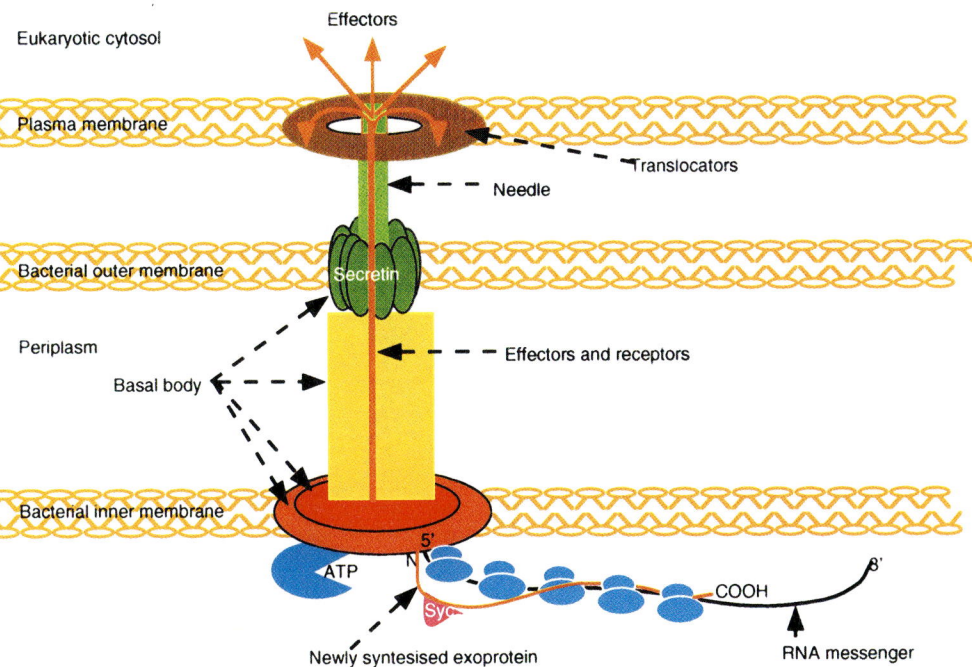

Chapter 7, Fig. 4. Type III secretion pathway.

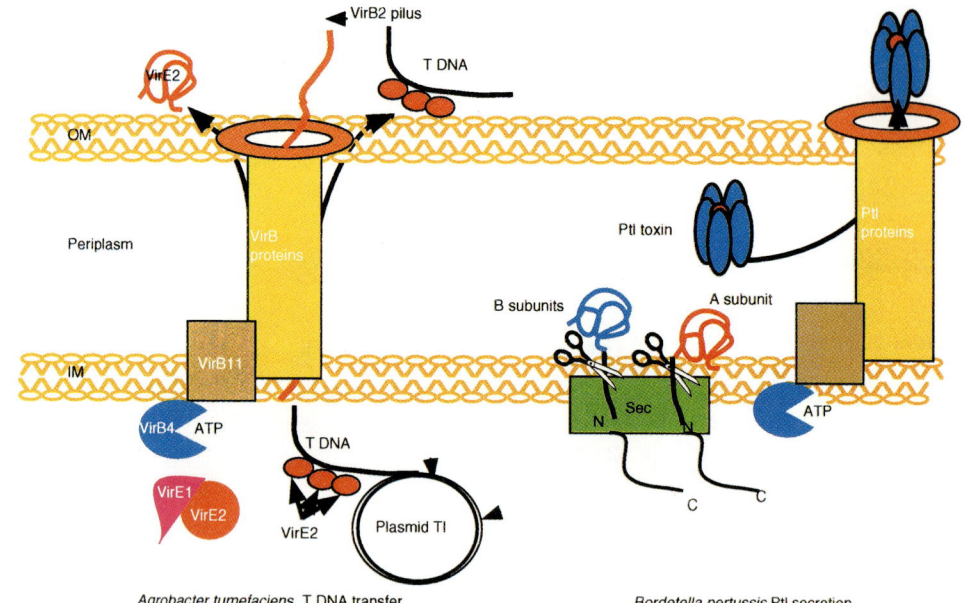

Chapter 7, Fig. 5. Type IV secretion pathways.

Color Plates | XXXIX

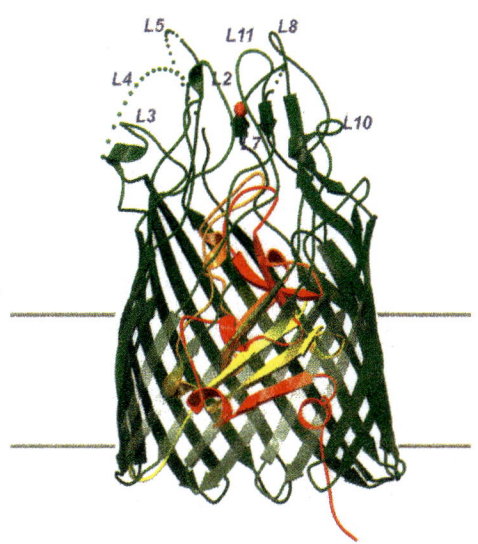

Chapter 11, Fig. 1. Ribbon diagram of FepA. The extracellular space is located at the top of the figure, and the periplasmic space is at the bottom. The position of the membrane bilayer is delineated by horizontal lines, as determined from the hydrophobic area found on the molecular surface. Part of the barrel has been rendered transparent to reveal the N-terminal domain located in the channel. The long extracellular loops are labeled; residues which could not be located in the loops are indicated by dotted lines. Reproduced with permission from S. Buchanan et al., *Nature Struct. Biol.*, **1999**, *6*, 56-63.

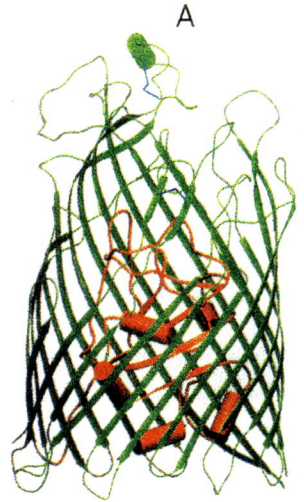

A

unliganded

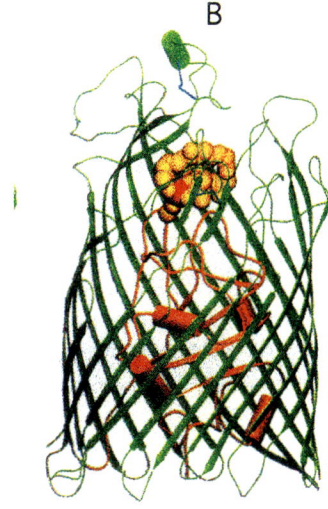

B

liganded

Chapter 11, Fig. 2. Structure of the free (**A**) and liganded (**B**) FhuA protein. In both conformations, the N-terminal domain is indicated by brown color located within the green color barrel formed by 22 antiparallel β-strands. In **B** the bound ferrichrome is depicted in yellow, with the complexed iron atom in red. In the liganded form, the $\alpha 1$ helix, at the bottom of the N-terminal domain, is unwound. Reproduced with permission from K. P. Locher et al., *Cell*, **1998**, *95*, 771-778.

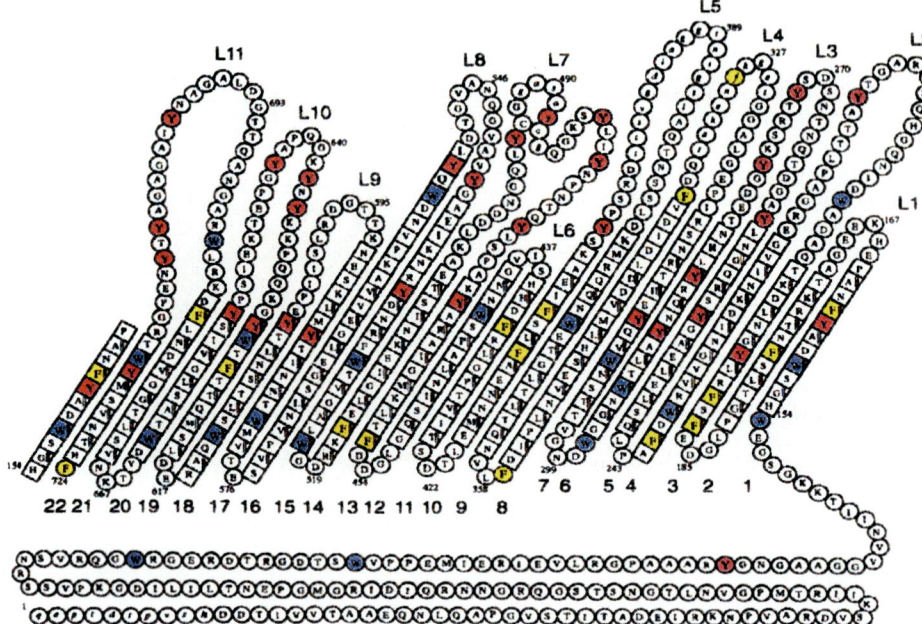

Chapter 11, Fig. 3. Topology of the FepA barrel, using amino acid sequence in one-letter code. The view is from the outside of the barrel, which is unrolled. Capital letters represent residues present in the model (Fig. 1), whereas small letters represent residues that have not been traced. Squares indicate residues in β-strand conformation. Side chains of residues marked with a double line point to the outside of the barrel. The topology of the N-terminal domain is not shown in this figure, although the residues are illustrated. Aromatic residues are colored magenta, blue and gold, for Tyr, Trp and Phe, respectively. Reproduced with permission from S. Buchanan et al., *Nature Struct. Biol.*, **1999**, *6*, 56-63.

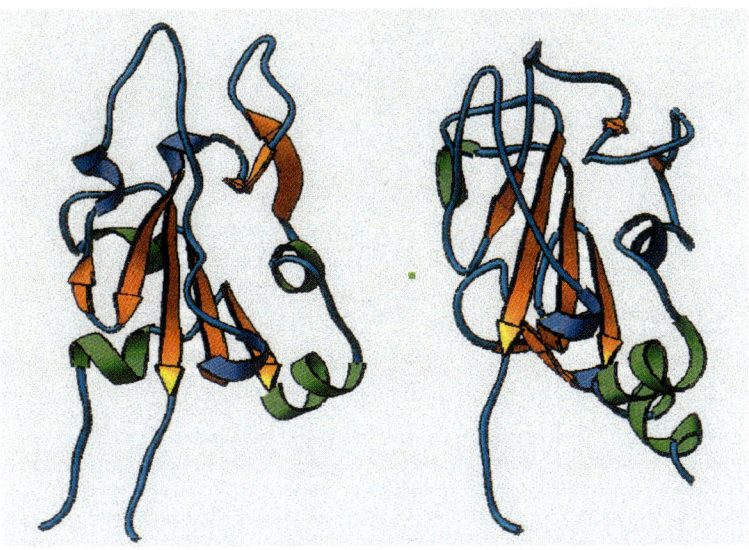

Chapter 11, Fig. 4. N-terminal domain of FepA (**A**) and FhuA (**B**). Short β-strands are indicated in brown color while the short α-helices are in green and blue color. Sensor loops are light blue in color at the top.

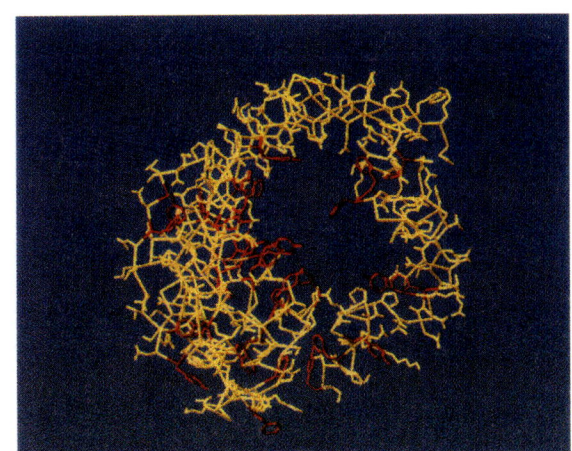

Chapter 11, Fig. 6. A view into the opening formed by the extracellular loops (yellow) in FhuA. This apex is lined by aromatic residues (red).

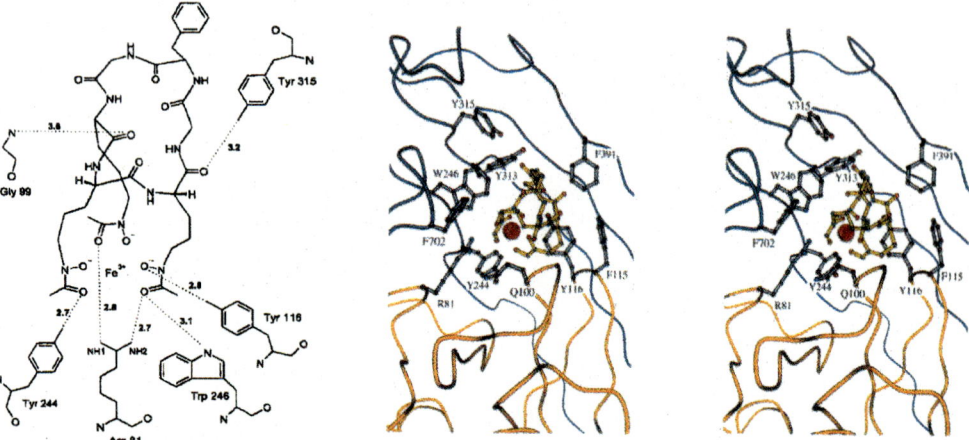

Chapter 11, Fig. 7. **A** Schematic representation of the hydrogen bonding pattern and electrostatic interaction of phenylferricrocin with FhuA side chain residues in the ligand binding site. Hydrogen bonds and charge interactions are indicated as dotted lines (distances given are in Å). **B** Stereoview of the ferrichrome-iron binding site in ribbon representation. The ferrichrome iron molecule is depicted as a green ball-and-stick model with red sphere as iron. Small red spheres are oxygen, small blue spheres are nitrogen atoms, and black spheres are carbon atoms of selected side chain residues of FhuA and small gray spheres are carbon atoms of the ferrichrome-iron molecule. The cork domain is shown yellow while barrel strands and loops are shown blue. Reproduced with permission from A. D. Ferguson et al., *Protein Sci.*, **2000**, *9*, 956-963 (**A**), and A. D. Ferguson et al., *Science*, **1998**, *282*, 2215-2220 (**B**).

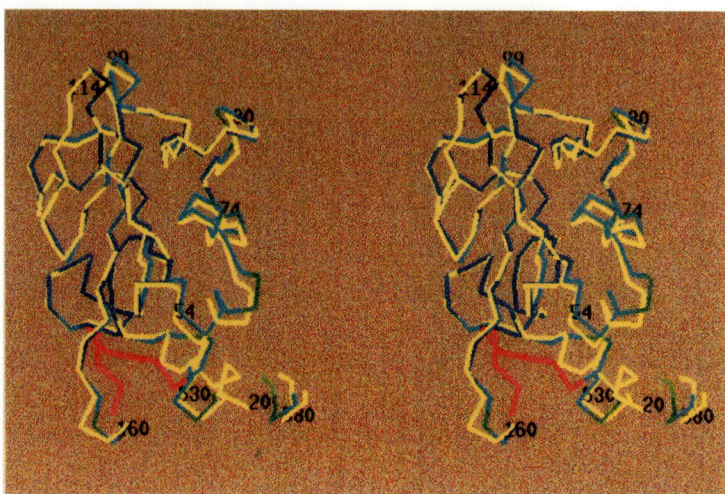

Chapter 11, Fig. 8. Structural changes in the N-terminal region of FhuA after ligand binding. Liganded FhuA is in blue (very little change) to light blue, green, and red indicating increasing translational or structural differences with unliganded FhuA (yellow). The $\alpha 1$ helix rests against the periplasmic loops P7 (530) and P8 (580). On ligand binding this helix unwinds (red).

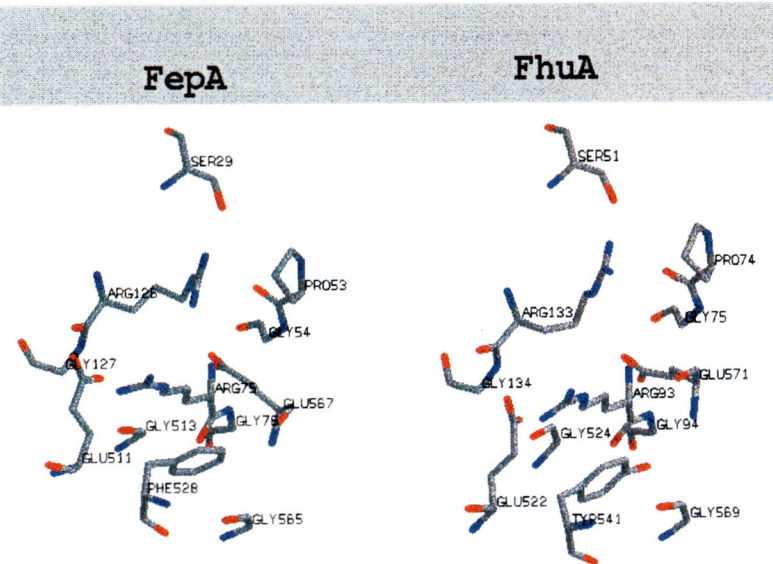

Chapter 11, Fig. 10. Cluster of 12 conserved residues in FepA and FhuA.

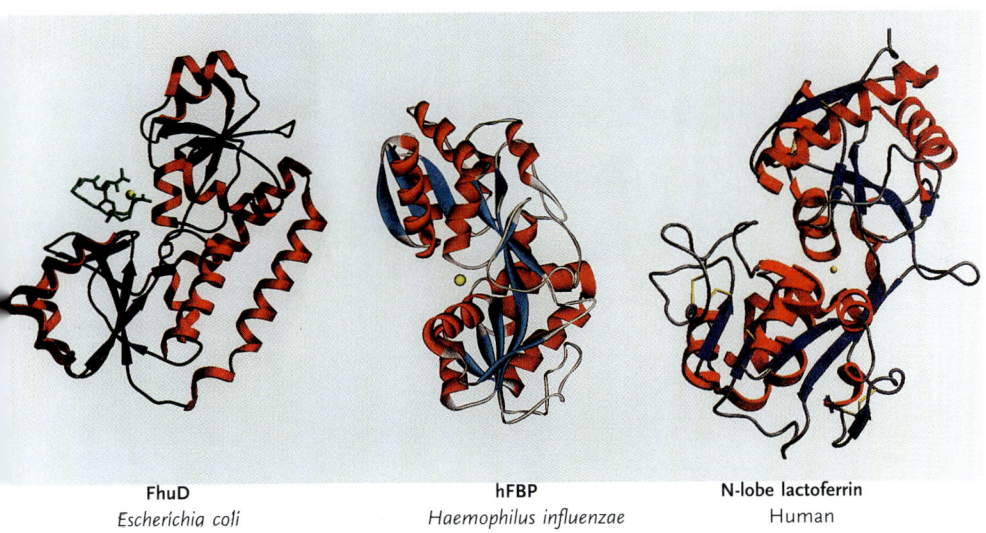

FhuD
Escherichia coli

hFBP
Haemophilus influenzae

N-lobe lactoferrin
Human

Chapter 12, Fig. 3. Comparison of the stereo-ribbon diagrams of the crystal structures of the *Escherichia coli* FhuD protein with bound gallichrome [56], the *Haemophilus influenzae* hFBP (HitA) protein with bound Fe^{3+} [58], and the N-lobe of human lactoferrin with bound Fe^{3+}.

1
Families of Transporters: A Phylogenetic Overview

Milton H. Saier, Jr.

1.1
Introduction

Over the past decade, work in our laboratory has led to the phylogenetic characterization of permease families [1–4]. This work has led us to formulate a novel classification system superficially similar to that implemented years ago for enzymes by the Enzyme Commission (EC). The Transporter Classification (TC) system has been reviewed and recommended for adoption by a panel of experts chaired by Prof. A. Kotyk of the International Union of Biochemistry and Molecular Biology. In contrast to the Enzyme Commission, which based its classification system solely on function, we have chosen to classify permeases on the basis of both function and phylogeny. In this chapter, I describe our proposal, point out some of its strengths, and emphasize its flexibility for the future inclusion of yet-to-be-discovered transporters. We hope that the TC classification system will prove to be as useful as is the EC system. Earlier treatises concerning the TC system [3–7] and transport protein evolution [8, 9] have appeared.

1.2
The TC System

A detailed description of the TC system can be found on our World Wide Web site (http://www-biology.ucsd.edu/~msaier/transport/). This site will be continuously updated as new relevant physiological, biochemical, genetic, biophysical and sequence data become available. Thanks to the participation of Can Tran, Yu-Feng Zhai, Nelson Yang and the Saier Lab bioinformatics group, the TC system will soon be available in database format. The system will provide a user-friendly search tool, so that the TC system can be readily accessed by keyword, TC#, gene name, protein name, sequence and sequence motif. These advances will render the TC system increasingly accessible to the entire scientific community worldwide.

In return, members of the scientific community are strongly encouraged to communicate novel findings and corrections to the author by e-mail, phone, fax, or snail mail.

1.3
The Value of Phylogenetic Classification

It is well recognized that any two proteins that can be shown to be homologous (i. e., that exhibit sufficient primary and/or secondary structural similarity to establish that they arose from a common evolutionary ancestor) will in general prove to exhibit strikingly similar three-dimensional structures, although a few exceptions have been noted [10]. Further, the degree of tertiary structural similarity correlates well with the degree of primary structural similarity. For this reason, phylogenetic analyses allow application of modeling techniques to a large number of related proteins and additionally allow reliable extrapolation from one protein member of a family of known structure to others of unknown structure. Thus, once three-dimensional structural data are available for any one family member, these data can be applied to all other members within limits dictated by their degrees of sequence similarity. The same can not be assumed for members of two independently evolving families, or for any two proteins for which common descent has not been established.

Similar arguments apply to mechanistic considerations. For example, the mechanism of solute transport is likely to be similar for all members of a permease family, and variations upon a specific mechanistic theme will be greatest when the sequence divergence is greatest. By contrast, for members of any two independently evolving permease families, the transport mechanisms may be strikingly different. Knowledge of these considerations allows unified mechanistic deductive approaches to be correctly applied to the largest numbers of transport systems, even when evidence is obtained piecemeal from the study of different systems.

The capacity to deduce and extrapolate structural and mechanistic information illustrates the value of phylogenetic data. However, another benefit that may result from the study of molecular phylogeny is to allow an understanding of the mechanistic restrictions that were imposed upon an evolving family due to architectural constraints. Specific architectural features may allow one family to diversify in function with respect to substrate specificity, substrate affinity, velocity of transport, polarity of transport, and even mechanism of energy coupling. By contrast, architectural constraints imposed on a second family may not allow functional diversification. Knowledge of the architectural constraints imposed on a permease family provides a clear clue as to the reliability of functional predictions for uncharacterized but related gene products revealed, e. g., by genome sequencing. Conversely, the functional diversity of the members of a permease family must be assumed to reflect architectural constraints, and thus phylogenetic/functional analyses lead to architectural predictions.

Finally, phylogenetic analyses provide valuable information about the evolutionary process itself. One can sometimes glean clues regarding the time of appearance of a family, the organismic type in which the family arose, and the pathway taken for the emergence of the family during evolutionary history. It is also sometimes possible to ascertain whether or not two distinct families arose independently of each other [1, 2].

1.4
Phylogeny as Applied to Transporters

Early studies revealed that transport proteins could be grouped into families based exclusively on the degrees of similarity observed for their amino acid sequences [10]. The significance of family assignment remained questionable until the study of internal gene duplications that had occurred during the evolution of some of these families established that these families had arisen independently of each other, at different times in evolutionary history, following different routes [1, 2]. In the next section we shall evaluate and utilize both function and molecular phylogeny for the purpose of conceptualizing transport protein characterization and classification.

1.5
The Basis for Classification in the TC System

According to the proposed classification system, now recommended by the transport nomenclature panel of the IUBMB, transporters are grouped on the basis of five criteria, and each of these criteria corresponds to one of the five entries within the Transporter Classification (TC) number for a particular permease. Thus, a permease-specific TC number has five components as follows: V.W.X.Y.Z. V corresponds to the transporter class while W corresponds to the subclass (see Tab. 1). X specifies the permease family (or superfamily) while Y represents the subfamily in a family (or the family in a superfamily) in which a particular permease is found. Finally, Z delineates the substrate or range of substrates transported as well as the polarity of transport (in or out). Any two transport proteins in the same subfamily of a permease family that transport the same substrate(s) using the same mechanism are given the same TC number, regardless of whether they are orthologs (i.e., arose in distinct organisms by speciation) or paralogs (i.e., arose within a single organism by gene duplication). Mode of regulation proves not to correlate with phylogeny and was probably superimposed on permeases late in the evolutionary process. Although regulation could conceivably be used as a final basis for classification, we have chosen not to do so.

1.6
Classes of Transporters

There are four recognized classes of transporters: (1) channels, (2) porters, (3) primary active transporters, and (4) group translocators (see Tab. 1). Sequenced homologs of unknown function or mechanism, and functionally characterized permeases for which sequence data are not available are included in a distinct class, class 9. Deficiencies in our knowledge will presumably be eliminated with time as more sequenced permeases become characterized biochemically and as sequences become available for the functionally but not molecularly characterized permeases. One additional class (class 8) is reserved for auxiliary transport proteins. It should be noted that each subclass of transporters has a two-digit

Tab. 1. Classes and subclasses of transporters according to the TC system[a]

1. Channels/pores
    1.A α-Type channels
    1.B β-Barrel porins
    1.C Pore-forming toxins (proteins and peptides)
    1.D Non-ribosomally synthesized channels

2. Electrochemical potential-driven transporters
    2.A Porters (uniporters, symporters, antiporters)
    2.B Non-ribosomally synthesized porters

3. Primary active transporters
    3.A P–P-bond hydrolysis-driven transporters
    3.B Decarboxylation-driven transporters
    3.C Methyltransfer-driven transporters
    3.D Oxidoreduction-driven transporters
    3.E Light absorption-driven transporters

4. Group translocators
    4.A Phosphotransfer-driven group translocators

8. Accessory factors involved in transport
    8.A Auxiliary transport proteins

9. Incompletely characterized transport systems
    9.A Recognized transporters of unknown biochemical mechanism
    9.B Putative but uncharacterized transport proteins
    9.C Functionally characterized transporters lacking identified sequences

[a]This system of classification was approved by the transporter nomenclature panel of the International Union of Biochemistry and Molecular Biology in Geneva, Nov. 28–30, 1999. No assignment has been made for categories 5–7. These will be reserved for novel types of transporters, yet to be discovered, that do not fall within categories 1–4.

TC# (V.W); each family has a three-digit TC# (V.W.X); each subfamily has a four-digit TC# (V.W.X.Y); and each permease-type has a five-digit TC# (V.W.X.Y.Z).).

The individual families of transporters included within each of the well characterized subclasses are listed in Tab. 2. Families within the large superfamilies are also specified [4].

Tab. 2. Families of transport proteins

1.A.	α-Type Channels	
	1.A.1	The Voltage-gated Ion Channel (VIC) Superfamily
	1.A.2	The Animal Inward Rectifier K^+ Channel (IRK-C) Family
	1.A.3	The Ryanodine-Inositol 1,4,5-triphosphate Receptor Ca^{2+} Channel (RIR-CaC) Family
	1.A.4	The Transient Receptor Potential Ca^{2+} Channel (TRP-CC) Family
	1.A.5	The Polycystin Cation Channel (PCC) Family
	1.A.6	The Epithelial Na^+ Channel (ENaC) Family
	1.A.7	ATP-gated Cation Channel (ACC) Family
	1.A.8	The Major Intrinsic Protein (MIP) Family
	1.A.9	The Ligand-gated Ion Channel (LIC) Family of Neurotransmitter Receptors
	1.A.10	The Glutamate-gated Ion Channel (GIC) Family of Neurotransmitter Receptors
	1.A.11	The Chloride Channel (ClC) Family
	1.A.12	The Organellar Chloride Channel (O-ClC) Family
	1.A.13	The Epithelial Chloride Channel (E-ClC) Family
	1.A.14	The Non-selective Cation Channel-1 (NSCC1) Family
	1.A.15	The Non-selective Cation Channel-2 (NSCC2) Family
	1.A.16	The Yeast Stretch-Activated, Cation-Selective, Ca^{2+} Channel, Mid1 (Mid1) Family
	1.A.17	The Chloroplast Outer Envelope Solute Channel (CSC) Family
	1.A.18	The Chloroplast Envelope Anion Channel-forming Tic110 (Tic110) Family
	1.A.19	The Influenza Virus Matrix-2 Channel (M2-C) Family
	1.A.20	The gp91^{phox} Phagocyte NADPH Oxidase-associated Cytochrome b$_{558}$ (CytB) H^+-channel Family
	1.A.21	The Bcl-2 (Bcl-2) Family
	1.A.22	The Large Conductance Mechano-sensitive Ion Channel (MscL) Family
	1.A.23	The Small Conductance Mechanosensitive Ion Channel (MscS) Family
	1.A.24	The Gap Junction-forming Connexin (Connexin) Family
	1.A.25	The Gap Junction-forming Innexin (Innexin) Family
	1.A.26	The Plant Plasmodesmata (PPD) Family
	1.A.27	The Phospholemman (PLM) Family
	1.A.28	The Urea Transporter (UT) Family
	1.A.29	The Urea/Amide Channel (UAC) Family
	1.A.30	The H^+- or Na^+-translocating Bacterial MotAB Flagellar Motor/ExbBD Outer Membrane Transport Energizer (Mot/Exb) Superfamily
	1.A.30.1	The H^+- or Na^+-translocating Bacterial Flagellar Motor (Mot) Family
	1.A.30.2	The TonB-ExbB-ExbD/TolA-TolQ-TolR (Exb) Family of Energizers for Outer Membrane Receptor (OMR)-Mediated Active Transport

Tab. 2. continued.

	1.A.31	The Annexin (Annexin) Family
	1.A.32	The Type B Influenza Virus NB Channel (NB-C) Family
	1.A.33	The Cation Channel-forming Heat Shock Protein-70 (Hsp70) Family
1.B.	**β-Barrel Porins**	
	1.B.1	The General Bacterial Porin (GBP) Family
	1.B.2	The Chlamydial Porin (CP) Family
	1.B.3	The Sugar Porin (SP) Family
	1.B.4	The *Brucella-Rhizobium* Porin (BRP) Family
	1.B.5	The *Pseudomonas* OprP Porin (POP) Family
	1.B.6	The OmpA-OmpF Porin (OOP) Family
	1.B.7	The *Rhodobacter* PorCa Porin (RPP) Family
	1.B.8	The Mitochondrial and Plastid Porin (MPP) Family
	1.B.9	The FadL Outer Membrane Protein (FadL) Family
	1.B.10	The Nucleoside-specific Channel-forming Outer Membrane Porin (Tsx) Family
	1.B.11	The Outer Membrane Fimbrial Usher Porin (FUP) Family
	1.B.12	The Autotransporter (AT) Family
	1.B.13	The Alginate Export Porin (AEP) Family
	1.B.14	The Outer Membrane Receptor (OMR) Family
	1.B.15	The Raffinose Porin (RafY) Family
	1.B.16	The Short Chain Amide and Urea Porin (SAP) Family
	1.B.17	The Outer Membrane Factor (OMF) Family
	1.B.18	The Outer Membrane Auxiliary (OMA) Protein Family
	1.B.19	The Glucose-selective OprB Porin (OprB) Family
	1.B.20	The Bacterial Toxin Export Channel (TEC) Family
	1.B.21	The OmpG Porin (OmpG) Family
	1.B.22	The Outer Bacterial Membrane Secretin (Secretin) Family
	1.B.23	The Cyanobacterial Porin (CBP) Family
	1.B.24	The Mycobacterial Porin (MBP) Family
	1.B.25	The Outermembrane Porin (Opr) Family
	1.B.26	The Cyclodextrin Porin (CDP) Family
	1.B.27	The *Helicobacter* Outer Membrane Porin (Hop) Family
	1.B.28	The Plastid Outer Envelope Porin of 24 kDa (OEP24) Family
	1.B.29	The Plastid Outer Envelope Porin of 21 kDa (OEP21) Family
	1.B.30	The Plastid Outer Envelope Porin of 16 kDa (OEP16) Family
	1.B.31	The *Campylobacter jejuni* Major Outer Membrane Porin (MomP) Family
	1.B.32	The Fusobacterial Outer Membrane Porin (FomP) Family
	1.B.33	The *Vibrio* Chitoporin/Neisserial Porin (VC/NP) Family
	1.B.34	The Corynebacterial Porin (PorA) Family
1.C.	**Pore-forming Toxins**	
	1.C.1	The Channel-forming Colicin (Colicin) Family
	1.C.2	The Channel-forming δ-Endotoxin Insecticidal Crystal Protein (ICP) Family
	1.C.3	The α-Hemolysin Channel-forming Toxin (αHL) Family
	1.C.4	The Aerolysin Channel-forming Toxin (Aerolysin) Family
	1.C.5	The Channel-forming ε-toxin (ε-toxin) Family

Tab. 2. continued.

1.C.6	The Yeast Killer Toxin K1 (YKT-K1) Family
1.C.7	The Diphtheria Toxin (DT) Family
1.C.8	The Botulinum and Tetanus Toxin (BTT) Family
1.C.9	The Vacuolating Cytotoxin (VacA) Family
1.C.10	The Pore-forming Haemolysin E (HlyE) Family
1.C.11	The Pore-forming RTX Toxin (RTX-toxin) Family
1.C.12	The Cholesterol-binding, Thiol-activated Cytolysin (TAC) Family
1.C.13	The Channel-forming Leukocidin Cytotoxin (Ctx) Family
1.C.14	The Cytohemolysin (CHL) Family
1.C.15	The Whipworm Stichosome Porin (WSP) Family
1.C.16	The Magainin (Magainin) Family
1.C.17	The Cecropin (Cecropin) Family
1.C.18	The Melittin (Melittin) Family
1.C.19	The Defensin (Defensin) Family
1.C.20	The Nisin (Nisin) Family
1.C.21	The Lacticin 481 (Lacticin 481) Family
1.C.22	The Lacticin 481 (Lacticin 481) Family
1.C.23	The Lactocin S (Lactocin S) Family
1.C.24	The Pediocin (Pediocin) Family
1.C.25	The Lactococcin G (Lactococcin G) Family
1.C.26	The Lactacin X (Lactacin X) Family
1.C.27	The Divergicin A (Divergicin A) Family
1.C.28	The AS-48 (AS-48) Family
1.C.29	The Plantaricin EF (Plantaricin EF) Family
1.C.30	The Plantaricin JK (Plantaricin JK) Family
1.C.31	The Channel-forming Colicin V (Colicin V) Family
1.C.32	The Amphipathic Peptide Mastoparan (Mastoparan) Family
1.C.33	The Cathilicidin (Cathilicidin) Family
1.C.34	The Tachyplesin (Tachyplesin) Family
1.C.35	The Amoebapore (Amoebapore) Family
1.C.36	The Bacterial Type III-Target Cell Pore (IIITCP) Family
1.C.37	The Lactococcin 972 (Lactococcin 972) Family
1.C.38	The Pore-forming Equinatoxin (Equinatoxin) Family
1.C.39	The Complement Protein C9 (CPC9) Family
1.C.40	The Bactericidal Permeability–Increasing Protein (BPIP) Family
1.C.41	The Tripartite Haemolysin BL (HBL) Family
1.C.42	The Channel-forming *Bacillus anthrax* Protective Antigen (BAPA) Family
1.C.43	The Earthworm Lysenin Toxin (Lysenin) Family
1.C.44	The Plant Thionine (PT) Family
1.C.45	The Plant Defensin (PD) Family
1.C.46	The C-type Natriuretic Peptide (CNP) Family
1.C.47	The Insect Defensin (Insect Defensin) Family
1.C.48	The Prion Peptide Fragment (PPF) Family
1.C.49	The Cytotoxic Amylin (Amylin) Family
1.C.50	The Amyloid β-Protein Peptide (AβPP) Family
1.C.51	The Pilosulin (Pilosulin) Family
1.C.52	The Dermaseptin (Dermaseptin) Family

Tab. 2. continued.

	1.C.53	The Bacteriocin AS-48 Cyclic Polypeptide (Bacteriocin AS-48) Family
	1.C.54	The Shiga Toxin B-Chain (ST-B) Family
	1.C.55	The Agrobacterial VirE2 Target Host Cell Membrane Anion Channel (VirE2) Family
	1.C.56	The *Pseudomanas syringae* HrpZ Target Host Cell Membrane Cation Channel (HrpZ) Family
	1.C.57	The Clostridial Cytotoxin (CCT) Family
1.D.	Nonribosomally-synthesized Channels	
	1.D.1	The Gramicidin A (Gramicidin A) Channel Family
	1.D.2	The Syringomycin Channel-forming (Syringomycin) Family
	1.D.3	The Syringomycin Channel-forming (Syringomycin) Family
	1.D.4	The Tolaasin Channel-forming (Tolaasin) Family
	1.D.5	The Alamethicin Channel-forming (Syringomycin) Family
	1.D.6	The Complexed Poly 3-Hydroxybutyrate Ca^{2+} Channel (cPHB-CC) Family
	1.D.7	The Beticolin (Beticolin) Family
	1.D.8	The Saponin (Saponin) Family
	1.D.9	The Polyglutamine Ion Channel (PG-IC) Family
	1.D.10	The Ceramide-forming Channel (Ceramide) Family
1.E.	Holins	
	1.E.1	The P21 Holin S (P21 Holin) Family
	1.E.2	The λ Holin S (λ Holin) Family
	1.E.3	The P2 Holin TM (P2 Holin) Family
	1.E.4	The LydA Holin (LydA Holin) Family
	1.E.5	The PRD1 Holin M (PRD1 Holin) Family
	1.E.6	The T7 Holin (T7 Holin) Family
	1.E.7	The HP1 Holin (HP1 Holin) Family
	1.E.8	The T4 Holin (T4 Holin) Family
	1.E.9	The T4 Immunity Holin (T4 Immunity Holin) Family
	1.E.10	The *Bacillus subtilis* φ29 Holin (φ29 Holin) Family
	1.E.11	The φ11 Holin (φ11 Holin) Family
	1.E.12	The φAdh Holin (φAdh Holin) Family
	1.E.13	The φU53 Holin (φU53 Holin) Family
	1.E.14	The LrgA Holin (LrgA Holin) Family
	1.E.15	The ArpQ Holin (ArpQ Holin) Family
	1.E.16	The Cph1 Holin (Cph1 Holin) Family
	1.E.17	The BlyA Holin (BlyA Holin) Family
	1.E.18	The *Lactococcus lactis* Phage r1t Holin (r1t holin) Family
2.A.	Porters (Uniporters, Symporters and Antiporters)	
	2.A.1	The Major Facilitator Superfamily (MFS)
	2.A.1.1	The Sugar Porter (SP) Family
	2.A.1.2	The Drug: H^+ Antiporter-1 (12 Spanner) (DHA1) Family
	2.A.1.3	The Drug: H^+ Antiporter-1 (14 Spanner) (DHA2) Family
	2.A.1.4	The Organophosphate: P_i Antiporter (OPA) Family
	2.A.1.5	The Oligosaccharide: H^+ Symporter (OHS) Family

Tab. 2. continued.

2.A.1.6	The Metabolite: H^+ Symporter (MHS) Family
2.A.1.7	The Fucose: H^+ Symporter (FHS) Family
2.A.1.8	The Nitrate/Nitrite Porter (NNP) family
2.A.1.9	The Phosphate: H^+ Symporter (PHS) Family
2.A.1.10	The Nucleoside: H^+ Symporter (NHS) Family
2.A.1.11	The Oxalate: Formate Antiporter (OFA) Family
2.A.1.12	The Sialate: H^+ Symporter (SHS) Family
2.A.1.13	The Monocarboxylate Porter (MCP) Family
2.A.1.14	The Anion: Cation Symporter (ACS) Family
2.A.1.15	Aromatic Acid: H^+ Symporter (AAHS) Family
2.A.1.16	The Siderophore-Iron Transporter (SIT) Family
2.A.1.17	The Cyanate Porter (CP) Family
2.A.1.18	The Polyol Porter (PP) Family
2.A.1.19	The Organic Cation Transporter (OCT) Family
2.A.1.20	The Sugar Efflux Transporter (SET) Family
2.A.1.21	The Drug: H^+ Antiporter-3 (12 Spanner) (DHA3) Family
2.A.1.22	The Vesicular Neurotransmitter Transporter (VNT) Family
2.A.1.23	The Conjugated Bile Salt Transporter (BST) Family
2.A.1.24	The Unknown Major Facilitator-1 (UMF1) Family
2.A.1.25	The Peptide-Acetyl-Coenzyme A Transporter (PAT) Family
2.A.1.26	The Unknown Major Facilitator-2 (UMF2) Family
2.A.1.27	The Phenyl Propionate Permease (PPP) Family
2.A.1.28	The Unknown Major Facilitator-3 (UMF3) Family
2.A.1.29	The Unknown Major Facilitator-4 (UMF4) Family
2.A.1.30	The Putative Abietane Diterpenoid Transporter (ADT) Family
2.A.1.31	The Nickel Resistance (Nre) Family
2.A.1.32	The Putative Aromatic Compound/Drug Exporter (ACDE) Family
2.A.1.33	The Putative YqgE Transporter (YqgE) Family
2.A.1.34	The Feline Leukemia Virus Subgroup C Receptor (FLVCR) Family
2.A.2	The Glycoside-Pentoside-Hexuronide (GPH): Cation Symporter Family
2.A.3	The Amino Acid-Polyamine-Organocation (APC) Family
2.A.3.1	The Amino Acid Transporter (AAT) Family
2.A.3.2	The Basic Amino Acid/Polyamine Antiporter (APA) Family
2.A.3.3	The Cationic Amino Acid Transporter (CAT) Family
2.A.3.4	The Amino Acid/Choline Transporter (ACT) Family
2.A.3.5	The Ethanolamine Transporter (EAT) Family
2.A.3.6	The Archaeal/Bacterial Transporter (ABT) Family
2.A.3.7	The Glutamate:GABA Antiporter (GGA) Family
2.A.3.8	The L-type Amino Acid Transporter (LAT) Family
2.A.3.9	The Spore Germination Protein (SGP) Family
2.A.3.10	The Yeast Amino Acid Transporter (YAT) Family
2.A.4	The Cation Diffusion Facilitator (CDF) Family
2.A.5	The Zinc (Zn^{2+})-Iron (Fe^{2+}) Permease (ZIP) Family
2.A.6	The Resistance-Nodulation-Cell Division (RND) Superfamily
2.A.6.1	The Heavy Metal Efflux (HME) Family
2.A.6.2	The (Largely Gram-negative Bacterial) Hydrophobe/Amphiphile Efflux-1 (HAE1) Family
2.A.6.3	The Putative Nodulation Factor Exporter (NFE) Family

Tab. 2. continued.

	2.A.6.4	The SecDF (SecDF) Family
	2.A.6.5	The (Gram-positive Bacterial Putative) Hydrophobe/Amphiphile Efflux-2 (HAE2) Family
	2.A.6.6	The Eukaryotic (Putative) Sterol Transporter (EST) Family
	2.A.6.7	The (Largely Archaeal Putative) Hydrophobe/Amphiphile Efflux-3 (HAE3) Family
2.A.7		The Drug/Metabolite Transporter (DMT) Superfamily
	2.A.7.1	The 4 TMS Small Multidrug Resistance (SMR) Family
	2.A.7.2	The 5 TMS Bacterial/Archaeal Transporter (BAT) Family
	2.A.7.3	The 10 TMS Drug/Metabolite Exporter (DME) Family
	2.A.7.4	The Plant Drug/Metabolite Exporter (P-DME) Family
	2.A.7.5	The Glucose/Ribose Porter (GRP) Family
	2.A.7.6	The L-Rhamnose Transporter (RhaT) Family
	2.A.7.7	The Chloramphenicol-Sensitivity Protein (RarD) Family
	2.A.7.8	The *Caenorhabditis elegans* ORF (CEO) Family
	2.A.7.9	The Triose-phosphate Transporter (TPT) Family
	2.A.7.10	The UDP-N-Acetylglucosamine: UMP Antiporter (UAA) Family
	2.A.7.11	The UDP-Galactose: UMP Antiporter (UGA) Family
	2.A.7.12	The CMP-Sialate: CMP Antiporter (CSA) Family
	2.A.7.13	The GDP-Mannose: GMP Antiporter (GMA) Family
	2.A.7.14	The Plant Organocation Permease (POP) Family
2.A.8		The Gluconate: H^+ Symporter (GntP) Family
2.A.9		
2.A.10		The 2-Keto-3-Deoxygluconate Transporter (KDGT) Family
2.A.11		The Citrate-Mg^{2+}: H^+ (CitM) – Citrate-Ca^{2+}: H^+ (CitH) Symporter (CitMHS) Family
2.A.12		The ATP: ADP Antiporter (AAA) Family
2.A.13		The C_4-Dicarboxylate Uptake (Dcu) Family
2.A.14		The Lactate Permease (LctP) Family
2.A.15		The Betaine/Carnitine/Choline Transporter (BCCT) Family
2.A.16		The Tellurite-resistance/Dicarboxylate Transporter (TDT) Family
2.A.17		The Proton-dependent Oligopeptide Transporter (POT) Family
2.A.18		The Amino Acid/Auxin Permease (AAAP) Family
2.A.19		The Ca^{2+}: Cation Antiporter (CaCA) Family
2.A.20		The Inorganic Phosphate Transporter (PiT) Family
2.A.21		The Solute: Sodium Symporter (SSS) Family
2.A.22		The Neurotransmitter: Sodium Symporter (NSS) Family
2.A.23		The Dicarboxylate/Amino Acid: Cation (Na^+ or H^+) Symporter (DAACS) Family
2.A.24		The Citrate: Cation Symporter (CCS) Family
2.A.25		The Alanine or Glycine: Cation Symporter (AGCS) Family
2.A.26		The Branched Chain Amino Acid: Cation Symporter (LIVCS) Family
2.A.27		The Glutamate: Na^+ Symporter (ESS) Family
2.A.28		The Bile Acid: Na^+ Symporter (BASS) Family
2.A.29		The Mitochondrial Carrier (MC) Family
2.A.30		The Cation-Chloride Cotransporter (CCC) Family
2.A.31		The Anion Exchanger (AE) Family
2.A.32		The Silicon Transporter (Sit) Family

Tab. 2. continued.

2.A.33	The NhaA Na$^+$: H$^+$ Antiporter (NhaA) Family
2.A.34	The NhaB Na$^+$: H$^+$ Antiporter (NhaB) Family
2.A.35	The NhaC Na$^+$: H$^+$ Antiporter (NhaC) Family
2.A.36	The Monovalent Cation: Proton Antiporter-1 (CPA1) Family
2.A.37	The Monovalent Cation: Proton Antiporter-2 (CPA2) Family
2.A.38	The K$^+$ Transporter (Trk) Family
2.A.39	The Nucleobase: Cation Symporter-1 (NCS1) Family
2.A.40	The Nucleobase: Cation Symporter-2 (NCS2) Family
2.A.41	The Concentrative Nucleoside Transporter (CNT) Family
2.A.42	The Hydroxy/Aromatic Amino Acid Permease (HAAAP) Family
2.A.43	The Lysosomal Cystine Transporter (LCT) Family
2.A.44	The Formate-Nitrite Transporter (FNT) Family
2.A.45	The Arsenite-Antimonite (ArsB) Efflux Family
2.A.46	The Benzoate: H$^+$ Symporter (BenE) Family
2.A.47	The Divalent Anion: Na$^+$ Symporter (DASS) Family
2.A.48	The Reduced Folate Carrier (RFC) Family
2.A.49	The Ammonium Transporter (Amt) Family
2.A.50	The Glycerol Uptake (GUP) Family
2.A.51	The Chromate Ion Transporter (CHR) Family
2.A.52	The Ni^{2+}-Co^{2+} Transporter (NiCoT) Family
2.A.53	The Sulfate Permease (SulP) Family
2.A.54	The Mitochondrial Tricarboxylate Carrier (MTC) Family
2.A.55	The Metal Ion (Mn^{2+}-iron) Transporter (Nramp) Family
2.A.56	The Tripartite ATP-independent Periplasmic Transporter (TRAP-T) Family
2.A.57	The Equilibrative Nucleoside Transporter (ENT) Family
2.A.58	The Phosphate: Na$^+$ Symporter (PNaS) Family
2.A.59	The Arsenical Resistance-3 (ACR3) Family
2.A.60	The Organo Anion Transporter (OAT) Family
2.A.61	The C$_4$-dicarboxylate Uptake C (DcuC) Family
2.A.62	The NhaD Na$^+$: H$^+$ Antiporter (NhaD) Family
2.A.63	The Monovalent Cation (K$^+$ or Na$^+$): Proton Antiporter-3 (CPA3) Family
2.A.64	The Twin Arginine Targeting (Type V Secretory Pathway) (Tat) Family
2.A.65	The Bilirubin Transporter (BRT) Family
2.A.66	The Multi Antimicrobial Extrusion (MATE) Family
2.A.67	The Oligopeptide Transporter (OPT) Family
2.A.68	The p-Aminobenzoyl-glutamate Transporter (AbgT) Family
2.A.69	The Auxin Efflux Carrier (AEC) Family
2.A.70	The Malonate: Na$^+$ Symporter (MSS) Family
2.A.71	The Folate-Biopterin Transporter (FBT) Family
2.A.72	The K$^+$ Uptake Permease (KUP) Family
2.A.73	The Inorganic Carbon (HCO$_3^-$) Transporter (ICT) Family
2.A.74	The 4 TMS Multidrug Endosomal Transporter (MET) Family
2.A.75	The L-Lysine Exporter (LysE) Family
2.A.76	The Resistance to Homoserine/Threonine (RhtB) Family
2.A.77	The Cadmium Resistance (CadD) Family
2.A.78	The Acetate Porter (AceP) Family
2.A.79	The Threonine/Serine Exporter (ThrE) Family

Tab. 2. continued.

2.B	**Non-ribosomally Synthesized Porters**	
	2.B.1	The Valinomycin Carrier (Valinomycin) Family
	2.B.2	The Monensin (Monensin) Family
	2.B.3	The Nigericin (Nigericin) Family
	2.B.4	The Macrotetrolide Antibiotic (MA) Family
	2.B.5	The Macrocyclic Polyether (MP) Family
	2.B.6	The Ionomycin (Ionomycin) Family
3.A	**P-P-bond-hydrolysis-driven transporters**	
	3.A.1	The ATP-binding Cassette (ABC) Superfamily

ABC-type Uptake Permeases (All from Prokaryotes (Bacteria and Archaea))

- 3.A.1.1 — The Carbohydrate Uptake Transporter-1 (CUT1) Family
- 3.A.1.2 — The Carbohydrate Uptake Transporter-2 (CUT2) Family
- 3.A.1.3 — The Polar Amino Acid Uptake Transporter (PAAT) Family
- 3.A.1.4 — The Hydrophobic Amino Acid Uptake Transporter (HAAT) Family
- 3.A.1.5 — The Peptide/Opine/Nickel Uptake Transporter (PepT) Family
- 3.A.1.6 — The Sulfate Uptake Transporter (SulT) Family
- 3.A.1.7 — The Phosphate Uptake Transporter (PhoT) Family
- 3.A.1.8 — The Molybdate Uptake Transporter (MolT) Family
- 3.A.1.9 — The Phosphonate Uptake Transporter (PhnT) Family
- 3.A.1.10 — The Ferric Iron Uptake Transporter (FeT) Family
- 3.A.1.11 — The Polyamine/Opine/Phosphonate Uptake Transporter (POPT) Family
- 3.A.1.12 — The Quaternary Amine Uptake Transporter (QAT) Family
- 3.A.1.13 — The Vitamin B_{12} Uptake Transporter ($VB_{12}T$) Family
- 3.A.1.14 — The Iron Chelate Uptake Transporter (FeCT) Family
- 3.A.1.15 — The Manganese/Zinc/Iron Chelate Uptake Transporter (MZT) Family
- 3.A.1.16 — The Nitrate/Nitrite/Cyanate Uptake Transporter (NitT) Family
- 3.A.1.17 — The Taurine Uptake Transporter (TauT) Family
- 3.A.1.18 — The Putative Cobalt Uptake Transporter (CoT) Family
- 3.A.1.19 — The Thiamin Uptake Transporter (ThiT) Family
- 3.A.1.20 — The *Brachyspira* Iron Transporter (BIT) Family
- 3.A.1.21 — The Yersiniabactin Fe^{3+} Uptake Transporter (YbtPQ) Family
- 3.A.1.22 — The Nickel Uptake Transporter (NiT) Family

ABC-type Efflux Permeases (Bacterial)

- 3.A.1.101 — The Capsular Polysaccharide Exporter (CPSE) Family
- 3.A.1.102 — The Lipooligosaccharide Exporter (LOSE) Family
- 3.A.1.103 — The Lipopolysaccharide Exporter (LPSE) Family
- 3.A.1.104 — The Teichoic Acid Exporter (TAE) Family
- 3.A.1.105 — The Drug Exporter (DrugE1) Family
- 3.A.1.106 — The Lipid Exporter (LipidE) Family
- 3.A.1.107 — The Putative Heme Exporter (HemeE) Family
- 3.A.1.108 — The β-Glucan Exporter (GlucanE) Family
- 3.A.1.109 — The Protein-1 Exporter (Prot1E) Family
- 3.A.1.110 — The Protein-2 Exporter (Prot2E) Family
- 3.A.1.111 — The Peptide-1 Exporter (Pep1E) Family
- 3.A.1.112 — The Peptide-2 Exporter (Pep2E) Family
- 3.A.1.113 — The Peptide-3 Exporter (Pep3E) Family

Tab. 2. continued.

	3.A.1.114	The Probable Glycolipid Exporter (DevE) Family
	3.A.1.115	The Na$^+$ Exporter (NatE) Family
	3.A.1.116	The Microcin B17 Exporter (McbE) Family
	3.A.1.117	The Drug Exporter-2 (DrugE2) Family
	3.A.1.118	The Microcin J25 Exporter (McjD) Family
	3.A.1.119	The Drug/Siderophore Exporter-3 (DrugE3) Family
	3.A.1.120	The Drug Resistance ATPase-1 (Drug RA1) Family
	3.A.1.121	The Drug Resistance ATPase-2 (Drug RA2) Family

ABC-type Efflux Permeases (Mostly Eukaryotic)

- 3.A.1.201 The Multidrug Resistance Exporter (MDR) Family
- 3.A.1.202 The Cystic Fibrosis Transmembrane Conductance Exporter (CFTR) Family
- 3.A.1.203 The Peroxysomal Fatty Acyl CoA Transporter (FAT) Family
- 3.A.1.204 The Eye Pigment Precursor Transporter (EPP) Family
- 3.A.1.205 The Pleiotropic Drug Resistance (PDR) Family
- 3.A.1.206 The a-Factor Sex Pheromone Exporter (STE) Family
- 3.A.1.207 The Conjugate Transporter-1 (CT1) Family
- 3.A.1.208 The Conjugate Transporter-2 (CT2) Family
- 3.A.1.209 The MHC Peptide Transporter (TAP) Family
- 3.A.1.210 The Heavy Metal Transporter (HMT) Family
- 3.A.1.211 The Cholesterol/Phospholipid/Retinal (CPR) Flippase Family
- 3.A.1.212 The Mitochondrial Peptide Exporter (MPE) Family
- 3.A.2   The H$^+$- or Na$^+$-translocating F-type, V-type and A-type ATPase (F-ATPase) Superfamily
- 3.A.3   The P-type ATPase (P-ATPase) Superfamily
- 3.A.4   The Arsenite-Antimonite (ArsAB) Efflux Family
- 3.A.5   The Type II (General) Secretory Pathway (IISP) Family
- 3.A.6   The Type III (Virulence-related) Secretory Pathway (IIISP) Family
- 3.A.7   The Type IV (Conjugal DNA-Protein Transfer or VirB) Secretory Pathway (IVSP) Family
- 3.A.8   The Mitochondrial Protein Translocase (MPT) Family
- 3.A.9   The Chloroplast Envelope Protein Translocase (CEPT or Tic-Toc) Family
- 3.A.10   The H$^+$-translocating Pyrophosphatase (H$^+$-PPase) Family
- 3.A.11   The Bacterial Competence-related DNA Transformation Transporter (DNA-T) Family
- 3.A.12   The Septal DNA Translocator (S-DNA-T) Family
- 3.A.13   The Filamentous Phage Exporter (FPhE) Family
- 3.A.14   The Fimbrilin/Protein Exporter (FPE) Family

3.B. Decarboxylation-driven Active Transporters
- 3.B.1   The Na$^+$-transporting Carboxylic Acid Decarboxylase (NaT-DC) Family

3.C. Methyl-transfer-driven Transporters
- 3.C.1   The Na$^+$-transporting Methyltetrahydromethanopterin: Coenzyme M Methyltransferase (NaT-MMM) Family

Tab. 2. continued.

3.D. Oxidoreduction-driven Active Transporters
- 3.D.1 The Proton-translocating NADH Dehydrogenase (NDH) Family
- 3.D.2 The Proton-translocating Transhydrogenase (PTH) Family
- 3.D.3 The Proton-translocating Quinol: Cytochrome c Reductase (QCR) Superfamily
- 3.D.4 The Proton-translocating Cytochrome Oxidase (COX) Superfamily
- 3.D.5 The Na^+-translocating NADH: Quinone Dehydrogenase (Na-NDH) Family
- 3.D.6 The Putative Ion (H^+ or Na^+)-translocating NADH: Ferredoxin Oxidoreductase (NFO) Family
- 3.D.7 The H_2: Heterodisulfide Oxidoreductase (HHO) Family
- 3.D.8 The Na^+- or H^+-Pumping Formyl Methanofuran Dehydrogenase (FMF-DH) Family
- 3.D.9 The H^+-translocating $F_{420}H_2$ Dehydrogenase ($F_{420}H_2$DH) Family

3.E. Light-driven Active Transporters
- 3.E.1 The Ion-translocating Microbial Rhodopsin (MR) Family
- 3.E.2 The Photosynthetic Reaction Center (PRC) Family

4.A Phosphoryl-transfer-driven Group Translocators
- 4.A.1 The PTS Glucose-Glucoside (Glc) Family
- 4.A.2 The PTS Fructose-Mannitol (Fru) Family
- 4.A.3 The PTS Lactose-N,N'-Diacetylchitobiose-β-glucoside (Lac) Family
- 4.A.4 The PTS Glucitol (Gut) Family
- 4.A.5 The PTS Galactitol (Gat) Family
- 4.A.6 The PTS Mannose-Fructose-Sorbose (Man) Family

5.A. Transmembrane Electron Transfer Carriers
- 5.A.1 The Disulfide Bond Oxidoreductase D (DsbD) Family
- 5.A.2 The Disulfide Bond Oxidoreductase B (DsbB) Family

8.A. Auxiliary Transport Proteins
- 8.A.1 The Membrane Fusion Protein (MFP) Family
- 8.A.2 The Secretin Auxiliary Lipoprotein (SAL) Family
- 8.A.3 The Cytoplasmic Membrane-Periplasmic Auxiliary-1 (MPA1) Protein with Cytoplasmic (C) Domain (MPA1-C or MPA1+C) Family
- 8.A.4 The Cytoplasmic Membrane-Periplasmic Auxiliary-2 (MPA2) Family
- 8.A.5 The Voltage-gated K^+ Channel β-subunit (VICβ) Family
- 8.A.6 The Auxiliary Nutrient Transporter (ANT) Family
- 8.A.7 The Phosphotransferase System Enzyme I (EI) Family
- 8.A.8 The Phosphotransferase System HPr (HPr) Family
- 8.A.9 The rBAT Transport Accessory Protein (rBAT) Family
- 8.A.10 The Slow Voltage-gated K^+ Channel Accessory Protein (MinK) Family
- 8.A.11 The Phospholamban (Ca^{2+}-ATPase Regulator) (PLB) Family
- 8.A.12 ABC Bacteriocin Exporter Accessory Protein (BEA) Family
- 8.A.13 The Tetratricopeptide Repeat (Tpr1) Family

Tab. 2. continued.

9.A. Transporters of Unknown Classification

9.A.1	The Polysaccharide Transport (PST) Family
9.A.2	The MerTP Mercuric Ion (Hg^{2+}) Permease (MerTP) Family
9.A.3	The MerC Mercuric Ion (Hg^{2+}) Uptake (MerC) Family
9.A.4	The Nicotinamide Mononucleotide (NMN) Uptake Permease (PnuC) Family
9.A.5	The Cytochrome Oxidase Biogenesis (Oxa1) Family
9.A.6	The Intracellular Nucleoside Transporter (INT) Family
9.A.7	The MerF Mercuric Ion (Hg^{2+}) Uptake (MerF) Family
9.A.8	The Ferrous Iron Uptake (FeoB) Family
9.A.9	The Low Affinity Fe^{2+} Transporter (FeT) Family
9.A.10	The Oxidase-dependent Fe^{2+} Transporter (OFeT) Family
9.A.11	The Copper Transporter-1 (Ctr1) Family
9.A.12	The Copper Transporter-2 (Ctr2) Family
9.A.13	The Short Chain Fatty Acid Transporter (scFAT) Family
9.A.14	The Nuclear Pore Complex Family
9.A.15	The YhaG Putative Tryptophan Uptake Permease (YhaG) Family
9.A.16	
9.A.17	The Metal Ion Transporter (MIT) Family
9.A.18	The Peptide Uptake Permease (PUP) Family
9.A.19	The Mg^{2+} Transporter-E (MgtE) Family
9.A.20	The Low Affinity Cation Transporter (LCT) Family
9.A.21	The ComC DNA Uptake Competence (ComC) Family
9.A.22	The NhaE $Na^+(K^+)$: H^+ Antiporter (NhaE) Family
9.A.23	The Ferroportin (FP) Family

9.B. Putative Uncharacterized Transporters

9.B.1	The Metal Homeostasis Protein (MHP) Family
9.B.2	The Ca^{2+} Homeostasis Protein (CHP) Family
9.B.3	The Putative Bacterial Murein Precursor Exporter (MPE) Family
9.B.4	The Putative Efflux Transporter (PET) Family
9.B.5	The KX Blood-group Antigen (KXA) Family
9.B.6	The Toxic Hok/Gef Protein (Hok/Gef) Family
9.B.7	The Putative Bacteriochlorophyll Delivery (BCD) Family
9.B.8	The Canalicular Bile Acid Transporter (C-BAT) Family
9.B.9	The Urate Transporter (UAT) Family
9.B.10	The 6TMS Putative MarC Transporter (MarC) Family
9.B.11	The Mitochondrial mRNA Splicing-2 Protein (MRS2) Family
9.B.12	The (Salt or Low Temperature) Stress-induced Hydrophobic Peptide (SHP) Family
9.B.13	The Putative Pore-forming Entericidin (ECN) Family
9.B.14	The Putative Heme Exporter Protein (HEP) Family
9.B.15	The Putative Chloroquine Resistance Na^+/H^+ Exchanger of *Plasmodium falciparum* (CQR) Family
9.B.16	The Putative Ductin Channel (Ductin) Family
9.B.17	The Putative Fatty Acid Transporter (FAT) Family
9.B.18	The SecDF-associated Single Transmembrane Protein (SSTP) Family
9.B.19	The Mn^{2+} Homeostasis Protein (MnHP) Family

Tab. 2. continued.

	9.B.20	The Putative Mg^{2+} Transporter-C (MgtC) Family
	9.B.21	The Frataxin (Frataxin) Family
	9.B.22	The Putative Permease (PerM) Family
	9.B.23	The Digestive Vacuole Transporter (DVT) Family
	9.B.24	The Testis-Enhanced Gene Transfer (TEGT) Family
	9.B.25	The YbbM (YbbM) Family
	9.B.26	The PF27 (PF27) Family
	9.B.27	The YdjX-Z (YdjX-Z) Family
	9.B.28	The YqaE (YqaE) Family
	9.B.29	The YebN (YebN) Family
	9.B.30	The Hly III (Hly III) Family
	9.B.31	The YqiH (YqiH) Family
	9.B.32	The Putative Vectorial Glycosyl Polymerization (VGP) Family
	9.B.33	The YaaH (YaaH) Family
	9.B.34	The Putative Membrane Peptide Cation Channel (PMP3) Family
	9.B.35	The Putative Thyronine-Transporting Transthyretin (Transthyretin) Family
	9.B.36	The Putative SgaT Transporter (SgaT) Family
	9.B.37	The HlyC/CorC (HCC) Family
9.C.		**Functionally Characterized Transporters With Unidentified Sequences**
	9.C.1	The Endosomal Oligosaccharide Transporter (EOT)
	9.C.2	Volume-sensitive Anion Channels (VAC)
	9.C.3	The *Rhodococcus erythropolis* Porin (REP) Family
	9.C.4	Nucleotide Sulfate (PAPS) Transporters (PAPS-T)
	9.C.5	The Endoplasmic Reticulum/Golgi ATP/ADP or AMP Antiport Transporters (ATP-T)
	9.C.6	T7 Phage DNA Uptake Tranlocator (T7-T)
	9.C.7	The Peroxisomal Protein Importers (PPI)

1.7
Class 1: Channels/Pores

1.A. *α-Type channels*. Transmembrane channel proteins of this class are ubiquitously found in the membranes of all types of organisms from bacteria to higher eukaryotes. Channel proteins usually catalyze movement of solutes by an energy-independent process, by passage through a transmembrane aqueous pore without evidence for a carrier-mediated mechanism. These channel proteins consist largely of α-helical spanners although β-strands may be present and may even contribute to the channel. Outer membrane porin-type channel proteins are excluded from this class and are instead included in class 1.B.

1.B. *β-Barrel porins*. These proteins form transmembrane pores that usually allow the energy-independent passage of solutes across a membrane. The transmembrane portions of these proteins consist exclusively of β-strands that usually form β-barrels. Porin-type proteins are found in the outer membranes of gram-negative bacteria, mitochondria, plant and algal plastids, and possibly acid-fast gram-positive bacteria.

1.C. *Pore-forming toxins*. These proteins/peptides are synthesized by one cell and secreted for insertion into the membrane of another cell where they form transmembrane pores. They may exert their toxic effects by allowing the free flow of electrolytes and other small molecules across the membrane, or they may allow entry into the target cell cytoplasm of a toxin protein that ultimately kills or controls the cell. Both large proteins and small ribosomally synthesized peptides are included in this category.

1.D. *Non-ribosomally synthesized channels*. These molecules, often chains of L- and D-amino acids as well as other small molecular building blocks such as hydroxy acids (i. e., lactate, β-hydroxybutyrate), form oligomeric transmembrane ion channels. Voltage may induce channel formation by promoting assembly of the oligomeric transmembrane pore-forming structure. These "depsipeptides" are often made by bacteria and fungi as agents of biological warfare. Other substances, completely lacking amino acids but capable of channel formation are also included in subclass 1.D.

1.8
Class 2: Electrochemical Potential-driven Porters

2.A. *Porters (uniporters, symporters, antiporters)*. Transport systems are included in this subclass if they utilize a carrier-mediated process to catalyze uniport (a single species is transported either by facilitated diffusion or in a membrane potential-dependent process if the solute is charged), antiport (two or more species are transported in opposite directions in a tightly coupled process involving only chemiosmotic energy), and/or symport (two or more species are transported together in the same direction in a tightly coupled process involving only chemiosmotic energy).

2.B. *Non-ribosomally synthesized porters.* These substances, like non-ribosomally synthesized channels, may be depsipeptides or non-peptide-like substances. Such a porter complexes a solute such as a cation in its hydrophilic interior and facilitates translocation of the complex across the membrane by exposing its hydrophobic exterior and moving from one side of the bilayer to the other. If the free porter can cross the membrane in the uncomplexed form, the transport process can be electrophoretic (the charged molecule moves down its electrochemical gradient), but if only the complex can cross the membrane, transport must be electroneutral. In electroneutral antiport, one charged substrate is exchanged for another.

1.9
Class 3: Primary Active Transporters

Primary active transporters use a "primary" source of energy to drive active transport of a solute against a concentration gradient. A "secondary" ion gradient is not considered a primary energy source because it is created by the expenditure of a primary energy source. Primary energy sources known to be coupled to transport are chemical, electrical, and solar.

3.A. *P–P-bond hydrolysis-driven transporters.* Transport systems are included in this subclass if they hydrolyze the diphosphate bond of inorganic pyrophosphate, ADP, ATP, or another nucleoside triphosphate, to drive the active uptake and/or extrusion of a solute or solutes. The transport protein may or may not be transiently phosphorylated, but the substrate is not phosphorylated. These transporters are found universally in all living organisms.

3.B. *Decarboxylation-driven transporters.* Transport systems that drive solute (e. g., ion) uptake or extrusion by decarboxylation of a cytoplasmic substrate are included in this subclass. These transporters are currently thought to be restricted to prokaryotes.

3.C. *Methyltransfer-driven transporters.* A single characterized multisubunit protein family currently falls into this subclass, the Na^+-transporting methyltetrahydromethanopterin:coenzyme M methyltransferase. These transporter complexes are currently thought to be restricted to archaea.

3.D. *Oxidoreduction-driven transporters.* Transport systems that drive transport of a solute (e. g., an ion) energized by the exothermic flow of electrons from a reduced substrate to an oxidized substrate are included in this subclass. These transporters are universal although some families are restricted to one domain or another.

3.E. *Light absorption-driven transporters.* Transport systems that utilize light energy to drive transport of a solute (e. g., an ion) are included in this subclass. One family (microbial rhodopsin) is found in archaea, bacteria and fungi, but the other (photosynthetic reaction center) is found only in bacteria and chloroplasts of eukaryotes.

1.10
Class 4: Group Translocators

4.A. *Phosphotransfer-driven group translocators.* Transport systems of the bacterial phosphoenolpyruvate:sugar phosphotransferase system (PTS) are included in this class. The product of the reaction, derived from extracellular sugar, is a cytoplasmic sugar phosphate. No porters of the PTS have been identified in the archaeal or eukaryotic domain.

1.11
Class 8: Accessory Factors Involved in Transport

8.A. *Auxiliary transport proteins.* Proteins that in some way facilitate transport across one or more biological membranes but do not themselves participate directly in transport are included in this class. These proteins always function in conjunction with one or more established transport systems. They may provide a function connected with energy coupling to transport, play a structural role in complex formation, serve a biogenic or stability function, or function in regulation.

1.12
Class 9: Incompletely Characterized Transport Proteins

9.A. *Transporters of unknown biochemical mechanism.* Transport protein families of unknown classification are grouped in this subclass and will be classified elsewhere when the transport mode and energy coupling mechanism are characterized. These families include at least one member for which a transport function has been established, but either the mode of transport or the energy coupling mechanism is not known.

9.B. *Putative but uncharacterized transport proteins.* Putative transport protein families are grouped in this subclass and will either be classified elsewhere when the transport function of a member becomes established, or will be eliminated from the TC classification system if the proposed transport function is disproved. These families include a member or members for which a transport function has been suggested, but evidence for such a function is not yet compelling.

9.C. *Functionally characterized transport proteins with unidentified sequences.* Transporters of particular physiological significance will be included in this category even though a family assignment cannot be made. When their sequences are identified, they will be assigned to an established family. This is the only protein subclass that includes individual proteins rather than protein families.

1.13
Transporters with Dual Modes of Energy Coupling

Members of a transporter family generally utilize a single mode of transport and energy coupling mechanism, thus justifying the use of functional categories as the primary basis for classification. However, a few exceptions have been noted. For example, the arsenite (ArsAB; TC #3.A.4) efflux permease of *E. coli* consists of two proteins, ArsA and ArsB. ArsB is an integral membrane protein that presumably provides the transport pathway for the extrusion of arsenite and antimonite [11–12]. ArsA is an ATPase that energizes ArsB-mediated transport. However, when ArsB alone is present, as in the case of the arsenical resistance pump of *Staphylococcus aureus*, transport is driven by the proton motive force (pmf) [13]. Expression of the *E. coli arsB* gene in the absence of the *arsA* gene similarly gives rise to pmf-driven transport. The presence or absence of the ArsA protein thus determines the mode of energy coupling [13].

The ArsB protein is a member of a large superfamily of ion transporters, the IT superfamily, in which at least two families exhibit the unusual capacity of being able to incorporate auxiliary constituents that alter the transport characteristics of the carrier [14]. Such promiscuous use of energy is exceptionally rare and has been documented in only a very few instances. When such an effect is reported, the TC system classifies the permease in accordance with the more complicated energy coupling mechanism [i.e., as an ATP-driven primary active transporter (Class 3), rather than as a secondary carrier (Class 2)]. However, in this unique case, the TC nomenclature panel of the IUBMB has recommended that a second family describing the pmf-driven ArsB homologs be included in the TC system (TC #2.A.45) as many ArsB homologs function by ATP-independent, ArsA-independent mechanisms.

1.14
Transporters Exhibiting More than One Mode of Transport

Examples of secondary carrier families in which promiscuous transport modes have been reported include the mitochondrial carrier (MC) family (TC #2.A.29) and the triose phosphate family of the Drug/Metabolite Transporter (DMT) superfamily (TC #2.A.7). Proteins of both families are apparently restricted to eukaryotic organelles. Members of these families normally catalyze carrier-mediated substrate:substrate antiport and are, therefore, classified as secondary carriers. However, treatment of certain mitochondrial carriers with chemical reagents, such as N-ethyl maleimide or Ca^{2+} [15–19], or imposition of a large membrane potential ($\Delta\Psi$) across a membrane into which a triose phosphate transporter has been incorporated [20–22] has been reported to convert these antiport-catalyzing carriers into anion-selective channels capable of functioning by uniport. Another secondary carrier that may be capable of exhibiting channel-like properties is the KefC protein of *E. coli* [23] which is a member of the CPA2 family (TC #2.A.37).

Again, the more complicated carrier-type mechanism, which appears to be relevant under most physiological conditions, provides the basis for classifying these proteins (i.e., as Class 2 carriers rather than Class 1 channels).

1.15
Conclusions and Perspectives

We believe that integral membrane transporters evolved independently of other classes of proteins such as enzymes, regulatory proteins, and receptors [8, 9]. If this is, in fact, true, then a system of classification for transporters distinct from those proposed for other types of proteins is fully justified. We have proposed a rational system of transporter classification (TC) which incorporates the most conserved features of transporters: first, mode of transport, second, energy coupling mechanism, third, phylogenetic family, and fourth, substrate specificity. Using this classification system we have been able to systematically extract tremendous amounts of information about transporters (see, for example, [4]). We expect that as increasing numbers of sequenced genomes become available for analysis, the experimental biologist will become increasingly dependent on theoretical approaches of computational biology. Moreover, scientists will rely on biosystematics to help render the reams of genomic data intelligible to the human brain. It is hoped that the TC system will prove effective in facilitating conceptualization of transporter structures, functions, mechanisms and evolution.

Acknowledgement

I would like to thank Yolanda Anglin for assistance in the preparation of this chapter.

References

1. M. H. Saier, Jr. *Microbiol. Rev.* **1994**, *58*, 71-93.
2. M. H. Saier, Jr. *Microb. Comp. Genom.* **1996**, *1*, 129-150.
3. M. H. Saier, Jr., in: *Advances in Microbial Physiology* (R. K. Poole, Ed.), Academic Press, San Diego, CA. **1998**, pp. 81-136.
4. M. H. Saier, Jr. *Microbiol. Mol. Biol. Rev.* **2000**, *64*, 354-411.
5. M. H. Saier, Jr., in: *Biomembrane Transport* (L. VanWinkle, Ed.), Academic Press, San Diego, CA. **1999**, pp. 265-276.
6. M. H. Saier, Jr. in: *International Review of Cytology: a Survey of Cell Biology* (K. W. Jeon, Ed.), Academic Press, San Diego, CA, **1999**, pp. 61-136.
7. M. H. Saier, Jr., Genome archeology leading to the characterization and classification of transport proteins, *Curr. Opin. Microbiol.* **1999**, *2*, 555-561.
8. M. H. Saier, Jr., T.-T. Tseng, Evolutionary origins of transmembrane transport systems. *SGM Symposium Series*, Leeds, UK, September 6-9, **1999**.
9. M. H. Saier, Jr., *J. Membr. Biol.* **2000**, *175*, 165-180.
10. R. F. Doolittle, *Of urfs and orfs: A Primer on how to Analyze Derived Amino Acid Sequences*. University Science Books, Mill Valley, CA, **1986**.
11. S. Bröer, G. Ji, A. Bröer, S. Silver,. *J. Bacteriol.* **1993**, *175*, 3480-3485.
12. S. Silver, G. Ji, S. Bröer, S. Dey, D. Dou, B. P. Rosen, *Mol. Microbiol.* **1993**, *8*, 637-642.
13. C. Xu, T. Zhou, M. Kuroda, B. P. Rosen, *J. Biochem.* **1998**, *123*, 16-23.
14. R. Rabus, D. L. Jack, D. J. Kelly, M. H. Saier, Jr., *Microbiology* **1999**, *145*, 3431-3445.
15. N. Brutovetsky, M. Klingenberg, *J. Biol. Chem.* **1994**, *269*, 27329-27336.
15. N. Brutovetsky, M. Klingenberg, *Biochemistry* **1996**, *35*, 8483-8488.
17. T. Dierks, A. Salentin, C. Heberger, R. Krämer. *Biochim. Biophys. Acta* **1990**, *1028*, 268-280.
18. T. Dierks, A. Salentin, C. Heberger, R. Krämer, *Biochim. Biophys. Acta* **1990**, *1028*, 281-288.
19. P. Jezek, D. E. Orosz, M. Modriansky, K. D. Garlid, *J. Biol. Chem.* **1994**, *269*, 26184-26190.
20. H. Wallmeier, A. Weber, A. Gross, U.-I. Flügge, in: *Transport and Receptor Proteins of Plant Membranes* (D. T. Cooke, D. T. Clarkson, Eds.), Plenum Press, New York. **1992**, pp. 77-89.
21. B. Schulz, W. B. Frommer, U.-I. Flügge, S. Hummel, K. Fischer, L. Willmitzer, *Mol. Gen. Genet.* **1993**, *238*, 357-361.
22. M. Schwarz, A. Gross, T. Steinkamp, U.-I. Flügge, R. Wagner, *J. Biol. Chem.* **1994**, *269*, 29481-29489.
23. G. P. Ferguson, Battista, J. R., Lee, A. T., Booth, I. R., *Mol. Microbiol.* **2000**, *35*, 113-122.

2
Energy-transducing Ion Pumps in Bacteria: Structure and Function of ATP Synthases

Jörg-Christian Greie, Gabriele Deckers-Hebestreit, and Karlheinz Altendorf

2.1
Introduction

One of the most frequently occurring reactions in biology is the synthesis of adenosine triphosphate (ATP). ATP acts as a carrier of energy in living organisms from bacteria and fungi to plants and animals including humans and has, therefore, been termed the cell's energy currency. The vast majority of ATP production is carried out by an almost ubiquitous multisubunit enzyme complex, the F_1F_O ATP synthase. Whether in mitochondria, chloroplasts, or eubacteria, F_1F_O ATP synthases catalyze the synthesis of ATP from ADP and inorganic phosphate (P_i) utilizing the energy of an electrochemical ion gradient ($\Delta\mu_{H^+}$ or $\Delta\mu_{Na^+}$) generated across the membrane by respiration or photosynthesis. The process of coupling vectorial ion transport to the synthesis of ATP may be explained by the same thermodynamics as ATP consuming ion pumping, such as in active ion transporters like the Na^+/K^+ ATPase (P-type ATPases). In fact, in case of low driving force, that is anaerobiosis, bacterial ATP synthases can also serve as primary ion pumps, thereby generating an ion gradient across the membrane at the expense of ATP, which may be derived from glycolysis. Therefore, ATP synthases are also termed F-type ATPases or just F-ATPases. However, structurally and mechanistically, F-type ATPases prove to be quite different from P-type ion pumps. Although of partially different subunit composition, all types of F-type ATPases share high homology with respect to the mechanism of catalysis and ion translocation as well as the mode of coupling.

2.2
Overview

Generally, the F_1F_O ATP synthase is built up of two different entities, the peripheral F_1 part and the membrane-embedded F_O complex (see Fig. 1). Despite a prevalent coarse organization comprising features like a headpiece, a basepiece, and two

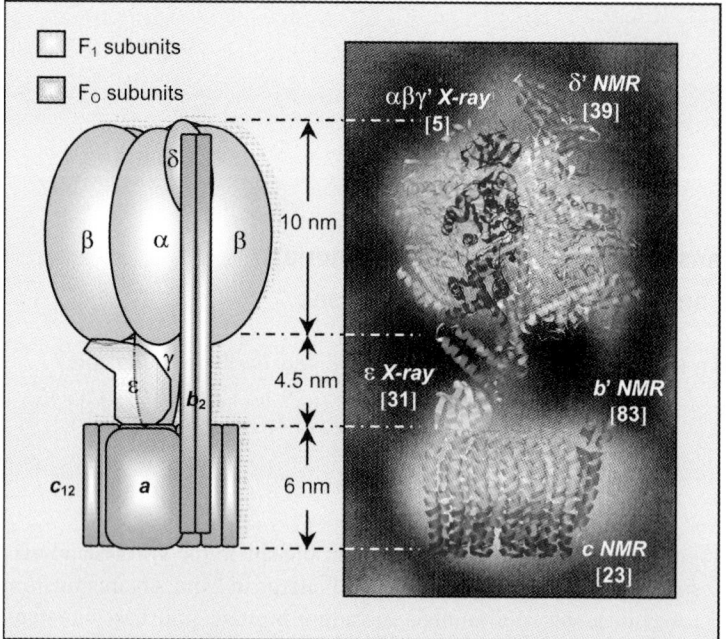

Fig. 1. Schematic presentation of the F_1F_O ATP synthase. Overview of subunit assembly and modeling of available structural information from either NMR spectroscopy or X-ray crystallographic analysis into the electron density map of the *E. coli* F_1F_O complex (taken from [7] with kind permission from *Nature*). Corresponding references are quoted in brackets; see color plates p. XXIII.

stalk-like structures connecting the head- and basepiece, there are some deviations in subunit composition between different species, especially within the F_O part of the enzyme. In the *Escherichia coli* ATP synthase, which is commonly used as a paradigm of structure and function due to its minimal constitution, three different F_O subunits are present in a stoichiometry of ab_2c_{12} [1, 2]. Whereas F_O complexes from eubacteria and chloroplasts prove to be quite similar, mitochondrial F_O complexes comprise at least five additional subunits as well as yet another seven F_O-associated polypeptides [3]. In contrast, the comparatively well conserved F_1 part consists of five different subunits with the same stoichiometry ($a_3\beta_3\gamma\delta\varepsilon$). Whereas isolated F_1 complexes from mitochondria and bacteria incessantly hydrolyze ATP with $a_3\beta_3\gamma$ resembling the minimal configuration for activity, the chloroplast F_1 is a latent ATPase that requires activation [4]. Within the F_1 complex, subunits α and β are alternately arranged in an $a_3\beta_3$ hexamer surrounding the centrally located subunit γ [5], which extends from the lower part of the hexamer to form a 4.5 nm stalk region, thereby connecting F_1 to F_O [6–8]. Subunit ε is in tight contact with subunit γ at the lower part of the mushroom-like F_1 complex [9–11], whereas subunit δ is supposed to be located at the upper periphery of the $a_3\beta_3$ hexagon [12–14]. During catalysis the three β-subunits carrying the catalytic sites are forced into asymmetry by eccentric rotation of the γ-subunit pivoted inside a hydrophobic sleeve within the $a_3\beta_3$ hexamer. Corresponding conformational changes result in different binding affinities for the nucleotides as described by the cooperative binding change mechanism, according to which each catalytic site is concomitantly occupied by a different nucleotide configuration [5, 15, 16].

Due to the rotational movement of a centrally located rotor part, a second stalk linking the F_1 and F_O entities is supposed to be necessary for the stabilization of the holoenzyme, which is built up at least of the two copies of subunit b of the F_O complex [7, 13, 14, 17–20]. Within the F_O part of the *E. coli* enzyme, subunits a and b are located outside a ring-like subunit c oligomer [21, 22]. Translocation of the coupling ion takes place via a highly conserved carboxylate (cD61) located in the second helix of subunit c, which folds in a hairpin-like structure connecting two transmembrane helices by a polar loop region [23]. Other amino acids essential for ion translocation are located within subunit a, especially the highly conserved aR210 in the penultimate transmembrane helix, thereby suggesting a direct interaction of subunits a and c via formation of a transient salt bridge [1, 24]. During coupled catalysis a $\gamma\varepsilon c_{12}$ subcomplex is supposed to rotate relative to the remainder of the F_1F_O complex propelled by the sequential protonation/deprotonation of residue cD61, thereby generating elastic torque, which is supposed to be stored in the two stalks, and then as a consequence drives ATP synthesis in F_1 or proton pumping in F_O [25–27].

2.3
Structure, Configuration, and Interaction of F_1 Subunits

In terms of enzyme structure, most information concerning the F_1 molecule is derived from X-ray crystallographic analysis of the bovine heart mitochondrial F_1 complex [5]. Due to its well conserved configuration, the structural information obtained can be considered as comprehensive. A main feature of the F_1 complex is the spherical arrangement of alternating a- and β-subunits extending from the top to the bottom side like segments of an orange (Figs. 1 and 2). On top of this hexameric assembly there is a central dimple extending about 1.5 nm toward the epicenter, below of which the interior surfaces of the a- and β-subunits form a hydrophobic sleeve with low surface potential, which is filled with two coiled helices comprising the N- and C-terminal portion of subunit γ. The helices protrude another 3.0 nm below the bottom side of the hexameric sphere, most likely being part of the 4.5 nm stalk region connecting F_1 to F_O. The remainder of the γ-subunit as well as subunits δ and ε were not sufficiently ordered within the crystals for structural determination. The domain structure of the a- and β-subunits proved to be quite similar, featuring an N-terminal scope of β-sheets, a central nucleotide binding domain containing Walker motifs A and B [28], and a C-terminal a-helical region comprising the well-conserved so-called DELSEED region, which contains numerous charged residues and is of important regulatory function [29]. The sites of nucleotide binding are located directly at the a/β interfaces, with residues of both subunit a and β participating in the formation of the catalytic sites. Within subunit β there is a water molecule hydrogen-bonded to a glutamic acid, thereby most probably attacking the γ-phosphate of the ATP molecule. The presence of a corresponding glutamine within the nucleotide binding site of subunit a could probably account for its lack of catalytic activity. However, although the domain structures of the a- and β-subunits superimpose quite well with deviations being less than 0.1 nm for the C_a backbone (except for the nucleotide binding domain of one β-subunit), there is a striking asymmetry in the overall structure of the hexamer, which principally resides in domain shifts arising from diverse orientations of the domains relative to the centrally located γ-subunit. The asymmetrically coiled helices of subunit γ provoke the synchronistic formation of different interactions with each domain, resulting in differential overall conformations of each of the a- and β-subunits.

In another recently published X-ray structural analysis of the bovine heart F_1 complex, also the residual part of subunit γ as well as the subunit ε analog could be resolved [30]. In subunit γ, the C-terminal a-helix extends further downward, followed by a so-called a/β domain again passing over to the N-terminal a-helix. The a/β domain packs aside the long C-terminal helix and features five β-strands arranged between another two extended a-helices, one of which is closely ordered to the DELSEED sequences of both subunit a and β at the a/β interface. The three-dimensional structure of subunit ε was obtained from both X-ray crystallographic analysis [30, 31] and NMR spectroscopy [32], featuring an N-terminal β-barrel domain tightly associated with an a-helical hairpin within the C-terminal region (see Fig. 1). In contrast, in the X-ray structural analysis of a $\gamma\varepsilon$ subcomplex from E. coli, the C-terminal pair of a-helices separate instead of forming a helix-turn-helix motif, thereby packing end-to-end at an angle of about 90° [33]. By use of cysteine-substituted F_1F_O complexes, subunit ε was shown to strongly interact with subunit γ [9–11, 34], with the interaction most likely occurring between the β-barrel domain and the N- and C-terminal helices of subunit γ protruding downward from the F_1 complex [30, 33]. Furthermore, in isolated F_1, residue εS108 of the helical domain could be cross-linked in a nucleotide-dependent manner with residues βE381 or aS411 located within the DELSEED region of subunits β and a at the lower part of the F_1 globe [35]. It should be denoted that this implicit interaction of subunit ε with the bottom part of the $a_3\beta_3$ hexagon could not be verified in the recently obtained crystal structures of either a yeast mitochondria F_1-c_{10} subcomplex [8] or bovine heart mitochondria F_1 [30], in both of which the location of subunit ε is shifted toward the subunit c oligomer with no contact sites whatsoever with the $a_3\beta_3$ hexamer. In contrast, chemical cross-linking of region εT26-G33 with residues cA40 to cD44 in cysteine-substituted F_1F_O complexes [34, 36, 37] connects the β-barrel domain of subunit ε to the hydrophilic loop region of the subunit c oligomer, which could be verified by X-ray structural analysis [8, 30, 33]. With respect to the spatial distance of about 4 nm between the bottom side of the F_1 sphere and the apex of the subunit c annulus, it seems likely that also subunit ε undergoes some spatio-temporal rearrangements during catalysis within the F_1F_O molecule in order to reach its contact sites in both F_1 and F_O, thereby possibly exerting a regulatory function [38]. This view is further supported by comparison of the X-ray structural analysis of the central stalk region from both the E. coli and the bovine heart enzyme, in which the C-terminal α-helical domains of subunit ε resume different positions with either orientation toward the F_1 sphere or to the subunit c oligomer, respectively [30, 33]. Although subunit δ could not be resolved in the crystal structure of the bovine heart F_1 molecule, structural information was obtained from NMR analysis of the N-terminal portion of subunit δ from E. coli comprising 134 of 176 residues, which was shown to be highly α-helical [39]. Subunit δ resides at the upper periphery of the F_1 sphere (see Fig. 1) as was derived from cross-linking experiments [12, 13] as well as from electron microscopic analysis of immunodecorated E. coli F_1F_O complexes [14].

2.4
Catalysis: Structural and Mechanistic Implications within the F_1 Complex

A closer view on the structure of the nucleotide binding sites within the bovine heart F_1 crystal structure essentially confirms a rather complicated as well as sophisticated mechanism of rotational catalysis (see Fig. 2). Crystallization was performed in the presence of ADP and Mg^{2+} together with the non-hydrolyzable ATP analog 5'-adenylylimidodiphosphate (AMP-PNP) in order to trap the enzyme in a well-defined conformation [5]. Whereas all three α-subunits were shown to bind AMP-PNP via their non-catalytic site, one of the catalytic sites on the β-subunits was filled with ADP, another with AMP-PNP (i.e., ATP), and the third site was found to be empty, strongly arguing for a different nucleotide occupancy of the catalytic sites at a time. According to the binding change mechanism postulated by Paul Boyer years before the elucidation of the enzyme structure (comprehensively reviewed in [15]), the three β-subunits are supposed to equivalently undergo sequential conformational changes induced by an asymmetric gyratory movement of the γ-subunit inside the $a_3\beta_3$ hexagon, thereby rotating different affinities for corresponding nucleotides within the catalytic sites. As a result, each of the three β-subunits concomitantly comprises another nucleotide occupancy during a catalytic cycle. Complying with these occupancies the different conformations of the β-subunits are commonly termed "tight", "loose", and "open" (Fig. 2). An adequate assignment of the three catalytic sites within the ADP-inhibited crystal structure of the bovine heart F_1 could be made with the site featuring bound ADP resembling the "tight" conformation in comparison to bound AMP-PNP (i.e., ATP) being assigned to "loose" (which should be vice versa in the non-inhibited state of the enzyme), whereas the third unoccupied site corresponds well with the "open" conformation. In the latter case, there is a notable difference in the superimposition of the C_a carbon trace of the surrounding hinge region with respect to the other subunits (see Fig. 2C and D), resulting in a kind of a champing movement on the bound nucleotide during the catalytic cycle. Whereas the formation of ATP from ADP and P_i occurs spontaneously within the catalytic site adopting the "tight" conformation, energy input by means of subunit γ

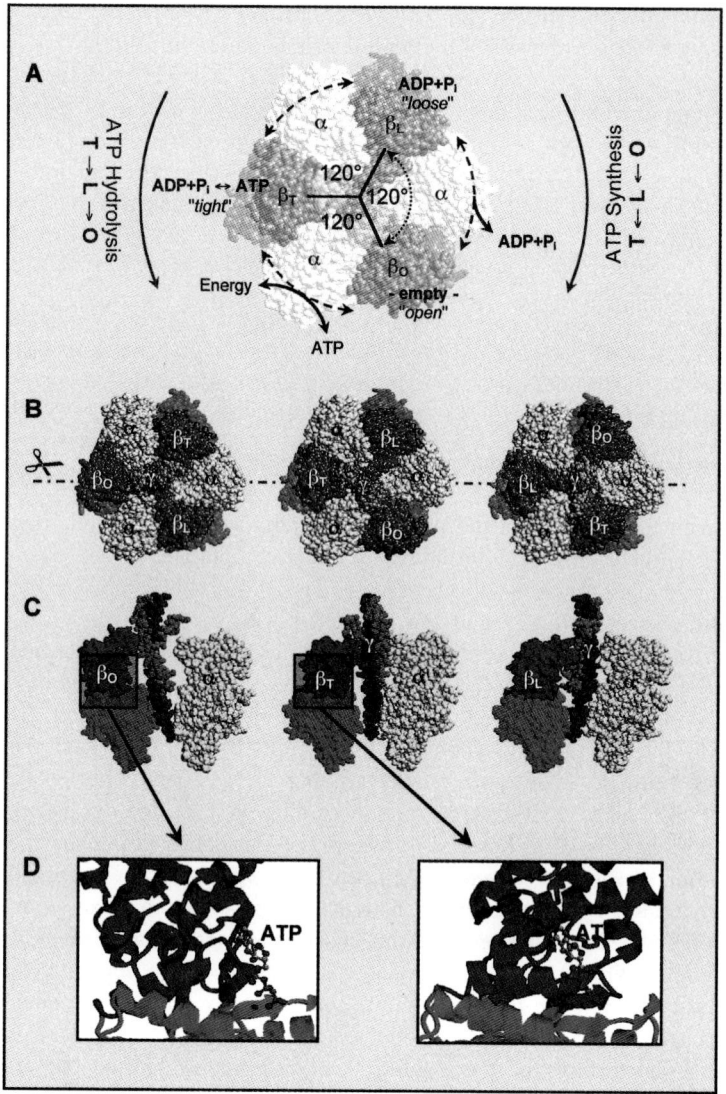

rotation is required for the release of the ATP molecule in transmission to the "open" state. It should be denoted at this point that despite these well matching structural data there is another convincing model of rotational catalysis, which is based on kinetic evaluation of enzyme activity under physiological conditions (i.e., steady state catalysis). According to this model, *all* catalytic sites are filled with nucleotides, including the "open" conformation instantly to be occupied by another ATP molecule, yielding no nucleotide-free intermediate "open" state whatsoever [40]. It should be denoted that there are also three-dimensional structural data of isolated F_1 subcomplexes, in which the three catalytic sites do not

Fig. 2. Catalysis within the F_1 complex – the binding change mechanism. **A** Different conformations assumed sequentially by each catalytic site during synthesis or hydrolysis of ATP as subunit γ rotates 120° within the $\alpha_3\beta_3$ hexamer. Sites are designated as "open" (β_O, no nucleotide bound), "loose" (β_L, ADP+P$_i$ bound), and "tight" (β_T, interconversion of bound ADP + P$_i$ and ATP). The sketch of the crystal structure from the bovine heart F_1 complex [5] is depicted as seen from the membrane. Clockwise rotation of subunit γ leads to ATP synthesis, whereas counter-clockwise rotation corresponds to ATP hydrolysis. Based on kinetic data it is likely that during steady state catalysis the "open" site is immediately occupied by another nucleotide. **B** Circulating conformational changes within the $\alpha_3\beta_3$ hexamer as subunit γ rotates stepwise at intervals of 120° each in counter-clockwise direction (i. e., ATP hydrolysis). **C** Cross-section through **B**. Nucleotide-dependent conformational changes within the C-terminal domain of the β-subunit during subunit γ rotation. Whereas the C-terminal domain undergoes spatio-temporal rearrangements during the catalytic cycle (red color), the N-terminal portion of subunit β (green) retains an approximately threefold symmetry around the rotational axis. The N- and C-terminal domain of subunit γ is depicted in gray and blue, respectively. **D** Clipping of the subunit β hinge region in either "open" (left) or "tight" (right) conformation. Refer to Sect. 4 for further details. Molecular sketches are kindly provided by Dr. G. Oster (Copyright © 2001, University of California, Berkeley); see color plates p. XXIV.

adopt different conformations [41, 42], probably representing different transition states of the enzyme. However, both models rely on a rotary movement of the centrally located subunit γ shaft, which simultaneously induces the conformational changes of all three catalytic sites when a 120° rotational step occurs. As a consequence, a full 360° turn is needed for each site to return to its original conformation.

The proposed gyratory movement of the γ-subunit sliding against the $\alpha_3\beta_3$ hexagon over infinite angles was variably approached by a whole set of experiments, finally leading to the assumption of a unidirectional rotation rather than a to-and-fro fluctuation. In two approaches, a particular residue on subunit γ was cross-linked to one individually labeled β-subunit. After disbanding the link generated in either isolated F_1 molecules [43] or F_1F_O complexes [44], the enzyme was allowed to carry out ATP hydrolysis and synthesis, respectively. After regeneration of the cross-link, subunit γ was shown to be connected to any of the three β-subunits, indicating equivalent interactions of γ with each β-subunit. Homologous experiments revealed that also subunit ε switches between the β-subunits during catalysis in assembled F_1F_O synthase [45], implying it to be part of the rotor, which was also deduced from the functional characterization of other $\gamma-\varepsilon$ cross-link products [9, 11, 34]. In another set of experiments a fluorescent dye covalently attached to the γ-subunit was shown to change its direction relative to the $\alpha_3\beta_3$ hexamer over a wide angle within isolated F_1 complexes [46], thereby virtually adopting three distinct orientations [47]. The unidirectional nature of subunit γ rotation was finally unequivocally shown by attaching a fluorescent actin filament to immobilized $\alpha_3\beta_3\gamma$ subcomplexes from the thermophilic bacterium PS3 (Fig. 3) The molecule was fixed on an appropriately coated surface via histidine residues engineered at the N-termini of the β-subunits, whereas the actin filament was attached to the protruding portion of subunit γ (Fig. 3A). ATP-driven rotation of the actin filament

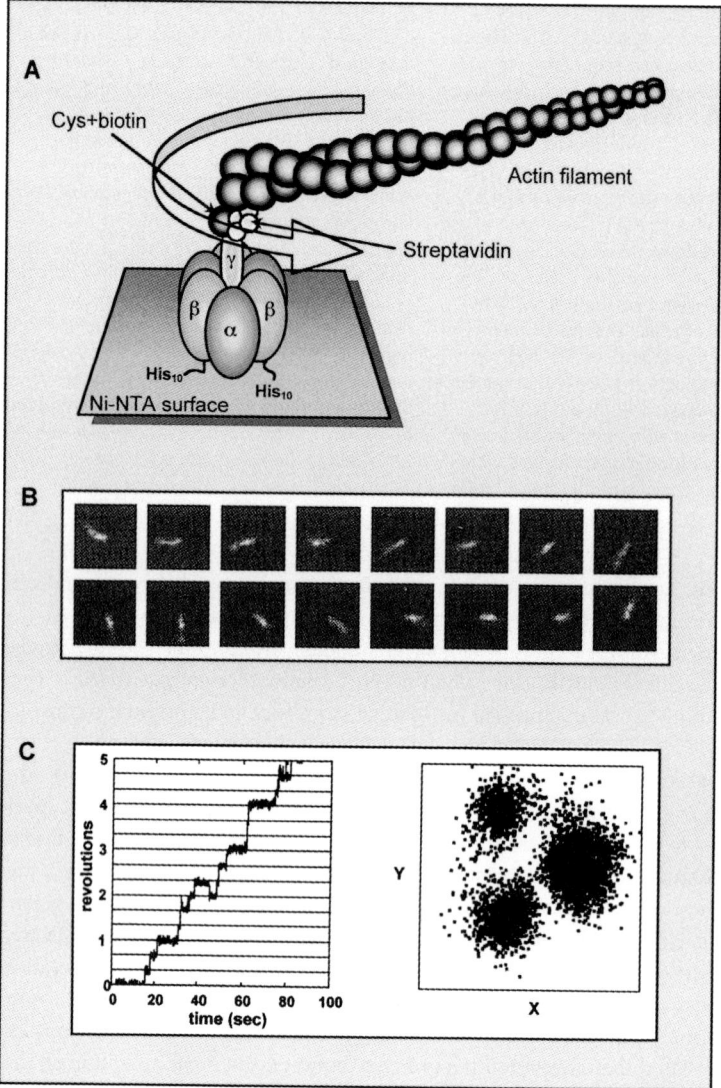

Fig. 3. Direct observation of subunit γ rotation. **A** Experimental setup. $\alpha_3\beta_3\gamma$ subcomplexes were immobilized onto a Ni^{2+}-nitrilotriacetic acid (Ni-NTA)-coated surface via poly-histidine affinity motifs genetically engineered at the N-termini of the β-subunits. A fluorescently labeled actin filament was covalently linked to cysteine-substituted subunit γ by use of streptavidin/biotin conjugates. **B** Sequential images of ATP-driven filament rotation at ATP concentrations of 2 mM. Unidirectional rotation occurs counter-clockwise as proposed by both the binding change mechanism and the X-ray structural data (see Fig. 1B) (reproduced from [48] with kind permission from *Nature*). **C** Stepwise rotation of subunit γ at low concentrations of ATP (20 nM). Left, revolutions versus time. Each revolution virtually divides in three unitary steps. Right, spatio-temporal trace of the centroid of the actin filament. The triangular distribution indicates a stepwise rotation of the γ-subunit in intervals of 120° each (reprinted from [125], Copyright © 2000, with kind permission from Elsevier Science).

could be observed with a fluorescence microscope [48] (see Fig. 3B). In case of low concentrations of ATP, rotation of the filament was shown to occur also unidirectional in a counter-clockwise manner when viewed from the membrane, but stepwise with three intervals of 120° each, both being perfectly in agreement with the threefold symmetry of the rotary binding change mechanism [49] (Fig. 2C). Analogous experiments demonstrating ATP-driven rotation of subunit γ were also performed using the chloroplast [50] as well as the *E. coli* enzyme [51, 52]. In addition, direct visualization of ATP-driven rotation was also obtained by attaching the actin filament to subunit ε from the PS3 F_1 complex, thereby clearly demonstrating that subunit ε is part of the rotor [53]. The stepwise rotation is likely to be triggered by periodically occurring polar interactions of subunit γ, in which residue γR75 exposed in a loop within the α/β domain interacts with the DELSEED sequences of neighboring α- and β-subunits, thereby generating stable conformational intermediates during the catalytic cycle via hydrogen bonding with neighboring carboxylates at the α/β interface [30].

Calculating the stall load due to hydrodynamic friction of the rotating actin filament imposed on subunit γ, the F_1 motor produces a constant torque of about 40 pN nm. This 40 pN nm times the angular displacement corresponding to one 120° step (i.e., $2\pi/3$ rad) results in about 80 pN nm of work done per ATP molecule. This is quite close to the free energy obtained from the hydrolysis of one molecule of ATP under intracellular conditions, which is about 90 pN nm, suggesting that the F_1 motor can work with nearly 100% efficiency [54]. In comparison, a common car engine works at only about 40% maximum efficiency, co-generating large amounts of caloric heat.

2.5
The F_1/F_O Interface: Contact Sites for Energy Transmission

Under conditions of steady state catalysis, the F_1F_O ATP synthase behaves like two counterrotating stepping motors mounted on a common shaft, one of which (F_1) is chemically driven by ATP hydrolysis, whereas the other (F_O) acts like a mechanical turbine powered by the constant flow of coupling ions. Thus, with respect to both the maintenance of structural integrity and the transmission of energy between the sites of ATP turnover and ion translocation, structural as well as functional links between the F_1 and F_O complex seem to be a prerequisite for coupled catalysis of the holoenzyme. From electron microscopic analyses of purified ATP synthase at least two structural domains can be derived linking the F_1 sphere to the F_O part [7, 14, 19, 20], one of which is located centrally at the long axis of the holoenzyme, whereas the other is placed at the outer periphery exhibiting a less prominent electron density (see Fig. 1). Whereas the so-called central stalk is likely to be composed of the tightly associated F_1 subunits γ and ε, the second stalk can be assigned to the subunit b dimer protruding from the F_O complex. Possible sites of interaction of the two b-subunits with the F_1 part of the enzyme can be derived from chemical cross-linking with both subunit α and β as well as with subunit δ

acting as counterparts in cross-link formation. By use of cysteine-substituted F_1F_O complexes, residue bI109 as well as bE110 could only be cross-linked to subunit a, whereas residue bA92 was shown to interact with both subunit β and region aI464-M483 of subunit a, thereby placing this region of the subunit b dimer at the a/β interface near a non-catalytic nucleotide binding site [55]. Due to the opposing face-to-face orientation of the polypeptide chains within the dimer, it seems likely that one copy of subunit b interacts with a, whereas the other makes contact with subunit β. The C-terminus of subunit b is supposed to interact with both the N-terminal region of subunit a and the C-terminal region of subunit δ [18, 56, 57], thereby bridging a distance of at least 11 nm from the membrane surface to almost the top of the F_1 molecule (see Fig. 1). Under conditions of steady state catalysis the stator part of the ATP synthase has to withstand a mechanical torque of 40 pN nm exerted by rotation of the γ-subunit. Thus, the interaction of F_O and F_1 subunits forming either the rotor or the stator part requires K_d values in the nM range. Whereas subunit δ exhibits a rather high affinity to the $a_3\beta_3$ hexagon with a K_d of less than 0.8 nM [58], the interaction of subunits b and δ was shown to be quite weak with a K_d of 5-10 µM, determined, however, for the hydrophilic domain of subunit b lacking the N-terminal membrane spanning region (b_{sol}) [55, 59]. Therefore, it seems reasonable that the subunit b dimer features multiple contact sites with the F_1 complex in the formation of the stator part of the enzyme.

In contrast, subunit interactions between F_1 and F_O within the rotor part (i.e., subunits γ, ε, and the subunit c annulus) seem to be less extended and rather restricted to just a few stretches of amino acids. Although both the ε- and the γ-subunit were shown to interact with the subunit c oligomer, the sites of interaction are restrained to the hydrophilic loop region located at the top of the ring-like structure. In subunit γ, residue γY205 as well as γY207 were shown to make contact with a stretch of residues including cQ42-D44 [10, 34, 60]. Since both γY205 and γY207 are located within a loop at the bottom of the C-terminal helix of subunit γ [33], this loop is supposed to pack between the front and back faces of the polar loop region of two adjacent c-subunits at the periphery of the ring. In another set of cross-linking experiments, a continuous stretch of residues εT26-G33 within subunit ε was also shown to make contact with the hydrophilic loop region (residues cA40-D44) at the surface of the subunit c oligomer [36, 37]. Residues εT26-G33 are supposed to form a turn connecting two antiparallel β-strands within the β-barrel domain at the very bottom of the ε-subunit [33]. Due to the close contact of subunits γ and ε, this turn is supposed to pack between the polar loop regions of a set of c-subunits adjacent to those interacting with subunit γ [33, 37]. This packing arrangement implies that not all of the c-subunits within the oligomer are involved in F_1 interaction. Although binding affinity studies with an engineered peptide comprising the hydrophilic loop region of subunit c revealed that under saturating conditions only one mol of peptide was bound per mol F_1 in the chloroplast system, the contribution of possible salt bridges by only a few polar residues within a single hydrophilic loop region is likely to be insufficient to withstand the torque generated during rotational catalysis [61]. Hence, with respect to energy coupling more than only one copy of subunit c should participate directly in F_1 binding. The packing arrangement of subunits γ and ε each between the front and back faces of neighboring polar loop regions involves at least three copies of subunit c in γ/ε interaction. Binding studies with monoclonal antibodies directed against the polar loop region of subunit c revealed that F_1 interaction was not affected by antibody binding, but occurred simultaneously, although the amino acid residues involved in antibody recognition overlap with those known to interact with subunits γ and ε [62]. Thus, even though the F_1 complex is present, at least a few copies of the polar loop regions of subunit c are accessible from the water phase and are not participating in F_1 interaction. Further proof can be derived from accessibility studies by use of F_1F_O complexes featuring a cQ42C substitution. Yet in the presence of F_1, at least 60% of the subunit c oligomer was shown to be readily accessible for the thiol-specific agent N-ethylmaleimide, whereas after F_1 removal all of the c-subunits could easily be labeled [63]. However, interaction with only a few copies of the subunit c oligomer places subunits γ and ε, and hence, the rotary shaft, at the periphery of the annulus. If the proposed rotation of the subunit c ring occurs along its central axis, this interaction would result in an eccentric co-rotation of the γ-subunit, thus adopting a more tilted or bent rather than parallel orientation relative to its rotational axis. On the other hand, in case of an incline of the subunit c ring relative to the rotational axis of a non-tilted γ-subunit, rotation would lead to an eccentric movement of the complete subunit c oligomer, rendering a close contact with the stator subunits within the F_O complex rather complicated. With respect to a functional F_1 interaction, neither the formation of the putative rotor part, nor the interaction of stator subunits can be held solely responsible for the formation of an F_1F_O complex stable under catalytic conditions. Reconstitution of purified F_O subunits or subcomplexes into artificial lipid vesicles resulted in the functional binding of isolated F_1 molecules only in the presence of all three F_O subunits a, b, and c [64–66]. Hence, a stable F_1 interaction can only be achieved in the presence of fully assembled F_O complexes providing all possible contact sites for both the maintenance of structural integrity and energy transduction (see [67]).

2.6
Structure, Configuration, and Interaction of F_O Subunits

Whereas detailed structural information on the F_1 part could be obtained from the three-dimensional crystallization of assembled protein complexes, purified F_O so far resisted the formation of highly ordered protein structures during crystallographic analysis possibly due to the strong hydrophobicity of its subunits. So far only monomeric subunit c and the membrane spanning domain of subunit b could be analyzed by NMR to straightly reveal three-dimensional structural features (Fig. 1). In case of subunit a, far less is known even on the topology of the polypeptide chain within the membrane. Especially for the N-terminal quarter of the protein, contradictory models exist concerning the orientation of hydrophilic loop regions and the number of transmembrane helices. Although most of the experiments concerning protein topology were performed by means of accessibility of hydrophilic loop regions from either side of the membrane, even the orientation of the N-terminus is controversially discussed (see [68]). Due to these discrepancies two different topological models emerged, one with the N-terminus being oriented toward the cytoplasm featuring six transmembrane helices [68–70], and another one with only five helices conditional upon the N-terminus facing the periplasmic side of the membrane [71–73]. Irrespective of this contradiction, both models are in agreement with former studies [74–76] concerning the topology of the C-terminal three-quarter of subunit a comprising the functionally important residues like aR210. Two cytoplasmic loop regions were found, ranging from around aK66 to aH95 and aK169 to aE196, respectively, and also two periplasmic domains including residues aV110 to aE131 and aL229 to aI246. In either model, the C-terminus is oriented toward the cytoplasm, resulting in four transmembrane helices beyond the first cytoplasmic loop region starting around residue aK66. Further evidence for the putative penultimate transmembrane helix of subunit a could be derived from cross-linking experiments, in which cysteine residues genetically engineered at one face of a putative a-helix between residues aL207 and aI225 were shown to cross-link with cysteines introduced into the second transmembrane helix of subunit c with positions ranging from cI55 to cY73 [77], thereby suggesting a parallel paired orientation of both helices (see also Fig. 5).

Subunit b is anchored in the membrane via an N-terminal stretch of hydrophobic amino acids. The remaining hydrophilic portion of the polypeptide chain is oriented toward the cytoplasm and participates in F_1 interaction and in the assembly of the F_1F_O complex [78, 79]. To elucidate the structure of subunit b, many experiments were carried out with both subunit b assembled within the F_1F_O complex and with b_{sol}, including extensive cross-linking studies and circular dichroism (CD) spectroscopy. From the results obtained a consistent model can be deduced, in which the two copies of subunit b are supposed to form an elongated dimer essential for F_1 binding, with predominantly a-helical conformation and parallel paired helices around residues bD53 to bK66, bS84, bQ104, bA105, bV124 to bS146, and the C-terminal residue bL156 [17, 18, 80–82]. Cross-linking experiments using cysteine residues genetically engineered within the N-terminal region of subunit b in assembled F_1F_O complexes together with the NMR analysis of the synthetic peptide b_{1-34} comprising the N-terminal membrane spanning domain revealed a possible direct interaction of the transmembrane helices around residues bT6 to bF17 at an angle of 23° [83]. In the NMR structure, residues bN4 to bM22 form a continuous a-helix, which is interrupted by a short bend around bK23 to bW26 with the helical structure resuming at residue bP27 at a 20° angle offset (see Fig. 1) most likely compensating the 23° tilt of the helices and thus allowing a re-orientation of the following helical segments at an angle more perpendicular to the membrane [83]. The helical stretches comprised by residues bN2 to bW26 are thought to span the core of the lipid bilayer, thereby anchoring the polypeptide in the membrane. With respect to characterization of the cross-link between bA92 and aI464-M483 obtained in cysteine-substituted F_1F_O complexes [55], the subunit b

dimer has to bridge a distance of about 6 nm emerging from the surface of the membrane in order to reach its contact site at the bottom part of the F_1 molecule. Provided that the stretch of residues between bW26 and bA92 adopts a rigid α-helical conformation, this would, however, contribute about 9.5 nm in length, so there seem to be some "extra" residues likely adopting a different conformation rather than a continuous α-helix. Secondary structure predictions suggest turn conformation around residues bN80 to bQ85 [84, 85]. The analysis of deletion mutants ranging from residues bA50 to bI75 revealed a tolerance of up to 11 amino acid residues in length, which corresponds to a distance of 1.6 nm [86], indicating an inherent stretching flexibility that cannot be explained by just one rigid rod-like α-helix protruding from the membrane to the top of F_1. The secondary structure composition of b_{sol} has been determined by CD spectroscopy to be highly α-helical [17, 80], whereas the spectroscopic analysis of membrane-embedded subunit b revealed an α-helical content of 80% together with 14% β-turn conformation, turning a proposed rigid α-helix into shorter α-helical stretches flexibly connected by β-turn motifs [66].

Within the F_O complex, the subunit b dimer was shown to contact both subunit a and c. Since subunits a and b are thought to participate in the formation of the putative stator part of the F_1F_O holoenzyme and are, therefore, supposed to withstand torque generated by rotation of the $\gamma\varepsilon c_{12}$ rotor part, the formation of sturdy subunit interactions within the stator subcomplex seems to be a prerequisite for the current theory of catalysis. First experimental indications for a possible proximity of subunits a and b were derived from non-specific cross-linking experiments [87, 88] as well as from characterization of the suppressor mutants bD9N/aP240A/L [89]. Cross-linking studies using specifically engineered cysteines revealed disulfide bond formation between residues bN2 and aG227 or aL228 [90], arguing for a close proximity of the N-terminal region of subunit b and the periplasmic part of the penultimate transmembrane helix of subunit a. The postulated stable interaction between subunit a and the subunit b dimer was unequivocally demonstrated by the purification and characterization of a stable and functionally assembled ab_2 subcomplex [67]. Since the formation of the subcomplex is not triggered by any cross-linking reagent and proved to be stoichiometric and functional, it most likely reflects subunit interactions occurring within the F_O complex in vivo. Besides the formation of the well-defined bN2/aG227-L228 cross-link product, further contact sites between subunit a and b can be assigned to residue bA32 [67] as well as bR36 [55], arguing in favor of both residues to be located at one face of an α-helical stretch, which is probably in contact with subunit a. The exact site of cross-link formation within subunit a remains to be elucidated, since cross-linking was performed by use of a semi-specific heterobifunctional agent instead of engineered disulfide bridging. Based on the NMR structure of b_{1-34}, residues bA32 and bR36 are supposed to be located about 1-2 nm above the membrane surface, which renders an interaction with residues of subunit a located directly at the lipid bilayer rather unlikely. The bend around bP27 amounts only up to 20°, which would not be enough for a re-orientation of the subsequent amino acid residues toward the membrane. Provided that the NMR structure of b_{1-34} reflects the structure of

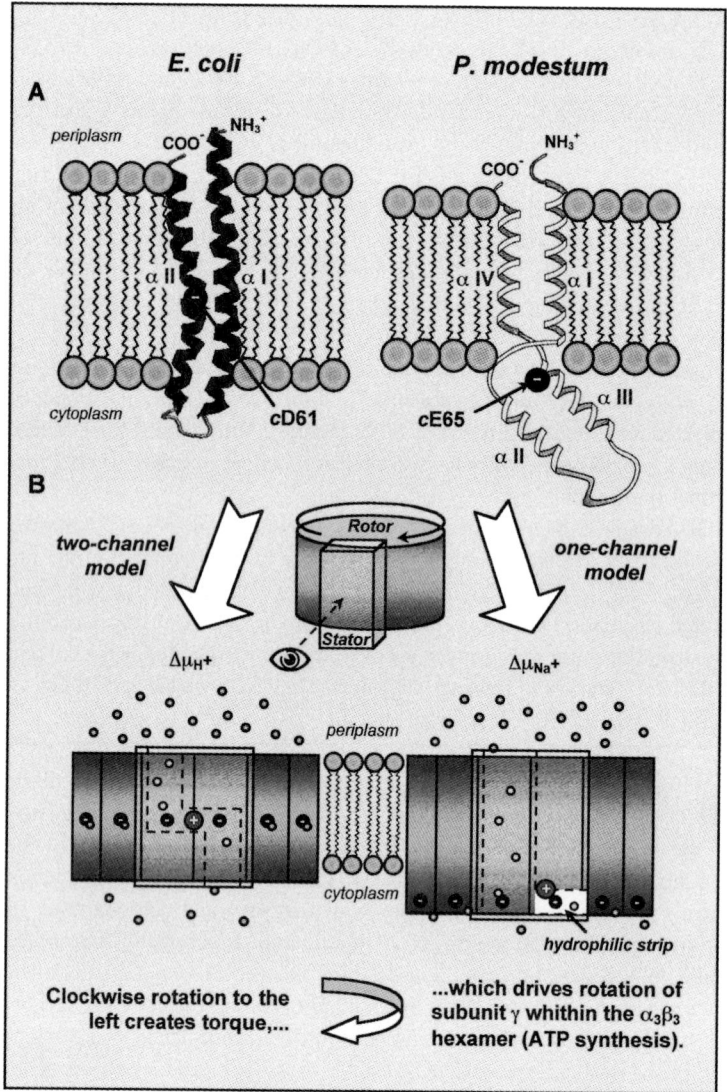

the N-terminal domain of subunit b *in vivo*, cross-link formation between bA32 or bR36 and subunit a can only occur with cytoplasmic loop regions of subunit a spanning the distance toward the site of cross-linking. In addition, both the N-terminus of subunit b and the penultimate transmembrane helix of subunit a can be cross-linked to the C-terminal region of subunit c [90, 91], suggesting a quaternary assembly of subunits a, b_2, and c within this region. Whether both b-subunits interact with subunit a or whether only one copy of the dimer is in contact with subunit a whereas the other interacts with the subunit c barrel remains to be established. The latter case seems to be attractive, since formation of the subunit b

Fig. 4. Structure of subunit c and channeling of the coupling ion within the F_O complex from *E. coli* and *P. modestum* ATP synthase. **A** Solution structure of monomeric subunit c solved by NMR spectroscopy in either mixed organic solvent (*E. coli*, left [23]) or SDS micelles (*P. modestum*, right [120]). Whereas in *E. coli* the reactive carboxylate cD61 is favored to be deeply embedded in the membrane, in case of *P. modestum* the corresponding cE65 residue is freely accessible from the aqueous phase (see Sect. 7). **B** Ion pathway as predicted by either the two-channel or the one-channel model. The rotor/stator assembly is presented as frontage view. The subunit c annulus is depicted in dark grey, whereas the stator is superimposed limpidly. In the two-channel model proposed for *E. coli* (left), protons enter the stator inlet channel connecting the reservoir of low pH with the reactive rotor site (i.e., cD61). Once protonated, these sites are supposed to rotate through the lipid bilayer for almost a complete revolution, rectified by the positive stator charge comprised by residue aR210. As soon as they reach the oppositely oriented outlet channel, protons dissociate into the high-pH environment. In the one-channel model as proposed for the sodium translocating *P. modestum* ATP synthase, all ion binding sites are freely accessible from the aqueous phase, which supercedes the outlet channel. Thus, sodium ions are directly transferred to the reactive cE65 via only one channel. Rectification is also ensured by a positive stator charge (i.e., residue aR227 in *P. modestum*), whereas an additional hydrophilic strip laterally connecting the channel with the cytoplasmic reservoir is thought to be necessary in order to permit charged (i.e., unoccupied) rotor sites to enter the rotor/stator interface from the right. In both models, rectified rotation of the subunit c annulus is supposed to create torque by intertwining helical domains of subunit γ, finally resulting in co-rotation of γ within the $\alpha_3\beta_3$ hexamer according to the binding change mechanism. Mechanistic sketches of the one- and two-channel model are depicted according to Oster and Wang [126] and Elston et al. [127], respectively.

dimer would then structurally link subunit a to the propelling subunit c annulus, thereby stabilizing possible dynamic interactions of helical interfaces between subunits a and c (see Fig. 5).

A high-resolution structure of monomeric subunit c from *E. coli* could be obtained by NMR analysis of the isolated proteolipid in mixed organic solvent [23], featuring a helical hairpin with the two transmembrane spanning α-helices connected by a short stretch of hydrophilic amino acids (see Fig. 4A). Both protein termini are oriented toward the periplasmic side of the membrane, whereas the hydrophilic loop region faces the cytoplasm, which is the F_1 binding site. cD61, most likely directly involved in the translocation of the coupling ion via salt bridging with residue aR210, is centered in the second transmembrane helix and was shown to be crucial for the conformation of the subunit c hairpin with respect to its protonated or deprotonated state [92] (see Fig. 5). The hairpin fold of monomeric subunit c in solution is likely to resemble the folding *in vivo*, since it features residue–residue interactions already predicted from the functional analysis of subunit c mutants, including a close proximity of residues cA24 and cI28 to cD61 [93, 94] as well as cA20 to cP64 [95, 96], either connecting the N- and C-terminal membrane spanning helices. The NMR model was further supported by extensive cross-linking studies using cysteine residues genetically introduced into subunit c in assembled F_1F_O complexes [97]. From the results obtained a packing arrangement could be derived, in which the radially ordered transmembrane helices of the subunit c monomers pack in two concentric rings with the N-terminal helix at the inside and the C-terminal helix oriented toward the outside of the annulus, thereby forming a hollow cylinder with an outer diameter of 5.5 to 6 nm and an inner space with a diameter of at least 1.1 nm [98]. Residue cD61 is thus thought to pack at the center of four transmembrane helices of two neighboring c-subunits, with a close proximity to residues cA24 and cI28 at the backface of the second c-subunit, which is perfectly in agreement with the analysis of a functional double mutant cA24D/D61G, in which the crucial carboxylate is shifted from position 61 to 24 [93]. In contrast, the analysis of genetically engineered subunit c mutants comprising tryptophan substitutions introduced in both the N- and C-terminal transmembrane domain led to steric conflicts in helix packing due to the bulky side chains in case of the C-terminal helix facing the outside of the ring [99, 100]. Furthermore, accessibility studies using monoclonal antibodies directed against the polar loop region also favor the N-terminal helix of subunit c to be packed at the outer periphery of the annulus [62].

The functional analysis of genetically engineered subunit c fusion proteins in *E. coli* were interpreted in a way that subunit c stoichiometry is likely to comprise a multiple of two and three, which is at least six but most likely twelve [101]. Chemical cross-linking of twofold cysteine-substituted subunit c monomers as well as engineered dimers and trimers within assembled F_1F_O complexes led to the formation of multiple cross-link products according to the number of potentially reticulating subunits within the ring-like oligomer, each of which comprising a multiple of the apparent molecular weight of the monomer. The finding of a maximum cross-link product equivalent to a 12-fold reticulated oligomer also argues in favor of 12 c-subunits being present in assembled F_1F_O complexes [97], which is in good accord with former stoichiometry analysis based on the quantification of biosynthetically incorporated radiolabel, resulting in 10±1 copies of subunit c per F_O complex [102]. It should clearly be denoted that the stoichiometry of subunit c may not be fixed, since at least among different organisms various stoichiometries have been reported, ranging from 10 c-subunits in yeast mitochondria F_O [8] over 12 for *E. coli* [101] to 14 in chloroplast ATP synthase [103]. In addition, in *E. coli* subunit c stoichiometry was shown to be dependent on the carbon source used for cultivation [104], thereby raising the possibility of some regulatory function in energy coupling during catalysis. In the structurally related eukaryotic V-type ATPases, the subunit c ring is supposed to consist of only six, but evolutionary duplicated copies of subunit c, each of which accordingly comprising four transmembrane helices, but only one carboxylate present in the ultimate membrane spanning domain [105]. Since V-type ATPases are solely capable of ATP-driven ion pumping, the presence of only one half of reactive carboxylates per ring in comparison to F-type ATPases might be the reason for their inability of coupling passive ion translocation to ATP synthesis *in vivo* [106, 107]. Further hints can be derived from the sodium translocating ATP synthase from *Acetobacterium woodii*. The concomitant presence of one type of duplicated c-subunit bearing only a single carboxylate like V-type subunit c together with two types of F-type-like c-subunits raises the possibility of a critical carboxylate ratio causing a possible switch from ATP-driven ion pumping to ATP synthesis [108]. Furthermore, the ATP synthase from the archaeon *Methanococcus jannaschii* contains a triplicated subunit c featuring six transmembrane helices together with two carboxylates potentially active in ion translocation [107, 109]. Hence, the determination of the exact stoichiometry of c-subunits in different organisms together with the number of carboxylates active in ion translocation constitutes a crucial challenge with respect to energy coupling.

2.7
Catalysis: Coupling Ion Translocation to ATP Synthesis

In the current model of energy transduction between the sites of ion translocation in F_O and ATP turnover in F_1, a $\gamma\varepsilon c_{12}$ rotor complex is supposed to propel against the static remainder of the F_1F_O ATP synthase functioning as a stator for the compensation of elastic torque generated during rotation [26, 27]. Stepwise movement of the subunit c ring triggered by the successive protonation/deprotonation of the conserved carboxylate within the C-terminal helix is thought to twist helical domains of subunit γ connected to the hydrophilic loop region of some of the c-subunits, thereby accumulating elastic tension within both the rotor and the stator part of the enzyme. Once the strain exerted transcends a critical threshold, tension is released by a 120° rotary step of the γ-subunit within the $\alpha_3\beta_3$ hexamer, thereby inducing the binding change of the catalytic sites as well as relapsing the system to its relaxed state. Relaxation at intervals of 120° each (i.e., stepping movement of the rotor part) could possibly be mediated through hydrogen bonding of subunit γ at the α/β interface via residue γR75 and corresponding carboxylates within the DELSEED sequence of neighboring α- and β-subunits [30]. The elastic coupling would also account for different subunit c stoichiometries eventually causing broken symmetries with respect to the three-stepped catalytic mechanism in F_1 [110]. Although this model is still a tempting hypothesis, there is some experimental evidence that the subunit c annulus co-rotates together with subunits γ and ε during catalysis. Mutations in the highly conserved loop region of subunit c prevent the formation of a coupled F_1F_O complex [111–114], most likely by interrupting energy transmission between F_1 and F_O. Subunit ε can be cross-linked to the subunit c oligomer without loss of enzyme activity [34], clearly arguing against a former model of rotary catalysis, in which a $\gamma\varepsilon$ rotor part slides along the surface of a static subunit c oligomer [115]. In contrast, cross-link formation between residues cV78 and bN2 abolishes ATP-driven proton pumping [91]. In addition, cross-linking also demonstrated that the c-subunit adjacent to the subunit b dimer exchanges with initially non-neighboring c-subunits during catalysis [91]. Finally, using an analogous approach as for rotation within the F_1 complex, a fluorescent actin filament covalently attached to the subunit c oligomer was shown to propel against immobilized F_1F_O complexes [116, 117]. The torque generated by the subunit c oligomer during rotation revealed to be almost identical to that exerted by subunit γ (i.e., 40 to 50 pN nm), indicating low-loss energy transmission perfectly in agreement with a direct structural link between both subunits. However, the structural and functional integrity of the analyzed enzyme complexes has not yet been unequivocally demonstrated [118].

Whereas the major principles of rotary energy transduction between F_1 and F_O have been clarified in different sets of experiments at least with respect to ATP-driven proton pumping, less consensus has yet been achieved on the exact mechanism of ion translocation within the F_O part of the enzyme. Especially the exact position and accessibility of the highly conserved carboxylate of subunit c revealed to be crucial for the current theory of enzyme function. Thus, two models emerged with either diverse location of the carboxylate together with an accordingly different channeling of the coupling ion (Fig. 4). In *E. coli*, the NMR spectroscopic data of the monomer accompanied by molecular modeling of the oligomeric structure by use of correspondingly generated cross-links argue in favor of an occluded location of residue cD61 near the middle of the membrane [98]. Thus, protonation/deprotonation of the carboxylate within the rotating oligomer is thought to occur via two half channels provided by the stator part of the F_O complex [27, 119]. Movement of the subunit c ring would basically occur by to-and-fro idling due to Brownian molecular movement. By involving the well-conserved positively charged residue aR210 of subunit a neighboring cD61 as a rectifying molecular ratchet, the to-and-fro fluctuation turns into unidirectional rotation (see Fig. 4B). In contrast, in the sodium translocating ATP synthase from *Propionigenium modestum*, subunit c features a slightly different structure. Based on NMR spectroscopic analysis, the helical hairpin was shown to be twofold interrupted, thereby comprising four helices instead of two [120]. The reactive carboxylate, that is cE65 in *P. modestum*, is situated at the C-terminal end of helix III (see Fig. 4A). Whereas there are only minor structural differences between the c-subunits from *P. modestum* and *E. coli*, strong discrepancies exist with respect to the accessibility of the ion binding site. In *E. coli*, residue cD61 is thought to be deeply embedded within the lipid bilayer. In contrast, the sodium binding site of *P. modestum* (i.e., cE65) was shown to be freely accessible from the aqueous phase, strongly arguing for an exposed position of the carboxylate [121, 122], thereby superceding the inlet channel with respect to the two-channel model mentioned above (see Fig. 4B). Furthermore, coupled occlusion of only one sodium ion per ATP synthase in assembled F_1F_O complexes from *P. modestum* favors only this subunit c to be occupied with the coupling ion, which directly interacts in channeling together with the stator subunits [122], whereas in the *E. coli* enzyme protonated c-subunits are supposed to rotate through the lipid bilayer for almost a complete revolution until their release at the outlet channel [119]. In addition, reconstituted F_1F_O complexes from *P. modestum* were shown to freely exchange sodium ions between the two compartments of the proteoliposome even in the absence of ATP, so that no rotation should occur except the to-and-fro idling of the ring caused by Brownian molecular motion. Since idling of the subunit c annulus along its axis can be considered as small angle rotation, ion translocation could take place involving just one rotor site to pick up and deliver a sodium ion to the nearby outlet channel without full-turn rotation [123, 124]. Hence, biochemical data clearly favor the one-channel model of ion translocation established for *P. modestum* versus the two-channel model proposed for *E. coli*, whereas structural data argue for a subunit *c* conformation, which places the carboxylate nearly at the center of the lipid bilayer. Whether either of these models is wrong or whether the subunit *c* ring from *P. modestum* and *E. coli* indeed features a different structure accompanied by a diverse mechanism of operation remains to be established. However, irrespective of the accessibility and exact position of the reactive carboxylate within the subunit *c* annulus, rotation of the ring past a static ab_2 stator subcomplex apparently poses some problems concerning the maintenance of structural integrity, since rotor and stator should be permanently fixed to each other in order to withstand the increasing torque during rotation. Especially a potentially eccentric rotation of the subunit *c* oligomer with subunits γ and ε attached to the periphery of the ring demands a bending of either subunit γ or of at least some stator components. Thus, the interaction between rotor and stator in F_O has to occur at more than only one site in order to maintain close contact and to secure the ion channel during rotation, which makes a rather dynamic and mechanical process more likely than simple rotational diffusion rectified by electrostatic forces. Correspondingly, a hand-over-hand mechanism as for kinesin has been proposed, according to which coordinate structural changes involving two or three *c*-subunits mechanically drive rotation of the ring, such that at least one of these *c*-subunits at a time is tightly bound to the stator [92] (Fig. 5). The structure of the subunit *c* monomer from *E. coli* was shown to drastically change upon protonation of residue *c*D61, resulting in a 140° rotation of the C-terminal helix with respect to the N-terminal domain accompanied by significant structural changes within the polar loop region. Helix rotation within the assembled *c* oligomer, in which a deprotonated *c* subunit is sandwiched by protonated counterparts, would thus bring a deprotonated *c*D61 close to its protonated complement, thereby resulting in proton transfer among the *c*-subunits and the essential *a*R210 residue [92].

No matter if all details of these mechanistic speculations prove to be right, more facts are required for the understanding of how the motor generates distinctive forces and how this is regulated. Although the proposed binding change mechanism could be verified by decisive experimental evidence, comparatively little is known about the mechanics within the F_O complex as on the atomic structure of its subunits. Both direct experimental evidence for proton-driven rotation and high-resolution structures seem to be a prerequisite for the understanding of F_1 and F_O interaction during rotary catalysis.

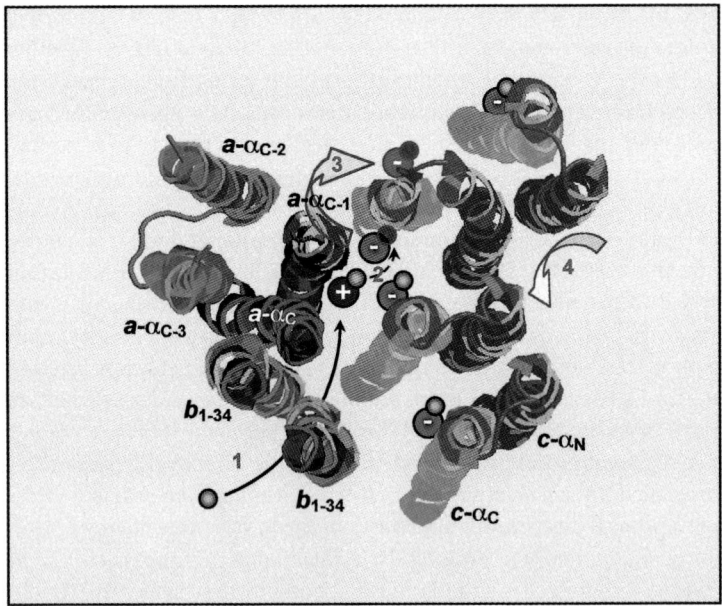

Fig. 5. Hand-over-hand pattern of the proton translocation pathway within the assembled F_O complex. Structural sketches are shown from four of the c-subunits (both the N- and C-terminal helix, c-a_N and c-a_C, respectively) as well as from the transmembrane domain of the subunit b dimer (b_{1-34}) and from the four C-terminal helices of subunit a (a-a_C – a-a_{C-3}) according to [92]. The assembly is presented as seen from the F_1 complex. The proposed functional cycle for the translocation of one proton is depicted according to the two-channel model established for the E. coli ATP synthase. The proton enters the complex via the inlet channel from the periplasmic side of the membrane, involving the positive stator charge aR210 (1). In the resting state, residue aR210 is sandwiched by both a protonated and a deprotonated cD61 side chain at the periphery of the subunit c oligomer. After proton transfer to cD61 (2), the C-terminal helix of the newly protonated monomer rotates 140° in order to adopt its protonated orientation (3), resulting in a fully protonated intermediate state of the oligomer. Simultaneously, by the interaction of cD61 and aR210 during helix rotation, the subunit c ring is pushed to rotate contrarily one step ahead (4), placing residue aR210 at the interface of the subsequent set of neighboring c-subunits. Concomitantly, residue cD61 of the next c-subunit loses its proton to the cytoplasmic side via the outlet channel (not shown), accompanied by rotation of the C-terminal helix in order to regenerate the deprotonated conformation of the resting state; see color plates p. XXVI.

References

1. G. Deckers-Hebestreit, K. Altendorf, *Annu. Rev. Microbiol.* **1996**, *50*, 791-824.
2. R. H. Fillingame, P. C. Jones, W. Jiang, F. I. Valiyaveetil, O. Y. Dmitriev, *Biochim. Biophys. Acta* **1998**, *1365*, 135-142.
3. J. Velours, P. Paumard, V. Soubannier, C. Spannagel, J. Vaillier, G. Arselin, P.-V. Graves, *Biochim. Biophys. Acta* **2000**, *1458*, 443-456.
4. M. L. Richter, F. Gao, *J. Bioenerg. Biomembr.* **1996**, *28*, 443-449.
5. J. P. Abrahams, A. G. W. Leslie, R. Lutter, J. E. Walker, *Nature* **1994**, *370*, 621-628.
6. E. P. Gogol, U. Lücken, R. A. Capaldi, *FEBS Lett.* **1987**, *219*, 274-278.
7. S. Wilkens, R. A. Capaldi, *Nature* **1998**, *393*, 29.
8. D. Stock, A. G. W. Leslie, J. E. Walker, *Science* **1999**, *286*, 1700-1705.
9. C. Tang, R. A. Capaldi, *J. Biol. Chem.* **1996**, *271*, 3018-3024.
10. S. D. Watts, C. Tang, R. A. Capaldi, *J. Biol. Chem.* **1996**, *271*, 28341-28347.
11. B. Schulenberg, F. Wellmer, H. Lill, W. Junge, S. Engelbrecht, *Eur. J. Biochem.* **1997**, *249*, 134-141.
12. H. Lill, F. Hensel, W. Junge, S. Engelbrecht, *J. Biol. Chem.* **1996**, *271*, 32737-32742.
13. I. Ogilvie, R. Aggeler, R. A. Capaldi, *J. Biol. Chem.* **1997**, *272*, 16652-16656.
14. S. Wilkens, J. Zhou, R. Nakayama, S. D. Dunn, R. A. Capaldi, *J. Mol. Biol.* **2000**, *295*, 387-391.
15. P. D. Boyer, *Annu. Rev. Biochem.* **1997**, *66*, 717-749.
16. T. Masaike, N. Mitome, H. Noji, E. Muneyuki, R. Yasuda, K. Kinosita Jr., M. Yoshida, *J. Exp. Biol.* **2000**, *203*, 1-8.
17. A. J. W. Rodgers, S. Wilkens, R. Aggeler, M. B. Morris, S. M. Howitt, R. A. Capaldi, *J. Biol. Chem.* **1997**, *272*, 31058-31064.
18. A. J. W. Rodgers, R. A. Capaldi, *J. Biol. Chem.* **1998**, *273*, 29406-29410.
19. B. Böttcher, L. Schwarz, P. Gräber, *J. Mol. Biol.* **1998**, *281*, 757-762.
20. B. Böttcher, I. Bertsche, R. Reuter, P. Gräber, *J. Mol. Biol.* **2000**, *296*, 449-457.
21. R. Birkenhäger, M. Hoppert, G. Deckers-Hebestreit, F. Mayer, K. Altendorf, *Eur. J. Biochem.* **1995**, *230*, 58-67.
22. S. Singh, P. Turina, C. J. Bustamante, D. J. Keller, R. A. Capaldi, *FEBS Lett.* **1996**, *397*, 30-34.
23. M. E. Girvin, V. K. Rastogi, F. Abildgaard, J. L. Markley, R. H. Fillingame, *Biochemistry* **1998**, *37*, 8817-8824.
24. R. H. Fillingame, in: *The Bacteria. A Treatise on Structure and Function* Vol. 12 (T. A. Krulwich, Ed.), Academic Press, New York, **1990**, pp. 345-391.
25. H. Wang, G. Oster, *Nature* **1998**, *396*, 279-282.
26. D. A. Cherepanov, A. Y. Mulkidjanian, W. Junge, *FEBS Lett.* **1999**, *449*, 1-6.
27. W. Junge, *Proc. Natl. Acad. Sci. USA* **1999**, *96*, 4735-4737.
28. J. E. Walker, M. Saraste, M. Runswick, N. J. Gay, *EMBO J.* **1982**, *1*, 945-951.
29. R. K. Nakamoto, C. J. Ketchum, M. K. Al-Shawi, *Annu. Rev. Biophys. Biomol. Struct.* **1999**, *28*, 205-234.
30. C. Gibbons, M. G. Montgomery, A. G. W. Leslie, J. E. Walker, *Nature Struct. Biol.* **2000**, *7*, 1055-1061.
31. U. Uhlin, G. B. Cox, J. M. Guss, *Structure* **1997**, *5*, 1219-1230.
32. S. Wilkens, F. W. Dahlquist, L. P. McIntosh, L. W. Donaldson, R. A. Capaldi, *Nature Struct. Biol.* **1995**, *2*, 961-967.
33. A. J. W. Rodgers, M. C. J. Wilce, *Nature Struct. Biol.* **2000**, *7*, 1051-1054.
34. B. Schulenberg, R. Aggeler, J. Murray, R. A. Capaldi, *J. Biol. Chem.* **1999**, *274*, 34233-34237.
35. R. Aggeler, R. A. Capaldi, *J. Biol. Chem.* **1996**, *271*, 13888-13891.
36. Y. Zhang, R. H. Fillingame, *J. Biol. Chem.* **1995**, *270*, 24609-24614.
37. J. Hermolin, O. Y. Dmitriev, Y. Zhang, R. H. Fillingame, *J. Biol. Chem.* **1999**, *274*, 17011-17016.
38. Y. Kato-Yamada, M. Yoshida, T. Hisabori, *J. Biol. Chem.* **2000**, *275*, 35746-35750.
39. S. Wilkens, S. D. Dunn, J. Chandler, F. W. Dahlquist, R. A. Capaldi, *Nature Struct. Biol.* **1997**, *4*, 198-201.
40. J. Weber, A. E. Senior, *Biochim. Biophys. Acta* **2000**, *1458*, 300-309.

41. M. A. Bianchet, J. Hullihen, P. L. Pedersen, L. M. Amzel, *Proc. Natl. Acad. Sci. USA* **1998**, *95*, 11065-11070.
42. G. Groth, E. Pohl, *J. Biol. Chem.* **2001**, *276*, 1345-1352.
43. T. M. Duncan, V. V. Bulygin, Y. Zhou, M. L. Hutcheon, R. L. Cross, *Proc. Natl. Acad. Sci. USA* **1995**, *92*, 10964-10968.
44. Y. Zhou, T. M. Duncan, R. L. Cross, *Proc. Natl. Acad. Sci. USA* **1997**, *94*, 10583-10587.
45. V. V. Bulygin, T. M. Duncan, R. L. Cross, *J. Biol. Chem.* **1998**, *273*, 31765-31769.
46. D. Sabbert, S. Engelbrecht, W. Junge, *Nature* **1996**, *381*, 623-625.
47. K. Häsler, S. Engelbrecht, W. Junge, *FEBS Lett.* **1998**, *426*, 301-304.
48. H. Noji, R. Yasuda, M. Yoshida, K. Kinosita Jr., *Nature* **1997**, *386*, 299-302.
49. R. Yasuda, H. Noji, K. Kinosita Jr., M. Yoshida, *Cell* **1998**, *93*, 1117-1124.
50. T. Hisabori, A. Kondoh, M. Yoshida, *FEBS Lett.* **1999**, *463*, 35-38.
51. H. Noji, K. Häsler, W. Junge, K. Kinosita Jr., M. Yoshida, S. Engelbrecht, *Biochem. Biophys. Res. Commun.* **1999**, *260*, 597-599.
52. H. Omote, N. Sambonmatsu, K. Saito, Y. Sambongi, A. Iwamoto-Kihara, T. Yanagida, Y. Wada, M. Futai, *Proc. Natl. Acad. Sci. USA* **1999**, *96*, 7780-7784.
53. Y. Kato-Yamada, H. Noji, R. Yasuda, K. Kinosita Jr., M., Yoshida, *J. Biol. Chem.* **1998**, *273*, 19375-19377.
54. K. Kinosita Jr., R. Yasuda, H. Noji, K. Adachi, *Phil. Trans. R. Soc. Lond.* B **2000**, *355*, 473-489.
55. D. T. McLachlin, A. M. Convey, S. M. Clark, S. D. Dunn, *J. Biol. Chem.* **2000**, *275*, 17571-17577.
56. D. T. McLachlin, J. A. Bestard, S. D. Dunn, *J. Biol. Chem.* **1998**, *273*, 15162-15168.
57. D. T. McLachlin, S. D. Dunn, *Biochemistry* **2000**, *39*, 3486-3490.
58. K. Häsler, O. Pänke, W. Junge, *Biochemistry* **1999**, *38*, 13759-13765.
59. S. D. Dunn, J. Chandler, *J. Biol. Chem.* **1998**, *273*, 8646-8651.
60. S. D. Watts, Y. Zhang, R. H. Fillingame, R. A. Capaldi, *FEBS Lett.* **1995**, *368*, 235-238.
61. T. Licher, E. Kellner, H. Lill, *FEBS Lett.* **1998**, *431*, 419-422.
62. R. Birkenhäger, J.-C. Greie, K. Altendorf, G. Deckers-Hebestreit, *Eur. J. Biochem.* **1999**, *264*, 385-396.
63. S. D. Watts, R. A. Capaldi, *J. Biol. Chem.* **1997**, *272*, 15065-15068.
64. E. Schneider, K. Altendorf, *Proc. Natl. Acad. Sci. USA* **1984**, *81*, 7279-7283.
65. E. Schneider, K. Altendorf, *EMBO J.* **1985**, *4*, 515-518.
66. J.-C. Greie, G. Deckers-Hebestreit, K. Altendorf, *Eur. J. Biochem.* **2000**, *267*, 3040-3048.
67. J.-C. Greie, G. Deckers-Hebestreit, K. Altendorf, *J. Bioenerg. Biomembr.* **2000**, *32*, 357-364.
68. G. Deckers-Hebestreit, J.-C. Greie, W.-D. Stalz, K. Altendorf, *Biochim. Biophys. Acta* **2000**, *1458*, 364-373.
69. H. Yamada, Y. Moriyama, M. Maeda, M. Futai, *FEBS Lett.* **1996**, *390*, 34-38.
70. H. Jäger, R. Birkenhäger, W.-D. Stalz, K. Altendorf, G. Deckers-Hebestreit, *Eur. J. Biochem.* **1998**, *251*, 122-132.
71. J. C. Long, S. Wang, S. B. Vik, *J. Biol. Chem.* **1998**, *273*, 16235-16240.
72. F. I. Valiyaveetil, R. H. Fillingame, *J. Biol. Chem.* **1998**, *273*, 16241-16247.
73. T. Wada, J. C. Long, D. Zhang, S. B. Vik, *J. Biol. Chem.* **1999**, *274*, 17353-17357.
74. C. Bjørbæk, V. Foërsom, O. Michelsen, *FEBS Lett.* **1990**, *260*, 31-34.
75. M. J. Lewis, J. A. Chang, R. D. Simoni, *J. Biol. Chem.* **1990**, *265*, 10541-10550.
76. G. Deckers-Hebestreit, K. Altendorf, *J. Exp. Biol.* **1992**, *172*, 451-459.
77. W. Jiang, R. H. Fillingame, *Proc. Natl. Acad. Sci. USA* **1998**, *95*, 6607-6612.
78. K. Steffens, E. Schneider, G. Deckers-Hebestreit, K. Altendorf, *J. Biol. Chem.* **1987**, *262*, 5866-5869.
79. M. Takeyama, T. Noumi, M. Maeda, M. Futai, *J. Biol. Chem.* **1988**, *263*, 16106-16112.
80. S. D. Dunn, *J. Biol. Chem.* **1992**, *267*, 7630-7636.
81. D. T. McLachlin, S. D. Dunn, *J. Biol. Chem.* **1997**, *272*, 21233-21239.
82. P. L. Sorgen, M. R. Bubb, K. A. McCormick, A. S. Edison, B. D. Cain, *Biochemistry* **1998**, *37*, 923-932.
83. O. Dmitriev, P. C. Jones, W. Jiang, R. H. Fillingame, *J. Biol. Chem.* **1999**, *274*, 15598-15604.
84. J. E. Walker, M. Saraste, N. J. Gay, *Nature* **1982**, *298*, 867-869.
85. A. E. Senior, *Biochim. Biophys. Acta* **1983**, *726*, 81-95.

86. P. L. Sorgen, T. L. Caviston, R. C. Perry, B. D. Cain, *J. Biol. Chem.* **1998**, *273*, 27873-27878.
87. J. P. Aris, R. D. Simoni, *J. Biol. Chem.* **1983**, *258*, 14599-14609.
88. J. Hermolin, J. Gallant, R. H. Fillingame, *J. Biol. Chem.* **1983**, *258*, 14550-14555.
89. C. A. Kumamoto, R. D. Simoni, *J. Biol. Chem.* **1986**, *261*, 10037-10042.
90. R. H. Fillingame, W. Jiang, O. Y. Dmitriev, P. C. Jones, *Biochim. Biophys. Acta* **2000**, *1458*, 387-403.
91. P. C. Jones, J. Hermolin, W. Jiang, R. H. Fillingame, *J. Biol. Chem.* **2000**, *275*, 31340-31346.
92. V. K. Rastogi, M. E. Girvin, *Nature* **1999**, *402*, 263-268.
93. M. J. Miller, M. Oldenburg, R. H. Fillingame, *Proc. Natl. Acad. Sci. USA* **1990**, *87*, 4900-4904.
94. R. H. Fillingame, M. Oldenburg, D. Fraga, *J. Biol. Chem.* **1991**, *266*, 20934-20939.
95. A. L. Fimmel, D. A. Jans, L. Langman, L. B. James, G. R. Ash, J. A. Downie, A. E. Senior, F. Gibson, G. B. Cox, *Biochem. J.* **1983**, *213*, 451-458.
96. Y. Zhang, R. H. Fillingame, *J. Biol. Chem.* **1995**, *270*, 87-93.
97. P. C. Jones, W. Jiang, R. H. Fillingame, *J. Biol. Chem.* **1998**, *273*, 17178-17185.
98. O. Y. Dmitriev, P. C. Jones, R. H. Fillingame, *Proc. Natl. Acad. Sci. USA* **1999**, *96*, 7785-7790.
99. G. Groth, Y. Tilg, K. Schirwitz, *J. Mol. Biol.* **1998**, *281*, 49-59.
100. C. Schnick, L. R. Forrest, M. S. P. Sansom, G. Groth, *Biochim. Biophys. Acta* **2000**, *1459*, 49-60.
101. P. C. Jones, R. H. Fillingame, *J. Biol. Chem.* **1998**, *273*, 29701-29705.
102. D. L. Foster, R. H. Fillingame, *J. Biol. Chem.* **1982**, *257*, 2009-2015.
103. H. Seelert, A. Poetsch, N. A. Dencher, A. Engel, H. Stahlberg, D. J. Müller, *Nature* **2000**, *405*, 418-419.
104. R. A. Schemidt, J. Qu, J. R. Williams, W. S. A. Brusilow, *J. Bacteriol.* **1998**, *180*, 3205-3208.
105. B. Powell, L. A. Graham, T. H. Stevens, *J. Biol. Chem.* **2000**, *275*, 23654-23660.
106. M. Mandel, Y. Moriyama, J. D. Hulmes, Y. C. Pan, H. Nelson, N. Nelson, *Proc. Natl. Acad. Sci. USA* **1988**, *85*, 5521-5524.
107. C. Ruppert, H. Kavermann, S. Wimmers, R. Schmid, J. Kellermann, F. Lottspeich, H. Huber, K. O. Stetter, V. Müller, *J. Biol. Chem.* **1999**, *274*, 25281-25284.
108. S. Aufurth, H. Schägger, V. Müller, *J. Biol. Chem.* **2000**, *275*, 33297-33301.
109. V. Müller, C. Ruppert, T. Lemker, *J. Bioenerg. Biomembr.* **1999**, *31*, 15-27.
110. S. J. Ferguson, *Curr. Biol.* **2000**, *10*, R804-R808.
111. M. E. Mosher, L. K. White, J. Hermolin, R. H. Fillingame, *J. Biol. Chem.* **1985**, *260*, 4807-4814.
112. M. J. Miller, D. Fraga, C. R. Paule, R. H. Fillingame, *J. Biol. Chem.* **1989**, *264*, 305-311.
113. D. Fraga, R. H. Fillingame, *J. Bacteriol.* **1991**, *173*, 2639-2643.
114. D. Fraga, J. Hermolin, M. Oldenburg, M. J. Miller, R. H. Fillingame, *J. Biol. Chem.* **1994**, *269*, 7532-7537.
115. R. H. Fillingame, *J. Exp. Biol.* **1997**, *200*, 217-224.
116. Y. Sambongi, Y. Iko, M. Tanabe, H. Omote, A. Iwamoto-Kihara, I. Ueda, T. Yanagida, Y. Wada, M. Futai, *Science* **1999**, *286*, 1722-1724.
117. O. Pänke, K. Gumbiowski, W. Junge, S. Engelbrecht, *FEBS Lett.* **2000**, *472*, 34-38.
118. S. P. Tsunoda, R. Aggeler, H. Noji, K. Kinosita Jr., M. Yoshida, R. A. Capaldi, *FEBS Lett.* **2000**, *470*, 244-248.
119. W. Junge, H. Lill, S. Engelbrecht, *Trends Biochem. Sci.* **1997**, *22*, 420-423.
120. U. Matthey, G. Kaim, D. Braun, K. Wüthrich, P. Dimroth, *Eur. J. Biochem.* **1999**, *261*, 459-467.
121. C. Kluge, P. Dimroth, *Biochemistry* **1993**, *32*, 10378-10386.
122. G. Kaim, U. Matthey, P. Dimroth, *EMBO J.* **1998**, *17*, 688-695.
123. G. Kaim, P. Dimroth, *EMBO J.* **1998**, *17*, 5887-5895.
124. P. Dimroth, H. Wang, M. Grabe, G. Oster, *Proc. Natl. Acad. Sci. USA* **1999**, *96*, 4924-4929.
125. E. Muneyuki, H. Noji, T. Amano, T. Masaike, M. Yoshida, *Biochim. Biophys. Acta* **2000**, *1458*, 467-481.
126. G. Oster, H. Wang, *Structure* **1999**, *7*, R67-R72.
127. T. Elston, H. Wang, G. Oster, *Nature* **1998**, *391*, 510-513.

3
Sodium/Substrate Transport

Heinrich Jung

3.1
Introduction

The major permeability barrier in bacteria is the cytoplasmic membrane. To cross this barrier solutes such as organic acids, peptides, sugars, and ions utilize specific passive or active transport systems localized in the membrane. Active transport systems are able to accumulate solutes against a concentration gradient, thereby an input of energy is required. According to a concept introduced by P. Mitchell [1] active transport can be classified based on the source of energy: (1) In primary active transport the energy comes from light or chemical reactions and is converted into electrochemical energy, e. g., a solute or ion concentration gradient. Examples are the light-driven H^+ pump bacteriorhodopsin, Na^+ transporting decarboxylases, and ATP-driven substrate uptake systems. (2) Group translocating transport systems are characterized by the fact that the transported solute becomes chemically modified during translocation. The only known examples are the phosphoenolpyruvate:sugar phosphotransferase systems (PTS) of bacteria. (3) In secondary transport energy stored in an existing electrochemical gradient of a solute is used to drive the accumulation of a second solute. Transport of the solutes can occur in the same direction or in opposite directions across the membrane. An example for movement in the same direction is given by Na^+-coupled sugar uptake in animal cells. Four decades ago R. K. Crane introduced the term cotransport for this type of process [2]. Coupled transport of solutes moving in opposite directions was called countertransport or exchange and is exemplified by Na^+/H^+ exchange. Based on P. Mitchell"s work and driven by the rapid development of molecular studies on bacterial transporters, particularly on the H^+/β-galactoside transporter (lactose permease, LacY) of *Escherichia coli* [3, 4], the terms symport and antiport were introduced for cotransport and countertransport (exchange), respectively [6]. In relation to these terms, uniport comprises the movement of a single solute in one direction across the membrane, thereby transport shows kinetic characteristics similar to symport or antiport. For a more detailed discussion of terms and definitions used to describe transport processes the reader is referred to [5–7]. Here, the terms symport, antiport, and uniport are used to discriminate between the different types of secondary transport.

In prokaryotes, H^+ and Na^+ are the most common coupling ions used by ion-coupled secondary transport systems. The necessary electrochemical ion gradients are built up by primary or secondary ion translocating processes. In many organisms, the proton motive force (pmf) generated by substrate oxidation via the respiratory chain is used to extrude Na^+ from the cell by a Na^+/H^+ antiport mechanism thereby establishing an electrochemical Na^+ gradient across the cytoplasmic membrane (sodium motive force, smf) [8]. Alternatively, a variety of prokaryotic cells possess primary Na^+ pumps such as membrane-bound decarboxylases, Na^+-translocating respiratory complexes, or ATPases to generate a smf [9–11]. Many prokaryotes maintain electrochemical gradients of both H^+ and Na^+ across the cytoplasmic membrane.

This chapter deals with Na^+-coupled substrate transport processes in prokaryotes thereby discussing their occurrence and physiological role and summarizing the knowledge on their structure, functional properties, and structure–function relationships. The well characterized bacterial transport systems MelB and PutP of *E. coli* and CitS of *Klebsiella pneumoniae* serve as model systems and are described in more detail. Different Na^+-dependent transporters are highly similar to H^+-dependent systems and both types of coupling can be found within one protein family. In some cases already a single amino acid substitution alters the ion selectivity. Therefore, results on H^+-dependent systems are included in the discussion where it appears appropriate.

3.2
Occurrence and Role of Na^+/Substrate Transport Systems

3.2.1
General Considerations

Transport of a given substrate can be coupled to H^+ in one organism and to Na^+ in another. There is no relation between substrate and ion specificity [12]. So, H^+- and Na^+-coupled transport systems exist within one organism, thereby overlapping substrate specificities can occur [13, 14]. There appear to be no major structural differences between H^+- and Na^+-dependent transporters (see Sects. 3.3.2 and 3.5.1). Furthermore, significant differences in the catalytic efficiency of H^+- and Na^+-coupled transport systems are not apparent. Therefore, the existence of H^+- and Na^+-coupled systems cannot be due to a need to optimize catalytic efficiency of secondary transporters during evolution but must be related to specific environmental conditions [12]. Obvious environmental parameters that influence the choice of the coupling ion are temperature, pH, and salinity [15–18]. In this context it has to be considered that high internal Na^+ concentrations are toxic and cells are required to keep the concentration of the ion at a low level [19]. Furthermore, many cellular functions require pH homeostasis, thereby Na^+/H^+ antiport plays a pivotal role

[20]. The choice of the transport system and thereby of the coupling ion for a given substrate can also be determined by metabolic requirements as they arise, e.g., upon a switch from aerobic to anaerobic growth conditions (e.g., [21–23]). Examples are discussed below.

3.2.2
Elevated Temperatures

H^+ and Na^+ cross a phospholipid bilayer most likely by different mechanisms which result in a significantly lower transport rate for Na^+ compared to H^+ at a given temperature [24, 25]. Besides alteration of the membrane lipid composition, some thermophiles use, therefore, Na^+ as sole energy coupling ion [25, 26]. *Caloramator fervidus* (previously *Clostridium fervidus*) is an example for a thermophilic anaerobe (optimum growth temperature 60 °C) that relies completely on Na^+ as the energy coupling ion [27]. A Na^+ electrochemical gradient across the cytoplasmic membrane is generated by a V-type Na^+-ATPase and is utilized, e.g., for the accumulation of neutral, acidic, and basic amino acids [28–30]. However, due to the high H^+ permeability of its membrane, *C. fervidus* is unable to maintain a constant intracellular pH thereby limiting growth to a narrow pH range (pH 6.3-7.7) [27].

Also the thermophilic aerobic bacterium *Bacillus stearothermophilus* (optimum growth temperature 63 °C) catalyzes the uptake of a variety of amino acids in a Na^+-dependent manner [31]. However, differing from *C. fervidus*, this organism does not contain a primary Na^+ pump. Instead, an electrochemical H^+ gradient generated by aerobic respiration is used to establish a smf via Na^+/H^+ antiport. It is speculated that the use of Na^+ as coupling ion may have the advantage that the Na^+ gradient is less sensitive to fluctuations in the environment of the bacterium [16, 32].

It should be pointed out that the completed genomes of archaea living under extreme conditions like the methanogenic anaerobes *Methanobacterium thermoautotrophicum* (optimum growth temperature 65 °C) and *Methanococcus jannaschii* (optimum growth temperature 85 °C) and the hyperthermophilic marine sulfate reducer *Archaeoglobus fulgidus* (optimum growth temperature 83 °C) reveal that these organisms contain a large number of ABC-type transporters while putative secondary transporters constitute only a small fraction of the total membrane transport proteins [33–35]. This might be explained by the higher binding affinities of ABC transporters ($K_d < 1$ μM), which are required to scavenge substrates available only at very low concentrations in the environment of the cells [26]. Among the remaining secondary transport systems is a putative Na^+/proline transporter (PutP) which is found, e.g., in *M. thermoautotrophicum*, *A. fulgidus*, and *Pyrococcus horikoshii* [34–37].

3.2.3
Na$^+$-rich Environments

A variety of bacteria living in Na$^+$-rich environments generate primary Na$^+$ cycles, thereby Na$^+$ becomes the main coupling ion for transport [21]. Marine bacteria live in a high-salt environment and need Na$^+$ for optimal growth. *Vibrio alginolyticus* is an example for a marine bacterium in which secondary substrate transport is obligatory coupled to Na$^+$. This gram-negative bacterium possesses a Na$^+$-translocating NADH-quinone oxidoreductase complex and is by this means able to directly generate a smf by respiratory chain activity [38, 39]. The smf is directly coupled to the active uptake of nutrients and to the rotation of polar flagella [38, 40, 41]. Thus, it is shown that Na$^+$ is essential for the active uptake of all amino acids of the organism [42]. Remarkably, it is also demonstrated that the activity of the high-affinity K$^+$ uptake system KtrAB of *V. alginolyticus* depends on Na$^+$ ions [43, 44]. Furthermore, protein export has been suggested to depend on a smf [45]. In *Vibrio parahaemolyticus*, one of the major causes of food poising, the smf is used for the uptake of a variety of solutes including galactose [46–49]. Recently, it was shown that the strain contains a Na$^+$-driven multidrug efflux pump (NorM) [50].

Also rumen bacteria live in a Na$^+$-rich habitat, and uptake of a variety of substrates in these bacteria is coupled to Na$^+$ [51]. Na$^+$-dependent transport has been observed, e.g., for neutral amino acids in *Streptococcus bovis* [52], and for glucose and probably also cellobiose in the cellulolytic anaerobe *Fibrobacter succinogenes* [51, 53]. Interestingly, *Selenomonas ruminantium* was shown to possess a Na$^+$/succinate transporter which is implicated in energy conservation via efflux of the fermentation product succinate in symport with Na$^+$ [54].

The halophilic archaea of the order Halobacteriales establish a smf at the expense of a primary generated pmf [17]. The underlying Na$^+$/H$^+$ antiport processes play an essential role in osmoregulation and pH homeostasis in the organisms [19, 20]. Na$^+$-coupled substrate transport (e.g., amino acid uptake) seems to be the main pathway of re-entry of Na$^+$ into the cells [17, 21]. The genome sequence of *Halobacterium* species NRC-1, an obligatory halophilic microorganism which is adapted to a salinity 10 times higher than seawater, reveals, e.g., the existence of a putative Na$^+$/proline transporter (PutP) [55]. However, most transport systems of the strain are putative ABC-type transporters.

3.2.4
High pH

Na$^+$-coupled uptake processes play an important role in alkaliphilic bacteria such as alkaliphilic *Bacillus* species [18]. These bacteria face the challenge to maintain a cytoplasmic pH that can be much more acidic than the external medium. Withstanding high external pH values by *Bacillus* species requires an active Na$^+$ cycle [56]. This cycle involves the outward movement of Na$^+$ through Na$^+$/H$^+$ antiporters and mechanisms that complete the cycle by catalyzing re-entry of

Na^+. Na^+-dependent substrate transport systems participate in the latter process [56]. The uptake of amino acids, carbohydrates, polyamines, and organic acids has been described as Na^+-dependent in alkaliphilic organisms [57–61]. At least under certain conditions the corresponding transport systems play an important role in pH homeostasis. So, at low Na^+ concentrations Na^+/substrate uptake can be rate limiting. Thus, the presence of a non-metabolizable amino acid analog whose uptake is coupled to Na^+ markedly improves pH homeostasis in an alkaliphilic *Bacillus* during an alkaline shift at low Na^+ concentrations [62]. However, appreciable pH homeostasis by alkaliphiles is also shown in the absence of solutes that enter with Na^+ [62]. As an explanation it has been suggested that the Na^+/substrate transporters themselves may have a mode that allows Na^+ influx in the absence of substrate above some threshold of cytoplasmic pH. The participation of specific pH-regulated Na^+ channels is also discussed [56].

3.2.5
Citrate Fermentation

Na^+/substrate transport can also be crucial for the energy metabolism of anaerobically growing mesophilic bacteria. This is exemplified by the citrate fermentation pathway of *K. pneumoniae* (Fig. 1). During anaerobic growth on citrate as sole source of carbon and energy the tricarboxylic acid is converted into acetate, formate, and CO_2. During this process 1 mol ATP mol^{-1} citrate is generated by substrate level phosphorylation via the acetate kinase reaction [23]. Use of the ATP to

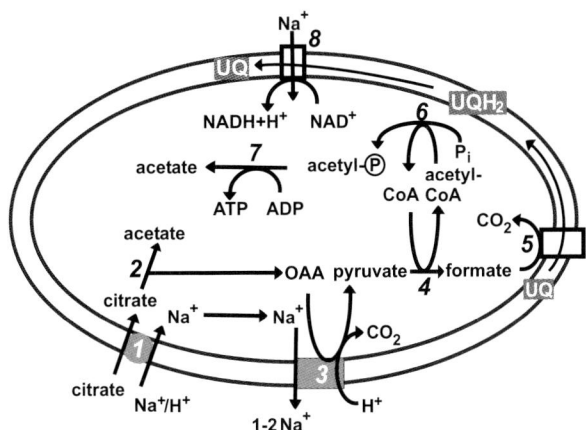

Fig. 1. Pathway of citrate fermentation in *K. pneumoniae* (according to [22]). Citrate is taken up in a Na^+-dependent manner catalyzed by CitS (1) and subsequently cleaved by citrate lyase (2) to acetate and oxaloacetate. The latter compound is decarboxylated by a Na^+ pumping decarboxylase (3) thereby generating a smf. The resulting pyruvate is further cleaved to formate and acetyl-CoA by pyruvate/formate lyase (4). After conversion of acetyl-CoA to acetyl-phosphate catalyzed by phosphotransacetylase (6) ATP is formed by the acetate kinase (7) reaction. Formate is oxidized to CO_2 by a membrane-bound formate dehydrogenase (5). Ubiquinol formed in this process is re-oxidized by smf-driven reverse electron transfer (8).

generate a pmf which subsequently drives citrate uptake would create a futile cycle. Instead, the citrate fermentation pathway is essentially coupled to a primary Na^+ cycle which involves a Na^+ pumping oxaloacetate decarboxylase acting as smf generator, the Na^+ pumping NADH:ubiquinone oxidoreductase operating in the direction of NADH formation, and the $Na^+/$citrate transporter CitS [22]. A similar Na^+-dependent pathway has been described for *Salmonella typhimurium* but not for *E. coli* [23, 63, 64].

3.2.6
$Na^+/$Substrate Transport in *Escherichia coli*

In *E. coli*, accumulation of proline, glutamate, serine/threonine, pantothenate, and melibiose by the $Na^+/$substrate transporters PutP, GltS, SstT, PanF, and MelB, respectively, occurs at the expense of an electrochemical Na^+ gradient [65–72] (Fig. 2). The transporters are part of a Na^+ cycle which is established secondarily to a primary existing H^+ cycle and involves the action of Na^+/H^+ antiporter(s) [12, 73, 74]. Interestingly, an *E. coli* mutant which lacks the Na^+/H^+ antiporter genes *nhaA* and *nhaB* has recently been shown to pump Na^+ by a primary mechanism involving the NADH:ubiquinone oxidoreductase (complex I) [75]. Nonetheless, secondary substrate transport in *E. coli* is mainly coupled to H^+, and the H^+-dependent systems can in part substitute for the Na^+-coupled transporters. Furthermore, the melibiose transporter (MelB) of *E. coli* can either use H^+ or Na^+ (or Li^+) to accumulate the sugar [71]. In this context, $Na^+/$substrate uptake does not appear to play a crucial role in *E. coli* metabolism and the corresponding transport systems may only represent the remains of earlier stages of evolution.

The Na^+-dependent transporters of *E. coli* are part of catabolic pathways and are used for the acquisition of the corresponding solute as carbon source [76, 77]. For

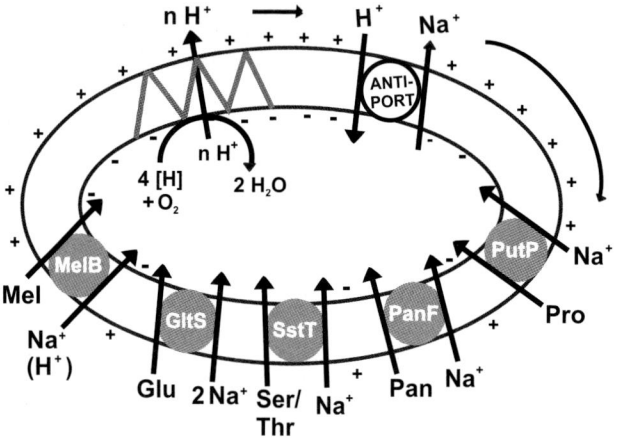

Fig. 2. Na^+-dependent substrate uptake systems of *E. coli*. The cells generate a primary H^+ cycle which is subsequently used to establish a Na^+ cycle which involves the action of Na^+/H^+ antiporter(s) and the indicated $Na^+/$substrate symport systems.

example, the *putP* gene encoding a Na$^+$/proline transporter of *E. coli* and *S. typhimurium* is part of the *put* operon which also encodes a proline dehydrogenase (product of *putA* gene) allowing the use of proline as a source of nitrogen and/or carbon [74, 76]. PanF, the Na$^+$/pantothenate transporter of *E. coli* and other prokaryotes, scavenges extracellular pantothenate for coenzyme A biosynthesis in the cells [78].

3.2.7
Osmotic Stress

In *E. coli* and *S. typhimurium*, increase of the osmolality of the medium leads to reduced Na$^+$/proline uptake by PutP for catabolic purposes [79]. Instead, two different transporters, ProP and ProU, are synthesized which accumulate proline but also glycine betaine and other betaines in response to osmotic stress [80–84]. ProP is a secondary, H$^+$-linked transport system and belongs to the major facilitator superfamily (TC 2.A.1) [85]. The isolated protein has recently been shown to act as an osmosensor and an osmoregulator [86]. ProU is a binding protein-dependent ABC transporter which preferentially transports glycine betaine and shows a relatively low affinity and uptake rate for proline compared to glycine betaine [80].

The situation is different in *Bacillus subtilis*. Here, the Na$^+$/proline transporter OpuE, a PutP homolog, participates in osmoadaptation and is apparently not involved in a catabolic pathway [87, 88]. OpuE activity is strongly enhanced in hypertonic media thereby highlighting the distinct physiological functions of the PutP and OpuE proteins [79, 88]. The osmoregulated proline transport activity of OpuE is entirely dependent on *de novo* protein synthesis and is controlled at the transcriptional level [87, 88]. Osmosensory properties as observed for the osmoprotectant uptake systems ProP of *E. coli* or BetP of *Corynebacterium glutamicum* could not be detected for OpuE [86, 88, 89]. The genome sequencing project of *B. subtilis* reveals that the bacterium probably contains a second proline transporter, YcgO, which is highly similar to OpuE and probably involved in proline catabolism [90] (E. Bremer, personal communication). Thus, two closely related proteins are obviously used for different physiological functions by putting their structural genes under the control of different regulatory schemes [88].

3.3
Functional Properties of Na$^+$/Substrate Transport Systems

3.3.1
General Considerations

The ever growing number of known transport proteins including Na$^+$-dependent transport systems is classified in different families based on their sequence similarities. There are mixed families of transporters that despite their sequence similarity accept either Na$^+$ or H$^+$ or both as coupling ion. Examples are the members of the glycoside–pentose–hexuronide (GPH) family (TC 2.A.2) (see Sect. 3.3.2), the cation/citrate family (CCS or 2-hydroxycarboxylate transporter family, TC 2.A.24) (see Sect. 3.3.4), and the dicarboxylate/amino acid-cation symporter (DAACS) family (TC 2.A.23) [92, 136–139, 160]. Other transporters fall into distinct families which either contain only Na^+- or H^+-dependent systems [92]. The Na^+/solute symporter (SSF) family (or SGLT/PutP family, TC 2.A.21) (see Sect. 3.3.3) and the Na^+/neurotransmitter symporter (NSS) family (TC 2.A.22) are examples for families of Na^+-dependent symporters [92, 119, 136, 139]. Also substrate uptake in symport with two different cations, H^+ and Na^+, has been observed [31, 32, 137]. The cation specificity can be altered by minor structural alterations as demonstrated by the influence of the lipid environment on the cation specificity of different glutamate transporters [69]. In the following, the functional properties of MelB and PutP of *E. coli* and of CitS of *K. pneumoniae* are summarized.

3.3.2
MelB

The melibiose transporter of *E. coli* (EcMelB) was discovered by Prestige and Pardee in 1965 [140]. The transporter is a member of the GPH family of transport proteins [138]. The MelB subfamily within this GPH family consists of melibiose transporters from *E. coli*, *S. typhimurium*, *K. pneumoniae*, and *Enterobacter aerogenes* [116, 138, 151]. Although there is a high degree of amino acid identity (78 to 85%) the melibiose transporters show distinct differences in ion selectivity. A special feature of EcMelB is its broad cation specificity, accepting Na^+, Li^+, and H^+ as coupling ions [109, 141, 142]. In the absence of Na^+ and Li^+ (< 20 to 50 µM) melibiose is taken up in symport with H^+. However, Na^+ and Li^+ readily compete with and displace H^+ and, therefore, it was suggested that Na^+ is the likely physiological coupling ion for melibiose uptake [142–146]. The melibiose transporter of *K. pneumoniae* (KnMelB) accepts H^+ and Li^+ but not Na^+ as coupling ions [152]. Other members of the GPH family like LacS of *Streptococcus thermophilus* catalyze sugar uptake only in symport with H^+ [153].

Since H^+ and Na^+ (and Li^+) probably share the same binding site in EcMelB, it appears likely that the interactions of the transporter with both H^+ and Na^+ are similar. Therefore, it has been suggested that H_3O^+ rather than H^+ may be the symported species [147]. If this is indeed the case, H^+ would not be translocated via a chain of protonable–deprotonable groups ("H^+ wire"), but H_3O^+ (similar as proposed for Na^+) would interact with the transporter through coordination to oxygen and/or nitrogen atoms thereby behaving as any other solute (see [14, 138, 146, 147] for further discussion of the subject).

Interestingly, it was shown that the ion selectivity profile of EcMelB is determined to some extent by the configuration of the transported sugar [71, 143, 145]. So, α-galactosides [melibiose, p-nitrophenyl-α-D galactopyranoside (NPG)] utilize Na^+, Li^+, or H^+ as coupling ion, whereas uptake of β-galactosides [lactose, methyl-β-D-thiogalactoside (TMG)] is preferentially coupled to Na^+ or Li^+. These data indicate a functional interaction between the sites of ion and sugar binding [138, 146].

Fig. 3. Ordered binding model of Na$^+$/substrate symport (based on [156–159]). According to this scheme, Na$^+$ binds to the empty transporter first thereby inducing a conformational alteration which increases the affinity of the transporter for the solute. The formation of the ternary complex induces another structural change that exposes Na$^+$ and substrate to the other site of the membrane. Substrate and then Na$^+$ are released and the empty transporter reorientates in the membrane allowing the cycle to start again.

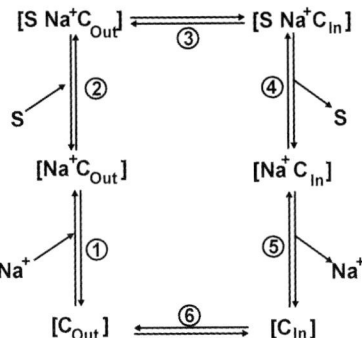

Sugar binding experiments indicate that Na$^+$ (or Li$^+$) increases the sugar affinity of EcMelB [148]. Furthermore, the mechanism of sugar transport by EcMelB was studied by analyzing the facilitated diffusion properties of the transporter in the presence of different coupling ions [149, 150]. The data show that the maximum rates (V_{max}) of melibiose efflux coupled to either Na$^+$, Li$^+$, or H$^+$ are by minimum one order of magnitude greater than the corresponding influx rates suggesting that the transporter functions asymmetrically. Exchange of melibiose by EcMelB in the presence of Na$^+$ proceeds without concomitant movement of the cation while influx and efflux of the sugar are stoichiometrically coupled to the transport of the ion. Obviously, exchange of melibiose proceeds without dissociation of the coupling ion. These and other kinetic properties of the different facilitated diffusion processes indicate that the rate of Na$^+$ release on the inner surface is several times slower than the release of melibiose and suggest that Na$^+$ release is a major rate limiting step of the transport process. Taken together the studies support a model according to which both binding and release of coupling ion and sugar are ordered processes (Fig. 3). Thereby, Na$^+$ binds first to the outer surface of the transporter followed by the sugar; on the inner surface sugar is released first and Na$^+$ last [149].

3.3.3
PutP

The Na$^+$/proline transporter (PutP) of *E. coli* is a member of the SSF which currently comprises more than 40 proteins from archaea, bacteria, yeast, insects, and mammals [37, 92, 119]. The known members of this family use Na$^+$ but not H$^+$ as coupling ion. PutP catalyzes the coupled translocation of proline and Na$^+$ according to a symport mechanism with a Na$^+$ to proline stoichiometry of 1:1 [65, 66, 154, 155]. Li$^+$ can substitute for Na$^+$, however, despite earlier assumptions H$^+$-driven proline uptake by PutP could not be demonstrated [66, 154]. An electrical potential ($\Delta \psi$) can drive proline accumulation if Na$^+$ ions are present. The K_m values for Na$^+$ and proline uptake by PutP into cells or right-side-out membrane vesicles are determined with 30 and 2 µM, respectively [66, 74, 156]. However, the apparent $K_m(Na^+)$ of PutP reconstituted into proteoliposomes in an inside-out orientation is calculated to be 730 µM [125]. The lower ion affinity may be explained by a functional asymmetry of the transporter which appears to have a lower cation affinity at the cytosolic side than on the outside. This idea is in general agreement with the finding that the apparent K_m for Na^+ uptake into intact cells or membrane vesicles is about 100 times lower than the K_d value for Na^+ representing ion binding to both sites of the membrane [156].

Binding of Na^+ to PutP causes a conformational change in the protein that results in an increase in its affinity for proline [156]. Further kinetic analyses of Na^+/solute symport catalyzed by different members of the SSF including PutP suggest that transport occurs according to an ordered binding mechanism [156–159] (Fig. 3).

3.3.4
CitS

The Na^+/citrate transporter (CitS) of *K. pneumoniae* belongs to the cation/citrate family (CCS or 2-hydroxycarboxylate transporter family, TC 2.A.24) [92, 160]. Most members of the CCS family catalyze substrate uptake with either Na^+ or H^+ and come from gram-negative [e.g., the Na^+/citrate transporters of *Salmonella pullorum* (CitC) and *Vibrio cholerae* (CitS)] and gram-positive bacteria (e.g., the putative Na^+/citrate transporter of *Bacillus halodurans*, the H^+-dependent citrate transporter of lactic acid bacteria (CitP), and the malate transporter of *Lactococcus lactis* (MleP) [63, 161–164]. The transporters differ in substrate specificity and physiological function. CitP of *L. lactis*, e.g., recognizes citrate, malate, citramalate, 2-hydroxybutyrate, 2-hydroxyisobutyrate, and lacate whereas CitS of *K. pneumoniae* only recognizes citrate [160, 164]. CitP and MleP, but not CitS, are involved in membrane potential generation via electrogenic citrate/lactate and malate/lactate exchange, respectively [164]. Instead, CitS accumulates citrate at the expense of a smf generated by the Na^+-pumping oxaloacetate decarboxylase [22] (Fig. 1, see also Sect. 3.2.5).

Analysis of the affinity constants of CitS of *K. pneumoniae* for the different citrate species at different pH values indicates that H-citrate^{2-} is the species transported by CitS [165]. Studies with intact cells and membrane vesicles prepared from *E. coli* expressing *citS* suggest that H-citrate^{2-} is electrogenic and can be driven by $\Delta\psi$ thereby implying that transport occurs in symport with at least three monovalent cations [165, 166]. Further studies revealed that increasing, non-saturating Na^+ concentrations cause a quadratic increase of the rate of citrate uptake suggesting that two Na^+ are translocated per transport cycle [167]. Taking into account that also a pH gradient is shown to drive citrate uptake, it is proposed that transport catalyzed by CitS has a stoichiometry of 2 Na^+ and 1 H^+/citrate [166, 167]. However, experiments using purified and reconstituted CitS question the electrogenic nature of the transport process [168]. In the reconstituted system ΔpH and ΔpNa$^+$ (chemical Na^+ concentration gradient) act as driving forces but $\Delta\psi$ proves to be unable to stimulate citrate uptake. Furthermore, Na^+ on the inside of the proteoliposomes stimulates citrate transport suggesting that Na^+ recycles during citrate transport. In addition, citrate counterflow kinetics reveal that CitS contains a high and a low binding site for citrate, which are exposed to opposite sides of the membrane thereby allowing a simultaneous type of transport [168].

3.4
Transporter Structure

3.4.1
General Features

Secondary transport proteins are believed to have a similar global secondary structure [91, 92]. Hydropathy profile analyses of the primary structures in combination with circular dichroic measurements, Laser Raman spectroscopy, and Fourier transform infrared (FTIR) spectroscopy predict that the transporters consist of hydrophobic segments in a-helical conformation (mean chain length about 20 amino acids) traversing the membrane in zigzag fashion (= putative transmembrane domains, TMs) and of hydrophilic loop regions connecting the TMs [14, 93, 94]. The 12 TM arrangement appears to be most common among secondary transporters including Na^+-dependent systems, thereby both the N- and C-termini are located at the cytoplasmic side of the membrane. The connecting hydrophilic loops are usually larger at the cytoplasmic side than at the periplasmic side of the membrane. Furthermore, the number of positively charged residues (Arg, Lys, His) is usually larger in cytoplasmic than in periplasmic loops (positive-inside rule) [95]. However, there are also exceptions that differ from these rules particularly with respect to the number of TMs and the location of the protein termini (see Sects. 3.4.3 and 3.4.4).

Understanding the molecular mechanism of function of the transport proteins requires detailed structural information. However, a major problem in the field of secondary transport is the difficulty inherent in obtaining structural information on the proteins involved at a relevant level of resolution. The first direct glimpse into the structure of a secondary transporter was provided by two-dimensional crystallization and electron cryo-microscopy of the Na^+/H^+ antiporter NhaA of E. coli [96]. Based on these studies a three-dimensional map of NhaA was produced that reveals 12 tilted, bilayer-spanning helices. A roughly linear arrangement of six helices is adjacent to a compact bundle of six helices, with the density for one helix in the bundle not continuous through the membrane [97]. Besides intensive trials to crystallize the transporters, cysteine cross-linking studies and site-directed labeling analyses in combination with spectroscopic methods are used to obtain information on helix packing in secondary transport proteins like the lactose permease of E. coli [98–100].

Much effort has been invested to elucidate the oligomeric state of secondary transporters (e.g., [101–104]). As to Na^+-coupled systems, the oligomeric state of the human Na^+/glucose transporter (SGLT1) was analyzed by freeze-fracture electron microscopic techniques [105]. The freeze-fracture particles were characterized by comparing the cross-sectional area of the SGLT1 particles with that of membrane proteins of known structure. The studies revealed that SGLT1 is an asymmetric monomer. Application of the method to the Na^+/galactose transporter of *V. parahaemolyticus* (vSGLT) yields a particle size consistent with the transporter existing as a monomer [106]. Recently, the quaternary structure of the LacS of *S. thermophilus*, like MelB of *E. coli* a member of the GPH family, was investigated using analytical ultracentrifugation and freeze-fracture particle analysis [107]. The study shows that LacS undergoes reversible self-association thereby existing as monomer or dimer. The functional relevance of the self-association process requires further investigation.

Information particularly on the secondary structure of the polypeptide chains of Na^+/substrate transporters has significantly improved in the last couple of years. The current state of knowledge is summarized below for the examples MelB and PutP of *E. coli* and CitS of *K. pneumoniae*.

3.4.2
MelB

The Na^+/melibiose transporter of *E. coli* is encoded by the *melB* gene which has been cloned and sequenced [108]. It consists of 473 amino acids (molecular mass of 53 kDa), 70% of which are apolar [108, 109]. FTIR spectroscopy suggests that the transporter is dominated by α-helical components (up to 50%) and contains β-structures (20%) and additional components assigned to turns, 3_{10} helix, and non-ordered structures (30%) [110]. Analysis of the hydropathy profile of the primary structure of EcMelB predicts 12 TMs [108, 111]. The prediction was tested by a gene fusion approach involving alkaline phosphatase (PhoA) as reporter protein. The activity pattern obtained with a series of MelB–PhoA hybrid proteins confirms the prediction of the 12-TM model and led to an assignment of TMs and loop regions to specific regions of the primary structure of EcMelB [111, 112] (Fig. 4). An immunological study shows that an antibody raised against the C-terminal decapeptide of EcMelB reacts with inside-out membrane vesicles indicating that the C-terminus and consequently also the N-terminus of the transporter are located in the cytoplasm [113].

Information on the packing of TMs of EcMelB has been gained from the analysis of second site suppressor mutations [114–117]. It is assumed that recovery of enzyme activity is most likely to happen if the sites of the primary and the secondary substitution are close together. From the studies the authors infer close proximity between (1) TM II and TMs IV, VII, and X, and the loop between TMs X and XI, (2) TMs IV and IX, and (3) TMs I and II. The arrangement is in agreement with studies on structure–function relationships of MelB (see Sect. 3.5.1). Furthermore, the arrangement is in accordance with recently published findings on interactions between TMs II, IV, VII, X, and XI of the homologous H^+/galactoside transporter LacS of *S. thermophilus* [118].

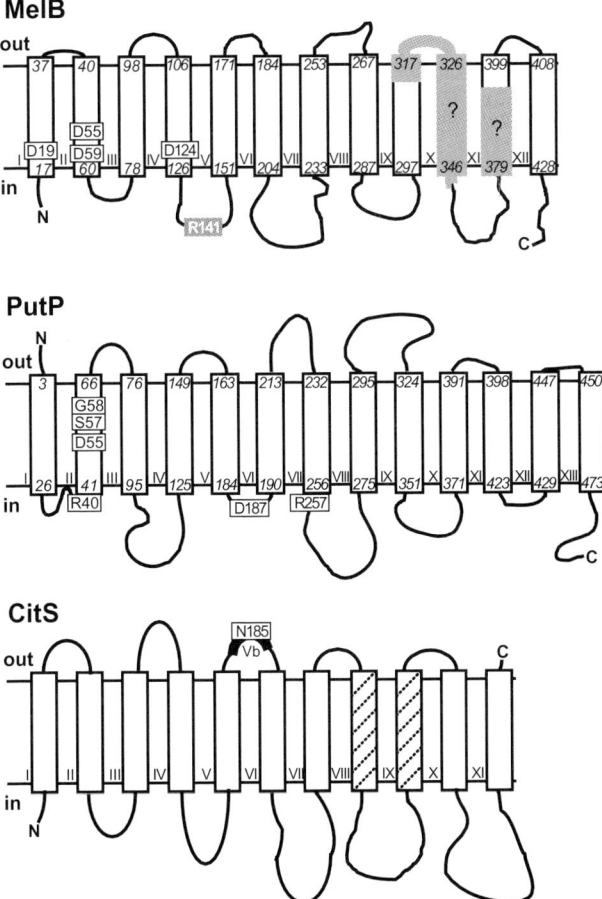

Fig. 4. Secondary structure models of MelB and PutP of *E. coli* and CitS of *K. pneumoniae*. The models are based on [111, 112] (MelB), [123] (PutP), and [134,135] (CitS). Putative transmembrane domains are represented as rectangles and numbered with Roman numerals. Italic Arabic numerals at the termini of each transmembrane domain represent the position of the corresponding amino acid in the primary structure of the proteins. Functional important amino acid residues are shown in boxes. The gray areas in the MelB model highlight regions which are photolabeled by the sugar analog *p*-azidophenyl α-D-galactopyranoside according to [192]. Other details are explained in the text (Sect. 3.4).

3.4.3
PutP and Other Members of the SSF

The average hydropathy plot for SSF proteins predicts 11 to 15 TMs in *α*-helical conformation [92]. Most experimental evidence on the topological arrangement comes from studies on the Na$^+$/proline transporter PutP of *E. coli* [37, 120]. PutP has originally been proposed to consist of a short hydrophilic N-terminal region, 12 TMs in α-helical conformation that traverse the membrane in zigzag fashion connected by hydrophilic loops, and a hydrophilic C-terminal tail [121]. According to the model 60 % of the 502 amino acids of PutP are found in hydrophobic domains with a mean length of 24 amino acids. Circular dichroic measurements of purified and reconstituted PutP indicate that the protein is about 63 % α-helical in conformation (T. Pirch, J. Greie, and H. Jung, unpublished data).

The 12-helix model of PutP was tested by applying a gene fusion approach, immunological and site-directed labeling techniques, and electron paramagnetic resonance (EPR) spectroscopy [122–124]. Analysis of a series of *putP–OphoA* and *putP–lacZ* fusions yields a reciprocal activity pattern of the reporter proteins alkaline phosphatase (PhoA) and β-galactosidase (LacZ) that is in agreement with the topology of TMs III to XII of the 12-helix model [123]. However, differing from the topology prediction, the fusion analysis suggests a shift of the boarders of putative TM II by 8 amino acids towards the C-terminus thereby creating a large preceding periplasmic loop. Placement of PutP–PhoA junction points into this loop did not yield conclusive results. Therefore, a cysteine accessibility analysis was performed to determine the topology of the N-terminal part of PutP [123]. Surprisingly, cysteine residues placed close to the N-terminus of PutP are highly accessible to membrane permeant and impermeant sulfhydryl reagents from the periplasmic space. In contrast, cysteine at the C-terminus is accessible only from the cytosolic side of the membrane as indicated also by immunological studies [122, 123]. These results contradict the 12-helix motif and indicate a periplasmic location of the N-terminus while the C-terminus faces the cytoplasm (Fig. 4). In addition, accessibility pattern of cysteine residues placed into the putative periplasmic loop (pL) preceding TM II (pL 2) suggests the formation of an additional TM formed by amino acids of this loop [123]. The results are in good agreement with site-specific proteolysis studies performed with unidirectionally reconstituted PutP [123, 125].

Finally site-directed spin labeling and electron paramagnetic resonance (EPR) spectroscopy were used to test the altered structural arrangement of PutP [124]. Information on spin label topography was obtained by analyzing the residual mobility of site-specifically attached nitroxide side chains and determination of collision frequencies of the nitroxide with non-polar oxygen and polar chromium oxalate The measurements indicate that TM II is composed of amino acids Ser41 to Gly66 while TM III (TM II in the 12-helix model) comprises residues Ser76 to Gly95 [124].

Based on these studies a revised secondary structure model of PutP is proposed according to which the protein consists of 13 TMs with the N-terminus on the periplasmic side of the membrane and the C-terminus facing the cytoplasm [123, 124] (Fig. 4). The new secondary structure model of PutP is in agreement with the recently proposed topological arrangement of TMs I to XIII of SGLT1, which is based on a N-glycosylation study [126]. The 13-helix feature was also proposed for the mammalian Na^+/I^- transporter [127]. Taken together, the results support the idea of a common topological motif for members of the SSF according to which, e.g., the bacterial transporters for proline (PutP) and pantothenate (PanF) and the mammalian Na^+/I^- transporter are composed of 13 TMs. Transporters with a C-terminal extension [e. g., the human SGLT1 and the Na$^+$/myoinositol transporter (SMIT1)] are proposed to have an additional 14th TM [119, 126]. As far as investigated, the 13-helix motif appears to be a special feature of the SSF.

Information on tertiary interactions within members of the SSF has recently been gained by chemical cross-linking of splits of the Na$^+$/galactose transporter of *V. parahaemolyticus* [128]. The studies suggest that the hydrophilic loops between TMs IV and V and between X and XI are within 8 Å of each other.

3.4.4
CitS

In *K. pneumoniae* the Na$^+$/citrate transporter CitS is induced upon anaerobic growth on citrate [129]. The gene encoding the transporter has been cloned and sequenced and shown to encode a highly hydrophobic protein (molecular mass 47.5 kDa) [130, 131]. The hydropathy profile suggests that the protein is very hydrophobic and contains 12 TMs. Contrary to MelB experimental evidence is not consistent with the prediction. Proteolysis experiments with CitS containing the biotin acceptor domain of the oxaloacetate decarboxylase of *K. pneumoniae* at its N- or C-terminus revealed that the N-terminus of CitS is located in the cytoplasm while the C-terminus faces the periplasm. Further investigation of the membrane topology by the *phoA* gene fusion approach suggested that the polypeptide is composed of 9 TMs leaving the C-terminal half of the transporter with two large hydrophobic periplasmic loops and two large hydrophilic cytoplasmic loops [132]. Later studies confirmed that the number of TMs is uneven, but corrected the total number of TMs. One experimental approach involved the insertion of C-terminally truncated CitS into the endoplasmic reticulum (ER) membrane by *in vitro* translation in the presence of dog pancreas microsomes thereby using leader peptidase as the insertion vehicle and the leader peptidase P2 domain as the topological reporter [133]. The study revealed that two hydrophobic segments of the larger hydrophobic loop (periplasmic loop between TMs VII and VIII of the 9 TM model) form a helical hairpin in the ER membrane. Based on these findings an 11 TM topology for CitS was suggested, thereby the two additional TMs form TMs VIII and IX (Fig. 4). The existence of the additional TMs was confirmed in *E. coli* by a cysteine accessibility analysis [134]. Furthermore, investigation of CitS–PhoA hybrid proteins by the latter method demonstrated that TM VIII is exported to the periplasm in the absence of the C-terminal CitS sequence, thus explaining why the *phoA* fusions used in the earlier study did not correctly predict the topology [134]. In another approach the biotin acceptor domain was inserted into different loop regions and the location of the domain relative to the membrane was determined by proteolysis experiments [135]. The results support the 11 TM model of CitS (Fig. 4).

3.5
Structure–Function Relationships

3.5.1
MelB

3.5.1.1 Site of Ion Binding

A variety of different experimental approaches have been employed to identify regions in EcMelB participating in cation binding. An approach used in early studies involves the isolation of Li$^+$-resistant *melB* mutants. In one of the mutants isolated by this technique Pro126[1)] is replaced with Ser (EcMelB-P126S)[2)] [169, 170]. In addition to being Li$^+$-resistant, this mutant can no longer use H$^+$ as coupling ion whereas Na$^+$-coupled sugar uptake is detected. Similar properties are found for mutants in which Pro146, Leu236, or Ala240 of EcMelB are substituted [171]. These studies provide a first indication of the role of the N-terminal domain of MelB in ion binding. A different approach utilizes the construction of chimeras between EcMelB and MelB of *K. pneumoniae* (KnMelB) [172]. The method is based on the fact that EcMelB can use Na$^+$ for sugar uptake whereas KnMelB is not able to use the ion. Accumulation of melibiose by chimeric proteins containing the N-terminal 81, 148, or 201 amino acids of EcMelB (the remaining C-terminal amino acids are from KnMelB) is strongly stimulated by Na$^+$ or Li$^+$ as it is the case with complete EcMelB. However, the activity of chimeras containing the N-terminal 302 or 353 amino acids of EcMelB is not significantly affected by Na$^+$ and Li$^+$. These results suggest that the N-terminal 81 residues of EcMelB are essential for Na$^+$-coupled melibiose uptake but additional residues from the C-terminal domain of EcMelB may also participate in Na$^+$ recognition. A drawback of these studies is that they do not discriminate between a direct effect on the Na$^+$ binding site and a secondary effect induced, e.g., by an alteration in the sugar binding site of the transporter. In fact, many manipulations affect both Na$^+$ and sugar binding and support the notion that the two ligand binding sites interact in a co-operative manner (see Sects. 3.3.2 and 3.5.2.1).

Other studies focus on acidic residues residing on TMs in the N-terminal domain of the transporter. Individual substitution of Asp at positions 19 (TM I), 55, 59 (both TM II), or 124 (TM IV) by a neutral amino acid inhibits Na$^+$-coupled sugar transport [173–177] (Fig. 4). Sugar binding to these EcMelB variants is comparable to wild-type EcMelB in the absence of Na$^+$, but binding is no longer stimulated by Na$^+$. Besides the effect on Na$^+$/sugar transport neutral substitution of Asp at positions 19, 59, or 124 inhibits also H$^+$-coupled transport. EcMelB in which Asp55 was replaced retained the capability to transport sugar with H$^+$ [175]. It

---

1) Prediction of the primary structure of EcMelB missed a stretch of N-terminal amino acids which was later identified by N-terminal sequencing [108, 109]. The amino acid positions given in the text refer to the revised sequence of EcMelB.

2) Amino acid replacements are designated as follows: The one-letter amino acid code is used followed by a number indicating the position of the native residue in the wild-type transporter. The sequence is followed by a second letter denoting the substitution at this position.

should be pointed out that the acidic residues of TM II (Asp55, Asp59) are highly conserved in all members of the GPH family [138]. Remarkably, neutral substitution of the equivalent of Asp55 in the H$^+$/galactoside transporter LacS of *S. thermophilus* (Glu67) results in a conditionally uncoupled phenotype but, similar to EcMelB-Asp55, is not essential for H$^+$-coupled sugar uptake. Replacement of Asp71 in LacS (corresponds to Asp59 in EcMelB) leads to a complete loss of H$^+$/sugar transport indicating that the carboxylate moiety is essential for ion coupled transport as observed upon substitution of EcMelB-Asp59 [138].

Interestingly, a single amino acid substitution in the H$^+$-dependent transporter KnMelB, A58N in TM II, is sufficient to enable Na$^+$-coupled sugar transport by the *K. pneumoniae* transporter thereby implicating Asn58 in Na$^+$ binding [178]. On the other hand, substitution of Asn58 in EcMelB reveals that the residue is important but not essential for Na$^+$ binding [179]. The findings fit the idea that Asp58 is located close to the ion binding site (e. g., to Asp59) but may not interact directly with the ion. Substitution of a series of other polar residues located close to Asp55, Asp59, or Asp124 of EcMelB confirms the notion that the carboxyl side chains of the acidic residues are of primary importance for cation binding [180]. Also a complete Cys scanning mutagenesis of TM II of MelB reinforces the unique role of the acidic residues in transporter function [181].

Besides the acidic and uncharged polar residues, Arg52 in TM II of EcMelB has been subjected to amino acid substitution analyses [116, 180]. Substitution of the residue with neutral polar or apolar residues leads to highly reduced rates of Na$^+$-coupled sugar transport and an uncoupling of H$^+$ sugar symport [116, 180, 181]. Based on the analysis of second-site revertants it is suggested that Arg52 is salt-bridged to the functional important Asp residues at positions 19 (TM I) and 55 (TM II) thereby playing a crucial role in cation-coupled sugar transport [116]. Furthermore, individual replacement of all seven His residues of EcMelB identified only His94 as important for melibiose uptake [182]. However, substitution of the latter residue alters expression and stability rather than the catalytic activity of the transporter [183]. In this respect, EcMelB clearly differs from LacY in which a His residue (His322) is proposed to play an essential role in coupling H$^+$ and sugar transport [184].

In conclusion, the results strongly support the idea that the N-terminal domain of EcMelB is responsible for Na$^+$ binding [138, 146]. Furthermore, the Asp residues at positions 19, 55, 59, and 124 are suggested to form a network for cation binding in EcMelB [138, 146, 185] (Fig. 4).

3.5.1.2 Sugar Binding and Functional Dynamics of MelB

Mutants exhibiting an altered sugar specificity were isolated to identify the sugar binding site of EcMelB. Mutants resistant to the non-metabolizable substrate TMG arose in a complex fashion from any of 23 unique amino acid substitutions mapping to 17 different positions in the N- (TMs I, II, IV) and C-terminal domain (TMs VII, X, XI, and the hydrophilic loop between TMs X and XI) of EcMelB [146, 186]. Strikingly, most of the mutations (including also P126S discussed above) affect both the ion and sugar specificity. Similar findings were made in other studies (e.g., [173, 175, 180, 181]). The observations are in agreement with the notion that binding of Na^+ and sugar occurs in a co-operative manner and suggest that the two binding sites are overlapping or interact with each other over a distance via conformational alterations [138, 146, 177] (see also Sects. 3.3.2 and 3.5.2.2).

To learn more about the structural organization of EcMelB including the ligand binding sites an approach combining site-directed mutagenesis and tryptophan fluorescence measurements was used [187–191]. The studies are based on the finding that ligand binding alters the fluorescence emission of EcMelB. So, binding of sugar leads to an increase of the tryptophan fluorescence signal with α-galactosides causing larger variations than β-galactosides, whereas Na^+ or Li^+ quench the signal. Strikingly, however, the ions potentiate the sugar-induced fluorescence increase thereby the ion effect increases in the order of $H^+ < Na^+ < Li^+$. The data suggest that the changes in fluorescence reflect conformational alterations occurring upon the formation of ternary cation/sugar/MelB complexes [187]. Substitution of Trp residues reveals that binding of β-galactosides affects the fluorescence of Trp residues in both the N- and C-terminal part (Trp299 in TM IX and Trp342 in TM X) of EcMelB proposing proximity and/or tight functional linkage between specific helices of both domains [188]. α-Galactosides alter only the fluorescence of two Trp residues in the C-terminal domain. Fluorescence energy transfer (FRET) studies using fluorescent sugar analogs confirm the proximity between N- and C-terminal helices [189,190]. Based on the FRET measurements it is calculated that the bound fluorescent galactoside is about 20 and 14 Å away from Trp64 (cytoplasmic loop between TMs II and III) and Trp299 (TM IX), respectively. A more detailed investigation of the six Trp residues of the N-terminal domain shows that individual replacement of Trp116 (TM IV) and Trp128 (cytoplasmic loop between TMs IV and V) impairs transport activity [191]. Furthermore it is demonstrated that the two residues are responsible for the β-galactoside-induced fluorescence attributed to the N-terminal domain and together with Trp54 also contribute to the ion-induced fluorescence alteration. The latter study proposes that TM IV plays a crucial role in connecting cation and sugar binding sites of EcMelB [191].

In addition, covalent photolabeling of EcMelB was performed [192] (Fig. 4). The study uses the sugar analog p-azidophenyl α-D-galactopyranoside (α-PAPG) which binds with high-affinity in a Na^+-dependent manner to the transporter. Labeling was completely prevented by an excess of melibiose or α-NPG. CNBr cleavage of labeled EcMelB yields different labeled peptides. One peptide comprises amino acids Asp124 to Met181 (compose the cytoplasmic loop connecting TMs IV and V, TM V, and the periplasmic loop between TMs V and VI). Arg141 of the cytoplasmic loop region was identified as the only labeled residue of the peptide (Fig. 4). The study supports the hypothesis that the loop between TMs IV and V plays a crucial role in sugar binding and transport by EcMelB [192]. Interestingly, also an Arg residue (Arg144) together with a Glu residue (Glu126) in the corresponding loop of LacY of E. coli have been shown to be essential for sugar binding and proposed to interact directly with the sugar substrate [193–195]. Although the primary structures of MelB and LacY show only a very limited sequence similarity, comparison of the structure–function relationships suggests that both transporters may have conserved domains involved in sugar recognition [192]. Analysis of other peptide fragments suggest that the sequences comprising amino acids Met310-Ile347 (TM X) and Met374-Gly393 (TM XI) are also labeling targets [192] (Fig. 4).

Finally, based on a functional EcMelB molecule devoid of all four native Cys residues [196], a complete Cys scanning mutagenesis of TMs II and XI was performed [181, 197]. Analysis of the sensitivity of the different single Cys MelB molecules to p-chloromercuribenzenesulfonic acid (PCMBS) treatment suggests that residues of these two TMs participate in the formation of a water-filled cleft or pore which could be part of the ion and/or sugar pathway through the membrane. The idea is supported by the fact that melibiose prevents the PSMBS effect at a variety of positions in both helices [181, 197]. Clearly, substrate protection could also be achieved by a sugar-induced conformational alteration that buries the sulfhydryl group in the protein.

In conclusion, the results provided by random and site-directed mutagenesis studies, spectroscopic analyses, and affinity labeling strongly support the contention that structural components both of the N- (TM IV-loop-TM V region) and C-terminal part (TM X-loop-TM XI region) of EcMelB contribute to sugar binding thereby suggesting that these regions are in close proximity. Taking into account that residues in the N-terminal part of the transporter (TMs I-IV) are involved in Na^+ binding, the studies suggest that the sites of Na^+ and sugar binding are located close to each other in the tertiary structure of the protein, a situation that would facilitate the observed functional interaction between the two sites.

3.5.2
PutP

3.5.2.1 Site of Na^+ Binding

Labeling experiments, random and site-directed mutagenesis have been employed to identify functionally important sites in PutP of *E. coli* and *S. typhimurium* (see also [37]).

To identify residues involved in cation binding, *putP* mutants that confer resistance to Li^+ during growth on proline were isolated. The location of each mutation was determined by deletion mapping: the mutations cluster in two small deletion intervals at the 5' and 3' termini of the *putP* gene [198]. Recent site-directed mutagenesis studies have established that amino acids of TM II of PutP are of particular functional importance thereby strongly supporting the idea that the N-terminal part of the transporter is responsible for Na^+ binding (Fig. 4). Thus, out of five acidic residues in the N-terminal domain analyzed, the carboxylate of Asp55 (TM II) proved to be essential for transport while Asp33, Asp34 [both putative cytoplasmic loop (cL) 2], and Glu75 (pL 3) are dispensable for function [199]. Significant albeit highly reduced activity was detected with Glu in place of Asp55 (PutP-D55E). Kinetic analysis of active transport catalyzed by the latter PutP derivative reveals a 50-fold decrease of the apparent affinity of the protein for Na^+ ions compared to the wild-type transporter. On the other hand, only a relatively small alteration of the apparent affinity for proline is observed [199]. Furthermore, Na$^+$ binding could not be detected for PutP missing a carboxylate at position 55. These results suggest that Asp55 is located at or close to a binding site of the coupling ion.

In addition, individual replacement of Ser54, Met56, Ser57, or Gly58 following Asp55 in TM II leads to significantly altered transport kinetics (Fig. 4). Replacement of Ser54 with Cys yields an about 8-fold decrease of the apparent affinity of PutP for Na$^+$ with no significant effect on the proline affinity. Substitution of Ser57 with Ala, Cys, Gly, or Thr increases the apparent K_m values for Na$^+$ (and Li$^+$) *and* proline by up to two orders of magnitude with little influence on V_{max} values [200]. Similarly, Cys in place of Met56 or Gly58 causes an increase of the apparent K_m values for Na$^+$ *and* proline although the extent of the effect is less dramatic as in case of Ser57 (M. Nietschke, S. Landmeier, M. Quick, and H. Jung, unpublished data). These results support the idea that the sites of Na$^+$ and proline binding are overlapping. However, Cys accessibility analyses challenge the idea of Ser57 and Gly58 being part of a Na$^+$ binding site. Thus, the reaction rate of a Cys placed at the positions of Ser57 and Gly58 of PutP with sulfhydryl-specific reagents is not blocked by Na$^+$ as expected for residues being directly involved in binding of the ligand but significantly stimulated by the ion (T. Pirch and H. Jung, unpublished data). Obviously, ion binding to a site different from Ser57 and Gly58 induces a conformational alteration that increases the accessibility of the amino acids at positions 57 and 58. These observations make a direct involvement of Ser57 and Gly58 in Na$^+$ binding unlikely. The accessibility of Cys at the positions of Ser54 and Met56 is not significantly influenced by Na$^+$.

Furthermore, the conserved residues Arg40 (cytoplasmic end of TM II), Arg187 (cL 6), and Arg257 (cL 8) were subjected to amino acid substitution analysis. The results suggest that Arg40 is located close to the site of ion binding and is important for the efficient coupling of ion and proline transport [201]. The data obtained upon substitution of Asp187 [202] indicate that electrostatic interactions of the amino acid side chain at position 187 in PutP with other parts of the transporter and/or the coupling ion are crucial for active proline transport. It is suggested that Asp187 is located close to the pathway of the coupling ion through the membrane and may be involved in the release of Na$^+$ on the cytoplasmic side of the membrane [203]. A reduced Na$^+$ dependence of proline binding is observed upon replacement of Arg257 with Cys [202]. The authors speculate that removal of the positive charge leads to an enhanced affinity of PutP for Na$^+$.

Finally, alignment of amino acid sequences of members of the SSF led to the proposal of a sodium binding (SOB) motif which has the following consensus sequence: GX$_{35-37}$AX$_3$(EQ)LX$_3$GR [92]. In PutP of *E. coli* the conserved amino acids are Gly328 (TM IX), Ala366, Glu370 (cL 10), Leu371, Gly375 and Arg376 (TM X). However, a substitution analysis of Arg376 reveals that the residue does not reside at the Na$^+$ binding site thereby suggesting that the proposed motif is not essential for function [204].

In conclusion, the results obtained so far suggest that the N-terminal domain of PutP, in particular amino acids of TM II are responsible for Na$^+$ binding similar as proposed for MelB of E. coli. But differing from the sugar transporter, the carboxylate at only one position (Asp55) is found to be essential for substrate transport by PutP while other acidic residues in the N-terminal domain prove to be dispensable for function. The participation of conserved polar residues of this domain in Na^+ liganding is currently investigated.

3.5.2.2 Regions Important for Proline Binding

Besides the discussed role of amino acids of TM II in Na^+ liganding, some residues in this segment are important for proline binding (Fig. 4). As already described above even conservative replacements for Ser57 and Gly58 cause a dramatic decrease of the apparent affinity of PutP for proline. In addition, substitution of Ser57 reduces the sensitivity of E. coli cells to the toxic proline analogs L-acetidine-2-carboxylate (AZT) and 3,4-dehydro-D,L-proline (DHP) [200]. Remarkably, while Na^+ increases the accessibility of Cys in place of Ser57 or Gly58, proline inhibits the reaction of the residue with sulfhydryl-specific reagents at both positions thereby reversing the Na^+ effect almost completely (T. Pirch and H. Jung, unpublished data). Taken together the findings propose a location of Ser57 and Gly58 at or close to the proline binding site of PutP. Clearly, it cannot be excluded that protection of Cys at positions 57 and 58 is achieved by a proline induced conformational alteration and not by direct steric hindering. In any case, it is very likely that Ser57 and Gly58 participate in the transmission of a Na^+-induced conformational alteration leading to high-affinity proline binding to a site that is at least close to these positions in the tertiary structure of the transporter. The observations confirm the functional importance of TM II described above.

The proposed participation of TM II in proline binding apparently contradicts results obtained with SGLT1 suggesting that the C-terminal domain is responsible for substrate binding and translocation. Thus, analysis of chimeras consisting of sequences of the high-affinity transporter SGLT1 and the low-affinity transporter SGLT2 suggest that recognition/transport of organic substrate is mediated by interactions distal to amino acid 380 [205]. In addition, it is shown that a SGLT1 fragment comprising only the last five TMs of the transporter is able to catalyze Na^+-independent facilitated diffusion of sugar [206, 207]. However, it has to be taken into account that the sugar affinity of the C-terminal fragments of SGLT1 tested was highly reduced thereby leaving the possibility that amino acid residues of the N-terminal part of the protein contribute to high-affinity binding of the organic substrate.

The idea that multiple regions of PutP contribute to solute binding is supported by the analysis of *putP* mutants with an altered specificity for proline, AZT, and DHP [208]. Deletion mapping of the mutants revealed that the mutations cluster in three distinct regions of the *putP* gene. One set of mutants was mapped at the distal end of the *putP* gene thereby implicating the C-terminal region of PutP in substrate recognition.

Originally, the native Cys residues at positions 281 (TM VIII) and 344 (TM IX) of PutP were identified as substrate-protectable residues and suggested to reside close to the binding site for the coupling ion and/or proline [209, 210]. However, replacement of these Cys residues neither affects proline uptake nor alters the sensitivity of *E. coli* cells to the toxic proline analog L-azetidine-2-carboxylate significantly thereby indicating that the two Cys residues may not be directly involved in binding [209, 210]. Finally, a fully functional PutP molecule devoid of all five native Cys residues was constructed demonstrating that Cys residues are not important for function of PutP similar as shown for other secondary transporters [123, 196, 211].

In conclusion, amino acids of TM II have been found to be crucial for high-affinity proline binding. In view of the findings on SGLT1 it appears likely that besides the N-terminal domain also selected regions of the C-terminal domain are involved in proline binding and translocation. Studies are underway to test this hypothesis. If the idea proves to be correct the general functional organization of PutP could resemble that of MelB.

3.5.2.3 Functional Dynamics of PutP

The ordered binding model implies that Na^+/solute transport is the result of a series of ligand-induced conformational changes in the protein. To explore the nature of these changes protein labeling and spectroscopic methods were used. The studies performed so far indicate that Na^+ binding leads to conformational changes involving TM II and the preceding loop. Subsequent binding of proline induces another conformational alteration thereby TMs II, VII, and IX and the adjoining loops are participating [37, 124].

Taken together with amino acid substitution analyses the studies support the following model of Na^+/proline transport (see also Fig. 3): In a first step Na^+ binds to the empty transporter at or close to Asp55 in TM II of PutP. Binding of the ion induces a conformational change that involves at least TM II and the adjoining loops. As part of this change Ser57 and Gly58 move from a buried into a highly accessible position (step 1). The structural rearrangement results in high-affinity proline binding to the transporter, thereby Ser57 and Gly58 either participate directly in proline binding or are located close to the binding site in the tertiary structure of the protein (step 2). The formed ternary complex undergoes another structural alteration that exposes the ligand binding sites to the other site of the membrane. This rearrangement involves at least TM II, VII, IX, and adjoining loops (step 3). Subsequently, proline and Na^+ are released, thereby Asp187 in cL 6 plays a critical role (steps 4 and 5). In the last step of the cycle, the empty transporter re-orientates in the membrane to allow the cycle to start again (step 6). Clearly, additional analyses are required to obtain a complete picture of the sites of Na^+ and proline binding and of the residues forming the translocation pathway.

3.5.3
CitS

An amino acid substitution analysis has been applied to investigate the role of region Vb of CitS of *K. pneumoniae* which is highly conserved within the CCS family [160, 164] (Fig. 4). Whereas a peptide comprising region Vb inserts into the membrane, it is shown to be located in the periplasmic space in the intact transporter [133, 135]. Replacement of Glu194 in the Vb domain of CitS with Gln has almost no effect on the kinetics of Na$^+$ and citrate transport. Interestingly, however, substitution of Asn185 by Val leads to a 9-fold increase of the apparent K_m of CitS for H-citrate^{2-} while the V_{max} and the Na$^+$ kinetics are only slightly altered. Based on these data it is proposed that Asn195 of CitS might be involved in the binding of citrate [160].

3.6
Concluding Remarks and Perspective

Na$^+$/substrate transport plays an important physiological role in prokaryotic cells growing in habitats characterized by elevated temperatures, high pH, or high Na$^+$ content. Furthermore, Na$^+$-coupled transport is crucial under certain anaerobic growth conditions. As shown at the examples of MelB and PutP of *E. coli* and CitS of *K. pneumoniae*, knowledge on the secondary structures of transport proteins has significantly improved over the last couple of years. The studies confirm the 12-helix arrangements as the most common structural motif of secondary transport proteins, thereby deviations from the number are possible. So, a 13-helix arrangement has been proposed as a common structural motif of members of the SSF. Studies on MelB and PutP reveal that acidic and polar residues in the N-terminal part of the proteins are crucial for Na$^+$ binding whereas regions in the N- and C-terminal domains are probably involved in substrate binding. There is functional interaction between the sites of ion and substrate binding. However, it is obvious that many results discussed above, particularly those related to the molecular mechanism of energy coupling, remain speculative without information on the tertiary structure. As demonstrated by the pioneering work on lactose permease, analysis of second site revertants, cysteine cross-linking, and a variety of spectroscopic approaches can yield valuable information on helix packing and tilts. These techniques are now also successfully applied to the investigation of Na$^+$-dependent transport systems as exemplified, e.g., by the fluorescence studies on MelB and PutP. Notwithstanding the power of these studies, a high-resolution structure obtained, e.g., by crystallographic means is indispensable for the understanding of interactions between amino acid side chains and between side chains, coupling ion, and substrates. Therefore, the development of strategies allowing crystallization of the polytopic membrane transport proteins is currently one of the major challenges in the field of secondary transport. These strategies may involve a decrease of the conformational flexibility of the transporters and/or an increase of the polar surface of the proteins [212–215]. Moreover, the results obtained with MelB, PutP, and a variety of other secondary transporters indicate that these proteins require a high degree of conformational flexibility in order to function, making it imperative to obtain dynamic information in order to fully understand function. Here, site-directed labeling in combination with spectroscopic measurements such as EPR and FTIR spectroscopy are very powerful techniques to obtain information on the nature and the extent of the changes in high time resolution.

Acknowledgement

This work was financially supported by the Deutsche Forschungsgemeinschaft (SFB 431-D4).

References

1. Mitchell, P., *Adv. Enzymol. Relat. Areas Mol. Biol.* 1967, *29*, 33-87.
2. Crane, R. K., *Fed. Proc. Fed. Am. Soc. Exp. Biol.* 1965, *24*, 1000-1006.
3. Kaback, H. R., *Science* 1974, *186*, 882-892.
4. Newman, M. J. Wilson, T. H., *J. Biol. Chem.* 1980, *255*, 10583-10586.
5. Harold, F. M., *The Vital Force: A Study of Bioenergetic*, Freeman, New York, 1986.
6. Harvey, W. R., Slayman, C. L., *J. Exp. Biol.* 1994, *196*, 1-4.
7. Wolfersberger, M. G., *J. Exp. Biol.* 1994, *196*, 5-6.
8. Padan, E., Schuldiner, S., *J. Bioenerg. Biomembr.* 1993, *25*, 647-669.
9. Dimroth, P., *Antonie Van Leeuwenhoek* 1994, *65*, 381-395.
10. Dimroth, P., Schink, B., *Arch. Microbiol.* 1998, *170*, 69-77.
11. Krebs, W., Steuber, J., Gemperli, A. C., Dimroth, P., *Mol. Microbiol.* 1999, *33*, 590-598.
12. Lolkema, J. S., Speelmans, G., Konings, W. N., *Biochim. Biophys. Acta* 1994, *1187*, 211-215.
13. Henderson, P. J., Baldwin, S. A., Cairns, M. T., Charalambous, B. M., Dent, H. C., Gunn, F., Liang, W. J., Lucas, V. A., Martin, G. E., McDonald, T. P., *Int. Rev. Cytol.* 1992, *137*, 149-208.
14. Poolman, B., Konings, W. N., *Biochim. Biophys. Acta* 1993, *1183*, 5-39.
15. Haney, S. A., Oxender, D. L., *Int. Rev. Cytol.* 1992, *137*, 37-95.
16. Tolner, B., Poolman, B., Konings, W. N., *Comp. Biochem. Physiol. A Physiol.* 1997, *118*, 423-428.
17. Oren, A., *Microbiol. Mol. Biol. Rev.* 1999, *63*, 334-348.
18. Ivey, D. M., Guffanti, A. A., Krulwich, T. A., in: *Alkali Cation Transport Systems in Prokaryotes* (E. P. Bakker, Ed.), CRC Press, Boca Raton, FL, 1993, pp. 101-124.
19. Padan, E., Krulwich, T. A., in: *Bacterial Stress Responses* (G. Storz, R. Hengge-Aronis, Eds.), ASM Press, Washington, DC, 2000, pp. 117-130.
20. Padan, E., in: *Microbiology and Biochemistry of Hypersaline Environments* (A. Oren, Ed.), CRC Press, Boca Raton, FL, 1999, pp 163-175.
21. Padan, E., Schuldiner, S., in: *Alkali Cation Systems in Prokaryotes* (E. P. Bakker, Ed.), CRC Press, Boca Raton, FL, 1993, pp. 3-24.
22. Dimroth, P., *Biochim. Biophys. Acta* 1997, *1318*, 11-51.
23. Bott, M., *Arch. Microbiol.* 1997, *167*, 78-88.
24. van de Vossenberg, J. L., Ubbink-Kok, T., Elferink, M. G., Driessen, A. J., Konings, W. N., *Mol. Microbiol.* 1995, *18*, 925-932.

25. Driessen, A. J., van de Vossenberg, J. L., Konings, W., *FEMS Microbiol. Rev.* **1996**, *18*, 139-148.
26. Albers, S. V., van de Vossenberg, J. L., Driessen, A. J., Konings, W. N., *Front Biosci.* **2000**, *5*, D813-D820.
27. Speelmans, G., Poolman, B., Abee, T., Konings, W. N., *Proc. Natl. Acad. Sci. USA* **1993**, *90*, 7975-7979.
28. Speelmans, G., de Vrij, W., Konings, W. N., *J. Bacteriol.* **1989**, *171*, 3788-3795.
29. Speelmans, G., Poolman, B., Konings, W. N., *J. Bacteriol.* **1993**, *175*, 2060-2066.
30. Honer zu Bentrup, K., Ubbink-Kok, T., Lolkema, J. S., Konings, W. N., *J. Bacteriol.* **1997**, *179*, 1274-1279.
31. Heyne, R. I., de Vrij, W., Crielaard, W., Konings, W. N., *J. Bacteriol.* **1991**, *173*, 791-800.
32. Tolner, B., van der Rest, M. E., Speelmans, G., Konings, W. N., in: *Molecular Mechanisms of Transport* (E. Quagliariello, F. Palmieri, Eds.), Elsevier Science Publishers, Amsterdam, **1992**, pp. 43-50.
33. Bult, C. J., White, O., Olsen, G. J., Zhou, L., Fleischmann, R. D., Sutton, G. G., Blake, J. A., FitzGerald, L. M., Clayton, R. A., Gocayne, J. D., Kerlavage, A. R., Dougherty, B. A., Tomb, J. F., Adams, M. D., Reich, C. I., Overbeek, R., Kirkness, E. F., Weinstock, K. G., Merrick, J. M., Glodek, A., Scott, J. L., Geoghagen, N. S. M., Venter, J. C., *Science* **1996**, *273*, 1058-1073.
34. Smith, D. R., Doucette-Stamm, L. A., Deloughery, C., Lee, H., Dubois, J., Aldredge, T., Bashirzadeh, R., Blakely, D., Cook, R., Gilbert, K., Harrison, D., Hoang, L., Keagle, P., Lumm, W., Pothier, B., Qiu, D., Spadafora, R., Vicaire, R., Wang, Y., Wierzbowski, J., Gibson, R., Jiwani, N., Caruso, A., Bush, D., Reeve, J. N., *J. Bacteriol.* **1997**, *179*, 7135-7155.
35. Klenk, H. P., Clayton, R. A., Tomb, J. F., White, O., Nelson, K. E., Ketchum, K. A., Dodson, R. J., Gwinn, M., Hickey, E. K., Peterson, J. D., Richardson, D. L., Kerlavage, A. R., Graham, D. E., Kyrpides, N. C., Fleischmann, R. D., Quackenbush, J., Lee, N. H., Sutton, G. G., Gill, S., Kirkness, E. F., Dougherty, B. A., McKenney, K., Adams, M. D., Loftus, B., Venter, J. C., *Nature* **1997**, *390*, 364-370.
36. Kawarabayasi, Y., Sawada, M., Horikawa, H., Haikawa, Y., Hino, Y., Yamamoto, S., Sekine, M., Baba, S., Kosugi, H., Hosoyama, A., Nagai, Y., Sakai, M., Ogura, K., Otsuka, R., Nakazawa, H., Takamiya, M., Ohfuku, Y., Funahashi, T., Tanaka, T., Kudoh, Y., Yamazaki, J., Kushida, N., Oguchi, A., Aoki, K., Kikuchi, H., *DNA Res.* **1998**, *5*, 55-76.
37. Jung, H., *Biochim. Biophys. Acta* **2001**, *1505*, 131-143.
38. Unemoto, T., Hayashi, M., *J. Bioenerg. Biomembr.* **1993**, *25*, 385-391.
39. Nakayama, Y., Yasui, M., Sugahara, K., Hayashi, M., Unemoto, T., *FEBS Lett.* **2000**, *474*, 165-168.
40. Tokuda, H., in: *Alkali Cation Transport Systems in Prokaryotes* (E. P. Bakker, Ed.), CRC Press, Boca Rota, FL, **1993**, pp. 125-138.
41. Sato, K., Homma, M., *J. Biol. Chem.* **2000**, *275*, 5718-5722.
42. Unemoto, T., *Yakugaku Zasshi* **2000**, *120*, 16-27.
43. Nakamura, T., Yuda, R., Unemoto, T., Bakker, E. P., *J. Bacteriol.* **1998**, *180*, 3491-3494.
44. Tholema, N., Bakker, E. P., Suzuki, A., Nakamura, T., *FEBS Lett.* **1999**, *450*, 217-220.
45. Tokuda, H., Kim, Y. J., Mizushima, S., *FEBS Lett.* **1990**, *264*, 10-12.
46. Tsuchiya, T., Shinoda, S., *J. Bacteriol.* **1985**, *162*, 794-798.
47. Sarker, R. I., Ogawa, W., Shimamoto, T., Tsuchiya, T., *J. Bacteriol.* **1997**, *179*, 1805-1808.
48. Sarker, R. I., Ogawa, W., Tsuda, M., Tanaka, S., Tsuchiya, T., *Biochim. Biophys. Acta* **1996**, *1279*, 149-156.
49. Sarker, R. I., Ogawa, W., Tsuda, M., Tanaka, S., Tsuchiya, T., *J. Bacteriol.* **1994**, *176*, 7378-7382.
50. Morita, Y., Kataoka, A., Shiota, S., Mizushima, T., Tsuchiya, T., *J. Bacteriol.* **2000**, *182*, 6694-6697.
51. Martin, S. A., *J. Anim. Sci.* **1994**, *72*, 3019-3031.
52. Russell, J. B., Strobel, H. J., Driessen, A. J., Konings, W. N., *J. Bacteriol.* **1988**, *170*, 3531-3536.
53. Maas, L. K., Glass, T. L., *Can. J. Microbiol.* **1991**, *37*, 141-147.
54. Michel, T. A., Macy, J. M., *J. Bacteriol.* **1990**, *172*, 1430-1435.
55. Ng, W. V., Kennedy, S. P., Mahairas, G. G., Berquist, B., Pan, M., Shukla, H. D., Lasky, S. R., Baliga, N. S., Thorsson, V., Sbrogna, J., Swartzell, S., Weir, D., Hall, J., Dahl, T. A.,

Welti, R., Goo, Y. A., Leithauser, B., Keller, K., Cruz, R., Danson, M. J., Hough, D. W., Maddocks, D. G., Jablonski, P. E., Krebs, M. P., Angevine, C. M., Dale, H., Isenbarger, T. A., Peck, R. F., Pohlschroder, M., Spudich, J. L., Jung, K. H., Alam, M., Freitas, T., Hou, S., Daniels, C. J., Dennis, P. P., Omer, A. D., Ebhardt, H., Lowe, T. M., Liang, P., Riley, M., Hood, L., DasSarma, S., *Proc. Natl. Acad. Sci. USA* **2000**, *97*, 12176-12181.

56. Krulwich, T. A., Ito, M., Gilmour, R., Hicks, D. B., Guffanti, A. A., *Adv. Microb. Physiol.* **1998**, *40*, 401-438.
57. Sugiyama, S., Matsukura, H., Imae, Y., *FEBS Lett.* **1985**, *182*, 265-268.
58. Chen, K. Y., Cheng, S., *Biochem. Biophys. Res. Commun.* **1988**, *150*, 185-191.
59. Guffanti, A. A., Susman, P., Blanco, R., Krulwich, T. A., *J. Biol. Chem.* **1978**, *253*, 708-715.
60. Ando, A., Kusaka, I., Fukui, S., *J. Gen. Microbiol.* **1982**, *128*, 1057-1063.
61. Wakabayashi, K., Koyama, N., Nosoh, Y., *Arch. Biochem. Biophys.* **1988**, *262*, 19-26.
62. Krulwich, T. A., Federbush, J. G., Guffanti, A. A., *J. Biol. Chem.* **1985**, *260*, 4055-4058.
63. Ishiguro, N., Izawa, H., Shinagawa, M., Shimamoto, T., Tsuchiya, T., *J. Biol. Chem.* **1992**, *267*, 9559-9564.
64. Woehlke, G., Wifling, K., Dimroth, P., *J. Biol. Chem.* **1992**, *267*, 22798-22803.
65. Stewart, L. M. D., Booth, I. R., *FEMS Microbiol. Lett.* **1983**, *19*, 161-164.
66. Chen, C. C., Tsuchiya, T., Yamane, Y., Wood, J. M., Wilson, T. H., *J. Membr. Biol.* **1985**, *84*, 157-164.
67. Hasan, S. M., Tsuchiya, T., *Biochem. Biophys. Res. Commun.* **1977**, *78*, 122-128.
68. Hama, H., Shimamoto, T., Tsuda, M., Tsuchiya, T., *Biochim. Biophys. Acta* **1987**, *905*, 231-239.
69. Tolner, B., Ubbink-Kok, T., Poolman, B., Konings, W. N., *Mol. Microbiol.* **1995**, *18*, 123-133.
70. Ogawa, W., Kim, Y. M., Mizushima, T., Tsuchiya, T., *J. Bacteriol.* **1998**, *180*, 6749-6752.
71. Wilson, D. M., Wilson, T. H., *Biochim. Biophys. Acta* **1987**, *904*, 191-200.
72. Tsuchiya, T., Wilson, T. H., *Membr. Biochem.* **1978**, *2*, 63-79.
73. Padan, E., Schuldiner, S., in: *Transport Processes in Eukaryotic and Prokaryotic Organisms* (W. N. Konings, H. R. Kaback, J. Lolkema, Eds.), Elsevier Science, Amsterdam, **1996**, pp. 501-531.
74. Yamato, I., Anraku, Y., in: *Alkali Cation Transport Systems in Prokaryotes* (E. P. Bakker, Ed.), CRC-Press, Boca Raton, FL, **1993**, pp. 53-76.
75. Steuber, J., Schmid, C., Rufibach, M., Dimroth, P., *Mol. Microbiol.* **2000**, *35*, 428-434.
76. McFall, E., Newman, E. B., in: *Escherichia coli and Salmonella: Cellular and Molecular Biology* (F. C. Neidhardt, Ed.), ASM Press, Washington, DC, **1996**, pp. 358-379.
77. Lin, E. C. C., in: *Escherichia coli and Salmonella: Cellular and Molecular Biology* (F. C. Neidhardt, Ed.), ASM Press, Washington, DC, **1996**, pp. 307-342.
78. Jackowski, S., Alix, J. H., *J. Bacteriol.* **1990**, *172*, 3842-3848.
79. Wood, J. M., *J. Membr. Biol.* **1988**, *106*, 183-202.
80. Csonka, L. N., Epstein, W., in: *Escherichia coli and Salmonella: Cellular and Molecular Biology* (F. C. Neidhardt, Ed.), ASM Press, Washington, DC, **1996**, pp. 1210-1223.
81. Wood, J. M., *Microbiol. Mol. Biol. Rev.* **1999**, *63*, 230-262.
82. Cairney, J., Booth, I. R., Higgins, C. F., *J. Bacteriol.* **1985**, *164*, 1218-1223.
83. Grothe, S., Krogsrud, R. L., McClellan, D. J., Milner, J. L., Wood, J. M., *J. Bacteriol.* **1986**, *166*, 253-259.
84. Culham, D. E., Lasby, B., Marangoni, A. G., Milner, J. L., Steer, B. A., van Nues, R. W., Wood, J. M., *J. Mol. Biol.* **1993**, *229*, 268-276.
85. Pao, S. S., Paulsen, I. T., Saier, M. H. J., *Microbiol. Mol. Biol. Rev.* **1998**, *62*, 1-34.
86. Racher, K. I., Voegele, R. T., Marshall, E. V., Culham, D. E., Wood, J. M., Jung, H., Bacon, M., Cairns, M. T., Ferguson, S. M., Liang, W. J., Henderson, P. J., White, G., Hallett, F. R., *Biochemistry* **1999**, *38*, 1676-1684.
87. Spiegelhalter, F., Bremer, E., *Mol. Microbiol.* **1998**, *29*, 285-296.
88. von Blohm, C., Kempf, B., Kappes, R. M., Bremer, E., *Mol. Microbiol.* **1997**, *25*, 175-187.
89. Rübenhagen, R., Ronsch, H., Jung, H., Kramer, R., Morbach, S., *J. Biol. Chem.* **2000**, *275*, 735-741.
90. Kunst, F., Ogasawara, N., Moszer, I., Albertini, A. M., Alloni, G., Azevedo, V., Bertero, M. G., Bessieres, P., Bolotin, A., Borchert, S., Borriss, R., Boursier, L., Brans, A., Braun, M., Brignell, S. C., Bron, S., Brouillet,

S., Bruschi, C. V., Caldwell, B., Capuano, V., Carter, N. M., Choi, S. K., Codani, J. J., Connerton, I. F., Danchin, A., *Nature* **1997**, *390*, 249-256.
91. Marger, M. D., Saier, M. H., Jr., *Trends Biochem. Sci.* **1993**, *18*, 13-20.
92. Reizer, J., Reizer, A., Saier, M. H. J., *Biochim. Biophys. Acta* **1994**, *1197*, 133-166.
93. Kramer, R., *Biochim. Biophys. Acta* **1994**, *1185*, 1-34.
94. van Geest, M., Lolkema, J. S., *Microbiol. Mol. Biol. Rev.* **2000**, *64*, 13-33.
95. von Heijne, G., *Nature* **1989**, *341*, 456-458.
96. Williams, K. A., Geldmacher-Kaufer, U., Padan, E., Schuldiner, S., Kuhlbrandt, W., *EMBO J.* **1999**, *18*, 3558-3563.
97. Williams, K. A., *Nature* **2000**, *403*, 112-115.
98. Jung, K., Jung, H., Wu, J., Prive, G. G., Kaback, H. R., *Biochemistry* **1993**, *32*, 12273-12278.
99. Sun, J., Voss, J., Hubbell, W. L., Kaback, H. R., *Biochemistry* **1999**, *38*, 3100-3105.
100. Wolin, C. D., Kaback, H. R., *Biochemistry* **2000**, *39*, 6130-6135.
101. Dornmair, K., Corin, A. F., Wright, J. K., Jahnig, F., *EMBO J.* **1985**, *4*, 3633-3638.
102. Costello, M. J., Escaig, J., Matsushita, K., Viitanen, P. V., Menick, D. R., Kaback, H. R., *J. Biol. Chem.* **1987**, *262*, 17072-17082.
103. Sahin-Toth, M., Lawrence, M. C., Kaback, H. R., *Proc. Natl. Acad. Sci. USA* **1994**, *91*, 5421-5425.
104. Schroers, A., Burkovski, A., Wohlrab, H., Kramer, R., *J. Biol. Chem.* **1998**, *273*, 14269-14276.
105. Eskandari, S., Wright, E. M., Kreman, M., Starace, D. M., Zampighi, G. A., *Proc. Natl. Acad. Sci. USA* **1998**, *95*, 11235-11240.
106. Turk, E., Kim, O., le Coutre, J., Whitelegge, J. P., Eskandari, S., Lam, J. T., Kreman, M., Zampighi, G., Faull, K. F., Wright, E. M., *J. Biol. Chem.* **2000**, *275*, 25711-25716.
107. Friesen, R. H., Knol, J., Poolman, B., *J. Biol. Chem.* **2000**, *275*, 33527-33535.
108. Yazyu, H., Shiota-Niiya, S., Shimamoto, T., Kanazawa, H., Futai, M., Tsuchiya, T., *J. Biol. Chem.* **1984**, *259*, 4320-4326.
109. Pourcher, T., Leclercq, S., Brandolin, G., Leblanc, G., *Biochemistry* **1995**, *34*, 4412-4420.
110. Dave, N., Troullier, A., Mus-Veteau, I., Dunach, M., Leblanc, G., Padros, E., *Biophys. J.* **2000**, *79*, 747-755.
111. Botfield, M. C., Naguchi, K., Tsuchiya, T., Wilson, T. H., *J. Biol. Chem.* **1992**, *267*, 1818-1822.
112. Pourcher, T., Bibi, E., Kaback, H. R., Leblanc, G., *Biochemistry* **1996**, *35*, 4161-4168.
113. Botfield, M. C. Wilson, T. H., *J. Biol. Chem* **1989**, *264*, 11649-11652.
114. Wilson, D. M., Hama, H., Wilson, T. H., *Biochem. Biophys. Res. Commun.* **1995**, *209*, 242-249.
115. Wilson, T. H. Wilson, D. M., *Biochim. Biophys. Acta* **1998**, *1374*, 77-82.
116. Franco, P. J. Wilson, T. H., *J. Bacteriol.* **1999**, *181*, 6377-6386.
117. Ding, P. Z., Wilson, T. H., *Biochem. Biophys. Res. Commun.* **2000**, *268*, 409-413.
118. Veenhoff, L. M., Geertsma, E. R., Knol, J., Poolman, B., *J. Biol. Chem.* **2000**, *275*, 23834-23840.
119. Turk, E., Wright, E. M., *J. Membr. Biol.* **1997**, *159*, 1-20.
120. Jung, H., *Biochim. Biophys. Acta* **1998**, *1365*, 60-64.
121. Nakao, T., Yamato, I., Anraku, Y., *Mol. Gen. Genet.* **1987**, *208*, 70-75.
122. Komeiji, Y., Hanada, K., Yamato, I., Anraku, Y., *FEBS Lett.* **1989**, *256*, 135-138.
123. Jung, H., Rübenhagen, R., Tebbe, S., Leifker, K., Tholema, N., Quick, M., Schmid, R., *J. Biol. Chem.* **1998**, *273*, 26400-26407.
124. Wegener, C., Tebbe, S., Steinhoff, H. J., Jung, H., *Biochemistry* **2000**, *39*, 4831-4837.
125. Jung, H., Tebbe, S., Schmid, R., Jung, K., *Biochemistry* **1998**, *37*, 11083-11088.
126. Turk, E., Kerner, C. J., Lostao, M. P., Wright, E. M., *J. Biol. Chem.* **1996**, *271*, 1925-1934.
127. Levy, O., De la Vieja, A., Ginter, C. S., Riedel, C., Dai, G., Carrasco, N., *J. Biol. Chem.* **1998**, *273*, 22657-22663.
128. Xie, Z., Turk, E., Wright, E. M., *J. Biol. Chem.* **2000**, *275*, 25959-25964.
129. Dimroth, P., Thomer, A., *Biol. Chem Hoppe Seyler* **1986**, *367*, 813-823.
130. Schwarz, E., Oesterhelt, D., *EMBO J.* **1985**, *4*, 1599-1603.
131. van der Rest, M. E., Schwarz, E., Oesterhelt, D., Konings, W. N., *Eur. J. Biochem.* **1990**, *189*, 401-407.
132. van Geest, M., Lolkema, J. S., *J. Biol. Chem.* **1996**, *271*, 25582-25589.
133. van Geest, M., Nilsson, I., von Heijne, G., Lolkema, J. S., *J. Biol. Chem.* **1999**, *274*, 2816-2823.

134. van Geest, M., Lolkema, J. S., *J. Biol. Chem.* **1999**, *274*, 29705-29711.
135. van Geest, M., Lolkema, J. S., *Biochim. Biophys. Acta* **2000**, *1466*, 328-338.
136. Saier, M. H., http://www-biology.ucsd.edu/~msaier/transport/ **2000**.
137. Slotboom, D. J., Konings, W. N., Lolkema, J. S., *Microbiol. Mol. Biol. Rev.* **1999**, *63*, 293-307.
138. Poolman, B., Knol, J., van der Does, C., Henderson, P. J., Liang, W. J., Leblanc, G., Pourcher, T., Mus-Veteau, I., *Mol. Microbiol.* **1996**, *19*, 911-922.
139. Saier, M. H. J., *Microbiol. Mol. Biol. Rev.* **2000**, *64*, 354-411.
140. Prestage, L. S., Pardee A. B., *Biochim. Biophys. Acta* **1965**, *100*, 591-593.
141. Bassilana, M., Damiano-Forano, E., Leblanc, G., *Biochem. Biophys. Res. Commun.* **1985**, *129*, 626-631.
142. Tsuchiya, T., Lopilato, J., Wilson, T. H., *J. Membr. Biol.* **1978**, *42*, 45-59.
143. Tsuchiya, T., Oho, M., Shiota-Niiya, S., *J. Biol. Chem.* **1983**, *258*, 12765-12767.
144. Tsuchiya, T., Wilson, D. M., Wilson, T. H., *Ann. N. Y. Acad. Sci.* **1985**, *456:342-9*, 342-349.
145. Tsuchiya, T., Raven, J., Wilson, T. H., *Biochem. Biophys. Res. Commun.* **1977**, *76*, 26-31.
146. Maloney, P. C., Wilson, T. H., in: *Escherichia coli and Salmonella: Cellular and Molecular Biology* (Neidhardt, F. C., Ed.), ASM Press, Washington, DC, **1996**, pp. 1130-1148.
147. Boyer, P. D., *Trends Biochem. Sci.* **1988**, *13*, 5-7.
148. Damiano-Forano, E., Bassilana, M., Leblanc, G., *J. Biol. Chem.* **1986**, *261*, 6893-6899.
149. Bassilana, M., Pourcher, T., Leblanc, G., *J. Biol. Chem.* **1988**, *263*, 9663-9667.
150. Bassilana, M., Pourcher, T., Leblanc, G., *J. Biol. Chem.* **1987**, *262*, 16865-16870.
151. Okazaki, N., Kuroda, M., Shimamoto, T., Tsuchiya, T., *Biochim. Biophys. Acta* **1997**, *1326*, 83-91.
152. Hama, H., Wilson, T. H., *J. Biol. Chem.* **1992**, *267*, 18371-18376.
153. Foucaud, C., Poolman, B., *J. Biol. Chem.* **1992**, *267*, 22087-22094.
154. Tsuchiya, T., Yamane, Y., Shiota, S., Kawasaki, T., *FEBS Lett.* **1984**, *168*, 327-330.
155. Cairney, J., Higgins, C. F., Booth, I. R., *J. Bacteriol.* **1984**, *160*, 22-27.
156. Yamato, I., Anraku, Y., *J. Membr. Biol.* **1990**, *114*, 143-151.
157. Yamato, I., *FEBS Lett.* **1992**, *298*, 1-5.
158. Eskandari, S., Loo, D. D., Dai, G., Levy, O., Wright, E. M., Carrasco, N., *J. Biol. Chem.* **1997**, *272*, 27230-27238.
159. Wright, E. M., Loo, D. D., Panayotova-Heiermann, M., Hirayama, B. A., Turk, E., Eskandari, S., Lam, J. T., *Acta Physiol. Scand. Suppl.* **1998**, *643*, 257-264.
160. Kastner, C. N., Dimroth, P., Pos, K. M., *Arch. Microbiol.* **2000**, *174*, 67-73.
161. Heidelberg, J. F., Eisen, J. A., Nelson, W. C., Clayton, R. A., Gwinn, M. L., Dodson, R. J., Haft, D. H., Hickey, E. K., Peterson, J. D., Umayam, L., Gill, S. R., Nelson, K. E., Read, T. D., Tettelin, H., Richardson, D., Ermolaeva, M. D., Vamathevan, J., Bass, S., Qin, H., Dragoi, I., Sellers, P., McDonald, L., Utterback, T., Fleishmann, R. D., Nierman, W. C., White, O., *Nature* **2000**, *406*, 477-483.
162. Takami, H., Horikoshi, K., *Extremophiles* **2000**, *4*, 99-108.
163. David, S., van der Rest, M. E., Driessen, A. J., Simons, G., de Vos, W. M., *J. Bacteriol.* **1990**, *172*, 5789-5794.
164. Bandell, M., Ansanay, V., Rachidi, N., Dequin, S., Lolkema, J. S., *J. Biol. Chem.* **1997**, *272*, 18140-18146.
165. van der Rest, M. E., Molenaar, D., Konings, W. N., *J. Bacteriol.* **1992**, *174*, 4893-4898.
166. van der Rest, M. E., Siewe, R. M., Abee, T., Schwarz, E., Oesterhelt, D., Konings, W. N., *J. Biol. Chem* **1992**, *267*, 8971-8976.
167. Lolkema, J. S., Enequist, H., van der Rest, M. E., *Eur. J. Biochem.* **1994**, *220*, 469-475.
168. Pos, K. M., Dimroth, P., *Biochemistry* **1996**, *35*, 1018-1026.
169. Niiya, S., Yamasaki, K., Wilson, T. H., Tsuchiya, T., *J. Biol. Chem.* **1982**, *257*, 8902-8906.
170. Yazyu, H., Shiota, S., Futai, M., Tsuchiya, T., *J. Bacteriol.* **1985**, *162*, 933-937.
171. Kawakami, T., Akizawa, Y., Ishikawa, T., Shimamoto, T., Tsuda, M., Tsuchiya, T., *J. Biol. Chem.* **1988**, *263*, 14276-14280.
172. Hama, H., Wilson, T. H., *J. Biol. Chem.* **1993**, *268*, 10060-10065.
173. Pourcher, T., Deckert, M., Bassilana, M., Leblanc, G., *Biochem. Biophys. Res. Commun.* **1991**, *178*, 1176-1181.
174. Pourcher, T., Zani, M. L., Leblanc, G., *J. Biol. Chem.* **1993**, *268*, 3209-3215.

175. Zani, M. L., Pourcher, T., Leblanc, G., *J. Biol. Chem.* **1993**, *268*, 3216-3221.
176. Wilson, D. M., Wilson, T. H., *J. Bacteriol.* **1992**, *174*, 3083-3086.
177. Wilson, D. M., Wilson, T. H., *Biochim. Biophys. Acta* **1994**, *1190*, 225-230.
178. Hama, H, Wilson, T. H., *J. Biol. Chem.* **1994**, *269*, 1063-1067.
179. Franco, P. J., Wilson, T. H., *Biochim. Biophys. Acta* **1996**, *1282*, 240-248.
180. Zani, M. L., Pourcher, T., Leblanc, G., *J. Biol. Chem.* **1994**, *269*, 24883-24889.
181. Matsuzaki, S., Weissborn, A. C., Tamai, E., Tsuchiya, T., Wilson, T. H., *Biochim. Biophys. Acta* **1999**, *1420*, 63-72.
182. Pourcher, T., Sarkar, H. K., Bassilana, M., Kaback, H. R., Leblanc, G., *Proc. Natl. Acad. Sci. USA* **1990**, *87*, 468-472.
183. Pourcher, T., Bassilana, M., Sarkar, H. K., Kaback, H. R., Leblanc, G., *Biochemistry* **1992**, *31*, 5225-5231.
184. Sahin-Toth, M., Karlin, A., Kaback, H. R., *Proc. Natl. Acad. Sci. USA* **2000**, *97*, 10729-10732.
185. Leblanc, G., Pourcher, T., Zani, M. L., in: *Molecular Biology and Function of Carrier Proteins* (Reuss, L., Russell, J. M., and Jennings, M. L., Eds.), The Rockefeller University Press, New York, **1993**, pp. 213-227.
186. Botfield, M. C., Wilson, T. H., *J. Biol. Chem* **1988**, *263*, 12909-12915.
187. Mus-Veteau, I., Pourcher, T., Leblanc, G., *Biochemistry* **1995**, *34*, 6775-6783.
188. Mus-Veteau, I., Leblanc, G., *Biochemistry* **1996**, *35*, 12053-12060.
189. Cordat, E., Mus-Veteau, I., Leblanc, G., *J. Biol. Chem.* **1998**, *273*, 33198-33202.
190. Maehrel, C., Cordat, E., Mus-Veteau, I., Leblanc, G., *J. Biol. Chem.* **1998**, *273*, 33192-33197.
191. Cordat, E., Leblanc, G., Mus-Veteau, I., *Biochemistry* **2000**, *39*, 4493-4499.
192. Ambroise, Y., Leblanc, G., Rousseau, B., *Biochemistry* **2000**, *39*, 1338-1345.
193. Frillingos, S., Gonzalez, A., Kaback, H. R., *Biochemistry* **1997**, *36*, 14284-14290.
194. Sahin-Toth, M., le Coutre, J., Kharabi, D., le Maire, G., Lee, J. C., Kaback, H. R., *Biochemistry* **1999**, *38*, 813-819.
195. Venkatesan, P. Kaback, H. R., *Proc. Natl. Acad. Sci. USA* **1998**, *95*, 9802-9807.
196. Weissborn, A. C., Botfield, M. C., Kuroda, M., Tsuchiya, T., Wilson, T. H., *Biochim. Biophys. Acta* **1997**, *1329*, 237-244.
197. Ding, P., Z. Wilson, T. H., *J. Membr. Biol.* **2000**, *174*, 135-140.
198. Myers, R., S. Maloy, S. R., *Mol. Microbiol.* **1988**, *2*, 749-755.
199. Quick, M., Jung, H., *Biochemistry* **1997**, *36*, 4631-4636.
200. Quick, M., Tebbe, S., Jung, H., *Eur. J. Biochem.* **1996**, *239*, 732-736.
201. Quick, M., Stolting, S., Jung, H., *Biochemistry* **1999**, *38*, 13523-13529.
202. Ohsawa, M., Mogi, T., Yamamoto, H., Yamato, I., Anraku, Y., *J. Bacteriol.* **1988**, *170*, 5185-5191.
203. Quick, M., Jung, H., *Biochemistry* **1998**, *37*, 13800-13806.
204. Yamato, I., Kotani, M., Oka, Y., Anraku, Y., *J. Biol. Chem.* **1994**, *269*, 5720-5724.
205. Panayotova-Heiermann, M., Loo, D. D., Kong, C. T., Lever, J. E., Wright, E. M., *J. Biol. Chem.* **1996**, *271*, 10029-10034.
206. Panayotova-Heiermann, M., Eskandari, S., Turk, E., Zampighi, G. A., Wright, E. M., *J. Biol. Chem.* **1997**, *272*, 20324-20327.
207. Panayotova-Heiermann, M., Leung, D. W., Hirayama, B. A., Wright, E. M., *FEBS Lett.* **1999**, *459*, 386-390.
208. Dila, D. K., Maloy, S. R., *J. Bacteriol.* **1986**, *168*, 590-594.
209. Hanada, K., Yoshida, T., Yamato, I., Anraku, Y., *Biochim. Biophys. Acta* **1992**, *1105*, 61-66.
210. Yamato, I., Anraku, Y., *J. Biol. Chem.* **1988**, *263*, 16055-16057.
211. van Iwaarden, P. R., Pastore, J. C., Konings, W. N., Kaback, H. R., *Biochemistry* **1991**, *30*, 9595-9600.
212. Frillingos, S., Sahin-Toth, M., Wu, J., Kaback, H. R., *FASEB J.* **1998**, *12*, 1281-1299.
213. Prive, G. G., Kaback, H. R., *J. Bioenerg. Biomembr.* **1996**, *28*, 29-34.
214. Padan, E., Venturi, M., Michel, H., Hunte, C., *FEBS Lett.* **1998**, *441*, 53-58.
215. Ostermeier, C., Michel, H., *Curr. Opin. Struct. Biol.* **1997**, *7*, 697-701.

4
Prokaryotic Binding Protein-dependent ABC Transporters

Winfried Boos and Tanja Eppler

Summary

Bacterial binding protein-dependent ABC transporters are multidomain transmembrane substrate translocators belonging to the largest superfamily of proteins found in Eubacteria, Eukarya, and Archaea. The common denominator of these systems are the well-conserved ABC (ATP Binding Cassette) domains or subunits that couple ATP hydrolysis to mechanical movement, generally to the active transport of small molecules across membranes. The substrate recognition sites are soluble high-affinity substrate binding proteins. The three-dimensional structure for many of these binding proteins has been solved revealing a common scheme of substrate binding and conformational change. The transmembrane domains or subunits, regarded as the transport channel proper, are the least understood. Structure and function of these proteins have been studied by mutational analysis. They reveal a common sequence motif that is involved in coupling the ABC domain to the transport domain. The last component invariably occurs as a dimer of nucleotide binding domains. Two members of this type of subunits have only recently become available for structural analysis. They may well contain the answer to the question of how ATP hydrolysis is coupled to active transport. In some cases the ABC subunit also contains the interaction site for active and passive regulation. In the first, metabolic activity (inducer exclusion) affects transport activity; in the second, transport or the lack of transport affects gene regulation. This chapter will focus on binding protein-dependent prokaryotic ABC importers.

4.1
A Brief History of ABC Systems

The origin of ABC transporters is going back to the sixties. At that time the major types of bacterial transport systems were discovered. The lactose system, eventually the standard proton motive force (PMF)-type transport system, was identified as a single membrane protein by Fox and Kennedy [1], the phosphotransferase system

(PTS) was first recognized as a novel type of PEP-dependent sugar phosphorylation system until mutants revealed the connection between phosphorylation and transport [2, 3]. The ABC systems (called as such only much later) made their appearance at the same time but in a rather roundabout way. The application of Heppel"s cold osmotic shock procedure for isolation of periplasmic proteins of gram-negative enteric bacteria [4] yielded among degradative enzymes (e. g., alkaline phosphatase) mysterious binding proteins of high affinity for small polar substrates such as sugars, amino acids, and ions. These proteins did not exhibit any known enzymatic activity but from the beginning on they were suspected to be part of transport systems [5]. The evidence was quite indirect and hotly debated. Cells treated by Heppel"s shock procedure lost their capability to take up certain substrates and the specificity as well as affinity for substrate of well studied binding proteins closely resembled uptake systems. Thus, the first name for an ABC transporter was as trivial as "shock sensitive" systems. The evidence for the essentiality of binding proteins for transport function came from the analysis of mutants in the respective binding protein that no longer were able to transport substrate while reversions to transport-plus brought back an active binding protein [6, 7]. The isolation of mutants in these shock sensitive transport systems that still contained fully active and apparently wild-type binding proteins revealed that additional components must be necessary. It was the use of the gene fusion technique that led the way to the identification of these additional proteins [8]. An important point in the discussion of the binding protein-dependent or shock sensitive systems was the energy source. Clearly aside from high substrate affinity these systems exhibited "active transport", that is, accumulation against the chemical gradient of the substrate across the membrane. After an interesting interlude of acetyl phosphate being a candidate for the energy donor it became clear, mainly by the work of Giovanna Ames and coworkers [9, 10], that ATP hydrolysis was the driving force characterizing these systems as primary pumps linking ATP hydrolysis directly to the movement of substrate across the membrane.

Major achievements in the understanding of bacterial ABC transporters were their purification and reconstitution in liposomes as well as the isolation and purification of the intact active complex from detergent solubilized membranes [10–15]. The advent of DNA sequencing connected bacterial binding protein-dependent transport systems to the big world of eukaryotic transport systems [16]. Invariably, one subunit of these systems contained a stretch of DNA encoding about 215 amino acids that could be found in close homology not only in all other bacterial transport systems of this type but also in a number of eukaryotic membrane proteins suspected of transport function [17, 18]. Since this domain encompasses a Walker A and B site typical of ATP-hydrolyzing proteins [19] this subunit or domain was coined the <u>A</u>TP <u>B</u>inding <u>C</u>assette or ABC subunit [18], a name which later on was also used as a synonym for the entire system.

A historical account of bacterial binding protein-dependent ABC transporters has to mention the progress that has been made in understanding the structure and function of periplasmic binding proteins, the substrate recognition site of the transport systems. The sulfate binding protein was the first to be crystallized [5].

Since then, mainly by the work of Florante Quiocho, Sherry Mowbray, and their respective coworkers, an overwhelming number of periplasmic binding proteins was crystallized and their three-dimensional structures were solved [20, 21]. These studies not only allowed to picture the common shape of these molecules but also to get a glimpse at the intrinsic mobility when they bind substrate.

How the different subunits of this multicomponent transport system interact with each other is still not well understood. The use of genetics (mutant and suppressor analysis) and the use of the *phoA* and *lacZ* protein fusion technique [22, 23] allowed to obtain a schematic view of the two-dimensional topology of the membrane spanning subunits and an estimation of possible next neighbors between membrane spanning helices [24]. Again, mainly the combination of genetics and structural studies [25–27] as well as genetics and biochemical techniques permitted to draw conclusions on the protein surfaces that interact with each other [28, 29]. The latest achievement in the understanding of ABC systems was the recent structure determination of two ABC subunits [30, 31] revealing mechanistic aspects of energy coupling as well as delivering the key for understanding regulatory phenomena associated with some of these systems.

4.2
What is an ABC System?

ABC systems in their most general definition are modularly composed mechanical machines which couple ATP hydrolysis to the physical movement of molecules. The ever increasing DNA sequence information reveals the ubiquitous appearance in all kingdoms of life of a conserved ABC module encompassing, on about 215 amino acids, among other highly conserved sequences, the Walker A and Walker B sites of ATPases. Most likely, proteins harboring the ABC module constitute the largest protein superfamily. The overwhelming majority of all ABC systems transports molecules into or out of cells and across membranes. Prominent examples are P-glycoprotein involved in multiple drug resistance, the gated chloride channel CFTR (cystic fibrosis conductance regulator) [32] of cystic fibrosis, eye pigment precursor importers, protein exporters, and many more [33]. ABC importers have been implicated in competence and virulence of pathogenic bacteria which require effective metal ion transporters during infection [34]. Sporulation in *Bacillus subtilis* is associated with the need to transport and with the recognition of certain peptides [35].

All ABC transporters are minimally composed of two hydrophobic transmembrane domains (TMDs) spanning the membrane several times, and two ABC domains attached to the TMDs at the inner face of the cytoplasmic membrane. As revealed by genome sequencing, ABC transporters are the prominent active transport systems in prokaryotes catalyzing high-affinity uptake of small polar substances such as sugars, amino acids, and ions across the membrane. Binding protein-dependent importers constitute a subfamily of ABC transporters. Prokaryotic export systems have less frequently been observed among ABC systems; they usually belong to the PMF-type transport systems. A well-studied exception of this rule is LmrA, recently identified in *Lactococcus lactis* as a multidrug-resistant pump [36] that even can confer typical multidrug resistance to human fibroblast cells [37]. These exporters do not contain substrate binding proteins. Other prokaryotic export systems of the ABC-type are actually protein and polysaccharide secretion systems. They include the specific exporter for hemolysine in *E. coli* [33] or the protease exporter of *Erwinia chrysanthemi* [38] as well as the exporter for capsular polysaccharides (for reviews, see [39–41]). According to a recent sequence study [42, 43] analyzing all ABC systems in *E. coli*, the following picture arises: there are 79 polypeptides with a total of 97 ABC cassettes. These belong to 69 independent ABC machineries. 57 of those are associated with TMDs. Of these, 44 are binding protein-associated importers and 13 are possibly exporters (even though only few of them have been identified in their function, e. g., CcmA for heme export and KpsT for export of capsular polysaccharide [44]). Of the remaining 11, one is UvrA involved in DNA repair [45], another is a system catalyzing the LolA-dependent release of lipoprotein from the cytoplasmic membrane [46], both clearly no classical transport functions. With nearly 5 % of the total sequence, ABC systems represent the largest protein family in *E. coli*. A similar analysis of the recently sequenced genome of *B. subtilis* reveals an even higher number of ABC exporters than in *E. coli*. In addition, a number of systems are lacking (on a genetic basis) the ABC subunit, indicating joint usage of a common ABC subunit [47]. The genome of *Mycobacterium tuberculosis* contains 2.5 % ABC encoding genes. These genes putatively encode more exporters than binding protein-dependent importers [48].

Binding protein-dependent prokaryotic ABC importers are considered as unidirectional primary pumps that directly couple the translocation step to the hydrolysis of ATP. With the use of mutants preventing the metabolism of the transported substrate, concentration gradients as high as $1:10^5$ can easily be observed [49]. So far, uncoupled mutants allowing diffusion in the absence of ATP hydrolysis (or in the absence of the ABC subunit) have not been reported. Several reviews on the different aspects of ABC transporters have been published: on structure and function [24, 33, 50–52], evolution [41, 42, 53, 54] (see also *http://www-alt.pasteur.fr/ ~eli.dassa/ABC_class.html*), multidrug resistance and lipid transport [55, 56], and secretion [40, 57, 58]. In this review, we will concentrate on binding protein-dependent ABC importers of prokaryotes.

4.3
The Composition of the Prokaryotic ABC Transporters

All ABC transporters are minimally composed of two ABC domains and two TMDs. As with another type of bacterial transport system, the phosphotransferase system (PTS), these domains can be arranged in nearly all possible combinations (Fig. 1 shows the situation in *E. coli* as an example). For instance, as four single individual polypeptides (the oligopeptide or the nickel transporter); heterodimeric

TMDs with homodimeric ABC domains (the maltose or the histidine system); a homodimeric TMD in combination with fused twin ABC domains (the Ara or the Rbs system); or fused twin TMDs associated with homodimeric ABC domains (the iron transporter Fhu). The fusion of TMD to the ABC domain also occurs, e.g., LmrA from *Lactococcus lactis* [59]. This combination seems to be reserved for systems lacking a binding protein (BP).

In *E. coli*, several systems have been recognized that are served by two binding proteins with different substrate specificities. For instance, the HisM/Q/P$_2$ ABC system transports histidine in conjunction with the HisJ binding protein specific for histidine, but it transports arginine in conjunction with LAO, the binding protein that is specific for lysine, arginine, and ornithine [60]. Similarly, different substrate specificities are endowed to the ABC transporter CysATW$_2$ by the two binding proteins SbpA and CysP [61], or to LivF/G/H/M by the two binding proteins LivK and LivJ [62]. Also, ABC subunits can be shared by more than one system (Fig. 1). In *Streptomyces* it has been observed that the ABC subunit MsiK can serve several TMDs with associated binding proteins [63, 64] (Fig. 1).

Without exception, the major substrate recognition sites of prokaryotic ABC importers are binding proteins. These are soluble proteins located at the outer surface of the cytoplasmic membrane. In gram-negative bacteria they are enclosed in the periplasm between cytoplasmic and outer membrane, in gram-positive bacteria they are flexibly anchored via a thioether between the N-terminal cysteine and a diacylated glycerol. The two fatty acids attached to the glycerol moiety, together with a third fatty acid attached to the amino group of the N-terminal cysteine, are embedded in the outer leaflet of the cytoplasmic membrane [64–66]. Nothing is known about the stoichiometry of binding proteins in gram-positive bacteria, supposedly surrounding in close proximity the protruding cognate TMDs. In gram-negative bacteria, the excess of soluble periplasmic binding protein over the TMDs must be substantial. In case of the *E. coli* maltose system it is in the order of 30–50-fold [67].

The trehalose/maltose binding protein of the archaeon *Thermococcus litoralis* is membrane bound as are the binding proteins of gram-positive bacteria and has a cysteine containing signal sequence typical for thio-glyceride lipidation [68]. This indicates the same type of membrane anchoring in archaea as in gram-positive bacteria. Yet, the glucose binding protein of *Sulfolobus solfataricus* is anchored in the membrane via an N-terminal hydrophobic a-helix revealing a novel type of membrane attachment for binding proteins [69]. Recently, a curious combination of subunits has been reported. The high-affinity betaine ABC transporter of *L. lactis* consists of a dimer of two heterodimers. The first, BusAB, combines the TMD with the covalently linked binding protein attached at the C-terminus of the TMD, whereas the other (BusAA) represents the ABC domain. Apparently this is not uncommon, since sequences resembling BusAB have been found in *Enterococcus faecalis* and *Streptococcus pyogenes* [70, 71].

Figure 2 schematically shows the different realizations for the high-affinity binding proteins of prokaryotic ABC transporters.

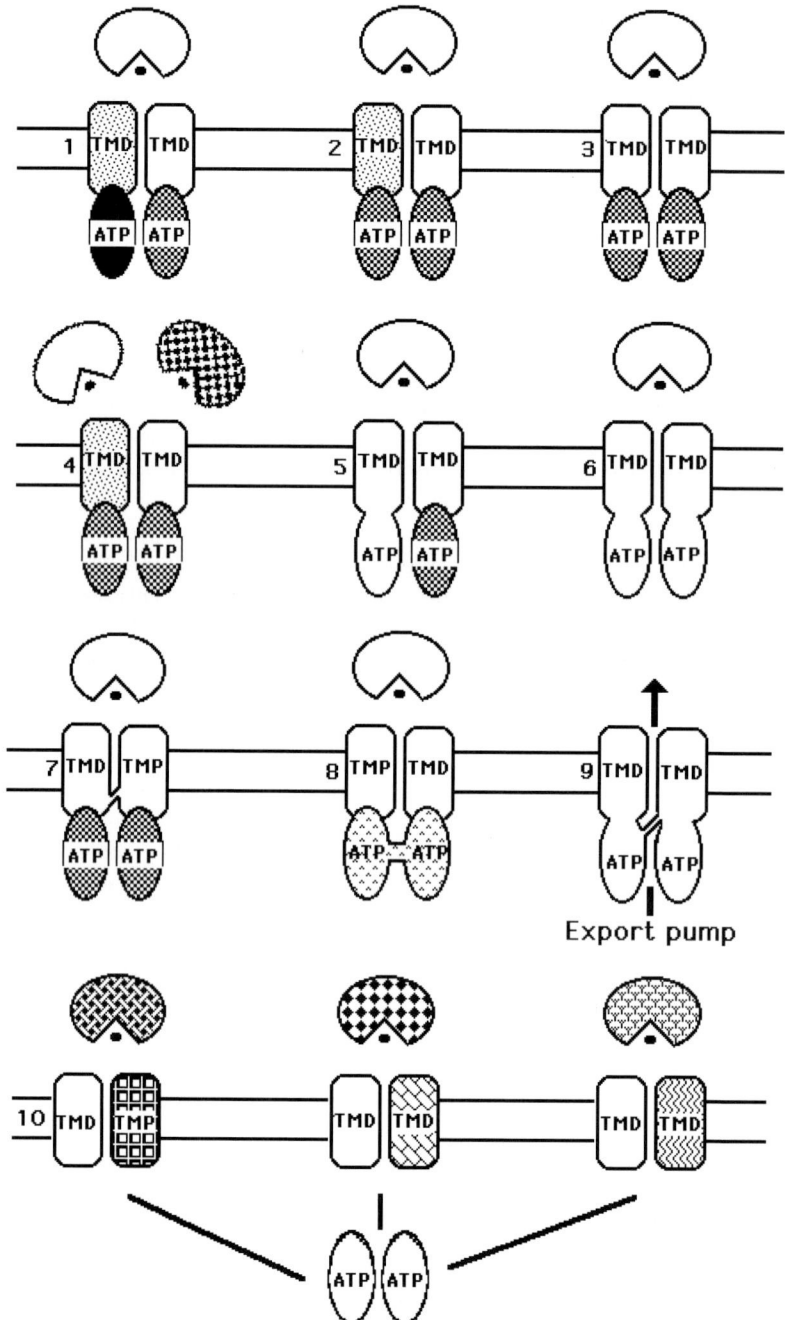

Fig. 1. Variations in the composition of the TMDs and ABCs. The archetypal ABC transporter has four domains, two transmembrane proteins (TMD) and two ATPase subunits (ABC), mostly with one binding protein [except (4) and (9)]. The domains (1–8 of *E. coli*) can be associated and fused in the following ways: (1) heterodimeric TMD and ABC, as in OppBCDF; (2) heterodimeric TMDs with homodimeric ABCs, e.g., MalFGK$_2$; (3) homodimeric TMDs and ABCs, as in GlnPQ; (4) with two binding proteins, e. g., the His system; (5) one TMD fused to one ABC as in YhiGH; (6) both TMDs fused to both ABCs, as in HlyB; (7) two separate ABCs with fused TMDs, e. g., FhuCB; (8) two separate TMDs with fused ABCs, as in RbsAC; (9) exporter of a single four-domain peptide with no binding protein, as in LmrA of *Lactococcus lactis*; (10) ABC transporters which share the same ATPase subunit (MsiK of *Streptomyces*).

There are examples of *E. coli* binding protein-dependent ABC transporters which, in the absence of the cognate protein, are able to functionally exchange their ABC subunits. For example, the Ugp system, specific for the uptake of glycerol-3-phosphate, and the maltose system exhibit high overall sequence homology in spite of their difference in substrate specificity, and are able to exchange their ABC domains [72]. Similarly, CymD, the ABC subunit of the cyclodextrin ABC transporter of *Klebsiella oxytoca* [73] can functionally substitute MalK, the ABC subunit of the *E. coli* maltose system [74]. Also, LacK, the ABC subunit of a lactose transporter of *Agrobacterium radiobacter* can, after mutational alteration, substitute for MalK in *Salmonella typhimurium* and *E. coli* [75]. For TMDs or for the specific binding proteins exchangeability between systems has not been demonstrated as yet.

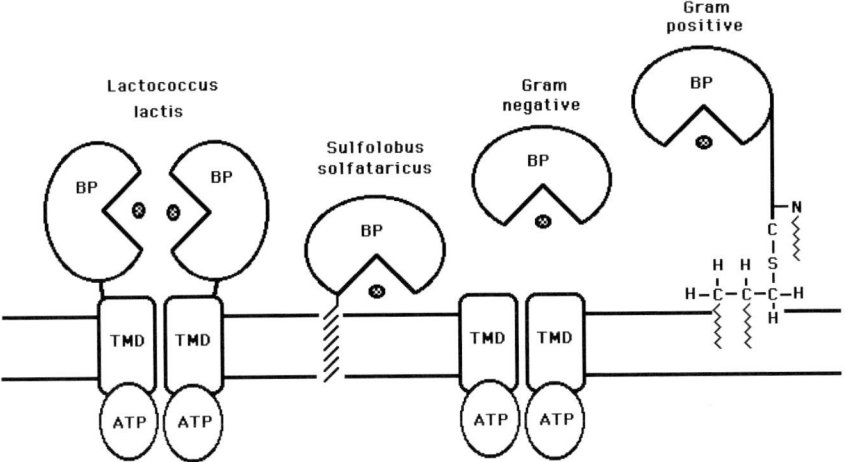

Fig. 2. Examples of the organization of different binding proteins (BP). Examples from *Lactococcus lactis* [the BPs are fused to the transmembrane protein (TMD)], from *Sulfolobus solfataricus* (anchored via an N-terminal hydrophobic α-helix), from gram-negative (with soluble BP located in the periplasm) and gram-positive bacteria (anchored to the cytoplasmic membrane via a lipid tail attached to the N-terminal cysteine).

4.4
Associated Proteins and Signal Transduction Pathways

Almost since the time of its discovery the galactose binding protein of E. coli has been associated with chemotaxis, functioning as a specific chemoreceptor [76, 77]. Now it is clear that only a few ABC-related binding proteins are in fact chemoreceptors, e. g., the maltose [78, 79], ribose [80], and dipeptide [81] binding proteins. They interact with their cognate TMDs to trigger ATP hydrolysis in the ABC subunit, thus initiating the transport step. In addition, they also bind to their cognate chemotaxis signal transducer to initiate the chemotaxis signal transduction pathway [21].

Some ABC transporters have been recognized as a part of signal transducing chains: the E. coli Pst system consisting of a classical binding protein-dependent high-affinity ABC transporter for inorganic phosphate (PstS/C/A/B) transfers by a still poorly understood mechanism the information of P_i transport or the lack of it via the mysterious PhoU protein to the membrane-bound sensor kinase PhoR, which in turn controls PhoB, the central gene activator (response regulator) of the *pho* regulon [82]. In *Salmonella typhimurium*, the SapB/C/D/F ABC transporter with its associated SapA binding protein (supposedly a peptide transporter) seems to regulate a K^+ channel [83]. The E. coli maltose ABC transporter uses MalK, its ABC subunit, to communicate with other systems. Under conditions of catabolite repression (PTS-mediated glucose transport) it interacts with dephosphorylated $EIIA^{Glc}$ of the PTS reducing (most likely by curbing ATPase activity) transport of maltose [84]. On the other hand, in the absence of maltose transport MalK associates with MalT, the central gene activator of the *mal* regulon curbing its activity (Fig. 4) [85] as will be discussed in detail below.

Binding protein-dependent ABC importers exhibit high affinity for their substrate (K_m usually 1 µM or below) and, in some cases, such as the E. coli maltose system, a high V_{max}. In gram-negative bacteria where the outer membrane poses a diffusion barrier, only partly overcome by unspecific pores, specific porins belonging to the same genetic regulon as the particular ABC transporter provide a sufficient rate of diffusion for the substrate into the periplasm. Well understood examples are maltoporin (the receptor for phage λ) of E. coli, specific for the diffusion of maltose and maltodextrins [86–90] or CymA, genetically associated with the Cym ABC system of *Klebsiella oxytoca* catalyzing the uptake of cyclomaltodextrins [91]. ABC transporters for amino acids or ions are not known to be associated with specific porins. This is probably due to the smaller size of their substrates and the much lower V_{max} of these systems compared to the sugar transport systems which are needed to ensure a high flow of carbon source.

A special case for ABC-associated proteins is the multitude of the ABC transporters for chelated Fe^{3+}. These systems all contain, in addition to the standard ABC complex in the cytoplasmic membrane (plus periplasmic binding protein), a high-affinity receptor in the outer membrane. The receptor, in combination with TonB, ExbB, and ExbD, couples the energy contained as PMF in the cytoplasmic membrane to the active transport of substrate into the periplasm [92]. Much has been learned by the recent elucidation of the three-dimensional structure of two of these receptors: FhuA [93, 94] and FepA [95]. As a rule, the affinity and specificity of these chelated Fe^{3+} transport systems are not determined by the periplasmic binding proteins but rather by the high-affinity receptor in the outer membrane. Signal transduction starting from substrate recognition at the outer membrane receptor and ending, via a specialized σ-factor, in gene expression has been recognized in one of these systems [96, 97].

4.5
The Components

4.5.1
The Binding Proteins

4.5.1.1 Substrate Recognition Sites are High-affinity Soluble Binding Proteins

Invariably, all prokaryotic ABC importers are associated with soluble high-affinity binding proteins. A listing of bacterial and archaeal ABC transporters with their respective binding proteins can be found on internet sites cited in Dassa et al. [43]. In 1993, over 50 solute binding proteins have been classified into 8 groups according to size, signature sequences, and relatedness [98]. Interestingly, one of these groups is related to classical transcriptional repressors of the LacI-type [99–101].

Binding proteins from gram-negative enteric organisms are secretory proteins equipped with a signal sequence that is recognized by the classical Sec system, and they are secreted into the periplasm after cleavage of the signal sequence by signal peptidase I [102–105]. From the periplasm they can easily be obtained through the classical osmotic shock procedure [4] preferably from cells harboring plasmid-encoded genes. As a rule they do not form inclusion bodies when overexpressed and represent the major protein among the periplasmic protein fraction from where they can easily be purified using standard biochemical techniques. They can be freed from naturally bound ligands by guanidinium hydrochloride denaturation followed by dialysis against buffer [106] where they easily renature. In their native form they are highly resistant against proteolysis and heat treatment. Binding proteins from gram-positive bacteria are always anchored to the cytoplasmic membrane via a lipid tail connected to the N-terminal cysteine. These proteins are also secretory proteins, recognized by the Sec system [105]. The N-terminal signal sequence of these proteins carries a particular motif that includes a cysteine [65] which, after cleavage of the signal sequence by signal peptidase II [107], is posttranslationally modified by lipidation leading to membrane attachment. These proteins when engineered free of their signal sequence or equipped with a corresponding signal sequence of gram-negative organisms can also be isolated and purified in soluble form. Then, they behave as binding proteins from gram-negative bacteria.

4.5.1.2 **The Binding Test**

There are several ways to test binding affinity. Aside from the classical equilibrium dialysis the most sensitive assay is based on the altered intrinsic tryptophan fluorescence of most binding proteins when the ligand is bound to it [108]. In a stopped flow setup this method is also able to measure the rate of dissociation and association of ligand [109]. From this type of assays it has become clear that the high affinity for substrate (K_d 1 µM and below) associated with the proteins is due to a slow rate of dissociation whereas the association rate is fast and close to diffusion controlled [110, 109]. A very convenient method to measure binding even in crude extracts is to precipitate the binding protein in the presence of radioactively labeled substrate by cold saturated ammonium sulfate followed by filtration and counting. Surprisingly, the protein-bound radioactivity can even be washed on the filter with a large excess of unlabeled substrate in saturated ammonium sulfate without loosing the signal [111]. Using the substrate retention phenomenon [112] the K_d of the protein can be determined in one dialysis (exit) experiment as long as the binding protein concentration is known [113].

4.5.1.3 **Special Examples**

A number of binding proteins have been isolated and characterized from *Bacillus subtilis*, some of which function in systems that are activated as well as induced by high osmolarity [114–116]. Transport-associated binding proteins from thermophilic prokaryotes have been characterized, e. g., a trehalose/maltose binding protein from the hyperthermophilic archaeon *Thermococcus litoralis* [68], a glucose binding protein from the archaeon *Sulfolobus solfataricus* [69], and a maltose binding protein from the bacterium *Thermotoga maritima* [117], or the maltose/trehalose binding protein from *Thermoanaerobacter ethanolicus* 39E [118] or probably from *Thermoplasma acidophilum* [119]. The glucose binding protein as well as several other binding proteins from the archaeon *Sulfolobus solfataricus* have been characterized as a glycoproteins [69, 119a]. Similarly, TMBP from *T. litoralis* also stains positively on SDS-PAGE with glycostain (Greller and Boos, unpublished data). It is interesting to note that maltose binding proteins from thermophilic organisms appear in two varieties: one exemplified by TMBP from *T. litoralis* or *T. ethanolicus* 39E that bind with equal affinity trehalose and maltose but not maltodextrins, and the other exemplified by the maltose binding protein from *T. maritima* [117] or the maltose binding protein from *Pyrococcus furiosus* [120] that bind maltose and maltodextrins but no trehalose. The difference in the binding pocket of these two types of maltose binders is the preference of hydrogen bonds in TMBP, but hydrophobic stacking in the maltose/maltodextrin binding protein from *P. furiosus* [121], the strength of which is increased at higher temperature.

Considering sugar binding sites it is noteworthy to mention an example of a multispecific sugar binding protein-dependent sugar ABC transporter of *Streptococcus mutans* that transports sugars as different as melibiose, raffinose, and isomaltotriose [122]. MsmE, the responsible binding protein, is a typical lipoprotein [123].

Up to a few years ago, the appearance of a soluble substrate binding protein was the sure sign for the presence of an ABC transporter. This dogma is no longer correct since it has been reported that the Na^+-dependent glutamate transport in *Rhodobacter sphaeroides* is mediated by a classical solute binding protein in conjunction with a secondary transporter [124].

4.5.1.4 Binding Proteins Undergo Conformational Changes upon Binding Substrate

The most prominent feature associated with binding proteins is their large conformational change upon binding substrate. One of the early observations attributed to a substrate-dependent conformational change was an altered electrical surface charge of the galactose binding protein as seen by a change in the electrophoretic mobility in non-denaturing polyacrylamide gels in the presence of galactose [108]. The change in the intrinsic fluorescence upon binding substrate was first seen in the glutamine [110], galactose [108] and maltose [49] binding proteins. The differential absorption in the spectrum corresponded to the absorption spectrum of a tryptophan model compound, and it became clear that binding-associated structural changes affected the chemical surrounding of tryptophan residues [125]. The introduction of 5-fluorotryptophan and ^{15}N-labeled amino acids made it possible to follow the conformational change by nuclear magnetic resonance [126–130]. Also labeling methionine residues in the galactose [131] and maltose [132] binding proteins by 5-iodoacetamidofluorescein demonstrated the large movement of binding proteins after substrate binding. The solution of the crystal structure of many binding proteins, in substrate-free as well as in substrate-bound form, revealed the underlying nature of the substrate-induced conformational change as a movement of two pseudosymmetrical lobes opening and closing around the substrate binding site established between the lobes [133–135]. Thus, in the case of the *E. coli* maltose binding protein, the two lobes rotate by an angle of 35° around a connecting hinge region and are twisted against each other by 8° [135]. In the case of the LAO binding protein, the angle is even 52° [133]. One may consider two modes in which the conformational change does occur. One would be a dynamic equilibrium of the open and closed form where binding of substrate would stabilize the closed form and strongly reduce the movement of the lobes. The other scenario would be that in the absence of substrate the open form prevails and is stable in solution. Binding of substrate would occur to one lobe of the open form closing the lobes (induced fit). Time-resolved tryptophan fluorescence anisotropy with the *E. coli* maltose binding protein has shown that the intrinsic mobility of the tryptophan residues in the absence of substrate is much higher than when substrate is bound demonstrating the dynamic nature (mobility) of the protein in the absence of substrate [136]. This observation and the fact that the closed but substrate-free form of the glucose/galactose binding protein can be crystallized [137] gives credit to the equilibrium model. The crystallization of multiple substrate-free forms of the *E. coli* ribose binding protein seems to draw a pathway in a series of conformational states leading to the fully closed protein [138].

In any case, binding nearly engulfs the substrate, preventing even the access of water. The closing of the two lobes is apparently an important feature for the successful function of the protein in delivering its substrate to the membrane components. In a series of publications Hall et al. [139–141] have demonstrated that the *E. coli* maltose binding protein can bind maltodextrins in two different modes, the end-on-mode (R) including the reducing end and the middle mode (B) excluding the reducing end. Only the R-mode leads to a complete closure of the two lobes and to transport whereas the B-mode (even though exhibiting potentially high affinity) remains partially open and does not allow transfer of substrate to the membrane components. In this context, it is interesting to refer to an oligopeptide transporter of *Lactococcus lactis* which can bind oligopeptides up to at least 35 residues [142]. In this study, it is shown that the first six amino acid residues are important for ligand binding (enclosed by the two lobes). The remainder of the oligopeptide ligand may interact in a non-opportunistic manner with the surface of OppA.

4.5.1.5 The Crystal Structure

Several periplasmic binding proteins have been crystallized and their structures have been determined by X-ray crystallography (see Tab. 1, for the relevant listing of references). The salient feature emerging from this analysis is the following: all binding proteins thus far analyzed follow the same pattern [80, 143–148]. Generally, two separate but similarly folded globular domains or lobes (the N-lobe and the C-lobe, as they contain the respective ends of the polypeptide chain) are connected by a hinge region made of two or three polypeptide chains. Even though the two lobes are made up from non-contiguous polypeptide segments both exhibit similar secondary structure. The two lobes forming a cleft or groove between them are always composed of a central β-pleated sheet of six or seven usually parallel strands (not all strands are parallel in the maltose and oligopeptide binding proteins) with two or three α-helices on each side. The three different and closely associated polypeptide segments that connect the domains (the hinge region) provide the base and the boundary for the deep cleft between the two domains that form the binding site. The substrate bound in the cleft is held primarily by hydrogen bonding, although hydrophobic and steric components are also important.

As an example, Fig. 3 shows the crystal structure of the maltose binding protein from *E. coli* [149] indicating the areas of putative interaction with the respective TMDs [27].

Apparently, also the open form will bind substrate on one half of the open cleft. This has been shown by crystallizing the substrate-free open form of the leucine/isoleucine/valine binding protein and soaking the crystals with substrate. Whereas the open structure remained as such, the substrate was exclusively bound to the binding cavity of one domain [150]. In the case of the sulfate binding protein, the closure of the lobes is aided by the formation of two salt bridges that span the cleft opening [151]. The engineered formation of disulfide bridges between the lobes greatly affects the binding kinetics of the protein [152]. The closed but substrate-free form of the *Salmonella typhimurium* galactose binding protein has

Tab. 1. Crystallized prokaryotic binding proteins

Substrate	Binding Protein	Organism[a]	Reference
Allose	ALLBP	E. C.	237
Arabinose (fucose, galactose)	AraF	E. C.	101, 147, 238–240
Dipeptides	DppA	E. C.	241, 242
Fe(III) hydroxamate	FhuD	E. C.	243
Glucose, galactose	GBP	S. T.	80, 137, 147, 154, 244, 245
Glucose, galactose	GBP	E. C.	246, 247
Glutamine	GlnBP	E. C.	248, 249
Heme	HbpA	H. I.	250
Histidine	HisJ	S. T.	148, 251
Leucine	LivK	E. C.	252
Leucine, isoleucine, valine, (threonine)	LivJ	E. C.	134, 150, 252
Lysine, arginine, ornithine	LAO	S. T.	133, 253
Maltose, maltodextrins	MBP (MalE)	E. C.	146, 149, 245, 254–257
Maltodextrins	MBP	P. F.	120
Molybdate, (tungstate)	ModA	E. C.	258
Oligopeptides	OppA	S. T.	157, 259, 260
Phosphate	PstS	E. C.	261–265
Polyamine	PotD	E. C.	266
Putrescine	PotF	E. C.	267
Ribose	RbsB	E. C.	80, 138, 147, 268–271
Sulfate	SBP	S. T.	151, 152, 263, 272–274
Trehalose, maltose	TMBP	T. L.	121

[a] **Abbreviations:**
E. C., *Escherichia coli*; S. T., *Salmonella typhimurium*; T. L., *Thermococcus litoralis*; P. F., *Pyrococcus furiosus*; H. I., *Haemophilus influenzae*.

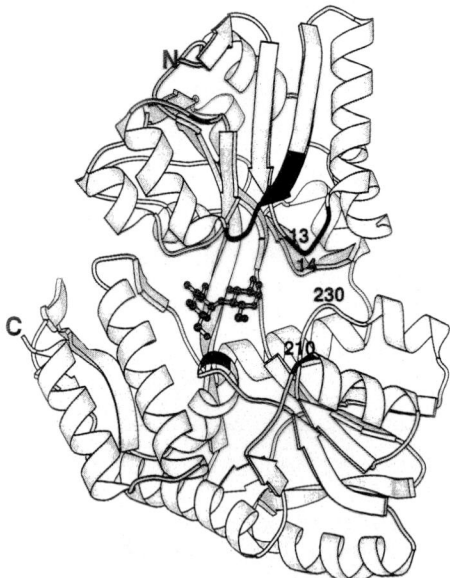

Fig. 3. Crystal structure of the maltose binding protein from *E. coli*. The protein is composed of two pseudo-symmetrical lobes that are connected by three polypeptides acting as a hinge. Shown is the conformation with enclosed maltose. The numbers indicate the positions of amino acids which, when mutated, exert a dominant negative phenotype and act as a suppressor for MalE-independent mutants in MalF or MalG indicating interaction sites with the TMDs [27]. Black areas indicate the location of mutations that affect transport. The figure was reproduced from [149] with permission from the authors and the publisher.

been determined [137] supporting the idea that even the substrate-free binding protein oscillates between the open and closed form.

One of the curiosities among binding proteins is the presence of a Ca^{2+} binding site in the galactose binding protein [137, 153, 154]. While the binding of Ca^{2+} does result in a local conformational change of the protein [155] it does not appear to be essential for the function of the protein in substrate binding [156]. The crystal structure of the oligopeptide binding protein of *S. typhimurium* is somewhat unusual [157]. The structure contains, beside the two domains found in all other binding proteins, a third domain whose function is unknown. The oligopeptide binding protein is larger than most periplasmic substrate binding proteins and binds peptides of two to five amino acid residues without regard to sequence. The dipeptide [81, 158], heme (from *Haemophilus influenzae* [159]), and nickel [160] binding proteins also share the property of the three-domain structure and, in addition, exhibit sequence homology. It is thus likely that these proteins are similar in structure and constitute a distinct novel class of binding proteins [98]. The corresponding proteins of *Streptococcus pneumoniae* [161] or *B. subtilis* [162], highly homologous to oligopeptide binding proteins of gram-negative enteric bacteria, have been implicated in the process of transformation competence suggesting the transport (or recognition) of peptides in this process. Recently, two structures for binding proteins from hyperthermophilic archaea have become available. One from *Thermococcus litoralis* specifically binds maltose and trehalose (TMBP) but not maltodextrins [121], the other from *Pyrococcus furiosus* is specific for maltodextrins [163, 120]. Both are nearly superimposable in their structure with the *E. coli* maltose binding protein. Increased thermostability of TMBP might be due to the 19 versus 3 surface-located salt bridges. Also, internal cavities are smaller in TMBP and helices tend to be more elongated in TMBP in comparison to the *E. coli* protein. Compared to the maltose binding protein from the mesophilic *E. coli*, the thermostable TMBP exhibits the same fast association of substrate at all temperatures (15–80°) but the dissociation of substrate is strongly reduced at room temperature [121]. Table 1 lists all citations reporting the crystal structure of prokaryotic and transport-related binding proteins found in MedLine up to November 2000.

4.5.2
The Integral Transmembrane Domains (TMDs)

4.5.2.1 Organization

The transmembrane domains represent the substrate channel of ABC transporters. In the majority of all binding protein-dependent import systems they are composed of homo- or heterodimers. Computer-aided analysis of the primary structure of these membrane proteins [23] reveals the occurrence of several uncharged, highly hydrophobic sequence stretches that are predicted to adopt an α-helical conformation spanning the lipid bilayer. These hydrophobic stretches are separated by more hydrophilic peptides of varying length forming loops at the aqueous phase of the cytoplasm or the periplasm. The construction of models for the orientation of these hydrophilic loops relative to the cytoplasmic membrane has been aided by application of the "positive-inside rule" that takes into account the typical distribution of charged amino acids found for integral membrane proteins [164, 165]. Most of the TMDs analyzed so far are predicted to comprise six transmembrane helices joined by hydrophilic sequence stretches that form three periplasmic and two cytoplasmic loops. The C- and N- termini of these proteins are located at the cytoplasmic face of the membrane [18]. Some proteins with this suggested topological organization include PstA and PstC [166, 167], PotH and PotI [168], PotB and PotC [169], and CysT [170]. The two-dimensional folding of the OppB and OppC proteins within the cytoplasmic membrane was analyzed by β-lactamase gene fusion and proteolysis experiments [171], and the topology of MalG was investigated with alkaline phosphatase fusions [22, 172]. The topology of these three proteins was shown to conform to the six membrane-spanner "consensus structure" [18], their C- and N-termini being located in the cytoplasm.

There are several exceptions to the six membrane-spanner organization. The MalF protein (514 aa) is relatively large in comparison to most other membrane proteins of this class, which usually consist of around 300 amino acids. MalF comprises eight membrane spanning segments including a large periplasmic loop between transmembrane helices 3 and 4, with both ends of the protein exposed to the cytoplasm [23, 173, 174]. The exceptional number of helices for the *E. coli* MalF can also be recognized in the corresponding protein of *Salmonella typhimurium, Enterobacter aerogenes* [175],as well as in CymF (cyclomaltodextrin) of *Klebsiella oxytoca* [176] and MalC of *Streptococcus pneumoniae* [177]. Yet, the large periplasmic loop between helices 3 and 4 is seen only in the MalF proteins of *E. coli, S. typhimurium,* and *E. aerogenes*. The trehalose/maltose system in the hyperthermophilic archaeon

Thermococcus litoralis, that is surprisingly homologous to the *E. coli* maltose system, lacks in its MalF homolog the large extracytoplasmic loop and has only six putative membrane spanning segments (MSS) [68].

Another scheme of transmembrane helices is established in ProW of *E. coli*. It comprises 7 membrane spanning segments, with the C-terminus located in the cytoplasm and a large, 99 aa N-terminal extension located entirely within the periplasmic space [178, 179]. This rather unusual structure may be correlated with the specialized function of the protein. ProW is involved in the uptake of osmoprotectants into the cell and its transport activity is regulated by the cell turgor [180]. The large N-terminal extension may be involved in measuring the cell turgor and transducing the mechanical stimulus into alterations of the transport system's activity.

The smallest number of transmembrane helices found so far for TMDs has been demonstrated for the HisQ (228 aa) and HisM (235 aa) proteins, that are rather short in comparison to most other proteins of this class. Here, only five such membrane spanners are present, with a periplasmic location of the N-termini and a cytoplasmic orientation of the C-termini. This structure has been confirmed experimentally with alkaline phosphatase fusions and proteolysis experiments [181].

This idea of a minimal core structure of 2×5 helices for any pair of TMDs is supported by the striking structural similarities observed when the topological models for TMDs with more than five transmembrane helices are compared with the "minimum" primary structure of the corresponding histidine transport proteins. The position and extent of five transmembrane segments and the connecting surface-exposed loops closely match that found for HisQ and HisM. In this structural alignment, the additional sequence stretches and transmembrane segments found in the longer polypeptides from other systems form an extension at the N-terminus, whereas the structures of the C-termini of all proteins can be aligned well with each other [181].

4.5.2.2 Composition and Structure

For the majority of binding protein-dependent ABC transporters, two separate TMDs are found and it is assumed that these proteins form a heterodimer within the transport complex. This has been directly shown for the histidine transporter from *S. typhimurium*, where the membrane-bound components were isolated and purified as a complex of one molecule each of HisM and HisQ – the integral membrane proteins – and two molecules of HisP, the membrane-associated ATP hydrolyzing module (see below) [182]. Likewise, the maltose transport complex from *E. coli* consists of one molecule each of the integral membrane proteins MalF and MalG, and two molecules of the MalK ATP binding subunit [12]. In about one-third of the systems analyzed just a single transmembrane protein is present, presumably as a homodimer. An interesting case is the *E. coli* FhuB protein. It is one of the largest of the integral membrane proteins (659 aa) Its N- and C-termini show strong sequence similarity to each other, suggesting that the *fhuB* gene arose by duplication of a primordial gene followed by fusion of the duplication products [183]. Both halves of the FhuB protein can be expressed as separate polypeptides without abolishing its function [183]. Thus, two integral membrane domains appear to be present, either as a homodimer of one protein or as heterodimers of two functionally related proteins [18]. There is no crystal structure available yet for any of the bacterial TMDs. Ehrmann et al. [24] have attempted to construct a model based on the *E. coli* maltose system that predicts the next neighbors of the transmembrane helices forming the translocation pathway. Some of the criteria were the established topology of MalF and MalG from the *E. coli* system as well as the computer-predicted topology of a number of homologous systems from different bacteria. The extent of periplasmic and cytoplasmic loops was then used as next neighbor criterion. Another criterion was the proposition of the proline content for helices being channel forming; still another consisted of the close positioning of mutations and intergenic suppressors. Thus, for instance, mutations in MSS 7 were suppressed by mutations in MSS 6 or 8, an indication that these helices are next neighbors. In addition, mutations in MSS 6 leading to altered substrate specificity occur only on one side of the helical wheel pointing to a participation of this surface in the substrate-specific transport channel [184]. The dispensability of the first transmembrane spanning segment of MalF [185] was a criterion for the minimal functional unit, as well as the head-to-tail arrangement of MalF and MalG as suggested by the observation that appropriately fused proteins were still functioning [186]. The outcome of these considerations was a minimal model in which the dimers were arranged in a symmetrical fashion such that only the three consecutive transmembrane helices of each monomer are participating in the transport channel. These helices are the fourth, the third, and the second to last helix in the linear arrangement of each polypeptide chain. This model proposes hydrophilic substrate recognition patches within the transport channel that are separated from each other by hydrophobic barriers. The movement of these would be connected to ATP hydrolysis by the ABC unit [24].

4.5.2.3 The Interaction of the TMDs with the Binding Protein

Even though representing the major substrate recognition site, the maltose binding protein (MBP) cannot be the only substrate binding site of the maltose transport system. MBP-independent mutants in MalF or MalG have been isolated that retain specificity for maltose indicating the presence of a latent substrate binding site in the MalF/MalG/MalK$_2$ membrane complex [25]. The V_{max} of these mutants for maltose transport can sometimes be similar to the wild type while the apparent K_m of uptake is always increased by a factor of about 10^3. Transport in these mutants is still against the concentration gradient and dependent on the MalK-mediated hydrolysis of ATP. Characteristically, these MBP-independent mutants in MalF always carry two mutations, one in MSS 5 near the periplasmic surface, in the p (proximal) region, and the other one deep within the cytoplasmic membrane, either in MSS 6, 7, or 8, in the d (distal) region [174]. It is interesting to note that substrate specificity mutations have been isolated in MSS 6 and that the next neighbor analysis predicts that MSS 6, 7, and 8 are close to each other [184] and at the same time form the d region.

The screening of mutations in *malF* and *malG* for the ability to transport other sugars revealed that one mutation, *malF*515, which changes Leu334 to Trp (L334W) on the periplasmic side of MSS 5, permitted the transport of lactose in addition to maltose [187]. MalK as well as MalG is required for lactose transport, as is MBP. The requirement for MBP is particularly intriguing because MBP clearly does not bind lactose. The position for such a "specificity" mutation on the periplasmic side of MalF is surprising since this domain is not expected to form the substrate recognition site of the transport channel. It may indicate that substrate specificity for lactose is already present in MalF or MalG but needs the proper opening of the gate for the substrate to enter. It is noteworthy that in this mutant lactose transport does not inhibit maltose transport as if the two sugars do not compete for a common binding site. In a different construct large portions of the MalF protein have been deleted (starting from the C-terminus and extending up to MSS 1). These deletions are able to transport lactose in a MalG-, MalK- and MBP-dependent manner without an additional mutation [188]. This may indicate that the substrate recognition site of the membrane complex for lactose resides in MalG and that a MalG dimer will mediate lactose transport. In any case binding of the unliganded MBP is apparently needed for lactose transport and can trigger ATP hydrolysis of the ABC subunit provided the TMDs are properly altered by mutation [189]. This is in line with the conclusion that not only the liganded but also the substrate-free form of the binding protein does interact with the membrane components when transporting its natural substrate [187], an observation that has also been made with the *S. typhimurium* histidine system [190].

The interaction of MBP with MalF and MalG has been studied using the genetic approach of mutant and suppressor analysis [26]. This study, in combination with the knowledge of the crystal structure of MBP, has led to the conclusion that one lobe of MBP interacts with MalF, the other with MalG [27]. The starting point for this genetic analysis was the isolation of mutants that transported maltose in the absence of MBP. Surprisingly, some of these MBP-independent mutations became maltose minus in the presence of wild-type MBP, a situation that allowed isolation of suppressor mutations in *malE* [26]. Most of these suppressor mutations in MBP were negative dominant in combination with a wild-type MalF/MalG/MalK$_2$ complex indicating a higher affinity towards the membrane components when in the non-ATP hydrolysis triggering mode [27, 149]. Similarly, after the introduction of two cysteines (G69C and S337C) by site-directed mutagenesis into each domain of MBP, the formation of an interdomain disulfide cross-link could be observed that holds the protein in a closed conformation. This mutant MBP confers a dominant negative phenotype for maltose transport. These mutant-suppressor analyses together with the properties of negative dominant mutations in MBP have led to the proposal that the MalF/MalG/MalK$_2$ complex must be able to attain at least two different conformations in its interaction with MBP, only one of which is able to trigger ATP hydrolysis by the MalK subunit [51, 149]. When the signal transduction pathway, starting with the recognition of the binding protein by the TMDs and ending in the ATPase activation of the ABC subunit, is interrupted (as in the MBP-independent mutants) ATPase activity becomes constitutive and uncoupled [15, 174].

The conclusion that one lobe of the binding protein interacts with TMD1 and the other lobe with TMD2 [27] tacitly assumes that the complete and operating transport system only contains one binding protein per complex, an assumption that is reflected in the numerous (non-atomic) models of ABC transporters. There is no compelling evidence for such a model. In contrast, the recent demonstration of binding proteins fused to the homodimeric TMDs [70, 71] indicates that a functional stoichiometry of more than one may be operating.

Recently, the structure of the ligand binding domain of mGluR1, a glutamate receptor of the mammalian central nervous system, has been crystallized and structurally analyzed [191]. The binding domain consists of two homodimers that are linked by a disulfide bridge. The monomers resemble periplasmic substrate binding proteins, in particular the LIV binding protein of *E. coli*. As in classical binding proteins, they consist of two flexible lobes in which the substrate is buried (see below). Structure determination was achieved in three different dimeric conformations: the substrate-bound and closed form and the two substrate-free forms. The latter appeared in two varieties: one was closed and resembled the substrate-bound form, the other represented the open form. The dimer interface is established between lobes 1 of the monomers. Changes in the dimer conformation upon binding substrate do not only occur in respect to the two lobes of the individual monomer but also within the interface of the dimer. However, the movements of the two lobes that are not part of the dimer interface are dramatic. They may well be the basis for the signal transmitted through the transmembranal portion of the receptor. One is tempted to think that such a dimeric arrangement may well also be the basis for the signal transduction in prokaryotic binding proteins when they interact with their TMDs.

The highly water-soluble character of solute binding proteins as well as the high K_d of the TMDs for their binding (90 μM estimated for the maltose binding protein [67]) have led to the idea that binding proteins are only loosely associated with their respective TMDs. Yet, recent experimental evidence using the ferrichrome iron ABC transporter Fhu points to an interaction of the binding protein FhuD deep within the structure of one of the cognate TMDs [192]. It was concluded that FhuD interacts with MSS 7 of FhuB including the cytoplasmic region that harbors the EAA signature sequence. Although FhuB consists of two homologous halves, FhuB(N) and FhuB(C), the sites identified for FhuD contact were asymmetrically arranged. This conclusion is based on the binding of various decapeptides representing the TMD sequence. If indeed binding of these peptides truthfully mimics the interaction with the corresponding site in the TMD instead of just involving the substrate binding site of the binding protein, this would alter our concept of the interaction between binding protein and TMDs. It would bring portions of the binding protein close to a site that is known to be in the vicinity of the signature sequence of the cognate ABC subunit and, thus, to where the action is. Surely, more corroborating evidence (e. g., cysteine cross-linking) is needed to accept such a dramatic conclusion.

4.5.2.4 The Sequence

The conservation of structure rather than actual amino acid sequence becomes apparent when the primary structure of the TMDs is compared. They show only limited sequence similarity, despite a high degree of conservation of their secondary structure [181]. Some heterodimeric TMDs are structurally more similar to each other than to components of other systems: CysT/CysW [170]) and PstA/PstC [166, 167] from the sulfate/thiosulfate and phosphate transporter, respectively, are examples for such pairs; it is likely that they arose by gene duplication [193]. Other TMDs are more closely related to corresponding proteins from other systems: PotB and PotC from the spermidine/putrescine transporter are more similar to PotH and PotI from the putrescine-specific transport system than to each other (37% and 36% amino acid identity, respectively [168]). A similar relation is shown by MalF and MalG from the maltose uptake system, that are more similar to UgpA and UgpE from the glycerol-3-phosphate transporter, respectively, than to each other [194]. In these cases, it is likely that the systems arose by a duplication and subsequent divergence of a primordial system that already possessed two transmembrane proteins. For detailed evolutionary aspects the reader is referred to the continually updated internet site of Milton Saier (*http://www-biology.ucsd. edu/~msaier/transport/titlepage.htm*).

A highly conserved sequence feature that is found in all TMDs is located near the C-terminus of these proteins, within the last cytoplasmic loop. This sequence motif was first noted in 1985 by Dassa and Hofnung [195] who defined the 20 amino acid consensus sequence $EAAX_3GX_9IXLP$ from an alignment of 7 proteins, and was later termed the EAA loop. When the primary structure of more proteins from this class became available, a similar segment was found in all cases, however, with certain variations. The amino acids of the conserved region are predominantly hydrophilic, consistent with its location in a cytoplasmic loop [181]. A glycine residue is found at an equivalent position in all EAA loop sequences, located between 94 and 115 aa from the C-terminus of the protein and invariably in the cytoplasmic loop between the second and third to last transmembrane helix. The amino acids at other positions within this stretch are less well conserved. A careful analysis of the sequence patterns within the conserved region showed that the proteins can be divided into several groups according to their sequence similarity [196, 197]. For example, the conserved segments of inner membrane proteins of uptake systems for chelated iron and vitamin B_{12} (Fhu, Fep, Fec, Btu) are quite similar to each other, but differ from the consensus sequence at several positions.

The high degree of conservation of the EAA loop [198] suggests an important common function of this sequence element for all TMDs. This idea is supported by the deleterious effect of mutations in this region. An insertion of four amino acids into the EAA loop region of MalG abolished maltose uptake completely [199]. Likewise, site-directed mutagenesis of the conserved glycine residue in both EAA loop regions from the internally duplicated FhuB protein strongly affected the function of FhuB [196]. A detailed analysis of the EAA loops of *E. coli* MalF and G was done by E. Dassa and coworkers [28]. Substitutions at the same positions in MalF and MalG have different phenotypes, indicating that the EAA loops in each subunit do not act symmetrically. Mutations in the EAA loop of MalF or MalG that only slightly affect transport can cause a completely defective phenotype when present together. This suggests that the EAA loops of MalF and MalG act in concert during transport. Suppressor mutations in MalK to mutations in the EAA loop of MalF and MalG were found predominantly in the so-called helical domain of MalK between the Walker A and B site indicating a close apposition (see Fig. 6 for the position in the MalK structure). The absence of the EAA loop in ABC exporters, in particular of eukaryotic systems, clearly indicates that EAA loops are involved in the functional connection between TMDs and ABC subunits.

4.5.3
The ABC Subunit

4.5.3.1 **The Sequence**

Sequence comparisons of the members of this protein class from all transport systems show a striking homology that extends over a region of about 215 amino acids and includes the two short Walker nucleotide binding fold motifs [19]. Hence, the ATP binding subunits of binding protein-dependent transporters have also been termed "conserved components" [193, 200]. Sequence similarities between these components are much higher then among the TMDs and the binding proteins. Moreover, strong similarity of this region within the ABC subunit is observed to a multitude of other pro- and eukaryotic proteins, most of which are involved in transport processes of substrate molecules across the cell membrane [201]. The close relation between the pro- and eukaryotic systems in the ABC transporter superfamily implies a very similar mechanism of transport and suggests that valuable mutual information concerning their function can be learned from experiments with both systems [32]. As an example, Fig. 6 shows the conserved sequences and motifs found in two prokaryotic ABC domains (MalK of *E. coli* and *Thermococcus litoralis*). The Walker A site is typically composed of a glycine-rich P-loop sandwiched between a helix and a β-strand (α1 and β3 in the structure shown in Fig. 7). The Walker B site consists of a β-strand harboring the highly conserved aspartate residue. The sequence between Walker A and B contains the helical region of relatively low sequence conservation. It contains a conserved glutamine residue within a stretch of amino acids that we termed the Lid region positioned within a loop between a helix and a β-strand (α2 and β7 in the structure shown in Fig. 7) [31]. Next to Walker B is the highly conserved ABC signature motif LSGGQQ(R,K)QR, previously also called linker peptide. Following the Walker B site one finds a conserved aspartate within the D-loop and the switch region harboring a reasonably conserved histidine residue. According to Linton and Higgins [42] this conserved histidine is always present at least in one copy of the two ABC subunits.

4.5.3.2 The Localization

In cell fractionation studies, ABC subunits are usually found in the cytoplasmic membrane fraction. For example, this has been shown for MalK [202, 203], HisP [204], OppF [205], and ProV [206], and it was concluded that these proteins are peripherally associated with the inner membrane. However, the mode of attachment to the membrane has been debated. HisP and OppF are found attached to the membrane even in strains that lack the integral inner membrane proteins [182, 205] and appear to have an affinity for the lipid bilayer. MalK, in contrast, can be found in the cytoplasm when its membrane partner proteins MalF and MalG are absent [203]. A stoichiometry of two for the ABC subunit within the transport complex has been experimentally determined in the case of the histidine system of *S. typhimurium* [14] and the maltose system of *E. coli* [12]. Also, the soluble protein in both systems has been determined as a dimer [207, 208].

A model claiming an intimate association of the ABC subunit with the membrane-bound translocation complex has recently emerged from a detailed analysis of the *S. typhimurium* HisP protein. Solubilization studies with strong detergents and chaotropic agents showed that HisP is attached more firmly to the membrane than a peripheral membrane protein, yet not as tightly as a typical integral membrane protein [182. 204]. HisP forms a complex with the integral membrane compounds, HisQ and HisM [182]. In this complex, HisP was shown to be accessible to proteases and a biotinylating reagent both from the cytoplasmic and the periplasmic aspect of the cytoplasmic membrane, suggesting that it traverses the lipid bilayer [209]. Since HisP does not comprise a potential hydrophobic transmembrane segment, it was suggested that HisP extends through a pore formed by the HisM and HisQ proteins [209]. HisP was no longer accessible to the periplasm in the absence of HisM and HisQ [209]. Such a transmembrane topology for HisP in the intact complex was suggested early on by the observation that a mutation in the histidine binding protein HisJ could be suppressed by a mutation in the ABC subunit HisP [210, 211], the classic genetic approach to identify a direct protein–protein interaction for a two-component system. A similar conclusion about the periplasmic accessibility of an ABC subunit by protease digestion experiments has been made for MalK of *S. typhimurium* [212]. Yet, the provocative conclusions about the transmembrane positioning of HisP and MalK have recently come under strong criticism [213]. The use of exceedingly high protease concentrations destroying the integrity of the cognate membrane components may have compromised the outcome of the experiments. The recently available crystal structures of HisP [30] and MalK [31] do not offer an obvious explanation for transmembrane positioning.

4.5.3.3 ATP Hydrolysis

The energy required for transport is provided by the hydrolysis of ATP [9, 12]. The first hint to the direct involvement of ATP hydrolysis in transport was the detection [16, 214] that these proteins comprise a consensus mononucleotide binding motif shared by a variety of other known nucleotide binding proteins (the Walker A and

B sequences) [19]. That this predicted nucleotide binding fold indeed is functional was shown by binding studies with ATP or structural analogs, e.g., for HisP [215, 216], OppD [214], FecE [217] or MalK [218]. In addition, it was shown that purified MglA [219] and MalK [220, 218] display ATPase activity *in vitro*. GTP is also bound by the energy transducing components [215, 220, 221] and can in fact energize transport [221] albeit with reduced efficiency. ATPase activity in the isolated MalK protein is high and not affected by the maltose binding protein or its substrate maltose. On the other hand, ATPase activity of the MalF/GK$_2$ complex [11, 12], either in membrane vesicles or reconstituted in liposomes, is strongly reduced, but it increases in the presence of substrate loaded maltose binding protein [222]. This apparently substrate transport coupled ATPase activity can still be observed in detergent solubilized MalF/GK$_2$ complex even though the basal activity (in the absence of maltose binding protein) is high. Presumably, detergent solubilization of the complex interferes to some extent with controlling the ATPase activity by the TMDs [15]. In line with coupling the ATPase activity of MalK by the recognition of substrate-loaded binding protein is the observation that mutations in MalF or MalG that show maltose binding protein-independent transport activity also exhibit uncoupled ATPase activity in liposome reconstituted or detergent solubilized MalF/GK$_2$ complex [222, 15]. For transport of substrate in the intact MalF/GK$_2$ complex to occur it appears that the ATPase activity of both MalK subunits has to be functioning [223]. This, and the observation that ATP hydrolysis in the intact complex exhibits cooperative behavior [224] shows the close interaction of the two nucleotide binding sites which is reflected in the structure discussed below.

ATPase activity in purified HisP of *S. typhimurium* has been analyzed. Results obtained in this system are somewhat different from those seen in the maltose system. With HisP, ATPase activity shows no cooperativity when using the soluble dimer but does so when using the membrane-bound complex [207], as in the maltose system. On the other hand, the cognate binding protein does not stimulate ATPase activity of the membrane-bound complex, a feature that is prominent in the *E. coli* and *S. typhimurium* maltose systems. In addition, the construction of heterodimers in which one monomer is defective in ATP hydrolysis still showed ATP hydrolysis and substrate translocation in the reconstituted complex, albeit at half the rate of the wild type [225]. In contrast, in the maltose system both ATP binding sites need to be functioning [223]. An alternating two-site binding and hydrolysis mechanism has been proposed for the homodimeric ABC exporter LmrA of *Lactococcus lactis* [226] that could very well be operating in the binding protein-dependent ABC importers such as the *E. coli* maltose system.

Binding protein-dependent ABC transport systems are regarded as unidirectional primary pumps that directly couple transport to ATP hydrolysis. Mutants that cannot metabolize the accumulated substrate will reach large and potentially toxic intracellular levels of it when they are exposed to the substrate in the medium. Indeed, *malQ* mutants lacking amylomaltase, the key enzyme in maltose metabolism, are highly sensitive to maltose or maltotriose. It would, therefore, be reasonable if ABC transporters were feedback inhibited at high internal substrate concentrations. This has been observed with the histidine transport system in recon-

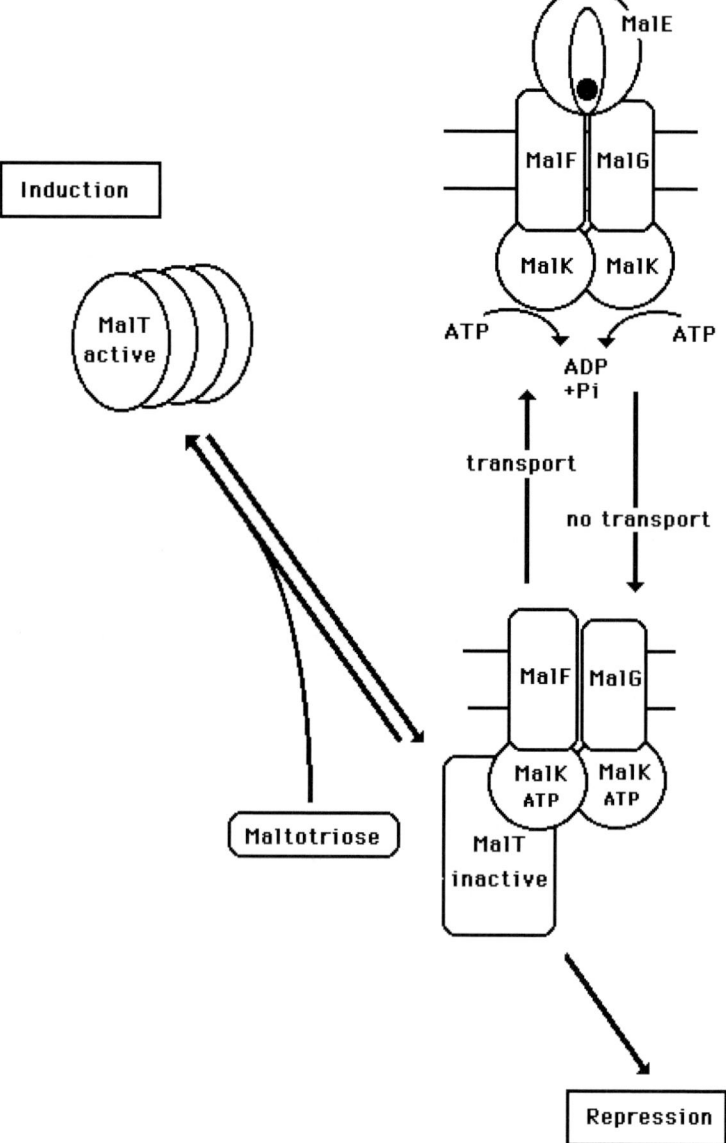

Fig. 4. A proposed model of transport and repression. When maltose (solid circle) is present in the medium, it is recognized by MBP (MalE) and transported via MalFGK$_2$. For transport, ATP is hydrolyzed by MalK. In this configuration, MalK has no affinity for MalT, and MalT can oligomerize in the presence of the inducer (maltotriose) acting as a *mal* gene transcriptional activator. In the absence of maltose transport the monomeric form of MalT interacts with MalK, keeping MalT in its inactive state. The key point in controlling the function of MalK is its association with ATP. When ATP is bound to MalK (but not hydrolyzed due to the absence of triggering substrate-loaded MBP) the protein interacts with and inhibits MalT. Triggering ATP hydrolysis by transport releases MalT from MalK.

stituted liposomes [13]. Transport and ATPase activity were inhibited by histidine. Also, in the case of the binding protein-dependent glycerol-3-phosphate (G3P) ABC transporter not the substrate itself, but P_i, a product of G3P metabolism, inhibits G3P uptake *in trans* [227].

ABC subunits have been recognized as targets for regulation. A well-studied example is MalK of *E. coli*. Mutants lacking MalK no longer transport maltose but, in addition, are constitutive for *mal* gene expression [228], whereas the overproduction of MalK results in strong repression [229, 230]. The basis of this phenomenon is a direct interaction of MalK with MalT [231], the transcriptional activator of all *mal* genes [232], inactivating its function (Richet et al., personal communication). Very clearly, transport activity (ATPase activity) determines the degree of repression: the MBP-independent mutation in *malF* leading to an unrestrained ATPase activity of the MalF/GK$_2$ complex is partially constitutive in spite of a wild-type MalK protein; the G137A exchange within the signature motif of MalK leading to loss of ATP hydrolysis but not ATP binding is a "super repressor" [231]. Thus, in the absence of transport and, therefore, in the absence of ATP hydrolysis (but not necessarily ATP binding), MalK has an affinity to MalT sequestrating it from its transcriptional activity and leading to *mal* gene repression. The model explaining our conclusions about the interaction between MalK and MalT is shown in Fig. 4.

All mutations in *malK* exhibiting a regulation-minus, transport-plus phenotype are located in the C-terminal portion of MalK [230, 233] (see Fig. 6). Looking at the structure of MalK from *Thermococcus litoralis* it is obvious that the C-terminal domain forms a β-barrel that is quite independent from the ATPase domain (Fig. 5). Nothing is known yet about the specific regulation of the *malEFG* gene cluster of *T. litoralis* except that expression is induced by trehalose and maltose. But since there are conserved regions within the C-termini of *E. coli* and *T. litoralis* MalK, and since also the length of the C-termini is nearly identical it appears likely that the C-terminal structure seen in Fig. 5 takes part in regulation, either actively (by interacting with a gene regulator) or passively (by being regulated via central metabolism). The C-terminal domain so apparent in the structure of *T. litoralis* MalK is evolutionarily conserved and must serve an important function other than transport (cluster 26 in ref. [41])

4.5.3.4 The Crystal Structure of MalK from *Thermococcus litoralis*

The hyperthermophilic archaeon *Thermococcus litoralis* contains a high-affinity binding protein-dependent ABC transporter exhibiting an unusual high affinity for its substrates trehalose and maltose [68, 234, 235]. Recently, it was possible to solve the structure of MalK, its ABC subunit, at a resolution of 1.9 Å [31]. The structure consists of two clearly separated domains. The amino terminal portion consisting of 223 residues contains the ATPase. It resembles an ellipsoidal planconvex lens of 55×40 Å, whereas the C-terminal residues 224-372 form a barrel with a diameter of 20 Å and a height of 45 Å that is made from β-strands. This C-terminal domain, by homology to its cousin from *E. coli*, most likely carries regulatory functions. Complete MalK is a dimer that connects via the flat surface of the ellipsoidal lens of two ATPases, forming an extended dumbbell-shaped molecule with a long axis of about 120 Å where the regulatory domains are attached at the opposite poles of the dimer (Fig. 5). The dimer interface covers a large surface area of 2,755 Å² that brings the conserved ABC motifs of the monomers in mutual contact. The contact interface extends along a twofold axis forming an angle of about 35° with the 120 Å long axis of the dimer. Fig. 5A shows the side view of the dimer where the interaction site with the TMDs would be expected to occur on the bottom. Fig. 5B represents the perpendicular view from the bottom.

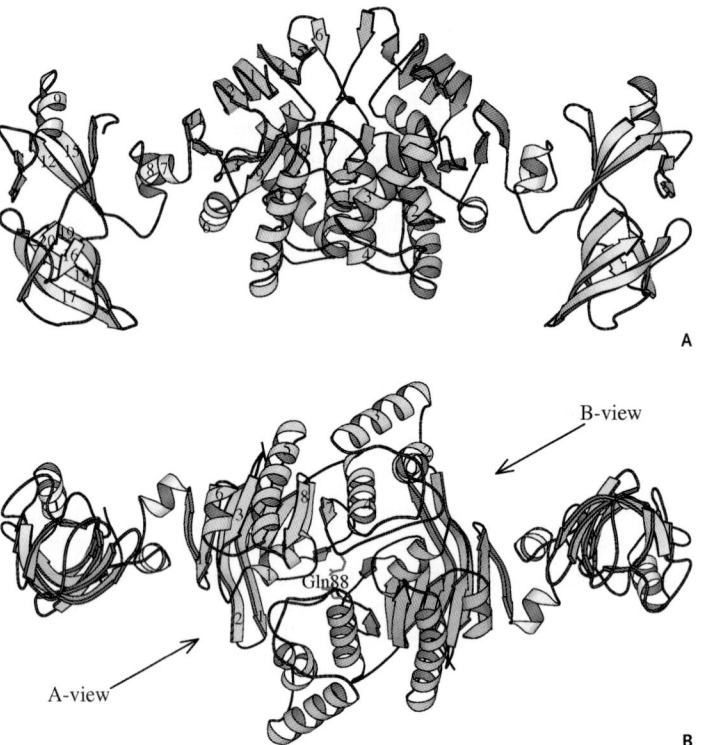

Fig. 5. Ribbon representation of the *Thermococcus litoralis* MalK dimer. The A- and B-molecules are colored yellow and blue, respectively, except for both regulatory domains which are gray. Labels indicate numbers of strands and helices according to the secondary structure assignment given in Fig. 6. **(A)** The side view shows the extended dumbbell shape resulting from the two regulatory domains on either end and the central ATPase domain dimer. The pseudo-twofold symmetry axis is oriented vertically and runs through the center of the dimer. The strong involvement of helices 2 and 4 in dimerization is seen. The bottom part of the dimer is supposed to interact with the TMDs MalFG. **(B)** The bottom view along the pseudo-twofold axis shows the deviation from twofold symmetry. The helical layer of one monomer is seen in contact with the two upper layers containing the nucleotide binding site of the other monomer. The symmetry axis between strands 6 of both monomers seems to provide a mechanical hinge for the dimer. Residues Gln88 from both monomers are shown to demonstrate their close apposition. The A- and B-viewing directions are indicated. Taken from [31] with permission from the author and the publisher; see color plates p. XXVII.

This view shows the close neighboring position of the two Gln88 residues of the "Lid" that is contained in the helical region between the Walker A and B site and is highly conserved among ABC transporters. Fig. 7 presents the "A" view on the dimeric ATPase as defined in Fig. 5. Again, the bottom part is expected to interact with the membrane components. Fig. 7 also depicts the position of all conserved sequences and motifs mostly within the interface between the monomers. From the cytoplasmic (top part) to the membrane attachment site one can differentiate three distinct layers: an antiparallel β-sheet (strands 2, 1, 4, 5, and 6) forms the top layer; the middle layer consists of a mixed β-sheet (strands 10, 3, 9, 8, and 7). Between the two sheets of upper and middle layer that approach each other at roughly right angles, the middle layer harbors the Walker A motif consisting of helix 1 followed by the P-loop and strand 3. Next to the P-loop, the pyrophosphate of ADP is seen at the dimer"s interface (only one pyrophosphate is shown). The last layer, supposedly next to the membrane components, is composed of helices 2–6 of which helix 2 and 4 form most of the interface between the dimer components. Close to the membrane interaction surface one can recognize the Lid region containing the conserved Gln 88, the D-loop, and the signature motif. Since the sequence of *T. litoralis* MalK in this region is highly conserved to the *E. coli* MalK, it is justified to project the mutations in *E. coli* or *S. typhimurium* MalK that suppress mutations within the EAA loop of MalF and MalG [28] to the now available structure of *T. litoralis* MalK. Fig. 6 shows an alignment of the two sequences including the positions of the suppressor mutations. They occupy one position on helix 3, two positions near Walker B (on β-strand 8) and one within the switch region on β-strand 9. Incidentally, all positions leading to suppression of EAA mutations are fully conserved in MalK from *E. coli* and *T. litoralis*. The conclusions from suppressor analysis about the interaction with the membrane components are corroborated by cross-link studies between MalK and MalF or MalG of *S. typhimurium* [29] which identify amino acids within the Lid region and helix 3 of MalK (Fig. 7) again indicating a close contact of these conserved structures with the membrane components.

It has been suggested that part of the MalK polypeptide becomes accessible to the periplasm, possibly in an alternating cycle of coupled substrate translocation and ATP hydrolysis. The so-called helical loop region between Walker A and B was proposed as a likely candidate for surface exposure [212]. The structure shown in Fig. 5 or 7 does not offer an obvious explanation for transmembranal surface exposure of the elements located between Walker A and B.

The crystal structure also explains well the phenotype of a number of mutants harboring insertions of 31 amino acids at various positions in the *E. coli* MalK protein [233]. Thus, insertions at positions 275, 291, 346, 364 in the regulatory domain (see Figs. 5, 6, and 7) abolish regulation but still allow transport, whereas of all insertions within the ATPase domain only one (at position 211) does not interfere with transport. Position 211 is within β-strand 11, in the outermost layer of the structure, far away from the center of action.

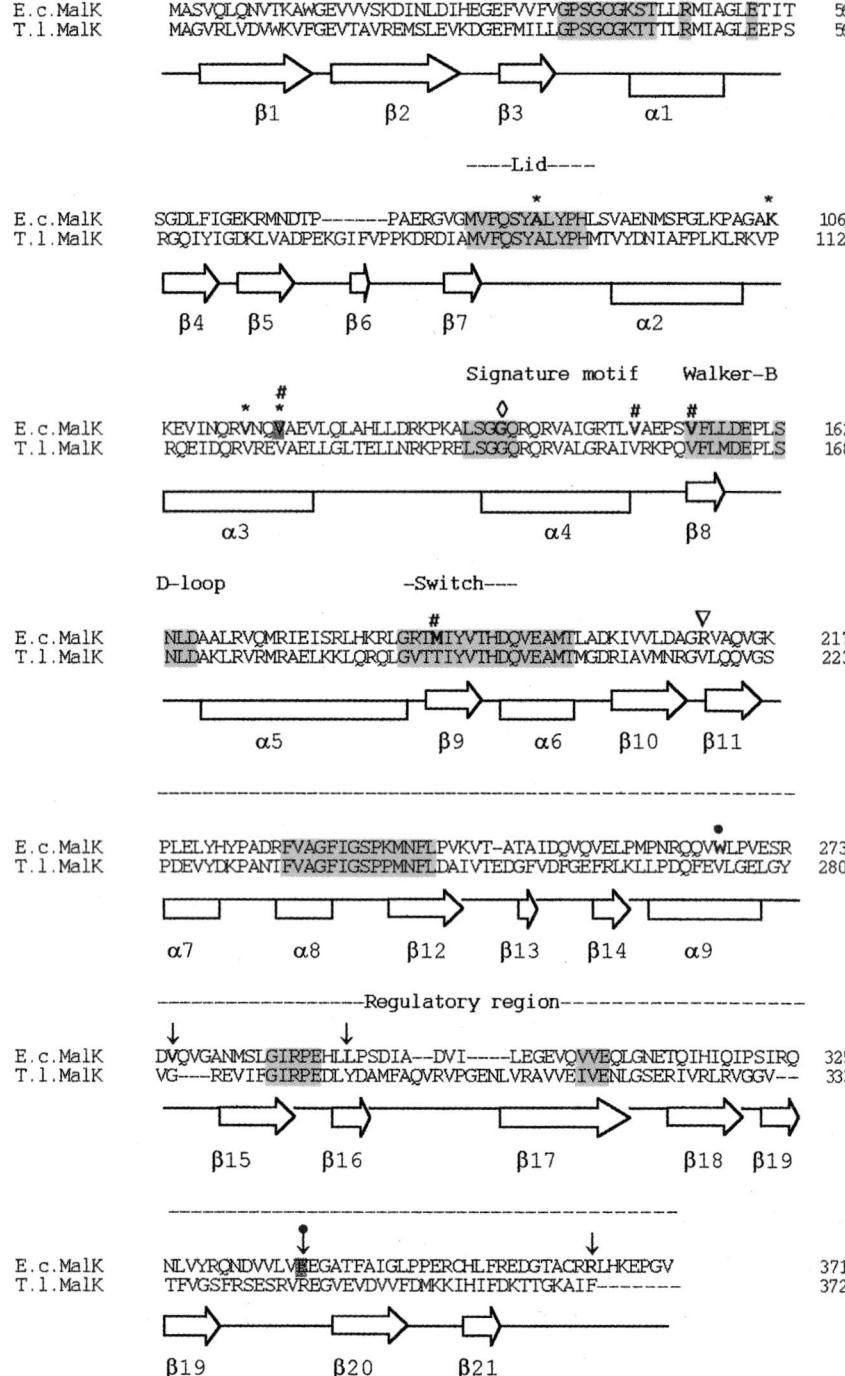

Fig. 6. Alignment of *T. litoralis* MalK and *E. coli* MalK, conserved sequences. Secondary structure elements of *T. litoralis* MalK are indicated and numbered: helices are shown as cylinders, strands as arrows. Conserved regions are primed yellow. All marked mutations, found in *S. typhimurium* or in *E. coli*, respectively, are at conserved sites. The following mutations are shown: ★ (blue), cross-link with EAA loop in TMD [29]; # (green), suppressor to mutations in EAA [28]; ◊ (red), mutation causing super-repression [231]; ↓ (lilac), insertion causing loss of regulation [233]; ∇ (magenta), insertion causing no alteration in transport [233]; ● (turquoise), point mutation causing loss of regulation [230]. Positions with double mutations are primed gray. With modification taken from [31] with permission from the author and the publisher; see color plates p. XXVIII.

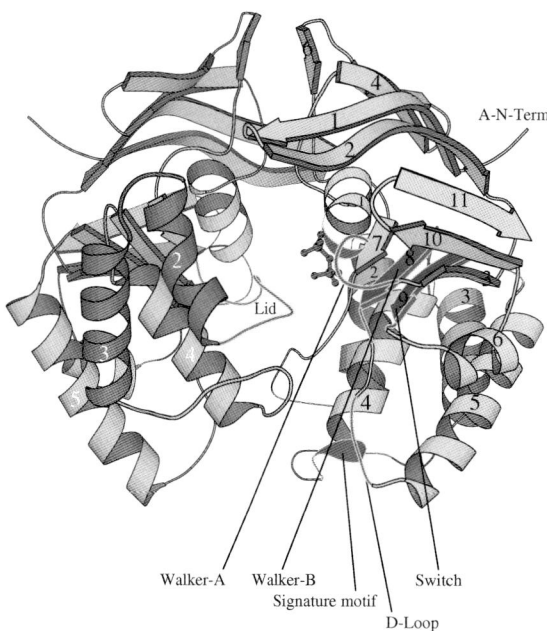

Fig. 7. A view along the interface perpendicular to the pseudo-symmetry axis of *T. litoralis* MalK. From top to bottom, the three layers are seen: antiparallel sheet, mixed sheet with P-loop and helix 1, and helical layer. Coloring is as in Fig. 5 except that the conserved regions (Walker A, Walker B, signature motif, D-loop, switch from monomer A and the Lid region from monomer B) are marked in red with yellow outlines. Labels indicate numbers of strands and helices according to the numbering given in Fig. 6. Taken from [31] with permission from the author and the publisher; see color plates p. XXVIII.

4.5.3.5 The Asymmetry within the MalK Dimer

Figure 5B shows the twofold axis between the two monomers of MalK. Close inspection reveals that the C-terminal regulatory domain of one monomer (A) can only be superimposed with the corresponding domain of the other monomer (B) by a rotation of 170°. In addition, the N-terminal structure comprising the nucleotide binding domain is not fully identical in the two monomers. This is best seen by following the A- and B-view as depicted in Fig. 5B. The resulting view is shown in Fig. 8. In the A-view, the interface appears narrower and His 95 of the B-monomer within the Lid approaches the pyrophosphate of the bound ADP in the A-monomer. In the B-view, the Lid of the A-monomer is shifted about 6 Å towards the bottom, opening the interface. This shift brings A-Tyr93 of the Lid into hydrogen bonding distance of another oxygen of the B-pyrophosphate. Helices 2 and 3 (Fig. 5B) are located in the peripheral part of the dimer interface and are among

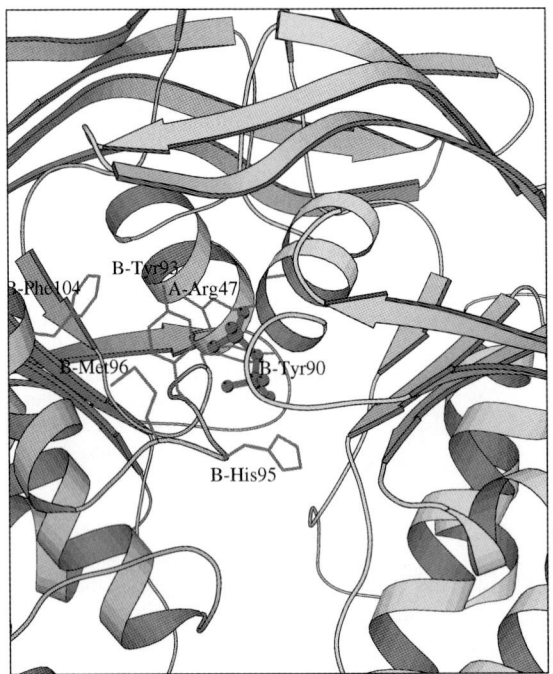

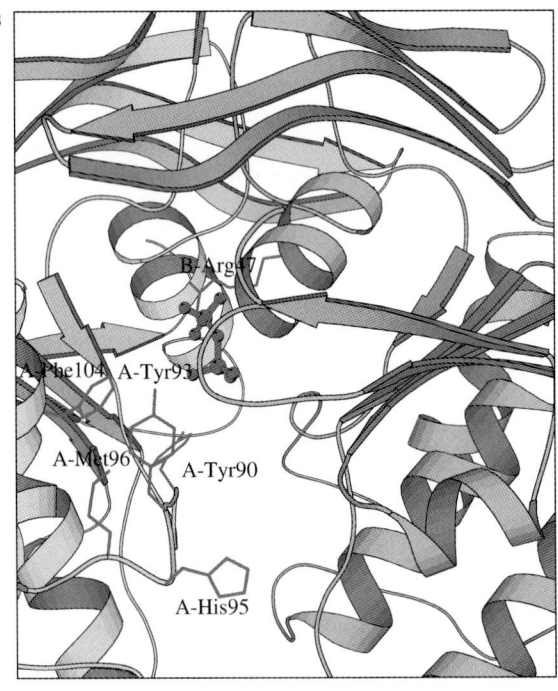

Fig. 8. The asymmetry within the MalK dimer. Comparison of the A- and B-views. In the A-view (**A**) the interface appears to be narrower than in the corresponding B-view (**B**) due to the upward shift of the loop containing the Lid region. This has been emphasized by including the side chains of residues 90, 93, 96, and 104. Notably B-His95 approaches the pyrophosphate to hydrogen bonding distance in the A-view, while A-Tyr93 plays this role in the B-view. Arg47 shows a difference in side chain conformation in the two views. The two helices 2 and 3 are shifted outwards in the B-monomer by approximately 3 as compared to the A-monomer. Taken from [31] with permission from the author and the publisher; see color plates p. XXX.

the most asymmetric elements in the dimeric structure of MalK whereas the antiparallel β-sheet on top of the ATPase domain (far away from the TMD interaction site) appears to be the most rigid part. At present, the functional relevance of the broken twofold symmetry in the MalK dimer is unclear. However, it is tempting to speculate that this observed asymmetry reflects the necessarily asymmetric interaction with MalF and MalG, the non-identical TMDs of the *in vivo* complex. Since suppressor mutations in *S. typhimurium* MalK to altered EAA loops of MalF and MalG can be found in the Lid region as well as in helix 3 of MalK, and since these regions are the most prominent asymmetric elements of the structure, one could imagine that a mechanical movement might be transmitted via these structures onto the TMDs.

It has long been proposed that two types of conformational changes have to occur in binding protein-dependent ABC transporters. The first must transmit the signal of the interaction between the loaded binding protein and the TMDs at the periplasmic side of the complex. This transmission through the TMDs will trigger ATP hydrolysis on the cytoplasmic side of the complex which in turn must be responsible for the second round of conformational changes opening and closing the transport channel within the TMDs. In the process, the attached and substrate-loaded binding protein must open up to release its substrate into the channel. It is unclear whether energy (from ATP hydrolysis) is needed for dissociation of substrate from the binding protein to occur. Estimates based on the *in vitro* rate of dissociation of substrate from the maltose binding protein in comparison to the maximal rate of transport suggest that little if any energy is needed for this process [51]. The structure of MalK cannot shed any light on the first set of conformational changes originating from the binding protein. However, the second set of conformational changes within the TMDs caused by ATP hydrolysis must have its origin in the structure that we are looking at. From the above discussion as well as from the observation of cooperativity [224] it is clear that hydrolysis at one site must affect the other. Thus, sequential ATP hydrolysis at these non-equivalent sites is likely to represent the complete transport cycle. An alternating two-site (two-cylinder engine) mechanism has been proposed for the multidrug LmrA ABC exporter of *Lactococcus lactis* [226]. The analysis of ATP hydrolysis in CFTR after its activation by phosphorylation revealed that ATP hydrolysis at one site causes the channel to open while ATP hydrolysis at the other site closes the channel [32]. Even though LmrA and more so the CFTR chloride channel are not strictly comparable to the situation of the supposedly stoichiometric transport (import; two ATPs hydrolyzed per one translocation cycle [236]) in the prokaryotic binding protein-dependent maltose system, the analogy to the two-cylinder engine is tempting and may well represent the basic underlying principle of energy coupling.

Recently, remarkable progress has been made in the understanding of the coupling of ATP hydrolysis to transport. With the use of the inhibitor vanadate a transition state complex in MalK of *E. coli* was recognized that is characterized by its ability to tightly bind MBP on the periplasmic side of the transport complex without its substrate maltose [236a]. In addition, fluorescence labeling of the cysteine in the Walker A site of MalK indicated that mutations in malF (rendering the transport system independent of MBP) behaved differently from the wild type, affecting not only ATPase activity but also binding of MBP [236b]. This demonstrates that MalK must undergo large conformational changes during the transport process that might involve a rearrangement of the MalK subunits towards each other.

References

1. C. F. Fox, E. P. Kennedy, *Proc. Natl. Acad. Sci. USA* **1965**, *54*, 891-899.
2. R. D. Simoni, M. Levinthal, F. D. Kundig, B. Anderson, H. P. E., S. Roseman, *Proc. Natl. Acad. Sci. USA* **1967**, *58*, 1963-1970.
3. S. Tanaka, E. C. C. Lin, *Proc. Natl. Acad. Sci. USA* **1967**, *57*, 913-919.
4. H. C. Neu, L. A. Heppel, *J. Biol. Chem.* **1965**, *240*, 3685-3692.
5. A. B. Pardee, *Science* **1968**, *162*, 632-637.
6. W. Boos, *J. Biol. Chem.* **1972**, *247*, 5414-5424.
7. G. F. Ames, J. E. Lever, *J. Biol. Chem.* **1972**, *247*, 4309-4316.
8. H. A. Shuman, T. J. Silhavy, J. R. Beckwith, *J. Biol. Chem.* **1980**, *255*, 168-174.
9. G. F. L. Ames, A. K. Joshi, *J. Bacteriol.* **1990**, *172*, 4133-4137.
10. l. Bishop, J. R. Agbayane, S. V. Ambudkar, P. C. Maloney, G. F.-L. Ames, *Proc. Natl. Acad. Sci. USA* **1989**, *86*, 6953-6957.
11. A. L. Davidson, H. Nikaido, *J. Biol. Chem.* **1990**, *265*, 4254-4260.
12. A. L. Davidson, H. Nikaido, *J. Biol. Chem* **1991**, *266*, 8946-8951.
13. C. E. Liu, G. F. L. Ames, *J. Biol. Chem.* **1997**, *272*, 859-866.
14. P.-Q. Liu, G. F.-L. Ames, *Proc. Natl. Acad. Sci. USA* **1998**, *95*, 3495-3500.
15. R. Reich-Slotky, C. Panagiotidis, M. Reyes, H. A. Shuman, *J. Bacteriol.* **2000**, *182*, 993-1000.
16. C. F. Higgins, I. D. Hiles, G. P. C. Salmond, D. R. Gill, J. A. Downie, I. J. Evans, I. B. Holland, L. Gray, S. D. Buckel, A. W. Bell, M. A. Hermodson, *Nature* **1986**, *323*, 448-450.
17. G. F.-L. Ames, C. S. Mimura, V. Shyamala, *FEMS Microbiol. Rev.* **1990**, *75*, 429-446.
18. C. F. Higgins, *Annu. Rev. Cell Biol.* **1992**, *8*, 67-113.
19. J. E. Walker, M. Saraste, J. M. Runswik, N. J. Gay, *EMBO J.* **1982**, *1*, 945-951.
20. F. A. Quiocho, P. S. Ledvina, *Mol. Microbiol.* **1996**, *20*, 17-25.
21. A. M. Stock, S. L. Mowbray, *Curr. Opin. Struct. Biol.* **1995**, *5*, 744-751.
22. D. Boyd, B. Traxler, J. Beckwith, *J. Bacteriol.* **1993**, *175*, 553-556.
23. B. Traxler, D. Boyd, J. Beckwith, *J. Membr. Biol.* **1993**, *132*, 1-11.
24. M. Ehrmann, R. Ehrle, E. Hofmann, W. Boos, A. Schlösser, *Mol. Microbiol.* **1998**, *29*, 685-694.
25. N. Treptow, H. Shuman, *J. Bacteriol.* **1985**, *163*, 654-660.
26. N. A. Treptow, H. A. Shuman, *J. Mol. Biol.* **1988**, *202*, 809-822.
27. L.-I. Hor, H. A. Shuman, *J. Mol. Biol.* **1993**, *233*, 659-670.
28. M. Mourez, M. Hofnung, E. Dassa, *EMBO J.* **1997**, *16*, 3066-3077.
29. S. Hunke, M. Mourez, M. Jehanno, E. Dassa, E. Schneider, *J. Biol. Chem.* **2000**, *275*, 15526-15534.
30. L.-W. Hung, I. X. Wang, K. Nikaido, P.-Q. Liu, G. F.-L. Ames, S.-H. Kim, *Nature* **1998**, *396*, 703-707.
31. K. Diederichs, J. Diez, G. Greller, C. Müller, J. Breed, C. Schnell, C. Vonrhein, W. Boos, W. Welte, *EMBO J.* **2000**, *19*, 5951-5961.
32. D. C. Gadsby, A. C. Nairn, *Physiol. Rev.* **1999**, *79*, S77-S107.
33. B. Holland, M. A. Blight, *J. Mol. Biol.* **1999**, *293*, 381-399.
34. A. Dinthilhac, G. Alloing, C. Granadel, J.-P. Claverys, *Mol. Microbiol.* **1997**, *25*, 727-739.
35. M. Perego, C. F. Higgins, S. R. Pearce, M. P. Gallagher, J. A. Hoch, *Mol. Microbiol.* **1991**, *5*, 173-185.
36. H. W. van Veen, M. Putman, A. Margolles, K. Sakamoto, W. N. Konings, *Biochim. Biophys. Acta* **1999**, *1461*, 201-206.

37. H. W. van Veen, R. Callaghan, L. Soceneantu, A. Sardini, W. N. Konings, C. F. Higgins, *Nature* **1998**, *391*, 291-295.
38. S. Létoffé, P. Delepelaire, C. Wandersman, *EMBO J.* **1996**, *15*, 5804-5811.
39. R. Binet, S. Létoffé, J. M. Ghigo, P. Delepelaire, C. Wandersman, *Gene* **1997**, *192*, 7-11.
40. M. J. Fath, R. Kolter, *Microbiol. Rev.* **1993**, *57*, 995-1017.
41. W. Saurin, M. Hofnung, E. Dassa, *J. Mol. Evol.* **1999**, *48*, 22-41.
42. K. J. Linton, C. F. Higgins, *Mol. Microbiol.* **1998**, *28*, 5-13.
43. E. Dassa, M. Hofnung, I. T. Paulsen, M. H. Saier, *Mol. Microbiol.* **1999**, *32*, 887-889.
44. M. S. Pavelka, S. F. Hayes, R. P. Silver, *J. Biol. Chem.* **1994**, *269*, 20149-20158.
45. R. F. Doolittle, M. S. Johnson, I. Husain, B. Van Houten, D. C. Thomas, A. Sancar, *Nature* **1986**, *323*, 451-453.
46. T. Yakushi, K. Masuda, S. Narita, S. Matsuyama, H. Tokuda, *Nature Cell Biol.* **2000**, *2*, 212-218.
47. Y. Quentin, G. Fichant, F. Denizot, *J. Mol. Biol.* **1999**, *287*, 467-484.
48. M. Braibant, P. Gilot, J. Content, *FEMS Microbiol. Rev.* **2000**, *24*, 449-467.
49. S. Szmelcman, M. Schwartz, T. J. Silhavy, W. Boos, *Eur. J. Biochem.* **1976**, *65*, 13-19.
50. E. Schneider, S. Hunke, *FEMS Microbiol. Rev.* **1998**, *22*, 1-20.
51. W. Boos, J. M. Lucht, *Book*, Vol. 1, 2nd Edn., (F. C. Neidhardt, R. Curtiss, J. L. Ingraham, E. C. C. Lin, K. B. Low, B. Magasanik, W. S. Reznikoff, M. Riley, M. Schaechter, H. E. Umbarger, Eds.), American Society of Microbiology, Washington, DC, **1996**, pp. 1175-1209.
52. P. M. Jones, A. M. George, *FEMS Microbiol. Lett.* **1999**, *179*, 187-202.
53. M. H. Saier, Jr., *J. Bacteriol.* **2000**, *182*, 5029-5035.
54. M. H. Saier, *Microbiol. Mol. Biol. Rev.* **2000**, *64*, 354-411.
55. C. A. Doige, G. F. L. Ames, *Annu. Rev. Microbiol.* **1993**, *47*, 291-319.
56. P. Borst, N. Zelcer, A. van Helvoort, *Biochim. Biophys. Acta* **2000**, *1486*, 128-144.
57. C. Wandersman, *Res. Microbiol.* **1998**, *149*, 163-170.
58. C. Wandersman, *Book*, Vol. 1 (F. C. Neidhardt, R. Curtiss, J. L. Ingraham, E. C. C. Lin, K. B. Low, B. Magasanik, W. S. Reznikoff, M. Riley, M. E. Schaechter, H. E. Umbarger, Eds.), American Society of Microbiology, Washington, DC, **1996**, pp. 955-966.
59. A. Margolles, M. Putman, H. W. van Veen, W. N. Konings, *Biochemistry* **1999**, *38*, 16298-16306.
60. C. F. Higgins, G. F. Ames, *Proc. Natl. Acad. Sci. USA* **1981**, *78*, 6038-6042.
61. A. Sirko, M. Zatyka, E. Sadowy, D. Hulanicka, *J. Bacteriol.* **1995**, *177*, 4134-4136.
62. M. D. Adams, L. M. Wagner, T. J. Graddis, R. Landick, T. K. Antonucci, A. L. Gibson, D. L. Oxender, *J. Biol. Chem.* **1990**, *265*, 11436-11443.
63. A. Schlösser, *FEMS Microbiol. Lett.* **1999**, *184*, 187-192.
64. A. Schlösser, T. Kampers, H. Schrempf, *J. Bacteriol.* **1997**, *179*, 2092-2095.
65. J. C. Sutcliffe, R. R. B. Russell, *J. Bacteriol.* **1995**, 1123-1129.
66. H. C. Wu, *Book*, Vol. 1 (F. C. Neidhardt, R. Curtiss, J. L. Ingraham, E. C. C. Lin, K. B. Low, B. Magasanik, W. S. Reznikoff, M. Riley, M. E. Schaechter, H. E. Umbarger, Eds.), American Society of Microbiology, Washington, DC, **1996**, pp. 1005-1014.
67. M. D. Manson, W. Boos, P. J. Bassford, B. A. Rasmussen, *J. Biol. Chem.* **1985**, *260*, 9727-9733.
68. R. Horlacher, K. B. Xavier, H. Santos, J. DiRuggiero, M. Kossmann, W. Boos, *J. Bacteriol.* **1998**, *180*, 680-689.
69. S. V. Albers, M. G. L. Elferink, R. L. Charlebois, C. W. Sensen, A. J. M. Driessen, W. N. Konings, *J. Bacteriol.* **1999**, *181*, 4285-4291.
70. D. Obis, A. Guillot, J.-K. Gripon, P. Renault, A. Bolotin, M.-Y. Mistou, *J. Bacteriol.* **1999**, *181*, 6238-6246.
71. T. van der Heide, B. Poolman, *Proc. Natl. Acad. Sci. USA* **2000**, *97*, 7102-7106.
72. D. Hekstra, J. Tommassen, *J. Bacteriol.* **1993**, *175*, 6546-6552.
73. M. Pajatsch, M. Gerhart, R. Peist, R. Horlacher, W. Boos, A. Böck, *J. Bacteriol.* **1998**, *180*, 2630-2635.
74. M. Pajatsch, Ph. D. Thesis, Ludwig-Maximilians Universität München, Germany, **1999**.
75. S. Wilken, G. Schmees, E. Schneider, *Mol. Microbiol.* **1996**, *22*, 655-666.
76. G. L. Hazelbauer, J. Adler, *Nature New Biol.* **1971**, *230*, 101-104.
77. H. M. Kalckar, *Science* **1971**, *174*, 557-565.
78. G. L. Hazelbauer, *J. Bacteriol.* **1975**, *122*, 206-214.

79. Y. Zhang, P. J. Gardina, A. S. Kuebler, H. S. Kang, J. A. Christopher, M. D. Manson, *Proc. Natl. Acad. Sci. USA* **1999**, *96*, 939-944.
80. S. L. Mowbray, *J. Mol. Biol.* **1992**, *227*, 418-440.
81. W. N. Abouhamad, M. Manson, M. M. Gibson, C. F. Higgins, *Mol. Microbiol.* **1991**, *5*, 1035-1047.
82. P. M. Steed, B. L. Wanner, *J. Bacteriol.* **1993**, *175*, 6797-6809.
83. C. Parra-Lopez, R. Lin, A. Aspedon, E. A. Groisman, *EMBO J.* **1994**, *13*, 3964-3972.
84. J. Van der Vlag, P. W. Postma, *Mol. Gen. Genet.* **1995**, *248*, 236-241.
85. W. Boos, A. Böhm, *Trends Genet.* **2000**, *16*, 404-409.
86. S. Freundlieb, U. Ehmann, W. Boos, *J. Biol. Chem.* **1988**, *263*, 314-320.
87. R. Benz, A. Schmid, G. H. Vos-Scheperkeuter, *J. Membr. Biol.* **1987**, *100*, 21-29.
88. M. Jordy, C. Andersen, K. Schülein, T. Ferenci, R. Benz, *J. Mol. Biol.* **1996**, *259*, 666-678.
89. T. Schirmer, T. A. Keller, Y. F. Wang, J. P. Rosenbusch, *Science* **1995**, *267*, 512-514.
90. Y.-F. Wang, R. Dutzler, P. J. Rizkallah, J. P. Rosenbusch, T. Schirmer, *J. Mol. Biol.* **1997**, *272*, 56-63.
91. M. Pajatsch, C. Andersen, A. Mathes, A. Böck, R. Benz, H. Engelhardt, *J. Biol. Chem.* **1999**, *274*, 25159-25166.
92. R. A. Larsen, M. G. Thomas, K. Postle, *Mol. Microbiol.* **1999**, *31*, 1809-1824.
93. K. P. Locher, B. Rees, R. Koebnik, A. Mitschler, L. Moulinier, J. P. Rosenbusch, D. Moras, *Cell* **1998**, *95*, 771-778.
94. A. D. Ferguson, E. Hofmann, J. W. Coulton, K. Diederichs, W. Welte, *Science* **1998**, *282*, 2215-2220.
95. S. K. Buchanan, B. S. Smith, L. Venkatramani, D. Xia, L. Esser, M. Palnitkar, R. Chakraborty, D. van der Helm, J. Deisenhofer, *Nature Struct. Biol.* **1999**, *6*, 56-63.
96. V. Braun, H. Killmann, *Trends Biochem. Sci.* **1999**, *24*, 104-109.
97. V. Braun, *Arch. Microbiol.* **1997**, *167*, 325-331.
98. R. Tam, M. H. Saier, *Microbiol. Rev.* **1993**, *57*, 320-346.
99. B. Müller-Hill, *Nature* **1983**, *302*, 163-164.
100. C. A. Mauzy, M. A. Hermodson, *Protein Sci.* **1992**, *1*, 843-849.
101. J. C. Nichols, N. K. Vyas, F. A. Quiocho, K. S. Matthews, *J. Biol. Chem.* **1993**, *268*, 17602-17612.
102. A. P. Pugsley, *Microbiol. Rev.* **1993**, *57*, 50-108.
103. C. K. Murphy, J. Beckwith, *Book*, Vol. 1 (F. C. Neidhardt, R. Curtiss, J. L. Ingraham, E. C. C. Lin, K. B. Low, B. Magasanik, W. S. Reznikoff, M. Riley, M. E. Schaechter, H. E. Umbarger, Eds.), American Society of Microbiology, Washington, DC, **1996**, pp. 967-978.
104. F. Duong, J. Eichler, A. Price, M. R. Leonard, W. Wickner, *Cell* **1997**, *91*, 567-573.
105. E. H. Manting, A. J. Driessen, *Mol. Microbiol.* **2000**, *37*, 226-238.
106. S. Y. Chun, S. Strobel, P. Bassford, L. L. Randall, *J. Biol. Chem.* **1993**, *268*, 20855-20862.
107. H. Tjalsma, G. Zanen, G. Venema, S. Bron, J. M. van Dijl, *J. Biol. Chem.* **1999**, *274*, 28191-28197.
108. W. Boos, A. S. Gordon, R. E. Hall, H. D. Price, *J. Biol. Chem.* **1972**, *247*, 917-924.
109. D. M. Miller, J. S. Olson, J. W. Pflugrath, F. A. Quiocho, *J. Biol. Chem.* **1983**, *258*, 13665-13672.
110. J. H. Weiner, C. E. Furlong, L. A. Heppel, *Arch. Biochem. Biophys.* **1971**, *142*, 715-717.
111. G. Richarme, A. Kepes, *Biochim. Biophys. Acta* **1983**, *742*, 16-24.
112. T. J. Silhavy, W. Boos, S. Szmelcman, M. Schwartz, *Proc. Natl. Acad. Sci. USA* **1975**, *72*, 2120-2124.
113. M. Argast, W. Boos, *J. Biol. Chem.* **1979**, *254*, 10931-10935.
114. B. Kempf, E. Bremer, *J. Biol. Chem.* **1995**, *270*, 16701-16713.
115. R. M. Kappes, B. Kempf, E. Bremer, *J. Bacteriol.* **1996**, *178*, 5071-5079.
116. R. M. Kappes, B. Kempf, S. Kneip, J. Boch, J. Gade, J. Meier-Wagner, E. Bremer, *Mol. Microbiol.* **1999**, *32*, 203-216.
117. D. Wassenberg, W. Liebl, R. Jaenicke, *J. Mol. Biol.* **2000**, *295*, 279-288.
118. C. R. Jones, M. Ray, K. A. Dawson, H. J. Strobel, *Appl. Environ. Microbiol.* **2000**, *66*, 995-1000.
119. A. Ruepp, W. Graml, M.-L. Santos-Martinez, K. K. Koretke, C. Volker, H. W. Mewes, D. Frishman, S. Stocker, A. N. Lupas, W. Baumeister, *Nature* **2000**, *407*, 508-513.

119a. M. G. L Elferink, S.-V. Albers, W. N. Konings, A. J. M. Driessen, *Mol. Microbiol.* **2000**, *39*, 1494-1503.
120. A. G. Evdokimov, D. E. Anderson, K. M. Routzahn, D. S. Waugh, *J. Mol. Biol.* **2001**, *305*, 891-904.
121. J. Diez, K. Diederichs, G. Greller, R. Horlacher, W. Boos, W. Welte, *J. Mol. Biol.* **2001**, *305*, 905-915.
122. R. R. B. Russell, J. Aduse-Opoku, I. C. Sutcliffe, L. Tao, J. J. Ferretti, *J. Biol. Chem.* **1992**, *267*, 4631-4637.
123. I. C. Sutcliffe, L. Tao, J. J. Ferretti, R. R. B. Russell, *J. Bacteriol.* **1993**, *175*, 1853-1855.
124. M. H. J. Jacobs, T. Van der Heide, A. J. M. Driessen, W. N. Konings, *Proc. Natl. Acad. Sci. USA* **1996**, *93*, 12786-12790.
125. E. B. McGowan, T. J. Silhavy, W. Boos, *Biochemistry* **1974**, *13*, 993-999.
126. S. D. Buckel, P. F. Cottam, V. Simplaceanu, C. Ho, *J. Mol. Biol.* **1989**, *208*, 477-489.
127. T. E. Cedel, P. F. Cottam, M. D. Meadows, C. Ho, *Biophys. Chem.* **1984**, *19*, 279-287.
128. J. F. Post, P. F. Cottam, V. Simplaceanu, C. Ho, *J. Mol. Biol.* **1984**, *179*, 729-743.
129. D. E. Robertson, P. A. Kroon, C. Ho, *Biochemistry* **1977**, *16*, 1443-1451.
130. Q. C. Shen, V. Simplaceanu, P. F. Cottam, C. Ho, *J. Mol. Biol.* **1989**, *210*, 849-857.
131. R. S. Zukin, P. R. Hartig, D. E. Koshland, *Proc. Natl. Acad. Sci. USA* **1977**, *74*, 1932-1936.
132. G. Giraldi, L. Q. Zhou, L. Hibbert, A. E. G. Cass, *Ann. Chem.* **1994**, *66*, 3840-3847.
133. B. H. Oh, J. Pandit, C. H. Kang, K. Nikaido, S. Gokcen, G. F.-L. Ames, S. H. Kim, *J. Biol. Chem.* **1993**, *268*, 17648-17649.
134. G. A. Olah, S. Trakhanov, J. Trewhella, F. A. Quiocho, *J. Biol. Chem.* **1993**, *268*, 16241-16247.
135. A. J. Sharff, L. E. Rodseth, J. C. Spurlino, F. A. Quiocho, *Biochemistry* **1992**, *31*, 10657-10663.
136. K. Döring, T. Surrey, P. Nollert, F. Jähnig, *Eur. J. Biochem.* **1999**, *266*, 477-483.
137. M. M. Flocco, S. L. Mowbray, *J. Biol. Chem.* **1994**, *269*, 8931-8936.
138. A. J. Björkman, S. L. Mowbray, *J. Mol. Biol.* **1998**, *279*, 651-664.
139. J. A. Hall, A. K. Ganesan, J. Chen, H. Nikaido, *J. Biol. Chem.* **1997**, *272*, 17615-17622.
140. J. A. Hall, K. Gehring, H. Nikaido, *J. Biol. Chem.* **1997**, *272*, 17605-17609.
141. J. A. Hall, T. E. Thorgeirsson, J. Liu, Y.-K. Shin, H. Nikaido, *J. Biol. Chem.* **1997**, *272*, 17610-17614.
142. F. J. M. Detmers, F. C. Laufermeijer, R. Abele, R. W. Jack, R. Tampé, W. N. Konings, B. Poolman, *Proc. Natl. Acad. Sci. USA* **2000**, *97*, 12487-12492.
143. P. Argos, W. C. Mahoney, M. A. Hermodson, M. Hanei, *J. Biol. Chem.* **1981**, *256*, 4357-4361.
144. F. A. Quiocho, *Book*, Vol. 55, **1986**, pp. 287-315.
145. F. A. Quiocho, *Philos. Trans. R. Soc. Lond.* **1990**, *B 326*, 341-351.
146. J. C. Spurlino, G.-Y. Lu, F. A. Quiocho, *J. Biol. Chem.* **1991**, *266*, 5202-5219.
147. N. K. Vyas, M. N. Vyas, F. A. Quiocho, *J. Biol. Chem.* **1991**, *266*, 5226-5237.
148. N. H. Yao, S. Trakhanov, F. A. Quiocho, *Biochemistry* **1994**, *33*, 4769-4779.
149. B. H. Shilton, H. A. Shuman, S. L. Mowbray, *J. Mol. Biol.* **1996**, *264*, 364-376.
150. J. S. Sack, M. A. Saper, F. A. Quiocho, *J. Mol. Biol.* **1989**, *206*, 171-191.
151. L. Jacobson, J. J. He, D. D. Lemon, F. A. Quiocho, *J. Mol. Biol.* **1992**, *223*, 27-30.
152. B. L. Jacobson, J. J. He, P. S. Vermersch, D. D. Lemon, F. A. Quiocho, *J. Biol. Chem.* **1991**, *266*, 5220-5225.
153. N. K. Vyas, M. N. Vyas, F. A. Quiocho, *Nature* **1987**, *327*, 635-638.
154. J. Y. Zou, M. M. Flocco, S. L. Mowbray, *J. Mol. Biol.* **1993**, *233*, 739-752.
155. L. A. Luck, J. J. Falke, *Biochemistry* **1991**, *30*, 4257-4261.
156. M. N. Vyas, B. L. Jacobson, F. A. Quiocho, *J. Biol. Chem.* **1989**, *264*, 20817-20821.
157. J. R. H. Tame, G. N. Murshudov, E. J. Dodson, T. K. Neil, G. G. Dodson, C. F. Higgins, A. J. Wilkinson, *Science* **1994**, *264*, 1578-1581.
158. E. R. Olson, D. S. Dunyak, L. M. Jurss, R. A. Poorman, *J. Bacteriol.* **1991**, *173*, 234-244.
159. M. S. Hanson, C. Slaughter, E. J. Hansen, *Infect. Immun.* **1992**, *60*, 2257-2266.
160. C. Navarro, L.-F. Wu, M.-A. Mandrand-Berthelot, *Mol. Microbiol.* **1993**, *9*, 1181-1191.
161. B. J. Pearce, A. M. Naughton, H. R. Masure, *Mol. Microbiol.* **1994**, *12*, 881-892.
162. D. Z. Rudner, J. R. Ledeaux, K. Ireton, A. D. Grossman, *J. Bacteriol.* **1991**, *173*, 1388-1398.

163. J. DiRuggiero, D. Dunn, D. L. Maeder, R. Holley-Shanks, J. Chatard, R. Horlacher, F. T. Robb, W. Boos, R. B. Weiss, *Mol. Microbiol.* **2000**, *38*, 684-693.
164. G. von Heijne, *EMBO J.* **1986**, *5*, 3021-3027.
165. G. von Heijne, *J. Mol. Biol.* **1992**, *225*, 487-494.
166. M. Amemura, K. Makino, H. Shinagawa, A. Kobayashi, A. Nakata, *J. Mol. Biol.* **1985**, *184*, 241-250.
167. B. P. Surin, H. Rosenberg, G. B. Cox, *J. Bacteriol.* **1985**, *161*, 189-198.
168. R. Pistocchi, K. Kashiwagi, S. Miyamoto, E. Nukui, Y. Sadakata, H. Kobayashi, K. Igarashi, *J. Biol. Chem.* **1993**, *268*, 146-152.
169. T. Furuchi, K. Kashiwagi, H. Kobayashi, K. Igarashi, *J. Biol. Chem.* **1991**, *266*, 20928-20933.
170. A. Sirko, M. Hryniewicz, D. Hulanicka, A. Böck, *J. Bacteriol.* **1990**, *172*, 3351-3357.
171. S. R. Pearce, M. L. Mimmack, M. P. Gallagher, U. Gileadi, S. C. Hyde, C. F. Higgins, *Mol. Microbiol.* **1992**, *6*, 47-57.
172. E. Dassa, S. Muir, *Mol. Microbiol.* **1993**, *7*, 29-38.
173. S. Froshauer, G. N. Green, D. Boyd, K. McGovern, J. Beckwith, *J. Mol. Biol.* **1988**, *200*, 501-511.
174. K. M. Y. Covitz, C. H. Panagiotidis, L. I. Hor, M. Reyes, N. A. Treptow, H. A. Shuman, *EMBO J.* **1994**, *13*, 1752-1759.
175. M. K. Dahl, E. Francoz, W. Saurin, W. Boos, M. D. Manson, M. Hofnung, *Mol. Gen. Genet.* **1989**, *218*, 199-207.
176. G. Fiedler, M. Pajatsch, A. Böck, *J. Mol. Biol.* **1996**, *256*, 279-291.
177. A. Puyet, M. Espinosa, *J. Mol. Biol.* **1993**, *230*, 800-811.
178. P. Whitley, T. Zander, M. Ehrmann, M. Haardt, E. Bremer, G. von Heijne, *EMBO J.* **1994**, *13*, 4653-4661.
179. M. Haardt, E. Bremer, *J. Bacteriol.* **1996**, *178*, 5370-5381.
180. E. Faatz, A. Middendorf, E. Bremer, *Mol. Microbiol.* **1988**, *2*, 265-279.
181. R. E. Kerppola, G. F. Ames, *J. Biol. Chem.* **1992**, *267*, 2329-2336.
182. R. E. Kerppola, V. K. Shyamala, P. Klebba, G. F.-L. Ames, *J. Biol. Chem.* **1991**, *266*, 9857-9865.
183. W. Köster, V. Braun, *Mol. Gen. Genet.* **1990**, *223*, 379-384.
184. R. Ehrle, C. Pick, R. Ulrich, E. Hofmann, M. Ehrmann, *J. Bacteriol.* **1996**, *178*, 2255-2262.
185. M. Ehrmann, J. Beckwith, *J. Biol. Chem.* **1991**, *266*, 16530-16533.
186. H. A. Shuman, C. H. Panagiotidis, *J. Bioenerg. Biomembr.* **1993**, *25*, 613-620.
187. G. Merino, W. Boos, H. A. Shuman, E. Bohl, *J. Theor. Biol.* **1995**, *177*, 171-179.
188. G. Merino, H. A. Shuman, *J. Biol. Chem.* **1998**, *273*, 2435-2444.
189. G. Merino, H. A. Shuman, *J. Bacteriol.* **1997**, *179*, 7687-7694.
190. G. F.-L. Ames, C. E. Liu, A. K. Joshi, K. Nikaido, *J. Biol. Chem.* **1996**, *271*, 14264-14270.
191. K. Kunishima, Y. Shimada, Y. Tsuji, T. Sato, M. Yamamoto, T. Kumasaka, S. Nakanishi, H. Jingami, K. Morikawa, *Nature* **2000**, *407*, 971-977.
192. A. Mademidis, H. Killmann, W. Kraas, I. Flechsler, G. Jung, V. Braun, *Mol. Microbiol.* **1997**, *26*, 1109-1123.
193. G. F.-L. Ames, *Annu. Rev. Biochem.* **1986**, *55*, 397-425.
194. P. Overduin, W. Boos, J. Tommassen, *Mol. Microbiol.* **1988**, *2*, 767-775.
195. E. Dassa, M. Hofnung, *EMBO J.* **1985**, *4*, 2287-2293.
196. W. Köster, B. Böhm, *Mol. Gen. Genet.* **1992**, *232*, 399-407.
197. W. Saurin, W. Köster, E. Dassa, *Mol. Microbiol.* **1994**, *12*, 993-1004.
198. W. Saurin, E. Dassa, *Protein Sci.* **1994**, *3*, 325-344.
199. E. Dassa, *Mol. Gen. Genet.* **1990**, *222*, 33-36.
200. C. S. Mimura, S. R. Holbrook, G. F.-L. Ames, *Proc. Natl. Acad. Sci. USA* **1991**, *88*, 84-88.
201. J. Young, I. B. Holland, *Biochim. Biophys. Acta* **1999**, *1461*, 177-200.
202. P. Bavoil, M. Hofnung, H. Nikaido, *J. Biol. Chem.* **1980**, *255*, 8366-8369.
203. H. A. Shuman, T. J. Silhavy, *J. Biol. Chem.* **1981**, *256*, 560-562.
204. G. F.-L. Ames, K. Nikaido, *Proc. Natl. Acad. Sci. USA* **1978**, *75*, 5447-5451.
205. M. P. Gallagher, S. R. Pearce, C. F. Higgins, *Eur. J. Biochem.* **1989**, *180*, 133-141.
206. G. May, E. Faatz, J. M. Lucht, M. Haardt, M. Bolliger, E. Bremer, *Mol. Microbiol.* **1989**, *3*, 1521-1531.

207. K. Nikaido, P.-Q. Liu, G. F.-L. Ames, *J. Biol. Chem.* **1997**, *272*, 27745-27752.
208. K. A. Kennedy, B. Traxler, *J. Biol. Chem.* **1999**, *274*, 6259-6264.
209. V. Baichwal, D. X. Liu, G. F. L. Ames, *Proc. Natl. Acad. Sci. USA* **1993**, *90*, 620-624.
210. G. F.-L. Ames, E. N. Spudich, *Proc. Natl. Acad. Sci. USA* **1976**, *73*, 1877-1881.
211. E. Prossnitz, *J. Biol. Chem.* **1991**, *266*, 9673-9677.
212. E. Schneider, S. Hunke, S. Tebbe, *J. Bacteriol.* **1995**, *177*, 5364-5367.
213. E. J. Blott, C. F. Higgins, K. J. Linton, *EMBO J.* **1999**, *18*, 6800-6808.
214. C. F. Higgins, I. D. Hiles, K. Whalley, D. J. Jamieson, *EMBO J.* **1985**, *4*, 1033-1040.
215. A. C. Hobson, R. Weatherwax, G. F. Ames, *Proc. Natl. Acad. Sci. USA* **1984**, *81*, 7333-7337.
216. C. S. Mimura, A. Admon, K. A. Hurt, G. F.-L. Ames, *J. Biol. Chem.* **1990**, *265*, 19535-19542.
217. G. Schultzhauser, B. Vanhove, V. Braun, *FEMS Microbiol. Lett.* **1992**, *95*, 231-234.
218. C. Walter, K. Höner zu Bentrup, E. Schneider, *J. Biol. Chem.* **1992**, *267*, 8863-8869.
219. G. Richarme, A. Elyaagoubi, M. Kohiyama, *J. Biol. Chem.* **1993**, *268*, 9473-9477.
220. S. Morbach, S. Tebbe, E. Schneider, *J. Biol. Chem.* **1993**, *268*, 18617-18621.
221. C. H. Panagiotidis, M. Reyes, A. Sievertsen, W. Boos, H. A. Shuman, *J. Biol. Chem.* **1993**, *268*, 23685-23696.
222. A. L. Davidson, H. A. Shuman, H. Nikaido, *Proc. Natl. Acad. Sci. USA* **1992**, *89*, 2360-2364.
223. A. L. Davidson, S. Sharma, *J. Bacteriol.* **1997**, *179*, 5458-5464.
224. A. L. Davidson, S. S. Laghaeian, D. E. Mannering, *J. Biol. Chem.* **1996**, *271*, 4858-4863.
225. K. Nikaido, G. F.-L. Ames, *J. Biol. Chem.* **1999**, *274*, 26727-26735.
226. H. W. van Veen, A. Margolles, M. Müller, C. F. Higgins, W. N. Konings, *EMBO J.* **2000**, *19*, 2503-2514.
227. P. Brzoska, M. Rimmele, K. Brzostek, W. Boos, *J. Bacteriol.* **1994**, *176*, 15-20.
228. B. Bukau, M. Ehrmann, W. Boos, *J. Bacteriol.* **1986**, *166*, 884-891.
229. M. Reyes, H. A. Shuman, *J. Bacteriol.* **1988**, *170*, 4598-4602.
230. S. Kühnau, M. Reyes, A. Sievertsen, H. A. Shuman, W. Boos, *J. Bacteriol.* **1991**, *173*, 2180-2186.
231. C. H. Panagiotidis, W. Boos, H. A. Shuman, *Mol. Microbiol.* **1998**, *30*, 535-546.
232. E. Richet, O. Raibaud, *EMBO J.* **1989**, *8*, 981-987.
233. J. Lippincott, B. Traxler, *J. Bacteriol.* **1997**, *179*, 1337-1343.
234. K. B. Xavier, L. O. Martins, R. Peist, M. Kossmann, W. Boos, H. Santos, *J. Bacteriol.* **1996**, *178*, 4773-4777.
235. G. Greller, R. Horlacher, J. DiRuggiero, W. Boos, *J. Biol. Chem.* **1999**, *274*, 20259-20264.
236. M. L. Mimmack, M. P. Gallagher, S. R. Pearce, S. C. Hyde, I. R. Booth, C. F. Higgins, *Proc. Natl. Acad. Sci. USA* **1989**, *86*, 8257-8261.
236a. J. Chen, S. Sharma, F. A. Quiocho, A. L. Davidson, *Proc. Natl. Acad. Sci. USA* **2001**, *98*, 1525-1530.
236b. D. E Mannering, S. Sharma, A. L. Davidson, *J. Biol. Chem.* **2001**, *276*, 12362-12368.
237. B. N. Chaudhuri, J. Ko, J. Park, S. L. Mowbray, *J. Mol. Biol.* **1999**, *286*, 1519-1531.
238. P. S. Vermersch, J. J. G. Tesmer, D. D. Lemon, F. A. Quiocho, *J. Biol. Chem.* **1990**, *265*, 16592-16603.
239. P. S. Vermersch, D. D. Lemon, J. J. G. Tesmer, F. A. Quiocho, *Biochemistry* **1991**, *30*, 6861-6866.
240. M. Zacharias, T. P. Straatsma, J. A. McCammon, F. A. Quiocho, *Biochemistry* **1993**, *32*, 7428-7434.
241. A. V. Nickitenko, S. Trakhanov, F. A. Quiocho, *Biochemistry* **1995**, *34*, 16585-16595.
242. P. Dunten, S. L. Mowbray, *Protein Sci.* **1995**, *4*, 2327-2334.
243. T. E. Clarke, S. Y. Ku, D. R. Dougan, H. J. Vogel, L. W. Tari, *Nature Struct. Biol.* **2000**, *7*, 287-291.
244. J. Åpvist, S. L. Mowbray, *J. Biol. Chem.* **1995**, *270*, 9978-9981.
245. B. H. Shilton, M. M. Flocco, M. Nilsson, S. L. Mowbray, *J. Mol. Biol.* **1996**, *264*, 350-363.
246. N. K. Vyas, M. N. Vyas, F. A. Quiocho, *Science* **1988**, *242*, 1290-1295.
247. M. N. Vyas, N. K. Vyas, F. A. Quiocho, *Biochemistry* **1994**, *33*, 4762-4768.
248. C. D. Hsiao, Y. J. Sun, J. Rose, P. F. Cottam, C. Ho, B. C. Wang, *J. Mol. Biol.* **1994**, *240*, 87-91.

249. C.-D. Hsiao, Y.-J. Sun, J. Rose, B.-C. Wang, *J. Mol. Biol.* **1996**, *262*, 225-242.
250. P. Dunten, S. L. Mowbray, *Protein Sci.* **1995**, *4*, 2335-2340.
251. B. H. Oh, C. H. Kang, H. Debondt, S. H. Kim, K. Nikaido, A. K. Joshi, G. F.-L. Ames, *J. Biol. Chem.* **1994**, *269*, 4135-4143.
252. J. S. Sack, S. D. Trakhanov, I. H. Tsigannik, F. A. Quiocho, *J. Mol. Biol.* **1989**, *206*, 193-207.
253. C. H. Kang, W. C. Shin, Y. Yamagata, S. Gokcen, G. F.-L. Ames, S. H. Kim, *J. Biol. Chem.* **1991**, *266*, 23893-23899.
254. L. Rodseth, F. A. Quiocho, *J. Mol. Biol* **1993**, *230*, 675-678.
255. A. J. Sharff, L. E. Rodseth, F. A. Quiocho, *Biochemistry* **1993**, *32*, 10553-10559.
256. A. J. Sharff, L. E. Rodseth, S. Szmelcman, M. Hofnung, F. A. Quiocho, *J. Mol. Biol.* **1995**, *246*, 8-13.
257. F. A. Quiocho, J. C. Spurlino, L. E. Rodseth, *Structure* **1997**, *5*, 997-1015.
258. Y. Hu, S. Rech, R. P. Gunsalus, D. C. Rees, *Nature Struct. Biol.* **1997**, *4*, 703-707.
259. J. R. H. Tame, E. J. Dodson, G. Murshudov, C. F. Higgins, A. J. Wilkinson, *Structure* **1995**, *3*, 1395-1406.
260. S. H. Sleigh, J. R. H. Tame, E. J. Dodson, A. J. Wilkinson, *Biochemistry* **1997**, *36*, 9747-9758.
261. B. D. Kubena, H. Luecke, H. Rosenberg, F. A. Quiocho, *J. Biol. Chem.* **1986**, *261*, 7995-7996.
262. Z. Wang, A. Choudhary, P. S. Ledvina, F. A. Quiocho, *J. Biol. Chem.* **1994**, *269*, 25091-25094.
263. F. A. Quiocho, *Kidney Int.* **1996**, *49*, 943-946.
264. N. Yao, P. S. Ledvina, A. Choudhary, F. A. Quiocho, *Biochemistry* **1996**, *35*, 2079-2085.
265. M. Hirshberg, K. Henrick, L. L. Haire, N. Vasisht, M. Brune, J. E. T. Corrie, M. R. Webb, *Biochemistry* **1998**, *37*, 10381-10385.
266. S. Sugiyama, D. G. Vassylyev, M. Matsushima, K. Kashiwagi, K. Igarashi, K. Morikawa, *J. Biol. Chem.* **1996**, *271*, 9519-9525.
267. D. G. Vassylyev, T. Kashiwagi, H. Tomitori, K. Kashiwagi, K. Igarashi, K. Morikawa, *Acta Cryst. D* **1998**, *54*, 132-134.
268. R. A. Binnie, H. Zhang, S. Mowbray, M. A. Hermodson, *Protein Sci.* **1992**, *1*, 1642-1651.
269. S. L. Mowbray, L. B. Cole, *J. Mol. Biol.* **1992**, *225*, 155-175.
270. A. J. Björkman, R. A. Binnie, H. Zhang, L. B. Cole, M. A. Hermodson, S. L. Mowbray, *J. Biol. Chem.* **1994**, *269*, 30206-30211.
271. S. L. Mowbray, A. J. Björkman, *J. Mol. Biol.* **1999**, *294*, 487-499.
272. J. W. Pflugrath, F. A. Quiocho, *J. Mol. Biol.* **1988**, *200*, 163-180.
273. J. J. He, F. A. Quiocho, *Science* **1991**, *251*, 1479-1481.
274. J. J. He, F. A. Quiocho, *Protein Sci.* **1993**, *2*, 1643-1647.

5
Glucose Transport by the Bacterial Phosphotransferase System (PTS): An Interface between Energy- and Signal Transduction

Bernhard Erni

5.1
Introduction

Phosphorylation of metabolites and proteins plays a central role in energy- and signal transduction. Sugars and sugar alcohols are primed for metabolic transformation by phosphorylation. The phosphorylated products are formed by kinases at the expense of ATP and by phosphorolytic cleavage of intracellular oligo- and polysaccharides. The equilibria of these reactions are far to the right and the intracellular concentration of free carbohydrates is kept low. In bacteria carbohydrates in addition can be phosphorylated at the expense of phosphoenolpyruvate (PEP). The reaction was discovered by Kundig, Gosh, and Roseman [1] while they were looking for bacterial kinases involved in the metabolism of sialic acid. Instead of a kinase, an activity was found which catalyzed the phosphorylation of N-acetyl-mannosamine and other sugars at the expense of PEP, hence the name *Phosphoenolpyruvate:Glycose Phosphotransferase System* (PTS). This novel activity originally comprised three protein fractions, a soluble fraction termed enzyme I, a membrane-bound, termed enzyme II, and a soluble heat-stable histidine containing protein (HPr). The subsequent biochemical and genetic dissection of the PTS in *Escherichia coli* and *Salmonella typhimurium* revealed that it was composed of a large number of components, in particular of several enzymes II of different specificity for hexoses, hexitols, and disaccharides (EII^{sugar} or II^{sugar}).

Whereas phosphorolytic and ATP-dependent phosphorylation entails the direct transfer of phosphate from donor to acceptor, phosphotransfer from PEP is mediated by a cascade of four proteins (Fig. 1). The components at the end of the cascade are transporters which couple translocation with phosphorylation of the substrate. This mechanism which couples uptake with chemical modification is termed *group translocation* [2].

Why is the PTS so complex (Fig. 1B), why is PEP the phosphoryldonor and not ATP, why is phosphotransfer mediated by a cascade of phosphoproteins? As often

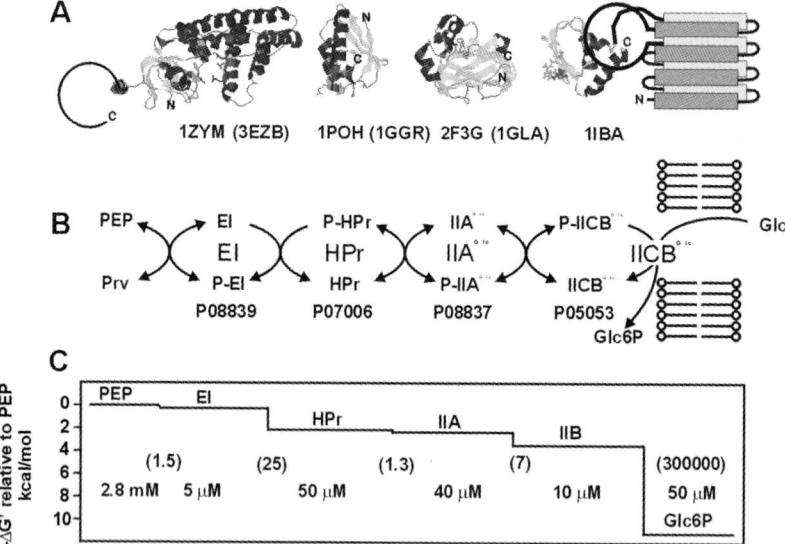

Fig. 1. Structural, biochemical and thermodynamic aspects of the glucose PTS. (**A**) Cartoons of EI, HPr, IIA^{Glc}, and IICB^{Glc}. Active site residues are represented as sticks. The PDB access codes are indicated. Codes in parentheses refer to structures of protein–protein complexes. Of EI, only the structure of the N-terminal domain is known. (**B**) Phosphotransfer chain of the glucose PTS. Arrowheads emphasize the reversibility of the phosphotransfer reaction. The SWISS-PROT access codes are indicated. (**C**) Free energy differences, equilibrium constants (in parentheses) of the phosphotransfer reactions between PTS proteins and intracellular concentrations of the components. Notice the small difference between HPr and IIA^{Glc}, which allows equilibrium distribution of phosphoryl groups between different IIA units via HPr. (Fig. 2). The values are from [24]; see color plates p. XXXI.

in biology, there is no answer to "why" but only to "how" a process works the way it does. ATP is the source of energy and phosphate in many biochemical processes and, therefore, one of the most highly connected metabolites [3]. PEP in contrast does not serve as a primary energy source for processes other than carbohydrate uptake, but is an important precursor for the biosynthesis of aromatic amino acids and the bacterial cell wall [4]. The PTS and glycolysis together form a biochemical cycle. From each sugar two PEP are generated of which one is consumed for transport of the next sugar. The phosphoprotein cascade, intercalated between the donor PEP and the acceptor sugar, presents a large and diverse surface for regulatory input and is itself the source of regulatory output (Sect. 5.4, Fig. 5). PTS proteins couple carbohydrate uptake chemically with glycolysis and allosterically or by protein phosphorylation with other metabolic systems (reviewed in [5; 6]). The regulatory activity varies with the degree of phosphorylation (ratio of non-phosphorylated to phosphorylated form), which in turn varies with the ratio of sugar-dependent dephosphorylation and PEP-dependent rephosphorylation (Sects. 5.4 and 5.5).

For more information on all aspects of the PTS readers are referred to the comprehensive reviews by Meadow, Fox, and Roseman [7]; Postma, Lengeler, and Jacobson [8, 9];, and the references given below in Sects. 5.2.1 and 5.2.2. This chapter describes the structure and the function of the transporter for glucose (Glc) of E. coli. A comprehensive review on the structure and function of the mannitol transporter and the other PTS transporters was published recently by Robillard and Broos [10].

5.2
The Components of the PTS and Their Function

5.2.1
Distribution of the PTS

The PTS is widely distributed but not universal among eubacteria. It does not occur in archaebacteria, animals, and plants. The comparative analysis of complete bacterial genomes indicates that gram-negative E. coli and S. typhimurium and gram-positive Bacillus subtilis have a very diverse PTS comprising up to 40 different genes ([11, 12] and Chap. 1). Other species have a much smaller PTS. Haemophilus influenzae has only one, Pseudomonas aeruginosa and Mycoplasma genitalium only two sugar-specific transporters. Interestingly, in all species the fructose transporter is always the first to be present. Genomes of some species, such as Treponema pallidum and Chlamydia trachomatis contain genes for proteins homologous to the soluble components of the PTS but none for sugar-specific transporters. These proteins may have regulatory functions. Other species, e.g., M. tuberculosis, have no PTS at all.

5.2.2
Modular Design and Classification

All PTS consist of at least two cytoplasmic proteins, EI (63.5 kDa) and HPr (9.1 kDa), and a variable number of sugar-specific transport complexes (II^{sugar}) (Figs. 1 and 2). EI transfers phosphoryl groups from PEP to the phosphocarrier protein HPr. HPr then transfers the phosphoryl groups to the different transport complexes. EI and HPr are usually unique. Inactivation of EI, therefore, paralyzes the entire PTS, inactivation of HPr most of it, because only the transport systems for Fru have an HPr-like unit (FruB) of their own [13] (Fig. 2). But there are exceptions. P. aeruginosa has two EI, one for N-acetyl-glucosamine (GlcNAc), the other for fructose (Fru) [14]. The two EI are domains of multiphosphotransfer proteins (MTP) [15] each dedicated to the transport of one substrate. The sugar-specific transporters consist of 3 to 4 functional units, IIA, IIB, and IIC and in the case of the mannose family also of IID. These units are either subunits of a complex or domains of a multidomain protein. A representative but by no means comprehensive selection of modular designs is shown in Fig. 2 (see also Chap. 1).

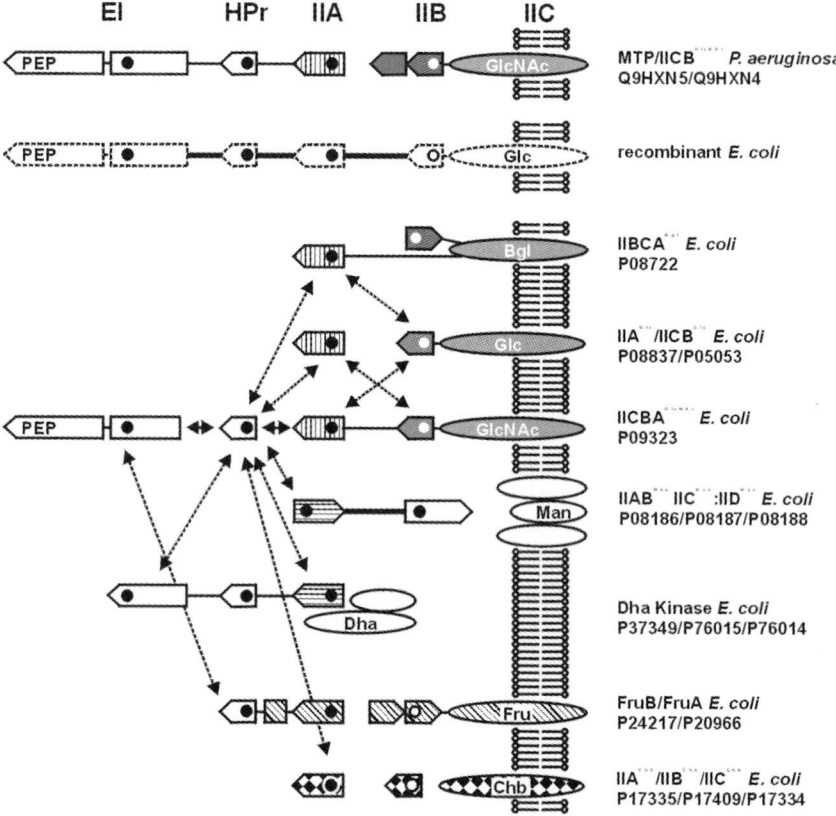

Fig. 2. Modular design of PTS proteins of the different families. Shown are representative examples of the glucose/β-glucoside, mannose, fructose/mannitol and lactose/chitobiose families from E. coli and P. aeruginosa. The functional units EI, HPr, IIA, IIB, IIC are vertically aligned. Units of the same family are filled identically. FruA and P. aeruginosa IICB^{GlcNAc} have duplicated B-domains. Solid and open circles indicate phosphorylated His and Cys. Dashed arrows indicate phosphotransfer between HPr and the sugar-specific IIA domains, and between homologous IIA and IIB domains on different transporters. Also shown is the heterotrimeric diyhydroxyacetone kinase (YcgC/YcgT/YcgS). The recombinant multi-domain protein was obtained by connecting the EI, HPr, IIA^{Glc}, and IICB^{Glc} units with the Ala−Pro rich linkers (thick line) taken from the IIAB^{Man} subunit of the mannose transporter [116]. The protein symbols are drawn to scale with the pointed end marking the C-terminus. SWISS-PROT accession codes are indicated.

EI and HPr are highly homologous between different species as well as within a species (e.g., *P. aeruginosa* and *E. coli*). In contrast, the amino acid sequences of the sugar-specific PTS transporters are very diverse with respect to sequence, fold, and oligomeric structure (reviewed in [10]). They can be grouped into five classes according to their amino acid sequences (glucose/sucrose/β-glucoside; mannose; mannitol/fructose; lactose/chitobiose; and not classified; for details see Chap. 1). A recent addition to this list of carbohydrate-specific phosphotrans-

ferases is the heterotrimeric dihydroxyacetone (Dha) kinase (YcgC, YcgT, YcgS) of
E. coli. What originally was characterized as a Dha-specific enzyme II [16] turned
out to be a "kinase" (YcgT,YcgS), which utilizes the multiphotransfer protein
YcgC instead of ATP as phosphoryl donor [17].

5.2.3
Active Sites

The EI, HPr, and the IIA domains are phosphorylated at a histidine, the
IIB domains at a cysteine [18, 19]. Exceptions are the B-domains of the mannose
family which are also phosphorylated at a histidine [20]. Phosphotransfer occurs
with inversion of the configuration at phosphorus [21], alternating between $N\varepsilon 2$
(EI and IIA) and $N\delta 1$ (HPr and IIB) of the imidazole rings. Interacting protein surfaces are complementary in shape and charge, concave and Asp/Glu-rich in EI and
IIA, convex and Arg-containing in HPr and IIB (reviewed in [22]).

Phospho-His and phospho-Cys have comparable phosphorylation potentials.
Whereas the overall drop of free energy from PEP to Glc-6P is 11 kcal mol^{-1},
the free energy between phospho-HPr, phospho-IIA, and phospho-IIB is small
(Fig. 1C). The flux of phosphate can be directed to and away from different transporters depending on demand (Fig. 2) [23]. Reversibility of phosphotransfer in the
PTS is important for the "propagation" and "distribution" of the regulatory signals
through the information transfer system ([24] and Sects. 4 and 5).

5.3
Structure and Function of the PTS Transporter for Glucose

E. coli utilizes two PTS transporters for uptake of Glc. The first consists of two subunits, IIA^{Glc} and IICB^{Glc}, which will be described in more detail in the remaining
sections. The second transporter consists of three subunits, IIAB^{Man}, IIC^{Man}, and
IID^{Man} [25, 26]. It has a broad specificity for Glc and derivatives modified at
carbon 2, and it is the only transporter for mannose (Man). It has no known regulatory function in E. coli but facilitates penetration of bacteriophage λ DNA across
the inner membrane [27, 28]. Finally, Glc is also taken up by the constitutively expressed galactose transporter but must then be phosphorylated by glucokinase.

Of all PTS transporters the mannitol transporter was the first to be purified [29]
and is by now the best characterized, mainly due to the work of Robillard and Jacobson (reviewed in [10]). The PTS transporters of E. coli for mannitol (Mtl), Glc,
Man, and GlcNAc can be purified in milligram amounts by metal chelate affinity
chromatography and they have been reconstituted into proteoliposomes [30–33].
Although dissimilar in structure, they share many similarities of function.

5.3.1
The Genes crr (IIA^{Glc}) and ptsG (IICB^{Glc})

The genes for IIA^{Glc} and IICB^{Glc} are located in two different regions (54.6 and 25.0 min) on the E. coli chromosome. IIA^{Glc} is encoded by crr (because inactivation of crr renders a cell catabolite repression resistant). Crr is the third gene of a tricistronic operon which also encodes HPr (ptsH) and EI (ptsI) of the PTS. The gene ptsG for IICB^{Glc} is in a monocistronic operon. The association of crr with ptsHI probably reflects the role played by IIA^{Glc} as central sensor of PTS activity. For comparison, the subunits of the other PTS transporters are usually encoded in a single operon (e.g., manXYZ, fruBKA, srlAEBD) and often associated with the genes for enzymes involved in the downstream metabolism of the transported substrate (e.g., kinase or dehydrogenase). Transcription of ptsG and ptsHIcrr is autoregulated. It is activated by the cAMP:CRP complex (cAMP receptor protein) and is repressed by the transcription repressor Mlc. The concentration of cAMP:CRP is itself controlled by the concentration of phospho-IIA^{Glc}, the activity of Mlc by the concentration of dephospho-IICB^{Glc} (Fig. 5A and Sect. 5.4).

5.3.2
The IIA^{Glc} Subunit

IIA^{Glc} (18.1 kDa) is a soluble β-sandwich protein consisting of two six-stranded antiparallel sheets and three short helical segments (Fig. 1A). It is phosphorylated at Nε2 of His90. The nearby invariant His75 stabilizes the phosphate in the transition state by a hydrogen bond from NHε2. The H75Q mutation decreases the phosphotransferase rate 200-fold [34]. IIA^{Glc} occurs in two forms, full-length and truncated at the N-terminus after Lys7. The truncated form is efficiently phosphorylated by HPr, but passes on the phosphate to IICB^{Glc} with only 2% of wild-type activity [35]. The N-terminus of IIA^{Glc} which, as judged by X-ray and NMR, [36, 37] is disordered in solution, assumes an amphipathic helical conformation when in contact with anionic membrane lipids [38]. It is proposed that the helical conformation of the N-terminus acts as a membrane anchor which facilitates the formation of a stable complex between IIA^{Glc} and membrane-bound IICB^{Glc}. Proteolytic processing of the N-terminus by a membrane associated protease [39] could change the distribution of IIA^{Glc} between the membrane and the cytoplasmic compartment and as a consequence modulate its function between transport and allosteric regulation (Fig. 5A and Sects. 5.4 and 5.5). The physiological relevance of this posttranslational modification remains to be demonstrated.

A number of PTS proteins contain IIA^{Glc}-like domains fused to HPr and/or IIB^{Glc}-like domains (Fig. 2). The transporter for GlcNAc (IICBA^{GlcNAc}) of E. coli has the IIA domain fused to the C-terminus of IICB. The GlcNAc transporter of P. aeruginosa has the IIA domain fused to the N-terminus of a soluble IIA-HPr-EI multiphosphotransfer protein [14]. It is likely that integration of IIA in a multidomain protein prioritizes its transport function and restricts its regulatory functions. Indeed, whereas IICBA^{GlcNAc} can functionally complement transport by

IICB^{Glc} in a IIA^{Glc} mutant [40, 41], the IIA domain of IICBA^{GlcNAc} must be expressed as a soluble protein in order to display and take over the regulatory functions of IIA^{Glc} [42].

As indicated (Fig. 5A), IIA^{Glc} must recognize a number of proteins of different structure. The complex of IIA^{Glc} with HPr was characterized by intermolecular nuclear Overhauser enhancement (NOE) [22], the complex with IIB^{Glc} by chemical shift mapping [43]. The IIA^{Glc}:glycerol kinase (GK) complex was analyzed by X-ray crystallography [44]. The backbone conformation of IIA^{Glc} does not change in the complexes. X-ray and NOE show changes of side chain conformation at the binding interfaces. The interfaces of IIA^{Glc} with HPr and IIB^{Glc} obviously surround the active sites histidines. The interface with GK contains a zinc binding site with Zn^{2+} coordinated by the active site His90 and His75 of IIA^{Glc}, water, and Glu478 of GK. In contrast to the two His residues of IIA^{Glc}, Glu478 is far from the active site of GK. This fits with the observation that GK can bind to IIA^{Glc} only in the presence of its substrate [45] (Sect. 5.4.1). Presence of zinc enhances inhibition of GK by IIA^{Glc} and replacement of Glu478 abolishes this enhancement. Phosphorylation at His90 prevents the cation-enhanced protein–protein regulatory interaction [46, 47].

5.3.3
The IICB^{Glc} Subunit

The IICB^{Glc} subunit (50.7 kDa) consists of two domains, the membrane-bound C- (41.1 kDa) and the cytoplasmic B-domain (9.6 kDa) (Fig. 3). The C-domain forms a stable homodimer [48–51], IIB is soluble [52]. The C-domain contains the sugar binding site, as concluded from the specificity of a chimeric protein between the IIB domain of the glucose transporter and the IIC domain of the homologous transporters for GlcNAc [53], and from the location of the point mutations conferring relaxed substrate specificity (see Sect. 5.3.3.4 and Tab. 1). The B-domain contains the phosphorylation site [19, 54]. IICB^{Glc} has good affinity for 1-thio-Glc, 5-thio-Glc, β- and α-methylglucoside (aMG), β-octylglucoside and low affinity for 6-F-Glc, 2-deoxy-Glc, and Man. Whether a low affinity sugar is classified as a "substrate" depends on how strongly the transporter is expressed and what substrate concentrations are considered physiological (see Sect. 5.3.3.4).

There exist over 30 sequences of orthologous and paralogous proteins with maximum affinity for Glc or GlcNAc, but not for both. Overall sequence similarities are stronger in the second half of the sequence (from cytoplasmic loop 2 to the end) than across the transmembrane segments 1 to 4 (Fig. 3). Two sequence motives of maximum similarity between the PTS transporters of the glucose family (domain order C–B) are LK**TPGRE**D in the linker region between IIC and IIB, and DA**CI**TRLR containing the active site Cys421.

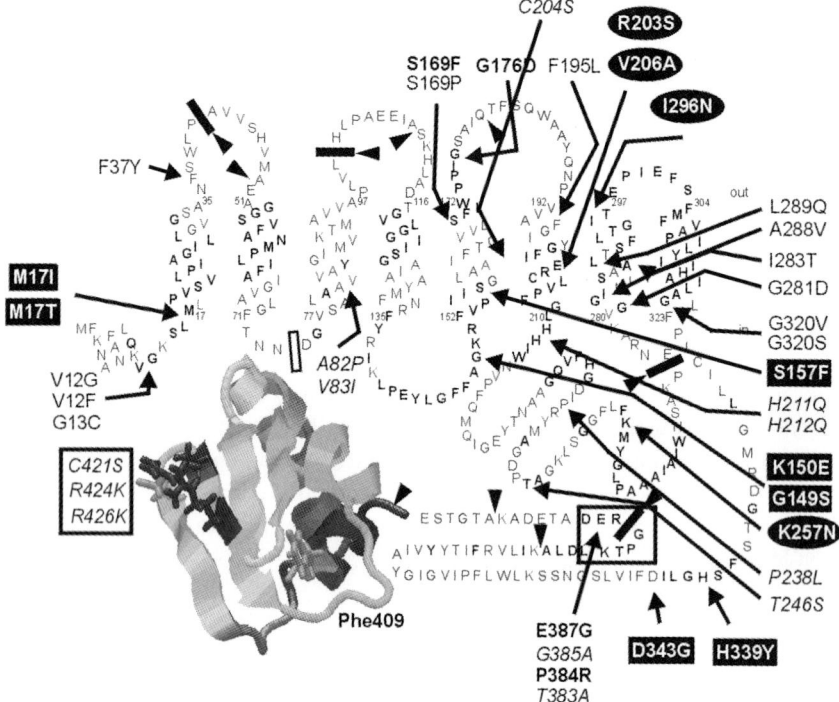

Fig. 3. Topology of IICB^{Glc} and structure of the IIB domain. Known mutations are labeled according to their phenotype as follows: *normal*, relaxed substrate specificity; **bold**, relaxed substrate specificity and alleviated repression of Mlc-regulated genes (sequestration of Mlc by IICB^{Glc}); *italics*, mutations which reduce or abolish IICB^{Glc} activity; boxed, phosphorylation normal, transport low; ovalled, uncoupled transport; solid wedges, sites of tolerated linker insertions and deletions; solid bars, site of new N- and C-termini in circularly permuted IICB^{Glc}; open bar, circularly permuted IICB^{Glc} with phosphorylation but without transport activity. Phe409, phosphorylation-sensitive chymotryptic cleavage site (Fig. 4). Indicated in **bold** are regions of strong amino acid similarity (according to Fig. 1 in [51]). Shown in frames are the conserved residues in the linker region between IIC and IIB and the phosphorylation site in IIB (PDB code 1IBA). See Tab. 1 and text for references.

5.3.3.1 Structure and Function of the IIC Domain

The C-domain consists of eight putative membrane spanning helices (TM1-8) with a large cytoplasmic domain between TM6 and 7. This topology was first derived from the activity of fusion proteins between IICB^{Glc} progressively truncated from the C-terminus and β-galactosidase or alkaline phosphatase [55]. It gained further support from random linker insertion mutagenesis. Twelve mutants with between 50% and 100% activity were found with inserts in the periplasmic loops 1 to 3, at the end of the cytoplasmic domain of IIC (loop 3), and in the linker between the IIC and IIB domains (Fig. 3). Split and circularly permuted variants of IICB^{Glc} with new N- and C-termini in the three periplasmic loops and in two cytoplasmic loops also display between 4% and 80% of wild-type activity. Only the two putative loops between TM3 and 4 and TM7 and 8, both in highly conserved regions, were refractive to any sort of manipulation [56, 57]. Whereas 8 TM are predicted for IICB^{Glc}, 6 are predicted for the mannitol transporter [58], 6 for IIC^{Man} [59], and at least 1 but more likely 3 for the IID^{Man} subunit of the mannose transporter [60]. Projection maps of 2-D crystals of IIC^{Mtl} [61] and of IIC^{Glc} [62] confirm the dimeric structure of these transporters. The 5 Å map of IIC^{Mtl} shows six domains of high electron density which most likely correspond to transmembrane α-helices. No X-ray structures of membrane spanning IIC domains have been published until January 2001. That PTS transporters of different classes have different topologies is not surprising in view of the great diversity of folds of their respective IIA and IIB domains [10].

5.3.3.2 Structure and Function of the IIB Domain

The IIB domain is a split α/β sandwich consisting of a four-stranded antiparallel β-sheet and three helices packed against one side of the sheet (Fig. 3) [63]. The active site Cys421 is in the loop between strands β1 and β2. This exposed position can result in disulfide cross-linking of the IICB^{Glc} subunits by oxidation [49]. Attachment is flexible and functional complementation between IIB and IIC domains on different subunits in a IICB^{Glc} dimer, or between different dimers is possible [51]. It is, however, not clear, whether and how fast subunits exchange between IICB^{Glc} dimers (for more about intersubunit complementation in IICBA^{Mtl} see [10]).

5.3.3.3 Structure and Function of the Linker Region

The linker region connecting the IIC with the IIB domain is surface exposed as judged from its susceptibility to proteolysis. Chymotrypsin and proteinase K cleave IICB^{Glc} after Phe409 and Asp392, respectively. Phosphorylation of IICB^{Glc} protects against protease cleavage, suggesting a conformational change of this region (Fig. 4). Phe409 is in helix 1, Asp392 in a region which is unstructured in the isolated IIB domain (Fig. 3) [63]. The linker, although invariant, is clearly not essential for transport for the following reasons. (1) The C- and B-domains can be split in the linker and expressed as individual protein subunits which, when recombined, retain *in vivo* transport and *in vitro* glucose phosphotransferase activity [52]. (2) A circularly permuted variant with the B-domain fused to the N-terminus of IIC has 70% of control activity [31]. (3) The structurally related transporters for β-glucoside and sucrose with the domain order IIB–IIC–IIA (Fig. 2) do not have this linker. (4) Mutations or a deletion of residues 387-394 in the linker can be more deleterious to transport and phosphorylation activity than its absence in split and circularly permuted forms [51]. (5) The linker region is the only crossover point for functional chimeric proteins between IICB^{Glc} and IICBA^{GlcNAc} (formed by *in vivo* circularization of a linearized plasmid containing *nagE* and *ptsG* at its ends). Other chimeras with fusion joints at Ser45, Ala68, Pro154 had only spurious activity ([53] and unpublished data). What function has the linker region? As described in

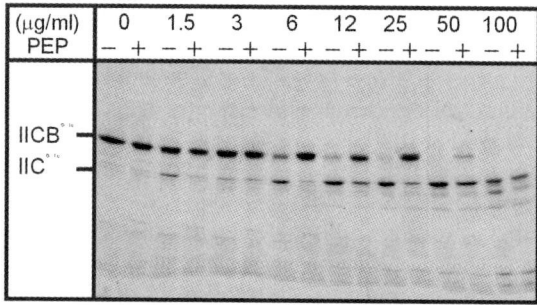

Fig. 4. Effect of phosphorylation on cleavage of IICB^{Glc} by chymotrypsin. Purified IICB^{Glc} was incubated with and without PEP in the presence of catalytic concentrations of EI, HPr, and IIA^{Glc}. After 10 min aliquots containing 0.5 mg mL^{-1} IICB^{Glc} were digested with the indicated concentrations of chymotrypsin at 25° for 30 min. The reactions were stopped with protease inhibitors and the products analyzed by polyacrylamide gel electrophoresis in the presence of sodium dodecylsulfate. Shown is a Coomassie blue stained gel.

Sect. 5.4, the linker may be involved in binding the transcriptional repressor Mlc in response to the phosphorylation state of IICB^{Glc} [64, 65].

5.3.3.4 Mutants of IICB^{Glc}

The activity of IICB^{Glc} conceptually can be dissected into (1) recognition, (2) transport and (3) phosphorylation of Glc, (4) phosphorylation of IICB^{Glc} by IIA^{Glc}, and (5) regulation of gene expression in response to the availability of Glc. Mutants exist which predominantly affect one or several of these functions (Fig. 3 and Tab. 1).

Recognition. Two classes of substrate specificity mutants were found. One displays low affinity for additional substrates (GlcN, Man) which, like Glc, are phosphorylated concomitantly with their uptake [65–67]. The second allows facilitated diffusion of substrates (ribose, Fru, Mtl) which, however, are not phosphorylated during transport and require an ATP-dependent kinase for further metabolism [68–70]. Mutations with relaxed substrate specificity are scattered over the entire IIC domain (Fig. 3) but are not found in the IIB domain. The mutated IICB^{Glc} have increased K_m and decreased V_{max} for the non-cognate substrates, whereas the catalytic constants for Glc remain close to normal. Because they were characterized *in vivo* by growth and uptake, it is not always clear whether the relaxed specificity is caused by the mutation, by upregulated expression of IICB^{Glc}, or by IICB^{Glc}-controlled expression of another transport system (see Sect. 5.4). The last mechanism could explain why competition for uptake between Glc and "new substrate" is usually weak. Perhaps not surprising for single residue changes, no mutants with altered specificity, that is increased affinity for a new substrate and decreased affinity for Glc, have been described.

Transport and phosphorylation. Four IICB^{Glc} mutants able to transport Glc by facilitated diffusion in the absence of EI have been isolated [68]. The mutants have a

Tab. 1. IICB^{Glc} mutants and their phenotypes. Mutations which occurred repeatedly in different laboratories and mutants which were tested in combinations are indicated bold

Mutation	Phenotype/Characteristic Properties		Comments	Reference
Specificity mutants	substrate specificity	Mlc-directed regulation		
V12F, V12G, G13C, **G176D**, P384R	growth on Glc, Man, GlcN; V_{max} of αMG uptake 3–10 x increased	constitutive transcription of ptsG	only 60% inhibition of αMG uptake by 10^4-fold excess of GlcN	67
A288V, **G320S**	growth on Glc, Man, GlcN; V_{max} of αMG uptake 3–10 x increased	not tested	only 60% inhibition of αMG uptake by 10^4-fold excess of GlcN	67
G320V	growth on Mtl, V_{max} in vitro 3 x increased	not tested		66
S169F, S169P	growth on Glc, GlcN, Man; V_{max} of αMG uptake 1–3 x increased	transcription of ptsG 4–5 x inducible, also by GlcN, Man	S169P has normal uptake (1 x)	65
E387G (umgC [104])	growth on Glc, GlcN, Man; V_{max} of αMG uptake 10 x increased	transcription of ptsG 7 x and constitutive	also identified as G176D/E472K [105]	65
F195L	growth on Glc, Man	transcription of ptsG 2 x and inducible by Man		65
F37Y, **G176D**, G281D, I283T, L289Q	growth on Rib; V_{max} of Rib uptake 5–12 x increased; ribokinase and RbsD are essential for growth and uptake, no PTS-dependent phosphorylation of Rib	experiments done in mlc background (constitutive high expression of ptsG)	80-95% inhibition of Rib uptake by 10^2 x excess of Glc; Glc uptake not affected (> 50%)	69
V12F	growth on Fru by facilitated diffusion; requires simultaneous overexpression of fructo(manno)-kinase (Mak)	not tested	possibly increased protein expression/stability; increased sensitivity to dGlc and αMG	70

Tab. 1. continued

Mutation	Phenotype/Characteristic Properties			Comments	Reference
Uncoupling of transport and phosphorylation	*phenotype*	*catalytic constants*			
R203S, V206A, K275N, I296N	growth on Glc by facilitated diffusion; requires active glucokinase	K_m of Glc oxidation 0.6–2.6 mM (wild type < 20 µM); V_{max} of uptake 50% of wt. low affinity for Glc. PT-activity *in vitro* 30%–100%		"conditionally uncoupled" K275N has no PT-activity because protein is unstable	68
M17T, M17I, G149S, K150E, S157F, H339Y, D343G	resistant to *a*MG but metabolize maltose; V_{max} of *a*MG uptake < 5%, of Glc uptake 15–40%	PT-activity *in vitro* V_{max} with *a*MG 20-60%, with Glc 45–140%		probably increased discrimination between Glc and *a*MG	71
Site-directed mutants	*protein phosphorylation*	*Glc phosphorylation*	*uptake*		
C204S	+ (unstable protein)	+	+++		54
C204I	+++	+++	+++		
C326S	+++	+++			
C421S	no	no	no (phosphorylation site)		
R424K, R426K	+++	No	no		54
H211Q, **H212Q**	+++	100%/15%		<10%/0, complementation between H212Q and **R424K** 10% of wt.	51
K382A, P384G, R386A, E387A, D388A	not tested	40%–120%		linker mutation	51
T383A, G385A	not tested	< 10%		linker mutation	51

Tab. 1. continued

Mutation	Phenotype/Characteristic Properties		Comments	Reference
	PTS activity	Mlc-directed regulation	comments	
Deletion mutants				
Δ218–232, Δ218–244, Δ218–254, Δ233–344, Δ233–254, Δ245–254	inactive, not negative dominant		deletions in cytoplasmic loop 3	unpubl.
Δ218–275, Δ245–275, Δ245–275, Δ255–275	inactive, no protein, not negative dominant		deletions in cytoplasmic loop 3	unpubl.
Δ387–477 (IIC)	2 % in presence of excess IIB (Δ1–390)	no upregulation of *ptsG* transcription, also not in the presence of IIB (Δ1–390)	no binding of Mlc	52, 64
Δ330–477 (TM1–8)	unstable protein	not tested		unpubl.
Δ321–477, Δ397–477, Δ460–477	growth arrest, unstable proteins	< 50 % upregulation of *ptsG* transcription		65
Δ80–320 (TM1,2, IIB) Δ1–320 (+IIB)		10 × induction of *ptsG* transcription (80 % of wt.)	binding of Mlc	64
Δ1–390 (IIB)	active, efficient complementation of Δ387–477 and C421S	no induction of *ptsG* transcription	used for NMR structure, no binding of Mlc observed	52
Δ387–394	stable, but only 2 % activity	not tested	linker deletion	unpubl.

Tab. 1. continued

Mutation	Phenotype/Characteristic Properties		Comments	Reference
	PTS activity	properties	comments	
Chimeric proteins				
R386/E370	Glc-specific 100 % active		IIC^{Glc}/IICBA^{GlcNAc}	53
S45/D34, V67/A61, P154/ I135, S204/ I183, I237/N217, S326/A306	inactive inactive inactive			unpubl.
IICB-IIA-HPr-EI	3-4 x more active than control	phosphate channeling	fusion protein (Fig. 2)	116

100 times higher K_m in the uncoupled mode, which is of a similar order as the K_m of natural facilitators. Uncoupling is conditional in so far as all mutants retain nearly normal activity when coupling between transport and phosphorylation of Glc is possible. It has not been tested, whether these mutants also have relaxed substrate specificity.

IICB^{Glc} mutants with the inverse properties, phosphorylation without transport, were not found. Mutants which approximate this phenotype have differentially inhibited transport and phosphorylation activity, less than 5% uptake, and between 20% and 70% phosphorylation of aMG. Uptake of aMG is 4–10 times more strongly inhibited than uptake of Glc. This increased selectivity for Glc and against aMG comes at the price of an only moderate (2- to 7-fold) reduction of V_{max} for Glc [71].

The C421S substitution in the active site is the only point mutation known to abolish phosphorylation of IICB^{Glc} by IIA^{Glc} [54]. Arg424 and Arg426 are essential for Glc phosphorylation and transport, but not for the phosphorylation of IICB^{Glc} by IIA^{Glc}. The H112Q mutation blocks transport but not phosphorylation. In H111Q both activities are moderately reduced. Noteworthy are the properties of mutations in the linker between IIC and IIB. Gly385A and the deletion of residues 387-394 reduce phosphorylation activity to below 1% and strongly compromise glucose fermentation which is contingent on glucose uptake. Alanine mutations of the other residues in the linker have either no effect or reduce PTS activity to between 10% and 60% ([51] and unpublished data). Short deletions in the large cytoplasmic loop between residues 218 and 275 afford inactive IICB^{Glc} or no protein at all (Tab. 1). This loop is of variable length between IICB homologs of different species and contains a 150-residue insertion in the IICB of *M. gentialium* (SWISS-PROT accession P47315). Uptake and phosphorylation of Glc are more sensitive to mutations than protein phosphorylation, because the former require many turnovers for detection, the latter only one. Moreover, given enough time even phage-displayed peptides containing either a Cys or a His can be phosphorylated by EI of the PTS, and there is no reason why IIA^{Glc} should not have similar capabilities [72].

5.4
Regulation by the PTS

Signaling by the PTS is based on the ratios of phosphorylated to non-phosphorylated proteins, signal propagation within the PTS on phosphotransfer between the components of the PTS. Generally, the steady state concentrations of the dephosphorylated form of a PTS protein will increase and its phosphorylated form decrease whenever a PTS substrate drains phosphoryl groups away and rephosphorylation is slowed down because of limitations in PEP, EI, or HPr (due to competition, gene expression, or mutations). Signal propagation in all directions of the network is possible, because the free energy difference between the phosphoproteins is rather small. The rate of phosphotransfer along the PTS, however, is kine-

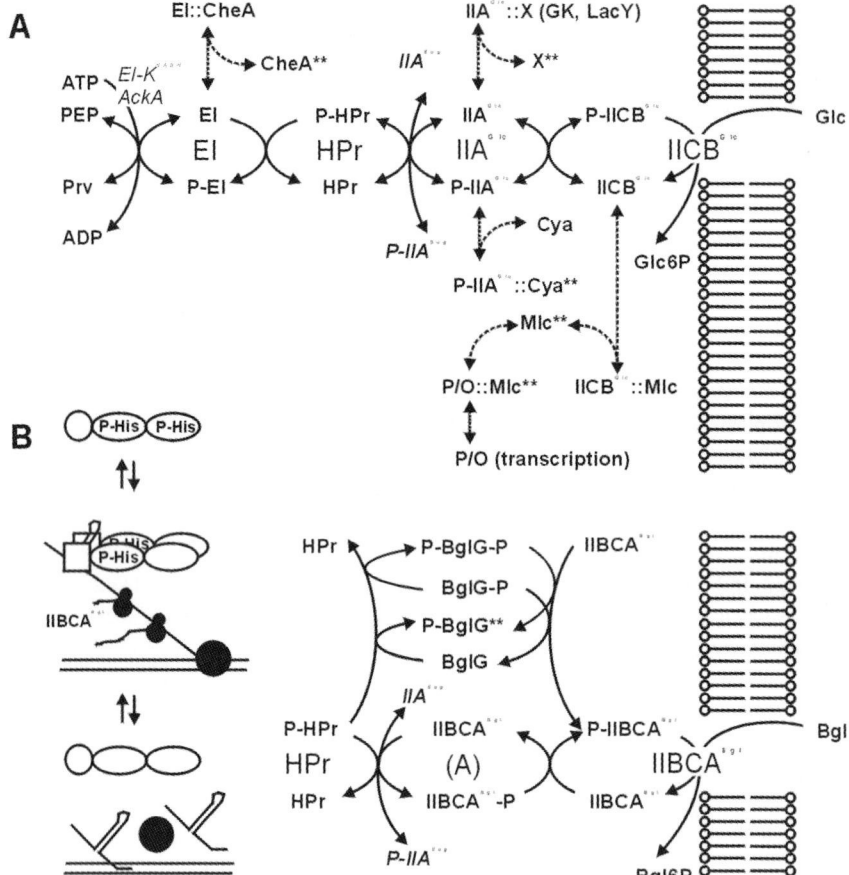

Fig. 5. Regulation of the PTS for Glc and β-glucosides. **(A)** Solid arrows indicate phosphotransfer between PTS proteins and between PTS proteins and metabolites. Dashed arrows indicate protein–protein interactions between PTS proteins and their targets of allosteric regulation. The (more) active form of these targets is labeled with two stars. EI-kinase (EI-K) and acetate kinase (AckA, P15046) can feed phosphate from ATP into the phosphotransfer chain. IIA^{Sug} are IIA subunits of other transporters which can compete with IIA^{Glc} for P-HPr. CheA, sensor kinase of the bacterial chemotaxis system (P07363). X, target enzymes (glycerol kinase, P08859; lactose permease, P02920), which are allosterically inhibited by IIA^{Glc}. Cya, adenylyl cyclase (P00936) is activated by phospho-IIA^{Glc}. Mlc (P50456), global transcription repressor of PTS operons (P/O), which is sequestered by binding to dephospho-$IICB^{Glc}$. **(B)** Regulation of BglG antiterminator protein (P11989) by multiple phosphorylation. The arrowheads indicate the flow of phosphoryl groups under the conditions of β-glucoside ("inducer") uptake by an active PTS. BglG is activated by dephosphorylation via $IIBCA^{Bgl}$ (P08722) and phosphorylation by P-HPr. The scheme on the left indicates that unphosphorylated and doubly phosphorylated BglG are less active than the singly phosphorylated form which dimerizes and by binding to the nascent mRNA allows continuation of transcription (adapted from [75, 117]).

tically controlled, and due to temporal delays the distribution of phosphoryl groups may never be in equilibrium.

The PTS regulates non-PTS systems and is itself regulated by non-PTS systems as well as by autoregulation. Regulation can be immediate involving allosteric interactions between PTS proteins and target enzymes and delayed via repression or induction of gene expression. In *E. coli*, the IIA^{Glc} and $IICB^{Glc}$ subunits are two major components mediating gene regulation. The dephosphorylated form of IIA^{Glc} inhibits expression of non-PTS genes, and dephosphorylated $IICB^{Glc}$ stimulates the expression of PTS genes. These activities will be described below, but three facts should be noticed first.

(1) IIA^{Glc} plays a central role in gram-negative bacteria only, whereas HPr has the analogous function in gram-positive bacteria (reviewed in [6]). The activity of HPr in gram-positive bacteria is modulated by an HPr-specific serine kinase and phosphatase [73]. HPr phosphorylated at Ser-46 has reduced phosphotransferase activity and acts as corepressor of the repressor protein CcpA [74]. (2) Several PTS transporters in addition to $IICB^{Glc}$ regulate gene expression by multiple phosphorylation and dephosphorylation of transcription antiterminators and repressor proteins [75]. (3) The discovery of PTS-related sequences in bacterial genomes devoid of genes for PTS transporters suggests additional as yet unknown regulatory functions [76]. For instance, it already has been shown for *E. coli* that EI plays a role in chemotaxis (Fig. 5A). Unphosphorylated EI inhibits the sensorkinase CheA [77–80]. Increasing concentrations of a PTS substrate, therefore, increase the ratio of smooth swimming to tumbling of a flagellated cell. The EI-like protein EI^{Ntr} (*ptsP*) is involved in linking carbon and nitrogen assimilation but not in sugar transport [76, 81], and was identified as a virulence factor in the *Caenorhabditis elegans–Pseudomonas aeruginosa* slow killing model [82].

5.4.1
Regulatory Role of IIA^{Glc}

The steady state concentrations of IIA^{Glc} increases and of P-IIA^{Glc} decreases (1) in the presence of Glc and (2) in the presence of other PTS substrates, which by competing for P-HPr direct the phosphoryl flow away from IIA^{Glc} [23]. Dephosphorylated IIA^{Glc} is an allosteric inhibitor of glycerol kinase (GK) [44, 47] and of non-PTS transporters for lactose (LacY), maltose (via binding to the MalK subunit of the maltose transport complex, see Chap. 5), and melibiose [42, 83, 84]. Dephosphorylated IIA^{Glc} binds to the target proteins only in the presence of their substrates, thus minimizing the sequestration of IIA^{Glc} by idling targets [45]. Inhibition by IIA^{Glc} always has the same effect: it prevents uptake and formation of metabolites which are inducers of gene expression. This mechanism is termed *inducer exclusion*. As long as Glc or other PTS substrates are available, the transcription of genes for alternative catabolic pathways is repressed. In addition, dephosphorylated IIA^{Glc} does *not* stimulate adenylyl cyclase, and the concentration of cAMP and of the cAMP binding protein (CRP; catabolite repression protein) remain low. Cyclic

AMP:CRP complex is, however, necessary for the expression of many catabolic genes and also of PTS genes as well as of CRP itself [85].

The concentration of phospho-IIA^{Glc} starts to increase only after the PTS substrates are consumed. Not only is inducer exclusion relieved but phospho-IIA^{Glc} now activates adenylyl cyclase [86–88] and the cAMP concentration increases. Genes and operons of which the transcription depends on the activation by the cAMP:CRP complex are transcribed, provided the specific inducer is also present (dual control by cAMP *and* inducer). Genes activated by the cAMP:CRP complex are termed catabolite repressed genes. Their promoter/operator sequences have different affinities for the cAMP:CRP complex, which allows for sequential and graded expression of these operons.

The importance of IIA^{Glc} for inducer exclusion is perhaps best demonstrated by the occurrences of a IIA-like domain in a non-PTS protein. In the lactose permease of *Streptococcus thermophilus* a IIA^{Glc}-like regulatory domain is fused to the C-terminus of the transporter domain. A protein subunit which acts as an allosteric inhibitor in *E. coli* has become an integral part of the lactose permease of *S. thermophilus* [89, 90].

5.4.2
Regulatory Role of IICB^{Glc}

IICB^{Glc} and other PTS transporters of the glucose and sucrose family can physically interact with transcriptional repressor, activator and antiterminator proteins (Fig. 5). Repressors and activators bind to operator DNA and prevent/stimulate initiation of transcription. Antiterminator proteins recognize stem-loop structures in nascent mRNA and prevent premature termination of transcription (Fig. 5B). Repressor activity can be regulated by different mechanisms. For instance, NagC, a repressor of genes *nagE* and *manXYZ* (encoding IICBA^{GlcNAc} and IIAB^{Man}/IIC^{Man}/IID^{Man}), is *inactivated* by binding of the inducer GlcNAc-6P [91]. Mlc is *inactivated* by binding to the dephosphorylated form of IICB^{Glc} [92]. Transcription activators (e.g., LevR of *B. subtilis* [93, 94] and antiterminators (e.g., BlgG of *E. coli* [95; 96]) are under dual control by HPr and IICB. They are *inactivated* by IICB-mediated phosphorylation and HPr-mediated dephosphorylation at multiple His (Fig. 5B) (reviewed in [75]). Only regulation by IICB^{Glc} will be discussed in more detail below.

The role of IICB^{Glc} as global regulator of PTS activity was elucidated mainly in the two laboratories of Plumbridge and Aiba. The key component is Mlc (44.6 kDa), a protein belonging to the ROK family of transcriptional repressors. Mlc mutants (*dgsA*) were first identified as cells with increased mannose transport activity during anaerobic growth [97]. Independently it was also found that overexpression of Mlc allowed prolonged growth of cells on Glc (<u>m</u>aking <u>l</u>arge <u>c</u>olonies) [98]. Mlc was then shown to inhibit transcription not only from the promoters of *manXYZ* (for the mannose transporter of the PTS), *ptsHIcrr* (EI, HPr, IIA^{Glc}), but also from *ptsG* encoding IICB^{Glc} [99–101]. Repression of chromosomal genes by Mlc was alleviated, if Mlc was titrated away with promoter sequences on high copy number plasmids. These results together indicate that Mlc is a transcriptional repressor with a broad specificity for several PTS genes. Repression by Mlc was relieved when cells were grown on Glc and to a varying degree also on other PTS substrates [100]. In spite of this finding, no low molecular weight inducer such as Glc or a metabolite of a PTS substrate could be identified by gel shift [101] and in *in vitro* transcription assays in presence of purified Mlc [102]. The riddle was solved based on the following key experiments [65, 92]: (1) Repression of genes by Mlc was relieved only in the presence of IICB^{Glc}, independently of whether Glc, or other PTS substrates which were transported by a different PTS transporter, were used as the inducer. (2) Deletion of *ptsG* led to a complete Mlc-mediated repression of all Mlc-controlled genes. (3) Overexpression of IICB^{Glc} from a multi-copy plasmid alleviated repression even in the absence of Glc. (4) Deletion of *ptsHIcrr* completely alleviated Mlc-mediated repression and indeed had the same effect as a deletion of *mlc* itself. (5) Deletion of both *ptsHIcrr* and *ptsG* again led to complete repression of Mlc-dependent genes. Based on these facts, it was concluded that the dephosphorylated form of IICB^{Glc} binds Mlc, and that IICB^{Glc} competes with the operator sequences for a limiting amount of Mlc (Fig. 5A). Binding of Mlc to dephosphorylated IICB^{Glc}, but not to phospho-IICB^{Glc}, has been demonstrated *in vitro* [64, 103]. Using deletion mutants of IICB^{Glc}, the junction region spanning residues 320-390 of the IIC domain could be identified as the potential binding site for Mlc [64, 65].

Some of the IICB^{Glc} mutants which grew faster on poor substrates (relaxed substrate specificity, see Sect. 5.3.3.4 and Tab. 1) also showed a constitutively increased expression of IICB^{Glc} [65, 67]. These mutants may have increased affinity for Mlc rather than for poor substrates, and by binding Mlc induce their own expression. This increased expression alone could account for fast growth on poor substrates, because, as shown in Sect. 5.5, the rates of solute uptake and growth are controlled by the concentration of IICB^{Glc} in the membrane. A particularly good example is the umgC mutation (uptake of aMG), which enables *E. coli*, lacking a functional mannose transporter, to grow on glucosamine and Man [104]. It has been mapped to *ptsG* and by sequencing was identified as E387G [65] or G176D/E472K [105] variant of IICB^{Glc}. In the presence of IICB^{Glc} (UmgC) Mlc-repressed reporter genes are expressed as strongly as in an Mlc deletion mutant, as expected if Mlc is sequestered to the membrane.

5.5
Kinetic Properties of the Phosphorylation Cascade

The PTS of *E. coli* is a network of interacting phosphoproteins. Certain proteins, such as IIA^{Glc} and IICB^{Glc}, make multiple contacts with PTS- and non-PTS proteins. How does such a complex system respond to concentration changes of metabolites (e.g., Glc, PEP) and PTS proteins (e.g., after induction or repression of their synthesis)? How does it respond to changes of phosphotransfer rate constants and of protein complex dissociation constants (e.g., in response to allosteric regulators)? How does it respond to the sequestration of its components by other proteins (e.g., sequestration of IIA^{Glc} by GK, sequestration of IICB^{Glc} by Mlc [64])?

Postma and coworkers analyzed the steady state activity of the glucose PTS in response to concentration changes of proteins and metabolites. They measured uptake of Glc, Glc oxidation, growth rates, IIA^{Glc}-dependent inhibition of GK, PEP/pyruvate and phospho/dephospho ratios of IIA^{Glc}, each as a function of the intracellular PTS protein concentrations and/or the extracellular concentrations of PTS and non-PTS substrates [84, 106–109]. Reduction of EI, HPr, and IIA^{Glc} concentrations, one at a time, down to 25 % of wild-type level had no effect on uptake, oxidation, and growth rates. Only when the intracellular concentrations dropped below 25 % the rates started to decrease. In contrast, variation of IICB^{Glc} caused an immediate and proportional change in the uptake rate. Reduction of EI and HPr also did not affect regulation of GK activity by IIA^{Glc}. GK activity was sigmoidally related to a reduction of IIA^{Glc} and attenuated when IICB^{Glc} dropped below 25 %.

Application of metabolic control theory to these data revealed that the sum of the enzyme flux *response* coefficients (relative change of uptake rate (J) to relative change of enzyme concentration $(\partial J/J)/(\partial [C]/[C])$ was around 1.5 and not 2 as would have been expected of hit and run collisions between freely diffusing enzymes [110, 111]. This result suggests that some PTS components are subunits of long living complexes. In these complexes phosphotransfer is intramolecular and, therefore, concentration-independent, leading to the observed reduction of the sum of the flux control coefficients. There is in fact biochemical evidence for the formation of such complexes between components of the glucose PTS [48, 112].

The sum of the enzyme flux *control* coefficients at physiological enzyme concentrations is 0.75, also less than the theoretical value of 1. The contribution of IICB^{Glc} is 0.73, indicating that IICB^{Glc} alone controls the uptake of Glc and the flux of phosphate through the glucose PTS. Variations of concentration and/or substrate specificity of IICB^{Glc}, therefore, are expected to have an immediate effect on substrate transport and cell growth (see Sect. 5.4.2). The sum total of less than 1 further suggests that there are sources other than PEP and sinks other than Glc which feed and consume phosphoryl residues. And in fact there are such sources. EI can be phosphorylated by acetate kinase and also by a NADH-regulated EI-specific kinase (EI-K) [113, 114]. And phosphate can be drained away from IICB^{Glc} by other PTS transporters via HPr and "cross talk" between complementary domains (Fig. 2).

The results obtained *in vivo* were further substantiated by *in vitro* experiments. The rate of glucose phosphorylation in crude extracts increased almost quadratically with protein concentration. The combined flux control coefficient (for all PTS proteins in the extract together) was close to 2, as expected, if the multisubunit complexes dissociate in dilution. The effect of dilution could be reversed by the addition of polyethylene glycol which reduced the flux control coefficient to close to 1. Polyethylene glycol induces macromolecular crowding and thus mimics the high protein concentration in intact cells [115].

Finally, a kinetic model was calculated which included 20 rate constants (4 for each phosphotransfer reactions AP+B↔APB↔A+PB) and 4 metabolite concentrations (PEP, Prv, Glc, Glc-6P) as constant parameters, and the concentrations of the 4 PTS proteins as variables [24]. The simulation reproduced the experimental data with astonishing accuracy, with only one exception concerning IIA^{Glc}. The simulation predicted 1.6 % of total IIA^{Glc} to be in the dephosphorylated form, not enough to bind and inhibit GK. However, when the experimentally determined GK concentration and GK:IIA^{Glc} equilibrium binding constant were included in the model, 60 % of dephospho-IIA^{Glc} was drawn into the complex, which is already closer to the measured 95 % of dephospho-IIA^{Glc} [108]. The remaining difference between simulation and experiment could be resolved by shifting of the PEP/pyruvate ratio from the steady state value to the value which was measured in response to a sudden increase of the glucose concentration.

Acknowledgement

I would like to thank Karin Flükiger for expert assistance, and the Swiss National Science Foundation (Grant 31-45838.95), ARPIDA AG Münchenstein, and Ciba-Geigy Jubiläumsstiftung, Basel for generous financial support.

References

1. W. Kundig, S. Gosh, S. Roseman, *Proc. Natl. Acad. Sci. USA* **1964**, *52*, 1067-1074.
2. J. B. Hays, in: *Bacterial Transport*, (B. Rosen, Ed.), Marcel Dekker, New York, **1978**, p. 43.
3. H. Jeong, B. Tombor, R. Albert, Z. N. Oltvai, A. L. Barabasi, *Nature* **2000**, *407*, 651-654.
4. C. T. Walsh, T. E. Benson, D. H. Kim, W. J. Lees, *Chem. Biol.* **1996**, *3*, 83-91.
5. M. H. Saier, T. M. Ramseier, J. Reizer, in: *Escherichia coli and Salmonella: Cellular and Molecular Biology*, (F. C. Neidhardt et al., Eds.), ASM Press, Washington, DC, **1996**, p. 1325.
6. M. H. Saier, S. Chauvaux, J. Deutscher, J. Reizer, J.-J. Ye, *Trends Biochem. Sci.* **1995**, *20*, 267-271.
7. N. D. Meadow, D. K. Fox, S. Roseman, *Annu. Rev. Biochem.* **1990**, *59*, 497-542.
8. P. W. Postma, J. W. Lengeler, G. R. Jacobson, *Microbiol. Rev.* **1993**, *57*, 543-594.
9. P. W. Postma, J. W. Lengeler, G. R. Jacobson, in: *Escherichia coli and Salmonella: Cellular and Molecular Biology*, (F. C. Neidhardt et al., Eds.) ASM Press, Washington, DC, **1996**, p. 1149.
10. G. T. Robillard, J. Broos, *Biochim. Biophys. Acta* **1999**, *1422*, 73-104.
11. J. Reizer, S. Bachem, A. Reizer, M. Arnaud, M. H. Saier, J. Stülke, *Microbiology* **1999**, *145*, 3419-3429.
12. J. Reizer, A. Reizer, *Res. Microbiol.* **1996**, *147*, 458-471.
13. R. H. Geerse, F. Izzo, P. W. Postma, *Mol. Gen. Genet.* **1989**, *216*, 517-525.
14. J. Reizer, A. Reizer, M. J. Lagrou, K. R. Folger, K. Stover, M. H. Saier, *J. Mol. Microbiol. Biotechnol.* **1999**, *1*, 289-293.
15. L. F. Wu, J. M. Tomich, M. H. Saier, *J. Mol. Biol.* **1990**, *213*, 687-703.
16. R. Z. Jin, E. C. C. Lin, *J. Gen. Microbiol.* **1984**, *130*, 83-88.
17. R. Gutknecht, R. Beutler, L. Garcia Alles, U. Baumann, B. Erni, *EMBO J.* **2001**, *20*, 2481-2486.
18. H. H. Pas, G. T. Robillard, *Biochemistry* **1988**, *27*, 5835-5839.
19. M. Meins, P. Jeno, D. Muller, W. J. Richter, J. P. Rosenbusch, B. Erni, *J. Biol. Chem.* **1993**, *268*, 11604-11609.

20. B. Erni, B. Zanolari, P. Graff, H. P. Kocher, *J. Biol. Chem.* **1989**, *264*, 18733-18741.
21. G. S. Begley, D. E. Hansen, G. R. Jacobson, J. R. Knowles, *Biochemistry* **1982**, *21*, 5552-5556.
22. G. S. Wang, J. M. Louis, M. Sondej, Y. J. Seok, A. Peterkofsky, G. M. Clore, *EMBO J.* **2000**, *19*, 5635-5649.
23. B. J. Scholte, P. W. Postma, *Eur. J. Biochem.* **1981**, *114*, 51-58.
24. J. M. Rohwer, N. D. Meadow, S. Roseman, H. V. Westerhoff, P. W. Postma, *J. Biol. Chem.* **2000**, *275*, 34909-34921.
25. B. Erni, B. Zanolari, H. P. Kocher, *J. Biol. Chem.* **1987**, *262*, 5238-5247.
26. R. Gutknecht, K. Flükiger, R. Lanz, B. Erni, *J. Biol. Chem.* **1999**, *274*, 6091-6096.
27. J. Elliott, W. Arber, *Mol. Gen. Genet.* **1978**, *161*, 1-8.
28. M. Esquinas-Rychen, B. Erni, *J. Mol. Microbiol. Biotechnol.* **2001**, *3*, 361-370.
29. G. R. Jacobson, C. A. Lee, M. H. Saier, *J. Biol. Chem.* **1979**, *254*, 249-252.
30. M. G. Elferink, A. J. Driessen, G. T. Robillard, *J. Bacteriol.* **1990**, *172*, 7119-7125.
31. R. Gutknecht, M. Manni, Q. C. Mao, B. Erni, *J. Biol. Chem.* **1998**, *273*, 25745-25750.
32. Q. Mao, T. Schunk, K. Flükiger, B. Erni, *J. Biol. Chem.* **1995**, *270*, 5258-5265.
33. S. Mukhija, B. Erni, *J. Biol. Chem.* **1996**, *271*, 14819-14824.
34. N. D. Meadow, S. Roseman, *J. Biol. Chem.* **1996**, *271*, 33440-33445.
35. N. D. Meadow, S. Roseman, *J. Biol. Chem.* **1982**, *257*, 14526-14537.
36. M. D. Feese, L. Comolli, N. D. Meadow, S. Roseman, S. J. Remington, *Biochemistry* **1997**, *36*, 16087-16096.
37. J. G. Pelton, D. A. Torchia, N. D. Meadow, S. Roseman, *Biochemistry* **1992**, *31*, 5215-5224.
38. G. Wang, A. Peterkofsky, G. M. Clore, *J. Biol. Chem.* **2000**, *275*, 39811-39814.
39. N. D. Meadow, P. Coyle, A. Komoryia, C. B. Anfinsen, S. Roseman, *J. Biol. Chem.* **1986**, *261*, 13504-13509.
40. A. P. Vogler, C. P. Broekhuizen, A. Schuitema, J. W. Lengeler, P. W. Postma, *Mol. Microbiol.* **1988**, *2*, 719-726.
41. A. P. Vogler, J. W. Lengeler, *Mol. Gen. Genet.* **1988**, *213*, 175-178.
42. J. Van der Vlag, P. W. Postma, *Mol. Gen. Genet.* **1995**, *248*, 236-241.
43. G. Gemmecker, M. Eberstadt, A. Buhr, R. Lanz, S. G. Grdadolnik, H. Kessler, B. Erni, *Biochemistry* **1997**, *36*, 7408-7417.
44. J. H. Hurley, H. R. Faber, D. Worthylake, N. D. Meadow, S. Roseman, D. W. Pettigrew, S. J. Remington, *Science* **1993**, *259*, 673-677.
45. S. O. Nelson, P. W. Postma, *Eur. J. Biochem.* **1984**, *139*, 29-34.
46. M. Feese, D. W. Pettigrew, N. D. Meadow, S. Roseman, S. J. Remington, *Proc. Natl. Acad. Sci. USA* **1994**, 3544-3548.
47. D. W. Pettigrew, N. D. Meadow, S. Roseman, S. J. Remington, *Biochemistry* **1998**, *37*, 4875-4883.
48. B. Erni, *Biochemistry* **1986**, *25*, 305-312.
49. M. Meins, B. Zanolari, J. P. Rosenbusch, B. Erni, *J. Biol. Chem.* **1988**, *263*, 12986-12993.
50. U. Waeber, A. Buhr, T. Schunk, B. Erni, *FEBS Lett.* **1993**, *324*, 109-112.
51. R. Lanz, B. Erni, *J. Biol. Chem.* **1998**, *273*, 12239-12243.
52. A. Buhr, K. Flükiger, B. Erni, *J. Biol. Chem.* **1994**, *269*, 23437-23443.
53. U. Hummel, C. Nuoffer, B. Zanolari, B. Erni, *Protein Sci.* **1992**, *1*, 356-362.
54. C. Nuoffer, B. Zanolari, B. Erni, *J. Biol. Chem.* **1988**, *263*, 6647-6655.
55. A. Buhr, B. Erni, *J. Biol. Chem.* **1993**, *268*, 11599-11603.
56. R. Beutler, M. Kaufmann, F. Ruggiero, B. Erni, *Biochemistry* **2000**, *39*, 3745-3750.
57. R. Beutler, F. Ruggiero, B. Erni, *Proc. Natl. Acad. Sci. USA* **2000**, *97*, 1477-1482.
58. J. E. Sugiyama, S. Mahmodian, G. R. Jacobson, *Proc. Natl. Acad. Sci. USA* **1991**, *88*, 9603-9607.
59. F. Huber, B. Erni, *Eur. J. Biochem.* **1996**, *239*, 810-817.
60. M. Kaufmann, *Thesis,* University of Bern, Switzerland, **1997**, 31-40
61. R. I. Koning, W. Keegstra, G. T. Oostergetel, G. Schuurman-Wolters, G. T. Robillard, A. Brisson, *J. Mol. Biol.* **1999**, *287*, 845-851.
62. J. P. Zhuang, R. Gutknecht, K. Flükiger, L. Hasler, B. Erni, A. Engel, *Arch. Biochem. Biophys.* **1999**, *372*, 89-96.
63. M. Eberstadt, S. G. Grdadolnik, G. Gemmecker, H. Kessler, A. Buhr, B. Erni, *Biochemistry* **1996**, *35*, 11286-11292.
64. S. J. Lee, W. Boos, J. P. Bouché, J. Plumbridge, *EMBO J.* **2000**, *19*, 5353-5361.
65. T. Zeppenfeld, C. Larisch, J. W. Lengeler, K. Jahreis, *J. Bacteriol.* **2000**, *182*, 4443-4452.

66. G. S. Begley, K. A. Warner, J. C. Arents, P. W. Postma, G. R. Jacobson, *J. Bacteriol.* **1996**, *178*, 940-942.
67. L. Notley-McRobb, T. Ferenci, *J. Bacteriol.* **2000**, *182*, 4437-4442.
68. G. J. G. Ruijter, G. Vanmeurs, M. A. Verwey, P. W. Postma, K. Vandam, *J. Bacteriol.* **1992**, *174*, 2843-2850.
69. H. Oh, Y. Park, C. Park, *J. Biol. Chem.* **1999**, *274*, 14006-14011.
70. H. L. Kornberg, L. T. Lambourne, A. A. Sproul, *Proc. Natl. Acad. Sci. USA* **2000**, *97*, 1808-1812.
71. A. Buhr, G. A. Daniels, B. Erni, *J. Biol. Chem.* **1992**, *267*, 3847-3851.
72. S. Mukhija, B. Erni, *Mol. Microbiol.* **1997**, *25*, 1159-1166.
73. A. Galinier, M. Kravanja, R. Engelmann, W. Hengstenberg, M. C. Kilhoffer, J. Deutscher, J. Haiech, *Proc. Natl. Acad. Sci. USA* **1998**, *95*, 1823-1828.
74. M. H. Saier, T. M. Ramseier, *J. Bacteriol.* **1996**, *178*, 3411-3417.
75. J. Stülke, M. Arnaud, G. Rapoport, I. Martin-Verstraete, *Mol. Microbiol.* **1998**, *28*, 865-874.
76. J. Reizer, A. Reizer, M. J. Merrick, G. Plunkett, D. J. Rose, M. H. Saier, *Gene* **1996**, *181*, 103-108.
77. M. S. Johnson, E. H. Rowsell, B. L. Taylor, *FEBS Lett.* **1995**, *374*, 161-164.
78. R. Lux, K. Jahreis, K. Bettenbrock, J. S. Parkinson, J. W. Lengeler, *Proc. Natl. Acad. Sci. USA* **1995**, *92*, 11583-11587.
79. R. Lux, V. R. Munasinghe, F. Castellano, J. W. Lengeler, J. E. Corrie, S. Khan, *Mol. Biol. Cell* **1999**, *10*, 1133-1146.
80. E. H. Rowsell, J. M. Smith, A. Wolfe, B. L. Taylor, *J. Bacteriol.* **1995**, *177*, 6011-6014.
81. R. Rabus, J. Reizer, I. Paulsen, M. H. Saier, *J. Biol. Chem.* **1999**, *274*, 26185-26191.
82. M. W. Tan, L. G. Rahme, J. A. Sternberg, R. G. Tompkins, F. M. Ausubel, *Proc. Natl. Acad. Sci. USA* **1999**, *96*, 2408-2413.
83. M. Kuroda, S. Dewaard, K. Mizushima, M. Tsuda, P. Postma, T. Tsuchiya, *J. Biol. Chem.* **1992**, *267*, 18336-18341.
84. B. M. Hogema, J. C. Arents, R. Bader, P. W. Postma, *Mol. Microbiol.* **1999**, *31*, 1825-1833.
85. H. Ishizuka, A. Hanamura, T. Inada, H. Aiba, *EMBO J.* **1994**, *13*, 3077-3082.
86. A. Peterkofsky, A. Reizer, J. Reizer, N. Gollop, P. P. Zhu, N. Amin, *Prog. Nucleic Acid Res. Mol. Biol.* **1994**, *44*, 31-65.
87. S. Levy, G. Q. Zeng, A. Danchin, *Gene* **1990**, *86*, 27-33.
88. P. Reddy, M. Kamireddi, *J. Bacteriol.* **1998**, *180*, 732-736.
89. B. Poolman, T. J. Royer, S. E. Mainzer, B. F. Schmidt, *J. Bacteriol.* **1989**, *171*, 244-253.
90. M. G. Gunnewijk, B. Poolman, *J. Biol. Chem.* **2000**, *275*, 34080-34085.
91. J. A. Plumbridge, *Mol. Microbiol.* **1991**, *5*, 2053-2062.
92. J. Plumbridge, *Mol. Microbiol.* **1999**, *33*, 260-273.
93. M. Arnaud, M. Débarbouillé, G. Rapoport, M. H. Saier, J. Reizer, *J. Biol. Chem.* **1996**, *271*, 18966-18972.
94. P. Tortosa, S. Aymerich, C. Lindner, M. H. Saier, J. Reizer, D. Le Coq, *J. Biol. Chem.* **1997**, *272*, 17230-17237.
95. O. Amster-Choder, F. Houman, A. Wright, *Cell* **1989**, *58*, 847-855.
96. K. Schnetz, B. Rak, *Proc. Natl. Acad. Sci. USA* **1990**, *87*, 5074-5078.
97. R. A. Roehl, R. T. Vinopal, *J. Bacteriol.* **1980**, *142*, 120-130.
98. K. Hosono, H. Kakuda, S. Ichihara, *Biosci. Biotechnol. Biochem.* **1995**, *59*, 256-261.
99. J. Plumbridge, *Mol. Microbiol.* **1998**, *27*, 369-380.
100. J. Plumbridge, *Mol. Microbiol.* **1998**, *29*, 1053-1063.
101. K. Kimata, T. Inada, H. Tagami, H. Aiba, *Mol. Microbiol.* **1998**, *29*, 1509-1519.
102. S. Y. Kim, T. W. Nam, D. Shin, B. M. Koo, Y. J. Seok, S. Ryu, *J. Biol. Chem.* **1999**, *274*, 25398-25402.
103. Y. Tanaka, K. Kimata, H. Aiba, *EMBO J.* **2000**, *19*, 5344-5352.
104. M. C. Jones-Mortimer, H. L. Kornberg, *J. Gen. Microbiol.* **1980**, *117*, 369-376.
105. J. Plumbridge, *Microbiology* **2000**, *146*, 2655-2663.
106. J. Van der Vlag, R. Van't Hof, K. Van Dam, P. W. Postma, *Eur. J. Biochem.* **1995**, *230*, 170-182.
107. B. M. Hogema, J. C. Arents, R. Bader, K. Eijkemans, T. Inada, H. Aiba, P. W. Postma, *Mol. Microbiol.* **1998**, *28*, 755-765.
108. B. M. Hogema, J. C. Arents, R. Bader, K. Eijkemans, H. Yoshida, H. Takahashi, H. Alba, P. W. Postma, *Mol. Microbiol.* **1998**, *30*, 487-498.
109. J. M. Rohwer, R. Bader, H. V. Westerhoff, P. W. Postma, *Mol. Microbiol.* **1998**, *29*, 641-652.

110. B. N. Kholodenko, J. M. Rohwer, M. Cascante, H. V. Westerhoff, *Mol. Cell. Biochem.* **1998**, *184*, 311-320.
111. B. N. Kholodenko, H. V. Westerhoff, *Trends Biochem. Sci.* **1995**, *20*, 52-54.
112. E. G. Jablonski, L. Brand, S. Roseman, *J. Biol. Chem.* **1983**, *258*, 9690-9699.
113. D. K. Fox, N. D. Meadow, S. Roseman, *J. Biol. Chem.* **1986**, *261*, 13498-13503.
114. H. K. Dannelly, S. Roseman, *J. Biol. Chem.* **1996**, *271*, 15285-15291.
115. J. M. Rohwer, P. W. Postma, B. N. Kholodenko, H. V. Westerhoff, *Proc. Natl. Acad. Sci. USA* **1998**, *95*, 10547-10552.
116. Q. Mao, T. Schunk, B. Gerber, B. Erni, *J. Biol. Chem.* **1995**, *270*, 18295-18300.
117. B. Görke, B. Rak, *EMBO. J.* **1999**, *18*, 3370-3379.

6
Peptide Transport

John W. Payne and Neil J. Marshall

6.1
Introduction

Over the last twenty-five years, periodic reviews of this topic have presented balanced and integrated accounts of the transport and utilization of peptides by microorganisms. This approach reflects the pre-eminent nutritional role of peptides, which act not only as preformed amino acids but also as important sources of nitrogen and carbon. Early reviews delineated the main types of transporters present in prokaryotic and eukaryotic microorganisms and described the fundamental characteristics of their substrate specificity, energy coupling, and regulation [1–5], whereas later ones concentrated on describing the molecular mechanisms and molecular genetics of the various types of peptide transporters and exploring their potential for delivering antimicrobial compounds [6–10].

Here we concentrate upon developments since the topic was last reviewed in detail [6]. Material covered then and in other reviews is only included when needed to clarify current discussion of a topic. We focus upon the following topics: classification and evolutionary relationships between generic types of transporters; the molecular genetics of peptide permeases and characterization of their protein components; the varied roles played by peptide transport systems; the structural basis for molecular recognition of substrates by peptide permeases; and exploitation of peptide transporters for targeted drug delivery.

6.2
Classification of Microbial Peptide Transport Systems

6.2.1
Classification Based upon Genome Sequencing

Coupled with the explosion in genome sequencing projects has come the necessary and timely classification of membrane transport systems, largely carried out by Saier, Paulsen, and coworkers [11–13], and the establishment of a comprehensive classification scheme known as the Transport Commission (TC) system [14] (see also Chap. 1), and the Web sites *http://www-biology.ucsd.edu/~msaier/transport/titlepage.html* and *http://www.biology.ucsd.edu/~ipaulsen/transport/*). The TC system is analogous to the Enzyme Commission (EC) scheme for enzymes, except that with the TC system, molecular phylogeny can now also be one of the criteria used within the classification scheme [14]. The families of transporters responsible for the transport of amino acids and their derivatives, including peptides, were recently reviewed [15] and divided into five categories (Tab. 1); one being a primary transporter (direct use of chemical energy) in the ATP-binding cassette (ABC) superfamily (traffic ATPases), the other four probably being secondary transporters (utilizing chemiosmotic energy generated by use of a primary energy source) driven, e. g., by a proton-motive force (pmf).

1. *The peptide/opine/nickel uptake transporter family (PepT) (TC 3.A.1.5)*

Two archetypal bacterial peptide permeases belong to this family within the ABC superfamily, the *Escherichia coli* dipeptide permease (DppABCDE; TC 3.A.1.5.2) and oligopeptide permease (OppABCDF; TC 3.A.1.5.1). ABC transporters are covered in detail in Chap. 4. These transporters are multicomponent systems comprising two relatively hydrophilic proteins (or domains) that bind and hydrolyze ATP [ATP-binding cassette (ABC) proteins] to energize the transport process (e. g., DppDE), two hydrophobic proteins (or domains) that generally contain 2 x 6 transmembrane segments (α-helices) (e. g., DppBC), and, for bacterial uptake systems, there is a substrate binding protein that is free and soluble in Gram-negative bacteria (e. g., DppA), or tethered to the cytoplasmic membrane by an N-terminal lipid moiety in Gram-positive species. The sequences of bacterial substrate binding proteins were found to cluster into eight groups [16], each with its own characteristic "signature sequence". Those for peptides (*E. coli* DppA and OppA) and nickel (*E. coli* NikA) grouped together in cluster 5 and were significantly larger (by 100 amino acid residues or more) than all comparable proteins analyzed, implying that they may contain an extra domain, which was later confirmed by the crystal structures of OppA and DppA (Sect. 6.3).

In *E. coli* there are a number of other paralogs (proteins derived from a common ancestor by gene duplication) that are putative peptide transporters, although some of these assignments are tentative and await experimental confirmation. For instance, the SapABCDF system (TC 3.A.1.5.5) transports peptides and peptide-based antibiotics and has been implicated as a contributing factor in the virulence of *Salmonella typhimurium*. In other bacterial species, there are ortho-

Tab. 1. Peptide transport systems. The PepT family (TC 3.A.1.5) are primary, active transporters driven by ATP hydrolysis, all other transporters are, or are presumed to be, active, secondary transporters driven by proton-motive force (pmf)

Transporter Family	TC Number[a]	Size Range[b]	TMDs[c]	Example
Peptide/opine/ nickel uptake transporter (PepT) family	3.A.1.5	BP[d] 490–550 CM[e] 275–340 ABC[f] 250–360	N/A 2 × 6 N/A	E. coli DppA E. coli DppBC E. coli DppDE
Proton-dependent oligopeptide transporter (POT) family	2.A.17	450–600	12	E. coli Tpp
Oligopeptide transporter (OPT) family	2.A.67	600–900	12–15	Candida albicans OPT1
Peptide-acetyl-CoA transporter (PAT) family	2.A.1.25	400–600	12	E. coli AmpG
Peptide-uptake permease (PUP) family	9.A.18	406–420	7	E. coli SbmA

[a] Transport commission number.
[b] Approximate number of amino acid residues in the proteins (including signal sequences for binding proteins).
[c] Number of transmembrane domains.
[d] Binding protein.
[e] Integral, cytoplasmic membrane protein.
[f] Peripheral, ATP-binding cassette (ABC) protein.

logs (proteins arising in different organisms by speciation) of the Dpp and Opp systems in *E. coli*. Thus, in *Bacillus subtilis*, there are the DppABCDE (also called DciAA, B, C, D, E) and OppABCDF systems (also called SpoOKA, B, C, D, F), as well as the AppABCDF system (TC numbers 3.A.1.5.2, 3.A.1.5.1, and 3.A.1.5.X, respectively). Interestingly, these ABC-type uptake permeases are only found in prokaryotes.

2. *The proton-dependent oligopeptide transporter family (POT) (TC 2.A.17)*

The POT family [17] is also known as the peptide transport (PTR) family [18]. These systems comprise single proteins that may contain 12 transmembrane domains, use pmf to drive the transport process and are found in bacteria and eukaryotes. In *E. coli* there are four POT paralogs; one of these is Tpp, one of the three best-characterized *E. coli* peptide transporters (Sect. 6.3.1.1.3) that contains the characteristic family motif [19]. Some members of the POT family show sequence similarity with members of the major facilitator superfamily (MFS), although, at present, the POT family is regarded as a separate family. This family is widespread, with members among bacteria, yeast, plants, and two clinically important mammalian members, intestinal PepT1 and renal PepT2. Although it is clear that these systems are energized by pmf, the stoichiometry of proton:substrate symport has been experimentally established for only a few members and appears to be variable with the nature of the substrate. These transporters appear to be specific for transporting only di- and tripeptides. The *Saccharomyces cerevisiae* transporter Ptr2 (TC 2.17), *B. subtilis* YclF, and *Lactococcus lactis* DtpT proteins are other POT family members.

3. *The oligopeptide transporter family (OPT) (TC 2.A.67)*

These transport oligopeptides comprising 4-6 amino acids and the best characterized members are from the yeasts, although there are plant homologs and there may be distantly related prokaryotic counterparts. The yeast proteins contain 12 transmembrane domains and are probably energized by pmf. So far, two *S. cerevisiae* OPT transporters (YJV2 and YGL4) have been functionally characterized, a third paralog having been identified from genome sequence analysis (Sect. 6.5).

4. *The peptide-acetyl-CoA transporter (PAT) family (TC 2.A.1.25)*

This family, of which one bacterial member, *E. coli* AmpG, has been partially characterized, are a family within the ubiquitous MFS [20]. *E. coli* AmpG transports peptides, including those derived from the cell wall (i.e., N-acetylglucosaminyl-β-1,4-anhydro-N-acetylmuramyl-tripeptide), glycopeptides, and inducers of β-lactamases (Sect. 6.4.5). Homologs of AmpG have been identified in *Haemophilus influenzae*, *Neisseria gonorrhoeae*, and *Ricksettia prowazekii*. Although it is not yet proved experimentally that these transporters operate as proton:substrate symporters, given their predicted 12 transmembrane domain topology and inclusion within the MFS, it seems highly probable. Whether further characterization of this family will substantiate its designation as a transporter of specialized peptides remains to be seen.

5. *The peptide-uptake permease (PUP) family (TC 9.A.18)*

This family is currently defined by only two proteins, *E. coli* SbmA and *Rhizobium meliloti* BacA, both of which have been functionally characterized in part (Sect. 6.4.3). SbmA transports the peptide antibiotics microcin B17 and microcin J25, as well as the non-peptide antibiotic bleomycin. SbmA and BacA share 64% sequence identity with each other, and are distant homologs of a few putative bacterial ABC-type permeases (TC 3.A.1). However, no ABC proteins have been found for the SbmA and BacA transporters, and their mechanism of energization remains unclear. The inclusion of this class is somewhat misleading here as their significance to true peptide transport is at best questionable.

6.2.2
Classification Based upon Substrate Specificity

Transporters can be classified using the above criteria, i.e., gene sequences, energy coupling, and protein number/architecture. However, peptide transporters may also conveniently be classified based upon their substrate specificity and this criterion can cut across the other classification schemes. For instance, the profile of substrates transported by *E. coli* Dpp, a multicomponent ABC transporter (TC 3.A.1.5.2), mirrors that of the human intestinal PepT1 and renal PepT2 transporters (TC 2.A.17) and members of the POT family, despite differences in their protein architecture [five versus one protein] and energization [ATP hydrolysis versus proton symport (pmf)] [21]. Using a combination of biochemical assays, molecular biology, and computer-based molecular modeling we have found that *E. coli* Dpp and Tpp recognize and transport distinct conformational forms of di-, and tripeptides [22, 23], while Opp prefers particular conformers of oligopeptides (3-6 residues) [24]. Because di-, and tripeptides all exist in distinct conformational forms, more than one transporter is required to ensure as complete an uptake of these peptides as possible. These studies also indicate that some dipeptides, e.g., those with N-terminal Asp or Glu residues, are recognized poorly by these generic transporters and may require the presence of specific systems tailored for such substrates. In this respect, it is noteworthy that *E. coli*, and other organisms, have additional, putative peptide transporters that have not yet been characterized, e.g., the three remaining *E. coli* POT paralogs. Thus, the peptide substrates transported by these systems do not necessarily depend upon particular protein architecture and mode of energization; indeed, these features may be considered as ancillary to those of substrate recognition itself, which has been driven by the repertoire of conformational forms universally found in peptide pools and available ubiquitously to these transporters. Aspects of the basis and evolution of molecular recognition by microbial peptide transporters is considered in greater detail in Sect. 6.6.

6.3
Peptide Transport in Prokaryotic Microorganisms

6.3.1
Gram-negative Bacteria

6.3.1.1 Enteric Bacteria
Extensive results from studies of the various peptide permeases found in *Escherichia coli* and *Salmonella typhimurium* have provided the basis for a general understanding of the physiological roles, structures and molecular mechanisms of such transporters in all microorganisms [6]. Three main peptide permeases are found in these species: the dipeptide (Dpp), oligopeptide (Opp) and tripeptide (Tpp) permeases. These show substrate specificities that at first sight are apparently over-

lapping, e.g., each can transport di- and tripeptides, yet recent studies of the structural basis of their molecular recognition reveals that they are complementary in several ways; firstly, each recognizes different conformational forms of a single peptide and secondly, acting together they are able to transport all small peptide components of the peptide pool.

6.3.1.1.1 The Dipeptide Permease (Dpp)

Initial studies of the dipeptide permease in *E. coli* showed it to be sensitive to osmotic shock and energized by phosphate-bond energy [25] and, in accord with this, the dipeptide binding protein (DppA) was subsequently isolated and shown to be also involved in peptide chemotaxis [26]. Dpp is a typical ABC transporter, which comprises an operon of five genes *dppABCDE* (Sect. 6.2.1) that shows growth-phase-dependent expression [27]. Its organization is analogous to the *opp* operon of *E. coli* and *S. typhimurium* (Sect. 6.3.1.1.2) and *spoOK* operon in *Bacillus subtilis* (Sect. 6.4.1). In certain *E. coli* mutants, DppA can be markedly overproduced when grown in minimal medium with no amino acids [6, 28]. DppA (and OppA) is normally repressed in cells growing in rich medium but high constitutive synthesis is observed in a mutant deleted for the *gcvB* gene, which encodes a small RNA transcript that is not translated *in vivo*, although the mechanism of this effect is not understood [29].

Substrate transport by Dpp (and other peptide permeases) is best determined by using several complementary assay procedures, e.g., radioactively labeled substrates and fluorescence-based techniques, because the rapid exodus of cleaved amino acid residues that can follow peptide uptake may lead to erroneous calculation of transport kinetics [6, 25]. Transport rates via Dpp have been measured for a range of natural and modified peptides using *E. coli* mutants defective in Opp and Tpp [6, 19]. Dpp can transport all normal dipeptides but with rates that vary up to about 10-fold (typically 20 nmol min^{-1} (mg bacterial protein)$^{-1}$; it can also transport tripeptides, although the rates are generally lower by about an order of magnitude than those for structurally related dipeptides [6, 19]. Using pure DppA from *E. coli* in a filter binding assay, a substrate:DppA stoichiometry of 1:1 was found with [^{14}C]AlaAla and Ala[^{14}C]Phe and substrate binding did not vary over the pH range pH 3-9.5. On isoelectric-focussing gels the native form of DppA migrates with a pI of 6.1, whereas binding of various, (neutral) dipeptides produces liganded DppA with pI values of 5.9 and 6.1, implying that DppA undergoes varied conformational changes to accommodate different substrates [6, 19]. For a range of dipeptides, their relative binding affinities for DppA, measured in competition assays with [^{125}I]Tyr dipeptides, were found to parallel their transport rates via Dpp, showing that DppA effectively determines the substrate specificity of Dpp [19]. *E. coli* DppA has been crystallized in the "open, unliganded" form [30] and in complex with GlyLeu [31]. This bilobate protein comprises three domains; domain I residues 1-33, 183-260 and 479-507, domain II residues 34-182, and domain III residues 261-478. Domains I and III are connected by two short polypeptides that form a "hinge", such that upon substrate binding domains I and II move relative to domain III by ~55°. The charged termini of dipeptides are bound by oppositely charged side chains within DppA (Asp^{408} and Arg^{355}), the peptide bond CO and NH groups H-bonding to DppA. The side chains of the dipeptides are accommodated within hydrated binding pockets. The characteristics and binding affinities for peptide binding to DppA measured by using isoelectric-focussing and filter binding assays [19], and isothermal titration calorimetry (ITC) [23] are in good agreement. ITC shows that dipeptide binding by DppA is exothermic (ΔH is negative), although binding entropies (ΔS) are generally small and there is entropy−enthalpy compensation across the series of peptides investigated. The structural basis for the molecular recognition of substrates by DppA can be explained in terms of the repertoire of conformers adopted by di-, and tripeptides (Sect. 6.6), the bound conformation of GlyLeu being one of the predominant torsional forms adopted by this dipeptide in solution [21, 22].

6.3.1.1.2 The Oligopeptide Permease (Opp)

The oligopeptide permeases of *E. coli* and *S. typhimurium* are paradigms for ABC transporters [32] (Sect. 6.2.1) comprising operons with five genes *oppABCDF*. Regulation of Opp is complex, being moderated at several levels including control by the leucine-responsive regulatory protein (Lrp) [6]. The oligopeptide binding protein (OppA) is normally the most abundant periplasmic protein, typically representing about 7–10% of total periplasmic proteins, but its synthesis is repressed in rich medium and by high phosphate levels [6], induced by presence of polyamines in the medium (Sect. 6.4.4.) and, as for DppA, it is enhanced in a mutant deleted for the *gcvB* gene (Sect. 6.3.1.1.1).

Models for the overall mechanism of substrate transport by Opp (and Dpp) have been proposed [6], involving substrate recognition and binding by OppA, translocation via OppBC with coupled ATP hydrolysis by OppDF that are broadly similar to ones proposed recently for the histidine permease [33, 34]. Further insight into how these proteins interact to effect transmembrane transport has come from studies in which a radioactive, photoaffinity-labeled substrate was attached to OppA (Marshall, DeUgarte, and Payne, unpublished data). Following derivatization of OppA, the protein was digested with trypsin and radioactively labeled peptide fragments were purified and sequenced; the most radioactively labeled fragment was found to be ...AspIleIleValAsnLys... (residues 300–305) associated not with amino acids at or around the ligand binding site but located about 30 Å distant at the protein surface in the entrance/exit of the cleft leading to the binding site (see below). Identical results were found with OppA from both *E. coli* and *S. typhimurium*. Single, site-directed mutations in this OppA sequence completely destroy the biological activity of Opp. However, ligand binding capability of the mutated OppA is essentially retained, with some mutations giving enhanced binding affinity and some decreased, and the synthesis of each mutant OppA is actually enhanced relative to synthesis of wild-type protein. It appears that these residues are involved in interactions with certain of the membrane-bound components (OppBCDF) that are essential for overall substrate translocation. This view was endorsed with isolation of compensative mutations in the membrane proteins that could restore transport with the mutated OppA proteins but did not function with wild-type OppA. Further analysis of these various mutants should clarify how these various proteins interact to transfer substrate from liganded binding protein to membrane complex.

Transport by Opp has been extensively studied with natural peptides and also natural and synthetic analogs [2, 5–7, 9, 10]. The system has a high affinity for small peptides, with the preferred substrates usually comprising $3 > 4 > 5 \cong 2 > 6$ residues, respectively; binding affinities are usually in the range 0.1–5 μM with typical uptake rates being 1-30 nmol min^{-1} (mg bacterial protein)$^{-1}$. Natural peptides from di-, up to and including hexapeptides, can be transported by whole cells, which reflects the overall size limit for passage through outer membrane porins and, consequently, the actual size limit for peptide uptake by Opp cannot be determined by measuring uptake of exogenous substrates and it remains a possibility that larger peptides generated by proteolysis within the cell envelope could be absorbed. Support for this possibility has come from filter binding assays with OppA, which have shown that peptides containing at least 16 residues can bind competitively with tripeptides to OppA; furthermore, [^{125}I]angiotensin (10-mer) showed good binding that was competitively inhibited by small substrates, e. g., LLL-Ala$_3$ but not by DDD-Ala$_3$ that is not itself a substrate [6, 9]. The biological significance of these observations has yet to be clarified but similar results have also been reported for a related protein from lactobacilli (Sect. 6.3.2.1). Comparison of results of transport studies with those on ligand binding to OppA, using equilibrium dialysis, filter binding assays, protein mobility shifts on isoelectric-focussing gels, ITC, and examination of crystal structures, all indicate that overall transport specificity is controlled by the recognition and binding properties of OppA [6, 9, 35, 36]. The native form of OppA from E. coli has a pI of 6.20, but it is commonly isolated with additional forms with pIs of 6.26 and 6.55 indicative of attached ligands; binding of various (neutral) oligopeptides can give rise to varied forms with different pIs implying that the protein may adopt subtly different conformational forms, as seen also with liganded DppA.

OppA from S. typhimurium has been crystallized in its free form [37] and also liganded with a variety of peptides [35–40]. OppA, like DppA, is a bilobate protein composed of three domains; domain I residues 1-44, 169-270 and 487-517, domain II residues 45-168, and domain III residues 271-486. Domains I and III are connected by two polypeptide stretches that form the "hinge" to allow domains I and II to move relative to domain III to engulf bound substrate, the calculated movement being only 26° [37]. The N-terminus of a bound peptide forms a salt bridge with the side chain of Asp^{419}, the C-terminal carboxylate of tri- and tetrapeptides forming salt bridges with the side chains of Arg^{413} or His^{371}, respectively. The CO and NH groups of the peptide bonds of the ligands form H-bonds with various OppA residues while the ligand side chains are contained within spacious, hydrated binding pockets. Upon binding, the different amino acid side chains displace variable numbers of water molecules from the binding pockets depending on their size, e. g., Trp displaces three more waters than Ala. As with DppA above, the bound conformation of peptides matches the predominant torsional form of oligopeptides in solution determined from modeling studies

[21, 24]. The thermodynamics of peptide binding have been investigated by using ITC of a series of Lys–X–Lys tripeptides where X = all of the 20 naturally occurring amino acids or 8 amino acid mimetics containing aliphatic chains, aliphatic amines, and ring structures [35, 36, 40]. Binding enthalpies (ΔH) were all positive (endothermic), the binding being driven by favorable (positive) entropy values (ΔS). However, even using such a closely related set of compounds, it was not possible to relate changes in the structure of the central side chain to changes in enthalpy and entropy. In consequence of the observed enthalpy/entropy compensation, the free energy values (ΔG) cover a narrow range but the binding affinities, which show correlation with ΔG, remain difficult to explain. The stability of OppA–oligopeptide complexes has also been investigated by using mass spectrometry (MS) [41] and results related to ITC measurements; however, at present the MS technique may be relatively insensitive, for two pentapeptides and two tripeptides with central D-residues were found not to bind, which is at odds with earlier reports [6]. Recently, the crystallization of E. coli OppA has been reported [42] and studies on its structure in complex with specifically selected ligands may provide additional and complementary information to that described above. Other aspects of Opp and OppA not exclusively concerned with peptide transport are considered later (Sect. 6.4). Many proteins interact with peptides, and this can be sequence-dependent or sequence-independent in manner [43]. Hubbard, Tame, and colleagues have used the OppA system as described above, in conjunction with other systems, to understand better the energetics of protein:ligand interactions with a view to being able to predict the affinities of such molecular recognition interactions, although, currently for OppA, the experimentally and computationally determined values differ markedly [44].

6.3.1.1.3 The Tripeptide Permease (Tpp)

Like Dpp, Tpp has specificity for di-, and tripeptides only but is distinguishable from Dpp both genetically and in terms of its precise substrate specificity. For a number of years, there have been conflicting reports about what type of transporter Tpp is and where it maps, mainly because most studies were complicated by being performed in strains that were not devoid of Opp and Dpp activities [6]. This situation has recently been clarified [19]. Starting with an E. coli strain (*dpp opp*), a *tpp* mutant was selected using resistance to alafosfalin, coupled with cross-resistance to ValGly and failure to utilize LeuTrp as a source of these auxotrophic amino acids. Using complementation analysis with episomes, the *tpp* mutation mapped to a gene at 36 min (p77304 in the SWISS-Prot database), homologous to the pmf transporter in lactobacilli (Sect. 6.3.2.1.) and containing the Ptr motif [19]. Recent studies have characterized differences between the substrate specificities of Dpp and Tpp [21–23].

6.3.1.2 Rumen Bacteria

These comprise a heterogeneous collection of mainly Gram-negative, anaerobic, proteolytic organisms for which peptides provide an important nutritional resource [45, 46]. *Provotella* (formerly *Bacteroides*) *ruminicola* is generally the most important bacterium and although few direct studies on peptide uptake have been reported [47], its complement of peptidases and proteases has been extensively studied [48, 49].

6.3.2
Gram-positive Bacteria

6.3.2.1 Lactic Acid Bacteria

The integrated systems involved in the utilization of nitrogen sources by lactococci and lactobacilli have been extensively studied. Peptides provide the main source of nutritional nitrogen, with uptake of free amino acids contributing only a minor proportion. Their proteolytic systems are all very similar, comprising extracellular, cell wall-bound proteases, transport systems specific for di-, tripeptides and oligopeptides and a large complement of intracellular peptidases [6, 50–53]. Detailed studies have been carried out with *Lactococcus lactis*, for which the protease PrtP and the oligopeptide transporter Opp are central components of the degradation pathway of exogenous proteins, for which no alternative activity is present; PrtP produces substrates only for Opp with no detectable di- or tripeptides being produced from casein [50, 54]. These bacteria occupy a special niche in their utilization of casein mixtures and it might be expected that to capitalize on this resource they may have evolved rather more specialized systems than would be the case, e.g., with enteric bacteria. Studies on peptide transporters have also mostly been carried out with *L. lactis*, in which three systems have been identified: the ABC transporters DtpP and Opp, specific for di-, tripeptides, and for oligopeptides, respectively; and a pmf-driven di-, tripeptide transporter, DtpT, that belongs to the POT family. Other organisms that can utilize casein, e.g., *Listeria monocytogenes*, possess transporters for di-, tripeptides and for oligopeptides [55]. In these Gram-positive organisms, which lack a periplasm, ABC transporters have a lipoprotein component anchored to the external face of the cytoplasmic membrane that plays the same role as a free, soluble periplasmic binding protein [56].

In an initial report [57], the specificity of DtpP was shown partially to overlap that of DtpT; both recognize only di- and tripeptides but DtpP was shown preferentially to transport peptides that are composed of hydrophobic residues whereas DtpT has a higher specificity for more hydrophilic and charged peptides. In this regard, these systems resemble *E. coli* and *S. typhimurium* Tpp and Dpp, respectively (see Sect. 6.3.1.). It was also proposed that DtpP was induced by the presence of di- and tripeptides containing branched-chain amino acids. Mutants of the DtpP transporter, isolated in a $\Delta dtpT$ strain using as a selective procedure resistance to the toxic dipeptide Phe-β-chloro-Ala, failed to transport di- and tripeptides but still showed uptake of oligopeptides. Using ionophores and metabolic inhibitors, it was concluded that DtpP-mediated transport was driven by ATP or a related energy-rich phosphorylated intermediate [57], and subsequently genome analysis confirmed it to be an ABC transporter [58]. Studies with its purified binding protein, DppA, showed its specificity profile to be broadly similar to that reported for the di-, tripeptide transporters in *E. coli*, most particularly to Tpp [58].

In contrast to the limited reports on DtpP, there has been a range of studies on DtpT [59–63]. DtpT is a member of the POT family (see Sect. 6.5) in comprising a single membrane-bound protein that is energized by a pmf. The gene for DptT has been cloned and functionally expressed in *E. coli* [60, 61] and conditions for optimizing protein purification and for expressing activity in artificial proteoliposomes have been defined [62, 63]. The substrate specificity of the transporter in these systems was studied using ProAla as a reporter peptide and found broadly to resemble that for Dpp in *E. coli* and the eukaryotic transporters PepT1 and PepT2.

Opp from *L. lactis* shows some similarities to the analogous transporters from *E. coli* and *S. typhimurium* but also significant differences. All consist of five proteins: the four membrane proteins, OppB,C,D,F and the substrate binding protein OppA. However, although the OppA of *L. lactis* is formally homologous to the oligopeptide binding proteins of *E. coli* and *S. typhimurium*, its amino acid similarity is not statistically significant (~ 20% identity). Furthermore, it is an "outsider" in the family of peptide binding proteins, with its sequence identity being low (~ 20%) even with peptide binding proteins from Gram-positive bacteria such as SpoOKA from *B. subtilis*. Opp can transport a range of peptides derived from cleavage of caseins by PrtP from 4 up to at least 18 residues [54, 64]. Significant peptide binding to OppA is only found for peptides with more than 5 residues. The dissociation constants for peptide binding to OppA varied from micromolar for dodecapeptides to millimolar for pentapeptides, and it was inferred that residues 6-12 of a peptide contribute to the binding affinity and must interact with the protein. It was suggested that variations in dissociation rate constants were mainly responsible for differences in binding constants and this was related to the kinetics of overall transport [65]. Subsequent studies indicated that the N-terminal amino group could be acylated and the C-terminal carboxylate amidated without markedly changing the dissociation constants of peptide substrates [66]. Using synthetic peptides containing Cys or azaTrp residues at various positions led to the conclusion that the first six amino acid residues were enclosed by the protein, whereas the remaining residues protruded and interacted with "the surface" of OppA [66]. Site-directed mutagenesis of specific residues in OppA, identified as putative binding sites, and study of their binding and transport properties showed a general correlation between decreased binding and lowered transport for larger peptides but little effect with a model tetrapeptide; the results were interpreted as implying that transport is determined to a large extent by donation of a peptide from OppA to the membrane transporter complex [67]. Recent studies, using combinatorial peptide libraries to probe binding by OppA, reveal that peptides up to 35 residues can be bound although optimum binding is for about 9 residues; the first 6 residues, which are enclosed by the protein, and the C-terminal 3 residues appear particularly important for binding [68]. Not surprisingly, a different mechanism of binding has needed to be invoked for *L. lactis* OppA compared with that of the archetypal enteric bacterial forms [68]. In summary, it seems this system may be better considered as an analog of a "protein transporter" rather than a peptide transporter and its recognition specificity better discussed in relation to the multiple sites normally considered for proteases, rather than the stricter specificities seen with binding proteins for small peptides and with peptidases. Furthermore, given the evidence that such binding proteins show properties characteristic of molecular chaperones (Sect. 6.4.6), it may be useful to look for mechanistic similarities between *L. lactis* OppA and proteins such as SecB that are components of protein translocation machinery.

6.3.2.2 Miscellaneous Organisms

Only a few of the organisms considered previously [6] have subsequently had their peptide transport systems characterized in any depth. In these bacteria, which lack an outer membrane and periplasm, their peptide binding proteins are attached to the outer surface of the cytoplasmic membrane by a lipid anchor. The first such system to be described in Gram-positive bacteria was the *ami* operon of the human pathogen *Streptococcus pneumoniae*, a fastidious, obligate parasite requiring several amino acids for growth and for which peptide utilization is important to its nutrition [69]. Mutations in the *ami* locus were found to increase transformation efficiency [70] and other pleiotropic effects, such as an influence on competence, accompany mutations in *ami* and the three homologous oligopeptide binding proteins AmiA, AliA, and AliB [71, 72]. Consequently, the suggestion has been made that in addition to a direct nutritional function, peptide transport systems may play a pivotal role in sensing environmental conditions and indirectly modulating the expression of several genes [73]. Cells of the oral bacterium *Streptococcus gordonii* express three membrane-bound lipoproteins that closely resemble the peptide binding proteins of *S. pneumoniae*; they are essential for uptake of penta- to heptapeptides and influence the organism's development of competence and adherence properties [74, 75]. Analogously, an oligopeptide binding protein component of a putative transport system acts as an adherence-associated lipoprotein in the cell wall-less bacterium *Mycoplasma hominis* [76]. In a strain of group A *Streptococcus*, operons coding for ABC-transporter sequences corresponding to Dpp and Opp systems have been identified; peptides in the growth medium enhanced the expression of these systems whereas mutations in these loci caused decreased expression of the virulence factor SpeB, a cysteine protease [77, 78]. In *Borrelia burgdorferi*, the causative agent of Lyme disease, an ABC-type oligopeptide permease has been identified [79, 80]. Unusually, it possesses three chromosomal copies of the lipoprotein binding protein, OppA, and also two plasmid-encoded copies, all of which are independently transcribed; increase in temperature induces one of the plasmid-encoded OppA, which appears to be an important cue for adaptive responses *in vivo* [79].

6.4
Bacterial Peptide Transport Systems with Specific Functions and Substrates

6.4.1
Role of Peptides and Peptide Transporters in Microbial Communication

Many microorganisms communicate by secreting and responding to extracellular peptides (pheromones). Some peptide pheromones act via cell surface receptors, which are often histidine protein kinases, whereas others are transported into the cell by an oligopeptide permease and interact with intracellular receptors to modulate gene expression [81, 82]. Microbial processes influenced in this way include sporulation, production of virulence factors, expression of gene transfer functions, and antibiotic production. Quorum sensing, which occurs at high cell density in many microorganisms, may be involved not only in regulation of the above processes but may also play a more central role in bacterial physiology by interacting with starvation-sensing pathways to regulate cell entry into stationary phase [83, 84].

The role of oligopeptide transport in sporulation in *B. subtilis* has been extensively researched. Oligopeptides are transported into *B. subtilis* by two ABC transport systems, Opp [85, 86] and App [87], and they may play both nutritional and regulatory roles. An extracellular peptide factor serves as a cell density signal for both competence development and sporulation [88, 89]. This CSF (competence and sporulation) peptide, Glu–Arg–Gly–Met–Thr (ERGMT) is transported into cells by the SpoOK oligopeptide permease (Opp) and stimulates expression of the surfactin synthesis operon at low concentration, while inhibiting competence gene expression and stimulating sporulation at high concentrations. A further signaling molecule, PhrA, is a pentapeptide that stimulates sporulation. Both peptides are produced by cleavage of secreted, precursor polypeptides and both must be transported back into the cell to be functional [88–90]. The *opp* operon is transcribed during exponential growth, whereas the *app* operon is induced at the outset of stationary phase; transcription of both operons is prevented by overproduction of the ScoC regulator, which is a negative regulator of sporulation and protease production that acts by binding directly to promoters of the genes it regulates [91]. In the filamentous bacterium *Streptomyces*, morphological differentiation involves a mechanism of extracellular signaling that culminates with the formation of an aerial mycelium. An oligopeptide permease is responsible for the import of an extracellular signal that acts at the first step in a cascade of developmental regulatory signals leading to production of aerial mycelium [92, 93]. In *Pseudomonas aeruginosa* and several other Gram-negative bacteria, a variety of diketopiperazines (cyclic dipeptides) have been found to interact with quorum sensing receptors, e.g., one for N-acylhomoserine lactones; the possible physiological role of this interaction has yet to be clarified [94].

6.4.2
Sap Genes and Resistance to Antimicrobial Cationic Peptides

Invertebrates, vertebrates, and plants all deploy cationic polypeptides such as defensins, melittin, and protamine to resist attack from bacterial pathogens. To withstand the antibiotic activity of these molecules, successful pathogens such as *S. typhimurium* and *Erwinia chrysanthenum* have evolved various mechanisms, with *sap* (sensitivity to antimicrobial peptides) genes playing an important role [95]. *Sap* genes are widely distributed among Gram-negative bacteria [96]. The *sapABCDF* operon resembles *opp* in enteric bacteria and has been implicated as a peptide transport system; mutations in this operon or in *sapG*, which encodes the NAD^+-binding protein TrkA, a component of a low-affinity K^+ transport system, confer hypersensitivity to cationic peptides [97]. The ways in which the various components are integrated remains unclear; virulence factors PhoP and PhoQ and the *lux* regulon have been implicated in regulation [96, 98]. *E. coli* also uses the outer-membrane protease OmpT to hydrolyze antimicrobial polypeptides such as protamine before they can exert their activity at its susceptible cytoplasmic membrane [99].

6.4.3
Uptake of Peptide Antibiotics

Given that a vast collection of peptide-based antibiotics are synthesized by a range of bacteria, fungi, and plants for targeting against the main generic peptide transporters, exemplified by *E.coli* Opp, Dpp and Tpp (see Sect. 6.3.1), it is ironic that a special family has been classified partly on the basis of peptide antibiotic uptake. This PUP (peptide-uptake permease) family comprises two proteins that exhibit 64% identity [15]. SbmA permits uptake by *E. coli* of microcins, which are thiazole ring-containing peptide antibiotics [100], whereas BacA is a nodulation protein that may take up peptidic compounds essential for bacterial development of symbiotic rhizobia with leguminous plants [101].

6.4.4
Polyamine Stimulation of OppA Synthesis and Sensitivity to Aminoglycoside Antibiotics

Addition of polyamines to growing cells of *E. coli* stimulates synthesis of a specific range of proteins, with increase in OppA being most marked; rapid synthesis of OppA on addition of a polyamine may contribute to polyamine-stimulated cell growth through enhancing the supply of nutrients. This stimulation occurs mainly at the level of translation and involves both the structure and location of the Shine Dalgarno sequence [102, 103]. In related studies, enhanced sensitivity to aminoglycoside antibiotics (gentamycin, isepamicin, kanamycin, neomycin, streptomycin) was observed in the presence of polyamines, which was related to the enhanced production of OppA; it was also shown that isepamicin could be bound to OppA, albeit poorly, and OppA enhanced its uptake into cells [104]. Spontaneous kanamycin-resistant mutants of *E. coli* were found to have either decreased or no OppA, to be cross-resistant to a range of other aminoglycosides and to show decreased uptake of isepamicin [105–107]. Many clinical isolates of *E. coli* showing resistance to aminoglycosides expressed reduced or undetectable levels of OppA, whereas other resistant strains were deficient in ornithine or arginine decarboxylases, or both, which might negatively affect OppA expression by decreasing polyamine synthesis [108]. The structural basis for the ability of OppA to bind aminoglycosides, albeit with affinities about 10^4 times less than typical peptides, has yet to be defined.

6.4.5
Role of MppA in Signaling Periplasmic Environmental Changes

Early studies implicated Opp in the uptake of cell-wall peptides during the recycling of murein in *E. coli* and *S. typhimurium* [109]. However, more recently, it has been shown that this involves not OppA but a different periplasmic binding protein, MppA, which is the same size as OppA with an amino acid sequence that is 47% identical but is present in the periplasm at a very much lower level than OppA [110]. MppA binds the murein tripeptide, L-alanyl-γ-D-glutamyl-*meso*-diaminopimelate, which is then transported into the cytoplasm via the membrane-bound complex OppBCDF, and it was speculated that the physiological function of MppA may be linked to signal transduction pathways that are involved in regulation of β-lactamase synthesis [110]. However, enigmatically in this regard, very little free murein tripeptide is actually transported by MppA, nearly all of it being absorbed by the pmf-driven AmpG permease in the form of N-acetylglucosaminyl-β-1,4-anhydro-N-acetylmuramyl-tripeptide [111]. AmpG is classified in the peptide-acetyl-CoA transporter (PAT) family of MFS (Sect. 6.2.1), homologs of which have been identified in the genomes of various microorganisms [15]. Alternatively, MppA may serve a signaling function, reporting on the changes in the periplasmic environment. Further studies showed that a null mutation in *mppA* conferred increased resistance to a wide spectrum of antibiotics; in the absence of MppA the cell exhibits all the properties associated with the multiple antibiotic resistance (MAR) phenotype, and it appears that MppA plays a central role in the regulation of a number of proteins including those responsible for MAR [112]. Whether these observations may also be linked to the role of OppA in aminoglycoside resistance has not been reported. MppA may also possess very low affinity for some α-linked oligopeptides, as ProPheLys was able to bind to MppA and satisfy the proline auxotrophy of an *E. coli oppA* mutant [110].

6.4.6
Periplasmic Substrate Binding Proteins as Molecular Chaperones

Protein folding in the periplasm is poorly understood and few general periplasmic chaperones have been found [113, 114]. However, proteins involved in other aspects of protein folding such as *cis–trans* isomerization of peptide bonds and formation of disulfide bridges have been identified and characterized [115]. Three periplasmic substrate binding proteins, *E. coli* OppA, maltose binding protein (MalE), and galactose binding protein (MglB), have been found to have chaperone properties comparable with those of the established chaperones DnaJ and DnaK [116]. For example, these three binding proteins promote the functional renaturation of citrate synthase *in vitro*, reduce its thermal aggregation, and also bind only to denatured proteins. These chaperone functions were unaffected by the addition of specific substrates, e. g., AlaAlaAla for OppA, implying that the conformation (open or closed) of the binding protein is not critical for this chaperone function. A 58 kDa protein isolated from *Rhodobacter sphaeroides* f. sp. *denitrificans* that was also able to act as a chaperone was found to be an *E. coli* DppA homolog [117], although at least two other periplasmic proteins are able to substitute for its function *in vivo* [118]. The binding of OppA by the homotetrameric SecB proteins, involved in the secretion of OppA and other proteins (e. g., MalE and MglB) across the cytoplasmic membrane, has been investigated *in vitro* [119, 120]. SecB binds to two non-contiguous regions of OppA, in contrast to the single regions with MalE and MglB, that span nearly the entire OppA protein, to form a complex of 1:1 stoichiometry (OppA monomer:SecB tetramer).

6.4.7
Transport of δ-Aminolevulinic Acid

The heme precursor δ-aminolevulinic acid, which resembles GlyGly without a peptide bond, was originally shown to enter *E. coli* and *S. typhimurium* via Dpp [121, 122]. In *E. coli*, it is also transported via Opp, this uptake being negatively regulated by Lrp [123]. Its transport in *E. coli* by Dpp and its competitive binding to DppA are less than 500-fold as effective as a typical dipeptide [19]. δ-Aminolevulinic acid not only lacks any capacity for the hydrogen-bond stabilization that occurs with a normal peptide bond but conformational analysis shows its nominal psi (ψ), phi (ϕ) and omega (ω) angles lie outside those optimal for recognition (J. W. Payne, unpublished data). In an example of an observation initially made in bacteria being applied to mammalian system, δ-aminolevulinic acid has also been shown to be absorbed by intestinal and renal peptide transporters PepT1 and PepT2, respectively; as the compound acts as a precursor in porphyrin synthesis it has potential therapeutic applications as an endogenous photosensitizer for photodynamic treatment of various tumors [124].

6.4.8
Transport of Glutathione

Glutathione transport has been described in a variety of Gram-positive species such as streptococci and lactococci, although such results appear difficult to demonstrate in all strains, perhaps indicating problems in establishing a suitable energy source for its uptake [125]. However, a specific transport system for glutathione has not been reported in Gram-negative bacteria such as *E. coli*. In *S. cerevisiae*, two kinetically distinguishable glutathione transport systems have been reported; a specific, high-affinity, ATP-driven system that is regulated and a low-affinity system [126]. A system resembling the former, high-affinity transporter has been cloned and characterized; uptake was inhibited by oxidized glutathione but was not sensitive to competition by simple di- or tripeptides [127].

6.5
Peptide Transport in Eukaryotic Microorganisms

Detailed characterization of eukaryotic microbial peptide transporters has been mainly confined to studies with the yeasts, *S. cerevisiae* and *Candida albicans*. Several general reviews on yeast transporters have appeared [128, 129]. On the basis of initial biochemical and genetic studies, evidence was found for at least two peptide transport systems in these species [8]. In *S. cerevisiae*, a gene (*Ptr2*) for a di-, tripeptide transporter was cloned and partially characterized, which was similar to a gene for a nitrate transporter from *Arabidopsis thaliana* [130]. Subsequent sequence analyses of genes for various transport systems, including the one present in *S. cerevisiae*, led to the classification of the PTR (Peptide Transport) group of peptide transporters, which possess a unique structural motif as well as conserved glycosylation and phosphorylation sites [18]. The PTR group has alternatively been called the POT (Proton-dependent Oligopeptide Transporters) family and representatives are also found in bacteria, plants, and animals (Sect. 6.2.1) [15, 17, 131]. A *C. albicans* peptide transport gene (*Ca Ptr2*) was cloned by functional complementation of a peptide transport-deficient mutant (strain ptr2-2) of *S. cerevisiae*; a high level of identity existed between this gene sequence and that of the above peptide transporter of *S. cerevisiae* [132]. Restoration of the CaPTR2 protein in *C. albicans* allowed transport of di-, tripeptides, and growth on dipeptides as a source of required amino acids together with sensitivity to toxic dipeptides.

A further *C. albicans* oligopeptide transport gene (*opt1*) was identified and cloned through heterologous expression in a di-, tripeptide-transport mutant of *S. cerevisiae* [133]. Presence of OPT1 protein conferred ability to transport tetra-, pentapeptides, to grow on these peptides as sources of required amino acids, and sensitivity to toxic tetra-, pentapeptides; the level of oligopeptide transport was influenced by the nitrogen source used for growth. Gene sequence comparisons revealed that *S. cerevisiae* and *Schizosaccharomyces pombe* possessed genes for similar proteins and that they comprise a group of transporters distinct from the ABC or PTR membrane transport families [133, 134]. This OPT (oligopeptide transporter) family appears to be specific for small oligopeptides (tetra-, pentapeptides) and confined to fungi and plants [15]. OPT1 in *S. cerevisiae* transports the pentapeptides, Leu- and Met-enkephalins, which is competitively inhibited by the opioid receptor antagonists naloxone and naltrexone [135].

Various studies have been made into the regulation of yeast peptide transporters. In *S. cerevisiae*, two genes, *ptr1* and *ptr3*, control the PTR2 transporter for di-, tripeptides [136]. *Ptr1* mutants fail to express PTR2 and are thus unable to transport di-, tripeptides. *Ptr1* was found to be identical to *Ubr1*, a gene previously described as encoding the recognition component of the N-end-rule pathway of the ubiquitin-dependent proteolytic system but the precise physiological role of this common component in these two processes awaits full clarification. *Ptr3* is required for amino acid-induced expression of PTR2 but is not needed for nitrogen catabolite repression of peptide import or PTR2 expression [137]. The *Ptr3* gene product functions similarly to Ssy1p, being implicated in relaying signals about the presence of extracellular amino acids that leads to the leucine-inducible transcription of the amino acid permease genes *Bap2*, *Bap3*, and *Tat1* [138]. Ssy1p and Ptr3p occur in the plasma membrane, where they act as components of a sensor system for extracellular amino acids; loss of either blocks induction of the PTR2 system together with extensive pleiotropic effects [139]. In a continuation of a more physiological approach to studies of the regulation of uptake and utilization of peptides by yeasts [6], continuous culture in chemostats has been used with a variety of nutritional sources and nutrient limitations [140, 141].

6.6
Structural Basis for Molecular Recognition of Substrates by Peptide Transporters

We address this topic using as examples the three model peptide permeases from *E. coli*: Dpp, Opp, and Tpp. The main features of their substrate specificities have been established using a variety of biological and biophysical assays for transport and binding and found to be common for all; additional information has come from the crystal structures of their liganded binding proteins. Optimally, substrates need a free, protonated N-terminal α-amino group, a free C-terminal α-carboxylate, all L-stereochemistry, and *trans*-peptide bonds, with the nature of the side chains being relatively unimportant [2, 3, 6]. However, despite having extensive, structure-activity data accumulated over many years, it remained impossible to explain why certain peptides were better substrates than others, let alone to be able to predict the relative binding abilities of different peptides or to design bioactive analogs that could be targeted to peptide transporters for delivery. Recently, a detailed explanation for the differential molecular recognition of substrates has come from application of computer-based molecular modeling [21–24].

The premise of our approach was that common structural patterns must be present among all small peptides that form the basis for the individual specificities of the different transporters. Thus, for each peptide transporter, a molecular recognition template (MRT) can be defined that includes the structural and electronic features upon which recognition and binding must depend. To define the MRT for each transporter we focussed upon the following features of their natural substrates: (1) N-terminal α-amino and C-terminal α-carboxylate groups, allowing hydrogen bonding and charge interactions; (2) backbone torsion angles psi (ψ), phi (ϕ), and omega (ω); (3) chiral centers at α-carbons; (4) N–C distance between terminal amino and carboxylate groups; (5) chi (χ) space torsions of side chains; (6) hydrogen bonding of peptide bond atoms; (7) charge fields around the terminal α-amino and α-carboxylate groups. To identify MRTs, we carried out computer-based, conformational analysis of extensive collections of representative di-, tri-, and higher oligopeptides [21–24]. Normal, zwitterionic peptides are in general flexible molecules that exist in aqueous solution as a collection of conformers. When dipeptides were modeled, each possessed its own repertoire of conformers but all existed with a common set of 9 combinations of backbone torsions defined by specific ψ and ϕ angles combined with a *trans*-ω peptide bond. To describe these results, ψ and ϕ torsional space was each divided into twelve 30° sectors, (A1-A12) and (B1-B12), respectively. The preferred ψ-values occur in sectors A7 (+150° to ±180°), A10 (+60° to +90°), and A4 (−60° to −90°) and these are combined with preferred ϕ-values in sectors B12 (−150° to ±180°), B9 (−60° to −90°), and B2 (+30° to +60°). For any peptide, the percentage of each of its conformers was calculated by comparing its energy with that of the minimum energy conformer using a Boltzmann distribution. The distribution of all conformers, weighted according to the percentage of each, can be visualized graphically and analyzed using a novel, three-dimensional pseudo-Ramachandran plot (Fig. 1).

Comparison of the different transport and binding specificities of di-, and tripeptides for Dpp and Tpp with the conformer profiles of the peptides revealed the MRT for each transporter [21–24]. Dpp recognizes dipeptides with A7B9 and A7B12 torsions, whereas Tpp recognizes A4B9, A4B12, A10B9 and A10B12 conformers; representations of these for AlaAla are shown in Fig. 2. Corroboration of these conclusions has come from various experimental sources including biological assays, measurement of thermodynamic binding parameters from ITC, and also from the crystal structure of DppA liganded with GlyLeu, which shows the peptide to be present in an A7B9 conformation [31]. In addition, Dpp and Tpp recognize different "folded" conformations of tripeptides in which backbone torsions and N–C distances match those of the relevant dipeptide substrates. Thus, for any peptide, the percentage of its conformers in a particular MRT correlates with its relative transport and binding by each transporter. The features that define the MRTs of these transporters are summarized in Tab. 2.

Conformational analysis of higher oligopeptides has allowed the important features of the MRT for Opp to be defined [22, 24]. Opp preferentially recognizes tri- and oligopeptides having A7B9 torsions with "extended" backbones in which N–C distances are typically > 6.5 . For tripeptides, these comprise a different set of conformers from those recognized by Dpp and Tpp, giving Opp complementary substrate specificity. For higher oligopeptides, A7B9 conformers become increasingly prevalent, underpinning the reason for Opp having optimized its specificity for such conformers. Further conclusions have been drawn from these studies. For example, certain types of di-, tripeptides are particularly poor substrates for Dpp and Tpp; e.g., with N-terminal Asp or Glu residues, many conformers cannot match an MRT because side chains with gauche$^+$ and gauche$^-$ torsions have their charged carboxylates stabilized with the N-terminal α-amino group: to overcome this difficulty there is evidence that a specific transporter for such peptides may be present [23].

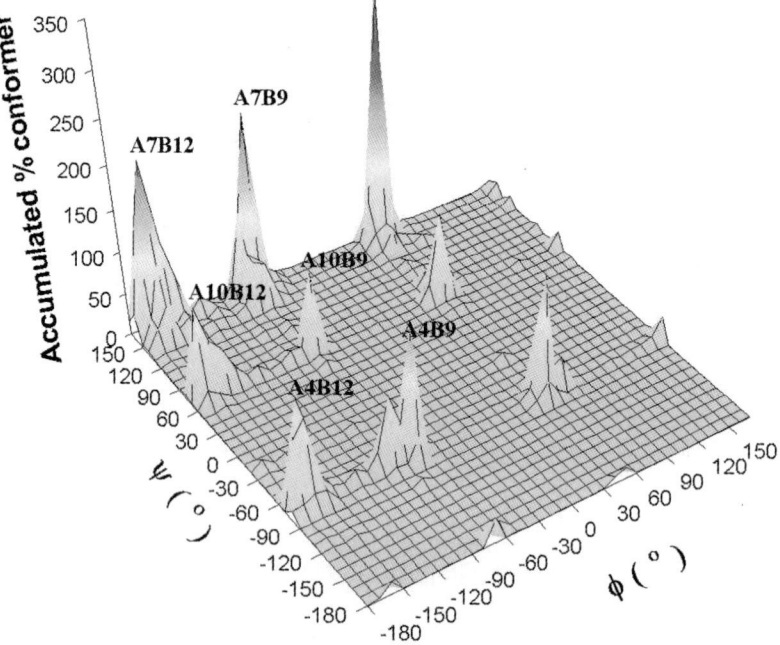

Fig. 1. 3-D pseudo-Ramachandran plot for 50 dipeptides. The conformational forms recognized by Dpp-type (A7B9 and A7B12) and Tpp-type (A4B9, A4B12, A10B9, and A10B12) transporters are indicated above the relevant peaks; see color plates p. XXXII.

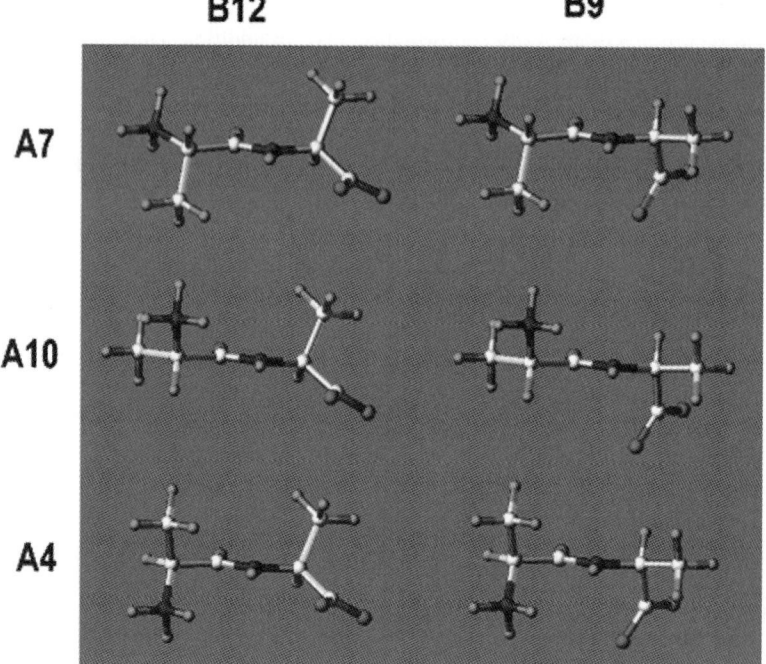

Tab. 2. Optimal features of the molecular recognition templates (MRTs) for Dpp-, Tpp- and Opp-type transporters

MRT Feature	Dpp-type	Tpp-type	Opp-type
Number of residues[a]	2 (3)	2 (3)	(2) 3–6
N-terminal α-amino group	positively charged amino group optimal; substitution with charge retention is acceptable		
C-terminal α-carboxyl group	negatively charged carboxylate is optimal		charged carboxylate is not essential
N-terminal ψ[b]	+140° to −175°	−50° to −85° and +50° to +85°	+140° to ±180°
C-terminal ϕ[c]	−50° to −95° and −130° to −175°		−60° to −90°
Peptide bonds	all trans (±180°) greatly preferred		
Residue chirality	all L-stereochemistry is optimal		
N–C distance[d]	5.1 Å to 6.4 Å	4.5 Å to 5.6 Å	> 6.5 Å
Side chains	all natural and some larger, unnatural side chains accepted		
Isopotential fields[e]	these should be symmetric about the N- and C-termini		
Stabilization by backbone atoms	unmodified peptide bond NH and CO groups are optimal to H-bond with appropriate groups on transporter proteins		

[a] Ideal number of peptide residues transported. The figures in brackets represent peptides transported less well.
[b] Psi torsion angle (ψ) of dipeptide before central peptide bond (or first peptide bond in tri- and higher oligopeptides).
[c] Phi torsion angle (ϕ) of dipeptide after central peptide bond (or first peptide bond in tri- and higher oligopeptides).
[d] Distance between N-terminal N- and C-terminal C-atoms.
[e] Distribution of positive or negative charge about the N- and C-terminus, respectively, of a peptide determined by using molecular modeling package SYBYL 6.5.

Fig. 2. Ball-and-stick representations of AlaAla in the conformational forms recognized by Dpp-type and Tpp-type transporters. Dpp-type transporters recognize A7B9 and A7B12 conformers, whereas Tpp-type transporters recognize A4B9, A4B12, A10B9, and A10B12 conformers. The relative orientations of the central peptide bond of the conformers are constant throughout; see color plates p. XXXIII.

In summary, these three archetypal transporters have evolved complementary specificities to optimize utilization of di-, tri- and higher oligopeptides. It is now apparent that the driving force for the evolution of the substrate specificities of these transporters has been the selective pressure represented by the conformer profiles of these peptides in solution. Although these conclusions have been derived from studies upon the main peptide permeases of *E. coli*, the fundamental principles of the relationship between conformer profiles and transporter specificities is such that we expect it to apply to all main peptide transporters, not only in microorganisms but also in all other organisms, including man, which have evolved in response to a common selection pressure present in the "universal" peptide pool. However, apart from a few notable exceptions [142], there exists a surprising reluctance among those who study analogous systems in mammalian cells and tissues to draw lessons from results of related studies in other species, and also to appreciate the fundamental principles of molecular recognition that underpin the common structural specificities found in all peptide transporters.

6.7
Exploitation of Peptide Transporters for Delivery of Therapeutic Compounds

Much of the stimulus for research into the characterization of peptide transport systems and their substrate specificities stems from the desire to capitalize upon the unrivalled opportunities they offer for transporting therapeutic compounds. The uptake of natural antimicrobial peptides and of synthetic peptidomimetic smuggling designed to exploit peptide permeases to transport impermeable compounds into microorganisms, has been discussed [6–10] together with wider discussions of peptide-based drug design and oral transport [143]. Based upon the information provided by MRTs, the potential of this rational approach will now be greatly enhanced by the ability to tailor the design of drug molecules and prodrugs precisely. Optimal analogs of novel compounds can be identified *in silico* at an early stage in a drug development program, thereby minimizing the need to synthesize and to test numerous compounds. In our laboratory, we have confirmed this expectation by conformational analysis of a range of orally available β-lactam antibiotics, of angiotensin-converting enzyme inhibitors and anti-viral prodrugs, which has clarified the basis for their recognition and transport by intestinal and renal peptide transporters. From these studies, it has proved possible to elaborate general principles for producing conformationally constrained analogs of the above classes of compounds and of novel anti-cancer compounds that offer good oral absorption (B. M. Grail, S. Gupta, N. J. Marshall, G. M. Payne, and J. W. Payne, unpublished data).

Acknowledgement

We thank B. M. Grail, S. Gupta, and G. M. Payne for their extensive help with the development of the ideas and results reported from our laboratory; this research was supported in part by grants from the Biotechnology and Biological Sciences Research Council and from the Research and Development Committee of the North West Wales NHS Trust.

Note added in proof: see p. 479.

References

1. J. W. Payne, *Adv. Microb. Physiol.* **1976**, *13*, 55-113.
2. J. W. Payne, Transport and utilization of peptides by bacteria, in: *Microorganisms and Nitrogen Sources: Transport and Utilization of Amino Acids, Peptides, Proteins and Related Substrates* (Payne, J. W., Ed.), John Wiley & Sons, Chichester, New York, **1980**, pp. 211-256.
3. D. M. Matthews, J. W. Payne, *Curr. Topics Membr. Trans.* **1980**, *14*, 331-425.
4. J. M. Becker, F. Naider, Transport and utilization of peptides by yeast, in: *Microorganisms and Nitrogen Sources: Transport and Utilization of Amino Acids, Peptides, Proteins and Related Substrates* (Payne, J. W., Ed.), John Wiley & Sons, Chichester, New York, **1980**, pp. 257-281.
5. C. F. Higgins, M. M. Gibson, *Methods Enzymol.* **1986**, *125*, 365-377.
6. J. W. Payne, M. W. Smith, *Adv. Microb. Physiol.* **1994**, *36*, 1-80.
7. J. W. Payne, Bacterial peptide permeases as a drug delivery target, in: *Peptide-Based Drug Design: Controlling Transport and Metabolism* (Taylor, M. D., Amidon, G. L., Eds.), American Chemical Society, Washington, DC, **1995**, pp. 341-367.
8. J. M. Becker, F. Naider, Fungal peptide transport as a drug delivery system, in: *Peptide-Based Drug Design: Controlling Transport and Metabolism* (Taylor, M. D., Amidon, G. L., Eds), American Chemical Society, Washington, DC, **1995**, pp. 369-386.
9. D. R. Tyreman, M. W. Smith, G. M. Payne, J. W. Payne, Exploitation of peptide transport systems in the design of antimicrobial agents, in: *Molecular Aspects of Chemotherapy* (Shugar, D., Rode, W., Borowski, E., Eds.), Springer-Verlag, Berlin, **1992**, pp. 127-142.
10. D. R. Tyreman, M. W. Smith, N. J. Marshall, G. M. Payne, C. M. Schuster, B. M. Grail, J. W. Payne, Peptides as prodrugs: the smugglin concept, in: *Peptides in Mammalian Protein Metabolism: Tissue Utilization and Clinical Targeting* (Grimble, G. M., Backwell, F. R. C., Eds.), Portland Press, London, **1998**, pp. 141-157.
11. I. T. Paulsen, M. K. Sliwinski, M. H. Saier, *J. Mol. Biol.* **1998**, *277*, 573-592.
12. I. T. Paulsen, L. Nguyen, M. K. Sliwinski, R. Rabus, M. H. Saier, *J. Mol. Biol.* **2000**, *301*, 75-100.
13. I. T. Paulsen, M. K. Sliwinski, B. Nelissen, A. Goffeau, M. H. Saier, *FEBS Lett.* **1998**, *430*, 116-125.
14. M. H. Saier, *Adv. Microb. Physiol.* **1998**, *40*, 81-136.
15. M. H. Saier, *Microbiology* **2000**, *146*, 1775-1795.
16. R. Tam, M. H. Saier, *Microbiol. Rev.* **1993**, *57*, 320-346.
17. I. T. Paulsen, R. A. Skurray, *Trends Biochem. Sci.* **1994**, *18*, 404.
18. H. Y. Steiner, F. Naider, J. M. Becker, *Mol. Microbiol.* **1995**, *16*, 825-834.
19. M. W. Smith, D. R. Tyreman, G. M. Payne, N. J. Marshall, J. W. Payne, *Microbiology* **1999**, *145*, 2891-2901.

20. S. S. Pao, I. T. Paulsen, M. H. Saier Jr., *Microbiol. Mol. Biol. Rev.* **1998**, *62*, 1-34.
21. J. W. Payne, B. M. Grail, N. J. Marshall, *Biochem. Biophys. Res. Commun.* **2000**, *267*, 283-289.
22. B. M. Grail, J. W. Payne, *J. Pept. Sci.* **2000**, *6*, 186-199.
23. J. W. Payne, B. M. Grail, S. Gupta, J. E. Ladbury, N. J. Marshall, R. O'Brien, G. M. Payne, *Arch. Biochem. Biophys.* **2000**, *384*, 9-23.
24. N. J. Marshall, B. M. Grail, J. W. Payne, *J. Peptide Sci.* **2001**, *7*, 175-189.
25. J. W. Payne, *Biochem. Soc. Trans.* **1983**, *19*, 794-798.
26. M. D. Manson, V. Blank, G. Brade, C. F. Higgins, *Nature* **1986**, *321*, 253-256.
27. W. N. Abouhamad, M. D. Manson, *Mol. Microbiol.* **1994**, *14*, 1077-1092.
28. E. R. Olson, D. S. Dunyak, L. N. Jurss, R. A. Poorman, *J. Bacteriol.* **1991**, *173*, 234-244.
29. M. L. Urbanowski, L. T. Stauffer, G. V. Stauffer, *Mol. Microbiol.* **2000**, *37*, 856-868.
30. A. V. Nickitenko, S. Trakhanov, F. A. Quiocho, *Biochemistry* **1995**, *34*, 16585-16595.
31. P. W. Dunten, S. L. Mowbray, *Protein Sci.* **1995**, *4*, 2327-2334.
32. K. J. Linton, C. F. Higgins, *Mol. Microbiol.* **1998**, *28*, 5-13.
33. C. E. Liu, P. Q. Liu, A. Wolg, E. Lin, G. F. L. Ames, *J. Biol. Chem.* **1999**, *274*, 739-747.
34. P. Q. Liu, C. E. Liu, G. F. L. Ames, *J. Biol. Chem.* **1999**, *274*, 18310-18318.
35. S. H. Sleigh, P. R. Seavers, A. J. Wilkinson, J. E. Ladbury, J. R. H. Tame, *J. Mol. Biol.* **1999**, *291*, 393-415.
36. T. G. Davies, R. E. Hubbard, J. R. H. Tame, *Protein Sci.* **1999**, *8*, 1432-1444.
37. S. H. Sleigh, J. R. H. Tame, E. J. Dodson, A. J. Wilkinson, *Biochemistry* **1997**, *36*, 9747-9758.
38. J. R. H. Tame, G. N. Murshudov, E. J. Dodson, T. K. Neil, G. G. Dodson, C. F. Higgins, A. J. Wilkinson, *Science* **1994**, *264*, 1578-1581.
39. J. R. H. Tame, E. J. Dodson, G. Murshudov, C. F. Higgins, A. J. Wilkinson, *Structure* **1995**, *3*, 1395-1406.
40. J. R. H. Tame, S. H. Sleigh, A. J. Wilkinson, J. E. Ladbury, *Nature Struct. Biol.* **1996**, *3*, 998-1001.
41. A. A. Rostom, J. R. H. Tame, J. E. Ladbury, C. V. Robinson, *J. Mol. Biol.* **2000**, *296*, 269-279.
42. Y. Papanikolau, R. Gessmann, K. Petratos, K. Igarashi, M. Kokkinidis, *J. Cryst. Growth* **2000**, *210*, 761-766.
43. R. L. Stanfield, I. A. Wilson, *Curr. Opin. Struct. Biol.* **1995**, *5*, 103-113.
44. T. G. Davies, J. R. H. Tame, R. E. Hubbard, *Perspect. Drug Discov. Design* **2000**, *20*, 29-42.
45. R. J. Wallace, *J. Nutrit.* **1996**, *126*, S1326-S1334.
46. R. J. Wallace, C. Atasoglu, C. J. Newbold, *Asian-Austral. J. Anim. Sci.* **1999**, *12*, 139-147.
47. J. R. Ling, I. P. Armstead, *J. Appl. Bacteriol.* **1995**, *78*, 116-124.
48. R. J. Wallace, N. McKain, G. A. Broderick, L. M. Rode, N. D. Walker, C. J. Newbold, J. Kopecny, *Anaerobe* **1997**, *3*, 35-42W.
49. K. E. Griswold, B. A. White, R. I. Mackie, *Curr. Microbiol.* **1999**, *39*, 187-194.
50. E. R. S. Kunji, I. Mierau, A. Hagting, B. Poolman, W. N. Konings, *Antonie van Leeuwenhoek* **1996**, *70*, 187-221.
51. I. Mierau, E. R. S. Kunji, K. J. Leenhouts, M. A. Hellendoorn, A. J. Haandrikman, B. Poolman, W. N. Konings, G. Venema, J. Kok, *J. Bacteriol.* **1996**, *178*, 2794-2803.
52. V. Juillard, C. Foucaud, M. Desmazeaud, J. Richard, *LAIT* **1996**, *76*, 13-24.
53. I. Mierau, E. R. S. Kunji, G. Venema, J. Kok, *Biotechnol. Genet. Eng. Rev.* **1997**, *14*, 279-301.
54. E. R. S. Kunji, G. Fang, C. M. Jeronimus-Stratingh, A. P. Bruins, B. Poolman, W. N. Konings, *Mol. Microbiol.* **1998**, *27*, 1107-1118.
55. A. Verheul, F. M. Rombouts, T. Abee, *Appl. Environ. Microbiol.* **1998**, *64*, 1059-1065.
56. I. C. Sutcliffe, R. R. Russell, *J. Bacteriol.* **1995**, *177*, 1123-1128.
57. C. Foucaud, E. R. S. Kunji, A. Hagting, J. Richard, W. N. Konings, M. Desmazeaud, B. Poolman, *J. Bacteriol.* **1995**, *177*, 46523-4657.
58. Y. Sanz, F. C. Lanfermeijer, W. N. Konings, B. Poolman, *Biochemistry* **2000**, *39*, 4855-4862.
59. A. Hagting, E. R. S. Kunji, K. J. Leenhouts, B. Poolman, W. N. Konings, *J. Biol. Chem.* **1994**, *269*, 11391-11399.
60. A. Hagting, J. Knol, B. Hasemeier, M. R. Streutker, G. Fang, B. Poolman, W. N. Konings, *Eur. J. Biochem.* **1997**, *247*, 581-587.
61. H. Nakajima, A. Hagting, E. R. S. Kunji, B. Poolman, W. N. Konings, *Appl. Environ. Microbiol.* **1997**, *63*, 2213-2217.
62. G. Fang, R. Friesen, F. Lanfermeijer, A. Hagting, B. Poolman, W. N. Konings, *Mol. Membr. Biol.* **1999**, *16*, 297-304.

63. G. Fang, W. N. Konings, B. Poolman, *J. Bacteriol.* **2000**, *182*, 2530-2535.
64. F. J. M. Detmers, E. R. S. Kunji, F. C. Lanfermeijer, B. Poolman, W. N. Konings, *Biochemistry* **1998**, *37*, 16671-16679.
65. F. C. Lanfermeijer, A. Picon, W. N. Konings, B. Poolman, *Biochemistry* **1999**, *38*, 14440-14450.
66. F. C. Lanfermeijer, F. J. M. Detmers, W. N. Konings, B. Poolman, *EMBO J.* **2000**, *19*, 3649-3656.
67. A. Picon, E. R. S. Kunji, F. C. Lanfermeijer, W. N. Konings, B. Poolman, *J. Bacteriol.* **2000**, *182*, 1600-1608.
68. F. J. M. Detmers, F. C. Lanfermeijer, R. Abele, R. W. Jack, R. Tampe, W. N. Konings, B. Poolman, *Proc. Natl. Acad. Sci. USA* **2000**, *97*, 12487-12492.
69. G. Alloing, P. DePhilip, J. P. Claverys, *J. Mol. Biol.* **1994**, *241*, 44-58.
70. B. J. Pearce, A. M. Naughton, H. R. Masure, *Mol. Microbiol.* **1994**, *12*, 881-892.
71. G. Alloing, C. Granadel, D. A. Morrison, J. P. Claverys, *Mol. Microbiol.* **1996**, *21*, 471-478.
72. G. Alloing, B. Martin, C. Granadel, J. P. Claverys, *Mol. Microbiol.* **1998**, *29*, 75-83.
73. J. P. Claverys, B. Grossiord, G. Alloing, *Res. Microbiol.* **2000**, *151*, 457-463.
74. R. McNab, H. F. Jenkinson, *Microbiology* **1998**, *144*, 127-136.
75. H. F. Jenkinson, R. A. Baker, G. W. Tannock, *J. Bacteriol.* **1996**, *178*, 68-77.
76. B. Henrich, M. Hopfe, A. Kitzerow, U. Hadding, *J. Bacteriol.* **1999**, *181*, 4873-4878.
77. A. Podbielski, B. Pohl, M. Woischnik, C. Korner, K. H. Schmidt, E. Rozdzinski, B. A. B. Leonard, *Mol. Microbiol.* **1996**, *21*, 1087-1099.
78. A. Podbielski, B. A. B. Leonard, *Mol. Microbiol.* **1998**, *28*, 1323-1334.
79. J. L. Bono, K. Tilly, B. Stevenson, D. Hogan, P. Rosa, *Microbiology* **1998**, *144*, 1033-1044.
80. J. A. Kornacki, D. N. Oliver, *Infect. Immun.* **1998**, *66*, 4115-4122.
81. B. A. Lazazzera, A. D. Grossman, *Trends Microbiol.* **1998**, *6*, 288-294.
82. G. M. Dunny, B. A. B. Leonard, *Ann. Rev. Microbiol.* **1997**, *51*, 527-564.
83. B. L. Bassler, *Curr. Opin. Microbiol.* **1999**, *2*, 582-587.
84. B. A. Lazazzera, *Curr. Opin. Microbiol.* **2000**, *3*, 177-182.
85. M. Perego, C. F. Higgins, S. R. Pearce, M. P. Gallagher, J. A. Hoch, *Mol. Microbiol.* **1991**, *5*, 173-185.
86. D. Z. Rudner, J. R. Ladeaux, K. Breton, A. D. Grossman, *J. Bacteriol.* **1991**, *173*, 1388-1398.
87. A. Koide, J. A. Hoch, *Mol. Microbiol.* **1994**, *13*, 417-426.
88. J. M. Solomon, B. A. Lazazzera, A. D. Grossman, *Genes Devel.* **1996**, *10*, 2014-2024.
89. B. A. Lazazzera, J. M. Solomon, A. D. Grossman, *Cell* **1997**, *89*, 917-925.
90. M. Perego, *Proc. Natl. Acad. Sci. USA* **1997**, *94*, 8612-8617.
91. A. Koide, M. Perego, J. A. Hoch, *J. Bacteriol.* **1999**, *181*, 4114-4117.
92. J. R. Nodwell, K. McGovern, R. Losick, *Mol. Microbiol.* **1996**, *22*, 881-893.
93. J. R. Nodwell, R. Losick, *J. Bacteriol.* **1998**, *180*, 1334-1337.
94. M. T. G. Holden, S. R. Chhabra, R. deNys, P. Stead, N. J. Bainton, P. J. Hill, M. Manefield, N. Kumar, M. Labatte, D. England, S. Rice, M. Givskov, G. P. C. Salmond, G. S. A. B. Stewart, B. W. Bycroft, S. A. Kjelleberg, P. Williams, *Mol. Microbiol.* **1999**, *33*, 1254-1266.
95. C. ParraLopez, M. T. Baer, E. A. Groisman, *EMBO J.* **1993**, *12*, 4053-4062.
96. H. Y. Chen, S. F. Weng, J. W. Lin, *Biochem. Biophys. Res. Commun.* **2000**, *269*, 743-748.
97. C. ParraLopez, R. Lin, A. Aspedon, E. A. Groisman, *EMBO J.* **1994**, *13*, 3964-3972.
98. J. S. Gunn, S. I. Miller, *J. Bacteriol.* **1996**, *178*, 6857-6864.
99. S. Stumpe, R. Schmid, D. L. Stephens, G. Georgiou, E. P. Bakker, *J. Bacteriol.* **1998**, *180*, 4002-4006.
100. R. A. Salomon, R. N. Farias, *J. Bacteriol.* **1995**, *177*, 3323-3325.
101. A. Ichige, G. C. Walker, *J. Bacteriol.* **1997**, *179*, 209-216.
102. K. Igarashi, T. Saisho, M. Yuguchi, K. Kashiwagi, *J. Biol. Chem.* **1997**, *272*, 4058-4064.
103. M. Yoshida, D. Meksuriyen, K. Kashiwagi, G. Kawai, K. Igarashi, *J. Biol. Chem.* **1999**, *274*, 22723-22728.
104. K. Kashiwagi, A. Miyaji, S. Ikeda, T. Tobe, C. Sasakawa, *J. Bacteriol.* **1992**, *174*, 4331-4337.
105. H. H. Tsuhako, L. C. S. Ferreira, S. O. P. da Costa, *Genet. Mol. Biol.* **1998**, *21*, 15-19.

106. K. Kashiwagi, M. H. Tsuhako, K. Sakata, T. Saisho, A. Igarashi, S. O. Pinto da Costa, K. Igarashi, *J. Bacteriol.* **1998**, *180*, 5484-5488.
107. M. B. Rodriguez, S. O. P. Costa, *Revista Microbiol.* **1999**, *30*, 153-156.
108. M. B. R. Acosta, R. C. C. Ferreira, G. Padilla, L. C. S. Ferreira, S. O. P. Costa, *J. Med. Microbiol.* **2000**, *49*, 409-413.
109. J. T. Park, *Mol. Microbiol.* **1995**, *17*, 421-426.
110. J. T. Park, D. Raychaudhuri, H. S. Li, S. Normark, D. Mengin-Lecreuix, *J. Bacteriol.* **1998**, *180*, 1215-1223.
111. N. D. Hanson, C. C. Sanders, *Curr. Pharm. Des.* **1999**, *5*, 881-894.
112. H. Li, J. T. Park, *J. Bacteriol.* **1999**, *181*, 4842-4847.
113. C. Wulfing, A. Pluckthun, *Mol. Microbiol.* **1994**, *12*, 685-692.
114. F. Shao, M. W. Bader, U. Jakob, J. C. A. Bardwell, *J. Biol. Chem.* **2000**, *275*, 13349-13352.
115. D. Missiakis, S. Raina, *J. Bacteriol.* **1997**, *179*, 2465-2471.
116. G. Richarme, T. D. Caldas, *J. Biol. Chem.* **1997**, *272*, 15607-12.
117. M. Matsuzaki, Y. Kiso, I. Yamamoto, T. Satoh, *J. Bacteriol.* **1998**, *180*, 2718-2722.
118. M. Matsuzaki, Y. Kiso, I. Yamamato, T. Satoh, *FEMS Microbiol. Lett.* **2000**, *193*, 223-229.
119. V. F. Smith, S. J. Hardy, L. L. Randall, *Protein Sci.* **1997**, *6*, 1746-1755.
120. J. E. Bruce, V. F. Smith, C. Liu, L. L. Randall, R. D. Smith, *Protein Sci.* **1998**, *7*, 1180-1185.
121. E. Verkamp, V. M. Bachman, J. M. Bjornsson, D. Soll, G. Eggertsson, *J. Bacteriol.* **1993**, *175*, 1452-1456.
122. T. Elliott, *J. Bacteriol.* **1993**, *175*, 325-331.
123. N. D. King, M. R. O'Brian, *J. Bacteriol.* **1997**, *179*, 1828-1831.
124. F. Doring, J. Walter, M. Focking, M. Boll, S. Amashesh, W. Clauss, H. Daniel, *J. Clin. Invest.* **1998**, *101*, 2761-2767.
125. C. Sherrill, R. C. Fahey, *J. Bacteriol.* **1998**, *180*, 1454-1459.
126. T. Miyake, T. Hazu, S. Yoshida, M. Kanayama, K. Tomochika, S. Shinoda, B. Ono, *Biosci. Biotech. Biochem.* **1998**, *62*, 1858-1864.
127. A. Bourbouloux, P. Shahi, A. Chakladar, S. Delrot, A. H. Bachhawat, *J. Biol. Chem.* **2000**, *275*, 13259-13265.
128. W. Tanner, T. Caspari, *Ann. Rev. Plant Physiol. Plant Mol. Biol.* **1996**, *47*, 595-626.
129. J. Horak, *Biochim. Biophys. Acta Rev. Biomembr.* **1997**, *1331*, 41-79.
130. J. R. Perry, M. A. Basrai, H. Y. Steiner, F. Naider, J. M. Becker, *Mol. Cell. Biol.* **1994**, *14*, 104-115.
131. D. Meredith, C. A. R. Boyd, *Cell. Mol. Life Sci.* **2000**, *57*, 754-778.
132. M. A. Basrai, M. A. Lubkowitz, J. R. Perry, D. Miller, E. Krainer, F. Naider, J. M. Becker, *Microbiology* **1995**, *141*, 1147-1156.
133. M. A. Lubkowitz, L. Hauser, M. Breslav, F. Naider, J. M. Becker, *Microbiology* **1997**, *143*, 387-396.
134. M. A. Lubkowitz, D. Barnes, M. Breslav, A. Burchfield, F. Naider, J. M. Becker, *Mol. Microbiol.* **1998**, *28*, 729-741.
135. M. Hauser, A. M. Donhardt, D. Barnes, F. Naider, J. M. Becker, *J. Biol. Chem.* **2000**, *275*, 3037-3041.
136. K. Alagramam, F. Naider, J. M. Becker, *Mol. Microbiol.* **1995**, *15*, 225-234.
137. D. Barnes, W. Lai, M. Breslav, F. Naider, J. M. Becker, *Mol. Microbiol.* **1998**, *29*, 297-310.
138. T. Didion, B. Regenberg, M. J. Jorgensen, M. C. Kielland-Brand, H. A. Andersen, *Mol. Microbiol.* **1998**, *27*, 643-650.
139. H. Klasson, G. R. Fink, P. O. Ljungdahl, *Mol. Cell. Biol.* **1999**, *19*, 5405-5416.
140. W. M. Ingledew, C. A. Patterson, *J. Am. Soc. Brew. Chem.* **1999**, *57*, 9-17.
141. C. A. Patterson, W. M. Ingledew, *J. Am. Soc. Brew. Chem.* **1999**, *57*, 1-8.
142. D. M. Matthews, *Protein Absorption: Development and Present State of the Subject*, Wiley-Liss, New York, **1991**.
143. M. D. Taylor, G. L. Amidon, *Peptide-Based Drug Design: Controlling Transport and Metabolism* (Taylor, M. D., Amidon, G. L., Eds.), American Chemical Society, Washington, DC, **1995**.

7
Protein Export and Secretion in Gram-negative Bacteria

Philippe Delepelaire and Cécile Wandersman

7.1
Introduction

Eukaryotic cells contain extensive sets of organelles, whereas bacteria have few distinct cellular compartments. Gram-positive bacteria have one cytoplasmic compartment, delimited by a single membrane. Gram-negative bacteria are more differentiated with two membrane delimited compartments: the periplasm and the cytoplasm. However, approximately one third of the total bacterial proteins do not remain in the cytoplasm where they are synthesized. Instead they are targeted to final extracytoplasmic locations, such as the inner membrane, the periplasmic space, the outer membrane, the cell surface (external part of the outer membrane), the extracellular medium, or into other recipient cells. Integral membrane proteins carry out many vital biochemical processes (respiration, transport, cell division, etc.). Although extracellular and surface exposed proteins are not usually essential for bacterial growth in laboratory culture conditions, they are often essential virulence determinants. Adhesins and invasins are necessary for pathogen–host interactions, colonization, and internalization by forced phagocytic uptake. Extracellular hydrolytic enzymes favor bacterial spread by degrading extracellular matrix components. They also provide nutrients by breaking down host polymers into small pieces, which can be internalized and metabolized. Exotoxins and pore-forming proteins are involved in host killing and the escape from endocytic vacuoles. Many secreted proteins are translocated into the eukaryotic cytoplasm, where they interfere with host signaling pathways and host defenses.

Most extracellular proteins would be ineffective or harmful if they were sent to the wrong site. This emphasizes the necessity for reliable cellular sorting. This cellular sorting involves multiple mechanisms.

One of the sorting mechanisms involved in the transport of proteins across the inner membrane is the universal signal peptide-dependent general export pathway. This mechanism was first described in eukaryotic cells by Blobel and Dobberstein in 1975 [1]. They established that protein precursors are initially synthesized with extra amino acids at their N-terminus, which act as a signal to initiate protein export. Gene fusions between signal sequences and genes encoding a cytoplasmic protein (β-galactosidase) showed that the signal peptide is involved in protein export [2]. Hybrid proteins allowed genetic tools to be devised to identify the bacterial secretion apparatus proteins: the Sec proteins, which are mostly encoded by essential genes. This pathway is now well characterized in gram-positive and gram-negative bacteria. It is also known as the *sec*-dependent pathway, and will be briefly described below.

Until recently it was believed that this was the only mechanism for exporting proteins across the inner membrane. Recently, a *sec*-independent mechanism, the double arginine pathway, was described. This pathway allows proteins to be exported to the periplasm, and involves a special signal peptide and a specific set of secretion proteins. It will be described in the second part of this review.

However, neither the general export pathway nor the double arginine pathway are sufficient to direct proteins beyond the outer membrane. This was first shown by expressing genes of different bacterial origins, encoding extracellular proteins, in *Escherichia coli*. The foreign proteins were synthesized in *E. coli*, but not secreted into the extracellular medium. Thus the secreted proteins do not contain sufficient information to allow their secretion, and the recipient strain probably lacks some secretion functions. Some foreign exoproteins with signal peptides were located in the periplasmic space when expressed in *E. coli*. This suggests that they were translocated through the inner membrane by the *sec*-dependent pathway and were unable to leave the periplasm in the absence of secretion functions. Consistent with this hypothesis, foreign proteins accumulated in the cytoplasm of *E. coli sec* mutants at non-permissive temperatures, indicating that their periplasmic localization was indeed dependent on the *sec* genes. Several viable secretion defective mutants were isolated from naturally secreting species. In these mutants polypeptides accumulated in the periplasm, thus translocation through the outer membrane may be carried out by non-essential mechanisms. Hence, polypeptides cross the two membranes by distinct mechanisms; the inner membrane by the general export pathway, and the outer membrane by a specific mechanism [3].

In other cases, the foreign exoproteins lacking a signal peptide, remained in the cytoplasm when expressed in *E. coli*; thus their secretion is signal peptide- and *sec*-independent.

Export beyond the outer membrane always requires specific systems, therefore, we use the term "protein export" to refer to the mechanisms by which proteins are transported through the inner membrane, and "protein secretion" for mechanisms by which proteins are transported beyond the outer membrane to their final locations.

Bacteria secrete many proteins which have diverse sequences and functions. However, only four secretion pathways have been described in gram-negative bacteria. Thus, unrelated proteins are often secreted by similar pathways. The systems are similar, but are often specific for one protein, for closely related proteins, or for unrelated proteins produced by the same species. This specificity is also reflected by the genetic organization of the secretion systems; the genes encoding the secretion apparatus and the secreted proteins are usually linked. The gene clusters may

Tab. 1. General features of the different secretion pathways found in gram-negative bacteria

Secretion Type	Sec Dependency	Supra-molecular Structure	Major Protein Function	Microorganisms
I	no	no	hemolysins protease lipases	*Bordetella* spp. *Neisseria* spp. *Escherichia coli*
ABC Transporter			toxins hemophores S-layers unknown	*Serratia marcescens* *Pseudomonas aeruginosa* *Pseudomonas fluorescens* *Yersinia pestis* *Erwinia chrysanthemi*
II	yes	no	IgA1 proteases pertactins proteases invasins	*Bordetella* spp. *Neisseria* spp. *Escherichia coli* *Shigella* ssp.
Auto-transporter			adhesins toxins S-layers unknown	*Haemophilus inflenzae* *Moraxella* spp. *Rickettsia* spp. *Serratia marcescens*
II One outer membran helper	yes	no	hemolysins hemagglutinins	*Bordetella* spp. *Proteus mirabilis* *Serratia marcescens*
II	yes	pilus-like structure	proteases lipases toxins pullulanases	*Pseudomonas aeruginosa*, *Vibrio cholerae* *Klebsiella oxytoca* *Erwinia*
GSP			unknown	*Xanthomonas* *Aeromonas* *Legionella pneumophila*
III	no	needle-like structure	protein tyrosin phosphatases protein kinases pore forming translocases toxins adhesins invasins unknown	*Shigella flexneri* *Yersinia* spp. *Salmonella* spp. *Escherichia coli* spp. *Pseudomonas aeruginosa* *Pseudomonas syringae* *Erwinia amylovora* *Ralstonia solanacearum.*
IV	a) yes	sex pilus	toxins unknown	*Bordetella pertussis* *Brucella suis* *Legionella pneumoniae*
	b) no	sex pilus	protein kinases SS DNA binding proteins nuclear targeting proteins	*Helicobacter pylori* *Agrobacterium tumefaciens*

have been maintained during evolution by horizontal transfer via conjugation, transformation, and bacteriophage infection. Indeed, many secretion systems (especially type III systems, see below) are located in pathogenicity islands (islands of conserved DNA sequences on distantly related genomes), on chromosomes, or on large virulence plasmids [4].

The four defined secretion pathways were numbered from one to four. Type I secretion pathways were identified in the early 1980s. Type II secretion pathways were identified in the mid-1980s. Type III and IV secretion pathways were identified within the last decade. Types I, III, and IV are quite homogeneous, whereas type II includes several unrelated mechanisms. Nevertheless, type II pathways are a defined entity, because they are the only secretion systems that require the general export pathway to the cross of the inner membrane. Type I and III pathways are independent of the *sec* system. However, some proteins secreted by pathway IV have a signal peptide, whereas others do not (Tab. 1).

This review will be divided into two parts. The first part will deal with *sec*-dependent and *sec*-independent protein export. The second part will describe protein secretion by the *sec*-dependent and *sec*-independent systems and the type IV pathway.

In gram-positive bacteria extracellular proteins are secreted by the signal peptide-dependent general export pathway. Translocation through the single cytoplasmic membrane and cleavage of the signal peptide usually release the mature polypeptide to the surrounding medium [5]. Assembly of surface organelles (pili and flagella) might be considered as a particular aspect of protein secretion, however, only the aspects of biosynthesis that are common or relevant for protein secretion will be addressed. Thus, this review will focus on protein export and secretion in gram-negative bacteria.

7.2
Protein Export

7.2.1
Sec Pathway

7.2.1.1 Introduction
Most proteins exported to the periplasm or the outer membrane use the Sec pathway, with a common signal peptide (SP). Extensive reviews have been written on the Sec translocase and its functioning [6–11]; thus we will review the recent developments on the targeting of proteins to the inner membrane translocase and the modularity of the different translocases found in the inner membrane of gram-negative bacteria.

After the signal hypothesis and initial experiments on eukaryotes, the development of genetic screens allowed experiments to be carried out on *E. coli*. These screens identified *prl* alleles, which can correctly localize proteins with defective signal sequences, and *sec* mutations which cause exported proteins to be mislocalized. Several *prl* genes were identical to *sec* genes, and the Sec nomenclature is now widely used. Biochemical approaches led to the functional reconstitution of the translocation reaction from purified components.

Efficient and reliable *in vitro* systems have been reconstituted in which SecY/SecE/SecG are the only integral membrane proteins associated in a complex and the SecA ATPase provides the driving energy for translocation via ATP hydrolysis [12–14]. The SecA ATPase is a major component of the Sec translocase [15]. It is a hydrophilic protein, which can be either soluble or membrane-bound due to its specific interaction with SecYE. It undergoes cycles of insertion/deinsertion in the membrane in an ATP-dependent manner, actively driving the insertion and translocation of segments of the polypeptide chain. This insertion/deinsertion cycle of SecA is coupled to the inversion of topology of SecG during the catalytic cycle of the translocase [16]. Another heterotrimeric Sec complex can also be isolated, SecD/SecF/YajC [17, 18]. SecYE is absolutely required for translocation across the cytoplasmic membrane, however, SecDFYajC plays a less essential role because viable null mutants can be obtained in the genes encoding those proteins, although their growth is cold-sensitive [19]. SecDFYajC enhances the proton-motive-force-dependent translocation of the periplasmic regions of translocated proteins, and aids the release of membrane-bound SecA during its catalytic cycle [17, 18]. The Sec translocase is required for the translocation of periplasmic and outer membrane proteins across the cytoplasmic membrane, and for the integration of several membrane proteins in the cytoplasmic membrane. The requirements of integral membrane proteins vary; some are Sec-independent, such as M13 procoat (see below), others are SecYE-dependent and SecA-independent [20, 21], and others are SecA- and SecYE-dependent [22]. Access to the translocase is a highly regulated process, involving several recognition steps before completion of translocation.

7.2.1.2 Targeting to the Sec translocase: SRP and Trigger Factor SecA/B Routes

At least two routes exist for the access to the Sec translocase: the SecA/B route for the translocation of protein precursors across the inner membrane, and the SRP (Signal Recognition Particle) route for the insertion of several membrane proteins. Bacterial SRP associates Ffh (analogous to the eukaryotic SRP54 component) and a 4.5S RNA (product of *ffs*). It then directs proteins to FtsY, which is analogous to the eukaryotic SRP receptor and is a peripheral membrane protein [23, 24]. Both Ffh and FtsY hydrolyze GTP. SecB is a cytoplasmic chaperone, dedicated to proteins export; it associates with the precursors, slows down their folding kinetics, and targets them to SecA via a high affinity SecB binding site on the C-terminus of SecA [25, 26]. The hydrophobicity of the sequence signal seems to determine whether the SecA/B or the SRP route is used [27]. The initial characterization of SRP function, based on *in vivo* depletion studies of Ffh or *ffs*, indicated that SRP was involved in the translocation of SecB-independent substrates, such as β-lactamase. The current view is that SRP cotranslationally targets proteins, with a highly hydrophobic signal sequence or signal anchor, to the membrane and is essential for the targeting of integral membrane proteins [28]. Ribosomes can also associate directly with the SecYEG translocon via the 23S RNA [29].

Beck and coworkers used ribosome-associated nascent chains from two distinct substrates, MtlA, a SRP-dependent polytopic inner membrane protein, and proOmpA, a SecA/B-dependent outer membrane protein to set up an *in vitro* system reproducing the early steps of these pathways [21, 30]. They found that Ffh was cross-linked to a 189 amino acid MtlA nascent chain and was tightly ribosome-bound. Addition of puromycin or of inner membrane vesicles plus GMP-PNP, a non-hydrolyzable GTP analog, abolishes the cross-linking to Ffh, compatible with the cotranslational binding of SRP to a nascent chain of a membrane protein and its release after membrane targeting. Other experiments have shown that the first signal anchor of MtlA is a recognition site for Ffh.

Nascent chains of proOmpA formed a major cross-link product with trigger factor. Trigger factor is a ribosome-bound peptidyl-prolyl isomerase, which also exists in the cytoplasm. It was isolated because it allows pro-OmpA to be in an efficient translocation competent state *in vitro*. Under identical experimental conditions trigger factor does not associate with nascent MtlA. This suggests that trigger factor which is tightly bound to ribosomes has a much lower affinity for the nascent chain of a polytopic membrane protein, MtlA, than for a secreted protein, pro-OmpA. Both SecA and B associate with nascent secretory proteins, but cross-linked adducts to SecA and SecB are only found after puromycin releases the pro-OmpA nascent chain from the ribosome, and are independent of each other. In ribosomal preparations from trigger factor minus strains both Ffh and SecA associate with nascent proOmpA chains. SecB binds after puromycin dissociates the nascent chain from the ribosome. Furthermore, the addition of trigger factor to nascent proOmpA chains completely displaces bound SecA and Ffh. This indicates that the affinity of trigger factor for nascent secretory proteins at native ribosomes is greater than that of SecA and Ffh.

SecB or SecA do not cross-link to MtlA after it is released from ribosomes by puromycin. No targeting factor, except Ffh, associates with MtlA, suggesting that membrane protein can only reach the membrane via SRP. Addition of inverted membrane vesicles to ribosome-associated nascent chains showed that MtlA is located close to SecY, similarly nascent proOmpA chains were clearly cross-linked to SecA and SecY.

It is known that the insertion of a membrane protein with large periplasmic loops or domains also depends on SecA [22, 31]. This process probably relies on the presence of sufficiently long periplasmic loops. SecA may be recruited at the translocase during the cotranslational insertion of the membrane protein and not before the nascent chain associates with the membrane. This is consistent with the existence of SecY mutants that affect the interaction with SecA and the translocation of pro-OmpA, but do not affect the integration of polytopic membrane proteins with short periplasmic loops [32, 33]. SecA is probably not required, *per se*, for the targeting of membrane proteins, irrespective of the presence of large periplasmic domains (Fig. 1).

Targeting pathways for membrane and secreted proteins are correlated with (1) the presence of a highly hydrophobic signal sequence, together with the absence of a trigger factor binding site. This allows exclusive recognition of polytopic mem-

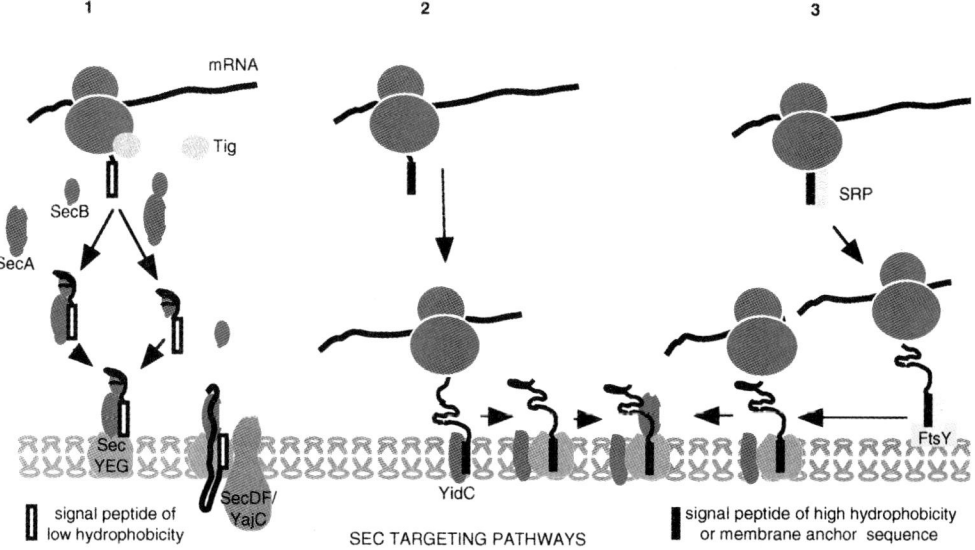

Fig. 1. Sec targeting pathways. The different targeting pathways are shown (1, 2, 3) according to the nature of the signal sequence, whether of low hydrophobicity (Tig/SecB targetin pathway) or of high hydrophobicity, as a membrane anchor (SRP/FtsY pathway). Some proteins (e.g., M13 procoat) might be directly targeted to YidC. See text for additional details; see color plates p. XXXIV.

brane proteins by Ffh, and (2) the sequence specific recognition of nascent chains by trigger factor prevents Ffh from binding proteins with N-terminal signal sequences with low hydrophobicity. In the absence of trigger factor, it would be interesting to test whether secretory proteins, such as pro-OmpA, are targeted to the Sec translocase via the SRP route or not and to determine whether other chaperones, such as DnaK whose synthetic lethality with trigger factor has been established [34], fulfill the functions of trigger factor.

All the studies described here have been carried out in *in vitro* reconstituted systems, thus they may not be completely reliable. For example, *in vivo* studies using depletion of both FtsY and SRP suggested that in the absence of SRP the targeting of the ribosome to the membrane is mediated by FtsY [35]. However, *in vitro* studies suggest that this is mediated by 23S RNA [29]. It might be that SRP operates after the ribosome binds to the membrane.

7.2.1.3 YidC, an Essential Component for Integration of Cytoplasmic Membrane Proteins

As mentioned above, some proteins, such as LacY permease and MltA, are also integrated into the inner membrane via Sec translocase. However, a few membrane proteins appear to be inserted into the cytoplasmic membrane independently of Sec, these include the M13 procoat protein, which is also spontaneously inserted into pure phospholipid vesicles [36]. It was initially thought that insertion of the M13 procoat did not require any other protein. However, it was recently shown that in mitochondria, which do not have Sec proteins, insertion of some of the respiratory chain components into the inner membrane was dependent upon Oxa1p [37]. Oxa1p homologs have also been found in bacteria and chloroplasts, and Oxa1p mutations cause defects in the assembly of some chlorophyll protein complexes [38].

In *E. coli* an unassigned reading frame, YidC, is an Oxa1p homolog. YidC is an essential gene in *E. coli*. Depletion studies have shown that it is required for the integration of all membrane proteins studied so far [39, 40]. The M13 procoat protein has an N-terminal signal sequence cleaved by leader peptidase, followed by a membrane anchor segment. Signal peptide processing of procoat does not exist in YidC depleted cells, whereas proOmpA processing is virtually unaffected. This shows that YidC is strictly required for the membrane insertion of a protein previously thought to be spontaneously inserted. The amino terminal domain of Lep (leader peptidase I) is inserted into the membrane in a Sec-independent manner. A fusion protein, comprising the Lep amino terminal domain fused to the N-terminal 18 amino acids of phage Pf3 procoat protein, also inserts in a YidC-dependent fashion, although it is less dependent upon YidC than M13 procoat. Similarly, the insertion of a chimeric ProW-Lep protein, consisting of the first three transmembrane segments (TMS) of ProW, whose insertion is Sec-dependent (but SecA-independent), fused to the P2 domain of Lep, was found to be partially YidC-dependent. Chemical cross-linking indicated that ProW-Lep is cross-linked to YidC. YidC can be copurified with the Sec translocase, composed of SecYEG. FtsQ, a membrane protein whose insertion is Sec-dependent, is cross-linked to both SecA and SecY via its hydrophilic domain, and to YidC via its membrane anchor. It thus appears that YidC functions alone in the case of M13 procoat, without any help from other identified factors. YidC may also function in association with SecYEG, possibly laterally releasing transmembrane segments (TMS) from integral membrane proteins, which would otherwise get stuck in the SecYEG translocon. The presence of large periplasmic loops, such as in FtsQ or AcrB, might require the action of SecA.

The Sec translocase appears to be a versatile structure, whose functions can be extended by the addition of the SecDFYajC, SecA, and YidC subunits. Likewise, several systems exist to carry protein precursors to the translocase. These systems depend on the nature of the protein being transported. The presence of a highly hydrophobic membrane anchor enables ribosome-bound proteins to be recognized by SRP and targets the complex to FtsY in the membrane. In the absence of such a hydrophobic anchor, proteins with cleavable signal sequences are targeted via ribosome-bound trigger factor to the SecB/SecA pathway, which prevents SRP binding.

7.2.1.4 Oligomeric State of the Sec Translocase

Further structural data is required to understand the molecular basis of the Sec-YEG/A translocase. Two studies used similar methods, with a trapped intermediate, to investigate the oligomeric state of the translocase, and gave contrasting results. Manting et al. used negative stain electron microscopy (EM) and scanning transmission EM to show that purified SecYEG exists as monomers and dimers [14]. They found that the addition of SecA, which is a homodimer, provokes the dimerization of SecYEG dimers in such a way that there is one SecA dimer for four SecYEG monomers. Conversely, Yahr and Wickner used an epitope tagged version of SecE and a similar translocation intermediate, and were unable to coprecipitate untagged SecE and tagged SecE, even though this should be possible if dimers and tetramers are formed [41]. The oligomeric state of the translocase is, therefore, unknown. By analogy with the eukaryotic translocase equivalent, which consists of a very large channel comprised of several Sec61 complexes [42], and with the SecYE from *Bacillus subtilis* [43], Manting and coworkers assumed that oligomerization of the SecYEG complex in the presence of the SecA protein is physiologically relevant. However, Yahr and Wickner have pointed out that the overproduction of the SecYEG proteins used by Manting et al. leads to artificial association. It is thus necessary to estimate the stoichiometry of the blocked intermediate and to ensure the functionality of the translocase. It is also possible that in the study by Yahr and Wickner the presence of excess blocking substrate may drive the translocon into a monomeric, but functional, state.

Studies on the Sec system have shown that the translocase is a highly dynamic structure. Subunits are recruited according to the nature of the local polypeptide chain and the precursor takes a specific route early in the synthesis of the protein, which might drive it to a specific translocase. We have presented an apparently simplistic view of the Sec translocase, mainly based on *in vitro* work and a few *in vivo* studies carried out either with mutants or using depletion of essential components. *In vivo* situations might not be as unambiguous due to either functional redundancy or back-up systems. Genetic screens should reveal new mutations in the essential genes of those pathways [44].

7.2.2
Tat Pathway

7.2.2.1 Introduction

The Tat (twin arginine transfer) pathway is a specialized translocation system for transporting folded proteins across the cytoplasmic membrane, independent of the Sec pathway. It takes its name from a highly conserved motif, (S/T)RRXFLK, found in the N-terminal region of the signal peptide of proteins exported by this pathway. Such a motif is not found in the equivalent region of the classical Sec signal peptides. Several recent reviews have been written on the subject [45–48]. These unusual signal peptides were first found in the proteins imported into the lumen of chloroplast thylakoids via the ΔpH-dependent pathway [49, 50]. Protein importation into the thylakoid lumen, which is topologically equivalent to the gram-negative periplasm, occurs by at least two independent pathways: a Sec-dependent pathway and the ΔpH-dependent pathway. In chloroplasts the signal sequence determines whether the Sec or the ΔpH pathway is used. It has also been shown that in one case transfer via the ΔpH pathway involves folded intermediates [51]. More recently Hcf106, a protein implicated in the ΔpH pathway of chloroplasts, was identified [52]. It has been proposed that several periplasmic proteins that bind a range of redox cofactors are exported by a similar specific pathway [53].

The export of a chimeric protein containing the signal peptide of the small subunit of *Desulfovibrio vulgaris* Ni−Fe hydrogenase, was studied in *E. coli*. This protein has a twin arginine motif signal peptide, fused to the mature region of β-lactamase. Its export to the periplasm was faster under anaerobic conditions. Mutations in the conserved twin arginine motif drastically reduced its export [54]. These two characteristics were not expected from the Sec pathway. Further studies were carried out on TMAO (trimethyl ammonium oxide) reductase, one of the enzymes induced in *E. coli* by TMAO under anaerobic conditions. TMAO is a terminal electron acceptor in the electron transfer chain. TMAO reductase, a periplasmic enzyme with a molybdopterin cofactor, is the *torA* gene product. The TorA precursor has a signal sequence containing a typical twin arginine motif, which is absent from the mature protein [55]. Export of TorA to the periplasm requires a functional signal sequence and the addition of the molybdopterin cofactor. Export of TorA is considerably slower than translocation by the Sec pathway ($t_{1/2}$ = 5 min compared with $t_{1/2}$ = 30 s, respectively). It is completely inhibited by CCCP (carbonyl cyanide m-chlorophenyl hydrazone) and only slightly inhibited by azide, which is a very strong inhibitor of SecA ATPase. Mutations in *secA*, *secB*, *secY*, and *secE* do not affect the translocation of TorA to the periplasm. Finally, Ffh depletion slightly affects TorA translocation. TorA is exported to the periplasm by a Sec-independent pathway.

7.2.2.2 Genetic and Genomic Evidence for the *tat* Pathway in *Escherichia coli*

DMSO (dimethyl sulfoxide) reductase, which is constitutively expressed under anaerobic conditions, allows bacteria to grow with glycerol as carbon source and DMSO as terminal electron acceptor. It consists of three polypeptides, encoded by the *dmsABC* genes; DmsC is an integral membrane protein; DmsA and B are peripheral membrane subunits which probably interact with each other and DmsC. DmsB is a Fe−S protein. DmsA is a molybdopterin enzyme whose precursor contains a twin arginine motif signal peptide [56]. A genetic screen identified one mutant deficient in anaerobic DMSO respiration, proficient in nitrate and fumarate respiration, and still able to express functional DMSO reductase in the cytoplasm. Other proteins containing twin arginine signal peptides did not reach their correct localizations in this mutant, whereas proteins with a classical signal peptide were correctly localized. In the absence of the DmsC subunit, DmsA and B accumulate in the cytoplasm in the mutant and in the periplasm in the wild type. This indicates that the presence of DmsC may force DmsA and B to be localized at the cytoplasmic side of the cytoplasmic membrane.

Complementation of this mutant with an E. coli chromosomal fragment led to the identification of the *mtt* (membrane targeting and translocation) locus, which is located at 88 min on the chromosome [57]. The *mtt* locus encodes three proteins one of which is analogous to Hcf106 [52]. The *mtt* locus was later sequenced, which revealed that there was a fourth polypeptide present. This work established that the *mtt* locus is essential for the correct localization of proteins with a twin arginine signal sequence. It also showed that, although devoid of a signal peptide, the DmsB subunit has the same localization as the DmsA subunit.

Three unassigned E. coli reading frames are homologous to the Hcf106 protein of maize, mutations in which disrupt the ΔpH translocation pathway across the thylakoid membrane. Two *tat* (twin arginine transfer) loci have been identified; *tatABCD* is analogous to the *mtt* locus identified by Weiner, and the *tatE* locus is not linked to *tatABCD* [58]. TatA is a 89 aa long protein with one putative TMS that is 53% identical to TatE (67 aa). TatB is a 171 aa long protein with one putative TMS and has region of similarity with TatA and TatE; TatC is a 258 aa long protein with six putative TMS. *tatD*, which is cotranscribed with *tatABC*, encodes a nuclease that is not involved in the Tat pathway [59]. The mutation described by Weiner et al. [57] is located in *tatB*, and results in the replacement of a proline with a serine. TatA, TatB, and TatE are all similar to Hcf106. Each of the *tat* genes was deleted by an in-frame deletion mutation, and their effect on the translocation of various substrates binding several types of cofactors measured.

7.2.2.3 Functions and Interactions of the Tat Proteins

Deletions of either *tatA* or *tatE* did not result in any growth defect when glycerol was the carbon source, and nitrate, fumarate, DMSO, or TMAO were the electron acceptor. However, a deletion of both *tatA* and *tatE* led to no growth in the presence of TMAO or DMSO. A double *tatE tatA* deletion can be fully complemented by either *tatA* or *tatE*, but not by *tatB*. Several substrates were tested [58] and the importance of both *tatA* and *tatE* and their functional redundancy in the Tat pathway shown.

Deletion of *tatC* strongly affects the correct localization of all substrates tested [60]. Deletion of *tatB* leads to a complete loss of translocation of all substrates tested [61, 62]. Whether the *tatB* deletion affects the activity of the non-translocated part or not depends on the substrate tested. Overproduction of TatB in an otherwise wild-type background leads to deficiencies in the export of Tat substrates. Finally, TatB stabilizes TatC but TatC does not stabilize TatB [62]. It is thus difficult to assign a specific function to either *tatB* or *tatC* in the *tatB* mutant.

Thus, all *tat* genes affect translocation to different extents, the strongest phenotypes were obtained with *tatB* and *tatC*. *tatA* and *tatE* were found to have partially overlapping functions. TatC homologs are also present in algae genomes and plant mitochondria. The *tat* mutants do not affect the Sec pathway or the assembly of cytoplasmic proteins which have the same cofactors as the proteins exported by the Tat pathway. However, the accumulation of inactive forms of some substrates in the mutants might indicate that a critical activation step is carried out by or at the translocase.

Supramolecular complexes have been found some proteins of the Tat system [63]. In *E. coli*, co-immunoprecipitation studies have shown that TatA forms a complex with TatB, or at least part of it. Furthermore, both TatA and TatB can form very large complexes by themselves (600 kDa). The stoichiometry of the complexes is currently unknown. It is not known whether other subunits (TatC or E) are present in the complex, because TatB is known to stabilize TatC. Similar high molecular weight complexes have been identified in chloroplasts: blocked intermediates of the ΔpH pathway associate with 500 to 600 kDa complexes [64].

A study on *B. subtilis* has elucidated the function of TatC [5]. In *B. subtilis*, there are two *tatC* homologs (*tatCd* and *tatCy*), each preceded by a *tatA* homolog [65]. Deletion of the two *tatC* genes is not deleterious to the cell, indicating that TatC function is not essential. A gene encoding PhoD, a secreted phosphodiesterase with a putative Tat SP, is located upstream of *tatCd*. Secretion of PhoD is strongly affected in *tatCd* mutants and double *tatCd*–*tatCy* mutants but not in *tatCy* single mutants. Therefore, TatC might be a specificity determinant for Tat substrates. It is not known whether *B. subtilis tat* functions can complement those of *E. coli*.

7.2.2.4 Role of the Tat Signal Peptide

Several studies have investigated what directs a Tat SP to its specific translocator and not to the Sec pathway and vice versa. The twin arginine leader peptide appears to be the main determinant for targeting precursors to Tat. Tat leader peptides also have other distinctive features that the Sec signal sequences lack: their hydrophobic domain is less hydrophobic, their N-terminal region is much longer, and their C-terminal region contains more K-and R-residues, corresponding to the Sec avoidance motif [66].

Mutation of the conserved twin arginine motif is deleterious to translocation by the Tat pathway [54, 67]. Mutation of the conserved F-residue of the twin arginine consensus motif also greatly affects translocation [68]. A chimeric protein comprising the Tat SP of TorA, fused to the second periplasmic domain of leader peptidase, was redirected to the Tat pathway. However, a Tat signal peptide fused to an integral membrane protein cannot direct this chimeric protein to the Tat translocase. It is, therefore, likely that the targeting information contained in the hydrophobic segments overrides the Tat signal peptide information. Finally, an increase in the mean hydrophobicity of the h-region is sufficient to convert a Tat SP into a Sec SP [67]. Removal of the residues associated with the Sec avoidance motif in the C-terminus region further increases the translocation efficiency by the Sec pathway.

The Tat machinery may also have other functions, such as the acquisition of the full functional activity for some Tat substrates. If the DmsA SP is exchanged for the TorA SP the resultant protein has only 20% of the normal specific activity. Mutations in the Tat SP from DmsA or deletion of Tat SP lead to absence of activity, whereas a TorA protein lacking the SP is fully active [69]. FdoG also forms part of a complex anaerobic formate dehydrogenase which is synthesized with a twin arginine leader peptide and is associated with the cytoplasmic side of the cytoplasmic membrane [70]. Therefore, targeting to the membrane via the Tat system without translocation might be a prerequisite for full functional activity of some substrates.

The Tat pathway can transport folded substrates with added cofactors. The absence of added cofactors leads to the absence of translocation. It is, however, not clear to what extent the folding of Tat substrates is complete before translocation. Furthermore, Tat may not require a folded molecule for transport *per se*. Instead Tat SP availability, as a function of folding, may regulate access to the Tat machinery. Hence, the low cross-complementation of the TorA SP with the DmsA SP is correlated with the strong sequence conservation of the Tat SP of a given protein in different bacteria. This suggests that evolutionary pressure is exerted on the SP by the mature protein.

The leader peptidase I is believed to cleave the Tat SP because it contains the consensus AXA leader peptide recognition site. However, a DmsA mutant, carrying a putative NHN cleavage site instead of an AHA cleavage site, which is predicted to be non-cleavable by leader peptidase I, yielded a fully active protein of the mature size [69].

The Tat translocase can also translocate multisubunit complexes by a "hitchhiking" mechanism, in which only one of the subunits has a functional Tat SP [71]. The *E. coli* NiFe hydrogenase 2 consists of two subunits, encoded by *hybO* and *hybC*. The large subunit, encoded by *hybC*, has no SP and has a C-terminal extension which is cleaved by a cytoplasmic protease after nickel acquisition. The small subunit has a typical Tat SP. In the absence of the large subunit, the small subunit is not translocated. Similarly, a defect in nickel incorporation or in the processing of the large subunit affects the targeting and/or the translocation of the small subunit. Conversely, the small subunit is required for both nickel acquisition and C-terminal processing of the large subunit. A HybO–HybC complex was identified in the cytoplasm. Exchange of the *hybO* Tat SP for that of a Sec substrate redirects the HybO–HybC complex to the Sec machinery, where it becomes blocked in the Sec translocon. It is thus very likely that the complex forms before translocation.

7.2.2.5 Open Questions

It is currently not known whether other Tat proteins exist, because only one mutant has been directly isolated by genetic screening and the *tat* loci have only been identified by sequence comparison. Tat proteins from different bacteria may or may not complement each other, depending on the bacteria. This might, in some cases, be related to the putative size of the Tat translocase pore. *E. coli* Tat translocase cannot translocate fructose oxidoreductase from *Zymomonas mobilis*, which is larger than any molecule normally transported by the *E. coli* Tat pathway [72].

The conserved consensus at the cleavage site suggests that leader peptidase I cleaves Tat SPs, but this has never been shown conclusively.

It is known that Tat translocase substrates are usually exported to the periplasm and that the Tat machinery has a translocation function. However, the data on the Dms and Fdo systems suggest that Tat may have a role in the targeting and membrane integration of those substrates, even though they are not integral membrane proteins. This requires further clarification. Furthermore, it is possible that some Tat substrates do not bind any cofactor [68] and it is questionable whether they are translocated in a folded conformation or they could use the Sec pathway.

As mentioned above Tat SP, might have other functions as well as targeting to the Tat translocase. Berks et al. [46] propose that the signal peptide is sequestered by specific protein–protein interactions until the cofactors are completely attached or the protein is correctly folded. They also proposed that the proteins involved in cofactor attachment mask the signal peptide until the reaction is completed.

Several groups have proposed the structure of the translocase, but its determination will probably require reconstitution of a reliable *in vitro* system.

7.3
Protein Secretion

7.3.1
***Sec*-Dependent Pathway:Type II Secretion Pathway**

7.3.1.1 Type II Secretion Pathway with a Helper Domain Encoded by the Secreted Protein: The Autotransporter Mechanism

The autotransporter secretion pathway is the simplest secretion system. It was first described for the IgA1 protease of *Neisseria gonorrhoeae* in the late 1980s [73]. The expression of the *iga* gene in *E. coli* K12 allowed the secretion of the IgA1 protease into the extracellular medium. As this foreign host does not normally secrete any proteins, the *iga* gene may contain all the information necessary for IgA1 protease secretion.

The IgA1 protease is synthesized as a large precursor of 170 kDa with an N-terminal signal peptide followed by a 106 kDa protease domain and a 45 kDa C-terminal domain. The signal peptide allows the precursor to be exported through the inner membrane via the *sec* pathway. The C-terminal domain sequence predicts that it can adopt a β-sheet structure with 14 membrane spanning antiparallel β-strands, each 9 residues long. β-Barrels are known to form stable secondary and tertiary structures in the presence of SDS, at room temperature. These SDS-resistant molecules migrate more slowly during SDS-PAGE. The C-terminal domain of AIDA, another autotransporter, was also SDS-resistant, which strongly supports the β-barrel sequence predictions [74]. Thus, this C-terminal domain appears to have a similar structure to many outer membrane proteins. The polypeptide terminates by a YSF motif that is also common in many outer membrane proteins [75]. It was also shown that the IgA β-domain inserts in the outer membrane. *In vitro* the β-domain forms channels in bilayers [74]. It is proposed that it forms a pore for the mature domain, which is translocated to the cell surface. However, it is not clear whether it forms trimers or oligomers like other outer membrane proteins. Release of the mature IgA1 protease domain occurs by autoproteolytic cleavage at a consensus site present on IgA1 immunoglobulins and on the protease precursor between mature and helper domains. The expected narrowness of the β-domain hydrophilic channel raised the question of the folding state of the mature domain during its secretion. As IgA1 protease does not contain any cysteine residues the effect of disulfide bridges formation could not be evaluated. However, contradictory results were obtained with passenger proteins fused to the IgA1 β-domain. The cholera toxin B subunit is only secreted to the surface when toxin folding was prevented by addition of 2-mercaptoethanol or by the presence of a *dsbA* mutation, both of which inhibit disulfide bond formation [76]. Conversely, a single chain antibody (scFv), made by fusing two variable domains from the heavy and light chains of an antibody, which is only fully active when it acquires two intramolecular disulfide bonds, was fused to the IgA1 β-domain. This chimeric protein was secreted by both a *dsbA+* strain and a *dsbA−* strain, albeit at reduced levels in the *dsbA+* background. However, the chimera secreted in *dsbA−* strain was totally inactive, which demonstrates that disulfide bond formation required for activity must occur prior to secretion, in the periplasm. Thus, folded proteins with S–S bonds can be translocated through the β-domain channel. The effect of *dsbA* mutation suggests that the chimera and probably the IgA1 protease also exist as periplasmic intermediates, that stable folding reduces secretion, and that this channel is compatible with some degree of folding, depending on the location of S–S bonds in the tertiary structure [77].

This secretion mechanism is not restricted to IgA1 protease. Many proteins are secreted by a similar mechanism: a signal peptide, followed by a mature domain, and a C-terminal β-sheeted helper domain [78]. These include proteases, adhesins, invasins, toxins, and several proteins of unknown function (see Tab. 1). Autoproteolytic cleavage only occurs for some proteases. Others proteins are cleaved by outer membrane non-specific proteases, such as the *E. coli* OmpT protease, or by a dedicated protease, such as the *Shigella* SopA protease [79]. Often no cleavage or partial cleavage occurs, hence the secreted protein remains bound to the cell surface by the β-helper domain.

These studies did not determine whether the C-terminal β-domain could be active *in trans*, functioning as an independent helper protein as described below.

7.3.1.2 Type II Secretion Pathway with one Helper Protein

The paradigm of this secretion mechanism is the *Serratia marcescens* hemolysin system [80]. The hemolytic activity is determined by two contiguous chromosomally encoded genes, *shlA* (the hemolysin structural gene) and *shlB*. Expression of these two genes in *E. coli* is sufficient for hemolysin secretion and activity. In the absence of ShlB in *E. coli* or in the natural host, ShlA accumulates in the periplasm, and is released during osmotic shock. However, this periplasmic protein,

ShlA, is totally devoid of hemolytic activity. ShlA can be activated *in vitro* by mixing pure ShlA with ShlB, in the presence of phosphatidylethanolamine [81]. Four ethanolamine molecules bind each ShlA molecule, even in the absence of ShlB. However, hemolytic activity requires both ShlB and phosphatidylethanolamine. ShlB has a signal peptide, is an integral outer membrane protein, and is incorporated into the outer membrane by the Sec-dependent pathway. DNA sequence analysis predicts that ShlB forms a β-barrel similar to many outer membrane proteins. The structure–function relationship of ShlB was studied by insertional mutagenesis with an eight amino acid long M2 epitope, which reacts with an anti-M2 monoclonal antibody [82]. Analysis of the surface exposed epitopes indicated that ShlB has a β-sheet structure, with several transmembrane segments. Unlike outer membrane porins, it appears to be a monomer and to form a narrow channel *in vitro*, as shown by measuring the conductance of artificial lipid bilayer membranes. This mutational insertion analysis allowed the secretion and activation functions of ShlB to be separated. Several ShlB mutant proteins could still secrete inactive ShlA molecules efficiently. However, because the mutations were scattered throughout *shlB* it was not possible to attribute a defined domain for the activation function on ShlB. Thus, ShlB probably forms a pore through which ShlA is transported. The folding state of ShlA in the periplasm that is required for secretion has not yet been determined.

Several proteins, some unrelated to ShlA, appear to be secreted by similar mechanisms, involving a helper outer membrane protein [83]. Activation by lipid incorporation has only been reported for ShlA and ShlA homologs [84].

7.3.1.3 Type II Secretion Pathway with 11 to 12 Helper Proteins

This secretion pathway is also known as the general secretory pathway, and the specific secretion proteins are named Gsp.

Proteins that utilize Gsp are synthesized with a classical signal peptide and are exported to the periplasm by the Sec-dependent pathway. In the absence of the specific secretion functions, the exoproteins accumulate in the periplasm.

There are 11 or 12 specific secretion proteins depending on the organisms. They are usually encoded by chromosomally clustered genes, in one or two operons. These secretion systems exist in many gram-negative bacteria, and are usually devoted to the secretion of hydrolytic enzymes (lipase, protease, pullulanase, etc.) (see Tab. 1). The Pul system of *Klebsiella oxytoca* was the first system identified and reconstituted in *E. coli* [85], but it cannot be considered as a model system because experimental data from various systems have converged to build our current vision of this secretion mechanism, which is far from being understood.

7.3.1.3.1 The Secretion Apparatus: A Pilus-like Structure?

Nine of the 11 or 12 Gsp proteins that constitute the secretion machinery are integral cytoplasmic membrane proteins. One or two are outer membrane proteins, and one is a peripheral inner membrane protein. This latter protein, GspE, has a nucleotide binding motif and was shown, in one case, to bind radiolabeled ATP, albeit inefficiently. ATP hydrolysis was not demonstrated [86]. GspE only binds to the membrane in the presence of the other Gsp components [87] and could provide the energy required for the transporter assembly and/or for the translocation process itself.

The functions of the nine inner membrane proteins are still unclear. Most of them have one, or several, large periplasmic domains, which might associate with the substrates and/or with the outer membrane components. Four of them (GspG, H, I, and J) have a short N-terminal signal peptide (six to nine residues) with several positively charged residues (arginine and/or lysine). This signal peptide is followed by a 16 to 18 amino acid hydrophobic transmembrane segment, which anchors the GspG, H, I, and J proteins in the inner membrane. The unusual signal peptide is cleaved by a specific peptidase (GspO) at a conserved glycine residue [88]. This peptidase has a second enzymatic activity; it methylates the new N-terminus resulting from signal peptide cleavage [89]. Insertion of these four proteins into the inner membrane is probably *sec*-independent. It also appears to be independent of the specific peptidase activity [90]. However, the secretion apparatus is non-functional in the absence of peptidase. Conversely, the study of a peptidase mutant which had lost its methylase activity, but had retained its peptidase activity, demonstrated that methylation of these four proteins is not required for efficient protein secretion. Nevertheless, it cannot be excluded that the missense mutant used retained some undetectable residual methylase activity, which would be sufficient to support secretion [91].

A very similar short signal peptide is found in PilA, the type IV pilin subunit. Moreover, the PilA precursor is also methylated at the +1 residue after cleavage of the short signal peptide. Both activities are carried out by a specific bifunctional enzyme, which is highly similar to GspO. This sequence and function homology between pilin and Gsp peptidases is even more striking in *Pseudomonas aeruginosa*, in which the same protein (PilD) processes this group of Gsp proteins and pilin [89]. Thus we can consider the GspG, H, I, and J proteins to be pseudopilins. However, apart from the signal peptide and the hydrophobic N-terminal transmembrane segment, pilin and Gsp proteins do not share sequence homology. If the entire *K. oxytoca* Gsp secretion machinery is overexpressed, electron microscopy shows the formation of GspG containing long pilus-like structures [92]. In *E. coli* overexpression mutants, in which a cryptic gene encoding type IV pilin, PpdD, is controlled by a constitutive promoter leads to the formation of pilus-like structures, only in the presence of the Gsp secretion machinery [93]. Immunogold labeling showed that this pilus contains both PpdD and GspG. Mutational analysis showed that pilus formation only requires a restricted set of Gsp proteins (GspH, GspJ, GspC, GspL, and GspM are not necessary). The role of this pilus-like structure in secretion is unknown, because secretion still occurs in the absence of these structures, e. g., the Gsp proteins are expressed at a low level in *E. coli* K12 in which the *ppdD* gene is not expressed.

7.3.1.3.2 An Outer Membrane Protein Belonging to the Secretin Family of Proteins Involved in Type IV Pilus Formation, Filamentous Phage Extrusion, Gsp and Type III Secretion Pathways

GspD is an outer membrane component present in all Gsp systems. It forms multimers of 12 to 14 β-sheet subunits, which are resistant to SDS dissociation. Electron microscopy showed a rosette superstructure with an internal diameter of 5 to 8 nm, which is large enough to allow the translocation of folded and oligomeric exoproteins [94]. GspD protein incorporated into bilayer lipids has a very low channel activity [95]. *In vivo* this gated channel is usually closed, opening may be induced by the substrate, as suggested by the observation that coexpression of *Erwinia chrysanthemi* GspD and its pectinase substrate in *E. coli* causes cell lysis, whereas expression of each protein alone is not harmful [96].

GspD belongs to a large family of outer membrane proteins, also called secretins, which are involved in type IV pilus formation, in the type III secretion pathway, and in the extrusion of filamentous phage. The C-terminal domain of secretins, which is predicted to adopt a β-sheet structure embedded in the outer membrane, is highly conserved. The N-terminal domain might be involved in substrate recognition, interacting with the other secretion components and channel gating [97].

Many secretins also form oligomers and structures that are visible under the electron microscope [98, 94]. Secretins are highly specific and are probably involved in substrate recognition as shown by the co-immunoprecipitation of *E. chrysanthemi* GspD and pectinase. Heterologous complementation between secretins is not usually observed [99]. However, there is one exception to this: the *Vibrio cholerae* Gsp and filamentous phage, Ctxϕ have a common secretin, which is located in the *V. cholerae gsp* locus [100]. This GspD protein is required for Ctxϕ phage release and for the secretion of several proteins (the phage-encoded cholera toxin, the chromosomal chitinase, and proteases). Hence, secretin plays a dual role in virulence, secreting virulence factors and phage-mediated horizontal transfer of the toxin gene. Yet, the molecular mechanism of this double function is puzzling because filamentous phages are assembled by a transmembrane complex comprising inner and outer membrane (secretin) proteins. In the absence of the phage secretin phage particles are not found in the periplasm, suggesting that it is a one-step secretion mechanism, without periplasmic intermediates [101]. In contrast, proteins secreted by the type II secretion pathway reach the periplasm before they are secreted through the outer membrane. This implies that in *V. cholerae* the same secretin may recognize substrates in two distinct cell sites.

A second outer membrane protein, GspS, is only required in a few cases. GspS is a lipoprotein, anchored to the outer membrane, which acts as a specific secretin chaperone [102]. It has been shown that the last 65 amino acids of GspD are the chaperone target in *K. oxytoca*. Addition of this domain to the end of an *E. coli* filamentous phage secretin renders the release of phage dependent on GpsS [103]. This extra C-terminal domain is only present in some secretins (Fig. 2).

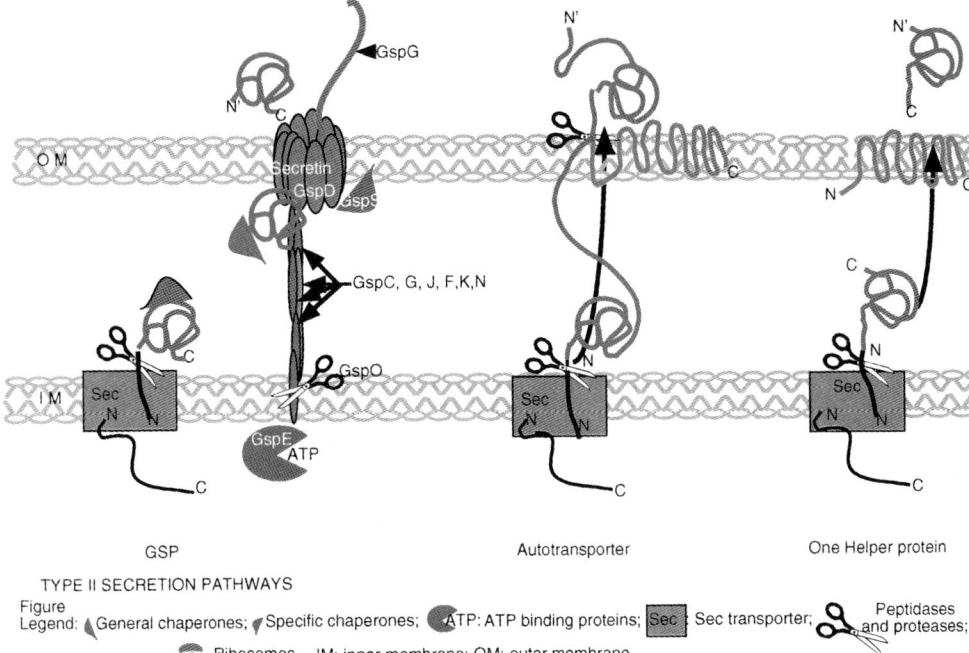

Fig. 2. Type II secretion pathways. In all cases the precursor is recognized by the sec machinery and the signal sequence cleaved on the periplasmic side of the cytoplasmic membrane; subsequent transport across the outer membrane may involve no other protein (autotransporter mechanism) or one helper protein or a set of 11-12 proteins with a specific secretin in the outer membrane (GSP pathway). See text for additional details; see color plates p. XXXV.

7.3.1.3.3 Exoprotein Recognition: Folding of the Periplasmic Intermediate is Essential for Secretion

Usually each species only has one Gsp system, which secretes several unrelated proteins, but does not secrete foreign proteins. Thus homologies were sought between the exoproteins which use the same apparatus. These studies did not reveal any obvious conclusions about the specificity of the domain(s). The study of chimeras between exoproteins with different specificity or between secreted and passenger proteins did not allow generally to identify a secretion signal. All the results suggest that the secretion signals and the specificity determinants do not share a consensus sequence, do not have the same localization on the polypeptides, require several non-adjacent regions, and are usually conformational. Furthermore, it is now well established that exoprotein folding in the periplasm, including oligomerization and disulfide bond formation, is usually required for secretion. Folding of lipases requires specific chaperones [104]. For example, the *Burkholderia glumae* lipase folding is achieved in the periplasm by the Lif product, a periplasmic protein anchored in the inner membrane by its N-terminal hydrophobic segment. In the absence of Lif, folding and secretion are prevented and there is a periplasmic accumulation of inactive lipase. Nevertheless, missense lipase mutants, lacking lipolytic activity, are secreted in normal amounts in the presence of Lif. This suggests that correct folding, but not lipase activity, is linked to secretion. As the chaperone anchored to the inner membrane is not co-secreted with the lipase, periplasmic folding must be retained during secretion [105].

Protein folding was also shown to be a prerequisite for the secretion of *P. aeruginosa* elastase. Elastase is synthesized as a preproprotein. After the signal peptide is processed the periplasmic zymogen is matured by autoproteolytic removal of the propeptide, which remains non-covalently bound to the mature protein. Both are secreted through the outer membrane [106, 107]. When expressed without the propeptide, the periplasmic elastase is inactive in *E. coli*. If the propeptide is provided *in trans* proteolytic activity is restored. This shows that elastase, like several other proteases, utilizes a propeptide as an intramolecular chaperone, but can also recognize and use it as an intermolecular chaperone. In *P. aeruginosa*, no elastase was found in the extracellular medium in the absence of the propeptide. Moreover, expression of the propeptide *in trans* restored elastase secretion. This shows that the propeptide is also required for secretion [108]. Proteins which are as large as the *Aeromonas* aerolysin dimer or the cholera toxin A−B hexamer are only secreted by the Gsp pathway after assembly in the periplasm [109−111]. Subunit A was not secreted in the absence of the B-pentamer, showing that it is not a Gsp substrate *per se*. Secretin pores seem to be large enough to allow the translocation of such folded molecules.

7.3.2
SEC-independent Pathways

This designation only refers to the substrates of these pathways, and not to the proteins constituting the secretion apparatus, which are usually translocated by the *sec*-dependent system.

7.3.2.1 Type I Secretion Pathway − ABC Protein Secretion in Gram-negative Bacteria

7.3.2.1.1 Introduction

ABC secretion across the gram-negative envelope was identified approximately 20 years ago in a model protein, *E. coli* α-hemolysin, secreted by some uropathogenic isolates [112]. It was later discovered in many other proteins secreted by gram-negative (see below) and gram-positive bacteria, and has several characteristic features.

The secretion apparatus consists of three (gram-negative), two or one (gram-positive) proteins located in the cell envelope [3]. It takes its name from one of the cytoplasmic membrane proteins, which belong to the ATP Binding Cassette class of proteins [113]; another cytoplasmic membrane protein belongs to the "membrane fusion proteins" class and is found in gram-negative bacteria and in some gram-positive bacteria [114]. The outer membrane component is exclusive to gram-negative bacteria. ABC proteins are almost exclusively associated with the transport of substrates across membranes. They are always composed of at least four modules, associating two integral membrane-bound domains and two ABC domains; these domains are part of either one or several polypeptide chains [115].

The proteins secreted by this pathway are secreted independently of the Sec system and, usually via an uncleaved C-terminal secretion signal. Secretion occurs in one step, from the cytoplasm to the extracellular medium with no periplasmic intermediate, unlike the type II system. All three secretion functions and the C-terminal secretion signal are strictly required for secretion [3].

7.3.2.1.2 Substrates Secreted by ABC and Putative Substrates of ABC – Genetic Organization

Some examples of proteins secreted by ABC systems in gram-negative bacteria are listed in Tab. 2; it is not exhaustive and is intended to show the variety of proteins secreted by this pathway and the variety of organisms it is found in.

1. *Proteins containing glycine-rich repeats*

The RTX (Repeats in Toxin) group, which includes the HlyA molecule and many other pore-forming toxins, are characterized by glycine-rich repeats with the consensus motif GGXGXDXXX (see below). Several of them are pore-forming toxins with specific target cells in the eukaryotic host [116]. Some of them are bifunctional, such as CyaA from *Bordetella pertussis* which is an adenylate cyclase-hemolysin [117], the hemolysin part of the molecule allows the cyclase moiety to be internalized in the target cell. *Vibrio cholerae* secretes the longest representative of the group, RtxA, which consists of over 4,000 residues. This is also internalized by the eukaryotic host, where it interacts with actin [118, 119].

The FrpA and FrpC proteins secreted by *Neisseria* under iron starvation conditions are of unknown function [138].

2. *Proteins without repeats*

The colicin V is unusual among the proteins secreted by ABC transporters: it possesses an N-terminal leader peptide which is essential for secretion and is cleaved during secretion. This leader peptide contains the typical double glycine sequence also found in some bacteriocins secreted by ABC transporters in gram-positive bacteria. The leader peptide is cleaved by the ABC protein, which contains an additional protease domain, during secretion [143].

During iron starvation *S. marcescens* secretes the HasA protein. This is a heme-binding protein, which can take up heme from hemoglobin and allows *S. marcescens* to use heme or hemoglobin as an iron source [144].

The structure of glycine-rich repeats is known for the *P. aeruginosa* alkaline protease. They form a "beta roll" which binds calcium ions very tightly [148]. Although the structures of other proteins belonging to these classes are unknown it is assumed that the glycine-rich repeats have the same structure in the other proteins. Several of these proteins require calcium for their activity and deletion of the repeats is harmful to their activity [149].

The number of repeats is highly variable, from 5 to 6 in the proteases, and up to 58 in oscillin. There is a loose correlation between the length of the secreted protein and the number of repeats. Although not all proteins with this kind of glycine-

Tab. 2. Diversity of proteins secreted by the ABC pathway

Protein Class	Organism	Secretion Functions	Reference
RTX toxins			
HlyA, hemolysin	*Escherichia coli*	yes	116
CyaA, adenylcyclase	*Bordetella pertussis*	yes	117
RtxA, actin binding	*Vibrio cholerae*	yes	118, 119
Metalloproteases			
PrtA, B, C, G,	*Pectobacterium chrysanthemi*	yes	120
AprA, alkaline protease	*Pseudomonas aeruginosa*	yes	121
	P. fluorescens	yes	122
PrtSM, metalloprotease	*Serratia marcescens*	yes	123
ZapA, metalloprotease	*Proteus mirabilis*	yes	124, 125
Lipases	*S. marcescens*	yes	126, 127
	P. fluorescens	yes	128
Mannuronan epimerases	*Azotobacter vinelandii*	no	129, 130
RzcA, bacteriocin	*Rhizobium leguminosarum*	yes	131
S-Layer proteins	*Caulobacter crescentus*	yes	132
	S. marcescens	yes	133
	Campylobacter fetus	yes	134
Oscillins, motility	*Phormidium uncinatum*	no	135, 136
	Synechococcus sp.	no	137
NodO, infection thread development	*Rhizobium leguminosarum*		139–141
FrpA, C	*Neisseria meningitidis*		138
Colicin V	*E. coli*	yes	142
Hemophores			
HasA$_{sm}$	*S. marcescens*	yes	144, 145
HasApa	*P. aeruginosa*	yes	146
HasApf	*P. fluorescens* *Yersinia pestis*	yes	147

rich repeat have been shown to be secreted by the ABC pathway (e.g., mannuronan epimerases from *Azotobacter vinelandi*), this is probably the case because they are often extracellular proteins without signal peptides.

Examination of the completed bacterial genomes shows that *P. aeruginosa* [150] contains two other potential ABC transporters which are probably involved in protein secretion, besides the HasApa and AprAX transporters. *Synechocystis* PCC6803 [151, 152] contains seven putative GR repeat proteins. *Xylella fastidiosa* [153] contains four putative proteins with GR repeats. *Aquifex aeolicus* [154] contains one putative GR repeat protein.

The genetic organization of ABC protein exporters is variable. The structural gene for the secreted protein is often associated with the genes for the secretion functions, e.g., the metalloprotease transporter from *Pectobacterium chrysanthemi* [155]. In several cases, the gene for one of the secretion functions is not contiguous with the others, e.g., hemolysin, where the hemolytic determinant comprises only the structural gene for the toxin (HlyA); the toxin activator (HlyC), which is not required for secretion; and the ABC and MFP components (HlyB and HlyD). The gene for the OMP component (*tolC*) [156] is not linked to the precedent ones. The other extreme is found in *Synechocystis*, in which none of the genes encoding a protein with glycine-rich repeats are close to a putative ABC, MFP, or OMP, and there is no link between the putative secretion functions. There is only one potential OMP component, four potential MFP components, and several ABC components which have similarities with ABC components from protein exporters.

The similarity between the secretion functions of proteins from different families is quite high; the similarity between secreted proteins from a given family is also high, and there is a good cross-complementation of secretion within a family, in all cases tested. However, complementation between families is quite low. The presence of several ABC protein exporters in a single bacterium or of many putative substrates with fewer potential secretion functions highlights the specificity of those systems.

7.3.2.1.3 Structure and Function of the Secretion Functions

Detailed studies of the roles of the secretion functions and the secretion signal have been carried out on several model systems, such as the HlyA secretion system from *E. coli*, the Prt secretion system from *P. chrysanthemi*, the HasA secretion system from *S. marcescens*, and the Lip system from *S. marcescens*. *In vivo, in vitro,* and *in vivo/in vitro* techniques have all provided important information on the role of the secretion functions.

In vivo studies reconstituted the Has, Prt and Lip systems and those of their various substrates in *E. coli*. The secretion functions of these systems are highly similar although they display specificity towards their substrates. The Has system is quite unique because it can secrete both HasA and metalloproteases in the reconstituted system, whereas the Prt system cannot secrete HasA. This, together with the finding that HasA can interact with the Prt transporter in an abortive manner stopping it from functioning as a Prt exporter, has led to the *in vivo* identification of specific interactions between the transporter subunits. The substrate is specifically recognized by the ABC protein [157]. Specific interactions also exist between the ABC protein and the MFP component, and between the MFP component and the OMP component [158]. Specific interactions are required for secretion to take place and most importantly the substrate specificity is provided by the ABC protein.

Two experimental approaches were used to test the functional organization of the secretion apparatus. One used the affinity of the native HasA for haemin-agarose and its properties towards the Prt and Has system or of the affinity of a GST–PrtC fusion protein for glutathione-agarose and its ability to block the Prt transporter [159]. The other one used a cross-linking approach on the Hly secretion system [160]. Both studies have yielded comparable, although not identical results. The whole transporter, consisting of the three secretion functions, is assembled when the C-terminal secretion signal interacts with the ABC protein. The ABC protein alone can interact with the substrate via the secretion signal, in the absence of the other components. In the Hly system, it was found that the MFP component alone can also interact with the substrate. The complex is assembled sequentially; the ABC assembles first, then the MFP, and the OMP. In the Has and Prt systems no pre-assembled ABC–MFP complex was found, whereas such a complex was found in the Hly system. This might be due to the difference in the experimental approach used, because the Hly study used cross-linking which might stabilize weak complexes or create artifactual associations, whereas the other approach relied on the stable interaction of the components in the absence of cross-linking. The second approach is more reliable, but unable to detect weaker interactions. The cross-linking approach also showed that the MFP and OMP components are probably trimeric. The nature of the modifications involved in the assembly of the functional complex is unknown, but differences in protease accessibility are seen when the complex assembles. The complex is transient and disassembles upon secretion. Assembly does not require a functional ATP-binding site on the ABC subunit. The C-terminal domain of HlyD, in which mutations have been found leading to an absence of secretion [161], is apparently not required for the assembly of the full translocator complex. This part of the molecule is probably involved in the later steps of the secretion process. Secretion is an ordered process, triggered by the specific association of the substrate and the ABC protein, followed by the association of the MFP and OMP components into a functional secretion complex.

The ABC protein of this type of system always consists of a membrane portion fused to an ABC domain. It is believed to function as a dimer, although there is no direct evidence. The ABC domain of the HlyB molecule was purified as a GST fusion and shown to possess unregulated ATPase activity [162]. The whole PrtD molecule was purified and also shown to display ATPase activity, which was specifically inhibited by the C-terminal secretion signal of one of the proteases and could be reconstituted in phospholipid vesicles, where it behaved similarly [163]. The ATPase activity is correlated with secretion efficiency [164].

There is a weak analogy between MFP proteins and viral proteins involved in membrane fusion [114]. They have one TM segment and most of the protein lies in the periplasm. Sequence analysis shows that they can form coiled coils. They are probably trimers. Many mutants have been isolated in HlyD [161], the MFP component of the Hly system, or the comparable CvaA protein of the colicin V secretion system [165]. These mutations are either located in the short N-terminal cytoplasmic region or in the C-terminal region and lead to a reduced hemolysin and colicin V secretion.

The OMP component is mostly located in the periplasm and is anchored in the outer membrane. The atomic structure of one of its representatives, TolC from *E. coli*, [166] has elucidated the functioning of this system. Its structure is very different from the known structures of the other outer membrane proteins. It is a trimer, each monomer comprises four β-strands in the outer membrane, and four very long α-helices which can span almost the entire periplasmic space because they are 140 Å long. The outer membrane pore is very wide (30-35 Å), as is the space limited by the α-helices in the periplasm. At the most distal part from the outer membrane the α-helices are arranged such that a tiny pore appears at this end of the molecule. This explains why TolC forms only very small channels in lipid bilayers [167]. Koronakis and coworkers proposed that C-terminal helices act as the diaphragm of a shutter controlling the pore size at the entrance of the tunnel of α-helices on the distal end of the molecule [166]. TolC is also the outer membrane component of many systems involved in drug efflux across the gram-negative envelope, associating either ABC and MFP components or RND and MFP components [168].

The structure of TolC and the length of the periplasmic helix bundles make it possible that there are also specific interactions between the ABC and the OMP, and that the MFP merely serves as a trigger for the conformational changes of the OMP and is not part of the actual protein conduit [168a]. The existence of MFP analogs in gram-positive bacteria that are required for the secretion of bacteriocins disfavors this possibility [169]. Furthermore, the TolC structure questions the putative fusion function of the MFP. As MFP are probably trimeric and can form coiled coils they may also be part of the conduit through the periplasmic space. However, the MFP component of the Acr system, which is involved in drug extrusion and functions with TolC as the OMP, has been shown *in vitro* to promote membrane fusion or hemifusion [170]. Finally, the TolC structure shows a large conduit open to the exterior, through which small hydrophilic molecules should be able to pass freely. In the presence of a TolC outer membrane component and a functional ABC protein transporter cells are more susceptible to vancomycin (MW 1,400 Da), a hydrophilic antibiotic whose target is in the periplasm. Mutants have been isolated in the ABC and MFP components of the transporter, which render the cells less susceptible to vancomycin under these conditions. This probably means that at least small molecules can pass through the TolC channel towards the periplasm. This is consistent with the hypothesis that part of the colicin V secreted by *E. coli* is located in the periplasm, although it has never been shown whether this is due to a leaky transporter or if periplasmic ColV can be secreted to the extracellular medium.

7.3.2.1.4 C-terminal Secretion Signal

Hemolysin was the first protein in which it was shown that the secretion signal is at the C-terminus of proteins secreted by an ABC transporter. N-terminally truncated forms of hemolysin are still secreted and a fusion of a portion of OmpA to the C-terminus of HlyA is specifically secreted to the extracellular medium [171]. The nature and precise function of the secretion signal are not completely clear. This is particularly obvious when one compares the C-terminal ends of several proteins secreted by one secretion system. There are few conserved residues in PrtSM, LipA, and SlaA, which are all secreted by the Lip system in *S. marcescens*. Likewise, there is little residue conservation between all the substrates that are efficiently secreted by the Hly system (HlyA, LktA, CyaA, and several hemolysins from *Morganella*, *Proteus*).

With a few exceptions the secretion signal is located at the C-terminus and is not cleaved during transport. It can be secreted autonomously by the cognate secretion apparatus. The shortest signal secreted by this kind of apparatus is the C-terminal secretion signal of the PrtG metalloprotease from *P. chrysanthemi*, which can be shortened to 29 residues [172].

The C-terminal secretion signals studied do not seem to adopt a preferential structure in aqueous solutions. They are mainly α-helices in membrane mimetic environments, with two main α-helices, and the very last residues are generally unstructured [173–175]. The C-terminal secretion signal of HlyA, which comprises 70 residues, can be divided into three zones; an amphiphilic α-helix, a linker region, and a second α-helix located near to the last amino acids. Extensive mutagenesis has indicated that the amphiphilic helix and the linker regions are both important, whereas the second helix is not very important [176]. There also appear to be a few scattered residues in the sequence which might be crucial for the recognition of the transporter [177].

The C-terminal secretion of PrtG was also studied [172]. It was found that a truncated non-secreted form of PrtG, devoid of the last 48 aa, was still specifically secreted after fusion to the last 15 residues of PrtG. Addition of one amino acid to the C-terminus drastically reduced or abolished secretion depending on the amino acid. This study also emphasized the role of a tetrapeptide, DVIV, at the extreme C-terminus of the protease and found that all the metalloproteases secreted by this pathway have the consensus DXXX (where X represents a hydrophobic amino acid). This kind of motif is also found 14 residues upstream of the C-terminal residue. Progressively truncated forms of PrtG re-exposing such a motif are again secreted. It is noteworthy that the DXXX motif is also found as part of the consensus of the glycine-rich repeat.

A genuine secretion signal should be able to direct secretion of foreign polypeptides, if the secretion functions are also provided. This has been carried out in several cases for this kind of secretion system and several conclusions can be drawn from these studies:

(1) Although significant, the secretion levels obtained with chimeric proteins are, with a few exceptions, quite low.

(2) The glycine-rich repeats enhance the secretion of the passenger proteins [178], either by enhancing the presentation of the C-terminal secretion signal to the secretion apparatus, by providing a flexible linker region, or by providing a direct interaction with the secretion apparatus because they function as secondary secretion signals in truncated versions of CyaA [179]. Thus, the very low level of cross-secretion between the different systems, particularly for the Hly system which can secrete several unrelated substrates at very low levels, might be due to recognition of the glycine-rich repeats found in those proteins.

7.3.2.1.5 Chaperones and ABC Secretion

The C-terminal position of the secretion signal implies that the protein is fully synthesized before specifically interacting with the ABC protein. The OMP component has a maximum pore size of 30 Å, which is incompatible with the secretion of a fully folded protein for many of the substrates identified. HasA which is one of the smallest proteins secreted by this system measures $40 \times 35 \times 35$ Å and thus could not accommodate fully folded in the TolC outer membrane channel. These constraints affect the folding status of the secreted protein during the early stages of the process.

Secretion of HasA in the reconstituted system in *E. coli* is almost strictly dependent upon the cytoplasmic SecB chaperone [180], which is the dedicated Sec pathway chaperone. The secretion of a C-terminal fragment of HasA, containing the

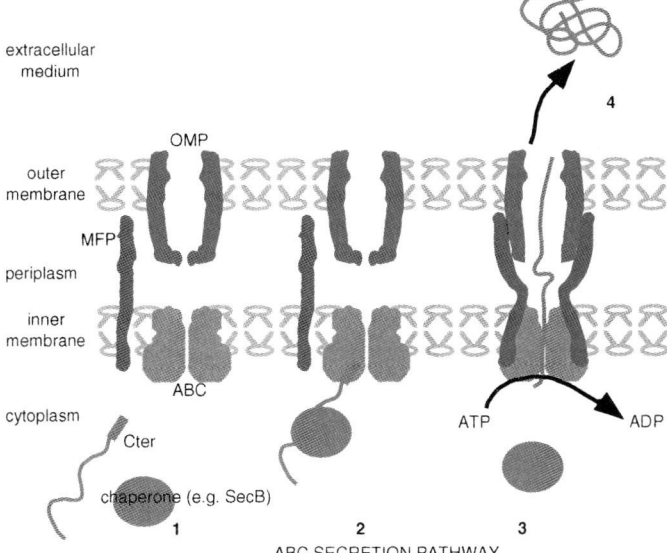

Fig. 3. ABC or type I secretion pathway. The secreted protein is synthesized in the cytosol with a C-terminal secretion signal (1); it might associate with chaperones and interact with the ABC protein via the C-terminal secretion signal (2), triggering the functional association of the MFP and OMP components (3); the protein is then secreted (4) and the transporter is ready for a new round. See text for additional details; see color plates p. XXXVI.

secretion signal of HasA, but which does not fold in the supernatant, is no longer SecB-dependent. SecB binds HasA *in vivo* and it probably maintains HasA in a "translocation competent" state. A SecB analog also exists in *S. marcescens* for HasA secretion, because a SecB interfering species inhibits HasA secretion in *S. marcescens*. Secretion of proteases is SecB-independent and HlyA secretion is SecB- and GroEL/ES-independent (Fig. 3).

Future work will determine the function of SecB in HasA secretion, whether it has unfolding and targeting activities like in the Sec pathway, and whether HasA is a unique case or other chaperones are required for the secretion of the other proteins secreted by this pathway.

7.3.2.1.6 Open Questions

Secretion by the ABC pathway is extremely fast, but a precise estimation of the secretion speed is still lacking, mostly due to technical reasons. In the Sec system, translocation is coupled to the cleavage of the precursor by the leader peptidase, which makes it easy to measure without separating the translocated and non-translocated species. This is not the case for the ABC system. There are probably two steps in the secretion process; one dependent upon the ATP, the other one dependent upon $\Delta\psi$.

The series of modifications that lead to the association of the secretion functions in a functional complex together with the substrate are unknown. It is difficult to progress without the reconstitution of some or all of the translocation processes *in vitro* and without knowing more about the structure of the secretion functions. Very little is known about the stoichiometry of the components of the ABC secretion pathway.

7.3.2.2 Type III Secretion Pathway

7.3.2.2.1 Introduction

Type III secretion systems are present in animal and plant pathogens, and also in opportunistic pathogens, such as *Pseudomonas aeruginosa*. The secretion apparatus is usually composed of approximately 20 proteins (secretion proteins), which are encoded by clustered genes located within chromosomal or plasmid-encoded pathogenicity islands. The GC content of these islands differs from that of the host DNA and they are bound by virtually complete transposable IS elements. Bacteria usually only possess one type III secretion system, except for *Salmonella typhimurium*, which has two. However, several unrelated proteins are secreted by unique apparatus in each species. Exoprotein structural genes are also often located on pathogenicity islands but are unlinked to each other and to the secretion cluster. Most of these exoproteins are virulence factors, also named effectors, which only exert their harmful functions inside the host cell where they interfere with host cell metabolism at various levels (see Tab. 1) [181]. Some of these proteins have various enzymatic activities in the host cytosol, such as phosphotyrosine phosphatase, serine threonine kinase, inositol phosphate phosphatase, ADP-ribosyltransferase, and adenylate cyclase. Other effectors are translocated into the host cell nucleus. Others bind to and modify the host cytoskeleton. Several effectors are inserted in the host plasma membrane and function as receptors for the entry of bacteria into host cells [182]. Thus, all these effectors are secreted out of the bacterial cells and injected into host cells where they are localized in various cell compartments. Moreover, secretion usually only occurs when bacteria are in contact with the host cells. This mechanism is still poorly understood. It clearly requires several levels of regulation, including transcription, translation, and secretion. We will only briefly describe the regulation effects associated with the secretion process. Both secretion and injection are carried out by type III machinery. The delivery of exoproteins into the host cell cytosol requires the crossing of the host plasma membrane and also thick cell walls in plant hosts.

7.3.2.2.2 The Secretion Apparatus: A Needle-like Structure with Some Similarities to Flagella?

The secretion proteins are highly conserved among species and there have been several reported cases of heterologous complementation, in which exoproteins from one species are secreted by a foreign secretion apparatus [183,184]. Moreover, a plant pathogen exoprotein encoded by *Pseudomonas syringae* can be secreted by the type III secretion pathway in *Yersinia enterolytica*. An *E. coli* strain carrying a cosmid encoding the *P. syringae* type III secretion pathway and one exoprotein allows the exoprotein to be secreted and injected into tobacco cells. This shows that the type III machinery is a universal mechanism operating in both animal and plant tissues [185].

Little is currently known about the individual functions of type III secretion proteins. Most of them are inner membrane proteins, sharing sequence homology with the flagellar export apparatus. One protein in each type III secretion apparatus is homologous to F_1 the cytoplasmic compound of the *E. coli* F_0/F_1 ATPase. In the *S. typhimurium* system, this purified protein hydrolyzes ATP [186]. As in the secretion pathways I and II, this protein might provide the energy for the secretion process and/or for the assembly of the secretion apparatus.

Only three secretion proteins are located in the outer membrane. One of them belongs to the secretin family and is involved in type IV pili formation, protein secretion by the type II pathway, and filamentous phage extrusion. It is noteworthy that pathway II proteins cross through the periplasm, whereas pathway III proteins do not have a signal peptide and are believed to be secreted directly from the cytoplasm (see below). This might be an indication that proteins belonging to the pathway III do have a periplasmic stage, but are bound to the secretion apparatus and cannot be released by osmotic shock.

The *S. typhimurium* and *Shigella* secretion proteins form multiprotein complexes which have been isolated and examined by electron microscopy. They assemble in supramolecular structures, with a 40 nm wide basal body and a 80 nm long needle. In osmotically shocked *S. typhimurium* or *Shigella* cells, the basal body spans the inner and outer membrane [187–190]. Biochemical analysis of purified *Shigella* and *S. typhimurium* needles demonstrated that the basal body consists of the secretin and two inner membrane lipoproteins. The *Shigella* needle is composed of at least one protein (MxiH) and overexpression of this protein leads to the formation of long needles and to an increase in *Shigella* invasiveness. When the gene encoding the *Shigella* putative ATPase was inactivated by a non-polar insertion, defective type III particles were assembled with a normal basal body, but lacking the needle [189]. Similar results were obtained with *S. typhimurium*, in which *invC* mutations (the ATPase) resulted in the complete absence of the needle segment without affecting the assembly of the base substructure [191]. Thus, ATP hydrolysis is not required for the basal body assembly, but for the assembly of the needle structure and the exoproteins secretion. However, the biological significance of the visualized superstructures is still questionable, because *Shigella* needles are present even in the absence of any inducer, such as eukaryotic cells or Congo red [190].

Although the structural components of needles and flagella do not share any sequence homology, flagella and needles have similar overall structures. Both have basal bodies with two rings and the needle-like portion corresponds to the flagella hook. However, there is no secretin-like protein in the flagella export apparatus. In flagella, new flagellin monomers are added at the tip of the filament. To reach the tip, the flagellin might pass through the secretion apparatus, which is composed by the flagellar basal body, and then be propelled through the hollow structure of the hook. It is now clearly established that the flagellar basal body forms a secretion apparatus that is related to type III secretion pathway (see below). Similarly, both components of the needle and exoproteins secreted by the pathway III could pass through the basal body and the needle canal. However, there is no experimental data supporting this hypothesis.

In plant pathogens, supramolecular structures have also been visualized by EM [192] and were shown to form thick pili that are very different from the needles.

Thus, exoproteins could pass through the needle and exit the bacteria. However, this does not explain how the exoproteins reach distinct extracellular locations. Indeed, type III secreted exoproteins are found in at least four compartments: bacterial outer membrane, extracellular medium, host plasma membrane, and host cytosol. Several exoproteins are hydrophobic and the effectors are not, however, they all require the same type III secretion apparatus. Moreover, unlike pore-forming proteins, purified type III effector proteins do not enter host cells *per se*. Several works have shown that cytotoxic effects require intimate contact between bacteria and host cells. This led to the concept that exoproteins use the same secretion apparatus, and have additional targeting signals which direct them to their final location.

7.3.2.2.3 How Many Signals on the Exoproteins?

A Signal Located at the 5′ End of the Exoprotein Messenger
Sequence comparison of the exoproteins secreted by type III secretion apparatus did not reveal any consensus signals. Deletion analysis and construction of hybrid proteins, carrying various lengths of the exoprotein N-terminus fused to reporter proteins, such as Npt (neomycin phosphotransferase) or CyaA (adenylate cyclase), reduced the minimal region necessary for secretion to the first 10 amino acids [193,

194]. Moreover, the open reading frame of this signal can be changed by frame shift mutations introduced just after the start codon, without inhibiting the secretion of the fusion proteins, if the correct reading frames were restored by a reciprocal change at the first codon of the reporter genes [195]. Such results were obtained with two *Y. enterolytica* (YopE and YopQ) and one *P. syringae* (AvrPto) exoproteins [184]. The AvrPto–Npt chimera, with a frame shift mutation which altered the reading frame of the first 15 codons, was secreted by its own transporter and also by the *Y. enterolytica* transporter. This shows unambiguously that the signal resides in the 5′end of the messenger and that this messenger signal might be conserved in various type III secretion systems. The implication of a secretion signal located in the messenger and not in the protein, is that translation and secretion must be coupled. Such coupling has been observed in YopQ [196]. There is no intracellular pool of YopQ. The 5′ end of YopQ is predicted to have a stem loop structure that buries the ribosome-binding site and the start codon. Hence, the translation of this messenger can only start when its 5′end interacts with the secretion machinery, which unwinds the translation initiation site. This mechanism is reminiscent of the eukaryotic signal recognition particle mechanism. SRP binds the signal peptide and stops translation until the SRP complex finds the docking protein on the ER membrane. However, there is currently no information about the nature of the secretion protein(s) which interact with the messenger.

A Second Signal within the Exoprotein

Contrastingly, many type III exoproteins accumulate in the cytoplasm before secretion, and are secreted when the bacteria enter into contact with host cells. This secretion is posttranslational, because inhibiting protein synthesis does not prevent secretion. This posttranslational secretion requires a second signal located within the first 100 amino acids of several exoproteins, and a dedicated chaperone binds to this second signal [197]. Secretion of mutants lacking the mRNA signal is entirely dependent on the presence of the specific chaperone and on the internal exoprotein signal [198]. Each chaperone binds and promotes the secretion of just one exoprotein. The exoprotein and its dedicated chaperone are usually encoded by linked genes. They do not share sequence homology, but have common features. Chaperones are small, negatively charged, a-helical proteins. Chaperones appear to have multiple functions. They act as secretion pilots, prevent the premature association of exoproteins in the bacterial cytoplasm, and prevent the degradation of exoproteins [199].

A Translocation Signal

The chaperone binding domain is also involved in translocation into the host cells. This was shown by using the 35 kDa adenylate cyclase domain of CyaA as a reporter gene. This bacterial enzyme is only active in the presence of calmodulin, a eukaryotic protein, which is present in the host cytosol. Increased synthesis of cyclic AMP reflects the translocation of the reporter protein into the host cells. This CyaA fusion protein approach showed that translocation requires the first 50 to 100 amino acids of various exoproteins. It was shown that if amino acids 49 to 159 were deleted from YopH, the CyaA reporter was secreted into the external medium due to the mRNA secretion signal, but translocation was prevented [193]. These results were largely confirmed by immunofluorescent detection, use of green fluorescent protein as reporter gene (GFP), and confocal laser scanning microscopy [200].

As well as the translocation domain and the chaperones, two proteins, themselves secreted by the type III secretion machinery, are required for translocation of the effectors into the host cytosol. Mutations inactivating these two proteins (*Yersinia* YopB and YopD, *Shigella* IpaB and IpaC) do not affect the secretion of the effectors, but abolish the injection in host cells. These two proteins (also called translocases) are targeted into the host plasma membrane forming the translocation machinery [201, 202].

A Scheduled Secretion?
These results suggest that timing is important in secretion. The contact between bacteria and host cells probably via the secretion needles, triggers the secretion of the translocases, which are secreted by a posttranslational chaperone-dependent mechanism. They pass to the tip of the needle and insert in the host plasma membrane. This, in turn, induces the cotranslational mRNA signal-dependent secretion of the effector proteins by the same apparatus. The various signals leading to this cascade are not yet identified. It is postulated that secretion might also regulate transcription by allowing the export of negative regulators.

7.3.2.2.4 The Flagellum Forms a Type III Secretion Apparatus and Mediates the Secretion of a Negative Regulator
The *S. typhimurium* flagellum is composed of a basal body structure located in the cell envelope, which spans both membranes, and an external hook to which the flagellar filament is attached. Thirteen proteins compose the flagellar secretion apparatus and allow the export of the basal body components, the hook components, and the flagellin subunits, which pass through the flagellar pore and are added at the tip. The ordered assembly of the flagella requires the sequential expression of the genes encoding these structures [203]. The genes are organized in operons, which are expressed from three temporal promoter classes. The class 1 promoter transcribes positive regulators, which are required for middle class gene expression. The class 2 operons encode the structural and assembly proteins that form the basal body and the hook. These operons also encode two antagonistic regulators: a σ-factor, FliA, and an anti-σ-protein, FlgM. Late class genes encode the flagellum subunit, hook-associated proteins, and the flagellar cap proteins. The FlgM protein binds directly to FliA preventing the transcription of class 3 genes. When the basal body and hook structures are assembled, they constitute a secretion apparatus through which the FlgM protein is secreted, allowing FliA to initiate class 3 transcription [204]. Very recently, it was shown that *flgM* gene is transcribed from two promoters; one class 2 promoter and one class 3 promoter. Most of the FlgM produced from class 2 promoters remains in the cytoplasm, whereas most of the FlgM produced from class 3 promoter is secreted. This suggests that the class 3

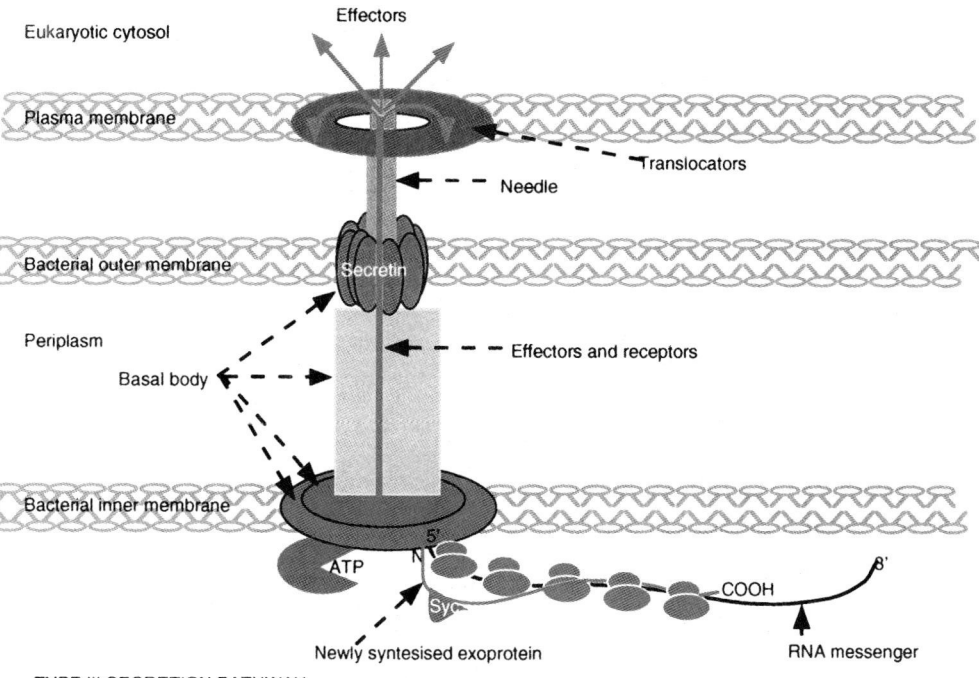

Fig. 4. Type III secretion pathway; see color plates p. XXXVII.

flgM transcript carries a secretion signal which directs the nascent FlgM protein to the membrane-bound ribosomes [205].

A phospholipase has also been reported to be secreted into the extracellular medium by the flagellar export apparatus. This suggests that this system might secrete proteins that are not directly linked to the flagella biosynthesis process [206].

A very similar mechanism was reported for the *Yersinia* type III secretion system. Using a *yopE–luxAB* operon fusion a large increase in transcription was observed upon host–pathogen interaction. A negative regulator of Yop operons (LcrQ) was recovered in the extracellular medium after host–pathogen interaction. When type III secretion mutant bacteria were used LcrQ was not recovered in the supernatant, strongly suggesting that the negative regulator is secreted by the type III machinery [207] (Fig. 4).

7.3.2.3 Type IV Secretion System

7.3.2.3.1 A Secretion Machinery for Conjugational DNA Transfer

Type IV secretion systems were first described as DNA delivery machines. Bacterial conjugation systems have been known for more than 50 years. They have provided invaluable tools for bacterial genetics, particularly before cloning was widely used. Self-transmissible plasmids (sex factors) can promote their own transfer, the transfer of integrated DNA, the transfer of co-resident non-autotransmissible plasmids, and the transfer of chromosomal DNA. Self-transmissible plasmids are highly infectious. They disseminate to other cells conferring the donor ability to them. They represent a major mechanism of horizontal gene transfer between bacterial species. Conjugation systems are also responsible for inter-kingdom gene transfer as shown by the early work on the crown gall tumor. Crown gall is a plant disease induced by a soil bacterium, *Agrobacterium tumefaciens*. In the early 1970s it was demonstrated that this plant tumor was induced by a bacterial plasmid. Bacterial oncogenicity was linked to the presence of a megaplasmid, Ti, which could be transferred from one *Agrobacterium* strain to another. It was subsequently shown that during plant infection only a segment of Ti (T-DNA) was transferred to the plant cell [208]. Wounded plants secrete phenolic compounds which are detected by the *Agrobacterium* Ti-encoded two-component regulatory system (VirA/VirG). This initiates the transcription of the other Ti *vir* genes, which are clustered in the Ti *virB* region. This allows two endonucleases (VirD1 and VirD2) to nick the edges of T-DNA resulting in the formation of a single-stranded T-DNA molecule. The single-stranded T-DNA remains covalently bound to one endonuclease at its 5' end (VirD2), and is entwined by a single-stranded DNA-binding protein (VirE2). This nucleoprotein complex is transported through the bacterial envelope, crosses the plant cell wall, the plasma, and the nuclear membranes before integrating in the host DNA. Both VirD2 and VirE2 possess nuclear localization signals (NLS) and thus drive the T-DNA to its final target. The host chromosome-integrated T-DNA directs the synthesis of proteins which interfere with the host signal transducing pathways leading to transformation of the plant cell. T-DNA also encodes enzymes involved in opine synthesis in plants, which are used by parasitic Agrobacteria as specific growth substances [209].

7.3.2.3.2 A DNA-binding Protein Secretion Apparatus: What is the Nature of the Signal, DNA or Protein?

Two decades ago, it was shown that mixed infections between two avirulent *Agrobacterium* mutants, one with a T-DNA deletion and the other with a *vir E2* mutation, lead to tumor in plants. This indicates that the VirE2 protein and the T-DNA–VirD2 complex can reach the plant independently, and can be assembled in the plant cytosol in a nucleoprotein complex, which is transported to the nucleus [210]. This secretion of each component is dependent on the *vir* genes. Moreover, a *vir E2* mutant elicits tumor formation in transgenic plants expressing VirE2. This demonstrates that VirE2 is only required in the plant [211]. Hence, VirE2 transfer to the plant is the first secretion system identified in gram-negative bacteria (see [212] for a review). VirE2 secretion requires a specific chaperone, VirE1, which can prevent VirE2 aggregation and premature binding to single-stranded DNA in *A. tumefaciens* [213]. Similar results were subsequently obtained with another Vir protein, VirF. However, VirE2, VirD2, and VirF are very inefficiently translocated to the periplasm and are secreted to the extracellular medium by an unidentified VirB-independent mechanism [214].

Extensive similarities exist between T-DNA transfer and bacterial conjugation. In both mechanisms, single-stranded DNA is generated by endonucleolytic cleavage at a defined site (oriT for transmissible plasmids and T-DNA border sequences for the Ti plasmid) and is transferred during close contact between the donor and recipient cells. Moreover, both mechanisms share important sequence similarities and have a similar genetic organization. Eleven Ti-encoded VirB proteins are homologous to Tra proteins of various self-transmissible plasmids. Self-transmissible plasmids direct the synthesis of pili composed of TraA subunits, which are essential during the intercellular contact between the donor and recipient cell. Pilus retraction leads to the formation of wall-to-wall contacts. The VirB2 protein shares sequence similarity with the F-pilus subunit (TraA) and is similarly processed. *A. tumefaciens* also makes VirB2-containing pilus-like structures, especially below 25 °C (see [215] for a review).

The T-DNA transport system can also deliver non-self-transmissible plasmids to both plants and bacterial recipient cells. Thus, this system can promote the transport of unrelated single-stranded DNA molecules, which raises the question of the recognition structure. As the system promotes the transport of both a nucleoprotein complex (VirD2-T-DNA) and single proteins (VirE2 and VirF), the specific recognition element may be on the proteins and not on the DNA.

7.3.2.3.3 A Secretion Machinery for Proteins with or without Signal Peptide

It was recently shown that the genetic determinant responsible for the secretion of *Bordetella pertussis* pertussis toxin PT is linked to the PT structural genes. Eight secretion proteins have been identified and seven show striking sequence similarities with the VirB and Tra proteins. PT is an oligomeric protein composed of five different subunits, known as S1 to S5. It belongs to the A–B group of toxins. S1, the ADP-ribosylating enzyme, is translocated into the host cytosol via the pentamer, which is composed of one S2, S3, and S5 molecule and two S4 subunits inserted in the host plasma membrane. Each subunit has a signal peptide and is transported to the periplasm by the *sec*-dependent pathway. None of the subunits is efficiently secreted alone. Thus, the oligomer probably has to be formed in the periplasm prior to secretion, like the A–B toxins secreted by the type II secretion pathway [216].

The *Helicobacter pylori* CagA protein is a 145 kDa protein which is injected into host cells where it is tyrosine-phosphorylated. CagA is secreted by a type IV secretion system, encoded by the *cag* pathogenicity island [217]. CagA does not have a signal peptide. In addition, the VirE2 and VirF proteins, which are secreted by the *A. tumefaciens* VirB machinery also lack signal peptides.

Several other putative type IV secretion systems have been identified in various pathogens, such as *Legionella pneumoniae* and *Brucella suis*. However, the exported substrates (DNA or proteins) have not yet been identified and the *Legionella pneumophila* Dot/Icm type IV system is only distantly related to the VirB system. In *L. pneumophila* the type IV system promotes non-self-transmissible plasmid conjugation, demonstrating that it is functional [218]. Furthermore, *L. pneumophila* and *B. suis* type IV secretion mutants are highly attenuated for multiplication inside host macrophages [219, 220].

7.3.2.3.4 The Secretion Apparatus

Nine of the eleven VirB proteins are associated with the cell envelope. VirB7, B9, and VirB10 form high molecular weight complexes. After being processed VirB2 polymerizes to form the T-pilus. VirB4 and VirB11 have nucleotide binding sites, and VirB11 displays ATPase activity [215]. They are associated with the inner membrane and may provide the energy required for secretion and/or the assembly of the secretion apparatus. It has been controversially proposed that the pilus structure acts as a conduit for protein and DNA. It is not clear whether the pilus forms a hollow structure. It is more likely that the pilus brings the donor and recipient cells together, allowing the formation of mating and secretion channels (Fig. 5).

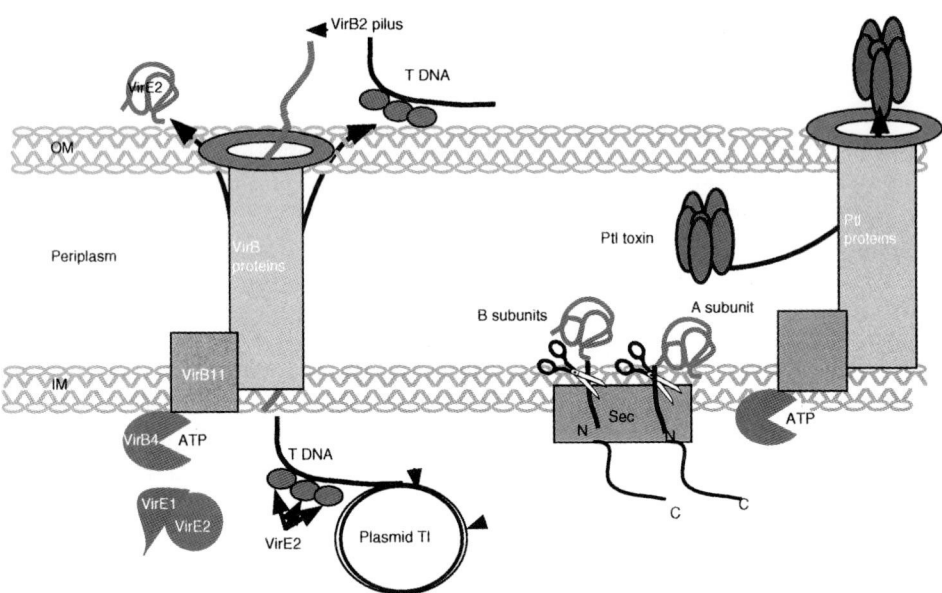

Fig. 5. Type IV secretion pathways; see color plates p. XXXVIII.

7.4
Concluding Remarks

Bacterial secretion pathways are currently separated into four distinct mechanisms. However, this classification is useful but schematic, because the mechanisms clearly share many common features, which suggests that they have common evolutionary origins. We will first emphasize the common features and then the most important divergences.

The most striking feature is that the secretion pathways are not dedicated to the secretion of one type of protein. Whereas hydrolytic enzymes are mostly secreted by type I and II pathways, toxins are secreted by all four pathways. Toxin entry into the host cytosol might occur after their secretion into the extracellular medium with the help of a pore-forming domain present on the toxins secreted by the type I apparatus (hemolysins, adenylate cyclases), the help of the B-pentamer secreted by type II or type IV apparatuses (A—B toxins). Contrastingly, secretion and injection can be coupled, as is the case for the type III injection device and probably for the type IV secretion pathway (for Cag protein) which occurs via by an uncharacterized mechanism.

All the systems (except for autotransporters and systems with one helper) have a specific ATPase, the function of which is clearly established in pathway I. It interacts with the substrates and modulates the formation of the multiprotein complex. In the other systems, it is not clear whether the ATPases interact with the substrate.

Another property shared by type II (GSP) and type III systems is the use of an outer membrane protein belonging to the secretin family. In GSP pathway, but not in pathway III, secretins interact directly with the exoproteins. It is not clear whether secretins play the same role in both systems or whether they interact with type III substrates. If they can interact with type III substrates it would imply that proteins secreted by pathway III must pass transiently by the periplasm if they are to interact with the secretins.

Similarly the E. coli outer membrane protein TolC is used by two distinct secretion pathways. We did not mention the E. coli heat-stable enterotoxin (STIp) secretion mechanism because the secretion apparatus has not been fully characterized. This small protein is synthesized as a precursor with a signal peptide, a propeptide, and an 18-amino acid mature domain [221]. The precursor is translocated across the inner membrane via the Sec pathway. Cleavage of the propeptide occurs in the periplasm, and the mature toxin is secreted from the periplasm to the extracellular medium in a TolC—dependent manner. TolC is an outer membrane protein belonging to type I transporters [222]. As type I substrates are thought to be translocated directly from the cytoplasm to the extracellular medium, it implies either that proteins belonging to the TolC family have different functions in protein secretion, or that type I substrates are transiently present in the periplasm where they can interact with the TolC-like proteins.

Another common feature to these secretion systems is that except for type I, type II autotransporters, and type II with one helper, they all form supramolecular structures resembling surface organelles. Type II GSP secretion apparatus can form structures resembling type IV pili. Type III secretion apparatuses form needles resembling flagella. Type IV secretion systems were first thought to be conjugation pili. These morphological similarities are strengthened by the sequence similarities between secretion proteins and proteins involved in organelle assembly and further by the sharing of common proteins, such as the prepillin peptidase. Some of these organelles are themselves secretion apparatuses, such as flagella which secrete a regulatory protein and a phospholipase. Sex pili and type IV pili may also form secretion machineries.

All secretion systems also use chaperones. However, these chaperones have different properties and are used in different manners.

The type I system uses general chaperones, such as SecB which maintain the substrates in a secretion-competent state. They might be also involved in the targeting of the substrate to the transporter. In type III secretion pathway, most of the substrates have dedicated chaperones which prevent aggregation and drive their exoproteins to the secretion apparatus. When the secretion is not coupled to translation (e. g., in mutants lacking the 5' mRNA secretion signal) secretion is entirely dependent on chaperones. Dedicated chaperones exist also in several type II GSP secretion systems. But they have opposite functions. They assist substrate folding in the periplasm, which is a prerequisite for this pathway. This is the major difference between these systems. Whereas GSP type II system recognizes and secretes folded substrates, the other systems probably secrete unfolded substrates. Another major difference between the systems is the dependency on the sec pathway. It was established that proteins with signal peptides do not use the same secretion pathway as those that lack a signal peptide. However, the discovery of pathway IV, which appears to have both types of substrates, demonstrates that this rule has exceptions.

References

1. Blobel, G., Dobberstein, B., *J. Cell Biol.* **1975**, *67*, 835-862.
2. Emr, S. D., Schwartz, M., Silhavy, T., *Proc. Natl. Acad. Sci. USA* **1978**, *75*, 5802-5806.
3. Wandersman, C., in: *Escherichia coli and Salmonella tiphymurium Cellular and Molecular Biology* (Neidhart, F. C., Ed.), ASM Press, Washington, DC, **1996**, pp. 955-967.
4. Hacker, J., Kaper J. B., *Annu. Rev. Microbiol.* **2000**, *54*, 641-679.
5. Tjalsma, H., Bolhuis, A., Jongbloed, J. D., Bron, S., van Dijl, J. M., *Microbiol. Mol. Biol. Rev.* **2000**, *64*(3), 515-547.
6. Danese, P. N., Silhavy, T. J., *Annu. Rev. Genet.* **1998**, *32*, 59-94.
7. Duong, F., Eichler, J., Price, A., Leonard, M. R., Wickner, W., *Cell* **1997**, *91*(5), 567-573.
8. Economou, A., *FEBS Lett.* **2000**, *476*(1-2), 18-21.
9. Fekkes, P., Driessen, A. J., *Microbiol. Mol. Biol. Rev.* **1999**, *63*(1), 161-173.
10. Manting, E. H., Driessen, A. J., *Mol. Microbiol.* **2000**, *37*(2), 226-238.
11. Wickner, W., Driessen, A. J., Hartl, F. U., *Annu. Rev. Biochem.* **1991**, *60*, 101-124.
12. Brundage, L., Hendrick, J. P., Schiebel, E., Driessen, A. J., Wickner, W., *Cell* **1990**, *62*(4), 649-657.

13. van der Does, C., Manting, E. H., Kaufmann, A., Lutz, M., Driessen, A. J., *Biochemistry* **1998**, *37*(1), 201-210.
14. Manting, E. H., van Der Does, C., Remigy, H., Engel, A., Driessen, A. J., *EMBO J.* **2000**, *19*(5), 852-861.
15. Hartl, F. U., Lecker, S., Schiebel, E., Hendrick, J. P., Wickner, W., *Cell* **1990**, *63*(2), 269-279.
16. Nishiyama, K., Suzuki, T., Tokuda, H., *Cell* **1996**, *85*(1), 71-81.
17. Duong, F., Wickner, W., *EMBO J.* **1997**, *16*(16), 4871-4879.
18. Duong, F., Wickner, W., *EMBO J.* **1997**, *16*(10), 2756-2768.
19. Pogliano, J. A., Beckwith, J., *EMBO J.* **1994**, *13*(3), 554-561.
20. Scotti, P. A., Valent, Q. A., Manting, E. H., Urbanus, M. L., Driessen, A. J., Oudega, B., Luirink, J., *J. Biol. Chem.* **1999**, *274*(42), 29883-29888.
21. Koch, H. G., Hengelage, T., Neumann-Haefelin, C., MacFarlane, J., Hoffschulte, H. K., Schimz, K. L., Mechler, B., Muller, M., *Mol. Biol. Cell* **1999**, *10*(7), 2163-2173.
22. Qi, H. Y., Bernstein, H. D., *J. Biol. Chem.* **1999**, *274*(13), 8993-8997.
23. Romisch, K., Webb, J., Herz, J., Prehn, S., Frank, R., Vingron, M., Dobberstein, B., *Nature* **1989**, *340*(6233), 478-482.
24. Poritz, M. A., Bernstein, H. D., Strub, K., Zopf, D., Wilhelm, H., Walter, P., *Science* **1990**, *250*(4984), 1111-1117.
25. Kumamoto, C. A., *Mol. Microbiol.* **1991**, *5*(1), 19-22.
26. Fekkes, P., van der Does, C., Driessen, A. J., *EMBO J.* **1997**, *16*(20), 6105-6113.
27. Valent, Q. A., de Gier, J.-W. L., von Heijne, G. Kendall, D. A., ten Hagen-Jongman, C. M., Oudega, B., Luirink, J., *Mol. Microbiol.* **1997**, *25*(1), 53-64.
28. Seluanov, A., Bibi, E., *J. Biol. Chem.* **1997**, *272*(4), 2053-2055.
29. Prinz, A., Behrens, C., Rapoport, T. A., Hartmann, E., Kalies, K. U., *EMBO J.* **2000**, *19*(8), 1900-1906.
30. Beck, K., Wu, L. F., Brunner, J., Muller, M., *EMBO J.* **2000**, *19*(1), 134-143.
31. Valent, Q. A., Scotti, P. A., High, S., de Gier, J. W., von Heijne, G., Lentzen, G., Wintermeyer, W., Oudega, B., Luirink, J., *EMBO J.* **1998**, *17*(9), 2504-2512.
32. Matsumoto, G., Yoshihisa, T., Ito, K., *EMBO J.* **1997**, *16*(21), 6384-6393.
33. Koch, H. G., Muller, M., *J. Cell Biol.* **2000**, *150*(3), 689-694.
34. Deuerling, E., Schulze-Specking, A., Tomoyasu, T., Mogk, A., Bukau, B., *Nature* **1999**, *400*(6745), 693-696.
35. Herskovits, A. A., Bibi, E., *Proc. Natl. Acad. Sci. USA* **2000**, *97*(9), 4621-4626.
36. Geller, B. L., Wickner, W., *J. Biol. Chem.* **1985**, *260*(24), 13281-13285.
37. Hell, K., Herrmann, J. M., Pratje, E., Neupert, W., Stuart, R. A., *Proc. Natl. Acad. Sci. USA* **1998**, *95*(5), 2250-2255.
38. Moore, M., Harrison, M. S., Peterson, E. C., Henry, R., *J. Biol. Chem.* **2000**, *275*(3), 1529-1532.
39. Samuelson, J. C., Chen, M., Jiang, F., Moller, I., Wiedmann, M., Kuhn, A., Phillips, G. J., Dalbey, R. E., *Nature* **2000**, *406*(6796), 637-641.
40. Scotti, P. A., Urbanus, M. L., Brunner, J., de Gier, J. W., von Heijne, G., van der Does, C., Driessen, A. J., Oudega, B., Luirink, J., *EMBO J.* **2000**, *19*(4), 542-549.
41. Yahr, T. L., Wickner, W. T., *EMBO J.* **2000**, *19*(16), 4393-4401.
42. Hamman, B. D., Chen, J. C., Johnson, E. E., and Johnson, A. E., *Cell* **1997**, *89*(4), 535-544.
43. Meyer, T. H., Menetret, J. F., Breitling, R., Miller, K. R., Akey, C. W., Rapoport, T. A., *J. Mol. Biol.* **1999**, *285*(4), 1789-1800.
44. Tian, H., Boyd, D., Beckwith, J., *Proc. Natl. Acad. Sci. USA* **2000**, *97*(9), 4730-4735
45. Berks, B. C., Sargent, F., Palmer, T., *Mol. Microbiol.* **2000**, *35*(2), 260-274.
46. Berks, B. C., Sargent, F., De Leeuw, E., Hinsley, A. P., Stanley, N. R., Jack, R. L., Buchanan, G., Tracy, P., *Biochim. Biophys. Acta* **2000**, *1459*(2-3), 325-330.
47. Voordouw, G., *Biophys. Chem.* **2000**, *86*(2-3), 131-140.
48. Wu, L. F., Ize, B., Chanal, A., Quentin, Y., Fichant, G., *J. Mol. Microbiol. Biotechnol.* **2000**, *2*(2), 179-189.
49. Cline, K., Henry, R., Li, C., Yuan, J., *EMBO J.* **1993**, *12*(11), 4105-4114.
50. Chaddock, A. M., Mant, A., Karnauchov, I., Brink, S., Herrmann, R. G., Klosgen, R. B., Robinson, C., *EMBO J.* **1995**, *14*(12), 2715-2722.
51. Creighton, A. M., Hulford, A., Mant, A., Robinson, D., Robinson, C., *J. Biol. Chem.* **1995**, *270*(4), 1663-1669.

52. Settles, A. M., Yonetani, A., Baron, A., Bush, D. R., Cline, K., Martienssen, R., *Science* **1997**, *278*(5342), 1467-1470.
53. Berks, B. C., *Mol. Microbiol.* **1996**, *22*(3), 393-404.
54. Niviere, V., Wong, S. L., Voordouw, G., *J. Gen. Microbiol.* **1992**, *138*(Pt 10), 2173-2183.
55. Santini, C. L., Ize, B., Chanal, A., Muller, M., Giordano, G., Wu, L. F., *EMBO J.* **1998**, *17*(1), 101-112.
56. Weiner, J. H., Rothery, R. A., Sambasivarao, D., Trieber, C. A., *Biochim. Biophys. Acta* **1992**, *1102*(1), 1-18.
57. Weiner, J. H., Bilous, P. T., Shaw, G. M., Lubitz, S. P., Frost, L., Thomas, G. H., Cole, J. A., Turner, R. J., *Cell* **1998**, *93*(1), 93-101.
58. Sargent, F., Bogsch, E. G., Stanley, N. R., Wexler, M., Robinson, C., Berks, B. C., Palmer, T., *EMBO J.* **1998**, *17*(13), 3640-3650.
59. Wexler, M., Sargent, F., Jack, R. L., Stanley, N. R., Bogsch, E. G., Robinson, C., Berks, B. C., Palmer, T., *J. Biol. Chem.* **2000**, *275*(22), 16717-16722.
60. Bogsch, E. G., Sargent, F., Stanley, N. R., Berks, B. C., Robinson, C., Palmer, T., *J. Biol. Chem.* **1998**, *273*(29), 18003-18006.
61. Chanal, A., Santini, C., Wu, L., *Mol. Microbiol.* **1998**, *30*(3), 674-676.
62. Sargent, F., Stanley, N. R., Berks, B. C., Palmer, T., *J. Biol. Chem.* **1999**, *274*(51), 36073-36082.
63. Bolhuis, A., Bogsch, E. G., Robinson, C., *FEBS Lett.* **2000**, *472*(1), 88-92.
64. Berghofer, J., Klosgen, R. B., *FEBS Lett.* **1999**, *460*(2), 328-332.
65. Jongbloed, J. D., Martin, U., Antelmann, H., Hecker, M., Tjalsma, H., Venema, G., Bron, S., Dijl, J. M., Muller, J., *J. Biol. Chem.* **2000**, g,
66. Bogsch, E., Brink, S., Robinson, C., *EMBO J.* **1997**, *16*(13), 3851-3859.
67. Cristobal, S., de Gier, J. W., Nielsen, H., von Heijne, G., *EMBO J.* **1999**, *18*(11), 2982-2990.
68. Stanley, N. R., Palmer, T., Berks, B. C., *J. Biol. Chem.* **2000**, *275*(16), 11591-11596.
69. Sambasivarao, D., Turner, R. J., Simala-Grant, J. L., Shaw, G., Hu, J., Weiner, J. H., *J. Biol. Chem.* **2000**, *275*(29), 22526-22531.
70. Benoit, S., Abaibou, H., Mandrand-Berthelot, M. A., *J. Bacteriol.* **1998**, *180*(24), 6625-6634.
71. Rodrigue, A., Chanal, A., Beck, K., Muller, M., Wu, L. F., *J. Biol. Chem.* **1999**, *274*(19), 13223-13228.
72. Wiegert, T., Sahm, H., Sprenger, G. A., *Eur. J. Biochem.* **1997**, *244*(1), 107-112.
73. Pohlner, J., Halter, R., Beyreuther, K., Meyer, T. F., *Nature* **1987**, *325*, 458-462.
74. Maurer, J., Jose, J., Meyer, T. F., *J. Bacteriol.* **1999**, *181*, 7014-7020.
75. Eppens, E., Nouwen, N., Tommassen, J., *EMBO J.* **1997**, *16*, 4295-4230.
76. Klauser, T., Pohlner J., T. F.g, M., *EMBO J.* **1990**, *6*, 1991-1999.
77. Veiga, E., de Lorenzo, V., Fernandez, L. A., *Mol. Microbiol.* **1999**, *33*, 1232-1243.
78. Henderson, I. R., Navarro-Garcia, F., Nataro, J. P., *Trends Microbiol.* **1998**, *6*, 370-378.
79. Egile, C., d'Hauteville, H., Parsot, C., Sansonetti, P., *Mol. Microbiol.* **1997**, g (23), 1063-1073.
80. Poole, K., Schiebel, E., Braun, V., *J. Bacteriol.* **1988**, *170*, 3177-3188.
81. Ondraczek, R., Hobbie, S., Braun, V., *J. Bacteriol.* **1992**, *174*, 5086-5094.
82. Konninger, U., Hobbie, S., Benz, R., Braun, V., *Mol. Microbiol.*, **1999**, *32*, 1212-1225.
83. Guedin, S., Willery, E., Tommassen, J., Fort, E., Drobecq, H., Locht, C., Jacob-Dubuisson, F., *J. Biol. Chem.* **2000**, *29*, 30202-30210.
84. Uphoff, T., Welch, R., *J Bacteriol.* **1990**, *172*, 1206-1216.
85. d'Enfert, C., A., R., Pugsley, A. P., *EMBO J.* **1987**, *11*, 3531-3538.
86. Sandkvist M, Bagdasarian M, Howard SP, VJ., D., *EMBO J.* **1995**, *14*, 1664-1673.
87. Ball G, Chapon-Herve V, Bleves S, Michel G, M., B., *J. Bacteriol.* **1999**, *181*, 382-388.
88. Nunn, D. N., Lory, S., *Proc Natl Acad Sci USA* **1992**, *89*, 47-51.
89. Nunn, D. N., Lory, S., *J. Bacteriol.* **1993**, *175*, 4375-4382.
90. Pugsley AP., *Mol. Microbiol.* **1993**, *9*, 295-308.
91. Pepe, J. C., Lory, S., *J. Biol. Chem.* **1998**, *273*, 19120-19129.
92. Sauvonnet, N., Vignon, G., Pugsley, A. P., Gounon, P., *EMBO J.* **2000**, *19*, 2221-2228.
93. Sauvonnet, N., Gounon, P., Pugsley, A. P., *J. Bacteriol.* **2000**, *182*, 848-854.
94. Bitter, W., Koster, M., Latijnhouwers, M., de Cock, H., Tommassen, J., *Mol. Microbiol.* **1998**, *27*, 209-219.
95. Nouwen, N., Ranson, N., Saibil, H., Wolpensinger, B., Engel, A., Ghazi, A., Pugsley, A. P., *Proc. Natl. Acad. Sci. USA* **1999**, *96*, 8173-8177.

96. Shevchik, V. E., Robert-Baudouy, J., Condemine, G., *EMBO J.* **1997**, *16*, 3007-3016.
97. Nouwen, N., Stahlberg, H., Pugsley, A. P., Engel, A., *EMBO J.* **2000**, *19*, 2229-2236.
98. Linderoth, N. A., Simon, M. N., Russel, M., *Science* **1997**, *278*, 1635-1638.
99. Russel M., *J. Mol. Biol.* **1998**, *279*, 485-499.
100. Davis, B. M., Lawson, E. H., Sandkvist, M., Ali, A., Sozhamannan, S., Waldor, M. K., *Science* **2000**, *288*, 333-335.
101. Feng, J. N., Model, P., Russel, M., *Mol. Microbiol.* **1999**, *34*, 745-755.
102. Hardie, K. R., Lory, S., Pugsley, A. P., *EMBO J.* **1996**, *15*, 978-988.
103. Daefler, S., Guilvout, I., Hardie, K. R., Pugsley, A. P., Russel, M., *Mol. Microbiol.* **1997**, *24*, 465-75.
104. El Khattabi, M., Ockhuijsen, C., Bitter, W., Jaeger, K. E., Tommassen, J., *Mol. Gen. Genet.* **1999**, *261*, 770-776.
105. El Khattabi, M., Van Gelder, P., Bitter, W., Tommassen, J., *J. Biol. Chem.* **2000**, *275*, 26885-26891.
106. Kessler, E., Safrin, M., Gustin, J. K., Ohman, D. E., **1998**, *273*, 30225-30231.
107. Braun, P., Bitter, W., Tommassen, J., *Microbiology* **2000**, *146*, 2565-2572.
108. Braun, P., Tommassen, J., Filloux, A., *Mol. Microbiol.* **1996**, *19*, 297-306.
109. Hirst, T. R., Holmgren J., *Proc. Natl. Acad. Sci. USA* **1987**, *84*, 7418-7422.
110. Peek, J. A., Taylor, R. K., *Proc. Natl. Acad. Sci. USA* **1992**, *89*, 6210-6214.
111. Hardie, K. R., Schulze, A., Parker, M. W., Buckley, J. T., *Mol. Microbiol.* **1995**, *17*(6), 1035-1044.
112. Goebel, W., Hedgpeth, J., *J. Bacteriol.* **1982**, *151*, 1290-1298.
113. Higgins, C. F., Hiles, I. D., Salmond, G. P., Gill, D. R., Downie, J. A. et al., *Nature* **1986**, *323*, 448-450.
114. Dinh, T., Paulsen, I. T., Saier, M. H. J., *J. Bacteriol.* **1994**, *176*(13), 3825-3831.
115. Higgins, C. F., *Annu. Rev. Cell. Biol.* **1992**, *8*, 67-113.
116. Welch, R. A., *Mol. Microbiol.* **1991**, *60*, 101-124.
117. Glaser, P., Sakamoto, H., Belladou, J., Ullman, A., Danchin, A., *EMBO J.* **1988**, *7*, 3997-4004.
118. Fullner, K. J., Mekalanos, J. J., *EMBO J.* **2000**, *19*(20), 5315-5323.
119. Lin, W., Fullner, K. J., Clayton, R., Sexton, J. A., Rogers, M. B., Calia, K. E., Calderwood, S. B., Fraser, C., Mekalanos, J. J., *Proc. Natl. Acad. Sci. USA* **1999**, *96*(3), 1071-1076.
120. Wandersman, C., Delepelaire, P., Létoffé, S., Schwartz, M., *J. Bact.* **1987**, *169*(11), 5046-5053.
121. Guzzo, J., Murgier, M., Filloux, A., Lazdunski, A., *J. Bacteriol.* **1990**, *172*, 942-948.
122. Kumeta, H., Hoshino, T., Goda, T., Okayama, T., Shimada, T., Ohgiya, S., Matsuyama, H., Ishizaki, K., *Biosci. Biotechnol. Biochem.* **1999**, *63*(7), 1165-1170.
123. Letoffe, S., Ghigo, J. M., Wandersman, C., *J. Bacteriol.* **1993**, *175*(22), 7321-7328.
124. Wassif, C., Cheek, D., Belas, R., *J. Bacteriol.* **1995**, *177*(20), 5790-5798.
125. Walker, K. E., Moghaddame-Jafari, S., Lockatell, C. V., Johnson, D., Belas, R., *Mol. Microbiol.* **1999**, *32*(4), 825-836.
126. Akatsuka, H., Kawai, E., Omori, K., Shibatani, T., *J. Bacteriol.* **1995**, *177*, 6381-6389.
127. Akatsuka, H., Kawai, E., Omori, K., Komatsubara, S., Shibatani, T., Tosa, T., *J. Bacteriol.* **1994**, *176*, 1949-1956.
128. Ahn, J. H., Pan, J. G., Rhee, J. S., *J. Bacteriol.* **1999**, *181*(6), 1847-1852.
129. Ertesvag, H., Hoidal, H. K., Hals, I. K., Rian, A., Doseth, B., Valla, S., *Mol. Microbiol.* **1995**, *16*(4), 719-731.
130. Ertesvag, H., Doseth, B., Larsen, B., Skjak-Braek, G., Valla, S., *J. Bacteriol.* **1994**, *176*(10), 2846-2853.
131. Oresnik, I. J., Twelker, S., Hynes, M. F., *Appl. Environ. Microbiol.* **1999**, *65*(7), 2833-2840.
132. Awram, P., Smit, J., *J. Bacteriol.* **1998**, *180*(12), 3062-3069.
133. Kawai, E., Akatsuka, H., Idei, A., Shibatani, T., Omori, K., *Mol. Microbiol.* **1998**, *27*(5), 941-952.
134. Thompson, S. A., Shedd, O. L., Ray, K. C., Beins, M. H., Jorgensen, J. P., Blaser, M. J., *J. Bacteriol.* **1998**, *180*(24), 6450-6458.
135. Hoiczyk, E., *J. Bacteriol.* *180*(15), **1998**, 3923-3932.
136. Hoiczyk, E., Baumeister, W., *Mol. Microbiol.* *26*(4), **1997**, 699-708.
137. Brahamsha, B., *Proc. Natl. Acad. Sci. USA* **1996**, *93*(13), 6504-6509.
138. Thompson, S. A., Wang, L. L., West, A., Sparling, P. F., *J. Bacteriol.* **1993**, *175*(3), 811-818.
139. de Maagd, R. A., Wijfjes, A. H., Spaink, H. P., Ruiz-Sainz, J. E., Wijffelman, C. A., Okker, R. J., Lugtenberg, B. J., *J. Bacteriol.* **1989**, *171*(12), 6764-6770.

140. Economou, A., Hamilton, W. D., Johnston, A. W., Downie, J. A., *EMBO J.* **1990**, *9*(2), 349-354
141. Walker, S. A., Downie, J. A., *Mol. Plant. Microb. Interact.* **2000**, *13*(7), 754-762.
142. Gilson, L., Mahanty, H. K., Kolter, R., *J. Bacteriol.* **1987**, *169*(6), 2466-2470.
143. Zhong, X., Kolter, R., Tai, P. C., *J. Biol. Chem.* **1996**, *271*(45), 28057-28063.
144. Letoffe, S., Ghigo, J. M., Wandersman, C., *Proc. Natl. Acad. Sci. USA* **1994**, *91*(21), 9876-9880.
145. Letoffe, S., Ghigo, J. M., Wandersman, C., *J. Bacteriol.* **1994**, *176*(17), 5372-5377.
146. Letoffe, S., Redeker, V., Wandersman, C., *Mol. Microbiol.* **1998**, *28*(6), 1223-1234.
147. Idei, A., Kawai, E., Akatsuka, H., Omori, K., *J. Bacteriol.* **1999**, *181*(24), 7545-7551.
148. Baumann, U., Wu, S., Flaherty, K. M., McKay, D. B., *EMBO J.* **1993**, *12*, 3357-3364.
149. Felmlee, T., Welch, R. A., *Proc. Natl. Acad. Sci. USA* **1988**, *85*(14), 5269-5273.
150. Stover, C. K., Pham, X. Q., Erwin, A. L., Mizoguchi, S. D., Warrener, P., Hickey, M. J., Brinkman, F. S., Hufnagle, W. O., Kowalik, D. J., Lagrou, M., Garber, R. L., Goltry, L., Tolentino, E., Westbrock-Wadman, S., Yuan, Y., Brody, L. L., Coulter, S. N., Folger, K. R., Kas, A., Larbig, K., Lim, R., Smith, K., Spencer, D., Wong, G. K., Wu, Z., Paulsen, I. T., *Nature* **2000**, *406*(6799), 959-964.
151. Kaneko, T., Sato, S., Kotani, H., Tanaka, A., Asamizu, E., Nakamura, Y., Miyajima, N., Hirosawa, M., Sugiura, M., Sasamoto, S., Kimura, T., Hosouchi, T., Matsuno, A., Muraki, A., Nakazaki, N., Naruo, K., Okumura, S., Shimpo, S., Takeuchi, C., Wada, T., Watanabe, A., Yamada, M., Yasuda, M., Tabata, S., *DNA Res.* **1996**, *3*(3), 185-209.
152. Kaneko, T., Sato, S., Kotani, H., Tanaka, A., Asamizu, E., Nakamura, Y., Miyajima, N., Hirosawa, M., Sugiura, M., Sasamoto, S., Kimura, T., Hosouchi, T., Matsuno, A., Muraki, A., Nakazaki, N., Naruo, K., Okumura, S., Shimpo, S., Takeuchi, C., Wada, T., Watanabe, A., Yamada, M., Yasuda, M., Tabata, S., *DNA Res.* **1996**, *3*(3), 109-136.
153. Simpson, A. J., Reinach, F. C., Arruda, P., Abreu, F. A., Acencio, M., Alvarenga, R., Alves, L. M., Araya, J. E., Baia, G. S., Baptista, C. S., Barros, M. H., Bonaccorsi, E. D., Bordin, S., Bove, J. M., Briones, M. R., Bueno, M. R., Camargo, A. A., Camargo, L. E., Carraro, D. M., Carrer, H., Colauto, N. B., Colombo, C., Costa, F. F., Costa, M. C., Costa-Neto, C. M., Coutinho, L. L., Cristofani, M., Dias-Neto, E., Docena, C., El-Dorry, H., Facincani, A. P., Ferreira, A. J., Ferreira, V. C., Ferro, J. A., Fraga, J. S., Franca, S. C., Franco, M. C., Frohme, M., Furlan, L. R., Garnier, M., Goldman, G. H., Goldman, M. H., Gomes, S. L., Gruber, A., Ho, P. L., Hoheisel, J. D., Junqueira, M. L., Kemper, E. L., Kitajima, J. P., Krieger, J. E., Kuramae, E. E., Laigret, F., Lambais, M. R., Leite, L. C., Lemos, E. G., Lemos, M. V., Lopes, S. A., Lopes, C. R., Machado, J. A., Machado, M. A., Madeira, A. M., Madeira, H. M., Marino, C. L., *Nature* **2000**, *406*(6792), 151-157.
154. Deckert, G., Warren, P. V., Gaasterland, T., Young, W. G., Lenox, A. L., Graham, D. E., Overbeek, R., Snead, M. A., Keller, M., Aujay, M., Huber, R., Feldman, R. A., Short, J. M., Olsen, G. J., Swanson, R. V., *Nature* **1998**, *392*(6674), 353-358.
155. Ghigo, J. M., Wandersman, C., *Res. Microbiol.* **1992**, *143*(9), 857-867.
156. Wandersman, C., Delepelaire, P., *Proc. Natl. Acad. Sci. USA* **1990**, *87*, 4776-4780.
157. Binet, R., Wandersman, C., *EMBO J.* **1995**, *14*(10), 2298-3206.
158. Akatsuka, H., Binet, R., Kawai, E., Wandersman, C., Omori, K., *J. Bacteriol.* **1997**, *179*(15), 4754-4760.
159. Letoffe, S., Delepelaire, P., Wandersman, C., *EMBO J.* 15(21), **1996**, 5804-5811.
160. Thanabalu, T., Koronakis, E., Hughes, C., Koronakis, V., *EMBO J.* **1998**, *17*(22), 6487-6496.
161. Schulein, R., Gentschev, I., Schlor, S., Gross, R., Goebel, W., *Mol. Gen. Genet.* **1994**, *245*(2), 203-211.
162. Koronakis, V., Hughes, C., Koronakis, E., *Mol. Microbiol.* **1993**, *8*(6), 1163-1175.
163. Delepelaire, P., *J. Biol. Chem.* **1994**, *269*, 27952-27957.
164. Koronakis, E., Hughes, C., Milisav, I., Koronakis, V., *Mol. Microbiol.* **1995**, *16*(1), 87-96.
165. Hwang, J., Tai, P. C., *Curr. Microbiol.* **1999**, *39*(4), 195-199.
166. Koronakis, V., Sharff, A., Koronakis, E., Luisi, B., Hughes, C., *Nature* **2000**, *405*(6789), 914-919.
167. Benz, R., Maier, E., Gentschev, I., *Zentralbl. Bakteriol.* **1993**, *278*(2-3), 187-196.
168. Nikaido, H., *J. Bacteriol.* **1996**, *178*(20), 5853-5859.

168a. Andersen, C., Hughes, C., Koronskio, V., EMBO Rep. **2000**, *1*, 313-318.
169. Venema, K., Dost, M. H., Beun, P. A., Haandrikman, A. J., Venema, G., Kok, J., Appl. Environ. Microbiol. **1996**, *62*(5), 1689-1692.
170. Zgurskaya, H. I., Nikaido, H., Proc. Natl. Acad. Sci. USA **1999**, *96*(13), 7190-7195.
171. Nicaud, J. M., Mackman, N., Gray, L., Holland, I. B., FEBS Lett. **1986**, *204*(2), 331-335.
172. Ghigo, J. M., Wandersman, C., J.Biol. Chem. **1994**, *269*, 8979-8985.
173. Yin, Y., Zhang, F., Ling, V., Arrowsmith, C. H., FEBS Lett. **1995** *366*(1),, 1-5.
174. Wolff, N., Ghigo, J. M., Delepelaire, P., Wandersman, C., Delepierre, M., Biochemistry **1994**, *33*(22), 6792-6801.
175. Wolff, N., Delepelaire, P., Ghigo, J. M., Delepierre, M., Eur. J. Biochem. **1997**, *243*(1-2), 400-407.
176. Hui, D., Morden, C., Zhang, F., Ling, V., J. Biol. Chem. **2000**, *275*(4), 2713-2720.
177. Kenny, B., Taylor, S., Holland, I. B., Mol. Microbiol. **1992**, *6*(11), 1477-1489.
178. Letoffe, S., Wandersman, C., J. Bacteriol. **1992**, *174*(15), 4920-4927.
179. Sebo, P., Ladant, D., Mol. Microbiol. **1993**, *9*(5), 999-1009.
180. Delepelaire, P., and Wandersman, C., EMBO J. **1998**, *17*(4), 936-944.
181. Hueck, C. J., Microbiol. Mol. Biol. Rev. **1998**, *62*(2), 379-433.
182. Kenny, B., DeVinney, R., Stein, M., Reinscheid, D. J., Frey, E. A., Finlay, B. B., Cell **1997**, *91*, 511-520.
183. Rosqvist, R., Hakansson, S., Forsberg, A., Wolf-Watz, H., EMBO J. **1995**, *14*, 4187-4195.
184. Anderson, D. M., Fouts, D. E., Collmer, A., Schneewind, O., Proc. Natl. Acad. Sci. USA. **1999**, *96*, 12839-12843.
185. Collmer, A., Badel, J. L., Charkowski, A. O., Deng, W. L., Fouts, D. E., Ramos, A. R., Rehm, A. H., Anderson, D. M., Schneewind, O., van Dijk, K., Alfano, J. R., Proc. Natl. Acad. Sci. USA. **2000**, *97*, 8770-8777.
186. Eichelberg, K., Ginocchio, C. C., Galan, J. E., J. Bacteriol. **1994**, *176*, 4501-4510.
187. Kubori, T., Matsushima, Y., Nakamura, D., Uralil, J., Lara-Tejero, M., Sukhan, A., Galan, J. E., Aizawa, S. I., Science **1998**, *280*, 602-605.
188. Galan, J. E., Collmer, A., Science. **1999**, *284*, 1322-1328.
189. Tamano, K., Aizawa, S., Katayama, E., Nonaka, T., Imajoh-Ohmi, S., Kuwae, A., Nagai, S., Sasakawa, C., EMBO J. **2000**, *19*, 3876-3887.
190. Blocker, A., Gounon, P., Larquet, E., Niebuhr, K., Cabiaux, V., Parsot, C., Sansonetti, P., J. Cell. Biol. **1999**, *147*, 683-693.
191. Kubori, T., Sukhan, A., Aizawa, S. I., Galan, J. E., Proc. Natl. Acad. Sci. USA **2000**, *97*, 10225-10230.
192. Roine, E., Wei, W., Yuan, J., Nurmiaho-Lassila, E. L., Kalkkinen, N., Romantschuk, M., He, S. Y., Proc. Natl. Acad. Sci. USA **1997**, *94*, 3459-3464.
193. Sory, M. P., Boland, A., Lambermont, I., Cornelis, G. R., Proc. Natl. Acad. Sci. USA **1995**, *92*, 11998-12002.
194. Schesser, K., Frithz-Lindsten, E., Wolf-Watz, H., J. Bacteriol. **1996**, *178*,7227-7233.
195. Anderson, D. M., Schneewind, O., Science. **1997**, *278*, 1140-1143.
196. Anderson, D. M., Schneewind, O., Mol. Microbiol. **1999**, *31*, 1139-1148.
197. Wattiau, P., Bernier, B., Deslee, P., Michiels, T., Cornelis, G. R., Proc. Natl. Acad. Sci. USA **1994**, *91*, 10493-10497.
198. Cheng, L. W., Anderson, D. M., Schneewind, O., Mol. Microbiol. **1997**, *24*, 757-765.
199. Menard, R., Sansonetti, P., Parsot, C., Vasselon, T., Cell **1994**, *79*, 515-525.
200. Cornelis, G. R., Van Gijsegem, F., Annu. Rev. Microbiol. **2000**, *54*, 735-774.
201. De Geyter, C., Wattiez, R., Sansonetti, P., Falmagne, P., Ruysschaert, J. M., Parsot, C., Cabiaux, V., Eur. J. Biochem. **2000**, *267*, 5769-5776.
202. Tardy, F., Homble, F., Neyt, C., Wattiez, R., Cornelis, G. R., Ruysschaert, J. M., Cabiaux, V., EMBO J. **1999**, *18*, 6793-6799.
203. Macnab, R. M., J. Bacteriol. **1999**, *181*, 7149-7153.
204. Hughes, K. T., Gillen, K. L., Semon, M. J., Karlinsey, J. E., Science **1993**, *262*, 1277-1280.
205. Karlinsey, J. E., Lonner, J., Brown, K. L., Hughes, K. T., Cell **2000**, *102*, 487-497
206. Young, G. M., Schmiel, D. H., Miller, V. L., Proc. Natl. Acad. Sci. USA **1999**, *96*, 6456-6461.
207. Pettersson, J., Nordfelth, R., Dubinina, E., Bergman, T., Gustafsson, M., Magnusson, K. E., Wolf-Watz, H., Science **1996**, *273*, 1231-1233.

208. Zambryski, P., Holsters, M., Kruger, K., Depicker, A., Schell, J., Van Montagu, M., Goodman, H. M., *Science* **1980**, *209*, 1385-1391.
209. Zambryski, P., Goodman, H. M., Van Montagu, M., Schell, J., in: *Mobile Genetic Elements* (Shapiro, J., Ed.), Academic Press, New York. **1983**, 505-535.
210. Otten, L., De Greve, H., Leemans, J., Hain, R., Hooykaas, P., Schell, J., *Mol. Gen. Genet.* **1984**, *195*, 159-163.
211. Citovsky, V., Zupan, J., Warnick, D., Zambryski, P., *Science* **1992**, *256*, 1802-1805.
212. Christie, P. J., *J Bacteriol* **1997**, *179*, 3085-3094.
213. Deng, W., Chen, L., Peng, W. T., Liang, X., Sekiguchi, S., Gordon, M. P., Comai, L., Nester, E. W., *Mol. Microbiol.* **1999**, *31*, 1795-1807.
214. Chen, L., Li, C. M., Nester, E. W., *Proc. Natl. Acad. Sci. USA* **2000**, *97*, 7545-7550.
215. Christie, P. J., Vogel, J. P., *Trends Microbiol.* **2000**, *8*, 354-360.
216. Weiss, A. A., Johnson, F. D., Burns, D. L., *Proc. Natl. Acad. Sci. USA* **1993**, *90*, 2970-2974.
217. Stein, M., Rappuoli, R., Covacci, A., *Proc. Natl. Acad. Sci. USA.* **2000**, *97*, 1263-1268.
218. Segal, G., Purcell, M., and Shuman, H. A., *Proc. Natl. Acad. Sci. USA* **1998**, *95*, 1669-1674
219. O'Callaghan, D., Cazevieille, C., Allardet-Servent, A., Boschiroli, M. L., Bourg, G., Foulongne, V., Frutos, P., Kulakov, Y., Ramuz, M., *Mol. Microbiol.* **1999**, *33*, 1210-1220.
220. Segal, G., Russo, J. J., Shuman, H. A., *Mol. Microbiol.* **1999**, *34*, 799-809.
221. Yang, Y., Gao, Z., Guzman-Verduzco, L. M., Tachias, K., Kupersztoch, Y. M., *Mol. Microbiol.* **1992**, *6*, 3521-3529.
222. Yamanaka, H., Nomura, T., Fujii, Y., Okamoto, K., *Microb. Pathog.*, **1998**, *25*, 111-120.

8
Bacterial Channel Forming Protein Toxins

J. Thomas Buckley

Channel forming proteins are found everywhere in nature. Many of them are integral components of cell membranes, facilitating the movement of molecules from one side of the membrane to the other. Others are water-soluble proteins that have the ability to form channels in target membranes. The latter proteins play offensive or defensive roles for a wide range of cell types, including animal and plant cells and bacteria. It is the bacterial channel forming protein toxins that are the subject of this review. They are a remarkably diverse family of molecules, comprised of a large number of seemingly unrelated toxins as well as a few groupings of proteins that are homologs, or at least analogs. The purpose of this chapter is not to make an exhaustive survey but rather to explore some of the general properties and common features of these channel forming proteins, using some of the best characterized toxins as examples. Only toxins for which knowledge is available about structure as well as function will be described. The reader will be referred to recent books and reviews for more detailed information about individual toxins, notably the recent comprehensive treatment edited by Alouf and Freer [1].

Nomenclature

Someone new to the field can quickly become confused by the vagaries of toxin nomenclature. Many toxins are given names associated with the cells they killed in the assay first used to identify them. Labels such as leukocidin or hemolysin are very common, leaving the impression that leukocytes or erythrocytes are primary toxin targets, when in many cases, if there is a primary target of a specific toxin, contributing to the virulence of the producing species, it has not been identified. This is especially true for toxins that are called hemolysins. Many of these toxins were so named because they were first identified by screening bacteria producing zones of hemolysis on blood agar plates; their activity is often quantitated using some variation of a hemolytic titer assay. In many cases of course, blood agar plates and hemolytic assays are only used because of their convenience. There is no reason to believe that erythrocytes should generally be favored targets for bacteria.

The words channel, pore and hole are all used rather indiscriminately to describe proteins that can breach the membrane permeability barrier, in spite of the fact that each word can imply a somewhat different connotation. The word pore brings to mind a hole of defined size with little or no selectivity. It might best describe toxins such as aerolysin of *Aeromonas* spp. and *Staphylococcus aureus* α-toxin, both of which produce extremely well defined, relatively non-selective pores. The word channel might be reserved for toxins such as the colicins and the *Bacillus thuringiensis* Cry toxins, which form well defined ion-selective channels. Finally, the word hole connotes a pore or channel that may be less uniform in properties. It might be best used with proteins such as the cholesterol-dependent toxins, which appear to form large holes of variable size. In the absence of an established convention, I will normally, but not exclusively, refer to all of the proteins in this review as channel forming proteins.

8.1
Toxins in Model Systems

A detailed discussion of the various model systems used to study channel forming toxins, such as lipid monolayers, planar lipid bilayers, and lipid vesicles, is beyond the scope of this chapter. Instead, the reader is referred to an excellent recent review of biophysical models and methods by Menestrina and Semjén [1a]. It is worth noting that the results of planar lipid studies have shown that the channels formed by many bacterial toxins exhibit some ion selectivity and many of them also exhibit some voltage gating. However, at least for many of the β-barrel toxins, ion selectivity and gating are of doubtful significance *in vivo*. Instead, voltage gating appears to be a fundamental physical property of the β-barrel proteins, including the porins [2].

8.2
Toxin Complexity

Bacterial protein toxins are typically rather large molecules. For example, α-toxin of *S. aureus* is 33 kDa and the *Escherichia coli* RTX toxin HlyA is more than 100 kDa. Since there are many examples of small peptides, such as valinomycin and mellitin, that can form defined channels in bilayers or disrupt bilayers by their detergent action, one must ask why there is a need for larger complex molecules. After all, if the only requirement for toxicity is that a channel be formed in a plasma membrane, then only a few amino acids would suffice. The answer to this question is that bacterial toxins have several important properties that require the complexity only a larger structure can provide. Among these properties are the following:

(1) Protein toxins do not normally kill the secreting bacteria. Often this is because they can exist in inactive conformations. Activation is typically accomplished by proteolytic processing, or by acylation, or by exposure to a change in environment, such as reduced pH.
(2) The channel forming proteins are polymorphic, existing in a soluble form (often the proform), which is released by the bacteria, as well as in an insoluble membrane-inserted form. The transformation from soluble to insoluble may be a direct result of a change in protein conformation, or it may be brought about by the oligomerization of monomers.
(3) Many channel forming proteins are primarily active against specific types of cells. This is possible because they contain separate domains that are capable of recognizing and binding unique receptors that are found on the cell surface.
(4) In addition to domains that are necessary for receptor binding and for their channel forming function, many toxins contain domains that have enzymatic activity.

8.3
Classification of Channel Forming Proteins

As already mentioned, bacterial channel forming proteins are a remarkably diverse family. They can be subdivided in a number of different ways, but perhaps the most useful first step is to divide them into two groups according to the means by which they generate the channel itself (Fig. 1). One group, which probably represents the majority of the known channel forming proteins, contains members that produce amphipathic β-barrels that can insert into target membranes producing relatively non-specific channels. These proteins are analogous to bacterial outer membrane porins and it is not unreasonable to speculate the porins and these toxins may have had common ancestors. The channels these toxins form may exhibit some anion or cation selectivity, and voltage gating may be observed in artificial lipid bilayers, however, as noted above, these characteristics probably have little significance *in situ*. The second group contains proteins that typically are transformed from soluble inserted states by conformational changes that expose α-helices that can then penetrate the bilayer. Members of this group can be more closely compared with the transmembrane proteins of the bacterial cytoplasmic membrane. Generation of the insertion-competent state is due to oligomerization in the case of the β-structure toxins, promoted by concentration of the protein on the cell surface. In contrast, many of the α-helical toxins appear to insert into the bilayer as monomers, the result of a conformational change in the protein that is often promoted by low-pH environment. A more detailed comparison of the strategies used by α-helical and β-barrel toxins to undergo soluble to membrane-associated conformations can be found in a recent discussion by Lesieur et al. [3].

Proteins in the two structural toxin groups are not only distinguished by the methods they use to form channels. None of the β-barrel toxins are known to be enzymes, whereas many toxins of the α-helix-type have enzyme activities that are lethal to target cells in addition to their channel forming activity. The β-barrel toxins tend to insert directly into the plasma membrane of target cells after oligomerizing (toxins produced by intracellular bacteria are special cases). These proteins do not need to be internalized. On the other hand, many of the toxins in the α-helix family are internalized after they have bound to the cell surface. Some of them require internalization to bring them to an acidic environment in an intracellular compartment where conformational changes can occur that result in the formation of a channel and the delivery of a toxic enzyme domain across a membrane to its substrate in another compartment.

8.4
Steps in Channel Formation

Bacterial protein toxins typically follow a series of common steps to achieve channel formation (Fig. 1). The order in which some of the steps occur may vary, depending upon the type of toxin and its final destination in the cell.

8.4.1
Binding to Target Cells

Not surprisingly, many toxins bind to specific receptors on target cells. Binding to a receptor on the cell surface effectively concentrates the toxin, and this can lead to an enormous increase in the rate of channel formation (compared to the rate of

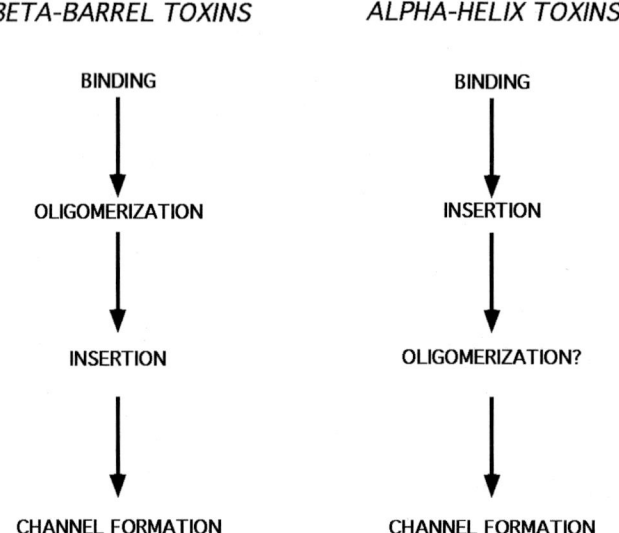

Fig. 1. Typical steps in channel formation by the β-barrel and α-helix classes of toxins.

channel formation in bilayers that lack receptors). This is especially true for the oligomeric toxins, where the high order of the oligomerization process (probably seventh order in the case of *Aeromonas hydrophila* aerolysin), leads to concentration dependence curves that become nearly vertical at some toxin concentration. Other toxins appear to lack specific receptors. These proteins tend to be active only at much higher concentrations than receptor-dependent toxins and many of them exhibit little or no discrimination, forming channels in virtually any lipid bilayer. Still other toxins may not use membrane surface molecules as high-affinity receptors, but still may exhibit cell specificity because they may only form channels in bilayers that have specific lipid components. For example, the cholesterol-dependent toxins are only active against membranes that contain cholesterol. Presumably the steroid confers some essential property on the bilayer that enables the oligomers formed by these proteins to insert. As a result, these toxins have no effect on bacterial membranes.

8.4.2
Activation

Although some toxins, such as *S. aureus* α-toxin and the cholesterol-dependent toxins, can form channels without the need for activation, many others must undergo an activation step before they can proceed to channel formation. For these toxins, activation is a strategy to control the transformation from the soluble to the insertion-competent state. Often activation is accomplished by proteolytic nicking, which may occur after the proform of the protein is bound to the surface of the target cell. Thus proaerolysin and the proform of anthrax protective antigen are both activated by furin or furin-like proteases after they have bound. In contrast, the Cry toxins of *Bacillus thuringiensis* are activated by proteases in the insect midgut before binding.

Proteolytic nicking is not the only method of toxin activation. Some proteins, such as anthrax protective antigen, which undergoes two activation steps, are activated by exposure to a low-pH environment, and members of one of the largest groups, the RTX toxins, including *E. coli* HlyA and the adenylate cyclase of *Bordetella pertussis*, undergo a unique form of activation that involves the acylation of one or two specific lysine residues.

8.4.3
Oligomerization

Toxins that form transmembrane β-barrels oligomerize before they insert into the membrane. This is consistent with the fact that these proteins do not contain any hydrophobic helices long enough to span the bilayer and, therefore, cannot stably interact with the hydrophobic interior of the membrane bilayer as monomers. Thus aerolysin, α-toxin, anthrax toxin protective antigen, and the cholesterol-dependent toxins are all thought to form "prepore" complexes. These are oligomers on the membrane surface that have not yet inserted. Toxins that penetrate lipid bilayers by inserting α-helices, such as the channel forming colicins and the *B. thuringiensis* Cry toxins, probably oligomerize after insertion, if they oligomerize at all.

8.4.4
Insertion

How does the insertion-competent form actually insert into the bilayer? Our knowledge of the chemistry of toxin action is most fuzzy in this area, particularly for the large prepore complexes assembled on the membrane surface before insertion by the β-barrel toxins. How these structures enter the membrane is largely a matter of speculation. Clearly a toxin approaching the bilayer surface will first encounter the polar head groups of the lipid and their water of hydration. This layer of the membrane will have to be penetrated in order to reach the hydrophobic interior which also has to be breached. It seems likely that the phospholipids in the region of the approaching toxin oligomer must at some stage form non-bilayer structures to allow penetration. In support of this, we have recently found that aerolysin insertion is promoted by the presence in the membrane of lipids favoring non-bilayer structures and we have speculated that aerolysin insertion may be comparable to some of the steps in bilayer fusion promoted by fusagens [4].

More is known about the insertion of some of the α-helical channel formers than insertion of the β-barrel toxins. A description of a current model of the insertion of colicin 1a can be found in Qui et al. [5].

8.5
Consequences of Channel Formation

Depending on the size and number of the channels and the type of target cell, there are several different possible consequences of channel formation. Erythrocytes are invariably lysed, because they are unable to repair any channels and they cannot withstand the osmotic imbalance resulting from the breached membrane permeability barrier. At high toxin concentrations, where many channels are formed, most other cell types are also lysed, however, when the number of channels is small, the cells may remain intact until death results from apoptosis. This is presumably because the cells are not killed immediately by channel formation, but nevertheless suffer changes in the intracellular concentrations of critical molecules and ions (such as K^+ and Ca^{2+}) that trigger apoptotic pathways.

8.6
Toxins that Oligomerize to Produce Amphipathic β-Barrels

This is probably the largest group of bacterial channel forming proteins. All of its members solve the problem of transformation from water-soluble to integral membrane forms by oligomerizing to produce insertion-competent amphipathic β-barrels. An outline of the steps in channel formation typically followed by these proteins is shown in Fig. 1. Often the first step is binding, which may be enhanced by the presence of a high-affinity receptor. Binding results in concentration of the protein on the membrane surface, favoring the next step, which is oligomerization, transforming the protein to an insertion-competent state. Before insertion, the oligomer is often referred to as a prepore complex. A conformational change in the protein (and probably in the membrane lipids), results in the insertion of the barrel into the bilayer, thus creating the channel. Within the β-barrel group, it is convenient to divide the proteins into two subgroups, those that form oligomers of up to seven monomers, producing rather small channels (in the order of 1 nm diameter), and those that form much larger oligomers and correspondingly larger channels or pores. The best characterized members of the first subgroup are *S. aureus* α-toxin, *A. hydrophila* aerolysin, and the protective antigen of *Bacillus anthracis*. Although there are no obvious similarities in the primary structures of these proteins, which is one of the characteristics of most of the members of this group, they share some remarkable similarities, perhaps most notably the fact that they all form heptameric oligomers. In the other subgroup is a family of homologous proteins. These are the cholesterol-dependent or oxygen-labile toxins such as perfringolysin O, listeriolysin O, and streptolysin O.

8.7
Toxins Forming Small β-Barrel Channels

8.7.1
Aerolysin

Aerolysin is secreted as an inactive precursor called proaerolysin by bacteria in the genus *Aeromonas* (see Fig. 2 and [6]). The protein utilizes the main terminal branch of the general secretion pathway in order to cross the inner and outer membranes of the bacteria and reach the exterior environment. In fact, studies of the protoxin's secretion by *A. hydrophila* and *A. salmonicida* have helped to characterize this pathway. One surprising observation has been that proaerolysin folds and even dimerizes in the periplasm before it is released from the cell: it is the dimeric form of the protein that crosses the outer membrane and this is the form found in solution and in the crystal. The proaerolysin dimer is extremely soluble (solutions of higher than 100 mg mL^{-1} can be made; unpublished data). Recent evidence from our laboratory has shown that the dimer has no tendency to dissociate, even at very low concentrations [7], contradicting an earlier report that the protein was monomeric at concentrations in the micromolar range [8]. Proaerolysin of *A. hydrophila* was the first channel forming protein to be solved by X-ray crystallography and the solution provided the clue that unraveled the mystery of how a water-soluble molecule might be transformed into an insertion-competent, channel forming state [9]. As mentioned, the protein is also a dimer in the crystal, and interactions between the two monomers are quite extensive, accounting for its stabi-

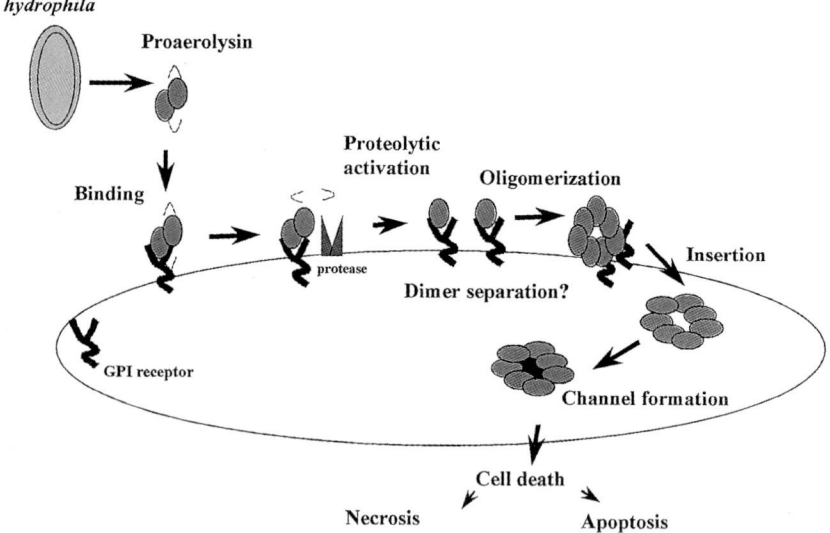

Fig. 2. Channel formation by aerolysin from *Aeromonas hydrophila*.

lity in solution. Each monomer is composed of two lobes, a small compact lobe that was later shown to be involved in receptor binding [10] and a large elongated lobe that contains extensive β-structure. This led to the proposal that some of the β-strands in the large lobe formed an amphipathic β-barrel in the oligomer and that, by analogy to porin proteins, it is the barrel that forms the channel.

Aerolysin is one of the few channel forming toxins for which a receptor has been identified. Both aerolysin and proaerolysin have the unique ability to bind to the glycosyl of glycosylphosphatidylinositol residue-anchored proteins on target cells [11]. Binding is selective and high affinity and as a result, cells with receptors are extremely sensitive to the action of aerolysin. For example, T-lymphocytes can be killed by toxin concentrations as low as 10^{-11} M [12]. Interestingly both lobes of the protein appear to be involved in binding, although the relative contributions of both are not known [10]. Presumably the binding site in the large lobe is the major recognition determinant for the anchor, as *Clostridium septicum* α-toxin, which is a homolog of the large lobe of aerolysin but lacks the small lobe, also binds glycosyl anchors [13]. The binding site in the small lobe may have a more general specificity for carbohydrates. We have found that a fold in this lobe is nearly identical to a fold in the S2 and S3 subunits of pertussis toxin, and similar to the carbohydrate domain of a variety of lectins [14]. Our evidence suggests that this site may facilitate interaction of aerolysin with cell surface proteins such as glycophorin, allowing the toxin to penetrate the glycocalyx. This may explain why the *C. septicum* toxin is far more active against erythrocytes when it is fused to the small lobe of aerolysin [15]. Proaerolysin is activated by proteolytic nicking in a region near the C-terminus of the protein. This can be accomplished by a variety of proteases, including furin on the surface of target cells as well as by soluble proteases such as proteinase K, trypsin, and chymotrypsin, and by at least one protease that is produced by the bacteria itself [16, 17]. Activation by furin may be the most important route *in vivo*, based on our recent observation that T-lymphocytes are not sensitive to proaerolysin when the furin recognition site in the protoxin is removed (Burr and Buckley, unpublished data).

Activation of proaerolysin by conversion to aerolysin is essential for the next step in channel formation, which is oligomerization to transform the protein from a water-soluble form to one that is insertion competent [16]. Early evidence obtained from analysis of 2-D crystalline arrays of the aerolysin oligomer indicated that it was heptameric. This was later confirmed by mass spectroscopy (the oligomer is stable enough for MALDI-TOF analysis), and by atomic force microscopy. The aerolysin oligomer has not been crystallized, however, the structure of the analogous oligomer from *S. aureus* has been solved and shown to be heptameric (see below). The concentration dependence of aerolysin oligomerization is extremely steep (Alonzo, Goni, and Buckley, unpublished data). The slope of the concentration dependence curve and the fact that no intermediates have been isolated suggest that the process may proceed in a single, seventh-order step. Oligomerization of other toxins, such as *S. aureus* α-toxin and anthrax protective antigen may also be single-step reactions, however, this is clearly impossible for the cholesterol-dependent toxins, which form oligomers with as many as 50 monomers.

Once formed, the aerolysin oligomer can spontaneously insert into a lipid bilayer. Neither accessory factors such as Ca^{2+} nor low pH are required for insertion. Whether or not the receptor plays any role in insertion is not known, in fact, as discussed above, very little is known about the insertion process for any toxin. The aerolysin channel is somewhat larger than 1 nm in diameter, voltage gated, and modestly anion ion-selective. Neither voltage gating nor ion selectivity is known to be important for the action of the toxin.

At relatively high aerolysin concentrations, erythrocytes are lysed and other mammalian cell types are rapidly killed by the changes in intracellular solutes caused by large numbers of toxin channels. However, at toxin concentrations where only a small number of channels are formed, cells, such as T-lymphocytes, die as a result of apoptosis [18]. Presumably these cells can withstand intracellular changes long enough so that the changes that do occur, such as a rise in intracellular Ca^{2+}, or a decline in intracellular K^+, trigger one or more apoptotic pathways. Similar observations have been made with other channel forming toxins, including *S. aureus* α-toxin.

8.7.2
α-Toxin

Most pathogenic strains of *Staphylococcus aureus* secrete α-toxin, a 33 kDa channel forming protein that is active against a broad range of cell types (reviewed in [19]). In contrast to aerolysin and anthrax protective antigen, *S. aureus* α-toxin is secreted in an active form. Most eukaryotic cells do not appear to have a specific high-affinity receptor for the toxin. Instead α-toxin can associate non-specifically with lipid bilayers, apparently preferring those containing phosphatidylcholine and cholesterol. However, some cells, such as rabbit erythrocytes [20] and human platelets and monocytes, have been shown to have a small number of high-affinity binding sites for the toxin [21]. The identity of these sites is not known, nor is it known whether or not they are primary targets for the toxin *in situ*, however, cells displaying these sites are lysed by nanomolar concentrations of the toxin, whereas concentrations in the micromolar range are required to lyse cells that lack them [21]. The structure of water-soluble α-toxin has not been solved, however, the structure of a related *S. aureus* protein LukF, which together with LukS forms a cation-selective pore, has been solved [22, 23]. In the LukF structure, the region that forms the channel in the oligomer is folded into an amphipathic three-stranded β-sheet.

Association of α-toxin with lipid bilayers promotes formation of a heptameric prepore complex that spontaneously inserts, producing channels approximately 1-1.5 nm in diameter. Like the aerolysin channel, the channel formed by α-toxin is voltage gated and slightly anion-selective in planar lipid bilayers, but as with aerolysin, these properties are not thought to contribute to channel function. The structure of the α-toxin heptamer has been solved [24]. Its overall morphology is quite similar to the structure that was predicted for the aerolysin heptamer by electron microscopic analysis of 2-D crystals [25]. The α-toxin heptamer has a mushroom shape, penetrated by a water-filled channel that runs along the sevenfold axis and ranges from 14 to 46 Å in diameter. The transmembrane domain is an antiparallel amphipathic β-barrel formed by seven amphipathic β-hairpin loops contributed by the seven monomers. Interestingly, hexameric oligomers of α-toxin have been observed by atomic force microscopy, evidence that the heptamer may not be the only stable form of the protein [26].

8.7.3
Anthrax Protective Antigen

Bacillus anthracis secretes three high-molecular-weight proteins that together are referred to as anthrax toxin (see [27] and [28] for recent reviews). One of them, anthrax protective antigen (PA), has some remarkable similarities to aerolysin and to a lesser extent to *S. aureus* α-toxin, in spite of the fact that the proteins bear no homology and that the defining function of PA is not channel formation. Like proaerolysin, PA binds with high-affinity to receptors on target cells [29]. Although the receptors have not been identified, the strength of binding is comparable to the strength of the proaerolysin—GPI-anchored protein interaction (K_d in the 10^{-9} M range). Once bound, just as in the case for proaerolysin, anthrax PA is processed by a furin-like enzyme on the cell surface and this converts the protein to a form that is able to oligomerize, producing heptamers just as aerolysin does. However, in contrast to the aerolysin oligomer, the PA oligomer must encounter a low-pH environment in order to insert. What is more, it seems likely that *in vivo* the primary purpose of the PA oligomer is not to form a channel *per se*, but to translocate one of the two other components of anthrax toxin, lethal factor (LF) or edema factor (EF) into the cytosol. Once the PA oligomer is formed on the cell surface it binds a molecule of LF or EF and the complex is internalized by receptor-mediated endocytosis. Exposure to the acid environment of an intracellular compartment triggers insertion of the PA heptamer, and the associated LF or EF is somehow translocated to the cell's cytosol. It is assumed that translocation is associated with the channel forming properties of the PA oligomers (in planar lipid bilayers PA forms channels approximately 1 nm in diameter [30]), however, the details of the translocation process are unknown. Once in the cytosol, edema factor acts as a calmodulin-dependent adenylate cyclase, whereas LF proteolytically inactivates mitogen-activated protein kinase kinases.

The structure of monomeric PA has been solved. As with proaerolysin, PA is organized mainly into β-sheets, and the protein has been divided into four domains. Some crystallographic information is also available about the heptamer, which is a ring-shaped structure with a negatively charged lumen. It is thought heptamerization results from the interaction of loops from the seven monomers, producing a 14-stranded amphipathic β-barrel, similar to the barrel of *S. aureus* α-toxin [31].

In summary, in spite of the fact that PA can oligomerize and insert to form discrete channels in lipid bilayers, promoting the release of small molecules, its primary function is to cause translocation of one of the associated proteins. Thus, although its channel forming properties are analogous to those of aerolysin, its translocation properties are analogous to those of the B-subunit of the AB-toxins such as diphtheria toxin. Aerolysin and α-toxin differ from PA because they do not require exposure to an acid environment to form channels, however, one has to wonder whether either or both of these proteins are also capable of effecting translocation of other proteins, and whether this may be their real function *in vivo*.

8.8
Toxins Forming Large β-Barrel Channels

8.8.1
The Cholesterol-dependent Toxins

The largest group of related channel forming toxins comprises the cholesterol-dependent (once called oxygen-labile) toxins produced by 19 or more gram-positive species in five genera. Many of these bacteria are important human pathogens and the toxins are thought to be virulence factors (see [32] and [33] for recent reviews). In terms of structure–function relationships and the mechanism of channel formation, the best studied of all of these toxins is perfringolysin O, which is produced by *Clostridium perfringens* as a soluble monomer. The structure of perfringolysin O has been solved [34]. Although it is different from the structures of aerolysin, α-toxin and anthrax protective antigen, perfringolysin O also contains a great deal of β-structure and it too has been divided into four domains. Recently it has been shown that channel formation is the result of a very large conformational change in domain 3 of the protein, in which 6 short helices are converted into two β-hairpins that form the transmembrane barrel in the inserted oligomer. The toxin oligomerizes on the surface of lipid bilayers to form prepore complexes containing up to 50 monomers. Hotze et al. [35] have been able to trap the perfringolysin O oligomer in the prepore state by introducing a cysteine into the transmembrane hairpin and another close by in domain 2 of the protein. The resulting disulfide bond prevents the prepore complex from undergoing conversion to the pore. This was an elegant demonstration that perfringolysin O oligomerizes before insertion, in a similar fashion to the three toxins described above.

Listeriolysin O is another member of the cholesterol-dependent toxin family. Its structure and the mechanism by which it forms channels are likely to be very similar to those of its homolog perfringolysin O, although they have been much less thoroughly studied. On the other hand, our knowledge of the contribution of listeriolysin O to bacterial virulence is perhaps the best understood of all of the cholesterol-dependent toxins. The toxin is produced by the intracellular pathogen *Listeria monocytogenes* and it is an essential virulence factor that enables the bacteria to escape from vacuoles in the host cell [36]. Interestingly, the toxin only disrupts the vacuolar membrane [37]. Recent evidence suggests that other membranes are not affected because the toxin is rapidly removed or degraded once it is released into the cytosol. Decatur and Portnoy [38] have recently shown that the protein contains a PEST sequence that appears to label the toxin for inactivation or degradation in the cytoplasm.

One question we might ask is why are there so many members in the cholesterol-dependent toxin family, when other toxins, except for the RTX toxins, appear to have few if any homologs. Perhaps the cholesterol-dependent toxins serve a specific common function to the species that produce them and this has led to the conservation of structure. A similar question might be asked about the other large group of related toxins, the RTX toxins discussed below.

8.9
The RTX Toxins

The RTX (repeat in toxin) toxins are a unique family of homologous proteins that are characterized by the presence of multiple repeats of a glycine- and aspartate-rich nonapeptide in their C-terminal domains (see [39] for a recent review). The two best characterized members of the family are HylA, an important virulence factor in *E. coli*, and CyaA, the adenylate cyclase of *Bordetella pertussis*. Related toxins are produced by *Actinobacillus, Pasteurella, Proteus, Morganella,* and *Moraxella* spp. All of these proteins are exported out of the cell by Type I secretion systems, indeed studies of the secretion of HylA by *E. coli* are largely responsible for our knowledge of the Type I pathway [40]. However, the most remarkable common property of these proteins is the method by which they are converted from their inactive forms to active toxins. This was first discovered in studies of *E. coli* HylA to be due to acylation of the protein by the cotranslated protein HlyC.

8.9.1
Escherichia coli HlyA

E. coli HlyA has a wide range of activity, lysing erythrocytes from a variety of species and killing many other mammalian cell types (see [41] for an unusually thorough recent review of HlyA). Other members of the RTX family are more specific. For example, the *Pasteurella hemolytica* leukotoxin lyses ruminant leukocytes and the *Actinobacillus actinomycetemcomitans* leukotoxin kills primate polymorphonuclear leukocytes. Most cells do not appear to have specific receptors for HlyA. Instead the first step in channel formation by the *E. coli* toxin is absorption to the cell surface. The toxin then inserts in a process associated with a change in conformation. Insertion results in the formation of 1 nm cation-selective channels. Whether oligomerization is necessary for channel formation has not been established, although it is clear that if oligomers are formed, they are much less stable than the oligomeric forms of the β-barrel toxins. Secondary structure analysis suggests that the membrane-spanning residues of the toxin are in a conserved region of the protein (residues 238 to 410) that can form four hydrophobic α-helices. It has been suggested that these helices as well as amphipathic helices formed by nearby residues, are responsible for channel formation.

The *E. coli* toxin is synthesized as an inactive 1,024-residue protoxin, which is activated before release by acylation at residues K564 and K690. In addition to acylation, calcium ions are absolutely required for the activity of the RTX toxins. It is thought that one Ca^{2+} ion binds to each glycine-rich repeat. The function of the two acyl chains in channel formation has not been established. Evidence concerning their importance in membrane association is conflicting. The idea that binding of HlyA is analogous to the binding of recoverin, recently suggested by Stanley et al. [41], has great appeal. Recoverin is also an acylated protein. In the presence of Ca^{2+}, its myristoyl chain is exposed and inserted into the lipid bilayer. Calcium may play a similar role in the binding of acylated HlyA.

8.9.2
Pertussis CyaA

The adenylate cyclase CyaA produced by *Bordetella pertussis* is one of the most remarkable of all bacterial toxins (reviewed recently in [42]) and the best characterized member of the RTX family of toxins after *E. coli* HlyA. The bacterium *B. pertussis* causes whooping cough in humans, and CyaA is an important virulence factor in the infection. As with HlyA, the protein is secreted as an inactive precursor which must be palmitolyated in a process that depends on the activity of an accessory protein, CyaC. The toxin itself is a large protein, consisting of 1,706 residues encoding two functional domains. The first 400 amino acids encode the adenylate cyclase activity, whereas the remaining residues form the domain responsible for the channel forming activity of the toxin. The two domains can function independently of each other.

Cya is often referred to as a bifunctional toxin because it is both a hemolysin and an intracellular activator of adenylate cyclase. The hemolysin domain is homologous to the comparable domain of HlyA and the other RTX toxins and the protein forms similar cation-selective channels in a calcium-dependent manner. More importantly, however, when added to a variety of different cell types, the hemolytic domain of Cya can deliver the toxin's catalytic domain into the host cell cytoplasm where it is activated by binding to calmodulin. Activation of the enzyme leads to a rapid rise in intracellular cyclic AMP levels and in the case of macrophages, this leads to apoptosis. Since calmodulin is not found in bacteria, the toxin has no effect on the secreting species.

Although it has been suggested that CyaA binds to gangliosides on target cells, a recent study of binding by flow cytometry found no evidence that binding was saturated at high toxin concentrations, suggesting that, as with HlyA, the protein may not have a specific receptor. Once bound to the cell surface, the enzyme domain of CyaA is transferred across the membrane by an unknown process, which depends on prior activation of the hemolytic domain by acylation, the presence of calcium ions, and a membrane potential, which may provide the driving force for transfer. Surprisingly, transfer does not appear to be directly related to formation of pores by the hemolytic domain of the toxin. Hence, in erythrocytes, adenylate cyclase appears inside the cells within seconds of addition of the CyaA, but hemolysis does not occur for nearly an hour, implying that channel formation by the hemolytic domain is secondary to its primary function in transferring the catalytic domain. Thus insertion of one or a few monomers may be enough to effect translocation, but channel formation and hemolysis may require oligomerization of the embedded toxin [43].

8.10
Ion Channel Forming Toxins

8.10.1
Channel Forming Colicins

Colicins are a family of plasmid-encoded proteins that are produced by and active against *E. coli* and related bacteria. One of the largest groups of colicins contains water-soluble proteins that can form channels in the cytoplasmic membranes of sensitive cells (reviewed in [44] and [45]). In order to do this, they must first bind to receptors on the surface of the outer membrane. This is followed by translocation across the outer membrane and insertion into the inner membrane. The sequences of a number of these colicins are known, and the crystal structures of several have been solved. The proteins are divided into three linearly organized domains, associated with binding, translocation, and insertion. In keeping with a common function in forming a voltage gated channel, the C-terminal insertion or C-domains of the colicins are homologous. The structure of the C-terminal domain of colicin A was the first to be solved, and it provided an explanation of the mechanism of transformation from a water-soluble to a membrane-inserted state [46]. This domain is comprised of 10 α-helices arranged in three layers. Eight of the helices are amphipathic and they surround the remaining two helices that together form a hydrophobic hairpin. Similar structures have been reported for colicin E1 and for colicin N. It is the hairpin that is exposed to transform these proteins into an insertion-competent state.

Colicins are divided into two groups depending upon how they are translocated across the outer membrane of the target cell. The group A toxins (including colicins A, E1 to E9, K, L, N, and S4) require specific porins as well as the Tol proteins. Group B colicins require TonB and its accompanying proteins, ExbB and ExbD. Colicin A is perhaps the best studied of all of the colicins. The protein binds to the vitamin B_{12} receptor BtuB on target cells via the toxin's central R-domain. This results in a substantial conformational change in the colicin that is followed by translocation, with the aid of OmpF and the Tol proteins, mediated by the colicin N-terminal T-domain. Once across the outer membrane, the C-domain inserts into the bacterial plasma membrane, driven by the transmembrane potential, producing discrete cation-selective voltage gated channels. This process is thought to begin with the insertion of the hydrophobic hairpin. The channel is then formed by the voltage-dependent insertion of two additional segments, so that portions of some helices span the membrane and other helices are transported completely across the membrane [5]. The T- and R-domains are thought to remain on the periplasmic side of the plasma membrane.

8.10.2
Bacillus thuringiensis Cry Toxins

The insecticidal crystal (Cry) proteins produced by *Bacillus thuringiensis* during sporulation have attracted a great deal of recent interest (see [47] for a recent review). More than 100 *cry* genes have been cloned and sequenced and the Cry proteins have been divided into four groups, based on their host range and primary structure. They are among the most effective of naturally occurring pesticides, and plant resistance to specific insects can be conferred on plant species using genetic engineering techniques. Once ingested by the insect, the toxins are solubilized in the alkaline midgut and activated by host proteases. The activated proteins then bind to specific receptors in the midgut epithelium and produce cation-selective channels that kill the cells.

The crystal structure of Cry3A, a member of the coleopteran-active group, was determined by Li et al. [48]. As with the colicins, there are three structurally distinct domains, one consists of a bundle of six α-helices surrounding a central helix, another of three antiparallel β-strands, and the third a sandwich of antiparallel β-sheets in a jelly-role configuration. The structure of lepidopteran-active Cry1Aa, which was later resolved by Grochulski et al. [49] is quite similar, in spite of the fact that the two toxins share only 25 % sequence homology.

The mechanism by which the Cry toxins form channels in target cells appears to be analogous to that described for the colicins above and for diphtheria toxin, which also contain three domains, including a helix bundle domain corresponding to domain 1 of the Cry proteins, which is responsible for insertion. The similarity to colicin A led to two models for the integration of Cry toxins into target membranes, a penknife mode and an umbrella model. Most recent evidence supports the umbrella model in which a hairpin formed by helices 4 and 5 inserts into the membrane, thereby promoting the spreading of the remaining helices in domain 1 over the membrane surface [48]. A tetrameric arrangement of four monomers is thought to be the minimal configuration required to produce a functional channel [50].

The other two domains of the Cry proteins are required for receptor binding and structure stabilization. Receptors clearly vary among the different Cry proteins, thereby providing target species specificity. One variant, *B. thurigiensis* cry toxin, Cry1Ac, has been shown to bind to the GPI-anchored protein aminopeptidase N on the surface of brush border cells in the *Manduca sexta* gut. Aminopeptidase N may not be the only receptor for the toxin in these cells, however, as site-directed mutagenesis of the toxin to destroy binding to the aminopeptidase only reduced its toxicity by twofold [51].

8.11
Other Channel Forming Toxins

No single chapter can do justice to all of the channel forming toxins that have been reported. I have chosen to restrict this survey to toxins for which there is both structural and functional information. As we gather more information about other channel forming proteins and as we discover new toxins, our understanding of the mechanisms and consequences of their action will continue to improve.

Acknowledgement

The author acknowledges the support of the Natural Sciences and Engineering Research Council of Canada and the Medical Research Council of Canada. I am grateful to Sarah Burr for her helpful comments.

References

1. J. E. Alouf, J. H. Freer (Eds.), *The Comprehensive Sourcebook of Bacterial Protein Toxins*, 2nd Edn., Academic Press, New York, **1999**, pp. 443-456.
1a. G. Menestrina, B. Semjén, in: *The Comprehensive Sourcebook of Bacterial Protein Toxins*, 2nd Edn. (Alouf, J. E., Freer, J. H., Eds.), Academic Press, New York, **1999**, pp. 287-309.
2. G. Bainbridge, I. Gokce, J. Lakey, *FEBS Lett.* **1998**, *431*, 305-308.
3. C. Lesieur, B. Vecsey-Semjen, L. Abrami, M. Fivaz, F. G. van der Goot, *Mol. Membr. Biol.* **1997**, *14*, 45-64.
4. A. Alonso, F. Goñi, J. T. Buckley, *Biochemistry* **2000**, *39*, 14019-14024.
5. X.-Q. Qui, K. Jakes, P. Kienker, A. Finkelstein, S. Slatin, *J. Gen. Physiol.* **1996**, *107*, 313-328.
6. J. T. Buckley, in: *The Comprehensive Sourcebook of Bacterial Protein Toxins*, 2nd Edn., (Alouf, J. E., Freer, J. H., Eds.), Academic Press, New York, **1999**, pp. 362-372.
7. R. Barry, S. Moore, A. Alonso, J. Ausio, J. T. Buckley, *J. Biol. Chem.* **2000**, *276*, 551-554.
8. M. Fivas, M. Velluz, F. G. van der Goot, *J. Biol. Chem.* **1999**, *274*, 37705-37708.
9. M. Parker, J. Buckley, J. Postma, A. Tucker, K. Leonard, F. Pattus, D. Tsernoglou, *Nature* **1994**, *367*, 292-295.
10. C. R. Mackenzie, T. Hirama, J. T Buckley, *J. Biol. Chem.* **1999**, *274*, 22604-22609.
11. D. B. Diep, K. L. Nelson, S. M. Raja, E. N. Pleshak, J. T. Buckley, *J. Biol. Chem.* **1998**, *273*, 2355-2360.
12. K. Nelson, S. M. Raja, J. T. Buckley, *J. Biol. Chem.* **1997**, *272*, 12170-12174.
13. V. M. Gordon, K. L. Nelson, J. T. Buckley, V. L. Stevens, R. K. Tweten, P. C. Elwood, S. H. Leppla, *J. Biol. Chem.* **1999**, *274*, 27274-27280.
14. J. Rossjohn, J. T. Buckley, B. Hazes, A. G. Murzin, R. J. Read, M. W. Parker, *EMBO J.* **1997**, *16*, 3426–3434.
15. D. B. Diep, K. L. Nelson, T. L. Lawrence, B. R. Sellman, R. K. Tweten, J. T. Buckley, *Mol. Microbiol.* **1999**, *31*, 785-794.
16. W. J. Garland, J. T Buckley, *Infect. Immun.* **1988**, *56*, 1249-53.
17. L. Abrami, M. Fivaz, E. Decroly, N. Seidah, F. Jean, G. Thomas, S. Leppla, J. Buckley, F. G. van der Goot, *J. Biol. Chem.* **1998**, *273*, 32656-32661.
18. K. Nelson, R. A. Brodsky, J. T. Buckley, *Cell. Microbiol.* **1999**, *1*, 69-74.
19. E. Gouaux, *J. Struct. Biol.* **1998**, *121*, 110-122.
20. P. Cassidy, S. Harshman, *Biochemistry* **1976**, *15*, 2348-2355.
21. A. Hildebrand, M. Pohl, S. Bhakdi, *J. Biol. Chem.* **1991**, *266*, 15195-2000.
22. R. Olson, H. Nariya, K. Yokota, Y. Kamio, E. Gouaux, *Nature Struct. Biol.* **1999**, *6*, 134-140.
23. J. Pédelacq, L. Maveyraud, G. Prévost, L. Baba-Moussa, A. Gonzâlez, E. Courcelle, J. Rossjohn, S. Feil. W. McKinstry, R. Tweten, M. Parker, *Cell* **1997**, *89*, 685-692.
24. L. Song, M. Hobaugh, C. Shustak, S. Cheley, H. Bayley, J. E. Gouaux, *Science* **1996**, *274*, 1859-1865.
25. H. U. Wilmsen, F. Pattus, W. Tichelar, T. J. Buckley, K. Leonard, *EMBO J.* **1992**, *11*, 2457-2463.
26. D. Czajkowski, S. Sheng, Z. Shao, *J. Mol. Biol.* **1998**, *276*, 325-330.
27. S. H. Leppla, in: *The Comprehensive Sourcebook of Bacterial Protein Toxins*, 2nd Edn. (Alouf, J. E., Freer, J. H., Eds.), Academic Press, New York, **1999**, pp. 243-263.
28. N. Duesbery, G. Vande Woude, *Cell. Mol. Life Sci.* **1999**, *55*, 1599-1609.
29. V. Escuyer, R. J. Collier, *Infect. Immun.* **1991**, *59*, 3381-3386.
30. A. Finkelstein, *Toxicology* **1994**, *87*, 29-41.
31. C. Petosa, R. J. Collier, K. Klimpel, S. Leppla, R. Liddington, *Nature* **1997**, *385*, 833-838.
32. R. K. Tweten, in: *Virulence Mechanisms of Bacterial Pathogens*, (Roth, J. A., Boloin, C. A., Brogden, K. A., Minion, C., Wannemuehler, M. J., Eds.), American Society for Microbiology, Washington, DC, **1995**, pp. 207-230.
33. J. E. Alouf, in: *The Comprehensive Sourcebook of Bacterial Protein Toxins*, 2nd Edn., (Alouf, J. E., Freer, J. H., Eds.), Academic Press, New York, **1999**, pp. 443-456.
34. J. Rossjohn, S. C. Feil, W. J. McKinstry, R. K. Tweten, M. W. Parker, *Cell* **1997**, *89*, 685-692.

35. E. Hotze, E. Wilson-Kubalek, J. Rossjohn, M. Parker, A. Johnson, R. Tweten, *J. Biol. Chem.* **2001**, *276*, 8261-8268.
36. D. Portnoy, T. Chakraborty, W. Goebel, P. Cossart, *Infect. Immun.* **1992**, *60*, 1263-1267.
37. M. Moors, B. Levitt, P. Youngman, D. Portnoy, *Infect. Immun.* **1999**, *67*, 131-139.
38. A. Decatur, D. Portnoy, *Science* **2000**, *290*, 992-995.
39. A. Ludwig, W. Goebel, in: *The Comprehensive Sourcebook of Bacterial Protein Toxins*, 2nd Edn. (Alouf, J. E., Freer, J. H., Eds.), Academic Press, New York, **1999**, pp. 330-348.
40. V. Koronakis, P. Stanley, E. Koronakis, C. Hughes, *FEMS Microbiol. Immunol.* **1992**, *5*, 45-53.
41. P. Stanley, V. Koronakis, C. Hughes, *Microbiol. Mol. Biol. Rev.* **1998**, *62*, 309-333.
42. D. Ladant, A. Ullmann, *Trends Microbiol.* **1999**, *7*, 172-176.
43. M. Gray, G. Szabo, A. Otero, L. Gray, E. Hewlett, *J. Biol. Chem.* **1998**, *273*, 18260-18267.
44. W. Cramer, J. Heymann, S. L. Schendel, B. Deri, F. Cohen, P. Elkins, C. Stauffacher, *Annu. Rev. Biophys. Biomol. Struct.* **1999**, *24*, 611-641.
45. C. Lazdunski, *Mol. Microbiol.* **1995**, *16*, 1059-1066.
46. M. Parker, J. Postma, F. Pattus, A. Tucker, D. Tsernoglou, *J. Mol. Biol.* **1992**, *224*, 639-657.
47. F. Rajamohan, M. Lee, D. Dear, *Prog. Nucleic Acids Res. Mol. Biol.* **1998**, *60*, 1-24.
48. J. Li, J. Carroll, D. J. Ellar, *Nature* **1991**, *353*, 815-821.
49. P. Grochulski, L. Masson, S. Borisova, M. Pusztai-Carey, J.-L. Schwarz, R. Brousseau, M. Cygler, *J. Mol. Biol.* **1995**, *254*, 447-464.
50. L. Masson, B. Tabashnik, Y.-B. Liu, R. Brousseau, J.-L. Schwartz, *J. Biol. Chem.* **1999**, *274*, 31996-32000.
51. J. Jenkins, M. Lee, S. Sangadala, M. Adang, D. Dean, *FEBS Lett.* **1999**, *462*, 373-376.

9
Porins – Structure and Function

Roland Benz

9.1
Introduction

The cell envelope of gram-negative bacteria consists of several layers, the outer membrane, the peptidoglycan, and the cytoplasmic membrane. The outer membrane serves as a molecular sieve for the passage of hydrophilic solutes. In enteric bacteria, the outer membrane also hinders the diffusion of hydrophilic solutes, such as the bile acids. The peptidoglycan layer is a large heteropolymer that is responsible for the maintenance of cell shape and for the ability of the cell to withstand the very high internal osmotic pressure in normal and dilute environments. It consists of a network of amino sugars and amino acids (see [1] for a review). The amino sugars (N-acetylglucosaminyl-N-acetylmuramyl dimers) form long linear strands, which are covalently linked between two muramyl residues by short tetrapeptides. Components of the outer membrane such as the lipoprotein or the porins are either covalently bound to the murein or interact with this macromolecule via ion bridges [2, 3]. The inner membrane acts as a real diffusion barrier and contains besides the energy converting devices a large number of transport systems for molecules in and out of the cell including many uptake systems for small hydrophilic substrates and different secretion systems for proteins (see Chaps. 1–7 and 13). The space between inner and outer membrane, the periplasmic space, appears to be isoosmolar to the cell interior (around 300 mM), which means that the osmotic pressure, normally around 3.5 bars and at maximum about 7 to 8 bars across the cell envelope, is maintained across the outer membrane and the attached peptidoglycan layer and not across the inner membrane [4]. The periplasmic space represents an additional compartment besides the cell interior and occupies between 5% and 20% of the total cell volume according to different estimations [4, 5]. It plays an important role in the metabolism of gram-negative bacteria because it contains binding proteins for uptake systems and many enzymes (see Chap. 5). The periplasmic space is strongly anionic compared to the external medium. This is partially due to anionic groups attached to the outer membrane and to the presence of anionic MDO (membrane-derived oligosaccharides) present in the periplasmic space to maintain part of the osmolarity. Both contribute to the Donnan potential across the outer membrane, which can be as large as -80 to -100 mV (inside negative) in media of low ionic strength [4, 6].

The outer membrane contains several classes of proteins that are either responsible for its molecular sieving properties, such as the porins, or the active uptake of nutrients, such as the receptors. The latter ones form single β-barrel cylinders in the outer membrane of gram-negative bacteria. Their structural and functional properties are described in Chaps. 12 and 13. Most porin channels sort according to the molecular mass of the solutes, which means that they possess more general than specific properties. Other porin channels contain binding sites for substrates (specific porins), which represents a striking advantage for solute transport in dilute media. These so-called specific porins are often induced under special growth conditions together with a variety of other proteins including components needed for fermentation and inner membrane transport. A third class of porins consists of the members of the TolC family [7, 8]. These outer membrane channels connect the outer membrane pathway with inner membrane export systems and link inner membrane transport in an elegant way with the surface of the cell thus shunting the periplasmic space [8]. The primary structure of many porin channels is known from the cloning and sequencing of their genes. It is noteworthy that the amino acid composition is similar to that of water-soluble proteins, which means that the arrangement of the amino acid sequence in the secondary and tertiary structure is responsible for the function of the porins as transmembrane channels. Several classes of porins including general diffusion pores, specific channels, and inner to outer membrane links have been crystallized and their 3-D structures are known from X-ray crystallography. The results allow interesting insight into the transmembrane structure of membrane channels that are formed by β-barrel cylinders. According to these studies general diffusion porins form trimers of three identical subunits of membrane channels with 16 β-strands. The carbohydrate-specific porins crystallized to date form also trimers but the monomeric channel contains 18 β-strands instead of 16 β-strands.

The studies of the porin channels are of special interest because of the excellent knowledge of the 3-D structure of a number of porin channels and the possibility to devise porin mutants. The stability of the porin oligomers towards denaturating detergents allows their isolation in sufficient large quantities to perform any type of structural studies including reconstitution experiments using liposomes and lipid bilayer membranes. This means that the porins allow excellent access to the investigation of the structure–function relationship. Consequently porin channels can serve as models of transmembrane diffusion of hydrophilic solutes.

9.2
Structure of the Outer Membrane of Gram-negative Bacteria and Isolation of Porin Proteins

The porins are deeply embedded into the outer membrane. This membrane is asymmetric regarding its lipid composition. The surface exposed monolayer contains lipopolysaccharide (LPS) as its major or exclusive (in enteric bacteria) lipid, while the inner leaflet contains phospholipids (mostly phosphatidylethanolamine [9]). LPS a is lipid-like amphiphilic molecule composed of lipid A, a diglucosamine phosphate dimer, to which 5 or 6 fatty acids are linked (also known as endotoxin [10]). Attached to the endotoxin is the rough oligosaccharide core and a variable number of repeated tri- to pentasaccharide units called O-antigen. Lipopolysaccharides carry net negative charges resulting in a strongly negatively charged cell surface [10]. LPS has a special role in assembly and maintenance of the outer membrane as permeability barrier for hydrophobic antibiotics, bile salts, detergents, proteases, lipases, and lysozyme and is important for the function of most outer membrane proteins because they interact with LPS through hydrophobic interactions and by non-covalent cross-bridging of adjacent LPS molecules with divalent cations in such a way that a tight network is produced. The treatment of gram-negative bacteria with ethylenediaminetetraacetate (EDTA) generally results in removal, by chelation, of divalent cations and consequent disruption of the outer membrane [3]. In the absence of such chelators, however, the ion bridges formed by negatively charged groups and divalent cations are responsible for many of the properties of the outer membrane of gram-negative bacteria.

Several different methods have been established for the isolation of porins from the cell envelope obtained by centrifugation of the homogenized bacterial cells by application of the French press, ultrasonification, or of glass-bead treatment. Some of them are tightly associated with the peptidoglycan layer [11]. In such a case the isolation is relatively simple because only a small number of outer membrane proteins are associated with the peptidoglycan layer. All the others are soluble in SDS-containing solutions. The SDS-insoluble material contains the peptidoglycan with the peptidoglycan-associated proteins including OmpA and some porins. An elegant way to separate the proteins from the peptidoglycan is the salt extraction method [11, 12]. The basic procedure is the treatment of the peptidoglycan porin complex with a buffer, which contains as essential components 1% SDS, 0.4 M NaCl, and 5 mM EDTA followed by chromatography according to standard biochemical procedures. One very elegant method to purify the carbohydrate-specific porins of the LamB family is the chromatography on a starch column (starch coupled to Sepharose 6B [13]).

Some porins are not tightly associated with the peptidoglycan layer. In this case a differential extraction of the cell envelope using different ionic and non-ionic detergents may help to isolate and purify the porins [14]. At low detergent concentration most of the components of the inner and outer membrane are soluble. At higher detergent concentration the porins can be found in the supernatant of the detergent wash steps. Ion-exchange columns are particularly useful for the purification of porins because the pore forming complexes may adsorb to the column material according to their net surface charge. The column is first washed with low ionic strength detergent buffer and then eluted with linear salt gradients ranging from 50 mM to 500 mM at a given pH. The salt and detergent content of the eluted protein can be decreased by dialysis procedures, but this is not essential for the channel forming activity of the porins, which remains constant for several months at 4° in a refrigerator or frozen at −20° in a freezer. The porin solutions can also be lyophilized in many cases without any loss of the pore forming activity. In the lyophilized form the protein remains active for at least 1 year stored in a freezer at −20°.

In some cases it may be essential to isolate the outer membrane from the cell envelope before the isolation and purification of the porins. The outer membrane fraction may be obtained from a sucrose step density centrifugation of the cell envelope. For this the cell envelope fraction is layered on top of a two- or three-step sucrose gradient [15]. After centrifugation in an ultracentrifuge for about 12 h at high speed the outer membrane is always found in the pellet or at higher sucrose concentration because of its higher density than the inner membrane. After isolation of the outer membrane a differential extraction of the outer membrane using increasing concentrations of different detergents may be used for the isolation of the porin as described above.

9.3
Model Membrane Studies with Porin Channels

The function of porin channels can be studied in their natural environment, i. e., in intact cells using the β-lactamase assay or the transport of radioactively labeled solutes in the cell by active transport processes [16, 17]. The latter type of investigation provides the unique advantage that the solutes can interact with the whole uptake system including periplasmic binding proteins and periplasmic enzymes. However, *in vivo* measurements only yield precise information on porin permeability if the flux of the radioactively labeled solutes or that of the β-lactam antibiotics across the outer membrane is the rate-limiting step, which is very often only the case when the concentrations of the solutes are very small [18]. Otherwise either the inner membrane transport or the enzyme activity is rate-limiting and the gradient of the solute transport is not established across the outer membrane. This means that the results of the studies do not provide any information about outer membrane permeability. The solute selectivity of porins can be measured *in vivo* by using the β-lactamase activity and positively or negatively charged cephalosporins [19, 20].

Model membrane studies provide the unique advantage of simple systems, which means that *in vitro* studies allow much better control of the experimental conditions than *in vivo* experiments. The first method for the identification of the pore forming proteins of the outer membrane of gram-negative bacteria was the vesicle permeability assay [21]. Liposomes are formed in a buffer solution containing two radiolabeled solutes (e.g., ^{14}C-sucrose and ^{3}H-dextran) from lipids in the presence of protein fractions or purified porin. After formation the liposomes are passed through a molecular sieve column and elute just after the void volume. The high molecular mass solute is retained in the liposomes, whereas ^{14}C-sucrose may leave the liposomes through the pores during the chromatography. Similar experiments can be performed with ^{14}C-solutes of different molecular masses. The ^{14}C- to ^{3}H-ratio within the eluted liposomes is a measure for the exclusion limit of the channel and for the effective diameter of the pore. It is noteworthy, however, that the vesicle permeability assay does not provide any information on the kinetics of permeation because it is an all or nothing process at the time scale of the experimental conditions.

The kinetics of solute permeation through porins may be achieved by using the liposome swelling assay [22]. Multilamellar liposomes are formed from lipid and protein in a similar way as described above in a buffer containing a large molecular mass osmolyte (around 40 mOsmol) that should be permeable through the channels. The liposomes are added under stirring to an isotonic solution of a test solute. When this solute can penetrate the pores, the osmolarity inside the liposomes increases because the high molecular mass solute is retained inside the liposomes. Water follows and the subsequent swelling of the liposomes leads to a decrease of the optical density. The swelling rate provides a measure of the penetration rate of the test solute through the channels. Only when the same liposome preparation is used for measurements of different solutes the rate of uptake of different solutes can be compared. For a general diffusion pore, i.e., a water-filled cylinder, the relative rate of permeation is given by the Renkin equation [23] times the aqueous diffusion coefficient of the solutes because the entry of the solutes into the channel represents the rate-limiting step, although the diffusion of the solutes in the aqueous phase plays also a certain role. The entry is according to the Renkin equation controlled by the size of the solutes and the effective diameter of the channel.

The liposome swelling assay has been used to compare the size of different *Escherichia coli* porins and to demonstrate the specificity of the LamB channel for maltose and maltooligosaccharides as compared to other carbohydrates [24, 25]. Furthermore, the properties of a variety of porin pores of different gram-negative bacteria were also investigated using this method [26, 27]. The pore diameter varied between 1.1 and 2.0 nm. Similar to the other methods for the study of porin function it has its advantages and disadvantages. Critical for the experiments is the time after the mixing of the liposomes with the test solute and the number of simplifications in the Renkin equation [23] that may influence the calculation of the pore diameter. For example, disaccharides of identical molecular weight caused a substantial variability of the swelling rates by 3- to 7-fold [24]. Furthermore, it is not possible to study the ion selectivity of the pores because membrane potentials occur in the liposome swelling assay when charged test solutes are used.

Experiments with lipid bilayer membranes allow much better access to the study of porin selectivity. Furthermore, only small amounts of membrane-active proteins are needed to obtain information about porin properties in planar lipid bilayer membranes. Two slightly different methods have been used for the reconstitution of porins into lipid bilayer membranes. The first one uses solvent-containing membranes formed according to the Mueller–Rudin method [28], while the other uses virtually solvent-free membranes [29]. Both have in common the electrical circuit. Two silver/silver chloride electrodes with salt bridges are switched in series with a voltage source and a highly sensitive current measuring device and allow the high-resolution detection of the ion flux through the channels reconstituted into the membrane. The porin solubilized in detergent solution is added in very small concentrations (1–100 ng mL^{-1}) to the aqueous phase bathing a painted (i. e., solvent containing) lipid bilayer membrane. Subsequently, the membrane conductance (the membrane current per unit voltage) increases in a step-like fashion as is demonstrated in Fig. 1 for the reconstitution of OprP from *Pseudomonas aeruginosa* outer membrane. The addition of the porin results in a stepwise increase of the membrane conductance. Almost all steps had a single-channel conductance of 150 pS, which means that the size of the channels inserted into the membrane was fairly homogeneous. Using the same approach, multichannel experiments (experiments with a large number of channels) can be performed. A maximum of between 10^4 to 10^6 pores per cm^2 can be incorporated into lipid bilayer membranes. This means that the reconstitution of porin pores is not a rare event. This is also an essential thing because porin preparations may contain contaminant proteins, which cannot be detected in SDS-PAGE, but can be observed in lipid bilayer experiments, and a few channels may be formed by these contaminants. On the other hand, it is possible to sort different porin channels using this method. It has to

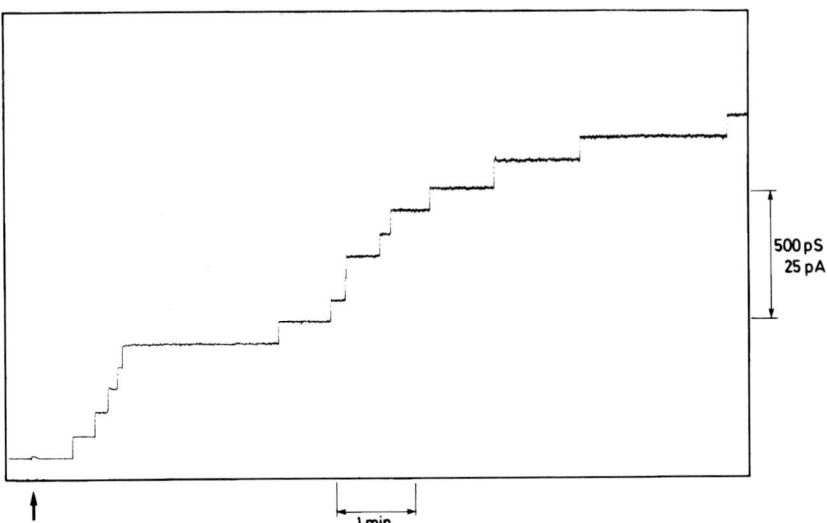

Fig. 1. Single-channel recording of a diphytanoyl phosphatidylcholine/n-decane membrane in the presence of 50 ng mL^{-1} of the phosphate-specific OprP from *P. aeruginosa* outer membrane. The aqueous phase contained 1 M KCl. $T = 20°$, membrane voltage 50 mV. Note that the single steps correspond to the reconstitution of single OprP trimers into the membrane. The arrow indicates the blackening of the membrane.

be noted that a similar sorting of channels is not possible by the liposome swelling assay and it is also not possible to check the integrity of porin channels because the results of the liposome swelling assay represent the average of all types of channels in the system.

The experiments with solvent-free membranes are performed in the following way. The basic difference is that the lipid is spread on the surface of the aqueous phase on both sides of thin Teflon foil using an organic solvent. The foil has a small circular hole (less than 100 µm diameter) above the initial water levels. The surfaces of both aqueous compartments are covered by lipid multilayers. The water levels on both sides of the membrane are now raised and a folded lipid bilayer membrane may be formed across the small hole when it was pre-treated with long-chain alkanes, such as hexadecane. The porins are also added to the aqueous phase, but the reconstitution rate is much lower than observed for solvent-containing membranes. So far it is not clear how the porin pores are incorporated into the lipid bilayers because very often channel formation has to be initiated using high voltage across the membrane. In other cases channels may be reconstituted in a similar way as described above for solvent-containing membranes. So far as single-channel conductance and channel selectivity are concerned, the results obtained from both types of lipid bilayer membranes are similar. Differences were only observed for the voltage dependence of the porin channels, which are voltage-dependent in solvent-depleted membranes, whereas porin channels incorporated into solvent-containing membrane show only minor or no voltage dependence [28–30]. The reason for the voltage dependence is not clear, but it has to be noted that the outer membrane permeability of *E. coli* is not dependent on the Donnan potential up to about 100 mV [6], which could mean that the voltage dependence of porins in solvent-depleted membranes does not own any physiological significance.

The lipid bilayer technique allows easy access to the determination of the ion selectivity of porins. For this type of measurement a salt gradient is established under zero current conditions across membranes which contain a sufficient number of channels [31]. Ions move now through the pores according to their permeability properties until the chemical concentration gradient is balanced by the membrane potential. From the measured zero current membrane potential V_m and the concentration gradient c''/c' across the membrane, the ratio P_c to P_a of the permeabilities of cation and anions may be calculated using the Goldman–Hodgkin–Katz equation [31]. The potential was found to be positive on the more dilute side for most porins from enteric bacteria and a variety of others. This indicates preferential movement of cations through the pores, i.e., they were cation-selective. NmpC and PhoE were anion-selective from the general diffusion porins of *E. coli*. It is noteworthy that the ion selectivity of the general diffusion pores was found to be dependent on the mobility of the permeant ions in the aqueous phase [31]. This means that their selectivity is not an absolute one as has been observed for the highly anion-selective channels (and phosphate-specific) OprP of *P. aeruginosa* and the outer membrane porin of *Pelobacter venetianus* [32, 33].

9.4
Structure and Function of the General Diffusion Porins

Many general diffusion porins have been characterized from gram-negative bacteria. Common to them is that they sort mainly according to the molecular mass of the solutes. Other properties, such as the charge of the solutes plays a minor role. Their diameters have been derived from the measurement of the exclusion limit or from the liposome swelling assay. They range between 1 and 1.5 nm. Other studies were performed with the lipid bilayer technique (see [34] for a list of bacteria). All experimental data suggest that the general diffusion porins form wide water-filled channels [31]. This means that a large number of ions are permeable through general diffusion porins without any detectable interaction with the channel interior. Tab. 1 shows the single-channel conductance, G, measured with OmpF of *E. coli* K12 [28] and *Rhodobacter capsulatus* porin [35]. The data suggest that the single-channel conductance is a linear function of the bulk aqueous concentration c. Succinylation of the *R. capsulatus* porin trimers changes this picture because the single-channel conductance of the chemically modified porin trimers is no longer a linear function of the bulk aqueous conductivity [35]. The difference between OmpF and the *R. capsulatus* porin probably reflects the larger diameter of the *R. capsulatus* channel that has also been found with the liposome swelling assay.

Tab. 1. Average single-channel conductance, G, of OmpF of *Escherichia coli*, of the native and the tetrasuccinyl porin of *Rhodobacter capsulatus* in different salt solutions[a]

Salt	c [M]	OmpF *E. coli* [28]	G [ns] Native Porin *R. capsulatus* [35]	Tetrasuccinyl Porin *R. capsulatus* [35]
LiCl	1.00	0.62	1.3	1.00
KCl	0.01	–	0.030	0.30
	0.03	0.06	0.095	0.53
	0.10	0.20	0.32	0.90
	0.30	0.58	1.00	1.45
	1.00	1.90	2.90	3.30
	3.00	4.90	7.40	6.50
K-acetate (pH 7)	1.00	1.10	2.00	2.70

[a] The membranes were formed of diphytanoyl phosphatidylcholine dissolved in *n*-decane. The aqueous solutions were unbuffered and had a pH of 6 unless otherwise indicated. The applied voltage was 20 mV, and the temperature was 20°. The average single-channel conductance, G, was calculated from at least 80 single events similar to those shown in Fig. 1. c is the concentration of the aqueous salt solutions.

Common to all general diffusion porins is the selectivity for one sort of charged solutes, either positively charged solutes, such as for most porins from enteric bacteria [31], or negatively charged ones for porins involved in the uptake of phosphate at phosphate limitations, such as PhoE of enteric bacteria [19, 36]. The cation selectivity of enteric bacterial porins has probably to do with the presence of bile acids and other anionic detergents in the intestine. Nevertheless, the selectivity is never ideal for general diffusion porins because oppositely charged solutes have also a certain permeability through the channels as it is shown in Tab. 1 for *R. capsulatus* porin. In particular, despite the cation selectivity of this porin potassium chloride and potassium acetate do not have the same single-channel conductance, which would be expected for a highly selective channel. This result suggests indeed that the channel lumen contains both negatively and positively charged amino acids with an excess of negative charges. The 3-D structure of *R. capsulatus* porin (see below) supports this observation because negatively and positively charged amino acids are localized opposite one another within the channel interior and create an internal electrical field inside the channel [37].

A limited number of porins has been investigated that are highly selective for charged solutes. Examples for this are the porins of *Pelobacter venetianus* [33] and *Comamonas acidovorans* [38] (formerly *Acidovorax delafildii* [39]). The presence of only one type of charged amino acids inside these channels results in Gouy–Chapman effects. As a consequence the single-channel conductance is not a linear function of the bulk aqueous ion concentration as in the case of general diffusion pores. Instead, a dependence on the square root of the concentration has been observed for the single-channel conductance similar as it is shown in Fig. 2 for the porin of *P. venetianus* [33]. Following an established procedure it is possible to correct the dependence of the single-channel conductance on the aqueous salt concentration. After correction for the charge effects a linear conductance–concentration relationship may be obtained and the size of the channel can be estimated from the correction of the single-channel conductance (see Fig. 2). It is noteworthy that a similar formalism can be used to correct the single-channel conductance of *R. capsulatus* porin after chemical modification with succinyl anhydride, which changes positively charged lysines into negatively charged groups in agreement with the 3-D structure of the modified porin (see Tab. 1 and [35]).

Figure 3 shows the 3-D structure of the *R. capsulatus* porin monomer as derived from X-ray crystallography [37, 40]. The general diffusion porins derived from different gram-negative bacteria form trimers of three identical subunits. The single subunit is formed by 16 antiparallel transmembrane β-strands, which are tilted by an angle of 45° to 60° regarding the outer membrane surface. They form a hollow cylinder with a length of approximately 4 nm and a diameter of approximately 2.4 nm (= 16×0.47 nm π^{-1} [37, 40]). The periplasmic side (bottom) of the porin trimer is almost flat and contains mostly β-turns, whereas the extracellular side (top) exhibits many long loops between two β-strands. Loop 3 from the N-terminal end is of special interest because it folds back into the lumen of the channel and restricts its cross-section diameter to approximately 0.8 x 1 nm (see Fig. 3B). This eyelet is also very important for the channel stability because its complete

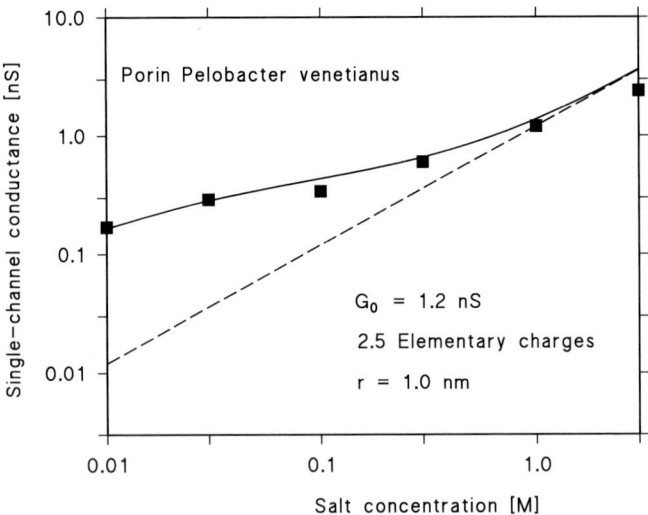

Fig. 2. Single-channel conductance of the *P. venetianus* outer membrane porin as a function of the KCl concentration in the aqueous phase (full squares). The solid line represents the fit of the single-channel conductance assuming the presence of positive point charges (2.5 positive charges; $q = 4 \times 10^{-19}$ As) at the channel mouth on both sides of the membrane and a channel diameter of 2 nm using equations (1) to (4) of reference [35]. Concentration of the KCl solution in M (molar); average single-channel conductance in nS (nano Siemens, 10^{-9} S). The straight (broken) line corresponds to a linear function between single-channel conductance and bulk aqueous concentration, which is obtained without point charges in or near the channel mouth.

deletion leads to complete inhibition of porin assembly in the outer membrane. Important is the eyelet also for the selectivity of the porin channels. Most negatively charged amino acids responsible for the cation selectivity are localized there, whereas the positively charged amino acids are found within the first β-strands from the N-terminal end [35, 37, 40].

It has been demonstrated by side-specific mutagenesis of PhoE of *E. coli* that Lys 125 (localized within this third external loop) plays a substantial role in the preference for anionic solutes and the interaction with phosphate and other anions [41]. The insertion of a 9 amino acid long stretch into the third extracellular loop of PhoE (the eyelet) results in a dramatic decrease of the cross-section of the channel and the single-channel conductance decreases by more than a factor of 10 [42]. The eyelet is on the other hand probably not responsible for the generation of a voltage dependence for certain porins when they are reconstituted into solvent-depleted lipid bilayer membranes [30]. This is the result of experiments with an OmpF mutant where the eyelet was linked through S—S bridges to the channel wall so that it could not move [43]. The structure of the porin channel itself should be very stable since it is stabilized by many hydrogen bonds [40], and a considerable energy would be required to break them down. This could mean that the vol-

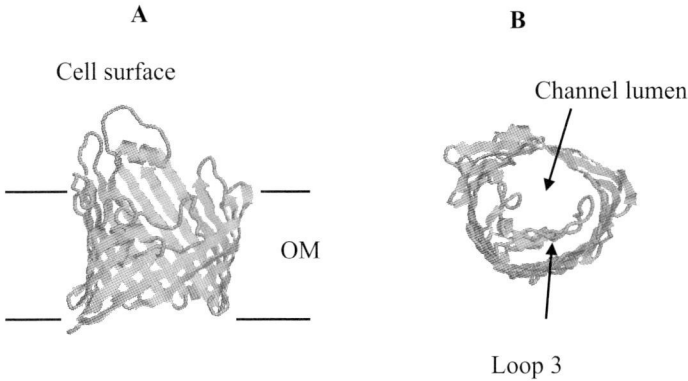

Fig. 3. Schematic view of the 3-D structure of the R. capsulatus porin monomer. The coordinates were taken from the crystallographic data of reference [37]. **(A)** Side view of the porin monomer. **(B)** View from the cell surface exposed side. The other two monomers are attached from the top side of the monomer, opposite to the location of loop 3.

tage dependence could be generated by the movement of surface exposed loops in and out of the membrane as it is discussed for the voltage dependence of the mitochondrial porins, which exhibit voltage-mediated closure already at very small transmembrane potentials of 30 mV [44].

9.5
Structure and Function of Specific Porins

The outer membrane of certain gram-negative bacteria contains, besides one or several of these general diffusion pores, channels which are specific for one class of solutes, such as carbohydrates (porins of the LamB [45], ScrY [46], or the BglH family [47]), cyclic dextrans (CymA of *Klebsiella oxytoca* [48]), nucleosides (Tsx of enteric bacteria [15]), phosphate (OprP of *P. aeruginosa* [32]), and polyphosphate (OprO of *P. aeruginosa* [49]). These specific porins have a very different effect on the permeation of molecules because they contain binding sites with a defined half-saturation concentration for the binding of the molecules. The carbohydrate-specific porins of enteric bacteria form a large family of proteins that have a common ancestor. Prototype of this family is LamB (maltoporin) of *E. coli*, whose properties are described in some detail here. The genes encoding for related proteins are present in a variety of enteric bacteria [45]. Transport of maltose and maltooligosaccharides in *E. coli* is mediated by a transport system that is composed of several components [50] including proteins for fermentation, a periplasmic binding protein, an inner membrane transport system, and an outer membrane protein, LamB (also known as "maltoporin"). This protein is the receptor for phage λ and is important for the uptake of maltose and maltooligosaccharides [51]. Reconstitution experiments have shown that LamB has a defined substrate specificity and that it is able to discriminate between disaccharides of identical molecular mass, e. g., between sucrose and maltose [25, 52]. Long-chain maltooligosaccharides are able to block permeation of glucose and ions through LamB [53, 54]. This possibility allows a meaningful investigation of carbohydrate transport through LamB *in vitro*. In particular, it is possible to study the mechanism and kinetics of maltooligosaccharide transport [25, 52]. Tab. 2 shows the stability constants for carbohydrate binding as derived from the measurements with lipid bilayer membranes [55] and the relative rate of permeation of the same carbohydrates through LamB [25]. The affinity of the carbohydrates for the binding site increases with increasing chain length of the maltooligosaccharides. For different disaccharides it is approximately the same despite a highly different permeability *in vitro* (see Tab. 2) [25] and *in vivo* [46]. This result suggests that the kinetics of carbohydrate permeation plays an important role for their transport through LamB. This has been confirmed by the analysis of the carbohydrate-induced current noise through LamB [52].

Tab. 2. Stability constants for the binding of different carbohydrates to the binding site inside the LamB channel [55] and relative rate of permeation of the same carbohydrates in the liposome swelling assay [25][a]

Carbohydrate	K [L mol^{-1}] Taken from [55]	K_S [mM] Taken from [55]	Relative Rate of Permeation Taken from [25]
D-Glucose	9.5	110	160
Maltose	100	10	100
Maltotriose	2,500	0.40	66
Maltotetraose	10,000	0.10	19
Maltopentaose	17,000	0.59	–
Maltohexaose	15,000	0.067	–
Maltoheptaose	15,000	0.067	2.5
Trehalose	46	22	76
Lactose	18	25	9
Sucrose	67	15	2.5

[a] The stability constants were derived from titration experiments where the ion current through LamB was inhibited by binding of the carbohydrates to the channel interior and assuming that the binding could be described by the Langmuir-adsorption isotherm [55]. The LamB-containing liposomes were prepared in 40 mM stachyose and diluted in 40 mM solution of the test solutes. The relative rate of permeation was derived from the swelling of the liposomes as measured from the decrease of the optical density of the liposomes at 500 nm and was set to 100 for maltose [25].

Some of the prototypes of carbohydrate-specific porins such as the LamB proteins of E. coli and Salmonella typhimurium or ScrY of the single copy plasmid pUR400 have been crystallized and their 3-D structure has been derived from X-ray crystallography [56, 57]. The trimers show a similar organization as the general diffusion porins with the exception that the monomers contain 18 instead of 16 antiparallel β-strands, tilted by an angle of about 45° to 60° regarding the membrane surface (see Fig. 4 for the 3-D structure of a LamB monomer). The periplasmic side (bottom) is again very smooth and contains mainly short β-turns, whereas the external loops (top) are very long and form a complicated architecture, which tends to collapse when single loops are removed [58]. The size of the channel is restricted by external loop 3 from the N-terminal end in a similar way as is the case in the 3-D structure of general diffusion porins (see Fig. 4B). The size of the channel itself is much smaller as compared to that of the general diffusion pores, which means that its permeability is also smaller for non-carbohydrate solutes. Moreover, its single-channel conductance is much smaller than that of a general diffusion pore such as OmpF of E. coli. LamB mutants have shown that the binding of the carbohydrates inside the channel is mediated by H^+-bridges. In particular, arginine R8 localized on the first β-strand from the N-terminal end has a strong influence on the binding affinity, and its mutation in a histidine lowers the stability constant for maltopentaose substantially (from 10,000 L mol^{-1} to 114 L mol^{-1} [59]). It seems that aromatic residues play also an important role in carbohydrate transport and form some sort of slide (greasy slide [60]), along which the carbohydrates move along on the way through the channel from the cell surface to the periplasmic space with the non-reducing end in advance [58]. The sucrose-specific ScrY (sucroseporin) combines features of a general diffusion pore and of carbohydrate-specific porins because it has a much higher permeability for ions [61]. Fig. 5 shows a comparison of the central constriction of both channels [56, 57]. LamB forms definitely a much smaller channel, which is obviously less favorable for the permeation of sucrose than ScrY [46, 62].

The expression of LamB in enteric bacteria is always induced with the expression of the periplasmic maltose binding protein MBP or MalE. MalE is present in the periplasmic space in a concentration in the millimolar range and confers this space into a sink for carbohydrates [50, 63]. This property is an essential part of carbohydrate uptake across the outer membrane because a carbohydrate bound from the cell surface to the binding site inside the LamB channel has still two possibilities with equal probabilities for further movement because the channel is symmetric with respect to its transport properties. It can move back to the cell surface or further on to the periplasmic space. In the latter case it is bound to MBP with a half-saturation constant of considerably less than 10 μM, which means that the carbohydrate is trapped within the periplasmic space, and as long as inner membrane transport functions it is unlikely that the carbohydrate can be lost through the outer membrane. Some of the uptake systems for carbohydrates, such as the plasmid pUR400, which encodes for the uptake of sucrose in enteric bacteria or the plasmid pRSD2 encoding for the uptake of raffinose do not encode for a periplasmic binding protein. In these cases the specific outer membrane

A	B
Cell surface	Loop 3

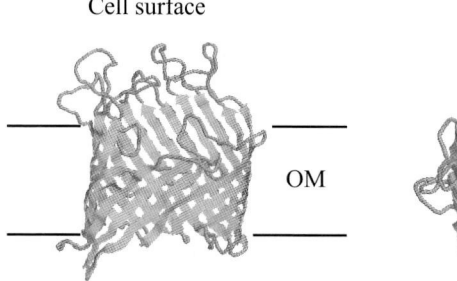
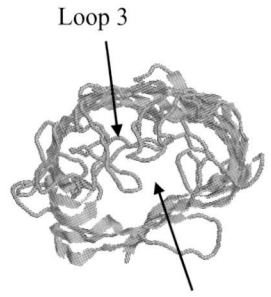

Periplasmic space   Channel lumen

Fig. 4. Schematic view of the 3-D structure of the *E. coli* LamB monomer. The coordinates were taken from the crystallographic data of reference [56]. **(A)** Side view of the LamB (maltoporin) monomer. **(B)** View from the cell surface exposed side. The other two monomers are attached from the bottom side of the monomer, opposite to the location of loop 3.

LamB	ScrY

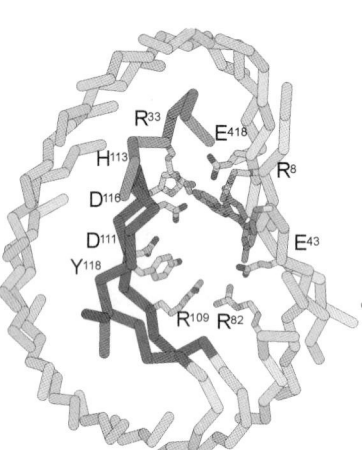
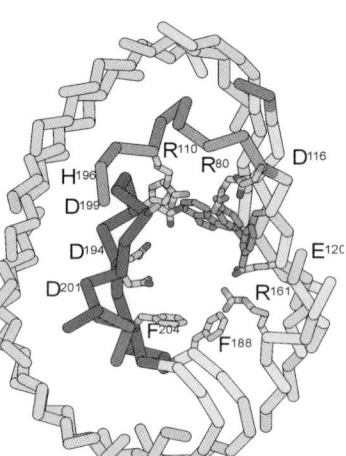

Fig. 5. Cross-sections of the *E. coli* LamB monomer (maltoporin, left) and the ScrY monomer (sucroseporin, right). Both panels show loop 3 (dark gray) and the amino acid residues (denoted with their numbers from the mature N-terminal end) that are relevant for passage of carbohydrates and ions through the central constriction of the channels. The β-strands of the β-barrel cylinder of both porins are given in light gray. The coordinates of maltoporin and sucroseporin were taken from the crystallographic data of references [56] and [57], respectively.

channel (ScrY) or the general diffusion pore (RafY) contain both N-terminal extensions with a length of about 80 amino acids, which may form according to structural predictions coiled-coil structures [46, 57, 64]. It is possible that these structures attached to the channels have a similar function as the periplasmic binding proteins because they bind carbohydrates [65], which may transform the periplasmic space in a sink. This does probably not mean that the N-terminal extensions influence directly the transport properties of the outer membrane channels.

9.6
The Inner and Outer Membrane Connector Channels

The export of E. coli α-hemolysin HlyA synthesized in the cytoplasm out of the cell requires besides the inner membrane transport system HlyB-HlyD also an outer membrane protein TolC that is not directly related to the expression of the inner membrane transport components (see Chap. 8). The same protein is also involved in a variety of other export and import processes, such as the export of toxic compounds, the export of the peptide antibiotics microcin J25, ColV, and Col10 , the import of Colicin E1, and the infection of E. coli cells by phage U3 [66]. In addition, TolC has been shown to be a channel forming outer membrane protein [67]. This means that TolC of is one of the most multifunctional proteins in E. coli, which interacts with a variety of export systems (Acr, Cva, Emr, and Hly, [68], see also Chap. 8). Furthermore, homologs of TolC are found in a large number of gram-negative bacteria [8]. These proteins are highly conserved and are also involved in outer membrane import and export processes, involving drug resistance-related transport processes important for the survival of bacteria in hostile environments [68].

TolC as the prototype of these important outer membrane proteins has recently been crystallized [7]. It represents a unique structure because its trimers contain two different parts (Fig. 6). The outer membrane part is a 4 nm long β-stranded cylinder with 12 β-strands that are not perpendicular to the membrane surface forming one single outer membrane channel. Four β-strands are provided by each monomer, which means that the three TolC monomers contribute to the outer membrane cylinder, in contrast to the situation of the general or the specific porins, where the three monomers form individual channels [37]. Attached to the single β-strands of the outer membrane cylinder are α-helical structures with a length of about 10 nm (see Fig. 6A). It seems that these α-helical structures span the periplasmic space in form of a bridge. Inner membrane transport systems, such as the HlyB/D or the Acr systems are likely to directly interact with the end of the α-helical structure of TolC. This view has an interesting impact on the export of molecules out of the bacterial cells, because it seems sufficient to transport them into the α-helical cage of TolC, which ends in the outer membrane channel. TolC homologous proteins in a variety of gram-negative bacteria have probably a similar structure as TolC and function in the same way as it has been shown for OprM of P. aeruginosa outer membrane [8, 69].

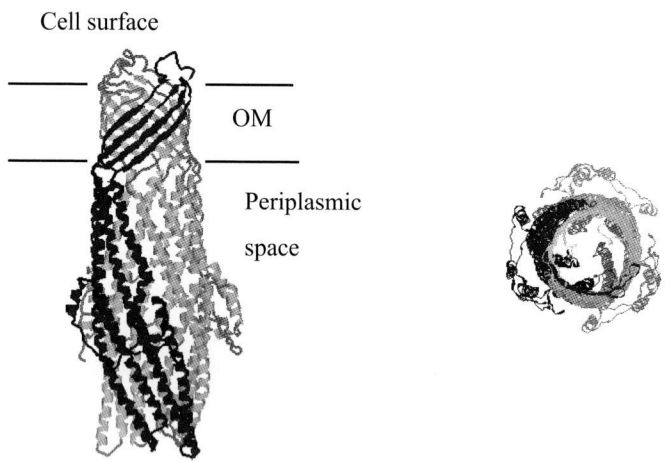

Fig. 6. Schematic view of the 3-D structure of the *E. coli* TolC trimer. The coordinates were taken from the crystallographic data of reference [7]. **(A)** Side view of TolC. The single monomers are given in black, dark gray, and light gray. **(B)** View from the cell surface exposed side in the TolC channel. Note that the α-helical structures limit the size of the TolC channel.

9.7 Conclusions

Gram-negative bacterial porins provide a unique tool for the study of channel permeability and substrate transport through molecular filters. The important advantage of the porin study is the knowledge of many of their structures at atomic resolution, which allows the understanding of substrate–channel interaction on a molecular level. This means that the study of porins represents an exciting object for the investigation of substrate translocation through membranes as model systems.

Acknowledgement

The author would like to thank Christian Andersen for the design of Fig. 5. The financial support of the authors own research by the Deutsche Forschungsgemeinschaft (project be 865/10) and the Fonds der Chemischen Industrie is gratefully acknowledged.

References

1. Schleifer, K. H., Kandler, O., Peptidoglycan types of bacterial cell walls and their taxonomic implications, *Bacteriol. Rev.* **1972**, *36*, 407-477.
2. Braun, V., Covalent lipoprotein from the outer membrane of *Escherichia coli*, *Biochim. Biophys. Acta* **1975**, *415*, 335-377.
3. Lugtenberg, B., Van Alphen, L., Molecular architecture and functioning of the outer membrane of *Escherichia coli* and other gram-negative bacteria, *Biochim. Biophys. Acta* **1983**, *737*, 51-115.
4. Stock, J. B., Rauch, B., Roseman, S., Periplasmic space in *Samonella typhimurium* and *Escherichia coli*, *J. Biol Chem.* **1977**, *252*, 7850-7861,
5. Nikaido, H., Vaara, M., Molecular basis of bacterial outer membrane permeability, *Microbiol. Rev.* **1985**, *49*, 1-32.
6. Sen, K., Hellman, J., Nikaido, H., Porin channels in intact cells of *Escherichia coli* are not affected by Donnan potentials across the outer membrane, *J. Biol. Chem.* **1988**, *263*, 1182-1187.
7. Koronakis, V., Sharff, A., Koronakis, E., Luisi, B., Hughes, C., Crystal structure of the bacterial membrane protein TolC central to multidrug efflux and protein export, *Nature* **2000**, *405*, 914-919.
8. Andersen, C., Hughes, C., Koronakis, V., Chunnel vision: Export and efflux through bacterial channel-tunnels, *EMBO Rep.* **2000**, *1*, 313-318.
9. Cronan, J. E., Vagelos, P. R., Metabolism and function of the membrane phospholipids of *Escherichia coli*, *Biochim. Biophys. Acta* **1972**, *265*, 25-60.
10. Galanos, C., Lüderitz, O., Rietschel, E. T., Westphal, O., Newer aspects of the chemistry and biology of bacterial lipopolysaccharides with special reference to their lipid A component, in: *International Review of Biochemistry*, Vol. 1: *Biochemistry of Lipids II* (Goodwin, T. W., Ed.), University Park Press, Baltimore, MD, **1977**. pp. 239-335.
11. Nikaido, H., Proteins forming large channels from bacterial and mitochondrial outer membranes: Porins and phage lambda receptor protein, *Methods Enzymol.* **1983**, *97*, 85-100.
12. Nakamura, K., Mizushima, S., Effects of heating in dodecyl sulfate solution on the conformation and electrophoretic mobility of isolated major outer membrane proteins from *Escherichia coli* K-12, *J. Biochem.* **1976**, *80*, 1411-1422.
13. Ferenci, T., Lee, K.-S., Directed evolution of the lambda receptor of *Escherichia coli* through affinity chromatographic selection, *J. Mol. Biol.* **1982**, *160*, 431-444.
14. Maier, E., Polleichtner, G., Boeck, B., Schinzel, R., Benz, R., Identification of the outer membrane porin of *Thermus thermophilus* HB8: the channel-forming complex has an unusually high molecular mass and an extremely large single-channel conductance, *J. Bacteriol.* **2001**, *183*, 800-803.
15. Maier, C., Bremer, E., Schmid, A., Benz, R., Pore-forming activity of the Tsx protein from the outer membrane of *Escherichia coli*. Demonstration of a nucleoside-specific binding site, *J. Biol. Chem.* **1987**, *263*, 2493-2499.
16. Zimmermann, W., Rosselet, A., Function of the outer membrane of *Escherichia coli* as a permeability barrier to beta-lactam antibiotics, *Antimicrob. Agents Chemother.* **1977**, *12*, 368-372.
17. Szmelcman, S., Hofnung, M., Maltose transport in *Escherichia coli* K-12: Involvement of the bacteriophage lambda receptor, *J. Bacteriol.* **1975**, *124*, 112-118.
18. West, I. C., Page, M. G. P., When is the outer membrane of *Escherichia coli* rate-limiting for uptake of galactosides? *J. Theor. Biol.* **1984**, *110*, 11-19.
19. Korteland, J., De Graff, P., Lugtenberg, B., PhoE protein pores in the outer membrane of *Escherichia coli* K-12 not only have a preference for Pi and Pi-containing solutes but are general anion-preferring channels, *Biochim. Biophys. Acta* **1984**, *778*, 311-316.
20. Nikaido, H., Rosenberg, E. Y., Foulds, J., Porin channels in *Escherichia coli*: studies with β-lactams in intact cells, *J. Bacteriol.* **1983**, *153*, 232-240.
21. Nakae, T., Outer membrane of *Salmonella typhimurium*: reconstitution of sucrose-permeable membrane vesicles, *Biochem. Biophys. Res. Commun.* **1975**, *64*, 1224-1230.

22. Nikaido, H., Rosenberg, E. Y., Effect of solute size on diffusion rates through the transmembrane pores of the outer membrane of *Escherichia coli*, *J. Gen. Physiol.* **1981**, *77*, 121-135.

23. Renkin, E. M., Filtration, diffusion and molecular sieving through porous cellulose membranes, *J. Gen. Physiol.* **1954**, *38*, 225-253.

24. Nikaido, H., Rosenberg, E. Y., Porin channels in *Escherichia coli*: studies with liposomes reconstituted from purified proteins, *J. Bacteriol.* **1983**, *153*, 241-252.

25. Luckey, M., Nikaido, H., Specificity of diffusion channels produced by λ-phage receptor protein of *Escherichia coli*, *Proc. Natl. Acad. Sci. USA* **1980**, *77*, 167-171.

26. Weckesser, J., Zalman, L. S., Nikaido, H., Porin from *Rhodopseudomoas sphaeroides*, *J. Bacteriol.* **1984**, *159*, 199-205.

27. Vachon, V., Laprade, R., Coulton, J. W., Properties of the porin of *Haemophilus influenzae* type b in planar lipid bilayer membranes, *Biochim. Biophys. Acta* **1986**, *861*, 74-82.

28. Benz, R., Janko, K., Boos, W., Läuger, P., Formation of large, ion-permeable membrane channels by the matrix protein (porin) of *Escherichia coli*, *Biochim. Biophys. Acta* **1978**, *511*, 305-319.

29. Schindler, H., Rosenbusch, J. P., Matrix protein from *Escherichia coli* outer membranes forms voltage-controlled channels in lipid bilayers, *Proc. Natl. Acad. Sci. USA* **1978**, *75*, 3751-3755.

30. Lakey, J. H., Voltage gating in porin channels, *FEBS Lett.* **1987**, *211*, 1-4.

31. Benz, R., Schmid, A., Hancock, R. E. W., Ion selectivity of Gram-negative bacterial porins, *J. Bacteriol.* **1985**, *162*, 722-727.

32. Benz, R., Hancock, R. E. W., Mechanism of ion transport through the anion-selective channel of *Pseudomonas aeruginosa* outer membrane, *J. Gen. Physiol.* **1987**, *89*, 275-295.

33. Schmid, A., Benz, R., Schink, B., Identification of two porins in *Pelobacter venetianus* fermenting high-molecular-mass polyethylene glycols, *J. Bacteriol.* **1991**, *173*, 4909-4913.

34. Benz, R., Bauer, K., Permeation of hydrophilic molecules through the outer membrane of Gram-negative bacteria: review on bacterial porins, *Eur. J. Biochem.* **1988**, *176*, 1-19.

35. M. Przybylski, M. O., Glocker, U., Nestel, V., Schnaible, K., Diederichs, J., Weckesser, M., Schad, A., Schmid, W., Welte, R., Benz X-ray crystallographic and mass spectrometric structure determination and functional characterization of succinylated porin from *Rhodobacter capsulatus*: Implications for ion selectivity and single-channel conductance, *Protein Sci.* **1996**, *5*, 1477-1489.

36. Benz, R., Schmid, A., van der Ley, P., Tommassen, J., Molecular basis of porin selectivity: membrane experiments with OmpC-PhoE and OmpF-PhoE hybrid proteins of *Escherichia coli* K-12, *Biochim. Biophys. Acta* **1989**, *981*, 8-14.

37. Weiss, M. S., Abele, U., Weckesser, J., Welte, W., Schiltz, E., Schulz, G. E., Molecular architecture and electrostatic properties of a bacterial porin, *Science* **1991**, *254*, 1627-1630.

38. Zeth, K., Diederichs, K., Welte, W., Engelhardt, H., Crystal structure of omp32, the anion-selective porin from *Comamonas acidovorans*, in complex with a periplasmic peptide at 2.1 Å resolution, *Structure Fold. Des.* **2000**, *8*, 981-992.

39. Brunen, M., Engelhardt, H., Schmid, A., Benz, R., The major outer membrane protein of *Acidovorax delafieldii* is an anion-selective porin, *J. Bacteriol.* **1991**, *173*, 4182-4187.

40. Weiss, M. S., Kreusch, A., Schiltz, E., Nestel, U., Welte, W., Weckesser, J., Schulz, G. E., The structure of porin from *Rhodobacter capsulatus* at 1.8 resolution, *FEBS Lett.* **1991**, *280*, 379-382.

41. K. Bauer, M., Struyvé, D., Bosch, R., Benz, J., Tommassen One single lysine-residue is responsible for the special interaction between polyphosphate and the outer membrane porin PhoE of *Escherichia coli*, *J. Biol. Chem.* **1989**, *264*, 16393-16398.

42. Struyvé, M., Visser, J., Adriaanse, H., Benz, R., Tommassen, J., Topology of PhoE porin: the "eyelet" region, *Mol. Microbiol.* **1993**, *7*, 131-140.

43. Bainbridge, G., Mobasheri, H., Armstrong, G. A., Lea, E. J., Lakey, J. H., Voltage-gating of *Escherichia coli* porin: a cystine-scanning mutagenesis study of loop 3, *J. Mol. Biol.* **1998**, *275*, 171-176.

44. Benz, R., Permeation of hydrophilic molecules through the mitochondrial outer membrane: review on mitochondrial porins, *Biochim. Biophys. Acta* **1994**, *1197*, 167-196.

45. Bloch, M. A., Desaymard, C., Antigenic polymorphism of the LamB protein among members of the family Enterobacteriaceae, *J. Bacteriol.* **1985**, *163*, 106-110.
46. Schmid, K., Ebner, R., Jahreis, K., Lengeler, J. W., Titgemeyer, F., A sugar-specific porin, ScrY, is involved in sucrose uptake in enteric bacteria, *Mol. Microbiol.* **1991**, *5*, 941-950.
47. Andersen, C., Rak, B., Benz, R., The gene *bglH* present in the *bgl*-operon of *Escherichia coli*, responsible for uptake and fermentation of *β*-glycosides encodes for a carbohydrate-specific outer membrane porin, *Mol. Microbiol.* **1999**, *31*, 499-510.
48. Pajatsch, M., Andersen, C., Mathes, A., Böck, A., Benz, R., Engelhardt, H., Properties of a cyclodextrin-specific, unusual porin from *Klebsiella oxytoca*, *J. Biol. Chem.* **1999**, *274*, 25159-25166.
49. Hancock, R. E. W., Egli, C., Benz, R., Siehnel, R. J., Overexpression in *Escherichia coli* and functional analysis of a novel pyrophosphate-selective porin OprO from *Pseudomonas aeruginosa*, *J. Bacteriol.* **1992**, *174*, 471-476.
50. Schwartz, M., The maoltose regulon, in: *Escherichia coli* and *Salmonella typhimurium* Vol. 2 (Neidhardt, F. C., Ed.), ASM, Washington, DC, **1987**, pp. 1482-1502.
51. Szmelcman, S., Hofnung, M., Maltose transport in *Escherichia coli* K-12: involvement of the bacteriophage lambda receptor, *J. Bacteriol.* **1975**, *124*, 112-118.
52. Andersen, C., Jordy, M., Benz, R., Evaluation of the rate constants of sugar transport through maltoporin (LamB) of *Escherichia coli* from the sugar-induced current noise, *J. Gen. Physiol.* **1995**, *105*, 385-401.
53. Luckey, M., Nikaido, H., Diffusion of solutes through channels produced by phage lambda receptor protein of *Escherichia coli*: Inhibition of glucose transport by higher oligosaccharides of maltose series, *Biochem. Biophys. Res. Commun.* **1980**, *93*, 166-171.
54. Benz, R., Schmid, A., Nakae, T., Vos-Scheperkeuter, G., Pore formation by LamB of *Escherichia coli* in lipid bilayer membranes, *J. Bacteriol.* **1986**, *165*, 978-986.
55. Benz, R., Schmid, A., Vos-Scheperkeuter, G. H., Mechanism of sugar transport through the sugar-specific LamB channel of *Escherichia coli* outer membrane, *J. Membrane Biol.* **1987**, *100*, 21-29.
56. Schirmer, T., Keller, T. A., Wang, Y. F., Rosenbusch, J. P., Structural basis for sugar translocation through maltoporin channels at 3.1 Å resolution, *Science* **1995**, *267*, 512-514.
57. Forst, D., Welte, W., Wacker, T., Diederichs, K., Structure of the sucrose-specific porin ScrY from *Salmonella typhimurium* and its complex with sucrose, *Nature Struct. Biol.* **1998**, *5*, 37-46.
58. Andersen, C., Bachmeyer, C., Täuber, H., Benz, R., Wang, J., Michel, V., Newton, S. M. C., Hofnung, M., Charbit, A., *In vivo* and *in vitro* studies of major surface loop deletion mutants of the *Escherichia coli* K12 maltoporin: contribution to maltose and maltooligosaccharide transport and binding, *Mol. Microbiol.* **1999**, *32*, 851-867.
59. Jordy, M., Andersen, C., Schülein, K., Ferenci, T., Benz, R., Rate constants of sugar transport through two lamB mutants of *Escherichia coli*: comparison to wild-type maltoporin and to LamB of *Salmonella typhimurium*, *J. Mol. Biol.* **1996**, *259*, 666-678.
60. Wang, Y. F., Dutzler, R., Rizkallah, P. J., Rosenbusch, J. P., Schirmer, T., Channel specificity: structural basis for sugar discrimination and differential flux rates in maltoporin, *J. Mol. Biol.* **1997**, *272*, 56-63.
61. Schülein, K., Schmid, K., Benz, R., The sugar specific outer membrane channel ScrY contains functional characteristics of general diffusion pores and substrate-specific porins, *Mol. Microbiol.* **1991**, *5*, 2233-2241.
62. Andersen, C., Cseh, R., Schülein, K., Benz, R., Study of sugar binding to the sucrose specific ScrY-channel of enteric bacteria using current noise analysis, *J. Membrane Biol.* **1998**, *164*, 263-274.
63. Brass, J. M., Bauer, K., Ehmann, U., Boos, W., Maltose-binding protein does not modulate the activity of maltoporin as ageneral porin in *Escherichia coli*, *J. Bacteriol.* **1985**, *161*, 720-726.
64. Ulmke, C., Lengeler, J. W., Schmid, K., Identification of a new porin, RafY, encoded by raffinose plasmid pRSD2 of *Escherichia coli*, *J. Bacteriol.* **1997**, *179*, 5783-5788.
65. Dumas, F., Frank, S., Koebnik, R., Maillet, E., Lustig, A., Van Gelder, P., Extended sugar slide function for the periplasmic coiled coil domain of ScrY, *J. Mol. Biol.* **2000**, *300*, 687-895.

66. Fath, M., Kolter, R., ABC transporters: bacterial exporters, *Microbiol. Rev.* **1993**, *57*, 995-1017.
67. Benz, R., Maier, E., Gentschev, I., TolC of *Escherichia coli* functions as an outer membrane channel, *Zbl. Bakteriol.* **1993**, *278*, 187-196.
68. Piddock, L. J., Mechanisms of fluoroquinolone resistance: an update 1994-1998, *Drugs* **1999**, *58* (Suppl. 2), 11-18.
69. Wong, K. K., Brinkman, F. S., Benz, R., Hancock, R. E. W., Evaluation of a structural model of *Pseudomonas aeruginosa* outer membrane protein OprM, an efflux component involved in intrinsic antibiotic resistance, *J. Bacteriol.* **2001**, *183*, 367-374.

10
Aquaporins

Ralf Kaldenhoff and Martin Eckert

10.1
Introduction

In all living cells water is the principal ingredient. Because of its dipolar character it is an ideal solvent for solutes of biologically important substances. For a single cell it is of crucial significance to manage and appropriately change the water content in accordance with the current physiological situation, which could change depending on medium or developmental conditions. In these cases, the biomembrane plays a pivotal role in water transport control. The lipid bilayer functions as a barrier and permits comparably smaller and slower perfusion of water and solutes. The molecular mechanisms modifying the influx and efflux of substances across the biomembrane bilayer include membrane-intrinsic proteins, which contribute to the specific features and functionality of the membrane. Representative examples are given by ATPases, ion channels, and metabolite transporters as well as the members of the Major Intrinsic Proteins (MIPs) superfamily. These operate as facilitators for the transport of water, glycerol, and other small uncharged solutes. The water channels, also named aquaporins as well as the glycerol conducting channels have a pore region which is responsible for a transmembrane water respectively solute flux [1]. The flux direction is dependent on the force driven by an osmotic gradient. Because of its energy-independent status, the transport process is classified as facilitated diffusion. By regulating the uptake, release, or production of osmotically active substances, the cell is able to alter the water potential of the cytosol (or vacuole) and thereby adjust the osmotic pressure and turgescence. Currently, three pathways are suggested to mediate water membrane movement (Fig. 1). Diffusion across the membrane, which does not need proteinaceous components, cotransport with charged molecules or ions, which pass the membrane in a hydrated configuration via a transporter protein, and thirdly, the aquaporin facilitated diffusion of water. The differences in the mechanism and efficiency of cotransporter, membrane diffusion and aquaporin-mediated water transport indicate the significance of aquaporins concerning cellular water management:

aquaporins

- selective for water or also permeable to small solutes
- water flux towards lower water potential
- independent of the gradient's chemical nature
- low activation energy
- linear kinetics

co-transporter

- specific for a substrate
- independent of gradient direction
- high activation energy
- non-linear kinetics

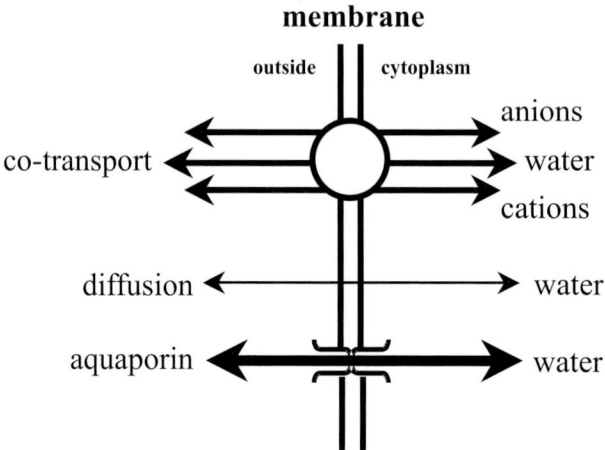

Fig. 1. Suggested pathways for water movement across biomembranes.

10.2
Diversity of Species with Aquaporin Genes

In the early 1990s, the first aquaporin AQP1 was characterized from red blood cells [2]. The protein was found to be related to MIP, the major intrinsic protein of the lens fiber [3]. Since then, the phenomena of a high water permeability of particular specialized cells turned out to correspond to the presence of MIP-related aquaporins [4]. Furthermore, the physiological function of aquaporins as channels selective for water became more and more evident. Today, the contribution of aquaporins in water transport mechanisms is widely accepted and provided improved insight in how organisms regulate their water relations [5–9].

In mammalians, 10 MIP-related proteins (AQP0-AQP9) have been described so far [10, 11]. They were characterized as functional aquaporins possessing significant specificity for water and in some cases also for small uncharged solutes like glycerol or urea. The occurrence of these aquaporins is strongly correlated with high water fluxes, e.g., in specialized kidney cells or with water homeostasis of the eye lens.

In plants, the number of aquaporin-like sequences in a single species seems to be generally higher. Computer-based sequence search in the completely sequenced genome of the small model plant *Arabidopsis thaliana* revealed about 30 MIP-related genes [12, 13]. However, many of the encoded proteins have not yet been functionally characterized. Studies of aquaporin expression in plants show a wide range in different cell types [13a–c]. High levels of promoter activity, mRNA or aquaporin protein were detected during processes of cell elongation [14], in and adjacent to stomata guard cells or vessels of roots, stems, or leaves. The physiological function of plant aquaporins is suggested to ensure an overall water supply during growth and development [15] as well as a cell-to-cell water movement and the intracellular water flux from the cytosol to the vacuole and vice versa [6]. MIP-like channels that transport water have also been identified in algae, mosses, ferns, and insect cells [16].

10.3
Microbial Aquaporins

In fungi and bacteria the single-cell organization results in a close interaction of cell and environment. Accordingly, the cell must be able to adapt to sudden changes in its growth conditions. To benefit from the rigidity of the bacteria murein or the cell wall of fungi, hyperosmolar conditions are important for maintaining the cell shape. Multiple strategies are conceivable, particularly the influx of osmotically active solutes like potassium or the extrusion of protons, e.g., a potassium influx via a K^+/H^+ exchanger. The question arises whether aquaporins are involved in the subsequent water movement or during water flux after cell division respectively expansion. However, increasing efforts led to the identification of a number of aquaporins in fungi like yeast, *Candida,* or *Aspergillus* and in gram-negative bacteria and archaea [8, 9]. The abundance of these genes is an argument in favor of the physiological relevance and importance of the encoded proteins.

10.4
Structural Properties of Aquaporins

Aquaporins and glycerol conducting channels are members of the MIP protein superfamily [17]. In general with a few exceptions, the proteins are rather small with a mass weight of approximately 30 kDa and characterized by six membrane spanning helices (Fig. 2). The helices are connected by five loops of which two (i.e., loop B and E) carry the highly conserved asparagine–proline–alanine (NPA) sequence motif [HNQA]-X-N-P-[STA]-[LIVMF]-[ST]-[LIVMF]-[GSTAFY] or parts of it. These elements dip into the membrane and form a single transmem-

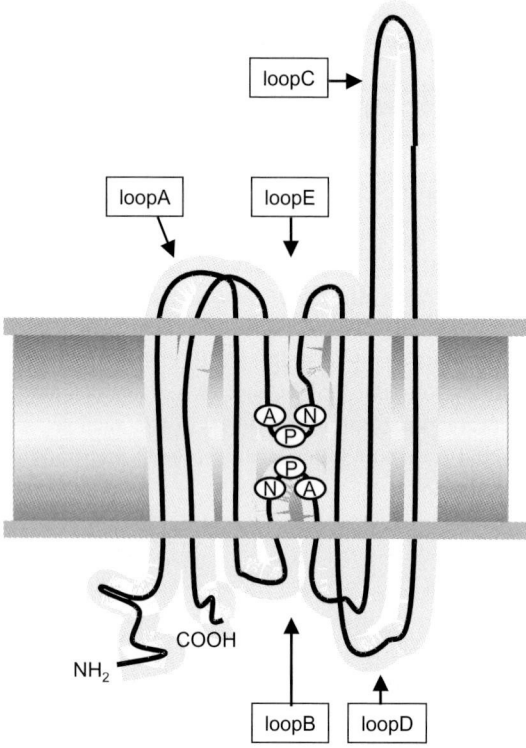

Fig. 2. Schematic drawing of the *Escherichia coli* aquaporin Z structure. Amino- (NH$_2$) and carboxy-terminus (COOH), the extramembrane loops (loopA to loopE), and the amino acids of the consensus motif (NPA) are indicated.

brane pore. Recent studies on the human aquaporin AQP1 respectively the *Escherichia coli* GlpF described a precise atomic model [18, 19]. According to these studies, a transport and selectivity mechanism was proposed (details will be discussed below).

10.5
Functional Analysis of Aquaporins

Once a gene or a cDNA is identified as a MIP-related sequence, a functional confirmation of the water transport capability is necessary to precisely classify the protein. One of the first and commonly used approaches is the heterologous expression in *Xenopus laevis* frog oocytes [20]. The cDNA is cloned into a specific expression vector between the upstream and downstream untranslated regions of the *Xenopus* β-globulin gene in order to maintain RNA stability and an efficient translation in the oocyte. Subsequently, the cloned cDNA is transcribed *in vitro* and the resulting copy RNA is injected into peeled oocytes. After an appropriate incubation time for protein expression and integration into the membrane, the cells are transferred from isotonic to hypoosmotic conditions. The resulting swelling is moni-

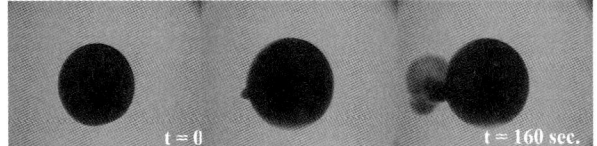

Fig. 3. *Xenopus* oocytes expressing an aquaporin in isoosmolar conditions ($t = 0$) or transferred to hypoosmotic conditions.

tored by a video system (Fig. 3), and changes in oocyte diameter per time and osmotic gradient are used to calculate the membrane water permeability coefficient (P_f). In case the expressed protein confers an increase in water permeability compared to water-injected oocytes, the protein function as an aquaporin can be confirmed in the heterologous oocyte system.

The characterization of microbial aquaporins is frequently also carried out in a prokaryotic system, e.g., *E. coli* [21]. Cells carrying the gene of interest were subjected to a hyperosmotic shock and the cell shape was documented by electron microscopy. If an aquaporin is expressed and functional located in the plasma membrane, the cytoplasm retracts due to rapid water loss and plasmolytic spaces become visible. Under the same conditions, no shrinkage was observed in control cells with a δ-aquaporin mutation.

In addition, a cell-free system has been established using isolated membrane vesicles, either from partitioning of native membranes or from reconstituted proteoliposomes [22, 23]. After changing the osmotic conditions, the size change of the vesicles is followed by specific light scattering properties, in a stopped flow device.

10.6
Unspecific Aquaporins

Although the members of the aquaporin family share high homologies at the amino acid sequence level, differences in transport specificity were observed. Uptake experiments with radioactively labeled compounds conducted with aquaporin-like proteins revealed that some transport also small uncharged molecules like urea or glycerol. In order to distinguish these aquaporins from proteins highly selective for water, Agre and coworkers created the term aquaglyceroporins [10]. Interestingly, aquaglyceroporins were not described for bacteria and fungi. Instead, the single cell organisms possess a high number of MIP-related proteins which facilitate glycerol transport and are mainly impermeable for water. The glycerol conducting channels are consequently called GlpFs. This supports the assumption that the employment of glycerol as a compatible solute in osmoregulation is a commonly used microbial feature.

10.7
Complexity of Microbial MIP-like Channel Genes

Today, more than 50 microbial genes or sequences are related to MIP-like water channels and glycerol conducting channels [8, 9]. Although only a few have already been characterized as functional aquaporins or GlpFs, the high sequence homology corroborates the assumption that a high number of microorganisms developed mechanisms for facilitated diffusion of water and glycerol. A phylogenetic sequence analysis uncovers an obvious categorization concerning the transport properties with regard to water or glycerol (Fig. 4). The glycerol conducting channels fall into three groups, clearly representing distinct clusters of mycobacteria, gram-negative and gram-positive bacteria. Focussing on aquaporins an obvious clustering becomes apparent, too. AQPs from fungi are phylogenetically distinct from those of gram-negative bacteria. Interestingly, MIPs from archaea form a se-

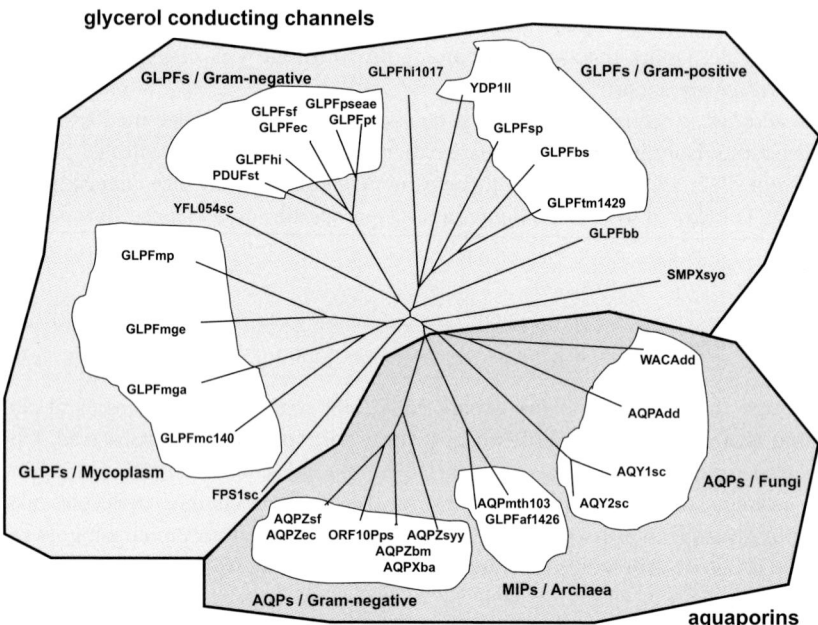

Fig. 4. Phylogenetic tree of microbial MIP-related sequences. Amino acid sequence data were retrieved from the National Center for Biotechnology Information (NCBI), MD, USA, http://www.ncbi.nlm.nih.gov. AQP, aquaporin; GLPF, glycerol conducting channel. Organism abbreviations: af = *Archaeoglobus fulgidus*, ba = *Brucella abortus*, bb = *Borrelia burgdorferi*, bm = *Brucella melitensis*, bs = *Bacillus subtilis*, dd = *Dictyostelium discoideum*, ec = *Escherichia coli*, hi = *Haemophilus influenzae*, ll = *Lactococcus lactis*, mc = *Mycoplasma capricolum*, mga = *Mycoplasma gallisepticum*, mge = *Mycoplasma genitalium*, mp = *Mycoplasma pneumoniae*, mth = *Methanobacterium thermoautotrophicum*, pa = *Pseudomonas aeruginosa*, ps = *Plesiomonas shigelloides*, pt = *Pseudomonas tolaasii*, sc = *Saccharomyces cervisiae*, sf = *Shigella flexneri*, sp = *Streptococcus pneumoniae*, st = *Salmonella typhimurium*, syo = *Synechococcus sp.*, syy = *Synechocystis sp.*, tm = *Thermotoga maritima*.

parate group containing both channel types AQP and GlpF. It must be kept in mind, however, that these classifications are sequence-related and not confirmed by functional characterization. The closest relative is the aqpZ family, a well characterized group of bacteria aquaporins and the group of fungal AQPs. Regarding the sequence relation, it is tempting to speculate that both archaea MIPs function as aquaporins. The phylogenetic classification of the two *Saccharomyces* glycerol conducting channels FPS1 and YFL054 is uncertain, since they do not fit into any phylogenetic cluster. Both sequences could, however, be assigned to the GlpF section.

10.8
Gene Structures

The genomic organization of aquaporins apparently differs from that of glycerol conducting channel proteins. It is reported that the *E. coli* aquaporin *aqpZ* displays a typical monocistronic organization [24]. A similar situation was described in the pathogenic bacterium *Brucella abortus* [25] and additional known bacterial aquaporin genes. In *B. abortus*, the *aqpX* is flanked by three copies of a palindromic sequence whose functions are still unknown. In contrast, glycerol conducting channels are often described to be part of an operon [8, 25]. This could also be demonstrated for GlpF genes in *E. coli* and *Pseudomonas aeruginosa*. A glycerol kinase was found to be transcriptionally associated with the glycerol conducting channel, indicating that glycerol transport and glycerol phosphorylation during uptake is correlated [26, 27]. This operon organization is also conserved in gram-positive and gram-negative bacteria [8].

10.9
Physiological Indications for Protein-mediated Membrane Water Transport

For a long time the molecular mechanism of water movement across membranes of living organisms has been a matter of contradicting scientific debates. On one hand, biomembranes are permeable for water and there seems to be no requirement for a facilitated cross membrane water transport in a normal environment, e.g., moderate osmotic conditions and sufficient water supply. With increasing knowledge of ion channel transport mechanisms, it became evident that water is also cotransported with ions, e.g., by chloride channels. However, the data also indicate that the contribution of ion cotransported water to the bulk membrane water flow is negligible. Arguments in favor of water conducting membrane pores were provided by the observation that the water permeability of membranes could be modified by addition of hormones in animal cells [28], and that individual tissue types exhibited extraordinary water permeability [29]. Some biophysical data also provide the hypothesis of a pore-mediated water flux. The activation energy for the movement of water across erythrocyte membranes, e.g., is about

4–6 kcal mol^{-1}, which is significantly less than that for water flow across a synthetic lipid bilayer (11–14 kcal mol^{-1}) [30]. The molecular basis for the increased water permeability of certain biomembranes in comparison to non-proteinaceous lipid bilayers was discovered in 1991 by Preston and Agre [2] in human blood cells. The corresponding protein initially named CHIP28 is today called aquaporin 1 (AQP1) and was the first functionally characterized water channel.

10.10
The Human Aquaporin 1 as a Model

Due to a high sequence conservation between all aquaporins detected so far, it is conceivable to transfer major features concerning structure–function relationships to other similar MIPs. In this regard the elucidation of AQP1 features could help to understand other aquaporins, e. g., those of microorganisms. AQP1 consists of 269 amino acid residues, which form a tandem repeat of three transmembrane a-helices. The amino and carboxy termini are located in the cytoplasm [31], and each of the two domains shows the highly conserved motif asparagine–proline–alanine in the cytoplasmic loop B as well as in the extracellular loop E (Fig. 2). From expression and function studies in Xenopus oocytes it became evident that these motifs are essential for the protein function. A mutation in or close to one NPA motif reduced the ability of water transport drastically. As a consequence, the so-called "hour glass" model was proposed, in which a cytoplasmic chamber formed by loop B connects with the extracellular loop E. By this, an amphiphilic pathway for water through the membrane is created [1]. Aquaporins seem to reside as tetramers in the membrane, although a monomer is able to form a functional water pore [32, 33]. Two-dimensional crystals of isolated AQP1 were used for a more detailed structure analysis and revealed that a monomer contains six tilted, membrane spanning a-helices, that form a right-handed bundle [33, 34]. The helices form a barrel that includes the water pore. The postulated "hour glass" model was furthermore confirmed by electron crystallographic data at 3.8 [18]. From these observation the nature of the water-selective pore could be determined. The selectivity for water resides in part on the pore size, which has a diameter of 3 (measured by van der Waals distance) at its narrowest site close to the NPA motifs. Regarding that the diameter of a water molecule is 2.8 and that of an ion is much larger a size selectivity is achieved. In addition the aquaporin does not contain a structure to liberate ions from their hydration shell. The shape of the pore shows a constriction at the center and a wide opening at the membrane surfaces, which result in a dielectric barrier that prevents the entry of ions and allows the penetration of neutral solutes [18]. The atomic model suggests a hydrogen bond interaction of the two pore asparagines with the oxygen atom in a water molecule, while the two hydrogen atoms are prevented from hydrogen bonding partners by hydrophobic residues at the opposite site of the pore. A H-bond isolation preventing protons to pass the pore is hypothesized to function through this mechanism, leading to a high specificity for water and impermeability for protons.

10.11
The *Escherichia coli* Aquaporin Z

The structure of the *E. coli* pendant named aquaporin Z (AqpZ) was deduced from 2-D crystals after overexpression in the homologous system. For this, the coding sequence was introduced into an expression vector with an IPTG-inducible promoter. The protein was expressed in fusion with a 10-histidine tag, which was used for purification on a nickel column. Reconstitution of the 10-His-aquaporin in proteoliposomes and subsequent water uptake measurements by light scattering in a stopped-flow apparatus indicated that the overexpressed protein was functionally active as an highly effective aquaporin [35]. It was found to be impermeable for glycerol, urea, or sorbitol. Sucrose gradient sedimentation analysis indicated that the naturally occurring AqpZ is a tetramer. The recombinant AqpZ was solubilized in octyl-β-D-glucopyranoside and the particles visualized by electron micrographs. The particle size and shape clearly confirmed the hypothesis of a tetrameric protein association. A three-dimensional reconstitution of negatively stained lattices, obtained from two-dimensional crystals, revealed the same packaging arrangement as with the human AQP1 described above. The striking resemblance of the 8 Å projection map further confirmed the high similarity of the two aquaporins originating from human or *E. coli* [36]. A further characterization was achieved by comparative atomic force microscopy [37] between the above mentioned recombinant His-AqpZ and a trypsin treated derivate without a His-tag at its amino terminus. While the cytoplasmic surface was drastically changed after proteolytic cleavage, the extracellular surface morphology remained unaltered, indicating that the N-terminus is located in the cytoplasm. A reversible force-induced conformational change obtained by atomic force microscopy could be related to the amino acid sequence by hydropathy analysis and was determined as the longest loop C, facing the extracellular space.

10.12
Physiological Relevance of Aquaporins

Two lines of evidences, gene expression and knockout mutants, strongly underline the importance of aquaporin Z for *E. coli* and possibly other bacteria. First, the expression of *aqpZ*, which is a single copy gene located at minute 19.7 [24], was analyzed and found to be regulated according to the growth phase. An *aqpZ::lacZ* gene fusion was introduced into *E. coli* and used to quantify the expression in a liquid culture. Highest values were obtained in the late logarithmic growth phase [38], indicating that aquaporin function is correlated to cell division or elongation. Besides growth phase regulation an influence of osmotic conditions was registered: it was increased in hypoosmotic (80 mosmol kg^{-1} H$_2$O) and decreased in hyperosmolar conditions (> 700 mosmol kg^{-1} H$_2$O). Regarding the function as water-selective pores and the possible association to the osmotic state, it is tempting to speculate that other proteins contributing to constant intracellular osmotic conditions during growth or change of extracellular osmolarity are coordinated with aquaporins. The discovery of mechanosensitive channels in all bacterial kingdoms [39, 40] and their role during transition into lower osmolarity [41] is one example that could complete the picture of physiological processes in bacteria under these conditions. The possible correlation between aquaporins and mechanosensitive channels is discussed in detail by Booth and Louis [42].

Since the *E. coli aqpZ* is a single copy gene and monocistronically expressed, a knockout mutant would be an efficient tool for elucidating aquaporin function. The mutant was generated by double allelic recombination with a *lacZ-kann*-disrupted *aqpZ* sequence [38]. When plated on LB-agar, the colonies of knockout mutants appear much smaller in comparison to those of wild type. Viability of *aqpZ* mutants at elevated temperatures was reduced and in accordance with the expression data also reduced in hypoosmotic medium. The mutant phenotype could be rescued by transformation with an aquaporin Z plasmid.

In a similar experiment, the aquaporin AqpX isolated from the gram-negative bacterium *Brucella abortus* was used to rescue an *E. coli* glycerol facilitator/aquaporin double mutant (*glpFaqpZ*). Hpyerosmotic shock treatment retained the ability for plasmolysis [25], and it was concluded that AqpX is responsible for outward directed water flux.

By database search the *Saccharomyces* homolog AQY1 could be identified [43]. In contrast to the results obtained with the *E. coli* mutant, *aqy1*$^-$ yeast strains show an improved viability in conditions inducing an osmotic stress. Taken together the data imply that AqpZ is required in osmoregulatory mechanisms and during rapid cell growth.

Whether the results concerning the physiological role of aquaporins could be transferred to other bacteria remains to be examined. Noticeably, aquaporin proteins could not be detected in most gram-positive bacteria [8]. It can be assumed that these genera developed different mechanisms to overcome osmotic stress [44, 45] and thus the function of an aquaporin is irrelevant.

10.13
Glycerol Conducting Channels

10.13.1
Structure

In contrast to aquaporins, which are almost completely absent in gram-positive genera, glycerol conducting channels like the *E. coli* glycerol facilitator (GlpF) seem to occur in diverse microorganism species. Due to a high sequence homology to aquaporins and a very similar protein structure, GlpFs belong to the same MIP protein family [46]. By isomorphous replacement and anomalous dispersion the *E. coli* GlpF structure was determined on frozen crystals [19]. A symmetric arrangement of four channels with three glycerol molecules in each was observed. According to the structural data, a molecular mechanism for the protein selectivity and glycerol transport mechanism was proposed. Although GlpF possess the conserved asparagines in the NPA motif at comparable positions to AQP1, and these residues could provide hydrogen bonds to glycerol during passage, a selectivity is achieved by different means. Close to the periplasmic site, the aromatic amino acids tryptophan (position 48) and phenylalanine (position 200) as well as arginine (position 206) form a constriction. The arginine residue forms hydrogen bonds with two hydroxyl groups of a glycerol molecule. By this, glycerol is oriented, facing the carbon backbone into the cavity built by the aromatic amino acids. As a result, glycerol is separated from certain other linear polyols and forced into a single file into the pore. This pore channel of 28 Å length is completely amphipathic, with polar residues opposing a hydrophobic wall [46].

10.13.2
Physiological Relevance of Glycerol Conducting Channels

Although glycerol conducting proteins are widespread among microorganisms, the most valuable information was obtained on those from *Saccharomyces* and *E. coli*. In addition to glycine [47–49], the yeast Fps1p and *E. coli* GlpF conduct the transport of some polyols, glyceraldehyde, glycine, and urea [50, 51]. Cotransport of water was found to be not important [52–53].

Because yeasts accumulate glycerol during growth in high osmolarity and release it after a hypoosmotic shock in an Fps1-dependent manner, it was concluded that the protein has its major function in glycerol export [47]. Fps1 deletion mutants are sensitive to hypoosmotic shock treatment and a deletion of the N-terminal domain results in dominant hyperosmosensitivity [49]. The findings suggest Fps1 as a glycerol export protein involved in osmoregulation [49, 54].

References

1. J. S. Jung, G. M. Preston, B. L. Smith, W. B. Guggino, P. Agre, Molecular structure of the water channel through aquaporin CHIP. The hourglass model, *J. Biol. Chem.* **1994**, *269*, 14648-14654.
2. G. M. Preston, P. Agre, Isolation of the cDNA for erythrocyte integral membrane protein of 28 kilodaltons: member of an ancient channel family, *Proc. Natl. Acad. Sci. USA* **1991**, *88*, 11110-11114.
3. M. B. Gorin, S. B. Yancey, J. Cline, J. P. Revel, J. Horwitz, The major intrinsic protein (MIP) of the bovine lens fiber membrane: characterization and structure based on cDNA cloning, *Cell* **1984**, *39*, 49-59.
4. A. S. Verkman, A. K. Mitra, Structure and function of aquaporin water channels, *Am. J. Physiol. Renal Physiol.* **2000**, *278*, F13-F28.

5. P. Agre, Aquaporin null phenotypes: The importance of classical physiology, *Proc. Natl. Acad. Sci. USA* **1998**, *95*, 9061-9063.
6. C. Maurel, Aquaporins and water permeability of plant membranes, *Ann. Rev. Plant Physiol. Plant Mol. Biol.* **1997**, *48*, 415-429.
7. R. Kaldenhoff, M. Eckert, Features and function of plant aquaporins, *J. Photochem. Photobiol. B-Biology* **1999**, *52*, 1-6.
8. S. Hohmann, R. M. Bill, G. Kayingo, B. A. Prior, Microbial MIP channels, *Trends Microbiol.* **2001**, *8*, 33-38.
9. G. Calamita, The Escherichia coli aquaporin-Z water channel, *Molecular Microbiology* **2000**, *37*, 254-262.
10. P. Agre, M. Bonhivers, M. J. Borgnia, The aquaporins, blueprints for cellular plumbing systems, *J. Biol. Chem.* **1998**, *273*, 14659-14662.
11. M. Borgnia, S. Nielsen, A. Engel, P. Agre, Cellular and molecular biology of the aquaporin water channels, *Ann. Rev. Biochem.* **1999**, *68*, 425-458.
12. A. Weig, C. Deswarte, M. J. Chrispeels, The major intrinsic protein family of *Arabidopsis* has 23 members that form three distinct groups with functional aquaporins in each group, *Plant Physiol.* **1997**, *114*, 1347-1357.
13. P. Kjellbom, C. Larsson, I. Johansson, M. Karlsson, U. Johanson, Aquaporins and water homeostasis in plants, *Trends Plant Sci.* **1999**, *4*, 308-314.
13a. M. Eckert, A. Biela, F. Siefritz, R. Kaldenhoff, New aspects of plant aquaporin regulation and specificity, *J. Exp. Bot.* **1999**, *50*, 1541-1545.
13b. R. Kaldenhoff, A. Kölling, G. Richter. Regulation of the *Arabidopsis thaliana* aquaporin gene AthH2 (PIP1b), *J. Photochem. Photobiol. B* **1996**, *36*, 351-354.
13c. D. Ludevid, H. Höfte, E. Himelblau, M. J. Chrispeels, The expression pattern of the tonoplast intrinsic protein g-TIP in *Arabidopsis thaliana* is correlated with cell enlargement. *Plant Physiol.* **1992**, *100*, 1633-1639.
14. R. Kaldenhoff, A. Kolling, J. Meyers, U. Karmann, G. Ruppel, G. Richter, The blue light-responsive AthH2 gene of *Arabidopsis thaliana* is primarily expressed in expanding as well as in differentiating cells and encodes a putative channel protein of the plasmalemma, *Plant J.* **1995**, *7*, 87-95.
15. R. Kaldenhoff, K. Grote, J. J. Zhu, U. Zimmermann, Significance of plasmalemma aquaporins for water-transport in *Arabidopsis thaliana*, *Plant J.* **1998**, *14*, 121-128.
16. M. Kuwahara, T. Asai, K. Sato, I. Shinbo, Y. Terada, F. Marumo, S. Sasaki, Functional characterization of a water channel of the nematode *Caenorhabditis elegans*, *Biochim. Biophys. Acta* **2000**, *1517*, 107-112.
17. J. Reizer, A. Reizer, M. H. Saier, Jr., The MIP family of integral membrane channel proteins: sequence comparisons, evolutionary relationships, reconstructed pathway of evolution, and proposed functional differentiation of the two repeated halves of the proteins, *Crit Rev. Biochem. Mol. Biol.* **1993**, *28*, 235-257.
18. K. Murata, K. Mitsuoka, T. Hirai, T. Walz, P. Agre, J. B. Heymann, A. Engel, Y. Fujiyoshi, Structural determinants of water permeation through aquaporin-1, *Nature* **2000**, *407*, 599-605.
19. D. X. Fu, A. Libson, L. J. W. Miercke, C. Weitzman, P. Nollert, J. Krucinski, R. M. Stroud, Structure of a glycerol-conducting channel and the basis for its selectivity, *Science* **2000**, *290*, 481-486.
20. R. Zhang, A. S. Verkman, Water and urea permeability properties of *Xenopus* oocytes: expression of mRNA from toad urinary bladder, *Am. J. Physiol.* **1991**, *260*, C26-C34.
21. C. Delamarche, D. Thomas, J. P. Rolland, A. Froger, J. Gouranton, M. Svelto, P. Agre, G. Calamita, Visualization of AqpZ-mediated water permeability in *Escherichia coli* by cryoelectron microscopy, *J. Bacteriol.* **1999**, *181*, 4193-4197.
22. A. Chang, T. G. Hammond, T. T. Sun, M. L. Zeidel, Permeability properties of the mammalian bladder apical membrane, *Am. J. Physiol.* **1994**, *267*, C1483-C1492.
23. R. Zhang, W. Skach, H. Hasegawa, A. N. van Hoek, A. S. Verkman, Cloning, functional analysis and cell localization of a kidney proximal tubule water transporter homologous to CHIP28, *J. Cell Biol.* **1993**, *120*, 359-369.
24. G. Calamita, B. Kempf, K. E. Rudd, M. Bonhivers, S. Kneip, W. R. Bishai, E. Bremer, P. Agre, The aquaporin-Z water channel gene of *Escherichia coli*: structure, organization and phylogeny, *Biol. Cell* **1997**, *89*, 321-329.

25. M. C. Rodriguez, A. Froger, J. P. Rolland, D. Thomas, J. Aguero, C. Delamarche, J. M. Garcia-Lobo, A functional water channel protein in the pathogenic bacterium Brucella abortus [In Process Citation], *Microbiology* **2000**, *146* (Pt 12), 3251-3257.

26. H. P. Schweizer, R. Jump, C. Po, Structure and gene-polypeptide relationships of the region encoding glycerol diffusion facilitator (glpF) and glycerol kinase (glpK) of *Pseudomonas aeruginosa*, *Microbiology* **1997**, *143* (Pt 4), 1287-1297.

27. V. Truniger, W. Boos, G. Sweet, Molecular analysis of the glpFKX regions of *Escherichia coli* and *Shigella flexneri*, *J. Bacteriol.* **1992**, *174*, 6981-6991.

28. V. Koefoed-Johnson, H. H. Ussing, The contributions of diffusion and flow to the passage of D2O through living membranes, *Acta Physiol. Scand.* **1953**, *28*, 60-76

29. D. A. Haydon, Some recent developments in the study of biomolecular lipid films, in: *Molecular Basis of Membrane Transport*, T. D. C., Editor, Prentice-Hall: Englewood Cliffs, NJ. **1969**, *28*, 111-132.

30. R. Fettiplace, D. A. Haydon, Water permeability of lipid membranes, *Physiol. Rev.* **1980**, *60*, 510-550.

31. G. M. Preston, J. S. Jung, W. B. Guggino, P. Agre, The mercury-sensitive residue at cysteine 189 in the CHIP28 water channel, *J. Biol. Chem.* **1993**, *268*, 17-20.

32. A. N. van Hoek, M. L. Hom, L. H. Luthjens, M. D. de Jong, J. A. Dempster, C. H. van Os, Functional unit of 30 kDa for proximal tubule water channels as revealed by radiation inactivation, *J. Biol. Chem.* **1991**, *266*, 16633-16635.

33. T. Walz, B. L. Smith, P. Agre, A. Engel, The three-dimensional structure of human erythrocyte aquaporin CHIP, *EMBO J.* **1994**, *13*, 2985-2993.

34. A. Cheng, A. N. van Hoek, M. Yeager, A. S. Verkman, A. K. Mitra, Three-dimensional organization of a human water channel, *Nature* **1997**, *387*, 627-630.

35. M. J. Borgnia, S. Kozono, G. Calamita, P. C. Maloney, P. Agre, Functional reconstitution and characterization of AqpZ, the *E. coli* water channel protein, *J. Mol. Biol.* **1999**, *291*, 1169-1179.

36. P. Ringler, M. J. Borgnia, H. Stahlberg, P. C. Maloney, P. Agre, Structure of the water channel AqpZ from *Escherichia coli* revealed by electron crystallography, *J. Mol. Biol.* **1999**, *291*, 1181-1190.

37. S. Scheuring, P. Ringler, M. Borgnia, H. Stahlberg, D. J. ller, P. Agre, A. Engel, High resolution AFM topographs of the *Escherichia coli* water channel aquaporin Z, *EMBO J.* **1999**, *18*, 4981-4987.

38. G. Calamita, B. Kempf, M. Bonhivers, W. Bishai, E. Bremer, P. Agre, Regulation of the *Escherichia coli* water channel gene aqpZ, *Proc. Natl. Acad. Sci. USA* **1998**, *95*, 3627-3631.

39. P. C. Moe, P. Blount, C. Kung, Functional and structural conservation in the mechanosensitive channel MscL implicates elements crucial for mechanosensation, *Mol. Microbiol.* **1998**, *28*, 583-592.

40. A. C. Le Dain, N. Saint, A. Kloda, A. Ghazi, B. Martinac, Mechanosensitive ion channels of the archaeon *Haloferax volcanii*, *J. Biol. Chem.* **1998**, *273*, 12116-12119.

41. P. Blount, S. I. Sukharev, P. C. Moe, S. K. Nagle, C. Kung, Towards an understanding of the structural and functional properties of MscL, a mechanosensitive channel in bacteria, *Biol. Cell* **1996**, *87*, 1-8.

42. I. R. Booth, P. Louis, Managing hypoosmotic stress: aquaporins and mechanosensitive channels in *Escherichia coli*, *Curr. Opin. Microbiol.* **1999**, *2*, 166-169.

43. M. Bonhivers, J. M. Carbrey, S. J. Gould, P. Agre, Aquaporins in *Saccharomyces*. Genetic and functional distinctions between laboratory and wild-type strains, *J. Biol. Chem.* **1998**, *273*, 27565-27572.

44. J. S. Lolkema, B. Poolman, W. N. Konings, Bacterial solute uptake and efflux systems, *Curr. Opin. Microbiol.* **1998**, *1*, 248-253.

45. B. Poolman, E. Glaasker, Regulation of compatible solute accumulation in bacteria, *Mol. Microbiol.* **1998**, *29*, 397-407.

46. V. M. Unger, Fraternal twins: AQP1 and GlpF, *Nature Struct. Biol.* **2000**, *7*, 1082-1084.

47. K. Luyten, J. Albertyn, W. F. Skibbe, B. A. Prior, J. Ramos, J. M. Thevelein, S. Hohmann, Fps1, a yeast member of the MIP family of channel proteins, is a facilitator for glycerol uptake and efflux and is inactive under osmotic stress, *EMBO J.* **1995**, *14*, 1360-1371.

48. F. C. W. Sutherland, F. Lages, C. Lucas, K. Luyten, J. Albertyn, S. Hohmann, B. A. Prior, S. G. Kilian, Characteristics of fps1-dependent and -independent glycerol transport in

Saccharomyces cerevisiae, *J. Bacteriol.* **1997**, *179*, 7790-7795.

49. M. J. Tamas, K. Luyten, F. C. Sutherland, A. Hernandez, J. Albertyn, H. Valadi, H. Li, B. A. Prior, S. G. Kilian, J. Ramos, L. Gustafsson, J. M. Thevelein, S. Hohmann, Fps1p controls the accumulation and release of the compatible solute glycerol in yeast osmoregulation, *Mol. Microbiol.* **1999**, *31*, 1087-1104.

50. K. B. Heller, E. C. Lin, T. H. Wilson, Substrate specificity and transport properties of the glycerol facilitator of *Escherichia coli*, *J. Bacteriol.* **1980**, *144*, 274-278.

51. A. H. Hofmann, A. C. Codon, C. Ivascu, V. E. A. Russo, C. Knight, D. Cove, D. G. Schaefer, M. Chakhparonian, J. P. Zryd, A specific member of the Cab multigene family can be efficiently targeted and disrupted in the moss *Physcomitrella patens*, *Mol. Gen. Genet.* **1999**, *261*, 92-99.

52. C. Maurel, J. Reizer, J. I. Schroeder, M. J. Chrispeels, M. H. Saier, Jr., Functional characterization of the *Escherichia coli* glycerol facilitator, GlpF, in *Xenopus* oocytes, *J. Biol. Chem.* **1994**, *269*, 11869-11872.

53. L. A. Coury, M. Hiller, J. C. Mathai, E. W. Jones, M. L. Zeidel, J. L. Brodsky, Water transport across yeast vacuolar and plasma membrane-targeted secretory vesicles occurs by passive diffusion, *J. Bacteriol.* **1999**, *181*, 4437-4440.

54. E. Nevoigt, U. Stahl, Osmoregulation and glycerol metabolism in the yeast *Saccharomyces cerevisiae*, *FEMS Microbiol. Rev.* **1997**, *21*, 231-241.

11
Structures of Siderophore Receptors

Dick van der Helm and Ranjan Chakraborty

11.1
Introduction

11.1.1
Iron Transport

Iron is an essential element required by all forms of life on earth with the exception of certain lactobacilli [1]. It is essential due to the fact that iron serves as a co-factor for many vital enzymes that facilitate electron transport, oxygen transport, reduction of oxygen for synthesis of ATP, reduction of ribotide precursors for DNA, and it is also essential for the activity of heme-containing compounds like cytochromes, ferridoxins, nitrogenase, and ribonucleotide reductase [2]. Its role in a wide range of electron transfer reactions is made possible by the existence of iron in two oxidation states spanning its redox potential between -300 mV to $+700$ mV [3]. Although iron is the fourth most abundant element in the earth"s crust, it is unavailable to the microorganism due to its insolubility at physiological pH of 7.0 (solubility 10^{-18} M). This is because at physiological pH and under aerobic conditions iron exists as Fe^{3+}, which has a tendency to hydrolyze and form insoluble $Fe(OH)_3$ polymers. To overcome the low solubility of Fe^{3+} (aerobically the most abundant form), microorganisms have evolved a high-affinity iron transport system. This is comprised of low molecular weight (500–1,500 Da) iron ligands produced by bacteria and fungi that are called siderophores and membrane receptor proteins recognizing specific siderophores. Siderophores bind ferric ion selectively and are transported actively into the periplasm via the outer membrane receptor proteins. Bacterial outer membranes are known to be devoid of any energy source. Therefore, the energy required for the active transport of an iron–ligand complex is presumably obtained through the physical interactions with the inner membrane protein TonB [4–7]. TonB functions in coordination with other inner membrane proteins ExbB and ExbD during energy transduction [8]. The membrane receptor protein component of the high-affinity system is the subject of this review.

11.1.2
Siderophores

Many structurally diverse forms of siderophores have been isolated and characterized from bacteria and fungi. Generally, siderophores are classified on the basis of the nature of their iron binding moieties, which is either hydroxamic acid, catechol, or a hydroxy carboxylic acid or they can be mixed ligands. Ferric siderophores are octahedral complexes, e.g., enterobactin and ferrichrome form hexadentate complexes with iron using catechol and hydroxamic acid, respectively, as a binding moiety. The coordination chemistry of siderophores has been reviewed [9–12]. Roles of siderophores in bacterial pathogenicity, as drug delivering agents and as antibiotic itself have also been recently reviewed [13–15].

11.1.3
Siderophore Receptors

Ferric siderophores are specifically recognized by membrane receptors present in the outer membranes of gram-negative bacteria. Many such siderophore receptor proteins have been identified [16] and crystal structures of two of them have recently been determined [17–19]. Biosynthesis of these receptor proteins is negatively controlled by iron [20] and in few cases it is also positively controlled by its cognate ligand [21, 22]. They can be classified into two groups as homologous and heterologous receptors: homologous receptors, for which the siderophore is produced by the same organism carrying the receptor, e.g., FepA, a receptor for ferric enterobactin, a siderophore produced by *Escherichia coli*. In case of heterologous receptors the siderophore involved is produced by different organism, e.g., FhuA of *E. coli* is a receptor for ferrichrome, a siderophore produced by fungi. Besides transporting ferric siderophore complexes across the outer membrane, these proteins also serve as receptor for bacteriophages, antibiotics and bactericidal proteins, colicins [20].

11.2
Biochemistry and Genetic Regulation of Siderophore Receptors

11.2.1
Chemistry

More than 25 different ferric siderophore receptor proteins (FSRPs) have been identified and sequenced from a variety of gram-negative bacteria including pathogens and facultative pathogens. The physical chemistry of outer membrane proteins has recently been reviewed [16]. All of them have been proved to be TonB-dependent and hence carry out energy-dependent transport of ferric siderophores.

Generally, gram-negative bacteria are reported to produce receptors for both homologous as well as heterologous siderophores. FepA and FhuA are examples of homologous and heterologous receptors produced by E. coli. In natural competitive environment, it is not unusual for gram-negative bacteria to express more than one ferric siderophore receptor each transporting distinctively different siderophores. This makes them more competent to acquire iron. For example, in E. coli K-12, there are six distinct siderophore-mediated Fe^{3+} transport systems, each recognizing a single type of siderophore while it synthesizes only one siderophore, enterobactin. Similarly, Pseudomonas aeruginosa has FptA and FpvA for the homologous pyochelin and pyoverdin but also PfeA for ferric enterobactin. There are many more examples of organisms producing FSRPs for homologous as well as heterologous siderophores [16]. On the other hand a specific ferric siderophore may be transported by more than one type of FSRPs found in different organism. PfeA and BfeA and FetA are the examples in P. aeruginosa, Bordetella pertussis, and Neisseria gonorrhoeae, respectively, which transport ferric enterobactin. These receptor proteins, like FepA, all transport ferric enterobactin but are not identical to FepA. PfeA and BfeA do share ~ 70% similarity to FepA but FetA shows very little similarity to FepA. Although FhuA and FcuA transport ferrichrome in E. coli and Yersinia enterocolitica, respectively, they share only 27% sequence homology. The same is true for the ferrioxamine B transporter FoxA in Y. enterocolitica and FhuE in E. coli. Siderophore receptors are of similar size ranging from 651 (FyaA) to 803 (AleB) amino acids in length and have similar molecular weights. Most of them have phenylalanine as C-terminal amino acid with either glutamine or alanine as the N-terminal. Some of them carry extended N-terminal sequences (e. g., FecA, BfeA, PupA, FpvA, PupX, etc.), a conclusion based on the location of the TonB box, a conserved region found in most, if not all, the FSRPs.

11.2.2
Genetic Regulation

The negative control of genes involved in iron transport as well as biosynthesis of siderophores is afforded by the Fur protein interacting with the consensus Fur box sequence adjacent to the promoter region [23, 24]. The Fur protein is only active when it is bound to Fe^{+2}. Recently it has been shown that the functional binding site is three repeats of the GATAAT motif rather than a palindromic sequence [25].

In some cases the transcription of the genes involved in iron metabolism is positively controlled by the ligand. Genetically the most studied of such systems are the ferric-dicitrate transport in E. coli [26] and ferric-pseudobactin BN7 and BN8 transport in Pseudomonas putida WCS 358 [27]. FecA in E. coli is induced by citrate and this involves two proteins FecI and R [28] while PupB is induced by either pseudobactin BN7 or BN8 with the help of regulatory proteins PupI and R [29]. Although, FecI and R share 43% and 37% sequence identity with PupI and PupR, respectively, they possibly differ in the mechanism by which they regulate induction of their respective operons [29].

11.3
Structures of FepA and FhuA

11.3.1
General

Within a period of one month two independent structures of FhuA [18, 19] and the structure of FepA [17] were published. For the FhuA protein it was possible to determine, in both cases, the structures of both the protein itself and the structures of the protein bound to its cognate ligand. The structures (Figs. 1 and 2) show two features which were not expected or predicted. Both proteins show a β-barrel formed from 22 β-strands, rather than the approximately 30 strands which had been proposed in model studies. In addition the structures show, unexpectedly, a N-terminal domain which is tugged within the β-barrel. This domain, sometimes called a plug or cork, closes the channel formed by the barrel. Even, once the ligand is bound, this domain prevents access to the periplasm. The similarity of the two structures allows the prediction that other proteins in this family have the same type of structure. Different aspects and comparisons of the structures will be discussed.

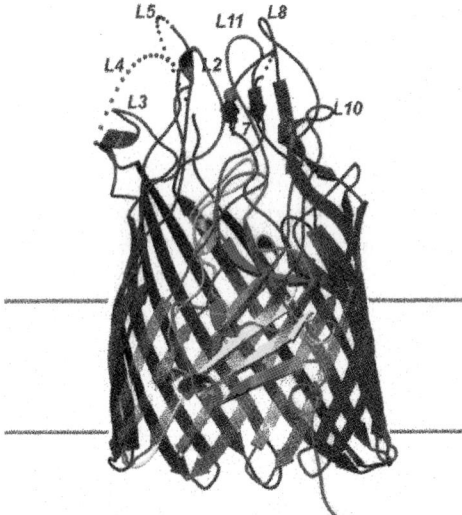

Fig. 1. Ribbon diagram of FepA. The extracellular space is located at the top of the figure, and the periplasmic space is at the bottom. The position of the membrane bilayer is delineated by horizontal lines, as determined from the hydrophobic area found on the molecular surface. Part of the barrel has been rendered transparent to reveal the N-terminal domain located in the channel. The long extracellular loops are labeled; residues which could not be located in the loops are indicated by dotted lines. Reproduced with permission from S. Buchanan et al., *Nature Struct. Biol.*, **1999**, *6*, 56-63; see color plates p. XXXIX.

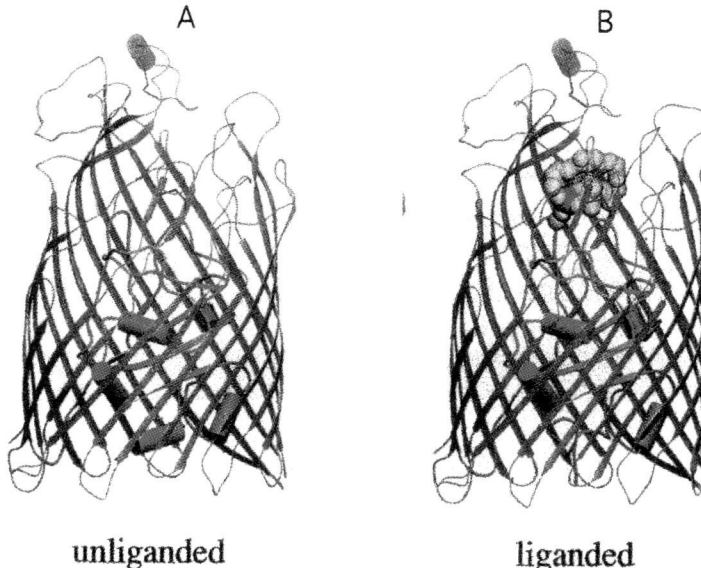

Fig. 2. Structure of the free (**A**) and liganded (**B**) FhuA protein. In both conformations, the N-terminal domain is indicated by brown color located within the green color barrel formed by 22 antiparallel β-strands. In **B** the bound ferrichrome is depicted in yellow, with the complexed iron atom in red. In the liganded form, the a1 helix, at the bottom of the N-terminal domain, is unwound. Reproduced with permission from K. P. Locher et al., *Cell*, **1998**, *95*, 771-778; see color plates p. XXXIX.

11.3.2
The β-Barrel and Periplasmic Loops

The topology of the FepA structure is shown in Fig. 3. The β-barrel is formed by antiparallel β-strands. The first strand is repeated to the left of strand 22. All strands are connected to adjacent strands by H-bonds of peptide groups as is common for antiparallel strands. Also strands 1 and 22 are connected in this manner, however, two additional H-bonds are formed for these strands between the side chains of Glu 165 in strand 1 and Trp 716 and Arg 714 in strand 22. The strand numbers are shown as are the external loop numbers. Capital letters indicate residues located in the model of FepA, whereas small letters represent residues that could not be traced from the electron density. Some of the β-strands continue above the membrane surface. Neighboring residues in adjacent strands share a side of the squares in the schematic (Fig. 3). The lengths of the strands located within the membrane consist of 10–11 residues. The shear number of the barrel is 24 which means that the strands are inclined with respect to the perpendicular to the membrane by an angle slightly more than 45° (about 47°; the angle between the strands and the membrane surface is thus 43°) (Figs. 1 and 2). The shear

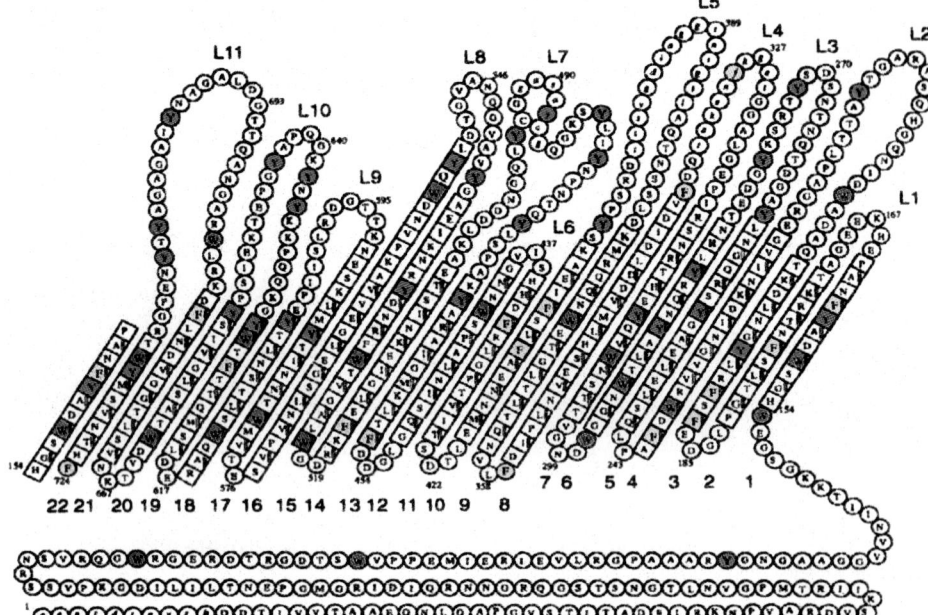

Fig. 3. Topology of the FepA barrel, using amino acid sequence in one-letter code. The view is from the outside of the barrel, which is unrolled. Capital letters represent residues present in the model (Fig. 1), whereas small letters represent residues that have not been traced. Squares indicate residues in β-strand conformation. Side chains of residues marked with a double line point to the outside of the barrel. The topology of the N-terminal domain is not shown in this figure, although the residues are illustrated. Aromatic residues are colored magenta, blue and gold, for Tyr, Trp and Phe, respectively. Reproduced with permission from S. Buchanan et al., *Nature Struct. Biol.*, **1999**, *6*, 56-63; see color plates p. XL.

number for the FhuA barrel is also 24[1]. Assuming that the distance between the Cα's of two residues n and $n + 2$ in a strand is 6.9 Å, this would give a membrane thickness of 24-26 Å. The strands have a right-handed twist forming the curvature of the barrel. The twists vary to some extent so that the barrel is not circular but slightly ellipsoidal. The distances between Cα's on opposite strands at the bottom of the barrel vary from 41–45 Å and at the top of the barrel between 38–47 Å. The barrel is thus more ellipsoidal at the top.

The barrel structures of FepA and FhuA are quite similar (a comparison of the two independent FhuA structures will be presented in Sect. 11.3.4). The barrel in FhuA is equally ellipsoidal at the bottom and the top of the barrel with axes of 37 and 48 Å (distances between Cα's on opposite strands). Fitting the Cα's of the barrels of FepA and FhuA by a L. S. procedure gives a RMS of 1.57 Å. There are thus

---

[1] The shear number, however, cannot be deduced from Fig. 2C in reference [19]. In this figure residues in adjacent strands in some cases share a corner vertically but in other cases horizontally.

significant differences in the structure of the barrels, due to differences in curvature of the barrels and differences in twist of equivalent strands.

The periplasmic loops are short (periplasmic loop P1 connects strand 2 and 3, etc.). The periplasmic loops in FepA are between 3 and 5 residues long while in FhuA they are between 1 and 7 residues in length (loops P9 and P10 are 6 and 7 residues, respectively). Although one might discard the significance of the periplasmic loops this may not be prudent (see Sect. 11.4).

The barrel structure is amphiphilic with a period of 2, that is for sequential residues in a strand the side chain points into the interior of the barrel while the next one points outward into the membrane. In Fig. 3 these latter residues are indicated by squares with a double vertical line. In FepA half of the side chains pointing into the membrane are from A, L, V, I, M, P, and G. However, polar and charged residues also occur in this group, although the latter are only present close to the membrane boundary. Lastly in this group one observes a large number (28) of aromatic residues, mostly, but not always, close to the membrane boundaries. This is a common observation for outer membrane proteins. The side chains pointing into the barrel are mostly polar or charged but there are a few hydrophobic residues and also four aromatic residues in this group. The distribution of residues in the FhuA barrel is similar.

11.3.3
The N-terminal Domain

The N-terminal domain closes the access of ligand from the extracellular space to the periplasm (Figs. 1 and 2). In FepA the N-terminal domain consists of residues 1-154 and in FhuA of residues 1–160. In FepA it was not possible to locate the first 10 residues [17], while in the two FhuA structures the first 18 [18] or 20 [19] could not be located. However, it is clear in both structures that the N-terminus extends into the periplasm, probably in extended form without secondary structure, causing the difficulty in locating these residues in the electron density maps. The domain does not fill the bottom of the barrel leaving a relatively empty space on the periplasmic side (Fig. 5 in reference [17]). The top of the domain extends far above the membrane surface (about 25–30 Å) on the extracellular side. However, the extracellular loops extend even farther (40–45 Å) leaving another space between the top of the extracellular loops and the top of the N-terminal domain. This space is readily accessible from the extracellular space (see Sect. 11.3.4).

If one compares the structures of the plug for FepA and FhuA it is not easy to superimpose the structures (Fig. 4), however, the topology for both is basically the same. The topology for FepA is shown in Fig. 5. The core of the domain is a four-stranded mixed β-sheet. If one starts from the N-terminus, the first recognizable secondary structure is a helix (α1) in both proteins (18-23 in FepA, 24-29 in FhuA). This is followed by a β-strand (β1) which is part of the four-stranded mixed β-sheet (27-33 in FepA and 49-55 in FhuA) visible in Fig. 4 on the right-hand side of the sheet. The number of residues between α1 and β1 is considerably larger in FhuA than in FepA, just 4 in FepA but 20 residues in FhuA. After β1 one finds two small helices followed by a β-strand (β2) leading to apex A in FhuA and NL1 in FepA (blue loop at right top of Fig. 4 for both proteins) consisting of residues 62-66 in FepA and 80-84 in FhuA. This is followed by a β-strand (β3). After this, one observes a significant structural difference between the two proteins. The subsequent loop rises again in FhuA forming another apex (apex B) (residue 81 in FhuA) while in FepA this loop stays well below apex A (residue 92 in FepA). After this, both proteins form the second strand of the mixed β-sheet (β4, fourth from the right, Fig. 4). Subsequently a third apex is formed (apex C, NL2), residue 102 in FepA and 116 in FhuA (the loops at left top in Fig. 4). The loop bends all the way downward to the bottom of strand β5 of the mixed β-sheet (residues 118-127 in FepA and 125-134 in FhuA) and after this, strand β6 is formed, completing the four-strand sheet (residues 139-147 in FepA and 146-154 in FhuA). After strand β6 the domain connects to the β-barrel at residue 154 in FepA and 160 in FhuA. The description shows that the topology is closely the same for the N-terminal domains of FepA and FhuA.

It is obvious from Fig. 4 that there are structural similarities and differences for the N-terminal domains of FepA and FhuA. However, the orientation and location with respect to the barrel is the same for both structures. For instance, the last residue in the helix which follows the first strand (β1) is close to the middle of strands 20 and 21 in both structures. Similarly, after a L. S. fit of the barrels of FepA and FhuA, the distances between the first and last residue of the first strand

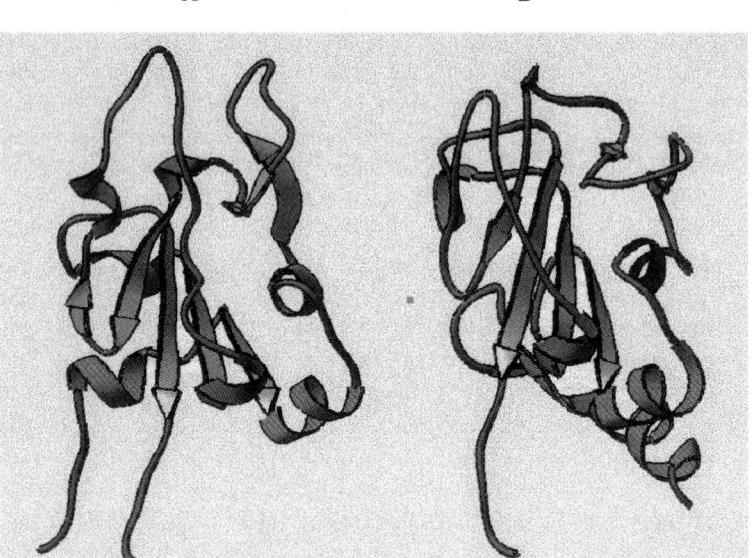

Fig. 4. N-terminal domain of FepA (**A**) and FhuA (**B**). Short β-strands are indicated in brown color while the short α-helices are in green and blue color. Sensor loops are light blue in color at the top; see color plates p. XLI.

Fig. 5. Topology diagram of the N-terminal domain in FepA. Reproduced with permission from S. Buchanan et al., *Nat. Struct. Biol.*, **1999**, 6, 56-63.

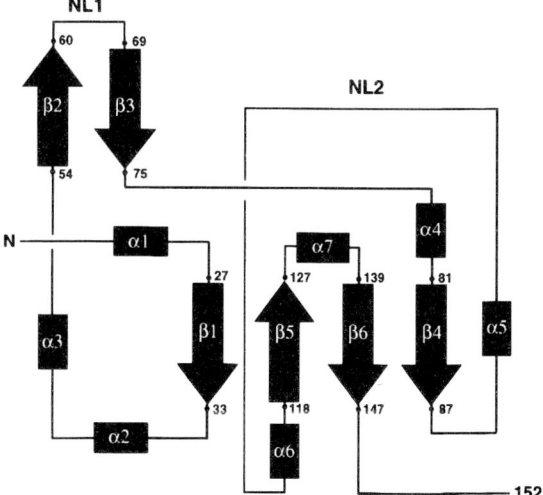

are only 1.4 and 1.1 Å, respectively. These are the distances thus between the Cα's of Gly 27 (FepA) and Ser 49 (FhuA) and the one between Ala 33 (FepA) and Ala 55 (FhuA). This is also true for β6, the third strand in the sheet where the distances after the L. S. fit of the barrels are 1.5 Å and 3.3 Å for the first and last residue. There is thus a remarkable structural similarity for the 4-stranded mixed β-sheets in FepA and FhuA both in the location and orientation of the sheet with respect to the barrel. There is only a general similarity in the structure of the loops, and there is a very distinct difference in the location of the first helix (α1). This can be seen in Fig. 4. In FepA this helix is at the bottom left and in FhuA at the bottom right. The distance between the first residues in this helix, that is Ala 18 in FepA and Pro 24 in FhuA, again after a L. S. fit of the barrels, is 25 Å. This will be discussed further in Sect. 11.4, in which the FhuA-ferrichrome structures are described.

The N-terminal domain is firmly anchored to the β-barrel. In FepA there are 46 interactions involving direct H-bonds or H-bonds mediated by a single water molecule between the N-terminal domain and the strands of the β-barrel. This involves 18 of the 22 strands. In addition there are 17 interactions between the domain and the extracellular loops. Upon association of the two domains there is a loss of 2,800 Å^2 in accessible surface area. The putative TonB box is of course a part of the N-terminal domain. In FepA it consists of residues 12-18 which could be located. In both structure determinations of FhuA, the TonB box could not be located (residues 7–13) but it is obvious that they are at the periplasmic side of the domain. In FepA these residues show an extended structure. Thr 362 in strand 9 is H-bonded to Val 15 and residues 11, 12, and 13 are below the bottom of the barrel. It is of course important that the TonB boxes in both proteins are easily accessible from the periplasm, in order to allow an interaction with TonB.

11.3.4
The Extracellular Loops

There are thus two independent structures for FhuA. In one case [19] the protein was purified by ion exchange and chromatofocussing, removing all LPS. In the other case an 11-residue His-tag was inserted between residue Pro 405 and Val 406, in L5 [18][1], and the protein was purified twice over a Ni^{2+}-charged column without chromatofocussing. The structure of the His-tagged protein shows an LPS molecule associated with the protein [18]. The two FhuA proteins crystallized in different space groups. The differences in the two structures are, however, minimal. A L. S. fit has been calculated, after excising the His-tag and three residues preceding and succeeding the insert. An RMSD of 0.50 Å is found and the maximum difference for the location of all $C\alpha$'s in the two proteins is only 2 Å. In various parts of the proteins, sequential residues differ by more than 1 Å between the two structures. These are 333-338 at the top of L4, residues before and after the His-tag in L5, also 464-466 in L6 and 553-555 in L8. On the periplasmic side there are differences between 1–2 Å in residues 185-187 in P1, 624-626 in P9, and 667-670 in P10. All together these differences are minor. Some are possibly caused by differences in intermolecular contacts while other differences are due to the insertion of the His-tag. In a subsequent publication [30] the complete structure of the LPS was determined using an improved electron density map and mass spectrometric analysis. In addition an LPS binding motif was identified.

We consider the extracellular loops to begin and end just above the membrane surface. In FepA it was not possible to locate 33 residues distributed in loops 4, 5, and 7. Even a cursory comparison shows that there is very little similarity in the loop topology or loop structure of FepA and FhuA. In FhuA the extension of several adjacent barrel β-strands continue this secondary structure far above the membrane surface, involving 10-14 residues. This is the case for strands 6-12. Even the intra-strand H-bonds are maintained and a sheet is formed by these six strands above the membrane surface resembling a wall. This structural feature is not present in FepA although in this protein some barrel β-strands also continue above the membrane surface, however, no significant sheet structure is formed in this case. Also the length of the 11 loops is totally different in FepA and FhuA structures (Tab. 1).

The overall appearance of the loop structure is conical, sloping towards the center of the molecule, still allowing, however, at the top, a wide opening for the entry of ferric siderophore ligands. The loops extend far above the surface of the membrane, about 35–40 Å, much farther than is observed for the porins. The extracellular loops are distinctly hydrophilic. Another common feature of the loop structures is the large number of Tyr and Phe residues. Many of these line the putative entrance for the ferric siderophore ligand into the apex of the

---

1) The numbering of the residues in the two FhuA proteins differ after residue 405, and one should subtract 11 from the numbering of the His-tagged protein for residues 406 and higher numbers.

Tab. 1. Extracellular loop lengths in FhuA and FepA

Loop	Number of Residues FhuA	FepA	Loop	Number of Residues FhuA	FepA
1	4	7	6	20	7
2	12	29	7	20	39
3	42	25	8	26	32
4	50	31	9	23	18
5	49	36	10	19	27
			11	23	37

loop structure as is shown in Fig. 6 for FhuA. This apex of the extracellular loop structure may actually be the initial binding site for the ferric siderophore (see below).

Due to the differences in structure of siderophores one would expect to see little homology in the extracellular loop region, in contrast to what may be expected for the barrel structure and N-terminal domain, when one compares proteins in this family.

The comparison of the FepA and FhuA structures described in the previous sections indicates that other proteins in this family can be expected to show the same type of structure. It is, however, important to attempt to locate the N-terminus, especially in those proteins which have an extension at the N-terminal for the purpose of signaling as established for FecA and PupB [28, 29].

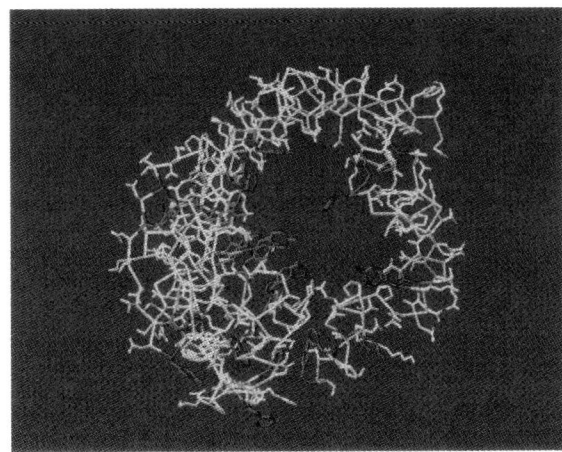

Fig. 6. A view into the opening formed by the extracellular loops (yellow) in FhuA. This apex is lined by aromatic residues (red); see color plates p. XLI.

11.4
The FhuA Structures with Ligand

The important difference between the structure determinations of FhuA and FepA is that in both FhuA structures it was possible to determine the structures of the proteins with bound ligand, ferrichrome. It has so far been impossible to obtain this result with FepA despite many efforts. In this case, besides ferric enterobactin, other metal chelates, such as Ga^{3+} and In^{3+} and synthetic analogs, e.g., Mecam and Trencam have been tried, but all without success. More likely than not the crystallization of FepA in a different space group will be required.

The ligand ferrichrome is a hexapeptide formed by three glycines and three ornithines. The ornithines are hydroxylated and acetylated to form three bidentate hydroxamate groups, which bind the ferric ion. The crystal structure of ferrichrome was determined [31] and was used to model the ligand in the protein ligand structures. The FhuA ligand structures were determined at 2.7 Å and they show closely similar results. The ligand is lying sideways resting on residues from apices A, B, and C of the N-terminal domain which were described earlier. Arg 81 (apex A) (Fig. 7) forms an H-bond with two hydroxamate groups of the ligand. There is an electrostatic interaction of a carbonyl group in the cyclic peptide of the ligand with Gly 99 (apex B), and Tyr 116 (apex C) forms an H-bond with a

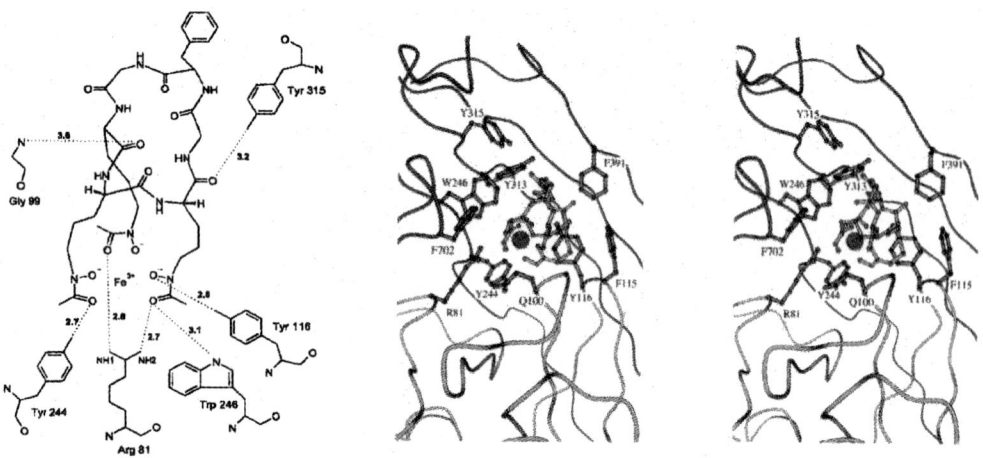

Fig. 7. A Schematic representation of the hydrogen bonding pattern and electrostatic interaction of phenylferricrocin with FhuA side chain residues in the ligand binding site. Hydrogen bonds and charge interactions are indicated as dotted lines (distances given are in Å). **B** Stereoview of the ferrichrome-iron binding site in ribbon representation. The ferrichrome iron molecule is depicted as a green ball-and-stick model with red sphere as iron. Small red spheres are oxygen, small blue spheres are nitrogen atoms, and black spheres are carbon atoms of selected side chain residues of FhuA and small gray spheres are carbon atoms of the ferrichrome-iron molecule. The cork domain is shown yellow while barrel strands and loops are shown blue. Reproduced with permission from A. D. Ferguson et al., *Protein Sci.*, **2000**, *9*, 956-963 (**A**), and A. D. Ferguson et al., *Science*, **1998**, *282*, 2215-2220 (**B**); see color plates p. XLII.

hydroxamate group of the ligand. In addition there are H-bonds of Tyr 244 and Trp 246, which are in loop 3, with two hydroxamate groups and of Tyr 315 (L4) with a carbonyl group of the cyclic peptide. The ligand tail is partially surrounded by aromatic side chains from L4, L5, and L11 involving van der Waals interactions (Fig. 7) but the other part of the tail remains solvent accessible. The ligand binding is the same in both FhuA-ligand structures.

What happens to the protein on the binding of ligand? One could have expected a possible closing of the loops around the ligand, closing off access from extracellular space. This is not the case at all and there is very little change in the loop structure between FhuA with and without ligand. However, many other changes do occur in other parts of the FhuA structure when the ligand is bound. We have compared the changes in both FhuA structures which occur on ligand binding and they agree with one another. Changes occur primarily in the N-terminal domain, some in the periplasmic loops and very few in the barrel and extracellular loops. The most visible change is the unwinding of the first helix (α1) in the N-terminal domain. This starts with a (φ Φ) change in residue 30 which causes the Cα of residue 29 to move 6 Å. The unwinding continues and the Cα of residue 20 is removed 20 Å from its original position in the non-liganded FhuA. This large structural change at the N-terminus of the protein can be related to the results of cross-linking and complex formation between FhuA and TonB, which increased on addition of ligand, ferricrocin [4]. This indicates that the structural changes observed in the N-terminal segment of FhuA in its complex with ligand facilitates the interaction with TonB. In unliganded FhuA the α1 helix rests against periplasmic loops P7 and P8. It has been suggested that the upwards motion of part of the N-terminal domain (see below) by 1–2 Å interferes with the hydrophobic interactions between the helix and periplasmic loops P7 and P8 and that this is the cause for the unwinding. This may be true, but it is interesting to note that as a result of the unwinding also periplasmic loops P7 and P8 move up by about 2 Å. This is, therefore, an indication that the hydrophobic interactions between the loops and the helix could be maintained in the liganded form of the protein and that other factors play a role in the unwinding of the α1 helix.

One may expect the binding of the ligand to signal the initiation of one or more interactions of TonB with the protein. The TonB box being one of those, but it is interesting to speculate that other interactions of TonB are initiated as well with, e.g., the periplasmic loops. It is doubtful, however, that the movement of P7 and P8 is involved. The reason for this statement is the fact that the α1 helix in FepA is located at a different place with respect to the barrel (as described earlier), and a similar change in the periplasmic loops P7 and P8 on ligand binding is, therefore, unlikely in FepA.

Another interesting structural change occurs when the ligand is bound to FhuA. It is the result of the induced fit in the binding of the ligand. Thr 80 in apex A is moved up by 2.5 Å and Gly 99 in apex B by 1.5 Å. However, apex C is left unchanged. As a result of these changes a part of the N-terminal domain is moved up by 1–2 Å (Fig. 11.8). It is clearly visible in the figure that this involves only a part of the domain. The significant movements (> 1 Å) occur for residue

53–73, 78–80, 98–100, and 157–158. Obviously this motion of the N-terminal domain relative to the β-barrel can be expected to disturb the H-bonding between the two units. An inspection shows this to be the case, however, the total number of direct H-bonds between residues 30-160 and the barrel and loops remains essentially unchanged. There are 49 such H-bonds in FhuA itself and 51 in the liganded form of FhuA.

The largest region of the N-terminal domain which is moved by more than 1 Å is the section between residues and 54 to 73. This region has only two H-bonds with the barrel in the unliganded form, but once the ligand is bound five H-bonds exist between this region and the barrel, in other words an apparent stabilization occurs due to the upward motion effected by ligand binding. From Fig. 8 it is also clear that the motion in the N-terminal domain is not a concerted one, in other words not like a piston in a cylinder. Neither does it seem to be a hinge motion, because in that case some of the changes would be downward. Instead, it is suggested that the positional changes are a result of two sequential effects, the ligand binding and the formation of new H-bonds. Many H-bonds between the N-terminal domain and the barrel and loops remain intact: 34 of the 49.

Periplasmic loops P9 and P10 move by 1–2 Å on binding of the ligand. Although no interactions are observed in the structures between these loops and those residues which could be located in the N-terminal domain, it is possible that loops P9 and P10 interact with some of the first 20 residues of the protein which could not be located.

Later on we will discuss the conserved residues Arg 93 and Arg 133 in the N-terminal domain and residues Glu 522 and Glu 571 in strand 14 and 16 of the bar-

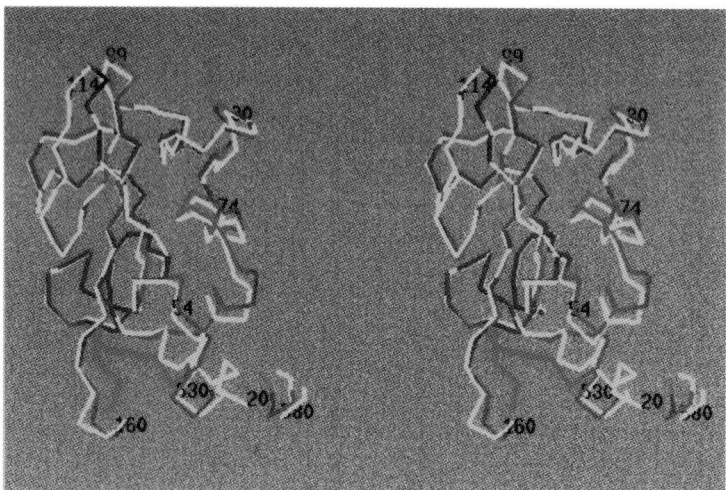

Fig. 8. Structural changes in the N-terminal region of FhuA after ligand binding. Liganded FhuA is in blue (very little change) to light blue, green, and red indicating increasing translational or structural differences with unliganded FhuA (yellow). The $\alpha1$ helix rests against the periplasmic loops P7 (530) and P8 (580). On ligand binding this helix unwinds (red); see color plates p. XLII.

rel. We believe these residues to be important for the transport process as delineated from the binding process. The four residues in the unliganded form [19] are located such that they form a quadrupole. Residue 93 has one H-bond with both 522 and 571, and 133 has a strong H-bond (2.6 Å) with 571. Although each of the four residues, on ligand binding, change position by only 0.2–0.3 Å, this H-bond pattern in one of the FhuA structures [19] changes, such that the H-bond between 133 and 571 disappears (distance: 3.3 Å) and four H-bonds are formed between residue 93 and 522 and 571. However, in the other FhuA structure [18] the H-bond pattern is retained on ligand binding. It is important to find out which of these two observations is correct, but this may require a higher resolution structure.

Recently [32] the structures of the albomycin–FhuA and phenylferricrocin–FhuA complexes were determined. Albomycin is a naturally produced antibiotic and uses the FhuA system for transport. It is a trihydroxy-N-acetyl-L-ornithine tripeptide, as exists in the natural ligand ferrichrome, which is covalently linked to a thioribosyl pyrimidine antibiotic group. The iron hydroxamate is bound closely similar to that found for the ferrichrome complex. The covalently linked antibiotic portion is located in the cavity below the apex of the extracellular loops. Also in this complex one observes an upward translation of apex A and B as described above for the ferrichrome complex with FhuA. FhuA is less specific in ligand recognition than FepA. Drug conjugates with dicatecholates, the chelating groups in enterobactin, are transported by Fiu rather than FepA. FhuA transports even antibiotics which are not conjugated to ferric hydroxamate. Examples are rifamycin [33] and microcin 25 [34].

11.5
Is the FepA Structure the Liganded or Unliganded Form of the Protein?

The crystals of FepA which were used for the structure determination were soaked for a short time with ferric enterobactin. Longer soaking invariably cracked the crystal. In the structure determination of FepA using anomalous dispersion a density for the ferric atom was observed. This was estimated to be 15–25 % of full occupancy. It was not possible to locate the organic ligand, enterobactin, in the electron density maps. The density of the iron atom was, however, not close to the sensor loops (NL1 and NL2), which correspond to apex A and C in FhuA. Instead, it was located close to the apex of the extracellular loops, the putative entrance of ligand from extracellular space to the cavity formed by the sensor loops and the apex of the loops. This entrance, both in FepA and FhuA, is lined by many Tyr residues and some Phe residues. This location may be the initial binding site for the ligand, rather than the ultimate binding site. It is interesting to note that the classical way to extract ferric siderophores involved $CHCl_3$–phenol or benzylalcohol. It is thus interesting to note that the protein with the use of the side chains of Tyr and Phe residues appears to follow a similar procedure, attracting and extracting the ligands from the extracellular space. If, therefore, the apex of the extracellular loop is the initial binding site and the apices A, B and C the ultimate binding site, this would agree with the biphasic binding kinetics which has been observed for ferric enterobactin [35]. The fact that the partial ferric ion signal in FepA is not observed close to the apices of the N-terminal domain and the fact that the $\alpha 1$ helix in this domain is not unwound in the FepA structure, leads to the conclusion that the FepA structure shows the unliganded form.

11.6
Biochemical and Genetic Experiments

With several structures known one would hope that many genetic and biochemical observations can now be explained. That is, however, quite often not the case. In general, from a structural point of view, large and even small deletions and insertions of the protein might alter the structure in an unknown manner and compromise the observations, disallowing a rationale for the results. Of course it is not practical to determine structures for all the mutagenized proteins.

FhuA has the advantage over FepA in the fact that both the structure of the protein and the complex with the ligand are determined. In FhuA it was observed that Lys 67 was protected from trypsinolysis in the complex compared to the free protein [36]. This indicates a conformational change which indeed occurs after binding (see Sect. 11.4). However, Lys 67 in both the protein and its complex has 0% accessibility (Swiss Protein PDB Viewer [37]). Ligand induced conformational changes for residues 1-20 and 21-59 were also observed. The section 21-59 contains the $\alpha 1$ helix which is unwound when the ligand is bound. Another observation with respect to ligand binding was that the addition of ferricrocin enhanced the interaction between FhuA and TonB [4]. The observed unwinding of the $\alpha 1$ helix in FhuA may well facilitate the interaction of TonB with the TonB box of the protein. This last observation gives hope that it might be possible to cocrystallize the proteins with segments of TonB. A natural mutant missing Asp 348 changed the character of the protein drastically in that ferrichrome was not taken up [38]. Asp 348 is a part of the β-sheet wall extending above the membrane surface (see Sect. 11.3.4). The residue is part of the binding pocket but not involved in the binding of the ligand. It is within the van der Waals distance of Tyr 116, involved in the binding of ferrichrome. A likely explanation is that deleting a residue in the β-sheet wall will cause a partial collapse of the sheet and, therefore, interfere with the binding of ligand. In this case it would be interesting to explore if any other deletions of a single amino acid in this sheet would give the same effect. A mutant V 347 C was constructed in order to attach a fluorophore [39]. Because ferrichrome inhibited the labeling of the residue, it was thought to be part of the binding pocket. However, Val 347 being adjacent to Asp 348, part of the β-sheet wall, points away from the binding pocket and the observation can, therefore, not be explained by the structure. On the other hand deletion of 236–248 resulted in loss of ferrichrome and albomycin transport [40]. The structures show that residues in this section form H-bonds with the ligand and the deletion will, therefore, directly affect the binding pocket. A large deletion $\Delta 322$–355 [41] and even $\Delta 335$–355 [42] causes a drastic change so that the protein behaves as a diffusion channel. One can speculate that the extracellular β-sheet collapses causing a disruption of many H-bonds with the N-terminal domain, and removing the domain from within the barrel. However, many other structural changes may occur. This is one of many cases where the structures cannot give an answer. A structure of the $\Delta 322-355$ protein would be necessary to give the proper answer. With insertion mutations [43] residues 321, 405 and 417 were correctly identified as surface accessible (40%). With MABs, determinants located between residues 1-20 and 21-59 were correctly identified as exposed to the periplasm [44]. On the other hand, the determinant located between residues 321-381, identified as surface exposed, involves two membrane β-strands as well as the larger part of L4. The mapping of determinants often involves sections of the protein which are too large to be able to make reliable conclusions, as in this case. Insertion derivatives in FhuA identified most (9 of 11) residues correctly as exposed to surface or periplasm [45]. Other experiments showed that small insertions in the N-terminal domain abolished receptor functions [46] while some deletions maintained growth promotion [47]. One cannot rationalize these results without knowing the mechanism although none involved residue Arg 93 which we propose to be essential for transport.

A similar discussion for FepA is more tenuous because only the structure of FepA is known and not its complex with ligand. The first two studies on the topology using MABs identified rather large epitopes [48, 49]. Several epitopes were correctly identified to be on the extracellular surface, however, one, as the structure showed, is accessible from the periplasm rather than the surface and another involves two β-strands embedded in the membrane. Two of these epitopes were combined and proposed to form a large loop and to act as gate for the channel formed by the remainder of the protein. Excision of this loop [49] which involves part of L2 and L4, and all of $\beta 4$, $\beta 5$, $\beta 6$ and $\beta 7$ and L3 gave a protein which showed passive, non-specific transport similar to the result of FhuA ($\Delta 335-355$). However, in FepA, the deletion is much larger and there is no way to predict the effect on the structure of such a large excision but for the fact that it will be a major change. The gate of course does not exist but instead the membrane channel is blocked by a different domain of the protein, the N-terminal domain, from the periplasmic side. It is a clear example that conclusions drawn from experiments using large epitopes, or large deletions and insertions should be made with greater care. The mechanism for transport is controlled by the N-terminal domain and not by extracellular loops. A study of limited scope predicted all of the $\beta 5$ strand correctly, both with respect to the location of residues in relation to the upper and lower membrane surface as well as their exposure to the channel or to the hydrophobic membrane phase [50]. This was an EPR study in which each sequential residue was replaced by Cys and subsequently linked to a spin label. A topological study using small epitope (2 residues) insertions [51] had several good results when compared to the structure. Two residues 55 and 142 were correctly placed on the periplasmic side (although their accessibility is calculated to be 0%) and two were properly placed on the cell surface. However, residue 359 was located in the membrane but in fact is on a periplasmic loop. Point mutants E280C and E310C were used to attach spin labels. An EPR study showed a structural change for these residues on ligand binding [52]. The side chains of the two residues are 20–25 Å from the sensor loops, the putative binding site of the ligand. They are also below the binding site. The observed effect must, therefore, be secondary. Another EPR study observed a burst of motion for the spin-labeled E280C mutant during transport [53]. Once the transport mechanism has been established this observation may be clarified. To draw conclusions from this single observation seems not warranted.

This overview indicates the importance of the biochemical and genetic studies, but it also shows the problems especially in the interpretation of the results. After the structures of FhuA and FepA were determined and, therefore, the existence of the N-terminal domain was known, FhuA Δ5–160 was constructed [54]. It excises the N-terminal domain as well as the TonB box. Cells that synthesized the deletion mutant displayed higher sensitivity to antibiotics. Non-specific diffusion of ferrichrome was observed in $tonB^-$ cells, but, most importantly, growth of a FhuA Δ5–160 $tonB^+$ strain occurred at low ferrichrome concentrations at about 45% of FhuA wild-type rate. The deletion mutant thus shows TonB-dependent transport despite missing the N-terminal domain and the TonB box. It is an unexpected result. It is possible (see Sect. 11.7) that other residues besides those on the TonB box interact with TonB. Secondly, as in all large deletions, one wonders which structural changes occur. However, the existence of the N-terminal domain, its participation in binding, and the apparent signaling which occurs after binding make it essential to investigate the binding and transport process for the complete protein. The results of these studies may give a clarification of the unexpected results for FhuA (Δ5–160).

11.7
Binding and Mechanism

It should be clearly stated that the structures are not an end in itself, instead, they are only a step in understanding the process of transport in which these proteins are involved. This has been the case with the evaluation of the catalytic mechanism of enzymes, where structures allowed to propose hypotheses which could be experimentally checked. In this case it is more difficult as well because the process which is being investigated is one for which there are no precedents in the literature while in addition it involves several steps and components. Neither the sequence of steps is well established nor are the interactions of the different components. Progress can be made by proposing hypotheses which can be tested by a combination of biochemical, genetic and structural methods. This is easier said than done. In this discussion it is important to realize the distinction between binding and transport.

The authors of one of the articles on the FhuA structure [18] suggested that, after binding, transport could occur by opening a channel below the binding site. The proposal involved residues on L5 (extension of $\beta 10$) and, successively lower, on $\beta 9$, $\beta 8$, and $\beta 7$. This is an attractive hypothesis but it has not been checked.

An earlier study into FepA using quantitative binding and transport measurements indicated that Arg 286 and 316, in combination, were important for both processes [55]. The experiments used R→A mutants for the residues. Residue 286 is on $\beta 6$ close to the membrane surface but not near the sensor loops. Residue 316, on L4, is 10–15 Å above the membrane surface and close to the putative binding site (residue 66). It was shown in Sect. 11.3.3 that the orientation of the N-terminal domain with respect to the barrel is the same for FepA and FhuA. It can also be expected that the transport mechanism is the same for both proteins. Therefore, the proposed channel in FhuA correlates quite well with the experiments on the 286, 316 mutants in FepA. Another effect of these two residues may involve the binding of ligand. Both interact with the beginning and end of sensor loop NL2, forming H-bonds with respectively Ser 93 and Asp 106 (although the former requires a small conformational change in Arg 286). Asp 106, in turn, affects sensor loop NL1 by an H-bond with Arg 66 close to the apex of the loop. However, in stark contrast is the observation that deletions in loops 3, 4, and 5 in FepA, had only slightly impaired uptake [56], compared to strongly impaired uptake due to deletions in L2, 10, and 11.

There is good kinetic evidence for two binding sites [35], one at the apex of the external loops, a site surrounded by aromatic residues in both FhuA and FepA, and the second at the apex of the N-terminal domain where the ligand has been located in both FhuA structures. Also this second site involves a number of aromatic residues. Some confirmation of the first binding site is obtained from the partial occupancy of the ligand in the FepA structure at this location. A number of aromatic residues in both sites were single- and double-substituted by Ala in FepA and the binding and transport of the mutants were quantitated [57]. It identified residues at both sites which, when mutated, are deficient in binding and transport. In addition R316 and E319, both on L4, and within 10–15 Å of the putative binding site were found to be important for both processes.

11.8
Proposed Mechanism

11.8.1
Overview

The molecular biological experiments prior to the determination of the crystal structures resulted in a proposed mechanism in which an external loop formed the gate for the proposed large porin-like channel formed by the β-barrel. The binding of ligand presumably signaled the TonB protein to open the gate after which the ligand could diffuse through the channel into the periplasm. The crystal structures showed, however, that the large channel was completely closed by the N-terminal domain of the proteins. In addition, binding of ligands has little or no effect on the extracellular loops, but instead a distinct effect on the N-terminal domain. Finally, transport to the periplasm cannot occur unless a channel is formed by the N-terminal domain, either by its removal or by an internal conformational change in the domain involving TonB and the proton motive force (PMF) in an energy-dependent process.

The crystal structures generated many questions which remain to be answered. What determines the specificity of binding of siderophore to its receptor? Which is the chain of events leading to the opening of the channel and movement of ferric siderophore complex into the periplasm? What is the interaction with TonB and how is the PMF involved? On the basis of the structures of FepA and FhuA one can assume that the overall structures of the proteins in this family should be the same. The binding specificity of the various siderophores will give differences in the binding pocket for the different proteins but one can assume that all proteins share a common mechanism of transport.

Transport of ferric siderophore is a multistep process involving multiple components which may be summarized as follows.

(1) $R + L \rightarrow RL$
(2) $RL \rightarrow R^*L$
(3) $R^*L \xrightarrow{TonB} R^{**}L$
(4) $R^{**}L + FhuD \rightarrow R + FhuD\ (L)$

R is the receptor without bound ligand; R* is the conformation observed in the FhuA ligand structures; R** is the unknown conformation as a result of the interaction with TonB, which, presumably, has a channel allowing access of the ligand to the periplasm; L is the ligand; FhuD is the periplasmic binding protein (FepB in the case of FepA).

11.8.2
Binding of Ligand to Receptor

Several observations described in the previous sections indicate that there are two sequential binding processes (steps 1 and 2). The first, a probable non-specific process involving aromatic residues at the apex of the extracellular loops where the ferric siderophore is in essence extracted from the medium, and a specific binding process (step 2) as observed in the FhuA ligand structures. It is only this second process which causes conformational changes observed in the FhuA structures. These allosteric conformational and structural changes are the same if either ferrichrome, albomycin, or ferricrocin are bound to FhuA [18, 32]. Comparing protein structures which transport ferrichrome one observes conservation for residues involved in this second binding pocket and the same is true when comparing proteins which transport ferric enterobactin. However, these homologies are lost when one compares both types of proteins. For instance, the sensor loops formed by the plug domain of FhuA which contribute two important residues Y116 and R81 for ferrichrome binding do not share any sequence homology with the sensor loops observed in FepA, although the overall topology of the plug domain is similar. This is not unexpected as ferric enterobactin is triple negatively charged compared to ferrichrome, which is neutral. Therefore, one finds several positively charged arginines present in the sensor loops of FepA [17], which probably participate in the high-affinity final binding of ferric enterobactin. Taking the sequence comparisons one step further by comparing many different proteins in this family which transport different ferric siderophores, one expects to loose all homologies due to binding but one may find instead homologies due to the next step in the process, the TonB-dependent transport of the ligands, indicated in step 3 (see below).

11.8.3
The TonB-dependent Transport

Step 3 is definitely a multistep process, and it is only written as a single step because it is not possible, at this moment, either to identify the events clearly or to establish the sequence of events which do occur. However, it is possible to discuss the probable events in this step.

In earlier sections it was argued that the first event is probably an interaction between the ligand-bound receptor protein (R*) and TonB, which is supported by the structural changes which occur in the receptor on binding of the ligand and by biochemical and genetic experiments. The last event of step 3 is probably a channel formation by either the dislodgment of the N-terminal domain or a conformational change within the domain (R**), for which there is, however, no structural or biochemical evidence but which seems necessary for transport. The ligand could then diffuse to the periplasm where it is bound by the periplasmic protein FhuD in the case of ferrichrome [58] or FepB in the case of ferric enterobactin [59], after which the protein reverts to its original state (R).

In between these two events are the one or more functions of TonB and also for these processes there is no structural information. It is evident, however, that the proton motive force (PMF) across the inner membrane mediated by TonB plays an important role in this event [5–8]. Recently, it has been shown that there are PMF-dependent conformational changes which take place in TonB and that these changes are also dependent on ligand binding to the receptor [8]. There is sufficient evidence suggesting the physical interaction between the periplasmic domain of TonB with the receptor protein. It seems most logical that this interaction allows a signal to the inner membrane to activate the PMF. It is, however, uncertain if the PMF, in turn, forces the TonB into a charged conformation, which interacts with the receptor protein to form the channel or if, as was suggested earlier [16], the PMF results in a localized pH change (periplasmic compartment), which induces conformational changes and channel formation in the receptor protein. There is no experimental data to allow answers to these questions. What is possible, however, is to identify residues in FepA and FhuA which are most likely involved in the transport process and those which may be involved in the interactions with TonB.

11.8.4
Homology

In order to target the amino acids which might be important in transport of the bound ligand, alignment studies on the TonB-dependent receptors belonging to this family were performed [60]. The receptors in these family do not share any homology with porins but their homology within the family allows the design of a phylogenetic tree [16]. Only 20% homology is observed when the sequences are aligned with one another individually except in case of FepA, PfeA, and BfeA where about 40–50% homology was observed. This is not unexpected as most of them transport structurally and chemically diverse siderophores while FepA, PfeA, and BfeA transport the same siderophore, ferric enterobactin.

It was, however, realized that by comparing all 19 proteins simultaneously (Fig. 9) it would be possible to determine domains that would be responsible for the common property of all these proteins, which is the mechanism of transport, and that this simultaneous comparison would eliminate homologies due to the binding specificity and recognition of the substrate. Simultaneous alignment of all the sequences resulted in only 1.4% (10 residues) identity and 3.6% (26 residues) similarities which do not include homologies due to Val, Leu, and Ile. However, the interesting feature of this alignment study was that a higher density of homologies occurs in the first 150 residues of the proteins and that in this portion two glycines exist at i and $i + 7$, which are conserved in all proteins. The original homology re-

```
FepA  IDIRGMGP---ENILILIDGKPVSSRNSVRQGWRGERDIRGDISWVPPEMIERIEVLRGP 128
PfeA  IDIRGMGP---ENILILVDGKPVSSRNSVRYGWRGERDSRGDINWVPADQVERIEVIRGP 130
BfeA  VDIRGMGP---ENILILIDGKPVTSRNAVRYGWNGDRDIRGDINWVPAEEVERIEVIRGP 134
RumA  VGIRGLPARLSPRSTILLDGIPLAAAPYGQPQLSMSPLSLG--------SISSIDVMRGA 126
FecA  FGIRGLNPRLASRSTVLMDGIPVPFAPYGQPQLSLAPVSLG--------NMDAIDVVRGG 198
FPVA  YYARGFSIN-----NFQYDGIPST---------ARNVGYSAGNILSDMAIYDRVEVLKGA 209
PupB  YWSRGFAIQ-----NYEVDGVPTS---------TRLDNYSQS-----MAMFDRVEIVRGA 209
PupA  IYSRGSAIN-----IYQFDGVTTY---------QDNQIRNMPSTIMDVGLYDRIEIVRGA 210
FhuE  YYSRGFQID-----NYMVDGIPTY---------FESR-WNLGDALSDMALFERVEVVRGA 128
FptA  YYVRGFKVD-----SFELDGVPAL---------LGN----TASSPQDMAIYERVEILRGS 124
PbuA  FYSRGFRMSG----QYQYDGVPLD---------IGSSYVQADSFNSDMAIYDRVEVLRGA 208
FatA  FKIRGFS--------SDIGDVM----------FNGLYGIAPYYRSSPEMYQRIDVLKGP 132
FcuA  YRIRGYN--------LDGDDIS----------FGGLFGVLPRQIVSTSMVERVEVFKGA 157
FhuA  LIIRGFAAEG----QSQNNYLNGLK-------LQGN--FYNDAVIDPYMLERAEIMRGP 135
FoxA  VALRGFHG-G----DVNNIFLDGLR-------LLSDGGSYNVLQVDPWFLERIDVIKGP 129
FyuA  ISLRGVSSAQDFYNPAVTLYVDGVP------------QLSTNTIQALIDVQSVELLRGP 114
AleB  QILRGRGML-----VLLDGIPLN---------INRDSARNLANIDPALVERVEVLRGS 215
IutA  MNVRGRPLV-----VLVDGVRLN---------SSRIDSRQLDSIDPFNMHHIEVIFGA 110
ViuA  PTIRGIDGSGPSVGGLASFAGTSPRLNMSIDG-RSLITYSEIAFGPRSLWDMQQVETYLGP 128
                **                                       :. : *
```

Fig. 9. An example of alignment of 19 OM receptor proteins. Alignment was done using clustal W [62].

gions identified by Lundrigan and Kadner are also confirmed [61] but these do not cover all identities and homologies observed.

Once the structures of FepA and FhuA were solved some of the homologies started to make some sense. For instance, (1) Arg 75 and 126 in the plug region form salt bridges with Glu 511 and 567 on the barrel (all four are conserved in the 19 proteins); (2) no homologies occur in the sensor loops of the plug region nor in the extracellular loops, as expected because they are involved in substrate binding and specific for the ferric siderophore (the sensor loops separate homology region III and IV in Lundrigan and Kadner's classification); (3) all homologies occur well below the sensor loops, that is between the binding site and the periplasmic side of the protein, a region which is expected to be involved in translocation related conformational changes; (4) the i and $i + 7$ Gly, which in FepA are Gly 127 and Gly 134, occur in a loop connecting strand $\beta 5$ and $\beta 6$ in the small 4-strand mixed β-sheet in the plug region. It is interesting to note that the N-terminal domain possesses 47 % of the conserved residues while the remaining 53 % are in the β-barrel.

11.8.5
Experimental Evidence

On the basis of the homologies described above one can identify at least six distinct areas including the plug and the barrel region. The first region in FepA involves a cluster around arginines 75/126 and glutamates 511/567, which actually consists of 12 conserved residues (Fig. 10). The figure shows the 12 residues in both FepA and FhuA, and the structural similarity is striking. In the homology alignment of 19 proteins, Arg 75 and 126 and Glu 511 and 567, correspond to Arg 93 and 133 and Glu 522 and 571 in FhuA, respectively. The arginine and glutamate residues form a quadrupole. Phe 528 supports the side chain of Arg 75, and Ser 29 keeps Arg 126 in place, the same holds true for Tyr 541 and Ser 51 in FhuA. The second cluster of homology is at the bottom of the small mixed β-sheet in the N-terminal domain, involving Thr 32, Glu 123, and Asn 143, while the third and fourth are small clusters of homologies on the β-barrel, one around Arg 431 (identical in 18 proteins) and the other, almost directly opposite, around Asn 677 (identical in 19 proteins). The former may be a part of the lining of the channel formed after a conformational change within the plug region. Another interesting aspect is the conservation of five glycines in the plug region (Gly 76, 88, 127, 134, and 54). Glycines can adopt many more conformations than other residues, and thus two or more could form the hinges in the assumed conformational change within the plug. Of course it is also possible that they are conserved because of structural requirements in forming bends. A recent publication showed that FhuA ($\Delta 5-160$) in which the complete plug region is excised (and, therefore, also the TonB box) still shows TonB-dependent transport [54]. This unexpected result might indicate that TonB interacts not only with the TonB box but also with the β-barrel. The C-terminal part of TonB is definitely basic. There are two acidic residues in the periplasmic bends, Asp 185 and Asp 422 (conserved only in 13-14 of

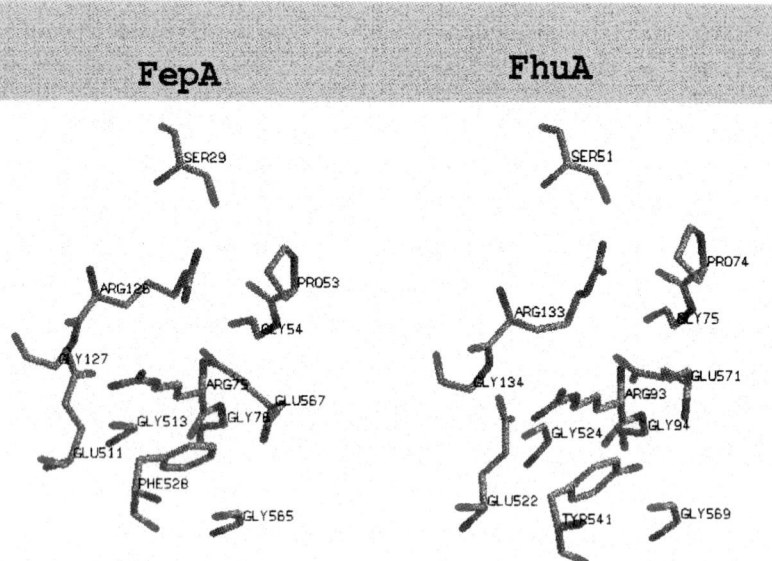

Fig. 10. Cluster of 12 conserved residues in FepA and FhuA; see color plates p. XLIII.

the 19 proteins), and these residues are possible candidates for interactions with TonB.

Several point mutants of the amino acids from these regions were made to study their effect on binding and transport of ^{59}Fe enterobactin under *in vivo* conditions [60]. Binding data were obtained at 0 °C to avoid further transport of the bound ferric enterobactin while transport studies were carried out at 37 °. The experimental data (K_D and K_M) clearly identify the mutants which show normal binding ($K_D \sim 0.2$) but are deficient in transport (3- to 10-fold increase in K_M value) (Tab. 2). The fact that K_M increases without an increase in K_D indicates a process which is different from a Michaelis–Menten mechanism, indicating thus the active participation of TonB. It is also encouraging to see the distinctive effects on transport of mutants of Arg 75, and Glu 511 and 567 without affecting binding. The explanation is only speculative. The disturbance of the quadrupole either prevents channel formation (R**) or disturbs the original protein conformation (R) or inhibits its restoration. The mutant F 528 A shows this residue not to be essential even though it is a component, albeit secondary, of the quadrupole structure. It was pointed out earlier that there is a discrepancy in the results of the two FhuA-ligand structures. The question is if this quadrupole region is or is not changed on ligand binding. The results on the mutant proteins indicate the importance to resolve this question. If indeed the H-bond for Arg 133 (in FhuA), equivalent to Arg 126 in FepA is broken, one can imagine strands β6 and β4 to be moved with respect to β1 and β5 in the mixed β-sheet of the N-terminal domain, a conformational change allowing the formation of a channel. The disruption of the mixed β-sheet would require a conformational change in the connection (127-139, in FepA) between

Tab. 2. Binding and transport properties of FepA mutants

Type of FepA	Binding (K_D)	Transport (K_M)
Wild type	0.19	0.25
R 75Q	0.11	2.58
G125A	0.11	0.84
E567A	0.09	0.91
E567Q/D511A	0.28	0.70

the two strands $\beta5$ and $\beta6$ and it is, therefore, interesting that the Gly 127 Ala mutant has a significant (4-fold) effect on transport. Such a mutation (Gly → Ala) can prevent conformational change. On the other hand, a similar mutation of Gly 76 and 134 has no effect on transport.

The earlier part of this section indicates many other residues which need to be mutated and investigated for transport and binding. In addition the results for FepA need to be confirmed by similar experiments with FhuA in order to assess their significance, and their importance for the mechanism of transport. Finally, structure determinations of the mutant proteins are required to properly interpret the results. The discussion in this section is, however, a clear example where knowledge of the structures can lead to verifiable hypotheses.

11.9
Conclusions

The crystal structures of FepA and FhuA provide a large step towards the understanding of the energy-dependent and specific transport of ferric siderophores across the outer membrane of gram-negative bacteria. Some, but certainly not all of the biochemical and genetic results can be rationalized from the structural information. In several cases these results indicate the need for structural data on previously genetically modified proteins. The structures certainly do not give a complete picture of the multistep transport process. Instead, the real significance of the structures is that they allow the framing of specific and directed questions and hypotheses which can be experimentally checked. A few of these were presented in this chapter and many others will be designed originating from various points of view. In many cases structures of the mutant proteins will be required in order to correlate the structural changes with the observed binding and transport properties or other experimental results. It will also be necessary to obtain structural information about the TonB interaction with the receptor proteins which may give the basis for an understanding about the energy dependence of the transport process.

Acknowledgement

We would like to thank Ms P. Singer for the preparation of the manuscript and Ms J. Keil for the help in scanning the figures. The research in our laboratory was supported by Grant GM 21822 from the National Institutes of Health.

References

1. Weinberg, E. D., *Physiol. Rev.* **1984**, *64*, 65-102.
2. Neilands, J. B., *J. Biol. Chem.* **1995**, *270*, 26723-26726.
3. Braun, V., Hantke, K., Köster, W., in: *Metal Ions in Biological Systems* (Sigel; A., Sigel, H., Eds.), Marcel Dekker, New York, **1998**, pp. 67-145.
4. Moeck, G. S., Coulton, J. W., Postle, K., *J. Biol. Chem.* **1997**, *272*, 28391-28397.
5. Bradbeer, C., *J. Bacteriol.* **1993**, *175*, 3146-3150.
6. Braun, V., *FEMS Microbiol. Rev.* **1995**, *16*, 295-307.
7. Cadieux, N., Bradbeer, C., Kadner, R. J., *J. Bacteriol.*, **2000**, *182*, 5954-5961.
8. Larsen, R. A., Thomas, M. G., Postle, K., *Mol. Microbiol.* **2000**, *31*, 1809-1824.
9. Hider, R. C., *Struct. Bonding* **1984**, *58*, 25-87.
10. van der Helm, D., Jalal, M. A. F., Hossain, M. B., in: *Iron Transport in Microbes, Plants and Animals* (Winkelmann, G., van der Helm, D., Neilands, J. B., Eds.), VCH, Weinheim, **1987**, pp. 135-165.
11. Matzanke, B. F., Matzanke, G. M., Raymond, K. N., in: *Iron Carriers and Iron Proteins* (Loehr, T. M., Ed.), VCH, New York, **1989**, pp. 1-121.
12. Matzanke, B. F., in: *Handbook of Microbial Iron Chelates* (Winkelmann, G., Ed.), CRC Press, Boca Raton, FL, **1991**, p. 61.
13. Roosenberg, II, J. M., Lin, Y. M., Lu, Y., Miller, M. J., *Curr. Med. Chem.* **2000**, *7*, 159-197.
14. Byers, B. R., Arceneaux, J. E. L., in: *Metal Ions in Biological Systems* (Sigel, A., Sigel, H., Eds.), Marcel Dekker, New York, **1998**, pp. 37-66.
15. Shanzer, A., Libman, J., in: *Metal Ions in Biological Systems* (Sigel, A., Sigel, H., Eds.), Marcel Dekker, New York, **1998**, pp. 329-354.
16. van der Helm, D., in: *Metal Ions in Biological Systems* (Sigel, A., Sigel, H., Eds.), Marcel Dekker, New York, **1998**, pp. 355-401.
17. Buchanan, S. K., Smith, B. S., Venkatramani, L., Xia, D., Esser, L., Palnitkar, M., Chakraborty, R., van der Helm, D., Deisenhofer, J., *Nature Struct. Biol.* **1999**, *6*, 56-63.
18. Ferguson, A. D., Hofmann, E., Coulton, J. W., Diederichs, K., Welte, W., *Science* **1998**, *282*, 2215-2220.
19. Locher, K. P., Rees, B., Koebnik, R., Mitschler, A., Moulinier, L., Rosenbusch, J. P., Moras, D. *Cell*, **1998**, *95*, 771-778.
20. Neilands, J. B., *Ann. Rev. Microbiol.* **1982**, *36*, 285-309.
21. Härle, C., Kim, I., Angerer, A., Braun, V., *EMBO J.* **1995**, *14*, 1430-1438.
22. Kim, I., Stiefel, A., Plantör, S., Angerer, A., Braun, V., *Mol. Microbiol.* **1997**, *23*, 333-344.
23. Hantke, K., *Mol. Gen. Genet.* **1981**, *182*, 288-292.
24. Hantke, K., *Mol. Gen. Genet.* **1984**, *197*, 337-341.
25. Escolar, L., Pérez-Martin, J., de Lorenzo, V., *J. Mol. Biol.* **1998**, *283*, 537-547.
26. Braun, V., *Arch. Microbiol.* **1997**, *167*, 325-331.
27. Koster, M., van de Vossenberg, J., Leong, J., Weisbeek, P. J., *Mol. Microbiol.* **1993**, *8*, 591-601.
28. van Hove, B., Staudenmaier, H., Braun, V., *J. Bacterol.* **1990**, *172*, 6749-6758.
29. Koster, M., van Klompenburg, W., Bitter, W., Leong, J., Weisbeck, P., *EMBO J.* **1994**, *13*, 2805-2813.
30. Ferguson, A. D., Welte, W., Hofmann, E., Lindner, B., Holst, O., Coulton, J. W., Diederichs, K., *Structure* **2000**, *8*, 585-592.
31. van der Helm, D., Baker, J. R., Wilmot, D. L. E., Hossain, M. B., Loghry, R. A., *J. Am. Chem. Soc.* **1980**, *102*, 4224-4231.

32. Ferguson, A. D., Braun, V., Fiedler, H.-P., Coulton, J. W., Diederichs, K., Welte, W., *Protein Sci.* **2000**, *9*, 956-963.
33. Pugsley, A. P., Zimmerman, W., Wehrli, W., *J. Gen. Microbiol.* **1987**, *133*, 3505-3511.
34. Salomón, R. A., Farías, R. N., *J. Bacteriol.* **1993**, *175*, 7741-7742.
35. Payne, M. A., Igo, J. D., Cao, Z., Foster, S. B., Newton, S. M. C., Klebba, P. E., *J. Biol. Chem.* **1997**, *272*, 21950-21955.
36. Moeck, G. S., Tawa, P., Xiang, H., Ismail, A. A., Turnbull, J. L., Coulton, J. W., *Mol. Microbiol.* **1996**, *22*, 459-471.
37. Schwede, T., Diemand, A., Guex, N., Peitsch, M. C., *Res. Microbiol.* **2000**, *151*, 107-112.
38. Killmann, H., Braun, V., *J. Bacteriol.* **1992**, *174*, 3479-3486.
39. Bös, C., Lorenzen, D., Braun, V., *J. Bacteriol.* **1998**, *180*, 605-613.
40. Killmann, H., Herrmann, C., Wolff, H., Braun, V., *J. Bacteriol.* **1998**, *180*, 3845-3852.
41. Killmann, H., Benz, R., Braun, V., *EMBO J.* **1993**, *12*, 3007-3016.
42. Killmann, H., Benz, R., Braun, V., *J. Bacteriol.* **1996**, *178*, 6913-6920.
43. Moeck, G. S., Bazzaz, B. S. F., Gras, M. F., Ravi, T. S., Ratcliffe, M. J. H., Coulton, J. W., *J. Bacteriol.* **1994**, *176*, 4250-4259.
44. Moeck, G. S., Ratcliffe, M. J. H., Coulton, J. W., *J. Bacteriol.* **1995**, *177*, 6118-6125.
45. Koebnik, R., Braun, V., *J. Bacteriol.* **1993**, *175*, 826-839.
46. Carmel, G., Hellstern, D., Henning, D., Coulton, J. W., *J. Bacteriol.* **1990**, *172*, 1861-1869.
47. Carmel, G., Coulton, J. W., *J. Bacteriol.* **1991**, *173*, 4394-4403.
48. Murphy, C. K., Kalve, V. I., Klebba, P. E., *J. Bacteriol.* **1990**, *172*, 2736-2746.
49. Rutz, J. M., Liu, J., Lyons, J. A., Goranson, J., Armstrong, S. K., McIntosh, M. A., Feix, J. B., Klebba, P. E., *Science* **1992**, *258*, 471-475.
50. Klug, C. S., Su, W., Feix, J. B., *Biochemistry* **1997**, *36*, 13027-13033.
51. Armstrong, S. K., McIntosh, M. A., *J. Biol. Chem.* **1995**, *270*, 2483-2488.
52. Liu, J., Rutz, J. M., Klebba, P. E., Feix, J. B., *Biochemistry* **1994**, *33*, 13274-13283.
53. Jiang, X., Payne, M. A., Cao, Z., Foster, S. B., Feix, J. B., Newton, S. M. C., Klebba, P. E., *Science* **1997**, *276*, 1261-1264.
54. Braun, M., Killmann, H., Braun, V., *Mol. Microbiol.* **1999**, *33*, 1037-1049.
55. Newton, S. M., Allen, J. S., Cao, Z., Qi, Z., Jiang, X., Sprencel, C., Igo, J. D., Foster, S. B., Payne, M. A., Klebba, P. E., *Proc. Natl. Acad. Sci. USA* **1997**, *94*, 4560-4565.
56. Newton, S. M., Igo, J. D., Scott, D. C., Klebba, P. E., *Mol. Microbiol.* **1999**, *32*, 1153-1165.
57. Cao, Z., Qi, Z., Sprencel, C., Newton, S. M. C., Klebba, P. E., *Mol. Microbiol.* **2000**, *37*, 1306-1317.
58. Clarke, T. E., Ku, S.-Y., Dougan, D. R., Vogel, H. J., Tari, L. W., *Nature Struct. Biol.* **2000**, *7*, 287-291.
59. Sprencel, C., Cao, Z., Qi, Z., Scott, D. C., Montague, M. A., Ivanoff, N., Xu, J., Raymond, K. N., Newton, S. M. C., Klebba, P. E., *J. Bacteriol.* **2000**, *182*, 5359-5364.
60. Chakraborty, R. N., Lemke, E., Cao, Z., Newton, S. M., Klebba, P. E. van der Helm, D., *J. Bacteriol.*, **2001** (in preparation).
61. Lundrigan, M. D., Kadner, R. J., *J. Biol. Chem.* **1986**, *261*, 10797-10801.
62. Thompson, J. D., Higgins, D. G., Gibson, T. J., *Nucleic Acids Res.* **1994**, *22*, 4673-4680.

12
Mechanisms of Bacterial Iron Transport

Volkmar Braun and Klaus Hantke

12.1
Introduction

Wherever bacteria live under oxic conditions, they are confronted with concentrations of available iron that are far below the level needed for multiplication ($\sim 10^{-7}$ M). In the soil and water, Fe^{3+} forms hydroxy polymers with a free iron concentration in the order of 10^{-9} M [1]. Organic acids such as citrate may solubilize a portion of the iron and deliver it to those bacteria that contain an iron uptake system. Most bacteria synthesize their own iron chelators, designated siderophores, which they secrete into their environment to form iron complexes that are then taken up into the cells.

Iron is an essential element for all bacteria, with the exception of a few species such as *Lactobacillus plantarum* [2] and *Borrelia burgdorferi* [3]. Lactobacilli usually grow under microaerophilic conditions and do not synthesize heme-containing proteins. *B. burgdorferi* grows normally in iron-limited medium, does not contain iron proteins commonly found in bacteria and iron-regulated genes, and contains less than 10 iron atoms per cell. *B. burgdorferi* apparently avoids iron limitation as a means of host defense.

Iron in bacteria is mostly contained in the reaction center of redox enzymes and directly participates in redox reactions by switching between the Fe^{2+} and Fe^{3+} states. The redox enzymes are membrane-bound and are also contained in the cytosol. Membrane-bound respiration cannot take place without iron. Aerobic respiration with oxygen as terminal electron acceptor and anaerobic respiration with nitrate, sulfate, sulfur, fumarate, carbon dioxide, or Fe^{3+} instead of oxygen occur in various bacteria. Soluble redox enzymes include ribonucleotide reductase, one of the key enzymes in DNA synthesis, and aconitase of the citric acid cycle. Iron-containing redox enzymes such as catalase and superoxide dismutase destroy oxygen radicals derived from iron-catalyzed reactions (Haber–Weiss cycle). Since iron is contained in so many essential enzymes, 10^5 ions are required per bacterial cell. Utilization of iron involves uptake from the culture medium and incorporation into the redox enzymes. Both processes are catalyzed by proteins that participate in intricate mechanisms [4].

Bacteria are confronted in the serum with iron bound to transferrin and in secretory fluids to lactoferrin. The free iron concentration in equilibrium with these iron-binding proteins in the extracellular fluids of the human host is in the order of 10^{-24}. Moreover, most of the body iron (78%) is contained in the heme of hemoglobin and myoglobin; either heme or iron have to be released by the bacteria to satisfy their iron requirement. Intracellular iron (15%) is also deposited in ferritin; little is known how ferritin iron is mobilized by bacteria.

Iron is the only nutrient known to be generally growth limiting. Therefore, iron supply systems are crucial in overcoming the iron shortage imposed by the host [5–11]. Recently, virulence was shown to be related to specific iron transport systems under defined *in vivo* experimental conditions. For example, in *Neisseria gonorrhoeae*, pili are essential for adhesion to human tissues. To circumvent immune reactions against pili, *Neisseria* contain a genetic mechanism that frequently changes the pili structure. The frequency of pilin antigenic variation increases five- to nine-fold under iron limitation, and the frequency of DNA recombination and DNA repair processes is also increased. As gonococci are faced with iron limitation during infection, iron limitation induces recombination events that are important for gonococcal pathogenesis [12]. *tonB*, *exbB*, and *exbD*, which are important for iron transport, are among 73 genes of *Neisseria meningitidis* found to be essential for bacteremic disease [13]. In *Actinobacillus pleuropneumoniae*, the etiologic agent of porcine pleuropneumoniae, an *exbB* deletion mutant elicits no detectable humoral immune response in pigs challenged with an aerosol containing the mutant bacteria, and no mutant bacteria were re-isolated [14]. *Neisseria* interfere with transferrin–iron homeostasis not only by using transferrin iron, which is crucial for bacterial colonization in the urethra of man [15], but also by affecting transferrin receptor cycling. Human epithelium cells infected by *N. gonorrhoeae* and *N. meningitidis* have reduced levels of transferrin receptor mRNA and cycling transferrin receptors. The ability of infected cells to internalize transferrin receptor is reduced, and the distribution between surface and cycling transferrin receptors is altered [16]. Iron uptake is important for the virulence of *Pseudomonas aeruginosa*. A *tonB* mutant devoid of Fe^{3+} uptake via the siderophores pyoverdin, pyochelin, and heme grows in the muscles and lungs of immunosuppressed mice, but does not kill the animals [17]. Pyoverdin- and pyochelin-negative double mutants multiply, but do not kill the mice; however, intranasal inoculation results in multiplication and killing [18], although growth in the lung is attenuated compared to wild-type bacteria. Iron withdrawal from bacterial pathogens might also involve the use of siderophore iron, in addition to binding iron to transferrin, lactoferrin, and ferritin. Human neutrophils, macrophages, and myeloid cell lines can acquire iron from pyoverdin and pyochelin of *P. aeruginosa* [19], and, therefore, may deplete *P. aeruginosa* of the essential iron. A *Mycobacterium tuberculosis* mutant devoid of mycobactin, a cell-bound siderophore, is impaired for growth in macrophage-like THP-1 cells [20]. Invasive *Escherichia coli* that synthesize the Fe^{3+}-aerobactin transport system grow in the serum of humans and animals [6, 21]. Virulence of the fish pathogen *Vibrio anguillarum* depends on the synthesis and transport of the strain-specific siderophore anguibactin [22]. *Yersinia enterocolitica* only kills mice when it synthesizes yersiniabactin and a related iron-repressible outer membrane transport protein [23]. *Y. enterocolitica* also contains a transport system for Desferal® [24], which explains the occasional occurrence of yersiniosis in Desferal®-treated patients. Iron acquisition via yersiniabactin is also important for the virulence of *Yersinia pestis* during the early stages of infection in mice [25]. In *Legionella pneumophila*, the *iraAB* locus is required for iron assimilation, intracellular infection, and virulence [26]. An *ira* mutant, defective in intracellular infection, yields 1,000-fold fewer bacteria from the lungs and spleen of inoculated guinea pigs compared to *ira* wild-type cells, in which a 50-fold lower dose was used. The mutant bacteria are subsequently cleared in contrast to wild-type bacteria, which multiply. In *Salmonella typhimurium* [now designated *Salmonella enterica* serovar *typhimurium*], the *sitABCD* genes encode a complete iron transport system. Transcription of the *sitABCD* genes is induced after invasion of the intestinal epithelium of BALB/c mice. A *sit* null mutant is attenuated in mice, which suggests that the *sit* iron transport system plays an important role in *S. typhimurium* virulence [27]. *Helicobacter pylori feoB* mutants do not colonize the stomach of mice after oral inoculation [28], showing that the major *in vitro* high-affinity iron transport system is also the most important one *in vivo*. An isogenic hemoglobin receptor-deficient mutant of *Haemophilus ducreyi*, the etiologic agent of the genital ulcer disease chancroid, is attenuated in a human model of experimental infection [29].

More comprehensive reviews than can be written – due to limited space – for this book on bacterial iron supply systems can be found in [4, 21, 30]. It is far beyond the scope of this chapter to describe the many siderophores, their synthesis and transport, and the many bacterial species in which they have been identified. This applies also to the heme and Fe^{3+} transport systems in which no siderophores are involved. Rather, well-studied examples will be discussed and common mechanisms will be emphasized.

12.2
Transport of Fe^{3+}-Siderophores

12.2.1
Transport of Fe^{3+}-Siderophores Across the Outer Membrane of Gram-negative Bacteria

Nutrients usually diffuse through water-filled pores formed by the porin proteins of the outer membrane. Most of the Fe^{3+}-siderophores are larger than 700 Da and, therefore, require specific transport proteins. Bacteria have turned the disadvantage of the diffusion barrier into an advantage by using the outer membrane to extract the scarce Fe^{3+} sources from the culture medium via highly specific and avid transport proteins with dissociation constants in the nanomolar range. Transport subsequent to binding is an active, energy-consuming process. Since there is no energy source in the outer membrane or in the periplasm, the proton motive force of the cytoplasmic membrane serves to drive active transport across the outer membrane. A protein complex, consisting of the proteins TonB, ExbB, and ExbD in *E. coli* and similar proteins in other gram-negative bacteria, transmits the energy from the cytoplasmic membrane into the outer membrane. Specific interaction of TonB through the region around residue 160 with the so-called TonB box of outer membrane transporters has been demonstrated (illustrated in Fig. 1) [31–33]. TonB is

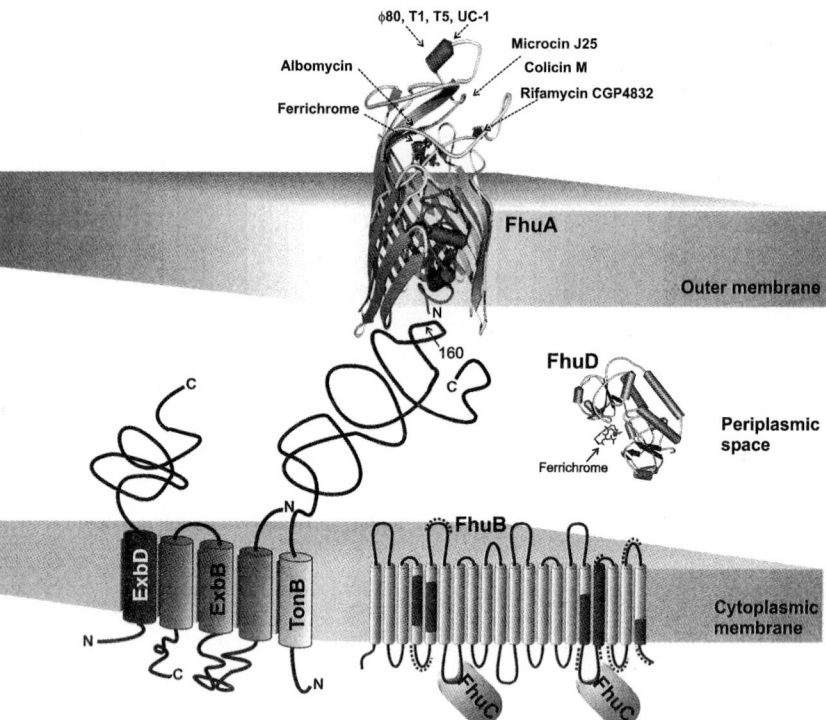

Fig. 1. Model of the FhuABCD activities of *E. coli*. The FhuA protein catalyzes transport of ferrichrome, albomycin, rifamycin CGP 4832, and uptake of colicin M and microcin 25 across the outer membrane of *E. coli*, and mediates infection by the phages T5, T1, ϕ80, and UC-1. The FhuBCD proteins catalyze transport of ferrichrome and albomycin across the cytoplasmic membrane. The crystal structure of FhuA is shown (39, 40) in which β-sheets in the front were removed to better visualize the cork domain. The binding sites of ferrichrome and albomycin [127] are indicated. The N-terminus of FhuA (marked N) contains the TonB box, which is thought to interact with the Gln-160 region of TonB (marked 160) in the periplasm. The crystal structure of FhuD with bound ferrichrome is shown [121]. The regions of FhuB that bind to FhuD, as revealed using synthetic FhuB peptides, are marked by dots in hydrophilic loops [the cytoplasmic loops may fold back into the predicted FhuB channel], and by shaded transmembrane segments [61]. It is assumed that two copies of FhuC, the tentative ATPase, might bind to the indicated FhuB regions, as revealed by mutation analysis [4]. The TonB, ExbB, ExbD protein complex involved in the transfer of energy stored in the proton motive force of the cytoplasmic membrane to FhuA indicates interaction of ExbB with TonB and ExbD [129], [132] and the transmembrane topology of TonB [131, 132], [125], ExbB [133], and ExbD [134].

anchored with the N-terminus in the cytoplasmic membrane and might assume an energized conformation in response to the proton motive force. The most likely mechanism, which has not been proved experimentally, is as follows. In the energized form, TonB interacts with the transport proteins, causes a structural transition of the transport proteins that releases the Fe^{3+}-siderophores from their transport protein binding sites and opens a channel in the transport proteins through which the Fe^{3+}-siderophores diffuse along certain amino acid side chains into the periplasm, where they are trapped by specific binding proteins.

Structural changes in FepA, the outer membrane transport protein of Fe^{3+}-enterobactin, in response to TonB and energy metabolism have been shown by electron spin resonance spectroscopy [34]. The nitroxide spin label was fixed to a genetically introduced cysteine at residue 280; in the FepA crystal structure, this residue is located in the L3 loop at the cell surface that connects the transmembrane β-pleated sheets 5 and 6 of the β-barrel ([34a]; see Chap. 11 for a detailed discussion of the FepA and FhuA crystal structures). This loop does not close the FepA channel, as was previously assumed from a FepA deletion derivative lacking residues 202 to 340 [35]. This deletion removed half of the L2 loop, the entire L3 and L4 loops, and the transmembrane segments 4 to 7, and converted FepA into an open channel. The resulting structure of the deletion derivative cannot be deduced from the crystal structure of the complete FepA and would require X-ray analysis. The electron spin resonance data suggest that loop 3 changes its conformation when FepA is energized. The same conclusion was drawn 1976 when it was demonstrated that irreversible adsorption of the phages T1 and ϕ80 to the FhuA protein of *E. coli* requires the proton motive force of the cytoplasmic membrane [36]. It was clear that the phage adsorption sites have to be at the cell surface. The phage adsorption site was then localized by competitive peptide mapping to a single region [37] which was localized at the cell surface by proteolytic cleavage of foreign peptides introduced at 34 sites along the entire FhuA protein [38]. The crystal structure of the FhuA protein indicates that the phage binding site is in fact located at the cell surface and involves loop 4 [39, 40].

The FhuA and FepA proteins consist of 22 antiparallel transmembrane β-strands that form a β-barrel [34a, 39, 40]. The β-barrel strands are interconnected by large loops at the cell surface and small turns in the periplasm. FhuA has been crystallized with and without bound ferrichrome, the Fe^{3+}-siderophore transported by FhuA. Ferrichrome is fixed above the cork well outside the membrane by hydrogen bonds and van der Waals contacts in a binding site formed by eight aromatic residues and one residue each of arginine, glutamine, and glycine. Compared to unloaded FhuA, apex B of the cork domain moves 1.7 Å towards ferrichrome. At the periplasmic side of FhuA, a short helix is completely unwound, and the resulting coil bends away by 180° from the previous helix axis, resulting in the transition of Glu-19 and Trp-22 17 Å away from their former α-carbon positions. A small structural change at the ferrichrome binding site at the cell surface is amplified to a large structural change in the periplasmic cavity. These results agree with biochemical data that demonstrate upon binding of ferrichrome conversion of isolated FhuA to trypsin resistance at Lys-67 [41] decrease of intrinsic tryptophan fluorescence of FhuA [42] *in vivo* fluorescence quenching of fluorescein-maleimide bound to the genetically introduced Cys-336 residue in loop 4 [43], and increase in the formation of a chemically cross-linked complex between FhuA and TonB [44]. Since Glu-19 is exposed to the periplasm, the TonB box (residues 7 to 11) is most likely located in the periplasm, where it interacts with TonB. Binding of ferrichrome enhances interaction of FhuA with TonB, so that preferentially substrate-loaded FhuA is coupled to the energy-providing Ton system. Site-directed spin labeling and electron paramagnetic resonance (EPR) studies carried out with BtuB, the vitamin B_{12} outer membrane transport protein with a transport mechanism highly similar to the iron transport mechanism, indicate that the TonB box of BtuB in the unliganded conformation might be located in a helix fixed to the β-barrel [45]. Binding of vitamin B_{12} converts this segment into an extended, disordered, and highly dynamic structure that probably extends into the periplasm to interact physically with TonB. A TonB-uncoupled TonB-box mutant of BtuB shows a strongly altered EPR spectrum and no longer responds to the addition of vitamin B_{12}. These results prove the functional interaction of the BtuB TonB box with the region around residue 160 of TonB.

The TonB box of FhuA is part of the cork domain. In theory, deletion of the cork should abolish interaction of TonB with FhuA and convert FhuA into an open channel through which ferrichrome diffuses in an energy-independent manner. A FhuA protein with no interaction with TonB should be catalytically inactive and no longer serve as binding site for the TonB-dependent phages T1 and ϕ80, uptake of colicin M, microcin J25, and albomycin. Only phage T5 could infect such cells by binding to loop 4 of the β-barrel because it occurs without the involvement of TonB. In contrast to these expectations, removal of the cork domain (residues 5–160) results in a protein (FhuAΔ5–160) through which substances only weakly diffuse, and, more surprisingly, the protein retains its TonB-dependent transport activity. In fact, all TonB-dependent activities of FhuA (Fig. 1) such as transport of the antibiotic albomycin, infection by phages T1 and ϕ80, and killing of cells by the bacterial protein toxin colicin M, are mediated by FhuAΔ5–160, with the exception of the uptake of microcin J25 [46]. These results point to other sites of interaction between FhuA and TonB in addition to the FhuA TonB box that contribute to the activity of FhuA. The additional sites in FhuA are presumably exposed to the periplasm as TonB is located in the periplasm and not firmly integrated into the outer membrane [47], which one would expect if it interacts with transmembrane segments of FhuA. However, there is no experimental evidence for an interaction of TonB with periplasmic regions of FhuA other than the TonB box. Mutants with point mutations in the TonB box are inactive, but insertions of peptides ranging from 4 to 16 amino acids at 5 out of 11 periplasmic turns of FhuA do not, or only slightly reduce the FhuA activities [38, 48, 49]. Since loop 4 of FhuA is the principal binding site for the phages [37], the structural transition in FhuAΔ5–160 upon binding to energized TonB presumably proceeds from the periplasm through the outer membrane to loop 4 at the cell surface. TonB-dependent activation of FhuA is mainly mediated through the β-barrel, although the crystal structure suggests a rather rigid β-barrel conformation, as evidenced by the low B-factors of the C_a-atoms.

FhuA displays multiple activities (Fig. 1). Overlapping regions are used by different ligands; the three unrelated phages T1, T5, and ϕ80 bind to loop 4, the iron chelate of albomycin is bound to the same site as ferrichrome to a region below loop 4, and most surprising, a structurally unrelated antibiotic, a chemically synthesized derivative of rifamycin, designated CGP 4832, binds to the same site as ferrichrome and is transported by FhuA [50]. These examples indicate that synthetic conjugates of FhuA substrates and antibiotics that consist of an iron-chelating siderophore, a peptide linker, and an antibiotic group may improve the efficacy of those antibiotics that diffuse too slowly into microorganisms to serve as therapeutic drugs [51].

Despite the many ligands, FhuA exerts a distinct specificity which is typical for all outer membrane proteins that transport Fe^{3+} from various sources. The exception is an outer membrane transport system of *Aeromonas hydrophila* that exhibits an extraordinarily broad specificity. In addition to the strain-specific siderophore amonabactin, four additional Fe^{3+}-siderophores with different molecular structures, iron-chelating functionalities, numbers of Fe^{3+} centers, and charges are transported by the same receptor protein [52]. A single chromium complex, which is not transported, inhibits transport of 4 Fe^{3+}–amonabactins and 10 additional Fe^{3+}-siderophores that are not structurally related to the chromium complex. The uptake of three other Fe^{3+}-siderophores is not inhibited by the chromium complex, thereby showing some sort of specificity and ruling out a toxic effect of the chromium complex. Although the dissociation constants are very high – up to 1,000-fold higher than that of ferric enterobactin with FepA – the transport rates are comparable. A shuttle mechanism has been proposed in which Fe^{3+} of the incoming Fe^{3+}-siderophore is transferred to a siderophore bound to the transport protein, which is then taken up into the cytoplasm as a Fe^{3+}-siderophore complex. The biochemical evidence for such a broad specificity should be complemented by the identification of the transport protein and the bound siderophore, isolation of knockout mutants in *A. hydrophila* and cloning of the gene in *E. coli*. The proposed transport mechanism may also apply to the uptake of pyoverdin by *P. aeruginosa*, where the highly purified outer membrane transport protein FpvA contains 1 mol of iron-free pyoverdin per mol of protein (53).

12.2.2
Transport of Fe^{3+}-Siderophores Across the Cytoplasmic Membrane by ABC Transporters

Once transported across the outer membrane, Fe^{3+}-siderophores are bound by a binding protein in the periplasm. Binding proteins are components of ABC (ATP-binding cassette) transporters (see Chap. 4 for a detailed discussion). The binding activity for Fe^{3+}-siderophores has been determined with FhuD, which belongs to the transport system of Fe^{3+}-hydroxamates of *E. coli* (see Fig. 1 and gene arrangement of the *E. coli fhu* operon in Fig. 2) FhuD binds ferrichrome, coprogen,

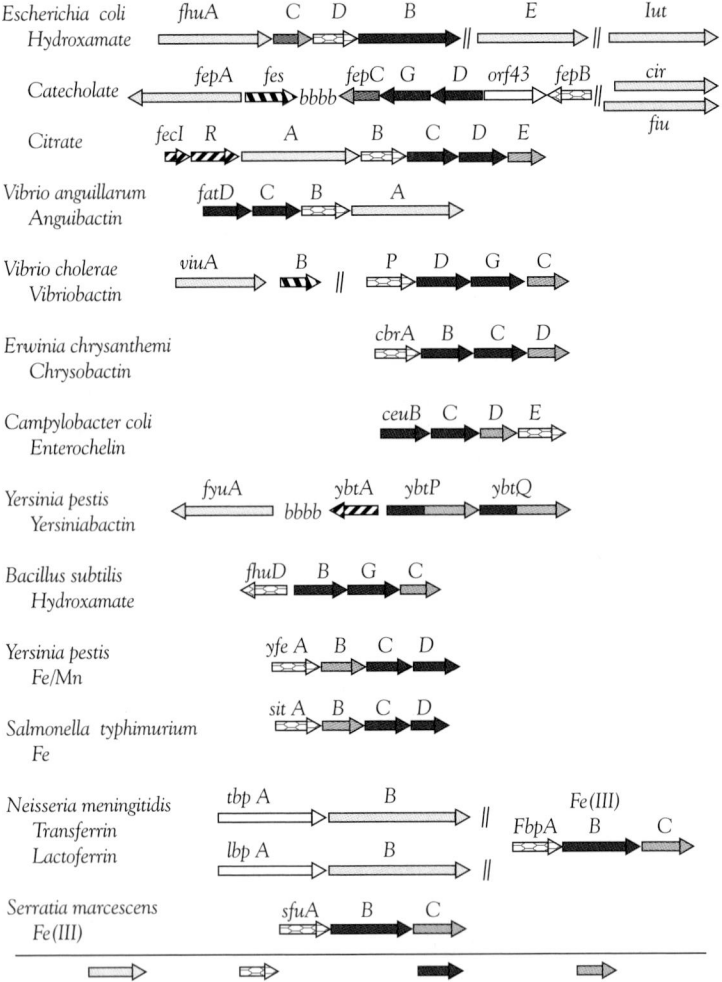

Fig. 2. Examples of genes that encode transport systems for Fe^{3+}-siderophores, Fe^{3+}, transferrin, and lactoferrin. In several systems not all participants have been identified: in *V. anguillarum* the ATPase component of the ABC transporter genes *fatDCB*; in *E. chrysanthemi* the outer membrane receptor for chrysobactin, in *C. coli* the outer membrane receptor for enterochelin, in *Y. pestis* the binding protein for yersiniabactin, in *B. subtilis* the binding protein for ferrioxamine; in *S. marcescens* the source of Fe^{3+} for the Sfu transporter is not known. For YfeA (*Y. pestis*) and SitA (*S. typhimurium*) the oxidation state of iron is not clear.

aerobactin, ferrioxamine B, shizokinen, rhodotorulic acid, and the antibiotic albomycin with dissociation constants ranging from 0.3 to 5.4 µM [54, 55]. FhuD displays a much broader specificity than the FhuA outer membrane transporter, which, of the Fe^{3+}-hydroxamates, transports only ferrichrome and albomycin. The broad substrate specificity can be explained by analysis of the crystal structure of FhuD with bound gallichrome (ferrichrome in which Fe^{3+} is replaced by Ga^{3+}) [56]. Gallichrome is not buried in a deep cleft like ferrichrome in FhuA, but is exposed at the protein surface (Fig. 3). Surface exposure provides fewer steric constraints to ligands than a deep cavity does. Only two residues are close enough to bind gallichrome to FhuD, and two residues bind gallichrome through an interconnected water molecule, whereas 10 residues of FhuA may bind ferrichrome.

Similarity between the crystal structures of the Fe^{3+}-binding N- and C-terminal domains of human transferrin and lactoferrin with the crystal structures of bacterial periplasmic binding proteins suggest a common evolution. This conclusion is particularly well supported by nearly identical secondary and tertiary structures of the *E. coli* sulfate binding protein and the lactoferrin N-domain [57]. If this applies to the bacterial sulfate binding protein, one would expect an even higher similarity in the folding pattern of the binding proteins of bacterial Fe^{3+} transport proteins and the Fe^{3+}-transporting transferrin. This, however, is not the case. The periplasmic Fe^{3+}-binding protein hFBP (also called HitA) of *Haemophilus influenzae* is composed of a two-domain structure connected by a hinge region similar to transferrin, but the folding pattern is entirely different from that of transferrin (Fig. 3) [58]. Moreover, the folding of FhuD differs from the folding of hFBP and transfer-

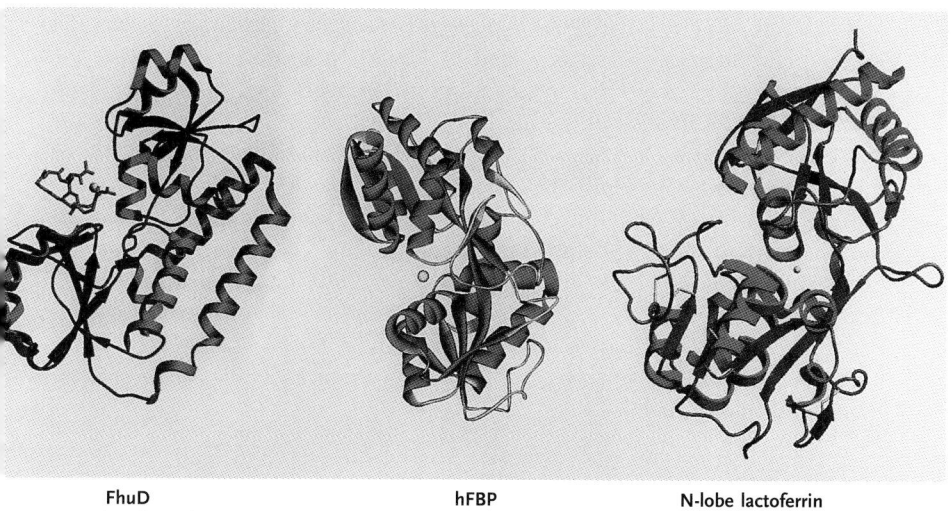

FhuD hFBP N-lobe lactoferrin
Escherichia coli *Haemophilus influenzae* Human

Fig. 3. Comparison of the stereo-ribbon diagrams of the crystal structures of the *Escherichia coli* FhuD protein with bound gallichrome [56], the *Haemophilus influenzae* hFBP (HitA) protein with bound Fe^{3+} [58], and the N-lobe of human lactoferrin with bound Fe^{3+}; see color plates p. XLIII.

rin, but results in an overall two-domain structure (Fig. 3). However, unlike in other binding proteins, transferrins, and lactoferrins, the two domains are interconnected by a rather rigid helix and the interface between the N- and C-terminal domains is predominantly hydrophobic. Therefore, it is unlikely that the two domains of FhuD close upon binding and open upon release of the substrates like a venus fly trap, as is usually observed in bacterial binding proteins, transferrins, and lactoferrins. hFBP, transferrin, and lactoferrin may have resulted from convergent evolution, but the evolution of FhuD resulted in a different structural solution. Sequence comparison of FhuD with other periplasmic binding proteins places FhuD in a group to which all Fe^{3+}-siderophore binding proteins belong, in contrast to hFBP, which forms a separate group together with other Fe^{3+}-binding proteins (see Sect. 12.3.2) [59].

FhuD delivers the Fe^{3+}-hydroxamates to the FhuB transport protein in the cytoplasmic membrane (Fig. 1). Interaction between FhuD and FhuB has been shown by protection of FhuB in spheroplasts against trypsin and proteinase K degradation through FhuD [55] cross-linking of His-tagged FhuD to FhuB, and binding of 10- and 20-amino acid residue peptides identical in sequence to segments of FhuB to FhuD [60, 61]. The FhuB peptides that bind in an ELISA assay to FhuD also inhibit ferrichrome transport when delivered into the periplasm through a FhuA derivative that forms an open channel, constructed by genetic excision of residues 322–355 [62]. Competitive peptide mapping not only has revealed sites of interaction between FhuD and FhuB, but also suggested a mechanism of FhuC activation that has hitherto not been envisioned for binding protein-dependent transport systems [60, 61]. According to a FhuB transmembrane topology model [63], three peptides that bind to FhuD are located in periplasmically exposed loops of FhuB. Unexpectedly, peptides of four transmembrane regions and even of four periplasmic regions bind to FhuD. These data suggest that the periplasmic segments of FhuB fold back into a channel formed by the transmembrane regions. FhuD inserts into the FhuB channel when it delivers ferrichrome and comes into close contact with one of the FhuC binding sites at FhuB, or into direct contact with FhuC. This model implies that ATP hydrolysis by FhuC is triggered without the need of a transmembrane signal through FhuB from the periplasm to the cytoplasm, which would be required if FhuD binds only to periplasmic loops of FhuB. Binding of FhuC to FhuB has been demonstrated [63a] and mutational analysis of FhuC [64] and FhuB [65, 66] has defined regions important for the activities of both proteins. These sites are equivalent to the activity-related MalFGK sites revealed by the well-advanced studies of the maltose transport system (Chap. 4).

12.3
Bacterial Use of Fe^{3+} Contained in Transferrin and Lactoferrin

Transferrin is contained in human serum (25–44 µM), and lactoferrin in mucosal secretions (6–13 µM) and leukocytes, from where it may be released at sites of microbial infections. These iron binding glycoproteins strongly reduce the free iron concentration for microbial invaders in extracellular fluids. Those bacteria that can use transferrin and lactoferrin iron have a great advantage over those that can not.

12.3.1
Bacterial Outer Membrane Proteins that Bind Transferrin and Lactoferrin and Transport Fe^{3+}

Gram-negative bacteria, such as members of the Neisseriaceae and Pasteurellaceae, synthesize two outer membrane proteins for binding of transferrin, TbpA and TbpB, and two for binding of lactoferrin, LbpA and LbpB [67]. Transferrin iron is mobilized and transported across the outer membrane by TbpA TbpB [68]. Binding of TbpA TbpB to transferrin may be sufficient for release of Fe^{3+} from transferrin if it induces opening of the two transferrin domains that enclose Fe^{3+} (see Sect. 12.2.2 and Fig. 3). Mutants lacking TbpB still take up transferrin iron, but much less efficiently than $TbpA^+$ $TbpB^+$ wild-type cells. TbpA contains the TonB box, through which TbpA receives the energy stored in the proton motive force of the cytoplasmic membrane – TonB-box mutants are transport inactive [15]. TbpA prevents TbpB proteolysis when energized via TonB, showing that TbpA functionally interacts with TbpB. TbpB is protease-sensitive in a TonB mutant, in a TbpA TonB-box mutant, and in de-energized cells. In *Actinobacillus pleuropneumoniae*, the genes of the energy-transducing proteins TonB, ExbB, and ExbD and the outer membrane transporters TpbA and TbpB are organized in the *tonB exbB exbD tbpB tpbA* operon [69]. Cotranscription of these genes under conditions of iron deprivation indicates that they form a functional unit in the transport of transferrin iron.

12.3.2
Transport of Fe^{3+} Across the Cytoplasmic Membrane

Fe^{3+} released from transferrin and lactoferrin and presumably transported as unchelated Fe^{3+} across the outer membrane by the TbpA TbpB and LbpA LbpB transport proteins is bound in the periplasm to binding proteins. Binding of a single Fe^{3+} to FbpA, the binding protein of *Neisseria gonorrhoeae*, and to HitA (also called hFBP; see Sect. 12.2.2), the binding protein of *Haemophilus influenzae*, has been demonstrated [70]. The crystal structure of HitA reveals an octahedral geometry of the six Fe^{3+} ligands with two tyrosine oxygens, one glutamate carboxylate, one histidine imidazole nitrogen, and two additional oxygens provided by a water molecule and a phosphate ion. In transferrin, phosphate is replaced by carbonate, which provides two oxygens and makes the water molecule dispensable. Fe^{3+} is more solvent-exposed in HitA than in transferrin [58].

Transport of Fe^{3+} across the cytoplasmic membrane was originally discovered in *Serratia marcescens* when a cloned *S. marcescens* DNA fragment conferred growth on iron-limited media to an *E. coli* mutant that did not synthesize its own siderophore enterobactin [71]. Various iron sources, except those that bound iron strongly, supported growth and iron transport. Sequence analysis revealed three genes, *sfuA*, *sfuB*, and *sfuC* (Fig. 2), cotranscribed in this order under the control of iron [72]. SfuA encodes a periplasmic protein that binds Fe^{3+} which was shown by protection against degradation by V8 protease. SfuC contains the Walker motifs found in ATPases, and SfuB is a highly hydrophobic protein typical for permeases that catalyze transport across the cytoplasmic membrane. No gene encoding an outer membrane protein was found near the *sfuABC* locus; this was also found later for the Fe^{3+} transport systems of *N. gonorrhoeae* and *H. influenzae* [70]. Under conditions of strong iron limitation, diffusion of Fe^{3+} across the outer membrane of *E. coli*, used as model, is no longer sufficient to support growth. Instead, when Fe^{3+}-dicitrate is provided as iron source, it is actively transported in a TonB-dependent manner by the outer membrane FecA protein. In the periplasm, Fe^{3+} in equilibrium with citrate is bound by SfuB and transported by SfuBC across the cytoplasmic membrane. Genes required for Fe^{3+} transport and gene arrangements analogous to *sfuABC* are found in *N. gonorrhoeae* and *H. influenzae* (Fig. 2).

Biochemical studies of the iron supply by transferrin to the gram-positive bacteria *Staphylococcus aureus* and *Staphylococcus epidermidis* have revealed the Tpn transferrin binding protein (42 kDa) in the cell wall. Tpn is synthesized under iron-limiting growth conditions and elicits antibody formation in the human serum and peritoneum upon staphylococcal infections [73]. The Tpn protein is the cell wall glyceraldehyde-3-phosphate dehydrogenase, which not only binds transferrin, but also plasmin [73, 74]. Transferrin binds to Tpn with the N-domain, where iron is first released before it is mobilized from the C-domain [74]. It is assumed that the released iron is taken up into the cytoplasm by ABC transporters. Two such ABC transporters, designated *sirABC* and *sstABCD*, have been partially characterized [75].

12.4
Bacterial Use of Heme

Heme is a generally used term and does not distinguish between the Fe^{2+} form (heme) as released from hemoglobin and the Fe^{3+}-form (hemin) as present in aqueous solution. For pathogenic and commensal bacteria, heme represents an iron source widely distributed throughout the human body in various forms, but is only available at low concentrations. Heme is very insoluble in aqueous solutions and is, therefore, bound to proteins. Free heme in the serum is bound to hemopexin (12 µM) with a dissociation constant below 1 pM. Hemoglobin in the serum (0.08–0.8 µM) is released from lysed erythrocytes, rapidly binds to serum haptoglobin (5–20 µM), and is removed by the liver. Heme is also bound to serum albumin (0.5 mM) and lipoproteins. Bacteria have developed transport systems for heme delivered as heme, hemoglobin, hemoglobin–haptoglobin, heme–hemopexin, and myoglobin. Iron is probably released from heme after transport into the cytoplasm. Intracellular release of iron from heme may occur by oxidative decomposition of heme, as has been found in *Corynebacterium diphtheriae* [76] and *N. meningitidis* [76a], in which heme oxygenase activities have been identified. To date, 19 heme delivery systems have been identified in various organisms and have been characterized to various extents [77, 78].

12.4.1
Bacterial Outer Membrane Transport Proteins for Heme

Complete heme transport systems have been genetically identified (Fig. 4), but only the binding to outer membrane transport proteins has been studied biochemically. Transport across the cytoplasmic membrane via binding protein-dependent ABC transporters similar to Fe^{3+}-siderophore transport has been inferred from sequence similarities, but no studies that support this hypothesis have been published.

Free heme directly binds to outer membrane transport proteins of gram-negative bacteria or to binding proteins of gram-positive bacteria [79]; the latter are anchored to the outside of the cytoplasmic membrane by a covalently linked lipid moiety of the murein-lipoprotein type [80]. Heme contained in hemoglobin is released by [1] outer membrane transport proteins, [2] secreted hemophores (see also Chap. 7), and [3] proteolytic degradation through proteases that may also bind the heme. Heme bound to hemopexin is released by specific outer membrane transport proteins.

The first and still one of the most complete molecular studies on heme transport was carried out with *Yersinia enterocolitica* [81]. The entire operon encoding the transport system has been cloned, and four genes and the encoded proteins have been identified, three of which are required for delivery of heme iron to an *E. coli* mutant defective in heme and enterobactin synthesis (Fig. 4). Regulation of transcription of the genes encoding the heme transport proteins by the Fur repressor has been shown. The fourth protein, HemS, is required for the iron supply, but not for heme supply, from which it was deduced that HemS is involved in iron mobilization from heme. Based on sequence similarities and SDS-PAGE analysis of the encoded proteins, the HemR protein was identified as an outer membrane transport protein, HemT as a putative periplasmic binding protein, HemU as a hydrophobic hemin transport protein in the cytoplasmic membrane, and HemV as a putative energy-providing cytoplasmic ATPase [82]. Functions in heme transport were assigned by growth stimulation tests on iron-limited media and [^{14}C]heme transport assays of genetically defined mutants and strains with a wild-type phenotype obtained by complementation of mutants with cloned genes on plasmids.

Heme transport systems of *Yersinia pestis* [83], *Pseudomonas aeruginosa* [84], *Shigella dysenteriae* [85], *Vibrio cholerae* [86], *Haemophilus influenzae* type b [87], and

Neisseria gonorrhoeae [88] have been subsequently well characterized. All systems are similarly composed of a single outer membrane transport protein and an ABC transport system across the cytoplasmic membrane (Fig. 4).

Another type of heme utilization system has been discovered in *Serratia marcescens* [89]. The HasA protein (<u>h</u>eme <u>a</u>cquisition <u>s</u>ystem) is secreted, releases heme from hemoglobin, and delivers the heme to the HasR outer membrane transporter. It is not known how HasA releases heme from hemoglobin, which binds heme very tightly (association constant 10^{12}–10^{15}). Proteins functionally similar to HasA, the so-called hemophores, have been identified in *P. aeruginosa* [89] and *Pseudomonas fluorescens* [90]. *H. influenzae* type b secretes the HxuA protein, which acquires heme from the heme–hemopexin complex and delivers it to the

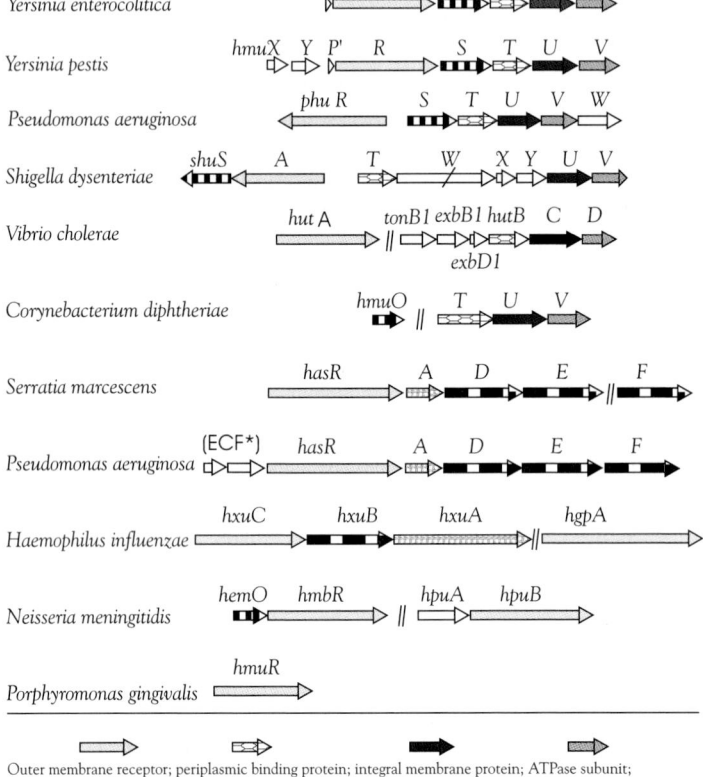

Fig. 4. Examples of genes that encode transport systems for heme from different heme sources. In several systems not all participants have been identified, e. g., in *S. marcescens*, *N. meningitidis*, *H. influenzae* and *P. gingivalis* the uptake system through the cytoplasmic membrane is not known. ECF* indicates that these two genes have similarity to *fecIR* and the FecI homolog may regulate as an <u>e</u>xtra<u>c</u>ytoplasmic-<u>f</u>unction σ-factor the *has* genes.

HxuC outer membrane transporter, from where it is taken up across the outer membrane by a TonB-dependent process [91, 92].

A third type of heme uptake has been identified in the human pathogenic *E. coli* strain EB1 [93]. A protease, Hbp, with a high affinity for hemoglobin is secreted, hydrolyzes hemoglobin, and binds heme. It is not known how the heme bound to the Hbp protease is further processed and finally transported into *E. coli*. A similar system seems to exist in *Porphyromonas gingivalis,* where the Kgp protease binds and degrades hemoglobin and delivers heme to the HmuR outer membrane transport protein [78].

12.4.2
More than one Ton System for Certain Heme Transport Systems

Transport of heme by the outer membrane transport proteins depends on the Ton system, as has been demonstrated with *tonB* mutants devoid of heme transport. The Ton system consists of the proteins TonB, ExbB, and ExbD (Fig. 1) [94]. Usually, a single TonB protein can interact with all outer membrane transport proteins of one strain and *tonB* mutants of one member of the Enterobacteriaceae can be complemented with *tonB* genes of other members of the Enterobacteriaceae. Even *tonB1* of *P. aeruginosa* [96], *tonB2* of *V. cholerae* [86], and *tonB* of *Bordetella bronchiseptica* [95] restore *tonB*-dependent activities of *E. coli tonB* mutants, in contrast to *tonB* of *Pseudomonas putida* [97], the linked *tonB1 exbB1 exbD1* genes of *V. cholerae* and *N. meningitidis* [98], and the linked *exbB exbD tonB* genes of *Haemophilus ducreyi* [99], which fail to complement *E. coli tonB* mutants. In the light of the rather promiscuous interaction of TonB with several outer membrane transport proteins, the identification of two Ton systems in *V. cholerae* came as a surprise [86]. The two *tonB exbB exbD* systems in *V. cholerae* are redundant in that only double mutations in both loci abolish growth on heme, ferrichrome, and Fe^{3+}-vibriobactin. The other unexpected finding is the linkage of *tonB1 exbB1 exbD1* with the heme transport operon (Fig. 4) and co-regulation of the six genes by the Fur repressor from a common promoter upstream of *tonB1*. This may indicate the acquisition of the complete operon by horizontal gene transfer resulting in a complete set of transport proteins and energy-providing proteins for heme import across the outer membrane and the cytoplasmic membrane.

P. aeruginosa also synthesizes two TonB proteins; TonB1 complements a *tonB* mutant of *E. coli* [96], while TonB2 fails to do so [100]. Interestingly, TonB1 is larger (341 residues) than *E. coli* TonB (239 residues) and contains an internal membrane anchor, in contrast to the N-proximal membrane anchor of all the other sequenced TonB proteins. The additional sequence is located in the N-proximal quarter, whereas the following sequence is rather similar to *E. coli* TonB (33% identity). *tonB2* forms an operon with *exbB exbD,* and the operon is not required for heme uptake; in contrast, *tonB1* mutants fail to grow on heme, hemoglobin, and Fe^{3+}-siderophores. These results suggest an additional set of *exbB exbD* genes which act together with *tonB1*.

Multiple *tonB exbB exbD* genes may also be found in iron transport systems unrelated to heme uptake. The chromosome of the plant pathogen *Xanthomonas campestris* contains the linked genes *tonB exbB exbD1 exbD2*; the protein products are anchored to the cytoplasmic membrane [101]. *tonB*, *exbB*, and *exbD1* are essential for iron uptake, whereas *exbD2* is dispensable. The only phenotype of an *exbD2* mutant that has been found is the failure to induce the hypersensitivity response and a multiplication rate higher than the *exbD2* wild-type strain on pepper leaf tissue.

12.5
Fe^{2+} Transport Systems

Under anoxic conditions, Fe^{2+} is stable and more soluble than Fe^{3+}. *E. coli* produces a ferrous iron transport system encoded by the three genes *feoABC* [102]. *feoB* encodes a 84 kDa protein located in the cytoplasmic membrane that presumably functions as transporter. FeoB contains a typical nucleotide binding motif whose function may be the energization of transport by nucleotide hydrolysis or the regulation of transport activity. *feoA* and *feoC* encode two small proteins of less than 10 kDa with unknown functions. *feo* genes have also been identified in *Salmonella enterica* serovar *typhimurium* [103]. In *Helicobacter pylori*, FeoB seems to be the major high-affinity iron acquisition system [28] since growth under iron-limiting conditions is impaired in a *feoB* mutant and cannot be restored by $FeSO_4$.

12.6
Regulation by Iron

12.6.1
Iron-dependent Repressors Regulate Iron Transport Systems

There are mainly two repressor families in bacteria: Fur in gram-negative bacteria and DtxR in gram-positive bacteria. The repressors regulate biosynthesis of siderophores and the transport proteins for the uptake of Fe^{3+}-siderophores, Fe^{3+}, and heme. Both repressors also regulate virulence factors of pathogenic bacteria, such as certain bacterial protein toxins, e.g., diphtheria toxin, shigella toxin, *P. aeruginosa* exotoxin A, and *S. marcescens* cytotoxin. Low-iron stress is taken as a signal for the host environment and induces genes required for multiplication in the host. By damaging cells of their eukaryotic hosts, these toxins make intracellular iron sources available for the bacteria. Another set of iron-regulated genes is involved with the oxidative-stress response, which is intimately connected to the iron-induced formation of cell damaging oxygen radicals. Since the oxidative burst is one of the weapons employed by mammalian macrophages to fight microbial invaders, this makes these genes all the more important for the survival of pathogenic bacteria.

The Fur protein loaded with Fe^{2+} represses transcription of genes that encode enzymes for the synthesis of siderophores and iron transport proteins [30]. The promoters of iron-regulated genes contain at least three copies of the hexamer GATAAT [104], called the Fur box. The DNA-binding portion of Fur is located in the N-terminal domain in an unusual helix-turn-helix motif [105, 106].

Some genes are positively regulated by Fur. The best-studied case is SodB, an iron-containing superoxide dismutase, whose activity is increased in the presence of iron and Fur. The *sodB* mRNA half-life is reduced in a *fur* mutant, indicating that Fur regulation is posttranscriptional and leads to stabilization of *sodB* mRNA [107]. A palindrome followed by an AT-rich stretch of bases in the first 40 transcribed bases is important for the Fur-dependent regulation of SodB, similar to the iron-responsive elements (IRE) that control iron metabolism in eukaryotic cells. In eukaryotes, the aconitase in the cytosol loses its enzyme activity when the iron–sulfur center is not formed or is disrupted by lack of iron. The iron-free form, called iron-dependent regulatory protein (IRP), binds to specific IREs in the mRNA 3′-region and stabilizes mRNA, which leads to enhanced translation, or binds to the mRNA 5′-end and inhibits translation [108]. In bacteria, e. g., in *Bacillus subtilis*, IREs exist, and binding of the aconitase protein has been demonstrated. An enzymatically inactive aconitase mutant still binds RNA and enhances sporulation 40-fold [109].

Diphtheria toxin is one of the first proteins whose synthesis was related to low iron in the growth medium. The fact that the toxin is formed upon infection and causes the disease proves the iron limitation of *C. diphtheriae* on the human throat. The Fe^{2+}-loaded DtxR protein encoded on the chromosome represses transcription of the phage-encoded toxin gene. Similar regulators found in mycobacteria (here called IdeR) and *Streptomyces* strains regulate siderophore biosynthesis and siderophore transport genes.

The crystal structures of DtxR and IdeR (the crystal structure of Fur has not been determined) reveal an N-terminal domain (residues 1–73) forming a helix-turn-helix motif that recognizes a nucleotide sequence of about 21 base pairs [110]. The central domain (74–40) has a function in dimerization; the role of the carboxy-terminal third domain (141–230) is uncertain. Although metal binding sites have been defined in these crystal structures, the mechanism by which metal binding causes the structural changes between apo- and holo-repressor is not clear.

Iron metabolism is not regulated by DtxR in all gram-positive bacteria. In *B. subtilis* (DNA with low G+C content), three Fur-like proteins have been characterized [111]. One, called Fur, regulates mainly iron uptake and siderophore biosynthesis. Another, called PerR, regulates peroxide stress response genes and acts with manganese as corepressor.

12.6.2
Regulation by Fe^{3+}

In a recently discovered Fe^{3+} regulatory mechanism, Fe^{3+} binds to the periplasmic segment of the PmrB transmembrane protein in the cytoplasmic membrane of S. enterica serovar *typhimurium* and activates PmrA, which results in a more than 16,000-fold increased resistance to the cationic antimicrobial peptide polymyxin [112]. Independent of the presence of iron, a *pmrA* mutant is 8,000-fold more sensitive to polymyxin than the wild type. The relationship to iron is supported by the killing of a *pmrA* mutant in a medium containing 1 µM $FeSO_4$, in contrast to the $pmrA^+$ wild-type strain that grows in the presence of 250 µM $FeSO_4$. The PmrA/PmrB two-component system does not regulate the intracellular iron level or affect adaptation to oxidative and nitric oxide stress. Rather, it has been suggested that the PmrA-controlled modification of LPS with 4-aminoarabinose reduces the negative charges at the cell surface to which Fe^{3+} binds. When administered orally to BALB/c mice, *pmrA* mutants display attenuated virulence [113].

12.6.3
Regulation by Fe^{3+}-siderophores

Synthesis of siderophores and synthesis of the Fe^{3+}-siderophore transport systems are not only controlled by repressors such as Fur and DtxR, but are also induced by Fe^{3+}-siderophores in the growth medium. This latter regulation has the advantage that the respective transport system is only formed when the cognate Fe^{3+}-siderophore is present in the growth medium. In contrast, iron limitation derepresses gene transcription of all iron transport systems regardless of which iron source is present. The Fe^{3+}-citrate transport system of E. coli, studied in most detail, is controlled by two regulatory processes. Transcription of the regulatory genes *fecI* and *fecR* is repressed by Fe^{2+}-Fur, and the genes are transcribed under iron limitation. The FecI and FecR proteins mediate the response of cells to Fe^{3+}-citrate in the medium, initiated by binding of Fe^{3+}-citrate to the outer membrane FecA protein [114]. This induces transcription of the *fecABCDE* transport genes (Fig. 2). FecR interacts with FecA and transmits the signal initiated by ferric–citrate-loaded FecA across the cytoplasmic membrane into the cytoplasm, where FecI is activated. FecI is a σ-factor that directs the RNA polymerase to the promoter upstream of the *fecABCDE* operon [114–118]. The novel mechanism of transcriptional control of the ferric–citrate transport system via transmembrane signaling is also observed in *Pseudomonas putida* and probably also occurs in *Pseudomonas aeruginosa*. Synthesis of the PupB outer membrane transporter for the uptake of Fe^{3+}-pseudobactin BN8 is induced by pseudobactin BN8, and PupB is required for induction. Two regulatory genes homologous to *fecIR* are encoded upstream of the *pupB* gene and are required for *pupB* transcription [119]. The genomes of P. putida and P. aeruginosa each contain approximately eight sets of genes homologous to *fecIR* [118], which suggests a number of FecIR-like regulatory devices in these organisms.

There are other mechanisms by which Fe^{3+}-siderophores induce the formation of their cognate transport systems. The reader is referred to recent reviews [10, 22, 114].

12.6.4
Regulation of Outer Membrane Transport Protein Synthesis by Phase Variation

Phase variation is a genetically determined high-frequency variation of gene expression designed to adapt bacteria to changing environmental conditions. By altering the outer membrane transport proteins, bacteria escape the host"s immune response to some of the strongest bacterial protein antigens; antibodies against these antigens are contained in the serum of patients. Changes in transporter proteins while maintaining transport activity also helps the microorganism to adapt to new iron sources since the specificity for the transported substrate is usually changed. Three genes of *H. influenzae* – *hgbA*, *hgbB*, and *hgbC*, which encode outer membrane receptor proteins for hemoglobin, hemoglobin–haptoglobin, or both compounds – contain multiple repeats of tetrameric CCAA units immediately after the sequence encoding the signal peptide. The same strain grown on heme or hemoglobin synthesizes HgbA (96 and 98 %) more frequently than HgbB (11 and 8 %). Cells that synthesize full-length HgbA contain 22 or 25 CCAA repeats, and cells that synthesize a prematurely terminated HgbA contain 23, 26, or 27 CCAA repeats. The results obtained with HgbB also relate protein synthesis to the number of CCAA repeats that give rise to in-frame or out-of-frame synthesis. HgbC is synthesized at a low level and has not been studied in detail [122].

In *Neisseria*, synthesis of HmbR, an outer membrane heme transport protein, correlates with a sequence of either 9 or 12 consecutive G nucleotides within the open reading frame, whereas a G track other than 9 or 12 nucleotides long abolishes HmbR synthesis. The rate of *hmbR* switching can differ in various serogroups from 10^{-2} to 10^{-6} [123]. Synthesis of HpuAB also depends on phase variation of a poly(G) tract [124]. An even more rapid phase variation between high and low expression levels has been found in the ferric enterobactin receptor FetA of *N. gonorrhoeae*, with a frequency of phase variation of approximately 1.3 % in the on and off directions [125]. The *fetA* transcription rate is determined by the number of C nucleotides in a poly(C) tract in the promoter region, with 12 cytosines in the high transcription state and 11 cytosines in the low transcription state.

12.7
Outlook

The more human, animal, and plant bacterial pathogens are identified and the genomes sequenced, the more iron transport systems will be found and characterized. To relate iron transport systems to virulence and pathogenicity, it will be necessary to use in experimental animal systems genetically well-defined strains that synthesize only one iron transport system. Bacteria isolated from humans and after experimental animal infections should be studied without further subculturing by microarray techniques to identify transcription of genes related to iron metabolism. These include genes that are controlled by iron, and that encode proteins that synthesize iron-chelating compounds, iron transport systems, protein toxins, and iron-related stress proteins, such as those that respond to oxygen radical formation and destruction, and genes of unknown functions. The genetic studies should be complemented by in-depth biochemical and biophysical studies on selected iron transport systems, as is done in the enzymological studies of siderophore biosynthesis [126]. Energized active transport across the outer membrane poses the most challenging problem since iron transport will reveal novel transport and energization mechanisms. Transport across the cytoplasmic membrane may also disclose novel aspects that differ from the partially understood ABC transporters of maltose and histidine, as the unique FhuD crystal structure (Fig. 3) demonstrates. Only the surface of iron-related regulation has been scratched and further studies may uncover novel transcription control mechanisms, as the examples of ferric–citrate transport and anguibactin synthesis indicate.

Acknowledgement

We thank Michael Braun for the computer drawing of Fig. 1 and Karen A. Brune for editing the manuscript. The authors" work was supported by the Deutsche Forschungsgemeinschaft (SFB 323, B1, B6, BR 330/19-1, BR 330/20-1, Ha 1186/3-1) and the Fonds der Chemischen Industrie.

References

1. Chipperfield, J. R., Ratledge, C., *Biometals* **2000**, *13*, 165-168.
2. Archibald, F., *FEMS Micobiol. Lett.* **1983**, *19*, 29-32.
3. Posey, J. E., Gherardini, F. C., *Science* **2000**, *288*, 1651-1653.
4. Braun, V., Hantke, K., Köster, W., in: *Metal Ions in Biological Systems* (Sigel, A., Sigel, H., Eds.), Marcel Dekker, New York, **1998**, pp. 67-145.
5. Bullen, J. J., Griffith, E., in: *Iron and infection, molecular, physiological and clinical aspects*, John Wiley & Sons, New York, **1999**.
6. Crosa, J. H., *Annu. Rev. Microbiol.* **1984**, *38*, 69-89.
7. Wooldrigde, K. G., Williams, P. H., *FEMS Microbiol. Rev.* **1993**, *12*, 325-348.
8. Payne, S. M., *Crit. Rev. Microbiol.* **1988**, *16*, 81-111.
9. Ratledge, C., Dover, L. G., *Annu. Rev. Microbiol.* **2000**, *54*, 881-9411.
10. Vasil, M. L., Ochsner, U. A., *Mol. Microbiol.* **1999**, *34*, 399-413.
11. Expert, D., Enard, C., Masclaux, C., *Trends Microbiol.* **1996**, *4*, 232-237.
12. Serkin, C. D., Seifert, H. S., *Mol. Microbiol.* **2000**, *37*, 1075-1086.
13. Sun, Y. H., Bakshi, S., Chalmers, R., Tang, C. M., *Nature Med.* **2000**, *6*, 1269-1273.
14. Baltes, N., Tonpitak, W., Gerlach, G. F., Hennig-Pauka, I., Hoffmann-Moujahid, A., Ganter, M., Rothkotter, H. J., *Infect. Immun.* **2001**, *69*, 472-478.

15. Cornelissen, C. N., Kelley, M., Hobbs, M. M., Anderson, J. E., Cannon, J. G., Cohen, M. S., Sparling, P. F., *Mol. Microbiol.* **1998**, *27*, 611-616.
16. Bonnah, R. A., Lee, S. W., Vasquez, B. L., Enns, C. A., So, M., *Cell. Microbiol.* **2000**, *2*, 207-218.
17. Takase, H., Nitanai, H., Hoshino, K., Otani, T., *Infect. Immun.* **2000**, *68*, 1834-1839.
18. Takase, H., Nitanai, H., Hoshino, K., Otani, T., *Infect. Immun.* **2000**, *68*, 4498-4504.
19. Britigan, B. E., Rasmussen, G. T., Olakanmi, O., Cox, C. D., *Infect. Immun.* **2000**, *68*, 1271-1275.
20. De Voss, J. J., Rutter, K., Schroeder, B. G., Su, H., Zhu, Y., Barry, C. E., *Proc. Natl. Acad. Sci. USA* **2000**, *97*, 1252-1257.
21. Crosa, J. H., *Microbiol. Mol. Biol. Rev.* **1997**, *61*, 319-336.
22. Crosa, J. H., in: *Iron and Infection* (Bullen J. J., Griffith, E., Eds.), John Wiley & Sons, Chichester, UK, **1999**, pp. 255-288.
23. Heesemann, J., Hantke, K., Vocke, T., Saken, E., Rakin, A., Stojiljkovic, I., Berner, R., *Mol. Microbiol.* **1993**, *8*, 397-408.
24. Bäumler, A. J., Hantke, K., *Mol. Microbiol.* **1992**, *6*, 1309-1321.
25. Bearden, S. W., Perry, R. D., *Mol. Microbiol.* **1999**, *32*, 403-414.
26. Viswanathan, V. K., Edelstein, P. H., Pope, C. D., Cianciotto, N. P., *Infect. Immun.* **2000**, *68*, 1069-1079.
27. Janakiraman, A., Slauch, J. M., *Mol. Microbiol.* **2000**, *35*, 1146-1155.
28. Velayudhan, J., Hughes, N. J., McColm, A. A., Bagshaw, J., Clayton, C. L., Andrews, S. C., Kelly, D. J., Venturi, V., Weisbeek, P., Koster, M., *Mol. Microbiol.* **1995**, *17*, 603-610.
29. Al-Tawfiq, J. A., Bauer, M. E., Fortney, K. R., Katz, B. P., Hood, A. F., Ketterer, M., Apicella, M. A., Spinola, S. M., *J. Infect. Dis.* **2000**, *181*, 1176-1179.
30. Hantke, K., Braun, V., in: *Bacterial Stress Responses* (Storz, G., Hengge-Aronis, R., Eds.), ASM Press, Washington, DC, **2000**, pp. 275-288.
31. Heller, K. J., Kadner, R. J., Günter, K., *Gene* **1988**, *64*, 147-153.
32. Schöffler, H., Braun, V., *Mol. Gen. Genet.* **1989**, *217*, 378-383.
33. Cadieux, N., Kadner, R. J., *Proc. Natl. Acad. Sci. USA* **1999**, *96*, 10673-10678.
34. Jiang, X., Payne, M. A., Cao, Z., Foster, S. B., Feix, J. B., Newton, S. M., Klebba, P. E., *Science* **1997**, *276*, 1261-1264.
34a. Buchanan, S. K., Venkatramani, L., Xia, D., Esser, L., Palnitkar, M., Chakraborty, R., van der Helm, D., Deisenhofer, J., *Nature Struct. Biol.* **1999**, *6*, 56-63.
35. Rutz, J. M., Liu, J., Lyons, J. A., Goranson, J., Armstrong, S. K., McIntosh, M. A., Feix, J. B., Klebba, P. E., *Science* **1992**, *258*, 471-475.
36. Hancock, R. W., Braun, V., *J. Bacteriol.* **1976**, *125*, 409-415.
37. Killmann, H., Videnov, G. Jung, G., Schwarz, H., Braun, V., *J. Bacteriol.* **1995**, *177*, 694-698.
38. Koebnik, R., Braun, V., *J. Bacteriol.* **1993**, *175*, 826-839.
39. Ferguson, A. D., Hofmann, E., Coulton, J. W., Diederichs, K., Welte, W., *Science* **1998**, *282*, 2215-2220.
40. Locher, K. P., Rees, B., Koebnik, R., Mitschler, A., Moulinier, L., Rosenbusch, J. P., Moras, D., *Cell* **1998**, *95*, 771-778.
41. Moeck, G. S., Coulton, J. W., *Mol. Microbiol.* **1998**, *28*, 675-681.
42. Locher, K. P., Rosenbusch, J. P., *Eur. J. Biochem.* **1997**, *247*, 770-775.
43. Bös, C., Lorenzen, D., Braun, V., *J. Bacteriol.* **1998**, *180*, 605-613.
44. Moeck, G. S., Coulton, J. W., Postle, K., *J. Biol. Chem.* **1997**, *272*, 28391-28397.
45. Merianos, H. J., Cadieux, N., Lin, C. H., Kadner, R. J., Cafiso, D. S., *Nature Struct. Biol.* **2000**, *7*, 205-209.
46. Braun, M., Killmann, H., Braun, V., *Mol. Microbiol.* **1999**, *33*, 1037-1049.
47. Letain, T. E., Postle, K., *Mol. Microbiol.* **1997**, *24*, 271-283.
48. Carmel, G., Hellstern, D., Henning, D., Coulton, J. W., *J. Bacteriol.* **1990**, *172*, 1861-1869.
49. Moeck, G. S., Bazzaz, B. S., Gras, M. F., Ravi, T. S., Ratcliffe, M. J., Coulton, J. W., *J. Bacteriol.* **1994**, *176*, 4250-4259.
50. Ferguson, A. D., Ködding, J., Walker, G., Bös, C., Coulton, J. W., Diederichs, K., Braun, V., Welte, W., *Structure* **2001**, in press.
51. Braun, V., *Drug Resistance Updates* **2000**, *2*, 363-369.
52. Stintzi, A., Barnes, C., Xu, J., Raymond, K. N., *Proc. Natl. Acad. Sci. USA* **2000**, *97*, 10691-61.

53. Schalk, I. J., Kyslik, P., Prome, D., van Dorsselaer, A., Poole, K., Abdallah, M. A., Pattus, F., *Biochemistry* **1999**, *38*, 9357-9365.
54. Köster, W., Braun, V., *J. Biol. Chem.* **1990**, *265*, 21407-21410.
55. Rohrbach, M. R., Braun, V., Köster, W., *J. Bacteriol.* **1995**, *177*, 7186-7193.
56. Clarke, T. E., Ku, S.-Y., Dougan, D. R., Vogel, H., Tari, L. W., *Nature Struct. Biol.* **2000**, *7*, 287-291.
57. Anderson, B. F., Baker, H. M., Norris, G. E., Rumball, S. V., Baker, E. N., *Nature* **1990**, *344*, 784-787.
58. Bruns, C. M., Nowalk, A. J., Arvai, A. S., McTigue, M. A., Vaughan, K. G., Mietzner, T. A., McRee, D. E., *Nature Struct. Biol* **1997**, *4*, 919-924.
59. Tam, R., Saier Jr., M. H., *Microbiol. Rev.* **1993**, *57*, 320-346.
60. Mademidis, A., Killmann, H., Kraas, W., Flechsner, I., Jung, G., Braun, V., *Mol. Microbiol.* **1997**, *26*, 1109-1123.
61. Braun, V., Killmann, H., *Trends Biochem. Sci.* **1999**, *24*, 104-109.
62. Killmann, H., Benz, R., Braun, V., *EMBO J.* **1993**, *12*, 3007-3016.
63. Groeger, W., Koster, W., *Microbiology* **1998**, *144*, 2759-2769.
63a. Schulz-Hauser, G., Köster, W., Schwarz, H., Braun, V., *J. Bacteriol.* **1992**, *174*, 2305-2311.
64. Becker, K., Köster, W., Braun, V., *Mol. Gen. Genet.* **1990**, *223*, 159-162.
65. Köster, W., Böhm, B., *Mol. Gen. Genet.* **1992**, *232*, 399-407.
66. Böhm, B., Boschert, H., Köster, W., *Mol. Microbiol.* **1996**, *20*, 223-232.
67. Schryvers, A. B., Stojiljkovic, I., *Mol. Microbiol.* **1999**, *32*, 1117-1123.
68. Gomez, J. A., Criado, M. T., Ferreiros, C. M., *Res. Microbiol.* **1998**, *149*, 381-387.
69. Tonpitak, W., Thiede, S., Oswald, W., Baltes, N., Gerlach, G. F., *Infect. Immun.* **2000**, *68*, 1164-1170.
70. Mietzner, T. A., Tencza, S. B., Adhikari, P., Vaughan, K. G., Nowalk, A. J., *Curr. Top. Microbiol. Immunol.* **1998**, *225:113-35*, 113-135.
71. Zimmermann, L., Angerer, A., Braun, V., *J. Bacteriol.* **1989**, *171*, 238-243.
72. Angerer, A., Klupp, B., Braun, V., *J. Bacteriol.* **1992**, *174*, 1378-1387.
73. Modun, B. J., Cockayne, A., Finch, R., Williams, P., *Microbiology* **1998**, *144*, 1005-1012.
74. Modun, B., Evans, R. W., Joannou, C. L., Williams, P., *Infect. Immun.* **1998**, *66*, 3591-3596.
75. Morrissey, J. A., Cockayne, A., Hill, P. J., Williams, P., *Infect. Immun.* **2000**, *68*, 6281-6288.
76. Chu, G. C., Park, S. Y., Shiro, Y., Yoshida, T., Ikeda-Saito, M., *J. Struct. Biol.* **1999**, *126*, 171-174.
76a. Zhu, W., Wilks, A., Stojiljkovic, I., *J. Bacteriol.* **2000**, *182*, 6783-6790.
77. Wandersman, C., Stojiljkovic, I., *Curr. Opin. Microbiol.* **2000**, *3*, 215-220.
78. Genco, C. A., Dixon, D. W., *Mol. Microbiol.* **2001**, *39*, 1-11.
79. Drazek, E. S., Hammack, C. A., Schmitt, M. P., *Mol. Microbiol.* **2000**, *36*, 68-84.
80. Hantke, K., Braun, V., *Eur. J. Biochem.* **1973**, *34*, 284-296.
81. Stojiljkovic, I., Hantke, K., *EMBO J.* **1992**, *11*, 4359-4367.
82. Stojiljkovic, I., Hantke, K., *Mol. Microbiol.* **1994**, *13*, 719-732.
83. Thompson, J. M., Jones, H. A., Perry, R. D., *Infect. Immun.* **1999**, *67*, 3879-3892.
84. Ochsner, U. A., Johnson, Z., Vasil, M. L., *Microbiology* **2000**, *146 (Pt 1)*, 185-198.
85. Wyckoff, E. E., Duncan, D., Torres, A. G., Mills, M., Maase, K., Payne, S. M., *Mol. Microbiol.* **1998**, *28*, 1139-1152.
86. Occhino, D. A., Wyckoff, E. E., Henderson, D. P., Wrona, T. J., Payne, S. M., *Mol. Microbiol.* **1998**, *29*, 1493-1507.
87. Morton, D. J., Whitby, P. W., Jin, H., Ren, Z., Stull, T. L., *Infect. Immun.* **1999**, *67*, 2729-2739.
88. Chen, C. J., Elkins, C., Sparling, P. F., *Infect. Immun.* **1998**, *66*, 987-993.
89. Letoffe, S., Nato, F., Goldberg, M. E., Wandersman, C., *Mol. Microbiol.* **1999**, *33*, 546-555.
90. Idei, A., Kawai, E., Akatsuka, H., Omori, K., *J. Bacteriol.* **1999**, *181*, 7545-7551.
91. Hanson, M. S., Pelzel, S. E., Latimer, J., Muller Eberhard, U., Hansen, E. J., *Proc. Natl. Acad. Sci. USA* **1992**, *89*, 1973-1977.
92. Jarosik, G. P., Maciver, I., Hansen, E. J., *Infect. Immun.* **1995**, *63*, 710-713.
93. Otto, B. R., van Dooren, S. J., Nuijens, J. H., Luirink, J., Oudega, B., *J. Exp. Med.* **1998**, *188*, 1091-1103.
94. Braun, V. ,*FEMS Microbiol. Rev.* **1995**, *16*, 295-307.

95. Nicholson, M. L., Beall, B., *Microbiology* **1999**, *145 (Pt 9)*, 2453-2461.
96. Poole, K., Zhao, Q., Neshat, S., Heinrichs, D. E., Dean, C. R., *Microbiology* **1996**, *142 (Pt 6)*, 1449-1458.
97. Bitter, W., Tommassen, J., Weisbeek, P. J., *Mol. Microbiol.* **1993**, *7*, 117-130.
98. Stojiljkovic, I., Srinivasan, N., *J. Bacteriol.* **1997**, *179*, 805-812.
99. Elkins, C., Totten, P. A., Olsen, B., Thomas, C. E., *Infect. Immun.* **1998**, *66*, 151-160.
100. Zhao, Q., Poole, K., *FEMS Microbiol. Lett.* **2000**, *184*, 127-132.
101. Wiggerich, H. G., Klauke, B., Koplin, R., Priefer, U. B., Pühler, A., *J. Bacteriol.* **1997**, *179*, 7103-7110.
102. Kammler, M., Schön, C., Hantke, K., *J. Bacteriol.* **1993**, *175*, 6212-6219.
103. Tsolis, R. M., Bäumler, A. J., Stojilkovic, I., Heffron, F., *J. Bacteriol.* **1995**, *177*, 4628-4637.
104. Escolar, L., Perez-Martin, J., De Lorenzo, V., *J. Mol. Biol.* **1998**, *283*, 537-547.
105. Holm, L., Sander, C., Rüterjans, H., Schnarr, M., Fogh, R., Boelens, R., Kaptein, R., *Protein Eng.* **1994**, *7*, 1449-1453.
106. Stojiljkovic, I., Hantke, K., *Mol. Gen. Genet.* **1995**, *247*, 199-205.
107. Dubrac, S., Touati, D., *J. Bacteriol.* **2000**, *182*, 3802-3808.
108. Hentze, M. W., Kuhn, L. C., *Proc. Natl. Acad. Sci. USA* **1996**, *93*, 8175-8182.
109. Alen, C., Sonenshein, A. L., *Proc. Natl. Acad. Sci. USA* **1999**, *96*, 10412-10417.
110. Pohl, E., Holmes, R. K., Hol, W. G., *J. Mol. Biol.* **1999**, *292*, 653-667.
111. Bsat, N., Herbig, A., Casillas-Martinez, L., Setlow, P., Helmann, J. D., *Mol. Microbiol.* **1998**, *29*, 189-198.
112. Wosten, M. M., Kox, L. F., Chamnongpol, S., Soncini, F. C., Groisman, E. A., *Cell* **2000**, *103*, 113-125.
113. Gunn, J. S., Ryan, S. S., Van Velkinburgh, J. C., Ernst, R. K., Miller, S. I., *Infect. Immun.* **2000**, *68*, 6139-6146.
114. Braun, V., *Arch.Microbiol.* **1997**, *167*, 325-331.
115. Kim, I., Stiefel, A., Plantör, S., Angerer, A., Braun, V., *Mol. Microbiol.* **1997**, *23*, 333-344.
116. Welz, D., Braun, V., *J.Bacteriol.* **1998**, *180*, 2387-2394.
117. Enz, S., Mahren, S., Stroeher, U. H., Braun, V., *J.Bacteriol..* **2000**, *182*, 637-646.
118. Stiefel, A., Mahren, S., Ochs, M., Schindler, P. T., Enz, S., Braun, V., *J Bacteriol* **2001**, *183*, 162-170.
119. Koster, M., van Klompenburg, W., Bitter, W., Leong, J., Weisbeek, P. J., *EMBO J.* **1994**, *13*, 2805-2813.
120. Venturi, V., Weisbeek, P., Koster, M., *Mol. Microbiol.* **1995**, *17*, 603-610.
121. Ren, Z., Jin, H., Whitby, P. W., Morton, D. J., Stull, T. L., *J. Bacteriol.* **1999**, *181*, 5865-5870.
122. Cope, L. D., Hrkal, Z., Hansen, E. J., *Infect. Immun.* **2000**, *68*, 4092-4101.
123. Richardson, A. R., Stojiljkovic, I., *J. Bacteriol.* **1999**, *181*, 2067-2074.
124. Lewis, L. A., Gipson, M., Hartman, K., Ownbey, T., Vaughn, J., Dyer, D. W., *Mol. Microbiol.* **1999**, *32*, 977-989.
125. Carson, S. D., Stone, B., Beucher, M., Fu, J., Sparling, P. F., *Mol. Microbiol.* **2000**, *36*, 585-593.
126. Cane, D. E., Walsh, C. T., *Chem. Biol.* **1999**, *6*, R319-R325.
127. Ferguson, A. D., Braun, V., Fiedler, H.-P., Coulton, J. W., Diederichs, K., Welte, W., *Protein Sci.* **2000**, *9*, 956-963.
128. Clarke, T. E., Ku, S. Y., Dougan, D. R., Vogel, H. J., Tari, L. W., *Nature Struct. Biol.* **2000**, *7*, 287-291.
129. Fischer, E., Günter, K., Braun, V., *J. Bacteriol.* **1989**, *171*, 5127-5134.
130. Braun, V., Gaisser, S., Herrmann, C., Kampfenkel, K., Killmann, H., Traub, I., *J. Bacteriol.* **1996**, *178*, 2836-2845.
131. Hannavy, K., Barr, G. C., Dorman, C. J., Adamson, J., Mazengera, L. R., Gallagher, M. P., Evans, J. S., Levine, B. A., Trayer, I. P., Higgins, C. F., *J. Mol. Biol.* **1990**, *216*, 897-910.
132. Roof, S. K., Allard, J. D., Bertrand, K. P., Postle, K., *J. Bacteriol.* **1991**, *173*, 5554-5557.
133. Kampfenkel, K., Braun, V., *J. Biol. Chem.* **1993**, *268*, 6050-6057.
134. Kampfenkel, K., Braun, V., *J. Bacteriol.* **1992**, *174*, 5485-5487.

13
Bacterial Zinc Transport

Klaus Hantke

13.1
Introduction

Zinc is, next to iron, the most important trace metal in bacteria. Zinc is found as a cofactor and as a structural element in many proteins. It is a "silent" metal in spectroscopic and analytical methods; therefore, its presence and importance may have been underestimated. In proteins, the metal is mainly tetrahedrally coordinated by Cys (C), His (H), and carboxylates of Asp (D) or Glu (E). The high affinity of thiolates for Zn^{2+} forces the cells to control the Zn^{2+} concentration in the cytoplasm carefully since too much Zn^{2+} may poison enzymes or the electron transport chain [1, 2]. The first transport systems for Zn^{2+} studied at the molecular level were for export of the toxic ion. Later, high-affinity uptake systems were identified that allow the cells to survive in Zn^{2+}-poor environments. Many ABC-type Me^{2+} permeases from pathogenic bacteria have been studied since they have a strong influence on virulence. Recently, this group of ABC transporters was excellently reviewed by Claverys [3].

13.2
Exporters of Toxic Zn^{2+}

13.2.1
RND Family of Exporters

Microorganisms living in zinc-rich environments have developed mechanisms to withstand the toxicity of this metal. In bacteria, mainly export systems protect the cytoplasm from high toxic levels of Zn^{2+}. The systems surveyed are given in Tab. 1.

The bacterium *Ralstonia metallidurans* (formerly *Alcaligenes eutrophus*) CH34 was isolated from a decantation tank of a zinc factory. Its minimal inhibitory concentration for Zn^{2+} is 12 mM in a Tris-based medium. The *czcNICBADRS* cluster of

Tab. 1. Genes and the function of the encoded proteins involved in Zn^{2+} export

Gene (Organism)	Function of the Encoded Proteins	Protein Family	Reference
czcA (Ralstonia metallidurans)	cation/proton antiporter for Cd^{2+}, Co^{2+}, and Zn^{2+}	RND (resistance, nodulation, division)	(6)
czcB/C (R. metallidurans)	CzcB/C couple CzcA to the outer membrane to allow extrusion of Me^{2+}	RND	(6)
czrA (Pseudomonas aeruginosa)	cation/proton antiporter	RND	(8)
czrB/C (P. aeruginosa)	CzrB/C couple CzrA to the outer membrane to allow extrusion of Me^{2+}	RND	(8)
czcD (R. metallidurans)	cation diffusion facilitator	CDF	(10)
zntA (Staphylococcus aureus)	cation diffusion facilitator	CDF	(11)
cadA (S. aureus)	$Cd^{2+} > Zn^{2+}$ exporter	P-type ATPase	(15)
cadA (Listeria monocytogenes)	Cd^{2+} exporter	P-type ATPase	(17)
zntA (Escherichia coli)	$Cd^{2+} > Zn^{2+}$ exporter	P-type ATPase	(18,19)

genes located on a megaplasmid is responsible for this high resistance. The products of the three genes czcABC form a cation export system across the inner and the outer membranes. The transporter mediates resistance to Co^{2+}, Cd^{2+}, and Zn^{2+}. The ions are exported with different kinetic parameters – Cd^{2+} with high affinity, Co^{2+} with low affinity – both following Michaelis–Menten kinetics. Zn^{2+} export is slow at low concentrations, but fast at high concentrations (sigmoidal substrate saturation kinetics), which seems to avoid Zn^{2+} depletion of the cell [4]. czcD encodes another zinc exporter (see below), the genes czcRS encode a Zn^{2+}-responsive two-component regulatory system, while the function of czcNI is unknown.

CzcA is a cation/proton antiporter that extrudes Zn^{2+} across the cytoplasmic membrane [5]. The special feature of this system is that CzcB in the periplasm and CzcC probably in the outer membrane form a channel that connects CzcA to the outside, thereby allowing extrusion of Zn^{2+} across two membranes [6]. CzcC is distantly related to TolC. The impressive crystal structure of TolC shows a porin-like barrel structure with 12 β-strands and a helical barrel structure that seems to form a tunnel spanning the periplasm. The β-barrel and the helical barrel are formed by three TolC monomers [7]. Further studies are necessary to determine whether CzcC has the expected structural similarity to TolC, which would nicely fit the model presented [6]. Similar Zn^{2+} exporters are found in *Pseudomonas aeruginosa* [8]. This type of export machinery is also involved in antibiotic and detergent export-resistance, nodulation, and division (RND) [9].

13.2.2
Cation Diffusion Facilitator

CzcD encoded as part of the Zn^{2+} resistance operon of *R. metallidurans* is a member of the cation diffusion facilitator family and exports Co^{2+}, Zn^{2+}, and Cd^{2+} across the cytoplasmic membrane [10]. A Zn^{2+} resistance determinant from *Staphylococcus aureus*, ZntA, shows 38% sequence identity to CzcD from *R. metallidurans* [11]. A *zntA* mutant is sensitive to 0.5 mM Zn^{2+}, which is lower than the 5 mM sensitivity of the parent strain. ZntR, a member of the ArsR family, regulates the expression of ZntA. The nomenclature is very unfortunate since these ZntA and ZntR proteins are not related to the ZntA and ZntR proteins of *Escherichia coli* and other bacteria, which are treated in the next sections. A few other members of the cation diffusion-facilitator protein family from eukaryotes have been characterized; in bacteria only sequence similarities indicate a wider distribution [11a].

13.2.3
P-Type ATPases Export Cd^{2+} and Zn^{2+}

P-type ATPases specific for Zn^{2+}, Cd^{2+}, and Pb^{2+} form a distinct group in this large protein family [12]. CadA from *S. aureus* was the first protein of this family studied intensely [13]. The 727-amino acid CadA protein is encoded on a plasmid and raises the Cd^{2+} resistance of *S. aureus* from 0.005 mM to 2.5 mM. Resistance to Zn^{2+} is only raised by a factor of three from 0.6 mM to 1.8 mM [14]. The small CadC protein is an ArsR/SmtB-type repressor that interacts with $Cd^{2+} > Zn^{2+}$ [15].

A nearly identical CadA/C transport system has been found in the gram-negative bacterium *Stenotrophomonas maltophila* (96% sequence identity to CadA/C from *S. aureus*) [16]. This indicates a recent horizontal gene transfer between these rather different bacteria.

In *Listeria monocytogenes*, a similar CadA/C Cd^{2+} export system with 66 and 48% identity to *S. aureus* CadA and CadC, respectively, has been identified [17]. The Zn^{2+} resistance level of a *Bacillus subtilis* strain transformed with this CadA/C export system is the same as that of the strain without CadA/C, possibly because of the high intrinsic Zn^{2+} resistance of the strain used. Results of similarity searches indicate a wide distribution of this type of export system in gram-positive bacteria.

Related P-type ATPases with a metal specificity for Cd^{2+}, Zn^{2+}, and Pb^{2+} are found in gram-negative bacteria. *zntA* [18,19] mutants of *E. coli* have been isolated that have a minimal inhibitory concentration (MIC) of 0.03 mM Cd^{2+}, compared to 1.5 mM in the parent strain. The MIC for Zn^{2+} of the *zntA* mutant is 0.5 mM, while the parent has an MIC of 2.0 mM. A ZntA- and ATP-dependent transport of $^{65}Zn^{2+}$ and $^{109}Cd^{2+}$ has been demonstrated using everted membrane vesicles. Vanadate inhibition of this transport (typical of P-type ATPases) has been shown *in vitro*. The export protein is regulated by the MerR-type regulator/activator ZntR, which activates the expression of *zntA* at Cd^{2+} concentration ≥ 0.019 mM and at Zn^{2+} concentrations ≥ 0.1 mM [20, 21].

Soft metal ion translocating P-type ATPases that transport Cu^+ and Ag^+ are very similar to transporters like ZntA, which is specific for the divalent ions Cd^{2+} and Zn^{2+}. In humans, mutations in the P-type ATPases for Cu^+ transport produce Menkes or Wilson's disease. ZntA mutants with a His475Gln or a Glu470Ala mutation have been constructed. The mutations are at the positions of the copper translocating P-type ATPase equivalent to those that cause Wilson's disease. Both ZntA mutants show a reduced metal ion-stimulated ATPase activity (about 30-40% of the wild-type activity) and are phosphorylated much less efficiently than the wild type. These results suggest that the mutations affect major stages in the transport process of both P-type ATPases [22].

13.3
High-affinity Uptake Systems for Zn^{2+} are ABC Transporters

13.3.1
Binding Protein-dependent Zn^{2+} Uptake in Gram-positive Bacteria

In certain environments, such as the serum where Zn^{2+} is tightly bound to albumin, Zn^{2+} is a scarce ion for bacteria. Therefore, most organisms have high-affinity Zn^{2+} uptake systems. Those discussed here are listed in Tab. 2.

Tab. 2. Binding protein-dependent metal ABC transporters

Gene (Organism)	Function of the Encoded Proteins	Reference
yfeA (*Yersinia pestis*)	Mn^{2+}/$Fe^?$ binding protein of an ABC transporter	(40)
yfeC and *yfeD*	integral membrane protein	
yfeB	ATP-binding protein	
sitA (*Salmonella typhimurium*)	$Fe^?$ binding protein of an ABC transporter	(41)
sitC and *sitD*	integral membrane proteins	
sitB	ATP-binding protein	
mntC (*Synechocystis* sp.)	Mn^{2+} binding protein of an ABC transporter	(42)
mntB	integral membrane protein	
mntA	ATP-binding protein	
psaA (*Streptococcus pneumoniae*)	Mn^{2+} binding protein of an ABC transporter	(23)
psaC	integral membrane protein	
psaB	ATP-binding protein	

Tab. 2. continued

Gene (Organism)	Function of the Encoded Proteins	Reference
scaA (Streptococcus gordonii) scaB scaC	Mn^{2+} binding protein of an ABC transporter integral membrane protein ATP-binding protein	(43)
mtsA (Streptococcus pyogenes) mtsC mtsB	Me^{2+} binding protein of an ABC transporter integral membrane protein ATP-binding protein	(27)
fimA (Streptococcus parasanguis) (ORF1) (ORF5)	putative Mn^{2+} binding protein of an ABC transporter putative integral membrane protein putative ATP-binding protein	(44)
troA (Treponema pallidum) troC troD troB	Mn^{2+} binding protein of an ABC transporter integral membrane proteins ATP-binding protein	(26)
ytgA (Bacillus subtilis) ytgC ytgD ytgB	Mn^{2+} binding protein of an ABC transporter integral membrane proteins ATP-binding protein	(45)
adcA (S. pneumoniae) adcB adcC	Zn^{2+} binding protein of an ABC transporter integral membrane proteins ATP-binding protein	(23)
ycdH (B. subtilis) yceA ycdI	Zn^{2+} binding protein of an ABC transporter integral membrane protein ATP-binding protein	(28)
znuA (Haemophilus ducreyi) ? ?	Zn^{2+} binding protein of an ABC transporter	(46)
pzp1 (Haemophilus influenzae) HI0407 HI0408	Zn^{2+} binding protein of an ABC transporter; not in an operon with the putative integral membrane protein putative ATP-binding protein	(29,47)
znuA (Escherichia coli) znuB znuC	Zn^{2+} binding protein of an ABC transporter integral membrane protein ATP-binding protein	(31)
ewlA (Erysipelothrix rhusopathiae) ORF 2	putative Zn^{2+} binding protein of an ABC transporter putative ATP-binding protein	
znuA (Thermotoga maritima) znuB znuC	putative Zn^{2+} binding protein of an ABC transporter putative integral membrane proteins putative ATP-binding protein	(48)

```
YFEA_YERPE  ---MLIKKKSPYLKMIERLNSPFLRAAALFTIVAFSSLIS--TAALAENNPSDTAKKFKVVTTFTI  Fe/Mn
SITA_SALEN  ------------MTNLHRLKT--LLIAGIVAILALS-------------PAYAKEKFKVITTFTV  Fe
MNTC_SYNSP  ---MATSFASRGGLLASGLAIAFWLTGCGTAEVTTSNAPSEEVTAVTTEVQGETEEKKKVLTTFTV  Mn
PSAA_STRPN  ---------------MKKLGTLLVFLSAIILVACAS---------GKKDT-TSGQKLKVVATNSI  Mn
SCAA_STRGO  ---------------MKKCRFLVLLLLAFVGLAACSS---------QKSSTDSSSSKLNVVATNSI  Mn
MTSA_STRPY  -------------MGKRMSLILGAFLSVFLLVACSS---------TGAKT-AESDKLKVVATNSI  Me
TroA_TREPA  ---------------MIRERICACVLALGMLTGF-THA---------FGSKDAAADGKPLVVTTIGM  Mn?
YTGA_BACSU  --------------MRQGLMAAVLFATFALTGCGTDS---------AG---KSADQQLQVTATTSQ  Mn
ADCA_STRPN  ---------------MKKIS-LLLASLCALFLVACSN---------QKQADG---KLNIVTTFYP  Zn
YCDH_BACSU  -------------MFKKWSGLFVIAACFLLVAACGNS-----S--TKGSADSKGDKLHVVTTFYP  Zn
ZNUA_HAEDU  ---------------------MFKKTVLTLAMLG---------------VTTVANADVLTSIKP  Zn
ZNUA_HAEIN  --------------MKKLLKISAISAALLS--------------APMMANADVLASVKP  Zn
ZNUA_ECOLI  MKCYNITLLIFITIIGRIMLHKKTLLFAALSAALWGG---------------ATQAADAAVVASLKP  Zn
EWLA_ERYRH  --------------MKKILAVLLVSLLVLTGCKSAP----------NNPSTENDGKINVVATTTM  Zn?
ZNUA_THEMA  -----------------MKKILLLLVLIVAVLNFG-------------------KTIVTTINP  Zn?
                                 :  ::

YFEA_YERPE  IQDIAQNIAGDVAVVESITKPGAEIHDYQPTPRDIVKAQSADLILWNGMNLER----WFEKFFESI  Fe/Mn
SITA_SALEN  IADMAKNVAGDAAEVSSITKPGAEIHEYQPTPGDIKRAQGAQLILANGLNLER----WFARFYQHL  Fe
MNTC_SYNSP  LADMVQNVAGDKLVVESITRIGAEIHGYEPTPSDIVKAQDADLILYNGMNLER----WFEQFLGNV  Mn
PSAA_STRPN  IADITKNIAGDKIDLHSIVPIGQDPHEYEPLPEDVKKTSEADLIFYNGINLETGGNAWFTKLVENA  Mn
SCAA_STRGO  IADITKNIAGDKINLHSIVPVGQDPHKYEPLPEDVKKTSKADLIFYNGINLETGGNAWFTKLVENA  Mn
MTSA_STRPY  IGDMTKVMAGDKIDLHSIVPIGQDPHEYEPLPEDVEKTSNADVIFYNGINLEDGGQAWFTKLVKNA  Me
TroA_TREPA  IADAVKNIAQGDVHLKGLMGPGVDPHLYTATAGDVEWLGNADLILYNGLHLET----KMGEVFSKL  Mn?
YTGA_BACSU  IADAAENIGGKHVKVTSLMGPGVDPHLYKASQGDTKKLMSADVVLYSGLHLEG----KMEDVLQKI  Mn
ADCA_STRPN  VYEFTKQVAGDTANVELLIGAGTEPHEYEPSAKAVAKIQDADTFVYENENMET----WVPKLLDTL  Zn
YCDH_BACSU  MYEFTKQIVKDKGDVDLLIPSSVEPHDWEPTPKDIANIQDADLFVYNSEYMET----WVPSAEKSM  Zn
ZNUA_HAEDU  LGFIANAITDGVTETKVLLPVTASPHDYSLKPSDIEKLKSAQLVVWVGDGLEA----FLEKSIDKL  Zn
ZNUA_HAEIN  LGFIVSSIADGVTGTQVLVPAGASPHDYNLKLSDIQKVKSADLVVWIGEDIDS----FLDKPISQI  Zn
ZNUA_ECOLI  VGFIASAIADGVTETEVLLPDGASEHDYSLRPSDVKRLQNADLVVWVGPEMEA----FMQKPVSKL  Zn
EWLA_ERYRH  IKDLVEIIGGDKVSVNGMMVAGVDPHLYKAKPSDVKAIQEADVVAFNGVHLEA----KLDDVLSGL  Zn?
ZNUA_THEMA  YYLIVSQLLGDTASVKLLVPPGANPHLFSLKPSDAKTLEEADLIVANGLGLEP----YLEKYREKT  Zn?
                ..  :         :        .#  .          *:.  .::    .

YFEA_YERPE  KDVPS---AVVTAGITPLPIR---E-----GPYS--------GIA--------NP-----------  Fe/Mn
SITA_SALEN  SGVPE---VVVSTGVKPMGIT---E-----GPYN--------GKP------NP-----------  Fe
MNTC_SYNSP  KDVPS---VVLTEGIEPIPIA---D-----GPYT--------DKP--------NP-----------  Mn
PSAA_STRPN  KKTENKDYFAVSDGVDVIYL----E-----GQNEK-------GKE------DP-----------  Mn
SCAA_STRGO  QKKENKDYYAVSEGVDVIYL----E-----GQNEK-------GKE------DP-----------  Mn
MTSA_STRPY  QKTKNKDYFAVSDGIDVIYL----E-----GASEK-------GKE------DP-----------  Me
TroA_TREPA  RGSRLV--VAVSETIPVSQR---------LSLEE-------A-E-----FDP-----------  Mn?
YTGA_BACSU  GEQKQS--AAVAEAIPKNKL----------IPAGE-------GKT-----FDP-----------  Mn
ADCA_STRPN  DKKKVK-TIKATGDMLLLPGGEEEE-----GDHDH-------GEEGHHHEFDP-----------  Zn
YCDH_BACSU  GQGHAV-FVNASKGIDLMEGSEEEH-----EEHDH-------GEHEHSHAMDP-----------  Zn
ZNUA_HAEDU  PKEKVLR-LEDVPGIKMIVDATKKK-----DHDH-------HDHDHDHDHDHDHEHIHGH---HH  Zn
ZNUA_HAEIN  ERKKVIT-IADLADVKPLLSKAHHEHFHEDGDHDHDHKHEHKHDHKHDHDHDHKHEHKHDHEHH  Zn
ZNUA_ECOLI  PGAKQVT-IAQLEDVKPLLMKSIHG------DDDD-------HDHAEKSDEDH-------H---HG  Zn
EWLA_ERYRH  EGSGKN-IIKLEDALEPSDIIN--D-----DEQ--------GGH------DP-----------  Zn?
ZNUA_THEMA  --------VFVSDFIPALLL----I-------DD-----------------NP-----------  Zn?
                  :                      .       :     :        .

YFEA_YERPE  ------------HAWMSPSNALIYIEN-IRKALVEHDPAHAETYNRNAQAYAEKIKALDAPLRERL  Fe/Mn
SITA_SALEN  ------------HAWMSAENALIYVDN-IRDALVKYDPDNAQIYKQNAERYKAKIRQMADPLRAEL  Fe
MNTC_SYNSP  ------------HAWMSPRNALVYVEN-IRQAFVELDPDNAKYYNANAAVYSEQLKAIDRQLGADL  Mn
PSAA_STRPN  ------------HAWLNLENGIIFAKN-IAKQLSAKDPNNKEFYEKNLKEYTDKLDKLDKESKDKF  Mn
SCAA_STRGO  ------------HAWLNLENGIIYAQN-IAKRLIEKDPDNKATYEKNLKAYIEKLTALDKEAKEKF  Mn
MTSA_STRPY  ------------HAWLNLENGIIYSKN-IAKQLIAKDPKNKETYEKNLKAYVAKLEKLDKEAKSKF  Me
TroA_TREPA  ------------HVWFDVKLWSYSVKA-VYESLCKLLPGKTREFTQRYQAYQQQLDKLDAYVRRKA  Mn?
YTGA_BACSU  ------------HVWFSIPLWIYAVDE-IEAQFSKAMPQHADAFRKNAKEYKEDLQYLDKWSRKEI  Mn
ADCA_STRPN  ------------HVWLSPVRAIKLVEH-IRDTLSADYPDKKETFEKNAAAYIEKLQSLDKAYAEGL  Zn
YCDH_BACSU  ------------HVWLSPVLAQKEVKN-ITAQIVKQDPDNKEYYEKNSKEYIAKLQDLDKLYRTTA  Zn
ZNUA_HAEDU  DKD--------WHIWFSPEASQLAAEQ-IAERLTAQLEPKKAKIAENLAAFKANLADKSNEITQQL  Zn
ZNUA_HAEIN  DHDHHEGLTTNWHVWYSPAISKIVAQK-VADKLTAQFPDKKALIAQNLSDFNRTLAEQSEKITAQL  Zn
ZNUA_ECOLI  DFN--------MHLWLSPEIARATAVA-IHGKLVELMPQSRAKLDANLKDFEAQLASTETQVGNEL  Zn
EWLA_ERYRH  ------------HIWFDVNLWKKSAQH-VADKLSEFDAENKDYYQQNAAQYVSELVDMDTYIKNRI  Zn?
ZNUA_THEMA  ------------HIWLDPFFLKYYIVPGLYQVLIEKFPEKQSEIKQKAEEIVSGLDTVIRDSFKAL  Zn?
              # *  .                :    :      .        .      :
```

```
YFEA_YERPE  SRIPAEQRWLVTSEGAFSYLAKDYGFKEVYLWPINAEQQGIPQQVRHVIDIIRENKIPVVFSESTI  Fe/Mn
SITA_SALEN  EKIPADQRWLVTSEGAFSYLARDNDMKELYLWPINADQQGTPKQVRKVIDTIKKHHIPAIFSESTV  Fe
MNTC_SYNSP  EQVPANQRFLVSCEGAFSYLARDYGMEEIYMWPINAEQQFTPKQVQTVIEEVKTNNVPTIFCESTV  Mn
PSAA_STRPN  NKIPAEKKLIVTSEGAFKYFSKAYGVPSAYIWEINTEEEGTPEQIKTLVEKLRQTKVPSLFVESSV  Mn
SCAA_STRGO  NNIPEEKKMIVTSEGCPKYFSKAYNVPSAYIWEINTEEEGTPDQIKSLVEKLRKTKVPSLFVESSV  Mn
MTSA_STRPY  DAIAENKKLIVTSEGCPKYFSKAYGVPSAYIWEINTEEEGTPDQISSLIEKLKVIKPSALFVESSV  Me
TroA_TREPA  QSLPAERRVLVTAHDAFGYFSRAYGFEVKGLQGVSTASEASAHDMQELAAFIAQRKLPAIFIESSI  Mn?
YTGA_BACSU  AHIPEKSRVLVTAHDAFAYFGNEYGFKVKGLQGLSTDSDYGLRDVQELVDLLTEKQIKAVFVESSV  Mn
ADCA_STRPN  SQAKEKS--FVTQHAAFNYLALDYGLKQVAISGLSPDAEPSAARLAELTEYVKKNKIAYIYFEENA  Zn
YCDH_BACSU  KKAEKKE--FITQHTAFGYLAKEYGLKQVPIAGLSPDQEPSAASLAKLKTYAKEHNVKVIYFEEIA  Zn
ZNUA_HAEDU  QAVKDKG--YYTFHDAYGYFERAYGLNSLGSFTINPTIAPGAKTLNAIKENIAAHKAQCLFAEPQF  Zn
ZNUA_HAEIN  ANVKDKG--FYVFHDAYGYFNDAYGLKQTGYFTINPLVAPGAKTLAHIKEEIDEHKVNCLFAEPQF  Zn
ZNUA_ECOLI  APLKGKG--YFVFHDAYGYFEKQFGLTPLGHFTVNPEIQPGAQRLHEIRTQLVEQKATCVFAEPQF  Zn
EWLA_ERYRH  AEIPEQQRVLVTAHDAFAYFGRYFGVHVEAIQGISTQSEAGIADINKVSDLIVDRKIKAIYTESSV  Zn?
ZNUA_THEMA  LPYTGKT--VVMAHPSFTYFFKEFGLELITLSSG-HEHSTSFSTIKEILRKKEQ--IVALFREPQQ  Zn?
                     .         # .    *:       ..               :   :       ::  *

YFEA_YERPE  SDKPAKQVSKETG-----AQYGGVLYVDSLSGEK-GPVPTYISLINMTVDTIAKGFGQ--------  Fe/Mn
SITA_SALEN  SDKPARQVARESG-----AHYGGVLYVDSLSAAD-GPVPTYLDLLRVTTETIVNGINDGLRSQQ--  Fe
MNTC_SYNSP  SDKGQKQVAQATG-----ARFGGNLYVDSLSTEE-GPVPTFLDLLEYDARVITNGLLAGTNAQQ--  Mn
PSAA_STRPN  DDRPMKTVSQDTN-----IPIYAQIFTDSIAEQG-KEGDSYYSMMKYNLDKIAEGLAK--------  Mn
SCAA_STRGO  DDRPMKTVSKEDN-----IPIYAKIFTDSIAEKG-EDGDSYYSMMKYNLDKISEGLAK--------  Mn
MTSA_STRPY  DRRPMETVSKDSG-----IPIYSEIFTDSIAKKG-KPGDSYYAMMKWNLDKISEGLAK--------  Me
TroA_TREPA  PHKNVEALRDAVQARGHVVQIGGELFSDAMGDAGTSEG-TYVGMVTHNIDTIVAALAR--------  Mn?
YTGA_BACSU  SEKSINAVVEGAKEKGHTVTIGGQLYSDAMGEKGTKEG-TYEGMFRHNINTITKALK--------  Mn
ADCA_STRPN  SQALANTLSKEAG-----VKTDVLNPLESLTEEDTKAGENYISVMEKNLKALKQTTDQEGPAIEPE  Zn
YCDH_BACSU  SSKVADTLASEIG-----AKTEVLNTLEGLSKEEQDKGLGYIDIMKQNLDALKDSLLVKS------  Zn
ZNUA_HAEDU  TPKVIDSLSKSTA-----VKVG---QLDPLGAKVKLSKTAYPQFLQAIADEFSQCLTQ--------  Zn
ZNUA_HAEIN  TPKVIESLAKNTK-----VNVG---QLDPIGDKVTLGKNSYATFLQSTADSYMECLAK--------  Zn
ZNUA_ECOLI  RPAVVESVARGTS-----VRMG---TLDPLGTNIKLGKTSYSEFLSQLANQYASCLKGD-------  Zn
EWLA_ERYRH  PKKTIESLQAAVKDRGFDVSIGGEIYSDSLKEDA-----SYIETYKINVDTIVDNLK---------  Zn?
ZNUA_THEMA  PAEILSSLEKELR--------MKSFVLDPLGVNG---EKTIVELLRKNLSVIQEALK--------  Zn?
                  :                    #  .
```

Fig. 1. Sequence comparison of periplasmic $Zn^{2+}/Mn^{2+}/Fe^?$ binding proteins that are part of Me^{2+} ABC transporter systems. An explanation of the abbreviations of the strain names (e.g., YERsinia PEstis) can be found in Tab. 13.2. The signal sequences are included, which may be the reason for the low similarity at the N-termini. The metal specificity of the binding proteins is not always clear; the existence of conflicting reports or a putative binding protein is indicated by ?. Identical residues are marked by an *, similar residues by : or . , and the positions involved in metal binding in the two known crystal structures of PsaA (25) and TroA (26) by #. Note the H-, D-, and E-rich domain in the Zn^{2+} binding proteins. AdcA has an additional C-terminal domain with unknown function; these 106 C-terminal residues are not shown in the comparison.

Mutants of several pathogenic streptococci that are deficient in interbacterial adhesion or adhesion to proteins of the host have been isolated. The mutations in different strains are in genes encoding surface-exposed lipoproteins with > 70 % sequence identity: *fimA* (fimbrial antigen), *psaA* (pneumococcal surface adhesin A), and *scaA* (streptococcal coaggregation-mediating adhesin). Dinthilhac et al. [23] found that the adhesin PsaA is in fact a binding protein of a Mn^{2+}-specific ABC transporter. In addition, the gene cluster *adcABC* encodes a binding protein-dependent ABC transporter for Zn^{2+}. Sequence comparisons have revealed that these lipoproteins belong to a new group of binding proteins not mentioned in the review by Tam and Saier [24]. Two subgroups have been defined, one with a predicted specificity for Mn^{2+} (or $Fe^?$) and one with a predicted specificity for Zn^{2+}. Some of these predictions have been verified since then. The proteins PsaA and TroA (a manganese binding protein from *Treponema pallidum*) have been crystallized, and their X-ray structures determined [25, 26]. For both proteins, the *in vivo* data on Mn^{2+} dependence and regulation by Mn^{2+} strongly argue for a function in Mn^{2+} uptake. However, in both crystal structures, the metal binding site is occupied by Zn^{2+} in a tetrahedral coordination, while Mn^{2+} is normally found in an octahedral coordination. Therefore, the biological data have been put to question, and for TroA, a function as a Zn^{2+} transporter has been postulated [26]. The metal is bound by the four amino acids H67, H139, E205, and D280 in PsaA and by H68, H133, H199, and D279 in TroA. The amino acids are in equivalent positions, as can be seen in Fig. 1. The sequence comparison of these Me^{2+} binding proteins also argues for an *in vivo* function as a Mn^{2+} binding protein. It is interesting to note that the Mn^{2+} binding residues in PsaA are well conserved in the other Mn^{2+} and Zn^{2+} binding proteins: position 67 only H, position 139 only H, position 205 E or H, and position 280 D or E.

The closely related protein MtsA from *Streptococcus pyogenes* has been shown *in vitro* to interact with Cu^{2+}, Zn^{2+}, and Fe^{3+}, and not with Mn^{2+}. An *mtsA* mutant has a 50% lower iron content and a 30% lower Zn^{2+} content than the parent strain [27]. From these results, it was postulated that the *mtsABC* genes encode a binding protein-dependent transport system with a very broad metal specificity for Cu^{2+}, Fe^{3+}, and Zn^{2+}. These results are unexpected since the binding protein MtsA shows more than 70% identity to PsaA, ScaA, and FimA, which have been shown to be manganese transporters. Uptake experiments with the *mtsA* mutant and the parent strain will verify the postulate.

Since the proteins from pathogenic streptococci were first described as adhesins or fimbrial antigens, many ORFs identified in various genome sequencing projects were annotated as putative adhesins. This is misleading since the loss of adhesive properties in the mutants was most likely a sign of their lower fitness.

The Zn^{2+} binding protein AdcA is roughly 100 residues longer than most of the other binding proteins of this group (Fig. 1). The function of this C-terminal domain is unknown. Small proteins with similarity to this domain, but with unknown function, are found in *E. coli* and *B. subtilis*.

In the gram-positive bacterium *B. subtilis*, the genes *ycdHIyceA* are regulated by Zn^{2+} and the repressor protein Zur, which is a member of the Fur (iron uptake regulation) protein family. Two lines of evidence suggest that the *ycdH* operon encodes a high-affinity Zn^{2+} transporter. First, a *ycdH* mutant is impaired in growth in low-Zn^{2+} medium. Second, mutation of *ycdH* alters the regulation of both the *yciC* (a gene regulated by Zn^{2+} and the repressor Zur) and *ycdH* operons such that much higher levels of exogenous Zn^{2+} are required for repression [28]. In addition, sequence similarity places YcdH into the family of Zn^{2+} binding proteins (Fig. 1).

13.3.2
Binding Protein-dependent Zn^{2+} Uptake in Gram-negative Bacteria

Members of this Me^{2+} binding protein family have also been found in gram-negative bacteria. The gene encoding the periplasmic Zn^{2+} binding protein Pzp1 from *Haemophilus influenzae* was cloned as a putative adhesin (similarity to PsaA, FimA, etc.). The isolated protein binds 2 to 5 Zn^{2+} ions per monomer [29]. A constructed *pzp1* mutant is unable to grow aerobically. Growth is recovered by addition of Zn^{2+}, which indicates that the binding protein Pzp1 is part of an ABC transporter. In the genome sequence, two genes encoding the putative integral membrane protein and the putative ATPase are found unlinked to *pzp1*.

Another Zn^{2+} transporter in *E. coli* has been described [31]. During the selection of recombinants with iron-regulated *lacZ* fusions with the transposing phage Mud1, fusions regulated by the availability of Zn^{2+} were obtained. These clones grow as red colonies on MacConkey agar with Zn^{2+} complexing chelators, indicating derepression of the *lacZ* fusion, while addition of $ZnCl_2$ leads to repression and growth of white colonies. On complex nutrient agar plates, growth of these mutants is inhibited by 5 mM EGTA or 0.4 mM EDTA. When $ZnCl_2$ is spotted on filter paper disks, a zone of growth around the disk is observed. Other metals, such as Ni^{2+}, Cu^{2+}, Mn^{2+}, and Fe^{2+} do not stimulate growth. A much smaller zone of growth was observed with Co^{2+}, which may substitute Zn^{2+} in some proteins and may lower the need for Zn^{2+}.

The Mud1 phage inserted into one of three genes encoding a binding protein-dependent ABC transporter. The *znuA* (zinc uptake) gene encodes a periplasmic binding protein, *znuB* encodes an integral membrane protein, and *znuC* encodes the ATPase component of the transporter.

$^{65}Zn^{2+}$ uptake by *znu* mutants and by the parent strain was measured. In a HEPES-buffered medium, the uptake of $^{65}Zn^{2+}$ was the same for the mutants and parent strain. Only when the cells were pre-grown in the presence of 5 mM EGTA did the mutants unexpectedly take up more $^{65}Zn^{2+}$ than the parent strain. Possibly another Zn^{2+} transporter was induced in the mutants. Addition of 0.5 mM EGTA to the transport medium lowered the uptake by a factor of 10 in the induced parent strain, whereas no uptake was observed in *znu* mutants regardless of the growth conditions.

With the use of *znu–lacZ* operon fusions as a reporter system, it was possible to identify the *E. coli* zinc-dependent repressor Zur, which binds with Zn^{2+} as corepressor to the promoter region of the *znu* genes [30, 31].

It is interesting to note that all unambiguous Zn^{2+} binding proteins contain histidine-, aspartate-, and glutamate-rich stretches in front of the conserved HXW motif (Fig. 1). The Pzp1 protein contains the largest extended region rich in H, D, and E (position 171/173 of Pzp1 in Fig. 1), which may function in binding of Zn^{2+} or in delivering Zn^{2+} to other proteins. H171 is in a Zn^{2+} binding position in the crystal structures of PsaA and TroA. Structural studies of Pzp1 in comparison to other ZnuA proteins with much smaller histidine-rich regions would be interesting.

13.4
Low-affinity Zn^{2+} Uptake Systems

Very little is known about the low-affinity uptake systems that provide Zn^{2+} to bacteria in metal-rich non-toxic media. A survey is given in Tab. 3. In *E. coli*, the *pit* genes encode a low-affinity phosphate transport system. It is known that the sub-

Tab. 3. Low-affinity transport systems for metals including Zn^{2+}

Gene (Organism)	Function of the Encoded Proteins	Reference
pitA (Eschericia coli)	metal-phosphate cotransport (Mg^{2+}, Ca^{2+}, Co^{2+}, or Mn^{2+}, or Zn^{2+})	(33)
citM (Bacillus subtilis)	cotransport of Mg^{2+}, Ni^{2+}, Mn^{2+}, Co^{2+}, or Zn^{2+} with citrate	(34)
citH (B. subtilis)	cotransport of Ca^{2+}, Ba^{2+}, or Sr^{2+} with citrate	(34)

strate of this transporter is $MeHPO_4$. The transporter is very unspecific since Mg^{2+}, Ca^{2+}, Co^{2+}, or Mn^{2+} is accepted [32]. Recently, it has been demonstrated that a *pit* mutant accumulates less Zn^{2+} than the parent strain in a Zn^{2+}-rich medium (0.5–2 mM) [33]. Interestingly, the PitA protein also catalyzes the exchange of metal ions between the cytoplasm and the outside, which may allow efflux of certain metals. Additional systems for low-affinity Zn^{2+} uptake may exist in *E. coli*.

Two metal–citrate cotransport systems in *B. subtilis* have been described; these systems have complementary substrate specificities: CitM transports Mg^{2+}, Ni^{2+}, Mn^{2+}, Co^{2+}, and Zn^{2+} with citrate, while CitH recognizes Ca^{2+}, Ba^{2+}, and Sr^{2+} together with citrate [34].

13.5
Concluding Remarks

Zinc export and uptake are regulated separately by individual regulators (Tab. 4), in contrast to iron uptake, where the global regulator Fur in *E. coli* regulates more than 50 genes not only involved in iron uptake, but also in the oxidative-stress response and in virulence [35]. Zinc metabolism in eukaryotes also seems to be linked to redox processes since mobilization of the tightly bound Zn^{2+} in metallothioneins may be accomplished in the cell by oxidation of zinc metallothioneins [36]. Among the bacteria, metallothioneins are only found in cyanobacteria. Storage of Zn^{2+} in other bacteria has not been observed. However, cloning of a eukaryotic metallothionein in *E. coli* enhances the Cd^{2+} and Zn^{2+} resistance of the bacterial cells [37].

As mentioned above, the metal specificity of the Zn^{2+}/Mn^{2+} binding proteins of pathogenic streptococci is not always clear [3]. Further research is certainly necessary to clarify the specificity question. However, pathogenic bacteria live in their host in an environment with only small concentration fluctuations for these metals. It may be that this allows a certain flexibility in metal ion specificity for these uptake systems, which would be deleterious for bacteria living in other environments.

Tab. 4. Regulators of Zn^{2+} transport systems

Gene (Organism)	Function of the Encoded Proteins	Reference
czcR czcS (Ralstonia metallidurans)	Zn^{2+}-dependent histidine protein kinase response regulator	(5)
zntR (Staphylococcus aureus)	ArsR/SmtB-like repressor; regulates zntA in S. aureus	(11)
cadC (S. aureus)	ArsR/SmtB-like repressor; regulates cadA in S. aureus; specificity: $Cd^{2+} > Zn^{2+}$	(49)
zntR (Escherichia coli)	MerR-like repressor/activator of zntA; specificity: $Cd^{2+} > Pb^{2+} > Zn^{2+}$	(21)
zur (E. coli)	Fur-like repressor protein	(30)
zur (Bacillus subtilis)	Fur-like repressor protein	(28)

Acknowledgement

I thank Volkmar Braun (this institute) for discussions, and Karen A. Brune for editing the manuscript. The BLAST [38] service of the NIH and the new SIB BLAST network service (ExPASy Molecular Biology Server) with the CLUSTALW [39] search possibility is gratefully acknowledged. The research of the author is supported by the Deutsche Forschungsgemeinschaft (HA 1186/3-1) and by the Fonds der Chemischen Industrie.

References

1. 1. Singh, A. P., Bragg, P. D., *FEBS Lett.* **1974**, *40*, 200-202.
2. Kasahara, M., Anraku, Y., *J. Biochem.* **1974**, *76*, 967-976.
3. Claverys, J.-P., *Res. Microbiol.* **2001**, *152*, 231-243.
4. Nies, D. H., *J. Bacteriol.* **1995**, *177*, 2707-2712.
5. Grosse, C., Grass, G., Anton, A., Franke, S., Santos, A. N., Lawley, B., Brown, N. L., Nies, D. H., *J. Bacteriol.* **1999**, *181*, 2385-2393.
6. Rensing, C., Pribyl, T., Nies, D. H., *J. Bacteriol.* **1997**, *179*, 6871-6879.
7. Koronakis, V., Sharff, A., Koronakis, E., Luisi, B., Hughes, C., *Nature* **2000**, *405*, 914-919.
8. Hassan, M. T., van der Lelie, D., Springael, D., Romling, U., Ahmed, N., Mergeay, M., *Gene* **1999**, *238*, 417-425.
9. Tseng, T. T., Gratwick, K. S., Kollman, J., Park, D., Nies, D. H., Goffeau, A., Saier, M. H. J., *J. Mol. Microbiol.Biotechnol.* **1999**, *1*, 107-125.
10. Anton, A., Grosse, C., Reissmann, J., Pribyl, T., Nies, D. H., *J. Bacteriol.* **1999**, *181*, 6876-6881.
11. Xiong, A., Jayaswal, R. K., *J. Bacteriol.* **1998**, *180*, 4024-4029.
11a. Paulsen, I.T., Saier, M. H. J., *J. Membr. Biol.* **1997**, *156*, 99-103.
12. Gatti, D., Mitra, B., Rosen, B. P., *J. Biol. Chem.* **2000**, *275*, 34009-34012.

13. Nucifora, G., Chu, L., Misra, T. K., Silver, S., *Proc. Natl. Acad. Sci. USA* **1989**, *86*, 3544-3548.
14. Yoon, K. P., Silver, S., *J. Bacteriol.* **1991**, *173*, 7636-7642.
15. Yoon, K. P., Misra, T. K., Silver, S., *J. Bacteriol.* **1991**, *173*, 7643-7649.
16. Alonso, A., Sanchez, P., Martinez, J. L., *Antimicrob. Agents Chemother.* **2000**, *44*, 1778-1782.
17. Lebrun, M., Audurier, A., Cossart, P., *J. Bacteriol.* **1994**, *176*, 3040-3048.
18. Rensing, C., Mitra, B., Rosen, B. P., *Proc. Natl. Acad. Sci. USA* **1997**, *94*, 14326-14331.
19. Beard, S. J., Hashim, R., Membrillo-Hernandez, J., Hughes, M. N., Poole, R. K., *Mol. Microbiol.* **1997**, *25*, 883-891.
20. Noll, M., Lutsenko, S., *IUBMB Life* **2000**, *49*, 297-302.
21. Binet, M. R., Poole, R. K., *FEBS Lett.* **2000**, *473*, 67-70.
22. Okkeri, J., Haltia, T., *Biochemistry* **1999**, *38*, 14109-14116.
23. Dinthilhac, A., Alloing, G., Granadel, C., Claverys, J.-P., *Mol. Microbiol.* **1997**, *25*, 727-739.
24. Tam, R., Saier Jr., M. H., *Microbiol. Rev.* **1993**, *57*, 320-346.
25. Lawrence, M. C., Pilling, P. A., Epa, V. C., Berry, A. M., Ogunniyi, A. D., Paton, J. C., *Structure* **1998**, *6*, 1553-1561.
26. Lee, Y. H., Deka, R. K., Norgard, M. V., Radolf, J. D., Hasemann, C. A., *Nature Struct. Biol.* **1999**, *6*, 628-633.
27. Janulczyk, R., Pallon, J., Bjorck, L., *Mol. Microbiol.* **1999**, *34*, 596-606.
28. Gaballa, A., Helmann, J. D., *J. Bacteriol.* **1998**, *180*, 5815-5821.
29. Lu, D., Boyd, B., Lingwood, C. A., *J. Biol. Chem.* **1997**, *272*, 29033-29038.
30. Patzer, S. I., Hantke, K., *J. Biol. Chem.* **2000**, *275*, 24321-24332.
31. Patzer, S. I., Hantke, K., *Mol. Microbiol.* **1998**, *28*, 1199-1210.
32. van Veen, H. W., Abee, T., Kortstee, G. J., Konings, W. N., Zehnder, A. J., *Biochemistry* **1994**, *33*, 1766-1770.
33. Beard, S. J., Hashim, R., Wu, G., Binet, M. R., Hughes, M. N., Poole, R. K., *FEMS Microbiol. Lett.* **2000**, *184*, 231-235.
34. Krom, B. P., Warner, J. B., Konings, W. N., Lolkema, J. S., *J. Bacteriol.* **2000**, *182*, 6374-6381.
35. Hantke, K., Braun, V., in: *Bacterial Stress Responses* (Storz, G., Hengge-Aronis, R., Eds.), ASM Press, Washington, DC, **2000**, pp. 275-288.
36. Maret, W., *J. Nutr.* **2000**, *130*, 1455S-1458S.
37. Hou, Y. M., Kim, R., Kim, S. H., *Biochim. Biophys. Acta* **1988**, *951*, 230-234.
38. Altschul, S. F., Madden, T. L., Schaffer, A. A., Zhang, J., Zhang, Z., Miller, W., Lipman, D. J., *Nucleic Acids Res.* **1997**, *25*, 3389-3402.
39. Thompson, J. D., Higgins, D. G., Gibson, T. J., *Nucleic Acids Res.* **1994**, *22*, 4673-4680.
40. Bearden, S. W., Perry, R. D., *Mol. Microbiol.* **1999**, *32*, 403-414.
41. Zhou, D., Hardt, W. D., Galan, J. E., *Infect. Immun.* **1999**, *67*, 1974-1981.
42. Bartsevich, V. V., Pakrasi, H. B., *EMBO J.* **1995**, *14*, 1845-1853.
43. Kolenbrander, P. E., Andersen, R. N., Baker, R. A., Jenkinson, H. F., *J. Bacteriol.* **1998**, *180*, 290-295.
44. Burnette-Curley, D., Wells, V., Viscount, H., Munro, C. L., Fenno, J. C., Fives-Taylor, P., Macrina, F. L., *Infect. Immun.* **1995**, *63*, 4669-4674.
45. Que, Q., Helmann, J. D., *Mol. Microbiol.* **2000**, *35*, 1454-1468.
46. Lewis, D. A., Klesney-Tait, J., Lumbley, S. R., Ward, C. K., Latimer, J. L., Ison, C. A., Hansen, E. J., *Infect. Immun.* **1999**, *67*, 5060-5068.
47. Tatusov, R. L., Mushegian, A. R., Bork, P., Brown, N. P., Hayes, W. S., Borodovsky, M., Rudd, K. E., Koonin, E. V., *Curr. Biol.* **1996**, *6*, 279-291.
48. Nelson, K. E., Clayton, R. A., Gill, S. R., Gwinn, M. L., Dodson, R. J., Haft, D. H., Hickey, E. K., Peterson, J. D., Nelson, W. C., Ketchum, K. A., McDonald, L., Utterback, T. R., Malek, J. A., Linher, K. D., Garrett, M. M., Stewart, A. M., Cotton, M. D., Pratt, M. S., Phillips, C. A., Richardson, D., Heidelberg, J., Sutton, G. G., Fleischmann, R. D., Eisen, J. A., Fraser, C. M., *Nature* **1999**, *399*, 323-329.
49. Endo, G., Silver, S., *J. Bacteriol.* **1995**, *177*, 4437-4441.

14
Bacterial Genes Controlling Manganese Accumulation

Mathieu Cellier

14.1
Introduction

The purpose of the chapter is to provide a brief overview of the biological properties of manganese, including its requirement for bacterial growth, and to briefly summarize recent literature on genes encoding membrane permeases and transcriptional regulators that control manganese uptake in bacteria. Taxonomic distribution and phylogenetic relationships of cloned genes are briefly mentioned. Emphasis was put on manganese transport and importance for bacterial growth, virulence, and pathogenesis.

14.1.1
Physicochemical Properties of Manganese

Manganese (Mn) is element 25 in the Periodic Table. It is a transition element, located in group 7 (formerly VIIA), that can exist in 11 oxidation states. The +2 valence [Mn(II)] is found in most biological systems, in aqueous solution in the form of sulfates, chlorides, and citrate. Mn(II) is resistant to oxidation, especially in acidic and neutral solutions. The redox potential of the Mn(III)/Mn(II) couple is ~ +1.5 V at 25°. Chelation and coordination substantially affect the redox potential of the Mn(III)/Mn(II) couple: ~ +1.0 V in the presence of excess of citrate ligand [1], and ~ +0.31 V in the active site of the manganese superoxide dismutase (Mn-SOD) [2]. In comparison, the maximal redox potential of the Fe(III)/Fe(II) couple is ~ +0.7 V. Mn(II) appears energetically quite stable, and is in fact oxidized 10^7 times slower than Fe(II) at pH 8 and 25° [1]. The Mn(III) state is subject to disproportionation to form Mn(II) and Mn(IV) compounds [1], unless it is stabilized by coordination, e.g., in Mn-SOD from bacteria and mitochondria [3]. Together with Ca(II) and Mg(II), Mn(II) is a hard Lewis acid, preferring ligands in the hard Lewis base category. O-donor moieties are preferred, such as the negatively charged oxygen atom in the carboxylate of aspartate, glutamate and tyrosinate, and the polar oxygen atom from solvent, H_2O, OH^-, O_2^-, and from asparagine and glutamine. Borderline hardness Lewis base category N-donor moieties, such as the imidazole nitrogen atom of histidine are also described as coordination ligands of Mn(II) [3–5]. Mn(II) has a ionic radius $r \sim 0.8$ Å, which is between Ca(II), $r \sim 0.99$ Å, and Mg(II), $r \sim 0.65$ Å, and in the range of ionic radii of other divalent transition metals, e.g., Fe(II) $r \sim 0.74$ Å. Mn(III) and Fe(III) exhibit $r \sim 0.66$ Å and $r \sim 0.64$ Å, respectively. Because of its higher electrostatic charge and smaller ionic radius, Mn(III) forms more stable complexes with electron donors, like Fe(III). In contrast, Mn(II) with a larger ionic radius and electronic configuration with five unpaired electrons forms rather weak complexes. Mn(II) complexes exhibit lower stability compared to those containing other transition metal divalent cations Fe(II), Cu(II), and Ni(II), but are still generally stronger than complexes involving Ca(II) or Mg(II) [6]. Mn(II) and the two latter are labile metal ions, which are not tightly bound in a metalloprotein and in dynamic equilibrium with their aqueous environment. Complexes involving Mn(II) as metal ion must, therefore, exhibit either very high affinity for Mn(II) or reside in a compartment with a very high concentration of Mn(II) [5]. Intracellular concentration of Mn(II) is generally estimated in the submicromolar range [4]. The protein ligands dictate Mn(II) coordination geometry according to catalytic requirements, and intact Mn-binding sites generally contribute to protein stability and catalytic function [4].

14.1.2
Physiological Role of Manganese in Bacteria

Mn is a critical element for plants and many living organisms including bacteria (e.g., *Lactobacillus*, *Treponema*, and *Borrelia*) for redox biochemistry, oxygen production, and protection against oxidative stress [4]. Redox cycling Mn(II) ↔ Mn(III) is crucial for aerobic life on earth, as the catalytic use of four Mn ions by photosystems of type II (PSII) for water oxidation, is considered to represent the major source of O_2 production in the atmosphere (Cyanobacteriaceae, algae, and plants) [7]. Mn redox cycling is also required for detoxification of oxygen-derived radicals. Disposing of superoxide radicals O_2^- by accelerating the conversion into H_2O_2 and O_2 is facilitated either directly by reduced Mn(II) contained in polyphosphate–protein aggregates present at high intracellular concentration (e.g., in *Lactobacillus*, *Deinococcus* and *Neisseria gonorrhoeae*) [8–10], or by Mn-based catalysis in the Mn-superoxide dismutase, SodA, the Sod of most aerobic bacteria, and mitochondria [2]. SodA is a functional equivalent of the bacterial Fe-Sod, SodB, predominant in anaerobes [11], and of the Cu, Zn-Sod, SodC, expressed in the bacterial periplasm and in eukaryotic cells cytoplasm [12]. Disposing of peroxides, including disproportionation of H_2O_2 to water and oxygen, is performed either directly by reduced Mn(II) in bicarbonate buffer [13] or following catalysis by Mn-catalase [14], and by lignin degrading Mn-peroxidase [15]. Redox cycling Mn(II) ↔ Mn(III) is thus an important catalytic mechanism to transform by-products of aerobic metabolism, and to resist to oxidative stress generated by host enzymes during infection (NADPH oxidase, myeloperoxidase, NO synthase). [16] Mn-based redox cycling is, however, not compatible with chain reactions involving electron transfer of lower redox potential (e. g., respiration). Since Mn(II) is much more resistant to oxidation by O_2 than Fe(II) or Cu(I), and does not produce Fenton reaction with H_2O_2 that yields the hydroxyl radical OH•, it thus plays a critical role in redox buffering of living cells [10]. Mn is also important for general metabolism and carbohydrate metabolism and glycosylation in particular, in both anabolic and catabolic functions in anaerobiosis and aerobiosis [5, 17]. A great variety of Mn-containing proteins are found in many organisms. In addition to oxido-reduction reactions (oxidation, dehydrogenation, decarboxylation, halogenation), Mn-containing enzymes perform transferase reactions (kinase, DNA and RNA polymerase, sulfo- and glycosyl transferase), hydrolysis reactions (C- and N-peptidase, GTPase, phosphatase, esterase, integrase, arginase), lyase reactions (carboxykinase and dehydratase, cyclase), sugar isomerase reactions, and ligase reactions [5, 17]. Some enzymes may require Mn in a particular taxonomic group and be more frequently found employing another transition metal (e. g., Fe, Cu, Zn, or Co). Other proteins containing bound Mn, including transcription factors and transport systems that use Mn as specific ligand, are presented in Sect. 14.2.

14.1.3
Effect of Manganese on Bacterial Growth

Most bacterial species show strong dependency on iron for growth and accumulate iron at 10- to 50-fold greater concentration than other transition metals including Mn, Cu, and Zn. Mn is probably important for growth of many bacterial species although at a very low level, to the extent that Mn requirements are generally not known. Iron plays a primordial catalytic role in metabolism, and is required for growth of most organisms, both anaerobic and aerobic. This is probably because Fe(II) and Fe(III) interconvert readily under physiological aerobic conditions, depending on the relative values of the redox potential and pH of the environment, so that iron can participate in electron exchange involving low-level free energy changes (e. g., respiratory chain intermediates). Also, iron is a borderline hardness Lewis acid that shows affinity for sulfur coordination ligands and forms various types of Fe–S cluster proteins. Iron–sulfur proteins are important for sensing redox potential and oxidative stress, e. g., the superoxide sensitive regulator SoxRS, which contains a [2Fe–2S] cluster [18], and for contributing directly to diminish bacterial sensitivity to oxidative stress, e. g., the [4Fe–4S]-containing dehydratases [19]. In turn, iron-based metabolisms ensure maximum energy production in aerobiosis, born iron scarcity and oxidative damage are problematic in aerobiosis [20]. In contrast, highly energetic oxidants such as UV radiation or the superoxide, peroxide are required to interconvert Mn(II) in Mn(III), and Mn(II) is a hard Lewis acid (see Sect. 14.1.1). Hence, Mn lacks the redox versatility, trivalence at physiological pH, and varied complexing ability under physiological aerobic conditions that are characteristic of iron. This probably explains why most organisms have retained iron as the key redox metal for the control of cellular growth. However, some facultative aerobic lactic acid bacteria have retained Mn(II) as a biological controller of metabolism instead of iron, and show strict Mn-dependent growth [10, 21]. Similarly, the Lyme disease pathogen has become iron-independent, by eliminating genes encoding iron-containing proteins, and by substituting Mn for Fe in the remaining metalloproteins [22]. Hence some bacteria may become Mn-dependent for growth as a result of adaptation to a particular niche, at the cost of elimination of iron-based metabolic pathways such as tricarboxylic acid cycle and respiratory electron transport chain, and biochemical mechanisms to avoid oxidative stress [10, 21, 22].

The molecular cloning and characterization of three different types of Mn(II) permeases has also revealed an important role for Mn in the control of growth of lactic acid bacteria and of several other bacterial species. The lactic acid bacteria are gram-positive non-spore-forming bacteria subdivided into two groups according to morphology as Lactobacillaceae (rods), or Streptococcaceae and Enterococcaceae (cocci). Because of heme and Fe−S proteins deficiency, fermentative glycolysis and substrate level phosphorylation are used as principal pathways for ATP production. Growth *in vitro* of *Lactobacillus plantarum* was shown to be proportional to total Mn(II) in the medium, reaching a plateau when intracellular Mn(II) was 30−35 mM [10, 23]. Mn(II)-dependent *L. plantarum* strains were obtained after chemical mutagenesis, which could only grow in a medium containing 20 mM Mn(II) in aerobiosis and around 2 mM Mn(II) in anaeobiosis, representing an increase of ∼ 5,000 times in Mn requirement for growth [24]. This Mn(II) high-affinity transporter that is required for growth of *L. plantarum* defines a group of Mn(II) primary transporter, that belongs to the P-type ATPase superfamily [25] and is presented in Sect. 14.2.2.1.1.

Some streptococcal species showed iron independence, and manganese dependency, for growth *in vitro* [21, 25, 26]. Streptococcaceae and Enterococcaceae apparently do not possess a homolog of *L. plantarum* MntA P-type ATPase Mn(II) permease. These species express another type of Mn(II) primary transporter, a multicomponent permease of the ATP-dependent binding cassette (ABC) superfamily, presented in Sect. 14.2.2.1.2. Inactivation of the gene coding for the surface metal binding-dependent component of this transport system imposes a Mn requirement for growth in defined medium on mutant strains of group C Streptococcaceae (GCS) species [27, 28]. Similarly, disruption of the homologous gene in Cyanobacteria results in lower growth rate in Mn(II)-deficient medium [29]. However, inactivation of the homologous genes in the other streptococcal species, group A Streptococcaceae (GAS) including *Streptococcus pyogenes*, which is known to require iron for growth, did not show growth defect in either rich or defined medium [30]. Apparent homologs of this group of primary Mn(II) ABC transporters were found to also influence growth in bacteria that are distant in evolutionary terms, e.g., Enterobacteriaceae. In *Yersinia pestis*, inactivation of the homologous multicomponent ABC permease results in growth reduction and defect in growth under iron-limiting conditions [31]. This family of Mn(II) ABC permease may contribute to bacterial pathogenesis by facilitating Mn(II) acquisition during host infection (e.g., *Streptococcus pneumoniae*, *S. gordonii*, *S. mutans*, *Enterococcus faecalis*, *Salmonella typhimurium*, *Y. pestis*, *Treponema palli-*

dum, Neisseria spp.). Some pathogenic species of Streptococcaceae, Enterococcaceae and Enterobacteriaceae also possess genes encoding another type of Mn permease (*S. mutans* plasmid pAM1, *E. faecalis*, and *S. typhimurium*, *Y. pestis*, see Sect. 14.2.2.2.1, MntH proteins).

The third group of bacterial Mn permeases that was shown to influence bacterial growth is represented by a family of proton-dependent secondary transporters. This family, denominated Natural Resistance-Associated Macrophage Protein (Nramp)/Divalent Metal Transporter (DMT), is evolutionary conserved from Bacteria to higher eukaryotes, including plants. Two genes were defined in mammals and associated to genetic disorders in mouse inbred strains [32]. The Natural Resistance-Associated Macrophage Protein (Nramp1) is expressed by professional phagocytes and mediates innate immunity towards some intracellular pathogens. Nramp1 controls the intracellular replication of certain bacterial parasites residing within a vacuole (phagosome), by extruding Mn(II) ions that are present soluble in the phagosome lumen [33]. Nramp2 protein, also denominated DMT1, is rather ubiquitously expressed and important for dietary iron absorption in the intestine [34, 35] and involved in the transferrin cycle for intracytoplasmic uptake of Fe(II) and Mn(II) [36]. Inactivation of *Nramp2* gene is responsible for microcytic anemia [37]. Yeast homologous SMF1-3 proteins are required for growth in the presence of metal chelator, and are known to control Mn and Cu homeostasis in *Saccharomyces cerevisiae* [38, 39], and to play a role in iron metabolism (see Chap. 20). Homologs in plant eudicots and monocots are also important for nutrition and development [40, 41]. The bacterial homologs of the Nramp/DMT family are denominated MntH for proton-dependent manganese transporter (presented in Sect. 14.2.2.2.1). The characteristics of *E. coli* MntH suggested that it corresponded to the single Mn(II) transporter known in this species [42]. Elimination of *Bacillus subtilis mntH* homolog imposes a requirement for 10 µM supplementary Mn(II) in defined medium *in vitro* to obtain growth that is similar to the wild-type isogenic strain [43].

Bacterial growth can be affected by regulators of gene expression that are sensitive to Mn(II). Some Mn-dependent transcriptional regulators which modulate the expression of Mn(II) permeases were recently described, and are presented in Sect. 14.2.3. They belong to the two major families of divalent metal-dependent transcriptional regulators of gene expression, the Ferric Uptake Repressor (Fur) and related factors, and the Diphteria Toxin Repressor (DtxR), and related factors (see Chap. 12). Alteration of Mn(II) permease expression, by disruption of defined Mn(II) permease genes or Mn(II)-dependent regulators of gene expression, allows to measure the impact of manganese acquisition on bacterial growth. Observation of a significant effect *in vitro* as with lactic acid bacteria, *B. subtilis*, and some Streptococcaceae, may reflect a substantial bias in Mn-based redox chemistry in the basic metabolism of these organisms. This could be due to specific adaptation, resulting in Mn(II) dependency for key regulatory functions and cell growth, and/or resistance to oxidative stress. Absence of a significant effect of the alteration of the Mn(II) permease expression *in vitro* may be indicative in contrast of bacterial species with a predominantly iron-based metabolism, and of a primary role of Mn(II) in oxidative defenses and/or could suggest a specific role for bacterial Mn(II) permease in pathogenesis.

14.2
Manganese Transport in Bacteria

14.2.1
Overview of Biochemical Studies with Whole Cells and Membrane Vesicles

A historical perspective on the biochemistry of Mn(II) transport in bacteria was recently provided by V.C. Culotta [39]. The biochemical characterization of Mn transport in whole cells was performed in the gram-negative bacteria *E. coli* and in the gram-positive *B. subtilis* and *Staphylococcus aureus*. A saturable secondary transporter was first identified in *E. coli* whole cells, showing temperature and energy dependency, pH-dependent inhibition by iron, slight inhibition by cobalt, and no inhibition by Ca(II) or Mg(II) [44, 45]. Parallel work with right-side-out isolated membrane vesicles showed independence from ATP of Mn(II) transport [46]. Co(II) was reported to inhibit Mn(II) uptake competitively, which was different from inhibition by Cd(II) [47]. Studies in *B. subtilis* demonstrated similar Mn uptake, that was regulated in response to metal bioavailability. The variations resulting from adaptation of *B. subtilis* to environmental conditions reflected alterations in V_{max} rather than in K_m values, suggesting regulation of the expression level of a single system [48, 49]. *B. subtilis* saturable Mn permease was examined in isolated membrane vesicles and shown to be energized by the membrane electrical potential and to transport Mn(II) and Cd(II), but not Ca(II) or Mg(II) [47, 50]. The Mn(II) secondary transporters corresponding to these similar Mn uptake activities measured in *E. coli* and *B. subtilis* belong to the MntH/Nramp family (see Sect. 14.2.2.2.1). Secondary transport of Cd(II) and Mn(II) was also demonstrated in *S. aureus* membrane vesicles, with each metal being a competitive inhibitor of the other [51]. Parallel studies demonstrated that Lactobacillaceae starved for Mn by lowering in the micromolar (µM) range the amount of Mn added in the medium, showed high-level Mn uptake upon Mn supplementation [52]. The calculated K_m value of 0.2 µM obtained in this study was in the range of values that were obtained for the other bacterial species (0.2–2.0 µM). Cd(II) was the sole metal competing for Mn(II) transport sites, and with higher affinity as was found for Mn(II) transport in *S. aureus* [51, 52]. However, V_{max} values obtained for *L. plantarum* were much higher than those for other bacteria, suggesting that to fulfill its extraordinary need for Mn(II), *L. plantarum* expresses a greater number of Mn(II) transport sites at the cell surface [10].

14.2.2
Genes Encoding Transport Systems for Manganese Acquisition

Permeases are generally classified according to transporter type, mode of energy coupling and molecular phylogeny (see Chap. 1). Mn(II) permeases characterized so far in bacteria belong to three different kinds of carrier versus channel type. Two types of primary transporters of Mn(II) were identified, that are believed to couple energy from pyrophosphate bond hydrolysis according to the mechanisms of transport described in their respective superfamilies. One family of secondary transporters of Mn(II) was shown to couple energy from the proton electrochemical gradient to the metal ion transport.

14.2.2.1 Primary Transport Systems

14.2.2.1.1 P-type ATPase, *Lactobacillus plantarum* MntA

A high-affinity, high-velocity Mn(II) uptake system expressed after induction by Mn(II) starvation was recently characterized in several *L. plantarum* strains [23]. This system also takes up Cd(II) following similar kinetics. The corresponding gene, denominated *mntA*, was cloned from *L. plantarum* ATCC 14917 into *E. coli* [24]. *L. plantarum mntA* gene is repressed in Mn(II)-sufficient cells. In response to Mn(II) starvation, the gene is transcribed and the MntA protein is expressed in the bacterial membrane [24]. Mn(II) uptake by Mn-starved cells at pH 5.5 required buffers containing metal chelators such as citrate or related tricarboxylic acids [52], while the same buffers were slight inhibitors of Cd(II) uptake, suggesting that the two ions Mn(II) and Cd(II) may be recognized in different forms by MntA [23]. Cd(II) uptake by Mn-starved *L. plantarum*, and by *E. coli* expressing recombinant MntA, was temperature-dependent. Uptake was inhibited by uncoupling agents known to increase membrane permeability to H^+, such as DNP and CCCP [23, 24]. These agents are believed to induce in bacteria a decrease in the total proton motive force, that is the sum of the membrane electrical potential and the pH gradient. In response, treated cells increase the rate of glycolysis and ATPase activity of the F_1F_0 proton pump [53]. The inhibition of MntA transport by proton uncouplers does not indicate which one from the proton motive force or ATP could directly provide the energy for MntA-dependent transport. Further studies using specific inhibitors of P-type pumps, such as vanadate, and using everted membrane vesicles would be required (e. g., [54, 55]). However, MntA predicted amino acid sequence demonstrates high similarity to the family of P-type (or E1–E2) ATPases, which are membrane pumps translocating monovalent cations H^+, Na^+, K^+ and Cu(I), as well as divalent metal cations, including Ca(II) and Mg(II), Zn(II), and Cd(II) (see Chap. 2). Sequence motifs corresponding to putative intracytoplasmic functional domains of P-type ATPase superfamily, the ion transduction domain, the phosphorylation domain, and the ATP-binding domain were identified in MntA sequence [24]. This would suggest that MntA couples ATP hydrolysis to mediate active intracellular accumulation of Mn(II) and

Cd(II). MntA protein sequences showed 20-30% identity to P-type translocating ATPases from both prokaryotes and eukaryotes [24].

Although the P-type ATPase superfamily is broadly conserved among Bacteria, Archaea, and Eukarya, amino acid sequence relationships enable to define a diversity of phylogenetic and functional groups or families [56]. The taxonomic distribution of bacterial sequences similar to *L. plantarum* MntA was evaluated by Blast analyses, and found to be very restricted. This suggests that MntA represents a novel family of P-type translocase specific for Mn(II) and Cd(II), and that it is involved in metal uptake. Closest related sequence to date is from an Archaea species, the Euryarcheota *Thermoplasma acidophilum* (Blast score of 429 bits of information, 34% identity). It shows a very high conservation with the presumed P-type ATPase functional domains in MntA. Another highly related sequence is from the γ-Proteobacteriaceae *Acidithiobacillus ferrooxidans* (Blast score of 394). A sequence from the archaean *Methanococcus jannachii*, and a second from the Bacteria species *A. ferrooxidans* gave Blast scores > 340. A sequence from *Chlorobium tepidum*, and one from a Bacteria species, the δ-Proteobacteriaceae *Geobacter sulfurreducens* show Blast scores > 300. A great variety of eukaryotic sequences, including green and red algae and plants, protozoans and fungi exhibit Blast scores > 270. Remarkably, a number of bacterial sequences, of which the previously characterized P-types involved in resistance to Cu, Zn [54, 55], and other heavy metal ions show Blast scores that are low in comparison < 200. Hence, *L. plantarum* sequence shows preferential conservation with sequences from archaeal and eukaryotic species. It is likely that the high-affinity Mn(II) transporter MntA discovered in *L. plantarum* defines a novel restricted family of Mn(II) and Cd(II) P-type translocase involved in metal uptake.

14.2.2.1.2 Metal Binding Protein-dependent ABC Transport System, Group PsaA

Another novel family of Mn(II) primary transporters was recently described, that belongs to the prokaryotic binding protein-dependent ABC transporter superfamily (see Chap. 1 and 4). These multicomponents permeases function with a specific ligand binding protein, which delivers the ligand to the permease, composed of one or two integral membrane proteins. Transport of metal ions across the cytoplasmic membrane is coupled to ATP hydrolysis by the cytoplasmic component (see Chap. 4). The novel family, denominated cluster 9, is specific for metal ions [27]. Homologs from cluster 9 of prokaryotic binding protein-dependent ABC transporters were recently described in distant bacteria including pathogens. The metal-binding protein component is either soluble in the periplasm as in the case of gram-negative bacteria, and of the TroA protein from *Treponema* [57], or a lipid-anchored protein exposed at the cell surface. This is the case with PsaA and related proteins in Streptococcaceae and Enterococcaceae [30, 58]. Comparison of the three-dimensional structure of PsaA and TroA proteins revealed a novel interdomain hinge unique to this metal binding permease family. In TroA, a Zn(II) atom was found coordinated by three His and one Asp, with two coordinating amino acid residues from the N- and C-globular domains [57]. Depending on the species, the metal binding component was characterized as part of a Mn(II),

Zn(II), or Fe(III) transport system during infection and, as an adhesin which is immunogenic.

The first example of a Mn(II)-binding protein-dependent ABC transport system was provided by Bartsevich and Pakrasi [59]. The prototype of this novel metal-specific group, MntABC, was discovered in the Cyanobacteria *Synecchocystis* 6803, in a genetic screen for photosynthesis-deficient mutants. A missense mutation in the ATP-binding component inactivated the transport system, and in turn prevented bacterial photosynthesis. The mutation could be compensated for by addition of µM Mn(II) which indicated another specific pathway for active Mn(II) uptake in *Synecchocystis* [59]. MntABC uptake system was induced by Mn starvation, displayed a K_m of 1-3 µM for Mn(II), and was competitively inhibited by Cd(II), Co(II), and Zn(II) [29]. Putative homologs were identified in several lactic acid bacteria [60, 61]. The so-called family of lipoprotein receptor-associated antigen I (LraI), important for virulence of Streptococcaceae and Enterococcaceae, corresponds in fact to the metal binding component of the transport system that is homologous to *Synecchocystis* MntABC [58, 62]. Observation of metal-dependent growth phenotypes in S. pneumoniae suggested that the lipoprotein member of the LraI family, PsaA, was the Mn(II)-binding component of the transport systems PsaABC, and that the related protein AdcA was similarly required for Zn(II) transport via the related system AdcABC [27]. Jenkinson and coworkers demonstrated that the homologous multicomponent transport system, ScaABC in S. gordonii, facilitates metal uptake through a similar Mn(II)-binding lipoprotein [28, 63]. ScaA is a protein lipid-modified at the N-terminus that is exposed at the cell surface. ScaA is expressed in response to Mn(II) starvation, after derepression of the *scaCBA* operon by the apo-repressor ScaR devoid of its co-repressor Mn(II) [26]. In S. *pyogenes* the cell surface component of the MtsABC uptake system displays when assayed *in vitro* multiple specificity in binding metal cations, with Zn(II) > Cu(II) > Mn(II). Fe(III) was apparently also bound to another specific site of the native MtsA protein [30]. A S. pyogenes strain devoid of *mtsABC* operon showed reduced Fe and Zn uptake [30]. S. gordonii ScaABC system was inhibited by Zn(II), but not by Mg(II), Ca(II), Cu(II), or Fe(II). It showed greater sensitivity toward 2-deoxyglucose treatment, which depletes intracellular stores of ATP, than toward the membrane protonophore CCCP, which collapses the proton electrochemical gradient [28]. The existence of another Mn(II) transport system in S. gordonii, presumably proton motive force dependent, was suggested in the same study, further supporting the notion of combinations of different high-affinity Mn(II) transporters in species that require Mn(II) for normal growth and function [27, 28, 59].

In addition to their role in transport as cell surface metal ion binding lipoproteins, the LraI proteins are involved in DNA transformation and bacterial adhesion, two other functions important for bacterial virulence and pathogenesis of lactic acid bacteria [64]. In particular streptococcal and enterococcal LraI proteins were directly implicated in bacterial co-aggregation, binding to fibrin [65], laminin [66], and adhesion to epithelial cells [62]. In Enterobacteriaceae Fe(III) acquisition systems using siderophore and host iron-containing proteins (e. g., hemin, hemoglobin, transferrin, and lactoferrin) are major determinants of virulence [67]. The search for factors facilitating iron acquisition in Yersinia pestis led to the cloning of a polycistronic operon encoding a four-components transport system (*yfeABCD*) homologous to those characterized in Cyanobacteria and lactic acid bacteria [68]. Expression of Yersinia *yfeABCD* operon is repressible by both Fe(III) and Mn(II) [68]. Y. pestis YfeABCD system requires cellular energy to transport Fe(III) and Mn(II). Uptake data suggested the existence of another high-affinity Mn(II) transporter [31], and in fact Y. pestis genome contains an *mntH* group A gene (see Sect. 14.2.2.2.1). In Salmonella, moderate attenuation of virulence resulting from inactivation of several iron uptake systems suggested extensive redundancy [69], and further stimulated the search for additional transport systems. A systematic effort aiming at the functional characterization of the Salmonella pathogenicity island 1, identified an apparent homolog of the transport systems characterized in Yersinia, Cyanobacteria, and lactic acid bacteria [70]. Both *yfeABCD* and *sitABCD* operons rescued E. coli growth in extreme iron depletion, in absence of siderophore and presence of iron chelator [68, 70]. The Salmonella SitA periplasmic protein is induced during infection of host intestinal epithelium, and required for full virulence in mice [71].

Binding protein-dependent ABC transport systems that are specific for metal ions compose the cluster 9 of bacterial ABC permeases [27]. Within cluster 9, three groups were distinguished: the PsaA group, the AdcA group, and another group represented by the Treponema sequence TroA, the Mycoplasma Erysipelothrix rhusiopathiae sequence [27], and the B. subtilis MntA sequence [43]. As summarized above, orthologs of PsaA were identifed in very distant species such as the Cyanobacteria MntA [59], the Enterobacteriaceae SitA [70] and YfeA proteins [68]. These homologs exhibit Blast scores in the range of ~ 205. S. pneumoniae PsaA sequence also exhibits moderate similarity with two parologous sequences (Blast scores 141 and 127), and with Treponema protein TroA (Blast score 145). One of the parologous sequences in S. pneumoniae is the proposed metal binding component AdcA of the Zn(II) transport system [27], and the other derives from a genomic segment so far anonymous. PsaA multigenic family is well conserved among Streptococcaceae. PsaA sequence relationships are in accordance with a phylogenetic tree established from analyses of *sodA* gene sequences in Streptococcaceae [72], except for two sequences. One sequence is from Streptococcus anginosus (Milleri group) and related to PsaA as if it were derived from the Mitis group (S. pneumoniae, S. mitis., S. oralis) with a Blast score of 594. The other is from E. faecalis and resembles S. peumoniae PsaA as if it were derived from group C Streptococcaceae (GCS; e. g., S. mitis, S. oralis, S. parasanguis, S. sanguis, and S. gordonii ScaA) or from GCA (e. g., S. pyogenes MtsA, and S. equi), with a Blast score of 490. Recent horizontal transfer is a possible explanation [70] for the peculiar relationships of these two PsaA-like sequences identified in E. faecalis and S. anginosus. Hence, the PsaA group of the cluster 9 familiy of solute binding ABC permeases is broadly conserved among distant prokaryotic species, partly due to some horizontal gene transfer, in possible relation to virulence and acquisition of Mn(II), Zn(II) and/or Fe(III) during host infection.

14.2.2.2 Secondary Transport Systems

14.2.2.2.1 pH-dependent Metal Ion Transporter: MntH Groups A, B, C

The Nramp/DMT family was defined by highly similar sequences identified in widely distant organisms including eukaryote and prokaryote species [73, 74]. The characterization of the yeast homologs Smf1p and Smf2p as Mn(II) transporters [75, 76], and the identification of mammalian Nramp2 as the intestinal secondary transporter mediating dietary iron absorption [34], prompted the search for functional homologs in several bacterial species, including pathogens. The secondary transporters so identified were denominated MntH [42, 43, 77], as they were likely to correspond to the Mn(II) transport activities previously described in whole cells and isolated vesicles (see Sect. 14.2.1). That *E. coli* K-12 Nramp homolog, a group A *mntH* gene, encodes a Mn(II) and Fe(II) secondary transporter was demonstrated by functional complementation of a divalent metal cation-dependent mutation affecting a cytoplasmic enzyme [78], and by direct measures in whole cells of ^{55}Fe(II) and ^{54}Mn(II) uptake at pH 6 and in the presence of 5 mM MgSO$_4$. In both cases uptake was saturable, temperature-dependent and energy- but not ATP-dependent [42]. Studies in a *uncBC E. coli* mutant lacking the F$_0$-F$_1$ ATPase showed that transport is strictly chemiosmotic (Cellier et al., unpublished data). Similar functional data was obtained under the same conditions with the "low GC" gram-positive *B. subtilis*, using a strain constitutively expressing the single Nramp homolog, as a result of inactivation of the Mn(II)-dependent transcriptional repressor MntR [43] (see also Sect. 14.2.3). This strain (*mntR*) is sensitive to Mn(II) and Cd(II), and incapable of sporulation in presence of μM concentration of Mn(II). The defects of *mntR* strain are corrected by inactivation of the *B. subtilis* gene encoding the group A MntH protein [43]. *B. subtilis* MntH secondary transporter showed a K_m for Mn(II) of 1.7 μM and, similarly to *E. coli* MntH was strongly inhibited by Cd(II) [42]. Further indirect evidence suggested the following substrate preference for *E. coli* MntH uptake: Mn(II) > Cd(II) > Zn(II) > Co(II) > Fe(II) > Ni(II) > Cu(II) [43]. The K_m for ^{54}Mn(II) of *S. typhimurium* MntH was 0.1 μM, and was pH-independent [77]. The most potent inhibitor was Cd(II) with a K_i of about 1 μM, followed by Co(II) with a K_i of 20 μM, and Fe(II) with a pH-dependent K_i, that was of minimal value, 10 μM at pH 5.5 [77]. In both *S. typhimurium* and *E. coli*, *mntH:lacZ* constructs were induced by hydrogen peroxide, and strains lacking a functional *mntH* gene were more susceptible to killing by hydrogen peroxide [77]. Expression of *mntH:lacZ* was induced 3 h after entry of *S. typhimurium* in macrophages, and *S. typhimurium mntH* mutants were reported to show slight attenuation of virulence in Balb/c mice [77]. Injection into *Xenopus laevis* oocytes of RNA encoding the single mycobacterial homolog induced an about 20-fold increase in ^{65}Zn(II) and ^{55}Fe(II) uptake, which was dependent on acidic extracellular pH, and maximal between pH 5.5 and 6.5. Transport was abolished by an excess of Mn(II) and Cu(II). Group A *mntH* mRNA level was up-regulated in response to increases in ambient Fe(II) and Cu(II) in *Mycobacterium tuberculosis* and expressed in *M. bovis* BCG cultured intracellularly [79]. It was also demonstrated by genomic

PCR that one *mntH* gene is present per mycobacterial genome, in both rapid- and slow-growing *Mycobacterium* species, and independently of pathogenicity [42].

All the MntH proteins characterized to date belong to one of three bacterial groups defined by sequence analyses, and tentatively denominated MntH groups A, B, and C [80]. Sequences from these groups exhibit a relatively low level of relationship (26-29% identity in pairwise comparisons between groups, Blast scores > 205). Phylogenetic analyses performed with 47 aligned Nramp and MntH sequences, including sequences from 13 bacterial genera, indicate that group C MntH proteins are more closely related to eukaryotic Nramp homologs. Yeast Smf3 and plant eudicot *Arabidopsis thaliana* atNramp1 and 6 are the eukaryotic protein sequences showing best conservation with group C prokaryotic sequences, and with one of *E. faecalis* sequences, EfMC1, in particular [80]. As summarized above group A MntH proteins function as pH-dependent secondary Mn(II) transporter in gram-negative and gram-positive bacteria. Sequence relationships between MntH proteins from group A are consistent with the phylogenetic relationships of the species from which sequences derive. Hence it is likely that an *mntH* group A gene existed in a common ancestor prior to the separation of gram-positive, gram-negative species and the group *Deinococcus* (*D. radiodurans*), and that an *mntH* group A gene was subsequently maintained in the genome of several eubacterial genera.

Group B MntH proteins are the most divergent in sequence. Two sequences encoded by the chromosomes of the green sulfur bacterium *Chlorobium tepidum* and the gram-positive bacterium *Clostridium acetobutylicum* have been identified so far. Group B sequences show different amino acid composition, and relatively low level sequence identity to each other. These sequences are separated by long branches on the Nramp tree [80]. However, group B sequences can be aligned without gap and show very similar hydrophobic profiles. The pattern of sequence and hydropathy profile conservation between group B MntH proteins is similar to those observed among other MntH groups, and suggests a predicted membrane topology with 11 transmembrane domains and the N-terminus intracellular. This topology is similar to that of yeast *smf1-3* gene products [81]. Interestingly, *Chlorobium* and *Clostridium* are strict anaerobes, whereas bacterial species expressing an MntH group A are facultative aerobes. *C. tepidum* carries out anoxygenic photosynthesis, which arose before the establishment of Mn-based oxygenic photosynthesis [82] (see Sect. 14.1.2). *C. tepidum* expresses a Fe^{2+}-based photosynthetic reaction center (PS), which is an evolutionary precursor of plant photosynthetic systems I and II (PSI and PSII) [83]. The uniqueness of *C. tepidum* MntH group B sequence is therefore interesting, in view of the peculiar PS sequence of this organism [84]. The prototypic PS from *C. tepidum* is believed to have emerged during primitive anaerobic atmospheric conditions (before 2.5 billion years) [85, 86], and to have further evolved without the selective pressure of oxygen toxicity, in an environment rich in soluble Fe^{2+} [84]. The separate clustering of MntH group B observed in phylogenetic analyses [80] may therefore reflect the early emergence of MntH group B, which predated the appearance of MntH group A. The gram-positive *Clostridium acetobutylicum* possesses a second gene that is carried on the megaplasmid pSOL1. This episomal sequence resembles group A MntH sequences from other species more than the chromosomal group B sequence from *Clostridium*. This suggests horizontal gene transfer (HGT) of a *mntH* group A gene via the plasmid pSOL1 into *C. acetobutylicum*. The apparent restriction of group B MntH sequences to strict anaerobes and the unique origin of MntH group B close to the root of Nramp tree [80], together suggest that group B *mntH* was close to the earliest ancestor of Nramp family, originating prior to the onset of aerobiosis and the appearance of group A. Co-existence of *mntH* genes from groups A and B in *Clostridium* may reflect HGT to acquire better resistance against environmental stress.

Group C *mntH* genes are restricted to animal and plant bacterial pathogens, and some species possess two group C genes: the lactic acid bacteria *E. faecalis*, and the γ-Proteobacteriaceae *Pseudomonas aeruginosa*. Both genes were found in different isolates of *P. aeruginosa* [42, 77]. The *E. faecalis* sequences EfMC1 appears to be among group C, the closest to eukaryotic sequences. One of *P. aeruginosa* sequences, PaMC1 is among group C the closest to gram-negative group A MntH proteins. Each remaining sequence from either *E. faecalis* or *P. aeruginosa* shows higher conservation with group C sequences from other species than with their respective paralogous sequence. The close sequence relationships between some MntH group C and eukaryotic sequences can be explained by a horizontal gene transfer from an eukaryotic species toward *E. faecalis*. This species is naturally competent for DNA transformation and can grow in presence of Mn instead of Fe. Further horizontal gene transfer between bacteria would have favored propagation among pathogens that form MntH group C [80].

A single group of MntH secondary Mn(II) transporters was so far characterized in gram-negative and gram-positive species. Metal ion transport by group A MntH proteins appears conserved with eukaryotic homologs, and therefore demonstrates the ancestral origin of the Nramp family. Substantial divergence among groups A, B, and C of MntH proteins, suggests, however, polyphyletic origins for bacterial *Nramp* homolog genes. A more recent origin for *mntH* group C genes could reflect bacterial adaptation for acquisition of redox metals at the pathogen–host interface [80].

14.2.3
Genes Encoding Transcription Factors Involved in Manganese Homeostasis

14.2.3.1 **Fur and Fur-related Factors**
The Ferric Uptake Repressor (Fur) is a transcriptional regulator, that uses predominantly Fe(II) as co-repressor [20]. Fur is active as a dimer with one Fe(II) per monomer, and a structural Zn(II) essential for activity per dimer [87]. All the genes that encode proteins involved in Fe(III) and Fe(II) cellular import are de-repressed under conditions of iron scarcity, including oxidative stress, that result in Fur inactivation [88]. Fur has also some positive effects on transcription, including of genes encoding iron storage functions [88]. Binding of Fur to DNA in the presence of Mn(II), and Co(II) as alternative divalent metal co-repressors, has been reported

in DNAse footprinting *in vitro* experiments [20]. *In vivo*, it was observed in *Y. pestis* that the *yfeABCD* operon encoding a multicomponent bacterial ABC permease for Fe(III) and Mn(II) (see Sect. 14.2.2.1.2) is regulated by Fur in response to Fe(II) and Mn(II) availability [68]. A plasmid *yfeA:phoA* fusion was used to show a 7-fold increase in expression under iron-deficient conditions, that was repressed 5.7-fold with 1 μM Fe, and 3.5-fold with 1 μM Mn, in the growth medium. This effect of Mn(II) as a Fur co-repressor was not observed for two other iron- and Fur-repressible promoters. It was thus concluded that iron and manganese repression of *yfeABCD* required a functional Fur protein [68]. A 19 bp palindrome was also identified in the promoter of the homologous operon in *S. typhimurium sitABCD*, and a *fur* mutation resulted in uncontrolled expression of a *sitB::lacZ* chromosomal fusion in presence of iron [70]. The Fur protein may thus play a role in the regulation of Mn(II) uptake transport systems especially in gram-negative bacteria.

Fur homologs were characterized functionally in more than 20 eubacterial genera [20, 88]. Fur family also includes the regulators of zinc transport, Zur (see Chap. 13) and peroxide stress resistance, PerR. A fur-like protein in the gram-positive species *B. subtilis* acts as an iron uptake repressor [89]. Another *B. subtilis* Fur-like protein, the peroxide stress regulator PerR, was described as a Mn(II)-dependent repressor of the transcription of peroxide stress regulon [90]. Of note, a *perR* gene is also required for iron repression of peroxide stress genes (*ahpC* and *katA*) in the gram-negative microaerophile *Campylobacter jejuni*, acting as a non-homologous substitution for the OxyR protein [91]. However, the presence of a single *perR* gene as a *fur* homolog in the Mn-dependent species *Borrelia burgdorferi* [22] further suggests that Fur and Fur-like proteins are important candidate transcriptional repressors of Mn(II) uptake functions.

14.2.3.2 DtxR and DtxR-related Factors

Factors more closely related to the DtxR/IdeR family play a central role in the regulation of expression of Mn(II) uptake in several species. DtxR/IdeR represent another family of regulators of iron acquisition, that is, however, less widespread than the Fur family, and so far described in gram-positive bacteria [92, 93], and in Archaea. Interestingly, two types of DtxR-related factors showing strict Mn(II) specificity were identified in the course of characterization of Mn(II) uptake systems.

A distant homolog of DtxR acts *in vivo* as a Mn(II)-dependent, transcriptional repressor of *B. subtilis mntH* group A gene (see Sect. 14.2.2.2.1). This factor is insensitive to iron and was denominated Manganese Transport Regulator (MntR). The apoprotein MntR also functions as a transcriptional activator of the operon encoding the multicomponent bacterial ABC permease *mntABCD* (see Sect. 14.2.2.1.2). *B. subtilis mntA* is transcriptionally activated at the end of logarithmic growth in Mn-limiting defined medium, and MntH may be the predominant uptake activity during logarithmic phase growth [43]. Bacterial species in which putative homologs of MntR are suggested by scores from Blast analyses with *B. subtilis* sequence as query include by decreasing order, *Bacillus halodurans*, several Euryarchaeoteae, the γ-Proteobacteriaceae *E. coli* and *Xylella fastidiosa*, and also *D. radiodurans*. The Archaea species *T. acidophilum, Methanococcus jannaschii* may also contain a MntA-like P-type ATPase (Sect. 14.2.2.1.1), while Pyrococcaceae, *Methanobacterium thermoautotrophicum, Archaeoglobus fulgidus, Halobacterium* sp. NRC-1 may possess a potential metal binding protein-dependent ATP permease (Sect. 14.2.2.1.2). The Bacteria species possess at least one *mntH* gene (Sect. 14.2.2.2.1), and *Bacillus* spp. and *D. radiodurans* also possess a binding protein-dependent ABC permease (Sect. 14.2.2.1.2). Co-existence of genes encoding some potential Mn(II) transporters and an MntR homolog in these species further supports the proposed existence of a conserved MntR family involved in Mn homeostasis [43].

Several gram-positive species, including Enterococcaceae, Streptococcaceae, *Staphylococcus epidermidis* and Mycobacteria, and also the gram-negative species *T. pallidum*, possess a gene encoding a protein with only low level similarity to DtxR and MntR, thus defining a novel DtxR-related group [26, 94, 95]. Both *T. pallidum* TroR and *S. gordonii* ScaR are Mn(II)-dependent transcriptional repressors that are insensitive to iron [95] and [26]. TroR binds to a 22-nt sequence within the −10 region of the promoter of the *troABCDR gpm* operon, and acts in presence of Mn(II) as a transcriptional repressor [95]. ScaR functions as a Mn(II)-dependent repressor of the expression of the *S. gordonii* virulence-related Mn(II) permease ScaABC [26, 28]. ScaR recognizes a 46 bp region in the promoter of the *scaCBA* operon that encompasses the −35 and −10 boxes. Palindromic elements were found around each box, and appeared conserved only in Streptococcaceae [26]. ScaA lipoprotein is expressed in response to Mn(II) starvation and exposure to human saliva and serum, suggesting that host body fluids have reduced available Mn(II), and constitute a signal for virulence expression [26]. It is interesting to note that the operon *sitABC* encoding a highly related metal binding protein ABC permease in Staphylococcaceae, which are iron-dependent organisms, is repressed by SirR, a ScaR/TroR-like regulator using Mn(II) or Fe(II) as co-repressors [94]. Hence, DtxR-like factors known as Mn(II)-specific regulators in bacteria that are Mn(II)-dependent for growth (e.g., *B. subtilis, T. pallidum*, and some Streptococcaceae) may function with another redox metal ion in species that are minimally affected by Mn(II) in growth.

14.3
Importance of Manganese Transport in Bacterial Pathogenesis

The host Nramp1 protein expressed in professional phagocytes was recently demonstrated to be a Mn(II)-withholding defense against potential microbial invaders [33]. A divalent cation-sensitive fluorescent probe was conjugated to zymosan particles, to analyze the ambient Mn(II) concentration within the phagosome of macrophages. The results suggest that Nramp1 protein extrudes Mn(II) from the lumen of the phagosome in a proton-dependent manner. It is likely that Fe(II) and other divalent metal cations are similarly removed from the phagosomal microenvironment so that macrophages impair the replication and intracellular survival of both Fe- and Mn-growth-limited bacteria [33].

It is possible that the importance of Mn(II) transport in bacterial pathogenesis is currently under-appreciated. The presence of multiple mechanisms employed by bacteria for obtaining redox metal ions, and described in great variety for iron (see Chap. 12), underscores the importance of metal ions assimilation during infection. Fur and DtxR-types of regulators are used by pathogenic bacteria to respond to host limitation of metal ions availability, by a coordinate regulation of key functions and virulence factors, including toxins, adhesins, and transport systems. Acquisition of iron is a dispensable feature only for bacterial pathogens that have adapted to live with an alternative redox metal ion, e.g., Mn [21, 22, 26, 95]. Bacteria showing Mn(II) dependency for growth (e.g., *S. suis, S. gordonii, S. mutans, T. pallidum, B. burgdorferi*) may thus coordinately regulate virulence expression to Mn(II) availability using Mn(II)-specific DtxR-related repressors. The ability to obtain and utilize Mn(II) for growth in place of iron would provide a selective advantage under conditions of iron restriction by host.

A level of redundancy in Mn(II) transport systems may be an indication of the metabolic importance of Mn(II) for a particular species, and may reflect the necessity to adapt to various environmental conditions for bacterial growth and/or pathogenicity. *E. faecalis* may also be primarily dependent on Mn for growth [96], and the extraordinary redundancy in putative Mn(II) transporters in this species (see Sects. 14.2.2.1.2 and 14.2.2.2.1) would support this view. Another lactic acid bacteria, *S. mutans*, is known to require Fe or Mn for growth [21]. *S. mutans* expresses a putative Mn(II)-binding ABC permease [97], a putative group C MntH encoded on the chromosome, and could possibly carry another group C *mntH* on plasmid, as was reported for the closely related, cariogenic species *S. cricetus* (GenBank # AB026123). This *mntH* group C gene found on plasmid pAM1 is linked to genes encoding Mn(II)-dependent glycosyl transferases. These enzymes produce exopolysaccharides that are required for infectivity and virulence of cariogenic species of Streptococcaceae. It is thus possible that Mn(II) transport could also affect adherence indirectly through the metabolic production of exopolysaccharides [98]. Multiple gene analyses of the regulation of Mn(II) transporter expression *in vivo* together with detailed metal transport measurements will increase our understanding of the importance of Mn(II) transport in bacterial pathogenesis.

In gram-positive species, the determination of the influence on bacterial pathogenesis of some multicomponent ABC Mn(II) permeases can be rendered more complicated, as the metal binding lipoprotein acts as a surface ligand for various host cell matrix proteins. Inactivation of genes encoding LraI family antigens can result in a defect not directly related to metal transport, but rather related to bacterial adhesion properties. In *S. mutans*, inactivation of SloC does not affect growth or DNA transformation in standard conditions, but selectively attenuates virulence in the rat model of endocarditis, and not cariogenicity in the gnotobiotic rat model [97]. In the group B Streptococcaceae (GBS) *S. agalactiae*, inactivation of *lmb*, which encodes a polypeptide showing equal similarity to *S. pneumoniae* Mn(II)- and Zn(II)-binding lipoproteins PsaA and AdcA, respectively, resulted specifically in reduced adherence to laminin [66]. No difference in growth rate or yield in regular medium and medium supplemented with Mn(II) was noted for the mutant strain. The defect in adherence to human laminin could not be circumvented by adding Mn(II) to growing bacteria or during adherence assay [66]. In the GCS *S. parasanguis*, insertion and deletion mutants of *fimA* exhibited reduced infectivity in the rat model of endocarditis, that was correlated to diminished adherence to fibrin [65]. A direct role in adherence was also demonstrated for the lipoprotein components PsaA and ScaA of the Mn(II) multicomponent permeases from GCS species *S. pneumoniae* and *S. gordonii*, respectively [27, 28, 62]. Multiple functions have been suggested for this novel class of lipoproteins [99, 100]. The studies indicate a direct role in adherence and virulence for the lipoproteins of Mn(II) ABC permeases, that is not dependent on Mn(II) transport.

In gram-negative species, periplasmic Fe(III)- and Mn(II)-binding protein-dependent uptake was measured (see Sect. 14.2.2.1.2). Inactivation of *Y. pestis yfeAB* genes resulted in considerable augmentation of the dose that is lethal for 50% of the mouse challenged by the subcutaneous route of infection. However, the mutant bacteria retained full virulence after infection via an intravenous route, and it is not yet clear whether attenuation results from a deficit in Mn or Fe acquisition [31]. The search for genes that are required for systemic infection by the related species *Yersinia entrocolitica* identified a system showing similarity to the *S. epidermidis* operon [101, 102]. Another study identified *S. typhimurium sitABCD* operon through an IVET screen for genes expressed in cultured murine hepatocytes [71]. A *sit* null mutant was significantly attenuated in Balb/c mice [71]. The results of these studies indicate an important role for virulence *in vivo* of Mn(II) and/or Fe(III) transport by the family of metal-specific multicomponent ABC permeases.

Mn(II) uptake should be beneficial for pathogens as they encounter aerobic environments, and oxidative burst, but the link is certainly not absolute. For example, *P. aeruginosa* MnSOD, SodA, may participate in the adaptation of mucoid strains to the stationary phase of growth in the lungs of cystic fibrosis patients [103]. A *Y. enterocolitica sodA* null mutant was strongly attenuated in a mouse intravenous infection model, but not in the orogastric infection model. Whether *Y. enterocolitica* SodA mediated resistance to oxygen radicals produced by bacterial endogenous metabolism or exogenously by phagocytes was not clear [104]. In *S. typhimurium*, resistance to the early oxygen-dependent microbicidal mechanisms of phagocytes involves the *sodA* gene product which, however, plays only a minor role in *Salmonella* pathogenesis [105]. In contrast *S. typhimurium sodC* mutant deficient in periplasmic Cu,Zn-superoxide dismutase had reduced survival in macrophages and attenuated virulence *in vivo* [12]. In addition, virulent *S. typhimurium* have acquired a second *sodC* gene by horizontal tranfer, and a *S. typhimurium* mutant devoid of *sodC* genes is less lethal [106]. Hence, Cu,Zn Sod may exert the predominant role of resisting to host oxidative burst in gram-negative bacteria, while gram-positive bacteria may rely primarily on Mn-based SodA protein, or on very high-level intracellular accumulation of Mn(II) in absence of SodA activity [107].

14.4
Concluding Remarks

Research on bacterial Mn transport in recent years has led to the molecular cloning and characterization of three different types of Mn(II) membrane permeases and of novel transcriptional regulators that are Mn(II)-dependent. Mn permeases show substantial evolutionary conservation, but the three protein families show quite distinct taxonomic distribution. It is likely that additional types of bacterial Mn transporters are yet to be discovered (e.g., in *B. burgdorferi*). Cloned genes and molecularly defined mutants have favored the appreciation of a somewhat overlooked role of Mn in bacterial growth, in addition to its protective role against oxidative stress. Genome-driven research has revealed that adaptation by reductive evolution, resulting in elimination of iron-based functions, was a viable approach for human pathogens. Hence the ability of bacterial pathogens to counteract host withholding defenses (e.g., Nramp1) and to acquire Mn(II) during infection deserves closer study. Further investigation of manganese transport and intracellular accumulation in bacteria in relation to expression of virulence and pathogenesis may reveal interesting knowledge to better fight some infectious diseases.

Acknowledgement

Work on this topic in my laboratory was supported in part by research grants from the Fonds pour la Recherche du Québec, the Natural Sciences and Engineering Research Council of Canada, and the Canadian Institutes of Health Research. I am grateful to Mireille Goetghebeur for her long-standing support. I also thank Philippe Gros and J. D. Helmann for their comments and suggestions on this manuscript.

References

1. Morgan, J. J., *Met. Ions Biol. Syst.* **2000**, *37*, 1-34.
2. Whittaker, J. W., *Met. Ions Biol. Syst.* **2000**, *37*, 587-611.
3. Reed, G. H., Poyner, R. R., *Met. Ions Biol. Syst.* **2000**, *37*, 183-207.
4. Christianson, D. W., *Prog. Biophys. Mol. Biol.* **1997**, *67*, 217-252.
5. Crowley, J. D., Traynor, D. A., Weatherburn, D. C., *Met. Ions Biol. Syst.* **2000**, *37*, 209-278.
6. Keen, C. L., Ensunsa, J. L., Clegg, M. S., *Met. Ions Biol. Syst.* **2000**, *37*, 89-121.
7. Debus, R. J., *Met. Ions Biol. Syst.* **2000**, *37*, 657-711.
8. Archibald, F. S., Fridovich, I., *Arch. Biochem. Biophys.* **1982**, *214*, 452-463.
9. Archibald, F. S., Fridovich, I., *Arch. Biochem. Biophys.* **1982**, *215*, 589-596.
10. Archibald, F. S., *Crit. Rev. Microbiol.* **1986**, *13*, 63-109.
11. Brioukhanov, A., Netrusov, A., Sordel, M., Thauer, R. K., Shima, S., *Arch. Microbiol.* **2000**, *174*, 213-216.
12. De Groote, M. A., Ochsner, U. A., Shiloh, M. U., Nathan, C., McCord, J. M., Dinauer, M. C., Libby, S. J., Vazquez-Torres, A., Xu, Y., Fang, F. C., *Proc. Natl Acad. Sci. USA* **1997**, *94*, 13997-14001.
13. Stadtman, E. R., Berlett, B. S., Chock, P. B., *Proc. Natl Acad. Sci. USA* **1990**, *87*, 384-388.
14. Yoder, D. W., Hwang, J., Penner-Hahn, J. E., *Met. Ions Biol. Syst.* **2000**, *37*, 527-557.
15. Gold, M. H., Youngs, H. L., Gelpke, M. D., *Met. Ions Biol. Syst.* **2000**, *37*, 559-586.
16. Babior, B. M., *Am. J. Med.* **2000**, *109*, 33-44.
17. F. C. Wedler, *Prog. Med. Chem.* **1993**, *30*, 89-133.
18. Storz, G., Imlay, J. A., *Curr. Opin. Microbiol.* **1999**, *2*, 188-194.
19. Benov, A. L., Fridovich, I., *J. Biol. Chem.* **1998**, *273*, 10313-10316.
20. Braun, V., Hantke, K., Koster, W., *Met. Ions Biol. Syst.* **1998**, *35*, 67-145.
21. Niven, D. F., Ekins, A., al-Samaurai, A. A., *Can. J. Microbiol.* **1999**, *45*, 1027-1032.
22. Posey, J. E., Gherardini, F. C., *Science* **2000**, *288*, 1651-1653.
23. Hao, Z., Reiske, H. R., Wilson, D. B., *Appl. Environ. Microbiol.* **1999**, *65*, 4741-4745.
24. Hao, Z., Chen, S., Wilson, D. B., *Appl. Environ. Microbiol.* **1999**, *65*, 4746-4752.
25. Drake, D., Taylor, K. G., Doyle, R. J., *Infect. Immun.* **1988**, *56*, 2205-2207.
26. Jakubovics, N. S., Smith, A. W., Jenkinson, H. F., *Mol. Microbiol.* **2000**, *38*, 140-153.
27. Dintilhac, A., Alloing, G., Granadel, C., Claverys, J. P., *Mol. Microbiol.* **1997**, *25*, 727-739.
28. Kolenbrander, P. E., Andersen, R. N., Baker, R. A., Jenkinson, H. F., *J. Bacteriol.* **1998**, *180*, 290-295.
29. Bartsevich, V. V., Pakrasi, H. B., *J Biol Chem.* **1996**, *271*, 26057-26061.
30. Janulczyk, R., Pallon, J., Bjorck, L., *Mol. Microbiol.* **1999**, *34*, 596-606.
31. Bearden, S. W., Perry, R. D., *Mol. Microbiol.* **1999**, *32*, 403-414.
32. Gruenheid, S., Gros P., *Curr. Opin. Microbiol.* **2000**, *3*, 43-48.
33. Jabado, N., Jankowski, A., Dougaparsad, S., Picard, V., Grinstein, S., Gros P., *J. Exp. Med.* **2000**, *192*, 1237-1248.
34. Gunshin, H., Mackenzie, B., Berger, U. V., Gunshin, Y., Romero, M. F., Boron, W. F., Nussberger, S., Gollan, J. L., Hediger, M. A., *Nature* **1997**, *388*, 482-488.
35. Canonne-Hergaux, F., Gruenheid, S., Ponka, P., Gros, P., *Blood* **1999**, *93*, 4406-4417.
36. Fleming, M. D., Romano, M. A., Su, M. A., Garrick, L. M., Garrick, M. D., Andrews, N. C., *Proc. Natl. Acad. Sci. USA* **1998**, *95*, 1148-1153.
37. Fleming, M. D., Trenor 3rd, C. C., Su, M. A., Foernzler, D., Beier, D. R., Dietrich, W. F., Andrews, N. C., *Nature Genet.* **1997**, *16*, 383-386.
38. Nelson, N., *EMBO J.* **1999**, *18*, 4361-4371.
39. Culotta, V. C., *Met. Ions Biol. Syst.* **2000**, *37*, 35-56.
40. Curie, C., Alonso, J. M., Le Jean, M., Ecker, J. R., Briat, J. F., *Biochem. J.* **2000**, *347*, 749-755.
41. Thomine, S., Wang, R., Ward, J. M., Crawford, N. M., Schroeder, J. I., *Proc. Natl Acad. Sci. USA* **2000**, *97*, 4991-4996.
42. Makui, H., Roig, E., Cole, S. T., Helmann, J. D., Gros, P., Cellier, M. F., *Mol. Microbiol.* **2000**, *35*, 1065-1078.

43. Que, Q., Helmann, J. D., *Mol. Microbiol.* **2000**, *35*, 1454-1468.
44. Silver, S., Kralovic, M. L., *Biochem. Biophys. Res. Commun.* **1969**, *34*, 640-645.
45. Silver, S., Johnseine, P., King, K., *J. Bacteriol.* **1970**, *104*, 1299-1306.
46. Bhattacharyya, P., *J. Bacteriol.* **1970**, *104*, 1307-1311.
47. Laddaga, R. A., Silver, S., *J. Bacteriol.* **1985**, *162*, 1100-1105.
48. Fisher, S., Buxbaum, L., Toth, K., Eisenstadt, E., Silver, S., *J. Bacteriol.* **1973**, *113*, 1373-1380.
49. Eisenstadt, E., Fisher, S., Der, C. L., Silver, S., *J. Bacteriol.* **1973**, *113*, 1363-1372.
50. Bhattacharyya, P., *J. Bacteriol.* **1975**, *123*, 123-127.
51. Perry, R. D., Silver, S., *J. Bacteriol.* **1982**, *150*, 973-976.
52. Archibald, F. S., Duong, M. N., *J. Bacteriol.* **1984**, *158*, 1-8.
53. Maloney, P. C., *FEMS Microbiol. Rev.* **1990**, *7*, 91-102.
54. Rensing, C., Mitra, B., Rosen, B. P., *Proc. Natl Acad. Sci. USA* **1997**, *94*, 14326-14331.
55. Rensing, C., Fan, B., Sharma, R., Mitra, B., Rosen, B. P., *Proc. Natl Acad. Sci. USA* **2000**, *97*, 652-656.
56. Moller, J. V., Juul, B., le Maire, M., *Biochim. Biophys. Acta* **1996**, *1286*, 1-51.
57. Lee, Y. H., Deka, R. K., Norgard, M. V., Radolf, J. D., Hasemann, C. A., *Nature Struct. Biol.* **1999**, *6*, 628-633.
58. Lowe, A. M., Lambert, P. A., Smith, A. W., *Infect. Immun.* **1995**, *63*, 703-706.
59. Bartsevich, V. V., Pakrasi, H. B., *EMBO J.* **1995**, *14*, 1845-1853.
60. Kolenbrander, P. E., Andersen, R. N., Ganeshkumar, N., *Infect. Immun.* **1994**, *62*, 4469-4480.
61. Sampson, J. S., O'Connor, S. P., Stinson, A. R., Tharpe, J. A., Russell, H., *Infect. Immun.* **1994**, *62*, 319-324.
62. Berry, A. M., Paton, J. C., *Infect. Immun.* **1996**, *64*, 5255-5262.
63. Jenkinson, H. F., *Trends Microbiol.* **1994**, *2*, 209-212.
64. Whittaker, C. J., Clemans, D. L., Kolenbrander, P. E., *Infect. Immun.* **1996**, *64*, 4137-4142.
65. Burnette-Curley, D., Wells, V., Viscount, H., Munro, C. L., Fenno, J. C., Fives-Taylor, P., Macrina, F. L., *Infect. Immun.* **1995**, *63*, 4669-4674.
66. Spellerberg, B., Rozdzinski, E., Martin, S., Weber-Heynemann, J., Schnitzler, N., Lutticken, R., Podbielski, A., *Infect. Immun.* **1999**, *67*, 871-878.
67. Byers, B. R., Arceneaux, J. E., *Met. Ions Biol. Syst.* **1998**, *35*, 37-66.
68. Bearden, S. W., Staggs, T. M., Perry, R. D., *J. Bacteriol.* **1998**, *180*, 1135-1147.
69. Tsolis, R. M., Baumler, A. J., Heffron, F., Stojiljkovic, I., *Infect. Immun.* **1996**, *64*, 4549-4556.
70. Zhou, D., Hardt, W. D., Galan, J. E., *Infect. Immun.* **1999**, *67*, 1974-1981.
71. Janakiraman, A., Slauch, J. M., *Mol. Microbiol.* **2000**, *35*, 1146-1155.
72. Poyart, C., Quesne, G., Coulon, S., Berche, P., Trieu-Cuot, P., *J. Clin. Microbiol.* **1998**, *36*, 41-47.
73. Cellier, M., Prive, G., Belouchi, A., Kwan, T., Rodrigues, V., Chia, W., Gros, P., *Proc. Natl. Acad. Sci. USA* **1995**, *92*, 10089-10093.
74. Cellier, M., Belouchi, A., Gros, P., *Trends Genet.* **1996**, *12*, 201-204.
75. Supek, F., Supekova, L., Nelson, H., Nelson, N., *Proc. Natl. Acad. Sci. USA* **1996**, *93*, 5105-5110.
76. Liu, X. F., Supek, F., Nelson, N., Culotta, V. C., *J. Biol. Chem.* **1997**, *272*, 11763-11769.
77. Kehres, D. G., Zaharik, M. L., Finlay, B. B., Maguire, M. E., *Mol. Microbiol.* **2000**, *36*, 1085-1100.
78. Herman, C., Lecat, S., D'Ari, R., Bouloc, P., *Mol. Microbiol.* **1995**, *18*, 247-255
79. Agranoff, D., Monahan, I. M., Mangan, J. A., Butcher, P. D., Krishna, S., *J. Exp. Med.* **1999**, *190*, 717-724.
80. Cellier, M., Bergevin, I., Boyer, E., Richer, E., *Trends Genet.* **2001**, *17*, 365-370.
81. Portnoy, M. E., Liu, X. F., Culotta, V. C., *Mol. Cell. Biol* **2000**, *20*, 7893-7902.
82. Des Marais, D. J., *Science* **2000**, *289*, 1703-1705.
83. Schubert, W. D., Klukas, O., Saenger, W., Witt, H. T., Fromme, P., Krauss, N., *J. Mol. Biol.* **1998**, *280*, 297-314.
84. Golbeck, J. H., *Proc. Natl Acad. Sci. USA* **1993**, *90*, 1642-1646.
85. Canfield, D. E., Habicht, K. S., Thamdrup, B.. *Science* **2000**, *288*, 658-661.
86. Xiong, J., Fischer, W. M., Inoue, K., Nakahara, M., Bauer, C. E., *Science* **2000**, *289*, 1724-1730.

87. Jacquamet, L., Aberdam, D., Adrait, A., Hazemann, J. L., Latour, J. M., Michaud-Soret I., *Biochemistry* **1998**, *37*, 2564-2571

88. Touati, D., *Arch. Biochem. Biophys.* **2000**, *373*, 1-6.

89. Bsat, N., Herbig, A., Casillas-Martinez, L., Setlow, P., Helmann, J. D., *Mol. Microbiol.* **1998**, *29*, 189-198.

90. Chen, L., Keramati, L., Helmann, J. D., *Proc. Natl. Acad. Sci. USA* **1995**, *92*, 8190-8194.

91. van Vliet, A. H., Baillon, M. L., Penn, C. W., Ketley, J. M., *J. Bacteriol.* **1999**, *181*, 6371-6376.

92. Schmitt, M. P., Holmes, R. K., *Infect. Immun.* **1991**, *59*, 1899-1904.

93. Dussurget, O., Rodriguez, M., Smith, I., *Mol. Microbiol.* **1996**, *22*, 535-544.

94. Hill, P. J., Cockayne, A., Landers, P., Morrissey, J. A., Sims, C. M., Williams, P., *Infect. Immun.* **1998**, *66*, 4123-4129.

95. Posey, J. E., Hardham, J. M., Norris, S. J., Gherardini, F. C., *Proc. Natl. Acad. Sci. USA* **1999**, *96*, 10887-10892.

96. Marcelis, J. H., den Daas-Slagt, H. J., Hoogkamp-Korstanje, J. A., *Antonie Van Leeuwenhoek*. **1978**, *44*, 257-267.

97. Kitten, T., Munro, C. L., Michalek, S. M., Macrina, F. L., *Infect. Immun.* **2000**, *68*, 4441-4451.

98. Munro, C., Michalek, S. M., Macrina, F. L., *Infect. Immun.* **1991**, *59*, 2316-2323.

99. Novak, R., Braun, J. S., Charpentier, E., Tuomanen, E., *Mol. Microbiol.* **1998**, *29*, 1285-1296.

100. Novak, R., Tuomanen, E., Charpentier, E., *Mol. Microbiol.* **2000**, *36*, 1505-1506

101. Cockayne, A., Hill, P. J., Powell, N. B., Bishop, K., Sims, C., Williams, P., *Infect. Immun.* **1998**, *66*, 3767-3774.

102. Gort, A. S., Miller, V. L., *Infect. Immun.* **2000**, *68*, 6633-6642.

103. Polack, B., Dacheux, D., Delic-Attree, I., Toussaint, B., Vignais, P. M., *Infect. Immun.* **1996**, *64*, 2216-2219.

104. Roggenkamp, A., Bittner, T., Leitritz, L., Sing, A., Heesemann, J., *Infect. Immun.* **1997**, *65*, 4705-4710.

105. Tsolis, R. M., Baumler, A. J., Heffron, F., *Infect. Immun.* **1995**, *63*, 1739-1744.

106. Fang, F. C., DeGroote, M. A., Foster, J. W., Baumler, A. J., Ochsner, U., Testerman, T., Bearson, S., Giard, J. C., Xu, Y., Campbell, G., Laessig, T., *Proc. Natl. Acad. Sci. USA* **1999**, *96*, 7502-7507.

107. Inaoka, T., Matsumura, Y., Tsuchido, T., *J. Bacteriol.* **1999**, *181*, 1939-1943.

15
The Unusual Nature of Magnesium Transporters

David G. Kehres and Michael E. Maguire

15.1
Introduction

Magnesium is different from other biological cations. Consequently, the proteins that transport it are unusual. While the kinetics and the regulation of Mg^{2+} *flux* across membranes have been studied in a number of mammalian systems (reviewed in [1]), only a single Mg^{2+} *transporter* has yet been cloned from any eukaryote [2]; as a consequence, all our molecular knowledge about Mg^{2+} transport proteins comes from the study of three bacterial Mg^{2+} transporter families: MgtE, CorA, and MgtA/B. This review will briefly summarize the chemical properties that make magnesium unique among biologically relevant metals. We will then concentrate on the three prokaryotic transporter types from multiple standpoints: genomics, physiology, structure, and mechanism.

15.2
The Properties of Mg^{2+}

15.2.1
Chemistry

Mg^{2+} represents an extreme case among biologically relevant cations both in the geometry and strength of its ionic interactions. Magnesium"s third row position in the periodic table and lack of *d*-electrons to participate in ligand coordination give it less bond angle flexibility than other common cations. It is almost invariably hexacoordinated, in a regular square bipyramidal geometry with all bond angles close to 90° [3–6]. Mg^{2+} strongly prefers oxygen as a ligand; it can frequently interact with nitrogen, as in chlorophyll, but interactions with sulfur are unknown in biological systems.

In terms of bond strength, magnesium has the highest charge density of all common cations [3]. However, in terms of bond dynamics, the absence of *d*-electron interactions apparently weighs more heavily than overall charge density in most cases because non-covalent bonds between Mg^{2+} and various ligands tend to be more labile than those formed by other cations [3, 7]. Nonetheless, charge density also affects binding to ligands indirectly, due to uncommonly strong interactions with water. Magnesium binds its inner hydration shell several orders of magnitude more tightly than do Na^+, K^+ or Ca^{2+}. Indeed, because of its avidity for water and rather rigid square bipyramidal structure, many of the activities of Mg^{2+} as an enzyme cofactor are mediated through a bound water molecule rather than by interaction with the metal ion directly [8–12]. Mg^{2+} has the largest hydrated radius of any common cation; its ionic radius, i.e., minus waters of hydration, is among the smallest seen with divalent cations. Because of these size considerations, the volume change between hydrated and ionic Mg^{2+} is almost 400-fold. Thus, any protein transporting Mg^{2+} must be capable of initially interacting with a rather large cation. Then, assuming that Mg^{2+} like other cations is transported in its ionic form, Mg^{2+} must pass through a pore that is quite small. Consequently, transport (or interaction with an enzyme) must, therefore, involve larger initial binding sites and/or more elaborate means of dehydration than are commonly found in other cation transporters. It is these atypical geometric and energetic features of magnesium chemistry that explain why magnesium transporters, as far as they have been characterized, tend to be novel types of proteins.

15.2.2
Association States of Magnesium

From the standpoint of transport thermodynamics, the vast majority of the magnesium in both intracellular and extracellular biological environments is complexed with other molecules; only a small fraction is present as the free solvated ion [13–20]. Much Mg^{2+} is "structural" in a relatively non-specific (i.e., rapidly exchanging) way, associated with macromolecules like nucleic acids, protein complexes, polysaccharides, and the polar head group regions of membranes. About half of cell Mg^{2+} is associated, more specifically (though not always more tightly), with small molecules – abundant ones such as ATP, or less abundant xenobiotics such as the drug tetracycline. A small fraction of Mg^{2+} is bound more tightly and specifically either as a stabilizing factor or as a catalytic factor to various protein and RNA enzymes, e.g., the G-proteins such as $G\alpha_s$, and hairpin and acyltransferase ribozymes. Overall, Mg^{2+} presents a variety of unique problems for the design of proteins that can transport it.

15.2.3
Technical Problems in Studying Magnesium

Mg^{2+} is devoid of convenient optical and magnetic properties that would allow its concentration in cells to be measured directly. The only useable radioactive isotope, $^{28}Mg^{2+}$, is of very high energy and has only a 21 h half-life; most importantly, it is not commercially available [21]. The stable isotopes ^{25}Mg and ^{26}Mg are available and can be measured but only by expensive and specialized mass spectrometry. In specialized cells such as large neurons, the free concentration of Mg^{2+} can also be measured by selective electrodes, but interference from other ions necessitates very careful controls [18, 22–24]. Magnesium concentration can also be measured indirectly, using fluorescent indicators such as mag-fura-2 [25–27]. However, such dyes are not completely selective for Mg^{2+} and indicators with a greater specificity as well as with a sensitivity over greater ranges of magnesium concentration still need to be developed. Because of this lack of probes and reporters for Mg^{2+}, transport of the known bacterial transporters is measured by uptake of other cations also transported, albeit non-physiologically, by the various systems. The MgtE class is measured using $^{57}Co^{2+}$, the MgtA/B class using $^{63}Ni^{2+}$, and the CorA class with either of these isotopes [21].

15.3
Prokaryotic Magnesium Transport

Three classes of magnesium transporters have been cloned from Bacteria and Archaea, MgtE, CorA, and MgtA/B class. CorA and MgtA transporters have been characterized physiologically and biochemically in some detail. Much of this work has previously been reviewed [28, 29]. Bacterial Mg^{2+} transporters have been cloned by complementation of the Mg^{2+} growth requirement of a *Salmonella typhimurium* strain engineered to lack all Mg^{2+} transport. This strain, MM281 [30–33], has insertions in its three Mg^{2+} transporters, CorA, MgtA, and MgtB. Wild-type *S. typhimurium* does not require addition of supplemental Mg^{2+} to the growth medium since the Mg^{2+} "contaminating" the medium is sufficient. A *S. typhimurium* strain lacking one or two of the three total transporters also does not require supplemental Mg^{2+} for growth. However, MM281, lacking all three transporters, requires addition of 100 mM Mg^{2+} for growth and shows no detectable $^{28}Mg^{2+}$ uptake. Introduction of any single Mg^{2+} transporter or of a Mg^{2+} transporter from another bacterium relieves both the growth and transport phenotype and hence can be used to identify other examples and classes of Mg^{2+} transporter.

The present review will discuss each of these three families first from the standpoint of genomics, since the explosion of microbial genome sequence projects makes this the fastest developing aspect of transporter studies, and then from the more molecular standpoints of physiology, structure, and mechanism. Since homologs have been found either in Archaea or in unicellular eukaryotes, these other microbes will be discussed here as well.

15.4
MgtE Magnesium Transporters

MgtE transporters were the last of the three known microbial types to be identified, and except for some basic transport studies have not been analyzed at the molecular level.

15.4.1
Genomics

MgtE transporters were serendipitously cloned from *Bacillus firmus* OF4 [34] and *Providencia stuarti* [35] in a search for additional CorA transporters, based on their ability to complement a magnesium transport deficiency in *Salmonella typhimurium*. At least 23 MgtE-like proteins can be identified currently in DNA sequence databases using a reasonably stringent BLAST search; MgtE transporters do not resemble any other known class of proteins. Twenty-two are in Bacteria, representing several deeply divergent branches, and one has been found in the Archaea. The phylogenetic relationships of their amino acid sequences follow quite closely the taxonomic relationships of the organisms containing them. No MgtE homolog has yet been found in any eukaryote.

15.4.2
Physiology

Biochemical data are only available for MgtEs from *Bacillus firmus* OF4 and *Providencia stuarti* [34, 35]. When expressed from multicopy plasmids in *Salmonella typhimurium*, these transporters have similar K_m values averaging about 70 µM and similar V_{max} values of about 0.50 nmol min^{-1} 10^8 cells^{-1} for ^{57}Co^{2+} uptake. The apparent K_i for Mg^{2+} inhibition is about 50 µM at 10 µM Co^{2+}, indicating that the K_m for Mg^{2+} uptake is 25–50 µM. Sr^{2+}, Mn^{2+}, Ca^{2+}, and Zn^{2+} inhibit with K_i values ranging from 80 µM down to 20 µM respectively, though it has not been determined if any of these cations is transported. Ni^{2+} does not inhibit nor is ^{63}Ni^{2+} taken up to any significant extent. There is no data whether MgtE is capable of cation efflux, nor have any regulatory or expression studies been done. Recently, however, an MgtE homolog has been cloned from *Aeromonas hydrophila* and shown to be necessary for biofilm formation and adherence in this opportunistic pathogen [36]. Thus, like CorA (*vide infra*), MgtE may be involved in bacteria pathogenesis.

15.4.3
Structure and Mechanism

Based on hydropathy analysis, MgtE transporters appear to have four or more likely five transmembrane domains (TM) following a large hydrophilic domain at the N-terminus; the latter is predicted to reside in the cytoplasm based on the positive-inside rule [34, 35]. Putative helical transmembrane segments contain a few modestly conserved charged residues and several well-conserved residues bearing hydroxyl side chains. There is no data whether MgtE transporters function as oligomers or monomers, and no evidence whether they associate with other types of proteins. MgtE sequences lack recognizable NTP binding motifs. It is accordingly presumed that they rely on some transmembrane electrochemical gradient to drive magnesium uptake.

15.5
CorA Magnesium Transporter

CorA was the first magnesium transporter to be isolated, being cloned from *Salmonella typhimurium* in 1985 [33] although the genetic locus had been identified by Silver much earlier [37]. Like the subsequently discovered MgtE, it represents a totally novel class of protein with no homology to any other type of transporter or membrane protein.

15.5.1
Genomics

Homologs of CorA are present in all classes of Bacteria for which there is reasonably complete genomic sequence. Similarly, all of the (relatively few) Archaea with genomic sequence exhibit a CorA homolog. Among the many bacterial genomes now extant, CorA is absent in only a few species, generally those with the smallest genomes [38]. (MgtE is usually present in these small genomes.) Thus, CorA's ubiquitous distribution indicates that it is the major Mg^{2+} transporter of both the Bacteria and the Archaea. As these two Kingdoms of life comprise the largest biomass on this planet, CorA is, therefore, the most abundant Mg^{2+} transporter on Earth.

When we analyzed the phylogeny of CorA [38, 39], we noted that some species had multiple CorA-like sequences, usually two and sometimes three. Alignment of these sequences indicated that there were likely two major branches of the CorA family. Sequences of many additional members of the CorA family have since emerged and have abundantly confirmed this initial suggestion. The "MPEL subclass", representing a signature sequence between TM2 and TM3 (see below for topology) comprises a coherent group of sequences closely related to the *S. typhimurium* CorA. The remaining sequences remain a "catchall" assortment at present, though possibly a distinct second "GGIP" subclass can be discerned. We term this group of sequences "CorA-II". Nothing convincing can yet be inferred about the function, origin, number and lineage(s) of this group except that it likely diverged from CorA relatively early in evolution.

In our initial analysis [38], there seemed a surprising lack of correlation between CorA molecular phylogeny and the underlying organismal phylogeny. This impression of a lack of phylogenetic correlation could have been due to the admixture of the two distinct subclasses of the CorA family in the multiple alignment. As already noted, many additional CorA family members have subsequently been found in genomic sequences. Nonetheless, even with these additional sequences and the separation of the CorA-II class from the analysis, there remains a lack of correlation of CorA molecular phylogeny with the generally accepted phylogeny derived from 16S rRNA sequence. This could imply that there have been multiple lateral transfers of *corA* between organisms, even possibly between Archaea and Bacteria as has recently been suggested for catalase−peroxidase genes [40]. The overall conclusion from genomics is that CorA is a very ancient protein in prokaryotes, probably predating the divergence of Bacteria and Archaea.

Although very faint homology has been claimed in some yeast proteins [41, 42], the relationship is generally confined to a small portion of the transmembrane domain and no direct evidence for Mg^{2+} transport has been presented. Overall, the amount of eukaryotic sequence now extant allows the conclusion that close homologs of the CorA Mg^{2+} transporter do not exist in eukaryotes.

15.5.2
Physiology

Salmonella and *Escherichia coli* CorAs are constitutive proteins, whose promoters do not respond to changes in magnesium concentration or to any other stimuli, as far as is known [43, 44]. Nothing is known about transcriptional regulation of other CorA homologs. Transport parameters have been established for CorA systems from the Bacteria *S. typhimurium*, *E. coli*, *Haemophilus influenzae*, and the Archaeon *Methanococcus jannaschii*. The CorAs of *S. typhimurium*, *E. coli* and *H. influenzae* mediate the influx of Mg^{2+}, Co^{2+}, and Ni^{2+} [31, 33]. CorA exhibits an affinity for Mg^{2+} of 15−20 µM, and affinities for Co^{2+} and Ni^{2+} of 20−40 and 200−400 µM, respectively. The affinities of the latter two cations are clearly within the toxic range for organisms such as *S. typhimurium* and so are highly unlikely to represent a physiologically meaningful uptake. The maximal rate of Mg^{2+} uptake by CorA is at least 1 nmol min^{-1} 10^8 $cells^{-1}$. Given the size of a bacterial cell and the cellular content of Mg^{2+}, this rate would double cell Mg^{2+} in less than 60 s if influx was unabated. This implies that there is some degree of cellular control of this transport system, but no data are available as to mechanism. The sequence of CorA contains no recognizable ATP binding site and since transport via CorA is sensitive to membrane potential, the likely mechanism of Mg^{2+} uptake is either through CorA functioning as a Mg^{2+} channel or as a Mg^{2+}/H^+ antiporter.

The *S. typhimurium* CorA can also mediate Mg^{2+} efflux but not of Ni^{2+} or Co^{2+}. The physiological significance of this capability is completely unknown as it does not occur under normal laboratory and probably environmental growth conditions [31, 45]. In the absence of a functional CorA protein, no Mg^{2+} efflux can be detected under a variety of conditions, thus demonstrating both that CorA is the only apparent Mg^{2+} efflux protein of *S. typhimurium* and that Mg^{2+} efflux is not essential to cell viability. Efflux via CorA requires relatively high extracellular Mg^{2+} concentrations, in the millimolar range, far above the K_m for influx. Thus the efflux does not represent a $Mg^{2+}-Mg^{2+}$ exchange process since the influx rate is already maximal before even minimal efflux can be detected.

The CorA of *M. jannaschii* exhibits transport properties remarkably like those of *S. typhimurium* when expressed in MM281, the Mg^{2+} transport deficient strain [46]. Normal conditions for this archaeal protein would be 85 °C, 250 atmospheres of pressure, a greatly different lipid milieu and an environment containing 55 mM Mg^{2+}. Yet the *M. jannaschii* CorA exhibits the same affinity for Mg^{2+} and other divalent cations. Its rate of influx is somewhat less than that of CorA, and, as would be expected, it is considerably more stable. Thus, this archaeal protein, only 12 % identical to the *S. typhimurium* CorA in the periplasmic domain and 19 % in the membrane domain, functions virtually identically.

Recently, this laboratory has identified selective and potent inhibitors of the CorA family of Mg^{2+} transporters [47]. As discussed above, the Mg^{2+} cation binds six waters which form a large hydration shell of almost 5 Å in diameter. Cations of similar size but with much greater stability exist. These consist of transition metals such as Co, Ru, Ni, and others with the waters of hydration replaced by *covalent* attachment of a variety of ligands, most notably ene-amines and ammines. Thus, Co(III)-hexaammine consists of a trivalent Co atom covalently bonded to six ammines (NH_3). The geometry of the complex is identical to the hydrated Mg^{2+} cation [48, 49] and can mimic hydrated Mg^{2+} in a variety of enzyme active sites [4, 9, 10]. We recently showed that Co(III)-, Ru(II)-, and Ru(III)-hexaammines were potent inhibitors of CorA but not of other Mg^{2+} transporters and other Mg^{2+} binding proteins. These hexaammines have affinities for CorA about 10-fold greater than Mg^{2+} itself. A variety of other similar compounds were either ineffective or less potent. This suggests that the initial binding site for Mg^{2+} on CorA, presumably in the periplasmic domain, is at least 5 Å in diameter and binds a fully hydrated cation. It is of interest that Co(III)-, Ru(II)-, and Ru(III)-hexaammines were also potent inhibitors of the *M. jannaschii* CorA with affinities identical to those for the *S. typhimurium* CorA. This implies, despite only 12 % identity in the soluble, periplasmic domain, that the structures of the archaeal and bacterial CorAs are very similar.

As noted above, many, perhaps most, Bacteria and Archaea carry paralogs of CorA. These may be divided into two major classes, a class most similar to CorA Mg^{2+} transporters already characterized with few or no charged residues in the sequence corresponding to TM1 of the three TM domains of *S. typhimurium* CorA (see below), and a CorA-II class possessing several (up to 9) charged residues in the sequence corresponding to TM1.

The first or CorA class seem to all be Mg^{2+} transporters. They contain three TM domains at the C-terminus with the initial 70 % of the protein residing in the periplasm. This class may be further divided into at least two groups, those CorAs carrying the "MPEL" sequence in the periplasmic loop between TM2 and TM3 and those with some other sequence in the TM2–TM3 loop. Members of both subclasses have been shown to be true Mg^{2+} transporters.

The CorA-II class, although homologous through their entire sequence to the CorA class, have a high degree of charge in the sequence corresponding to TM1 in the CorA Mg^{2+} transporters. Some sequences have as many as nine changes in the putative transmembrane domain. Clearly, this degree of charge cannot be accommodated within the lipid bilayer. This implies that this class has only two transmembrane segments, corresponding to TM2 and TM3 of CorA itself. This conclusion has the interesting corollary that the large soluble domain, rather than being in the periplasm, would be in the cytosol. We predict that the CorA-II branch are also cation transporters but instead of Mg^{2+} most likely mediate efflux of transition metal cations, possibly including a number of toxic cations.

15.5.3
Structure

CorAs have a large and less well-conserved N-terminal hydrophilic domain followed by a better conserved hydrophobic domain. The large majority of CorAs have only a short soluble sequence at the C-terminus. The membrane domain of *S. typhimurium* CorA has been shown experimentally to consist of three TM domains, separated by very short loops [50]. This places the C-terminus in the cytosol and the N-terminus in the periplasm. This topology is unique among membrane transporters. Thus overall, CorA can be considered as a two-domain protein, with a large N-terminal domain in the periplasm and a C-terminal membrane domain. The small number of transmembrane segments suggests that CorA must function as some sort of oligomer. Preliminary data (M. A. Szegedy and M. E. Maguire, unpublished data) suggest that CorA is pentameric.

At over 25 kDa (about 240 amino acids in length) the soluble domain of CorA is the largest known N-terminal sequence to be translocated across the plasma membrane without a signal peptide. It contains an unusually high percentage of charged amino acids and this soluble domain is predicted to have a pI of about 4. A truncated protein consisting of the entire soluble domain has been purified using a 6X His tag. The purified protein appears to retain structure, and, as predicted by various computer algorithms, is virtually all α-helix as measured by circular dichroism, and crystallization efforts are in progress.

Only one CorA homolog out of the over 50 now known contains even a single charge in TM2 and TM3, though the presence of multiple hydroxyl bearing residues renders both of these TM domains amphipathic. There is only a single, non-conserved Glu residue in TM1 of the *S. typhimurium* CorA. This latter residue can be mutated to alanine without detectable effect on CorA transport [51]. This lack of negatively charged residues within the membrane domain sets CorA quite apart from virtually all other cation transporters which require multiple Glu and Asp residues within the membrane domain for transport. In contrast, CorA mediates the influx of the most charge dense of the biological cations without use of a single negatively charged residue within the membrane.

What residues within the membrane are involved? Site-directed mutagenesis [51, 52] suggests that three conserved residues on a single face of the α-helix of both TM2 and TM3 are important to transport and seem to participate directly in substrate binding. These mutagenesis studies also indicate that all the residues in a highly conserved "YGMNF" sequence near the C-terminal end of TM2 are essential for transport. This region appears to play a role in maintaining a proper critical loop conformation between TM2 and TM3 rather than participating in direct substrate binding.

15.6
MgtA/MgtB Mg^{2+} Transporters

15.6.1
Genomics

The MgtA/B class transporters are an eclectic and not yet logically classified set of proteins from a genomic standpoint. *S. typhimurium* MgtA and MgtB are siblings and seemingly interchangeable P-type ATPases whose regulation and transport properties differ only in nuances; however, their phylogenetic distributions differ markedly [53, 54]. Southern blot experiments and available genomic sequences suggest that MgtA may be ubiquitous in the Enterobacteriaceae, but MgtB was only found in three of the nine species tested [55]. Unlike either *E. coli* or *S. typhimurium* MgtA, MgtB resides on a pathogenicity island (SPI-3) specific to only one branch of the *Salmonella enterica* lineage [55]. What this means in terms of its function and whether it generalizes to other Enterobacteria is not yet known.

15.6.2
Structure

Based on sequence alignments, the MgtA/B class of Mg^{2+} transporter belong to the P-type ATPase superfamily [56, 57]. They are most similar to the yeast H^+-ATPases and the mammalian Ca^{2+}-ATPases of the sarco(endo)plasmic reticulum. MgtB was the first P-type ATPase, bacterial or mammalian, whose complete membrane topology was unequivocally determined [58](58). Like most eukaryotic P-type ATPases, MgtB has 10 TM domains with both its N- and C-termini in the cytoplasm, consistent with more recent electron diffraction images of the H^+- and Ca^{2+}-ATPases [59–61]. This is in contrast to other prokaryotic P-type ATPases which are about 300 amino acids shorter and lack the 4 C-terminal TM domains. With the exception of the Kdp K^+-ATPase of *E. coli*, MgtA and MgtB are also different from all other P-type ATPases, prokaryotic or eukaryotic, in that their apparent primary function is to mediate the *influx* of a cation[62]. Other P-type ATPases mediate the *efflux* of their primary substrate, only occasionally mediating the influx of a secondary substrate, e.g., the Na^+, K^+-ATPase. The mechanistic basis for this difference is not apparent.

P-type ATPases all have a conserved Asp in a cytosolic loop between TM4 and TM5 that is phosphorylated during the reaction cycle. Mutation of this residue to Ala in the *S. typhimurium* MgtB abolishes function. The mammalian Na^+, K^+-

and Ca^{2+}-P-type ATPases have been shown to possess six highly conserved residues within the membrane domains that are apparently responsible for binding cation during membrane passage [63, 64]. These residues are also conserved in MgtA and MgtB; however, mutagenesis of *S. typhimurium* MgtB suggests there are puzzling differences in addition to intriguing similarities. Indeed, only two of the six conserved residues appear to have any role in transport (D. G. Kehres, L. M. Kucharski, and M. E. Maguire, unpublished data).

15.6.3
Physiology

The MgtA/B class of Mg^{2+} transporters are fundamentally different from other Mg^{2+} transporters in that their expression is regulated while the CorA and MgtE classes are all apparently constitutively expressed. Neither MgtA nor MgtB of *S. typhimurium* is expressed to any significant extent in the presence of millimolar $[Mg^{2+}]$ in the growth medium, but both are induced to an enormous extent (MgtB at least 1,000-fold) upon Mg^{2+} deprivation [43, 54]. This induction is mediated by the PhoPQ two-component regulatory system [65–67]. Correspondingly, MgtA and MgtB are also induced upon *S. typhimurium* invasion of macrophages as are other PhoPQ-regulated genes, many of which have apparent virulence function. Nonetheless, despite their large induction during residence within the macrophage, neither MgtA nor MgtB appear to have a major role in *S. typhimurium* pathogenesis [55]. For a more detailed discussion of the regulation of the *mgtA* and *mgtB* loci see [29].

MgtA and MgtB both transport magnesium with K_m values comparable to that of CorA. Their true V_{max} values are hard to measure because of their regulation [31]. The apparent V_{max} of each transporter is perhaps 10-fold lower than the "channel"-like throughput of CorA. The two proteins differ very slightly in substrate and inhibitor specificity and in pH and temperature dependence. For example, MgtB is extremely temperature sensitive, being fully active at 37° but completely inactive in terms of transport at 20°. No physiological basis for these differences is yet apparent.

Several puzzles remain regarding the actual physiological role of MgtA/B transporters. Their only known transport activity is cation uptake, while their closest homologs clearly export some cation from the cytoplasm. Do MgtA transporters export some undetermined substrate? With a K_m similar to that of the primary CorA Mg^{2+} uptake system, they apparently do not fulfill the role of a scavenger transporter, expressed only to transport an important nutrient in less than optimal environments. Phylogenetically, they are relatively sparse. Perhaps Mg^{2+} uptake is not the primary role of the MgtA/B class of proteins in Bacteria? A hint that *S. typhimurium* MgtB may perform some additional function comes from the phenotype of a E337A mutant. When transformed into the Mg^{2+} transport deficient *S. typhimurium* strain MM281, this mutant will not support growth on LB agar plates in the absence of 100 mM magnesium supplementation, yet it transports Mg^{2+} with kinetics indistinguishable from the wild-type protein (D. G. Kehres, L.

Kucharski, and M. E. Maguire, unpublished data). Other mutants of MgtB, even some with very low transport capacity, fully complement the growth requirement for supplemental Mg^{2+}. Could the E337A mutation be compromising the export half of some bidirectional transport cycle?

15.6.4
The MgtC Protein

In *S. typhimurium*, the *mgtB* gene is part of a 3-gene operon, the only current example of an MgtA/B class Mg^{2+} transporter residing within such a genetic structure. The operon comprises the right end of *Salmonella* Pathogenicity Island 3 [55, 68]. It is apparently a separate addition or insertion in SPI-3 since unlike the other genes in SPI-3, the genes of the operon have normal GC content and normal codon usage for *Salmonella enterica* lineage [53]. The third gene of this operon is of unknown function and has no homology to other known genes. However, the first gene of this operon, *mgtC*, is of some relevance. Since it is part of the same operon, it is also subject to regulation by extracellular Mg^{2+} via the PhoPQ two-component system [32, 54]. A few other bacterial species have apparent homologs of MgtC, but their phylogenetic distribution does not allow for any conclusions about the origin, and the function of these homologs is not known [55]. The expressed protein is about 23 kDa in size and is very hydrophobic with 5 or 6 TM segments.

Using intraperitoneal injection into mice, Blanc-Potard and Groisman [55] showed that *mgtC* is an essential virulence gene for *S. typhimurium* and suggested that it may be a fourth Mg^{2+} transporter in *S. typhimurium*. They also demonstrated that the *mgtC* homolog of *Mycobacterium tuberculosis* has a similar phenotype [69]. However, we subsequently showed that expression of MgtC alone in the Mg^{2+} transport deficient MM281 strain of *S. typhimurium* does not give detectable Mg^{2+} transport and does not relieve the requirement for Mg^{2+} supplementation in the growth medium [70], although the latter is slightly reduced by such expression. This suggests that MgtC is not a Mg^{2+} transporter. Curiously, although Mg^{2+} deprivation markedly and rapidly induces transcription of both *mgtC* and *mgtB*, only MgtB protein can be detected for several hours after transcription of the operon. The reason for this marked delay in MgtC translation and indeed its function remain a mystery.

15.7
Conclusions and Perspective

Knowledge of prokaryotic magnesium transporters to date is quite consistent with the unique chemistry of their substrate. CorA has no homology to other known transporters and mediates influx of Mg^{2+} without use of charged residues in the membrane domain, again unlike other cation transporters. Although the MgtA/B class of Mg^{2+} transporters are clear members of the P-type ATPase superfamily, they are highly unusual members since they mediate influx of cation rather than efflux, are phylogenetically much closer to eukaryotic than prokaryotic P-type ATPases, and do not appear to transport cation using the same intramembrane residues as transporters of their class. Finally, the MgtE class of Mg^{2+} transporter, like CorA, has no homology to other known transport proteins. These results support our hypothesis that Mg^{2+} transporters will most likely be unique transport proteins or at least highly unusual members of known classes of transport proteins [71, 72].

What about the relationship of these prokaryotic Mg^{2+} transporters to eukaryotic systems? There are no clear homologs of either CorA or MgtE in any eukaryote. Of the multitudinous P-type ATPases described in eukaryotes, none have been shown to transport Mg^{2+}. Given the extent of microbial genomic sequence now available from a wide phylogenetic distribution, it is highly unlikely that another widely distributed class of Mg^{2+} transporter exists among the prokaryotes. Thus, the Eukarya appear to have evolved quite different Mg^{2+} transport systems than the Bacteria and Archaea. Why this has occurred is completely unknown.

What do eukaryotic Mg^{2+} transporters look like? This is almost a complete mystery. A Na^+/Mg^{2+} transporter has recently been cloned from *Arabidopsis thaliana* (2), and some variant of the mammalian Na^+/Mg^{2+} antiporter will likely be cloned soon. How these transporters will relate to other major classes of eukaryotic transporters is unknown. However, based on our current knowledge of prokaryotic Mg^{2+} transport, eukaryotic Mg^{2+} transport systems are likely to be as diverse and novel as their prokaryotic neighbors.

Acknowledgement

Research from this laboratory has been supported by grants GM39447 and HL18708 from the National Institutes of Health.

References

1. Romani, A. M., Scarpa, A., *Front. Biosci.* **2000**, *5*, D720-D734.
2. Shaul, O., Hilgemann, D. W., de-Almeida-Engler, J., Van Montagu, M., Inzé, D., Galili, G., *EMBO J.* **99 A. D.**, *18*, 3973-3980.
3. Diebler, H., Eigen, M., Ilgenfritz, G., Maass, G., Winkler, R., *Pure Appl. Chem.* **1969**, *20*, 93-115.
4. Huang, H.-W., Cowan, J. A., *Eur. J. Biochem.* **1994**, *219*, 253-260.
5. Martin, R. B., *Met. Ions Biol.* **1990**, *26*, 1-13.
6. Cowan, J. A., *Inorg. Chem.* **1991**, *30*, 2740-2747.
7. Eigen, M., *Pure Appl. Chem.* **1963**, *6*, 97-115.
8. Jou R., Cowan, J. A., *J. Am. Chem. Soc.* **1991**, *113*, 6685-6686.
9. Black, C. B, Cowan, J. A., *Eur. J. Biochem.* **1997**, *243*, 684-689.
10. Cowan, J. A., *J. Inorg. Biochem.* **1993**, *49*, 171-175.
11. Suga, H., Cowan, J. A., Szostak, J. W., *Biochemistry* **1998**, *37*, 10118-10125.
12. Cowan, J. A., *Chem. Rev.* **1998**, *98*, 1067-1087.
13. Murphy, E., Steenbergen, C., Levy, L. A., Raju, B., London, R. E., *J. Biol. Chem.* **1989**, *264*, 5622-5627.
14. Gupta, R. K., Moore, R. D., *J. Biol. Chem.* **1980**, *255*, 3987-3993.
15. Hess, P., Metzger, P., Weingart, R., *J. Physiol.* **1982**, *333*, 173-188.
16. Rink, T. J., Tsien, R. Y., Pozzan, T., *J. Cell Biol.* **1982**, *95*, 189-196.
17. Quamme, G. A., *Am. J. Physiol. Gastrointest. Liver Physiol.* **1993**, *264*, G383-G389.
18. Hall, S. K., Fry, C. H., Buri, A., McGuigan, J. A. S., *Magnes. Trace Elem.* **1992**, *10*, 80-89.
19. Quamme, G. A., Dai, L.-J., Rabkin, S. W., *Am. J. Physiol. Heart Circ. Physiol.* **1993**, *265*, H281-H288.
20. Preston, R. R., *J. Membr. Biol.* **1998**, *164*, 11-24.
21. Grubbs, R. D., Snavely, M. D., Hmiel, S. P., Maguire, M. E., *Methods Enzymol.* **1989**, *173*, 546-563.
22. Luthi, D., Spichiger, U., Forster, I., McGuigan, J. A., *Exp. Physiol.* **1997**, *82*, 453-467.
23. Luthi, D., Gunzel, D., McGuigan, J. A., *Exp. Physiol.* **1999**, *84*, 231-252.
24. Hintz, K., Günzel, D., Schlue, W. R., *Pflügers Arch.* **1999**, *437*, 354-362.
25. Csernoch, L., Bernengo, J. C., Szentesi, P., Jacquemond, V., *Biophys. J.* **1998**, *75*, 957-967.
26. Konishi, M., *Jpn. J. Physiol.* **1998**, *48*, 421-438.
27. Sun, H., Jacquey, F., Bernengo, J. C., *Biochem. Biophys. Acta* **1998**, *1403*, 57-71.
28. Smith, R. L., Maguire, M. E., in: *The Biological Chemistry of Magnesium* (J. A. Cowan, Ed.), John Wiley & Sons, Chichester, **1995**, pp. 211-234.
29. Smith, R. L., Maguire, M. E., *Mol. Microbiol.* **1998**, *28*, 217-226.
30. Hmiel, S. P., Snavely, M. D., Florer, J. B., Maguire, M. E., Miller, C. G., *J Bacteriol.* **1989**, *171*, 4742-4751.
31. Snavely, M. D., Florer, J. B., Miller, C. G., Maguire, M. E., *J. Bacteriol.* **1989**, *171*, 4761-4766.
32. Snavely, M. D., Gravina, S. A., Cheung, T. T., Miller, C. G., Maguire, M. E., *J. Biol. Chem.* **1991**, *266*, 824-829.
33. Hmiel, S. P., Snavely, M. D., Miller, C. G., Maguire, M. E., *J. Bacteriol.* **1986**, *168*, 1444-1450.
34. Smith, R. L., Thompson, L. J., Maguire, M. E., *J. Bacteriol.* **1995**, *177*, 1233-1238.
35. Townsend, D. E., Esenwine, A. J., George III, J., Bross, D., Maguire, M. E., Smith, R. L., *J. Bacteriol.* **1995**, *177*, 5350-5354.
36. Merino, S., Gavin, R., Altarriba, M., Izquierdo, L., Maguire, M. E., Tomás, J. M., *FEMS Microbiol. Lett.* **2001**, *198*, 189-195.
37. Silver, S., *Proc. Natl. Acad. Sci. USA* **1969**, *62*, 764-771.
38. Kehres, D. G., Lawyer, C. H., Maguire, M. E., *Microb. Comp. Genomics* **1998**, *43*, 151-169.
39. Smith, R. L., Maguire, M. E., *J. Bacteriol.* **1995**, *177*, 1638-1640.
40. Faguy, D. M., Doolittle, R. F., *Trends Genet.* **2000**, *16*, 196-197.
41. MacDiarmid, C. W., Gardner, R. C., *J. Biol. Chem.* **1998**, *273*, 1727-1732.
42. Bui, D. M., Gregan, J., Jarosch, E., Ragnini, A., Schweyen, R. J., *J. Biol. Chem.* **1999**, *274*, 20438-20443.

43. Tao, T., Grulich, P. F., Kucharski, L. M., Smith, R. L., Maguire, M. E., *Microbiology* **1998**, *144*, 655-664.
44. Smith, R. L., Kaczmarek, M. L., Kucharski, L. M., Maguire, M. E., *Microbiology* **1998**, *144*, 1835-1843.
45. Gibson, M. M., Bagga, D. A., Miller, C. G., Maguire, M. E., *Mol. Microbiol.* **1991**, *5*, 2753-2762.
46. Smith, R. L., Gottlieb, E., Kucharski, L. M., Maguire, M. E., *J. Bacteriol.* **1998**, *180*, 2788-2791.
47. Kucharski, L. M., Lubbe, W. J., Maguire, M. E., *J. Biol. Chem.* **2000**, *275*, 16767-16773.
48. Basolo, F., Pearson, R. G., *Mechanisms of Inorganic Reactions* 2nd Edn. **1967**, pp. 158-170, John Wiley & Sons, New York.
49. Meek, D. W., Ibers, J. A., *Inorg. Chem.* **1970**, *9*, 465-470.
50. Smith, R. L., Banks, J. L., Snavely, M. D., Maguire, M. E., *J. Biol. Chem.* **1993**, *268*, 14071-14080.
51. Smith, R. L., Szegedy, M. A., Walker, C., Wiet, R. M., Redpath, A., Kaczmarek, M. L., Kucharski, L. M., Maguire, M. E., *J. Biol. Chem.* **1998**, *273*, 28663-28669.
52. Szegedy, M. A., Maguire, M. E., *J. Biol. Chem.* **1999**, *274*, 36973-36979.
53. Snavely, M. D., Miller, C. G., Maguire, M. E., *J. Biol. Chem.* **1991**, *266*, 815-823.
54. Tao, T., Snavely, M. D., Farr, S. G., Maguire, M. E., *J. Bacteriol.* **1995**, *177*, 2654-2662.
55. Blanc-Potard, A. B., Groisman, E. A., *EMBO J.* **1997**, *16*, 5376-5385.
56. Carafoli, E., Brini, M., *Curr. Opin. Chem. Biol.* **2000**, *4*, 152-161.
57. Scarborough, G. A., *Curr. Opin. Cell Biol.* **1999**, *11*, 517-522.
58. Smith, D. L., Tao, T., Maguire, M. E., *J. Biol. Chem.* **1993**, *268*, 22469-22479.
59. Toyoshima, C., Nakasako, M., Nomura, H., Ogawa, H., *Nature* **2000**, *405*, 647-655.
60. Scarborough, G. A., *J. Exp. Biol.* **2000**, *203 Pt 1:147-54*, 147-154.
61. Zhang, P., Toyoshima, C., Yonekura, K., Green, N. M., Stokes, D. L., *Nature* **1998**, *392*, 835-839.
62. Maguire, M. E., Snavely, M. D., Leizman, J. B., Gura, S., Bagga, D., Tao, T., Smith, D. L., *Ann. NY Acad. Sci.* **1992**, *671*, 244-256.
63. Clarke D. M., Loo T. W., Inesi G., MacLennan D. H., *Nature*, **1989**, *339*, 476-478.
64. MacLennan D. H., Rice W. J., Green N. M., *J. Biol. Chem.*, **1997**, *272*, 28815-28818.
65. Miller, S. I., Kukral, A. M., Mekalanos, J. J., *Proc. Natl. Acad. Sci. USA* **1989**, *86*, 5054-5058.
66. Groisman, E. A., Chiao, E., Lipps, C. J., Heffron, F., *Proc. Natl. Acad. Sci. USA* **1989**, *86*, 7077-7081.
67. Miller, S. I., *Mol. Microbiol.* **1991**, *5*, 2073-2078.
68. Blanc-Potard, A. B., Solomon, F., Kayser, J., Groisman, E. A., *J. Bacteriol.* **1999**, *181*, 998-1004.
69. Buchmeier, N., Blanc-Potard, A., Ehrt, S., Piddington, D., Riley, L., Groisman, E. A., *Mol. Microbiol.* **2000**, *35*, 1375-1382.
70. Moncrief, M. B. C., Maguire, M. E., *Infect. Immun.* **1998**, *66*, 3802-3809.
71. Grubbs, R. D., Maguire, M. E., *Magnesium* **1987**, *6*, 113-127.
72. Maguire, M. E., *Met. Ions Biol.* **1990**, *26*, 135-153.

16
Bacterial Copper Transport

Marc Solioz

Copper in biological systems presents a formidable problem: it is essential for life, yet highly reactive and a potential source of cell damage. The recent discovery of ATP-driven copper pumps in bacteria, fungi, plants and mammals and the association of such human enzymes with the copper metabolic disorders, Menkes disease and Wilson disease, have yielded new insight into trace metal transport. In bacteria, copper ATPases have a cytoplasmic or thylakoid membrane localization and function in copper uptake or extrusion. Copper and other heavy metal ATPases form a subclass of the classical P-type ATPases involved in non-heavy metal ion transport. Other copper transport system have been described, but their mechanisms remain elusive.

16.1
Introduction

The discovery of the oldest known microfossils in deep-sea volcanic rock suggests that hydrothermal vents may have hosted the first living systems on earth [1]. The hot, acidic seawater encountered at these vents releases metals like iron, manganese, zinc and copper from the volcanic rock [2] and resistance to these metal ions may have been an evolutionary priority for the first life forms. However, copper may not have been an essential trace element in early cells. Rather, it may have become a cellular constituent with the advent of a more oxidized biosphere and could thus be considered to be a "modern" bioelement [3] Cuproenzymes function almost exclusively in the metabolism of O_2, N_2O or NO_2^-, which only became necessary with the advent of an oxidizing environment 3×10^9 years ago. The corresponding need for a redox-active metal with potentials between 0 and 0.8 V is ideally fulfilled by the Cu(I)/Cu(II) redox pair.

While cuproenzymes that catalyze oxygen transport or redox reactions have been under investigation for several decades, how cells take up copper, route it intracellularly to sites of utilization, and secrete it when in excess has remained unknown until a few years ago. With the advent of the discovery of copper pumping ATPases in *Enterococcus hirae* in 1992 [4], the field of copper homeostasis has been virtually exploding. Today, we have a fairly detailed, although not yet complete, picture of copper homeostasis in eukaryotic and prokaryotic cells. Some of the key elements of copper homeostasis and transport will be summarized in this chapter.

16.2
The New Subclass of Heavy Metal CPx-type ATPases

P-type ATPases (earlier called E_1E_2-type ATPases) get their name from the fact that they form a phosphorylated intermediate in the course of the reaction cycle [5]. P-type ATPases are classically represented by the Na^+,K^+-ATPase of the plasma membrane and of the Ca^{2+}-ATPases of the sarcoplasmic reticulum. Other metal ions transported by P-type ATPases are Mg^{2+} in bacteria and H^+ in plants and fungi [6, 7]. With the discovery of cadmium and copper transporting P-type ATPases, it has become clear that there exists a subclass of the P-type ATPases involved in the transport of heavy metal ions. Members of this subclass differ from the classical P-type ATPases not only in their transport specificities, but also in their membrane topology. Accordingly, they form a distinct evolutionary branch. Figure 1 shows an unrooted phylogenetic tree with representative members of heavy metal and non-heavy metal P-type ATPases. Based on this divergence, it has been proposed to call the heavy metal ATPases P_1-type or CPx-type ATPases, respectively [8, 9]. The division into heavy metal and non-heavy metal ATPases probably took place before the division into prokaryotes and eukaryotes.

It has recently been proposed that the evolution of early life forms took place near thermal vents [1]. The hot and often acidic waters in these environments can be highly enriched in heavy metal ions [2]. An organism thriving in these conditions would have be endowed with mechanisms to deal with toxic metal ions and it is conceivable that efflux mechanisms for heavy metal ions evolved before or concomitantly with their use as cofactors. In line with such a hypothesis, the CPx-type ATPases encompass a wider spectrum of ion specificities than the non-heavy metal ATPases, now including Cu^+, Ag^+, Zn^{2+}, Cd^{2+}, and Pb^{2+}. It is to be expected that other metal ions will be added to this list. ATPases transporting silver, zinc, cadmium and lead are involved in bacterial resistance to these toxic metal ions, while copper transporting ATPases have a role in both, copper uptake to meet cellular demands, and copper extrusion when ambient copper is excessive. Some of the key features of heavy metal ATPases will be discussed below.

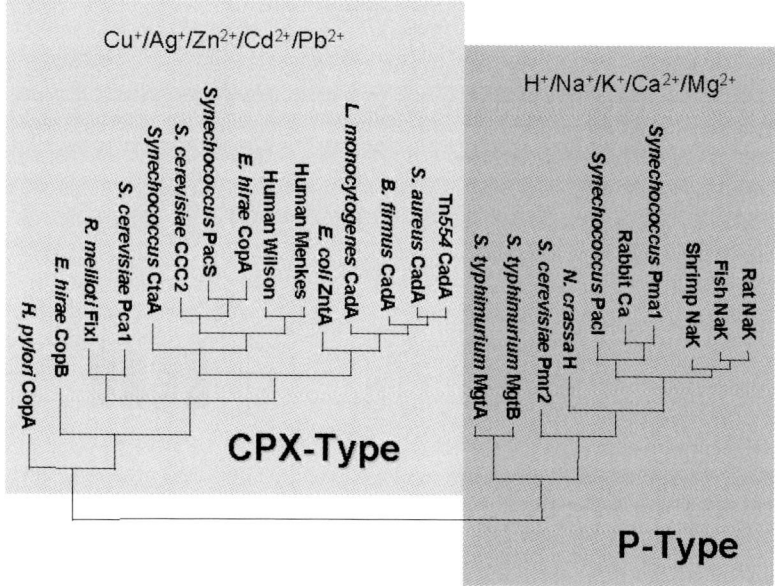

Fig. 1. Phylogram of ATPases. Divergence was scored for a selected sample of P-type and CPx-type ATPases by the Jukes–Cantor method [60]. Relationships between distant branches are not reliable. The shaded areas delineate the subgroups of heavy metal transporting CPx-type ATPases and the non-heavy metal transporting P-type ATPases. The ion specificities known for each group are indicated in the respective fields.

16.2.1
Membrane Topology of CPx-type ATPases

Figure 2 shows a comparison of the membrane topology of the *E. hirae* CopB copper ATPase and the Ca^{2+}-ATPase of rabbit sarcoplasmic reticulum. The three-dimensional structure of the latter has recently been derived to a resolution of 2.6 Å [10]. The residue that has been demonstrated to be phosphorylated is the aspartic acid in the conserved sequence DKTGT (given in the one-letter amino acid code used throughout this chapter). This sequence, also called the aspartyl kinase domain, is present in all of the more than 100 P-type ATPases sequenced today. Other motifs common to P-type ATPases are the so-called phosphatase domain of consensus sequence TGES, and the ATP-binding domains of consensus GCGINDAP, discussed in Chap. 2 [11]. However, the copper ATPases as well as the cadmium ATPases exhibit several striking features not found in any of the other P-type ATPases: (1) they have putative heavy metal binding sites in the polar N-terminal region, (2) they have a conserved intramembranous CPC, CPH or CPSH motif, (3) they have a conserved HP motif 34 to 43 amino acids C-terminal to the CPC motif, (4) they have two additional predicted transmembranous helices on the N-terminal end of the protein, and (5) they only have two membrane

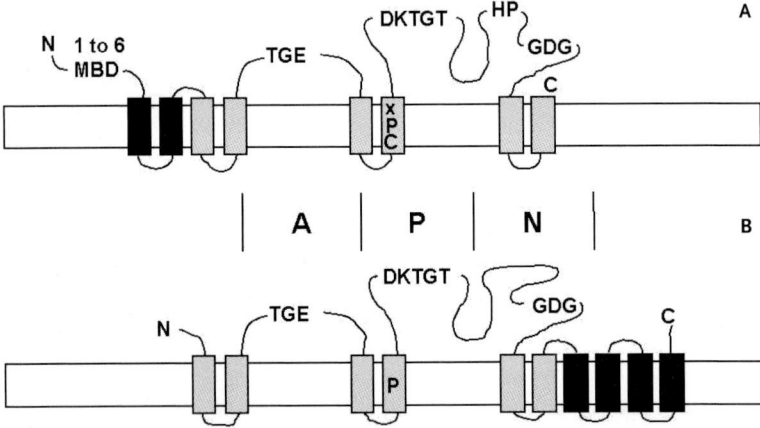

Fig. 2. Comparison of the membrane topology of a CPx-type ATPase and a non-heavy metal ATPase. Shown are CopB (**A**) of *E. hirae* and the Ca^{2+}-ATPase of sarcoplasmic reticulum (**B**). Helices common to both types of ATPases are in gray and helices unique to one type of ATPase are in black. Key sequence motifs are indicated in the one-letter amino acid code. In the center of the figure, the approximate locations of the three cytoplasmic domains A, P, and N are indicated. MBD, metal binding domain containing repeat metal binding sites; TGE, conserved site in transduction domain (A); CPx, putative copper binding site; DKTGT, phosphorylation site in domain P; HP, motif of unknown function, probably in domain N; GDG, nucleotide binding site residues in domain N.

spans at the C-terminal end, thus lacking four membrane spans present at this position in other P-type ATPases. Clearly, these ATPases form a subgroup of the P-type ATPases that is quite distinct.

16.2.2
Role of the CPx Motif

The CPx motif, which is CPC in Menkes and Wilson ATPase and CPH in CopB, has been postulated to be part of the ion channel through the membrane, chiefly based on site-directed mutagenesis studies with the Ca^{2+}-ATPase [9;12]. With the recent advent of a three-dimensional structure for the Ca^{2+}-ATPase of sarcoplasmic reticulum, the role of this domain in ion binding is now clear [10]. In this ATPase, the type II calcium binding site is made up primarily by residues in the transmembranous helix four (TM4). The key residues forming the calcium binding site are VAAIPE-309. The main-chain carbonyl oxygen atoms of V304, A305, and I307, and a side-chain oxygen atom of E309 (plus oxygens from N796 and D800) contribute to the site, requiring some unwinding of helix TM4. Table 1 shows a comparison of the corresponding motifs in relation to the transport specificities of ion-motive ATPases. Due to the different membrane topologies of heavy metal and non-heavy metal ATPases, the critical residues are located in helix TM4 in the calcium

Tab. 1 Sequence motifs in the ion channels of P-type and CPx-type ATPases. The sequence of the purported ion binding domains in transmembranous helices (TM) 4 or 6 are listed for non-heavy metal P-type ATPases and heavy metal CPx-type ATPases. See Fig. 2 for the location of the transmembranous helices in the overall structure

Organism	ATPase	Ions	TM	Motif
Rat	NaK-ATPase	Na^+K^+	4	VANVPE
Rabbit	SERCA1	Ca^{2+}	4	VAAIPE
E. coli	MgtA	Mg^{2+}	4	VGLTPE
N. crassa	H^+-ATPase	H^+	4	IIGVPV
E. hirae	CopB	Cu^+/Ag^+	6	IIACPH
Salmonella	SilP	Ag^+	6	IIACPC
E. hirae	CopA	Cu^+/Ag^+	6	VIACPC
Human	Menkes	Cu^+	6	CIACPC
E. coli	ZntA	$Cd^{2+}/Zn^{2+}/Pb^{2+}$	6	LIGCPC
S. aureus	CadA	Cd^{2+}	6	VVGCPC

ATPase and other non-heavy metal ATPases, but in helix TM6 in heavy metal ATPases (see Fig. 2).

Since nitrogen and sulfur atoms are much better copper ligands than oxygen, it is apt that the residues corresponding to I307 and E309 are cysteines or histidines in heavy metal ATPases. In agreement with the proposed key role of the CPx motif in ATPase function, the C369S mutation of CopB showed no function *in vivo* and the purified enzyme had no detectable ATPase activity [12a]. Also, the corresponding Menkes disease mutation C1000R changing the conserved CPC motif to RPC has been described as causing a severe phenotype, although with a long survival [29]. Direct binding measurements would be required to demonstrate the involvement of the CPC motif in high-affinity copper binding. However, preliminary studies in our laboratory indicated that the binding affinity of CopB for copper is in the low nanomolar range. Current instrumentation does not allow to measure such low copper concentrations. A detailed understanding of the CPx motif in copper binding and translocation will thus have to await further technical developments.

16.2.3
N-Terminal Heavy Metal Binding Sites

A conspicuous feature of CPx-type ATPases is the occurrence of one to six copies of conserved metal binding domains in the polar N-terminus preceding the first predicted membrane span. These metal binding sites are of two types. Usually, they feature a CxxC motif in a conserved domain encompassing 40 to 60 amino acids, but in some instances (e.g., CopB of *E. hirae*) an exceptionally histidine-rich N-terminus is present instead. A role of the CxxC motif in heavy metal binding has first been proposed by Silver et al. [13]. They pointed to the presence of this motif in the cadmium ATPase of *Staphylococcus aureus*, in the periplasmic mercury binding protein MerP, and in three different MerA mercuric reductases. In the meantime, new proteins containing this sequence element have been found, notably the copper ATPases and the copper chaperones. The latter are proteins of around 70 amino acids that bind copper and route it intracellularly [14]. They are members of a homologous family that includes representatives from humans (HAH1), *Caenorhabditis elegans* (CUC-1), *Arabidopsis* (CCH), yeast (Atx1), and bacteria (CopZ, [15–19]). A related bacterial chaperone is MerP, which routes mercury in the intermembrane space to the uptake system of mercury-resistant bacteria [20].

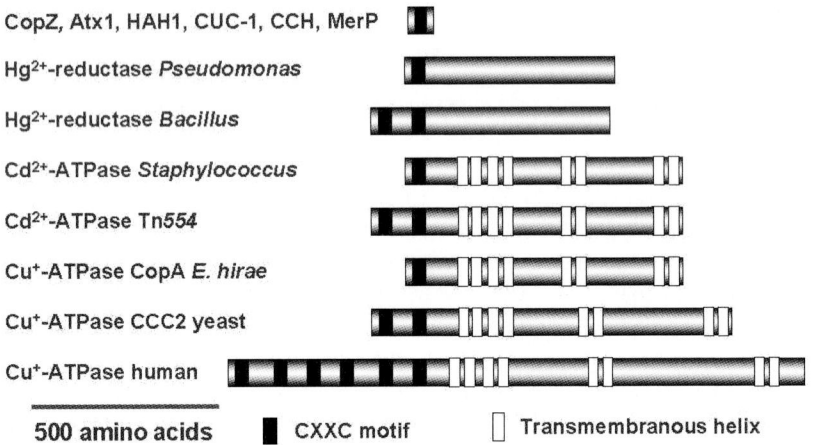

Fig. 3. Schematic representation of the occurrence of CxxC motifs in various proteins. The polypeptide chains are drawn to scale as boxes. Transmembranous helices are indicated by empty rectangles and CxxC motifs by filled rectangles.

Fig. 4. Ribbon Model of the Structure of CopZ. The position of the copper ion is inferred.

Figure 3 depicts a schematic overview of the occurrence of CxxC motifs, suggesting that the sequence is a feature of proteins which interact with heavy metal ions. Detailed structural studies have shown that members of this homologous family of chaperones, and MNKr4, the fourth metal binding domain of the Menkes copper ATPase, all have the same fold. It consists of four β-strands forming an antiparallel β-sheet, situated below two α-helices (Fig. 4). The $\beta\alpha\beta\beta\alpha\beta$ arrangement of secondary structure elements is characteristic of the ferredoxin-like proteins and is known as an "open-faced β-sandwich" [21]. The metal binding sequence motif CxxC occurs on the mobile loop between the first β-strand and the first α-helix. The function of copper chaperones is discussed in more detail below.

16.2.4
The HP Locus

A HP dipeptide motif is universally present in the CPX-type ATPases but absent in other P-type ATPases [9]. It is located 34 to 43 amino acids C-terminal to the phosphorylated aspartic acid residue. In the Ca^{2+}-ATPase, this region is divided into two clearly separated domains, the phosphorylation domain (P), extending roughly 8 amino acids beyond the DKTGT phosphorylation site, and the nucleotide binding domain (N), formed by the remainder of the large cytoplasmic loop (see Fig. 2). By analogy, the HP motif would be located in the N-domain near the ATP-binding site, but there is no recognizable sequence similarity between the Ca^{2+}-ATPase and copper ATPases in the region of the HP motif. In Wilson copper ATPase (ATP7B), which is mainly expressed in the liver and required for copper secretion via the bile, more than 100 point mutations have been identified, but mutation of this histidine, H1069Q, accounts for 30%–40% of the patients in North America and Northern Europe [22, 23]. Presumably, this renders the Wilson copper ATPase non-functional, thus causing the accumulation of copper in the liver and subsequent liver damage as well as neurological symptoms.

The impact of the H1069Q mutation on Wilson ATPase function has previously been tested by functional complementation, but the findings remained contradictory. When expressed in fibroblast of the mottled mouse, a model for Menkes disease, the Wilson ATPase gene carrying the H1069Q mutation could not rescue the mottled phenotype, while a wild-type Wilson gene did. The mutant enzyme mis-localized to the endoplasmic reticulum at normal growth temperatures and was degraded more rapidly than wild-type Wilson ATPase [24]. In contrast to this, several groups have shown that H1069Q could rescue the iron uptake-deficient phenotype of a yeast *Ccc2* knockout strain [23, 25]. It had been speculated that this was due to overexpression of the mutated, yet slightly active protein [23]. Since expression was determined with antibodies on Western blots, it is inherently not possible to compare expression of complementing ATPase to normal expression of the endogenous Ccc2 copper ATPase in this system.

In CopB of *E. hirae*, it was shown that H480Q CopB, corresponding to the H1069Q Wilson mutation, did not complement a CopB knockout strain. *In vitro*, H480Q CopB exhibited residual ATPase activity. The mutation did not significantly affect the K_m for ATP, but reduced V_{max} over 40-fold [12a]. This suggests that the HP motif is not involved in ATP binding, but is essential in a later step of the pump cycle, such as in coupling ATP hydrolysis to copper transport. Interestingly, an HP locus is also conserved between a group of related copper proteins, including CopA from *Pseudomonas syringae*, a similar protein from *Xanthomonas campestris*, laccase from four different fungi, and ascorbate oxidase from cucumber [26]. The function of this motif in these proteins has so far not been investigated.

16.3
Copper Homeostasis in *Enterococcus hirae*

The *E. hirae cop* operon is located on the chromosome and consists of four closely spaced genes in the order: *cop*Y, *cop*Z, *cop*A, and *cop*B. CopY and *cop*Z encode regulatory proteins, whose function is described below, while *cop*A and *cop*B encode CPx-type ATPases of 727 and 745 amino acids, respectively [27]. Figure 5 provides a summary of the current understanding of copper homeostasis in *E. hirae*. CopA and CopB were the first copper ATPases to be described [4] and were cloned fortuitously while trying to clone a potassium ATPase using an antibody of low specificity. The sequence similarity of the histidine-rich N-terminus of CopB to a 120 amino acid periplasmic copper binding protein of *P. syringae*, CopP, initially gave the clue to an involvement of CopB in copper homeostasis.

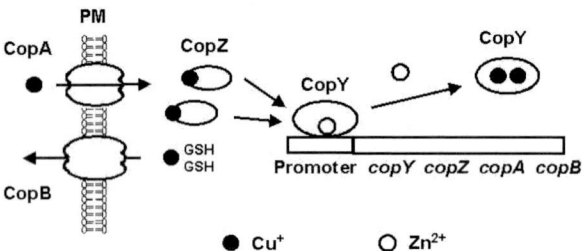

Fig. 5. Copper homeostasis in *E. hirae*. Under copper limiting conditions, copper is pumped into the cell by CopA. The CopZ copper chaperone picks up copper at this site of entry. Under physiological copper conditions, Zn(II)CopY binds to the promoter and represses transcription of the *cop* operon. In conditions of copper excess Cu–CopZ donates Cu(I) to CopY which leads to the replacement of the Zn(II), loss of DNA binding affinity and ultimately synthesis of the operon products. Excess copper is secreted by the CopB efflux pump. The substrate for this pump may be a copper–glutathione (GSH) complex, rather than Cu–CopZ.

16.3.1
Function of CopA in Copper Uptake

CopA of E. hirae exhibits 43 % sequence identity with the human Menkes and Wilson ATPases; in the transduction domain, sequence identity between these enzymes is even 92 %. This suggests that CopA is a representative model of a copper ATPase. Based on indirect evidence, CopA appears to function in copper uptake, although the evidence for this is still indirect. Cells disrupted in *cop*A cease to grow in media in which the copper has been complexed with 8-hydroxyquinoline or *o*-phenanthroline. This growth inhibition could be overcome by adding copper to the growth media. Null mutants in *cop*A could grow in the presence of 5 µM AgNO$_3$, which fully inhibit the growth of wild-type cells. The CopA ATPase thus appears to be a route for the entry of copper as well as silver into the cell [27]. Silver transport by CopA is probably fortuitous as silver has no known biological role. The transport of Ag(I) by CopA is an indication that Cu(I) rather than Cu(II) is transported by CopA of E. hirae.

Interestingly, a screen for virulence genes in *Staphylococcus aureus* revealed a gene, *ivi44*, which encodes a protein with 50 % sequence identity to CopA of E. hirae [28]. This suggests that copper may be a limiting nutrient in pathogenesis and copper import becomes a cellular priority when bacteria infect a host. It would be interesting to test a *copA* knockout strain of E. hirae for its ability to infect a host.

CopA could be expressed in *Escherichia coli* and purified to homogeneity by Ni-NTA affinity chromatography by means of an added histidine tag [28a]. Purified CopA had a pH optimum of 6.3 and a K_m for ATP of 0.2 mM. The enzyme formed an acylphosphate intermediate, which is a hallmark of P- and CPx-type ATPases [5]. Purified CopA can now serve to analyze mechanistic aspects of copper transport and to characterize structure–function relationships.

Using purified CopA and CopZ, it could be shown by surface plasmon resonance analysis that the two proteins interact directly with each other (Multhaup and Solioz, unpublished data). This interaction was significantly reduced, but not abolished, when the CxxC copper binding motif of CopA was mutated to SxxS. Thus, the interaction of CopA and CopZ does not only take place by virtue of bound copper, but also involves direct protein–protein interaction between CopA and CopZ. This allows to refine the model of copper circulation in E. hirae (see Fig. 5)

16.3.2
Function of CopB in Copper Excretion

For the CopB copper ATPase copper transport has been demonstrated directly. CopB was shown to catalyze ATP-driven accumulation of copper(I) and silver(I) in native membrane vesicles of E. hirae. These vesicles showed ATP-dependent accumulation of radiolabeled copper, but only under strongly reducing conditions. ^{64}Cu(I) was thus the transported species. Uptake of copper by these vesicles would correspond to copper extrusion in whole cells. Use of null mutants in either *cop*A, *cop*B, or *cop*A and *cop*B made it possible to attribute the observed transport to the activity of the CopB ATPase. Copper transport exhibited an apparent K_m for Cu^+ of 1 µM and a V_{max} of 0.07 nmol min^{-1} mg^{-1} of membrane protein. $^{110m}Ag^+$ was transported with similar affinity and rate [29]. Since Cu^+ and Ag^+ are complexed to Tris-buffer and dithiothreitol under the experimental conditions, the K_m values must be considered as relative only. The results obtained with membrane vesicles were further supported by evidence of $^{110m}Ag^+$ extrusion from whole cells pre-loaded with this isotope. Again, transport depended on the presence of functional CopB [30]. These findings suggest that CopB functions as a Cu^+/Ag^+-ATPase for the export of Cu^+ and Ag^+ *in vivo*.

Vanadate, a diagnostic inhibitor of P-type ATPases, showed an interesting biphasic pattern of inhibition of ATP-driven copper and silver transport: maximal inhibition of Cu^+ transport was observed at 40 µM VO_4^{3-} and of Ag^+ transport at 60 µM VO_4^{3-}. Higher concentrations relieved the inhibition of transport. This behavior is unexplained at present, but may relate to the complex chemistry of vanadate involving many oxidation states [31].

16.3.3
Regulation of Expression by Copper

The two copper ATPases of *E. hirae* exhibit biphasic regulation: induction of the genes is lowest in standard growth media (average copper content = 10 µM). If media copper was increased, expression is increased 50-fold at 2 mM extracellular copper. Induction was also observed by 5 µM Ag^+ or 5 µM Cd^{2+}. The induction by silver and cadmium is in all likelihood fortuitous, since *E. hirae* does not exhibit significant resistance to these highly toxic metal ions. Interestingly, high induction was also observed when copper was depleted from the media. Since CopA serves in copper uptake and CopB in its extrusion, this co-induction of CopA and CopB by high and low copper seems puzzling. It could be a safety mechanism: if cells express, under copper limiting conditions, only the import ATPase, they would become highly vulnerable to copper poisoning in the event of a sudden increase in ambient free copper, such as by acidification of the ambient.

Regulation of the *cop* operon is accomplished by *CopY*, a repressor protein of 145 amino acids [32]. The N-terminal half of CopY exhibits around 30 % sequence identity to the bacterial repressors of β-lactamases, MecI, PenI, and BlaI [33–35]. In the best studied of these, PenI, this N-terminal portion appears to be the domain that recognizes the operator [36]. In the C-terminal half of CopY, there are multiple cysteine residues, arranged as $CXCX_4CXC$. The consensus motif $CXCX_{4-5}CXC$ is also found in the three yeast copper responsive transcriptional activators, ACE1, AMT1 [37, 38], and MAC1 [39]. Disruption of the *E. hirae copY* gene resulted in constitutive overexpression of the *cop* operon [32]. Binding of CopY to an inverted repeat sequence upstream of the *copY* gene has been demonstrated *in vitro* [40]. CopY binds to DNA as a Zn(II)CopY complex. For the release of CopY from the DNA and induction of the operon, Cu(I)CopZ donates copper to CopY, thereby displacing the bound Zn(II) and releasing CopY from the DNA [41]. The displacement of Zn(II) by copper suggests that the copper ions bind to the same thiolate binding site that can be occupied by zinc. The Cu(I)CopY complex exhibited luminescence, implying that Cu(I) was sequestered in an environment where it was protected from solvent. It is plausible that the Cu(I) ions are being sequestered in a Cu(I)-thiolate cluster as found in the Cu(I)-regulated transcription factor ACE1 and the metallothioneins [38]. The inability of the displaced Zn(II) to bind to CopZ indicated that the metal binding site in CopZ is specific for Cu(I).

The transfer of Cu(I) from the CxxC metal binding site of CopZ is probably driven by the higher affinity of the more cysteine-rich CxCxxxxCxC metal binding site in CopY. The failure of the structural analog MNKr2, the second N-terminal metal binding domain of the Menkes ATPase, to deliver its copper to CopY suggests that the presence of the conserved metal binding site and global fold is not sufficient to effect copper transfer. Rather, the mechanism probably involves the specific docking of the chaperone to the recipient protein, with subsequent transfer of the metal ion. A two-step mechanism of this type would help protect the cell from the toxic effects of copper ions by preventing their non-specific release to inappropriate sites.

16.4
Copper Resistance in *Escherichia coli*

ORF *f834* of *Escherichia coli* encodes an 834-residue P-type ATPase that exhibits 36% identity with CopA from *E. hirae*. Since the gene product of *f834* could be shown to catalyze copper export, the gene was renamed *copA* and the gene product CopA (it will be called ecCopA herein to differentiate it from *E. hirae* CopA). EcCopA exhibits all the structural features of CPx-type ATPases. Interestingly, ecCopA possesses two N-terminal CxxC motifs. *E. coli* cells with a disrupted *copA* gene exhibited decreased resistance to copper. This apparent copper sensitivity could be complemented by introduction of a plasmid expressing ecCopA. *CopA*-disrupted strains were still relatively resistant to copper salts, which can be attributed to other genes involved in copper tolerance in *E. coli* [42]. However, their function in copper resistance is still largely unclear.

Uptake of copper into everted membrane vesicles from cells expressing ecCopA could be demonstrated, but only when ATP was present as an energy source. Transport was inhibited by the classical P-type ATPase inhibitor vanadate. Dithiothreitol, a strong reductant, was required for CopA catalyzed ^{64}Cu uptake suggesting that the substrate of CopA is Cu(I). So the function of ecCopA resembles that of the *E. hirae* CopB ATPase by functioning as a copper efflux mechanism when copper is in excess in the cytoplasm [43].

A role in copper transport in *E. coli* has also been ascribed to the products of at least six chromosomal genes, *cutA*, *cutB*, *cutC*, *cutD*, *cutE*, and *cutF* [44]. Mutation of one or more of these genes resulted in an increased copper sensitivity. Only the *cutC* and *cutF* genes were cloned and sequenced. The *cutC* gene encodes a cytoplasmic protein of 146 amino acids, and the *cutF* gene, which is identical to the *nlpE* gene, encodes a putative outer membrane protein with metal binding capacity [45]. No further information on the function of these proteins is currently available.

16.4.1
Regulation of the *Escherichia coli* Copper ATPase

It was found that ecCopA, the principal copper efflux ATPase of *E. coli*, is induced by elevated copper in the medium. A regulatory protein homologous to MerR was recently identified, encoded by the *ybbI* gene [46]. The *ybbI* gene was thus designated *cueR* for Cu efflux regulator. Inspection of the copA promoter revealed signature elements of promoters controlled by metalloregulatory proteins in the MerR family. These same elements are also present upstream of *yacK*, which encodes a putative multi-copper oxidase. Homologs of YacK are found in copper resistance determinants that facilitate copper efflux. CopA and YacK expression are apparently regulated in a copper responsive manner by CueR, with increased copper levels leading to increased expression.

16.5
Synechococcal Copper ATPases

Synechococcus PCC7942 is a Gram-positive bacterium that harbors a photosynthetic apparatus (thylakoid) similar to that of chloroplasts. From *Synechococcus*, two CPx-type ATPases, CtaA and PacS, have recently been cloned. The *ctaA* gene was found fortuitously while attempting to clone the biotin-carboxyl carrier protein from this organism with a DNA probe from the corresponding *E. coli* gene. This probe had, by coincidence, a perfect 17-nucleotide match with the *ctaA* gene. CtaA encodes a CPx-type ATPase of 790 amino acids with an intramembranous CPC sequence and an N-terminal CxxC metal binding motif [47]. Disruption of the *ctaA* gene by cassette mutagenesis resulted in a strain that still showed some growth in 10 µM Cu^{2+} while wild-type cells were completely inhibited. It also retained better viability in the presence of copper while other cations tested had no effect. From these data, it was concluded that *ctaA* encodes a copper export ATPase. However, the change in copper resistance was marginal, with wild type and mutant showing nearly the same growth behavior in 3 µM Cu^{2+}. The sequence similarity of CtaA, which is greatest to the CCC2 copper ATPase of yeast, does, however, suggests that CtaA is a copper pump.

The *pacS* gene was found in a systematic search for P-type ATPases in *Synechococcus*. With degenerate primers corresponding to the phosphorylation domain and the ATP-binding region, respectively, of P-type ATPases, PCR fragments were generated from total *Synechococcus* DNA and these fragments then used to screen a library [48]. One of the genes so cloned was *pacS*. It encodes a CPx-type ATPase of 747 amino acids with an N-terminal CxxC motif. Several lines of evidence suggest that PacS is a copper translocating ATPase located in the thylakoid membrane.

PacS mRNA was induced 20- to 30-fold by 5 µM Cu^{2+} or 40 µM Ag^+, but not by metal-depleted growth media. Deletion of the *pacS* gene resulted in increased copper sensitivity of growth. Growth was inhibited by 5 µM Cu^{2+} or 25 µM Ag^+, conditions that did not significantly affect the wild type. With an antibody against PacS, the protein was localized in the thylakoid membranes. Taken together, these results point to a role of PacS in intracellular copper distribution in *Synechococcus*.

16.6
The *Helicobacter pylori* Copper ATPases

Helicobacter pylori is a curved, microaerophilic Gram-negative bacterium that currently receives attention because of its association with chronic active type B gastritis in humans [49]. Efficient treatment of *H. pylori* infections can be accomplished with a combination of the antibiotic roxithromycin and omeprazole [50]. Omeprazole is a prodrug that is converted to the active form in the acid environment of the stomach. When activated, it reacts with sulfhydryls and strongly inhibits the gastric P-type K^+H^+-ATPase, but also the growth of *H. pylori*. Because of this dual effect, it seemed reasonable to search for an essential, omeprazole-sensitive ATPase in *H. pylori* and many laboratories set out to clone P-type ATPases from this organism.

A *H. pylori* CPx-type copper ATPase designated hpCopA here, was cloned independently by two groups [51;52]. Knockout mutants in the *hpcopA* gene were more sensitive to Cu^{2+}. N-terminal peptides of CopA exhibited affinity for Cu^{2+} in support of a role of this enzyme in copper transport. The membrane topology of hpCopA was experimentally demonstrated by an *in vitro* transcription/translation/glycosylation system [53]. HpCopA was found to have eight membrane spans, in line with previous models. This is so far the only data available on the membrane topology of CPx-type ATPases.

A second gene downstream of *hpcopA*, termed *hpcopP*, encodes a CopZ-like protein of 66 amino acids. hpCopP has the most extensive amino acid similarity to *E. hirae* CopZ protein but expression of this protein or a function similar to that of CopZ in intracellular copper routing has not been demonstrated [51].

16.7
The Copper ATPase of *Listeria monocytogenes*

CtpA of *Listeria monocytogenes* appears to be a CPx-type ATPase involved in copper homeostasis. Growth of ctpA insertion mutants was inhibited by the copper-chelating agent 8-hydroxyquinoline. Expression levels of *ctpA* mRNA were increased following growth in media containing low and high copper concentrations. This biphasic regulation resembles that observed for the *cop* operon of *E. hirae* (see before). CtpA lacks N-terminal metal binding sites common to the CPx-type ATPases, but possesses all the other typical features, such as the CPC motif in membrane helix 6 and the HP motif in the second cytoplasmic loop. The metal binding sites may have been lost during cloning. Francis and Thomas [54] proposed a membrane topology for CtpA with six transmembranous helices only, but a membrane helix assignment in accord with the eight membrane spans exhibited by other CPx-type ATPases is possible.

A mutant strain of *L. monocytogenes* with an insertion of an antibiotic resistance cartridge in the *ctpA* gene was compared to the wild type in tissue culture invasion assays and mouse infection studies [55]. Mutants in CtpA were unaltered for intracellular growth in J774 and HeLa cell lines but recovery of mutants from tissue of infected mice was dramatically reduced compared to wild type. Also, *in vivo* persistence in mixed-infection competition experiments was impaired. These results demonstrate a role of CtpA in establishing an *in vivo* infection by *L. monocytogenes*. Possibly, CtpA has a role in copper accretion by *L. monocytogenes*, which is expected to be more difficult during infection of a host than under culture conditions. Such an interpretation would be in line with the identification of a virulence gene homologous to *copA* of *E. hirae* in *Staphylococcus aureus* (see above).

16.8
Other Copper Resistance Systems

Plasmid-encoded copper resistance determinants have been identified in *Pseudomonas syringae*, *E. coli*, and *Xanthomonas campestris* [42]. The systems of *X. campestris* [26] and *E. coli* (*pco* genes) are essentially equivalent to that identified in *P. syringae* (*cop* genes). They consist of an operon with four structural genes, *pcopABCD* (*E. coli* nomenclature), and two regulatory genes, *pcoSR*. The two regulatory proteins are thought to act as membrane sensor and DNA-binding regulator, respectively. This conclusion is based on sequence similarities to the well-studied two-component ATP-kinase sensor-regulatory systems [56].

Three of the structural gene products of *P. syringae* have been purified [57]. CopA is a periplasmic protein with copper binding sites and was shown to bind 11 copper per monomer, resulting in a blue-green color of bacterial colonies grown in the presence of high copper concentrations [58]. CopC and CopD of *P. syringae* appear to function in copper uptake into the cytoplasm [58]. Copper transport studies in *E. coli* revealed that cells containing a copper resistance plasmid had reduced copper accumulation during the log phase of growth, while increased accumulation had previously been observed during stationary phase. Some data indicate that this plasmid-borne copper resistance is linked with chromosomal systems for copper management [59], but mechanistic details are still lacking.

16.9
Conclusion

Today, copper homeostasis is a research area of intense interest, and work in this field has recently uncovered several surprising new concepts of trace metal homeostasis – and more are likely to emerge. The molecular defects in the inherited disorders of copper metabolism, Menkes disease and Wilson disease, have been elucidated and clinical treatment can now be approached or improved. Study of the *E. hirae* model system has significantly contributed to the current understanding. It has shown modes of copper entry into and out of the cell by the action of copper ATPases, transcriptional control of copper homeostatic genes by a copper responsive repressor, and intracellular copper routing by a copper chaperone. However, there is still a lack of information on copper ATPase structure and function. For some copper resistance systems like those encoded by the *Pseudomonas cop* genes or the *E. coli cut* and *pco* genes, the mechanisms are still entirely unclear and further work is needed.

Acknowledgement

Part of the work described here was supported by grant 32-56716.99 from the Swiss National Foundation and a grant from the Novartis Foundation.

References

1. Rasmussen, B., *Nature* **2000**, *405*, 676-679.
2. Zierenberg, R. A., Adams, M. W., Arp, A. J., *Proc. Natl. Acad. Sci. USA*, **2000**, *97*, 12961-12962.
3. Kaim, W., Rall, J., *Angew. Chem. Int. Ed. Engl.* **1996**, *35*, 43-60.
4. Odermatt, A., Suter, H., Krapf, R., Solioz, M., *Ann. NY Acad. Sci.* **1992**, *671*, 484-486.
5. Pedersen, P. L., Carafoli, E., *Trends Biochem. Sci.* **1987**, *12*, 186-189.
6. Maguire, M. E., *J. Bioenerg. Biomembr.* **1992**, *24*, 319-328.
7. Fagan, M. J., Saier, M. H., Jr., *J. Mol. Evol.* **1994**, *38*, 57-99.
8. Lutsenko, S., Kaplan, J. H., *Biochemistry* **1995**, *34*, 15607-15613.
9. Solioz, M., Vulpe, C., *Trends Biochem. Sci.* **1996**, *21*, 237-241.
10. Toyoshima, C., Nakasako, M., Nomura, H., Ogawa, H., *Nature* **2000**, *405*, 647-655.
11. MacLennan, D. H., Rice, W. J., Green, N. M., *J. Biol. Chem.* **1997**, *272*, 28815-28818.
12. Vilsen, B., Andersen, J. P., Clarke, D. M., MacLennan, D. H., *J. Biol. Chem.* **1989**, *264*, 21024-21030.
12a. Bissig, K.-D., Wunderli-Ye, H., Duda, P., Solioz, M., *Biochem. J.* **2001**, *357*, 217-223.
13. Silver, S., Nucifora, G., Chu, L., Misra, T. K., *Trends Biochem. Sci.* **1989**, *14*, 76-80.
14. Harrison, M. D., Jones, C. E., Solioz, M., Dameron, C. T., *Trends Biochem. Sci.* **2000**, *25*, 29-32.
15. Klomp, L. W., Lin, S. J., Yuan, D. S., Klausner, R. D., Culotta, V. C., Gitlin, J. D., *J. Biol. Chem.* **1997**, *272*, 9221-9226.
16. Wakabayashi, T., Nakamura, N., Sambongi, Y., Wada, Y., Oka, T., Futai, M., *FEBS Lett.* **1998**, *440*, 141-146.
17. Himelblau, E., Mira, H., Lin, S. J., Cizewski Culotta, V., Penarrubia, L., Amasino, R. M., *Plant Physiol.* **1998**, *117*, 1227-1234.
18. Lin, S. J., Culotta, V. C., *Proc. Natl. Acad. Sci. USA* **1995**, *92*, 3784-3788.

19. Wimmer, R., Herrmann, T., Solioz, M., Wüthrich, K., *J. Biol. Chem.* **1999**, *274*, 22597-22603.
20. Steele, R. A., Opella, S. J., *Biochemistry* **1997**, *36*, 6885-6895.
21. Gitschier, J., Moffat, B., Reilly, D., Wood, W. I., Fairbrother, W. J., *Nature Struct. Biol.* **1998**, *5*, 47-54.
22. Shah, A. B., Chernov, I., Zhang, H. T., Ross, B. M., Das, K., Lutsenko, S., Parano, E., Pavone, L., Evgrafov, O., Ivanova-Smolenskaya, I. A., Anneren, G., Westermark, K., Urrutia, F. H., Penchaszadeh, G. K., Sternlieb, I., Scheinberg, I. H., Gilliam, T. C., Petrukhin, K., *Am. J. Hum. Genet.* **1997**, *61*, 317-328.
23. Forbes, J. R., Cox, D. W., *Am. J. Hum. Genet.* **1998**, *63*, 1663-1674.
24. Payne, A. S., Kelly, E. J., Gitlin, J. D., *Proc. Natl. Acad. Sci. USA* **1998**, *95*, 10854-10859.
25. Iida, M., Terada, K., Sambongi, Y., Wakabayashi, T., Miura, N., Koyama, K., Futai, M., Sugiyama, T., *FEBS Lett.* **1998**, *428*, 281-285.
26. Lee, Y. A., Hendson, M., Panopoulos, N. J., Schroth, M. N., *J. Bacteriol.* **1994**, *176*, 173-188.
27. Odermatt, A., Suter, H., Krapf, R., Solioz, M., *J. Biol. Chem.* **1993**, *268*, 12775-12779.
28. Lowe, A. M., Beattie, D. T., Deresiewicz, R. L., *Mol. Microbiol.* **1998**, *27*, 967-976.
28a. Wunderli-Ye, H., Solioz, M., *Biochem. Biophys. Res. Commun.* **2001**, *280*, 713-719.
29. Solioz, M., Odermatt, A., *J. Biol. Chem.* **1995**, *270*, 9217-9221.
30. Odermatt, A., Krapf, R., Solioz, M., *Biochem. Biophys. Res. Commun.* **1994**, *202*, 44-48.
31. Pope, M. T., Dale, B. W., *Rev. Chem. Soc.* **1968**, *22*, 527-545.
32. Odermatt, A., Solioz, M., *J. Biol. Chem.* **1995**, *270*, 4349-4354.
33. Himeno, T., Imanaka, T., Aiba, S., *J. Bacteriol.* **1986**, *168*, 1128-1132.
34. Suzuki, E., Kuwahara Arai, K., Richardson, J. F., Hiramatsu, K., *Antimicrob. Agents Chemother.* **1993**, *37*, 1219-1226.
35. Hackbarth, C. J., Chambers, H. F., *Antimicrob. Agents Chemother.* **1993**, *37*, 1144-1149.
36. Wittman, V., Wong, H. C., *J. Bacteriol.* **1988**, *170*, 3206-3212.
37. Zhou, P. B., Thiele, D. J., *Proc. Natl. Acad. Sci. USA* **1991**, *88*, 6112-6116.
38. Dobi, A., Dameron, C. T., Hu, S., Hamer, D., Winge, D. R., *J. Biol. Chem.* **1995**, *270*, 10171-10178.
39. Jungmann, J., Reins, H. A., Lee, J. W., Romeo, A., Hassett, R., Kosman, D., Jentsch, S., *EMBO J.* **1993**, *12*, 5051-5056.
40. Strausak, D. Solioz, M., *J. Biol. Chem.* **1997**, *272*, 8932-8936.
41. Cobine, P., Wickramasinghe, W. A., Harrison, M. D., Weber, T., Solioz, M., Dameron, C. T., *FEBS Lett.* **1999**, *445*, 27-30.
42. Silver, S., Phung, L. T., *Annu. Rev. Microbiol.* **1996**, *50*, 753-789.
43. Rensing, C., Fan, B., Sharma, R., Mitra, B., Rosen, B. P., *Proc. Natl. Acad. Sci. USA* **2000**, *97*, 652-656.
44. Brown, N. L., Camakaris, J., Lee, B. T., Williams, T., Morby, A. P., Parkhill, J., Rouch, D. A., *J. Cell Biochem.* **1991**, *46*, 106-114.
45. Gupta, S. D., Lee, B. T., Camakaris, J., Wu, H. C., *J. Bacteriol.* **1995**, *177*, 4207-4215.
46. Outten, F. W., Outten, C. E., Hale, J., O'Halloran, T. V., *J. Biol. Chem.* **2000**, *275*, 31024-31029.
47. Phung, L. T., Ajlani, G., Haselkorn, R., *Proc. Natl. Acad. Sci. USA* **1994**, *91*, 9651-9654.
48. Kanamaru, K., Kashiwagi, S., Mizuno, T., *Mol. Microbiol.* **1994**, *13*, 369-377.
49. Dick, J. D., *Annu. Rev. Microbiol.* **1990**, *44*, 249-269.
50. Cellini, L., Marzio, L., Di Girolamo, A., Allocati, N., Grossi, L., Dainelli, B., *FEMS Microbiol. Lett.* **1991**, *68*, 255-257.
51. Ge, Z. and Taylor, D. E., *FEMS Microbiol. Lett.* **1996**, *145*, 181-188.
52. Melchers, K., Herrmann, L., Mauch, F., Bayle, D., Heuermann, D., Weitzenegger, T., Schuhmacher, A., Sachs, G., Haas, R., Bode, G., Bensch, K., Schafer, K. P., *Acta Physiol. Scand. Suppl.* **1998**, *643*, 123-135.
53. Bamberg, K., Sachs, G., *J. Biol. Chem.* **1994**, *269*, 16909-16919.
54. Francis, M. S., Thomas, C. J., *Mol. Gen. Genet.* **1997**, *253*, 484-491.
55. Francis, M. S., Thomas, C. J., *Microb. Pathog.* **1997**, *22*, 67-78.
56. Parkinson, J. S., Kofoid, E. C., *Annu. Rev. Genet.* **1992**, *26*, 71-112.
57. Cha, J. S., Cooksey, D. A., *Proc. Natl. Acad. Sci. USA* **1991**, *88*, 8915-8919.
58. Cooksey, D. A., *FEMS Microbiol. Rev.* **1994**, *14*, 381-386.
59. Brown, N. L., Barrett, S. R., Camakaris, J., Lee, B. T. O., Rouch, D. A., *Mol. Microbiol.* **1995**, *17*, 1153-1166.
60. Swofford, D. L., Olson, P. D., in: *Molecular Systematics* (Hillis, D. M., Moritz, C., Eds.), Sinauer, Associates, **1990**, pp. 21-132.

17
Microbial Arsenite and Antimonite Transporters

Parjit Kaur

17.1
Introduction

17.1.1
Why Arsenic Transporters?

Why do the arsenic resistance mechanisms exist? Arsenic is prevalent in our environment naturally as well as a result of human activity [1]. Arsenic compounds have been used in the past as antimicrobial agents in paints and wood-treating agents and as active ingredients of insecticides. In addition, arsenic is also used as a catalyst in the gold mining industry, which creates a major environmental problem in the abandoned gold mine areas. Hence, it is likely that resistance mechanisms evolved in microorganisms to deal with the presence of the toxic compounds in the environment. Molecular analysis of arsenic resistance mechanisms from different organisms has given important clues about the evolution of resistance. It seems that over time and with increasing selective pressures the arsenic resistance mechanisms have become sophisticated and more efficient. Hence, arsenic resistance systems exhibiting different levels of complexity, thus conferring different levels of resistance in the host organism, have been isolated. Several of these arsenic resistance mechanisms will be discussed in this chapter.

17.1.2
Efflux as a Mechanism for Resistance

Microorganisms are known to adopt many different strategies/mechanisms for resistance to toxic compounds. One of the commonly found mechanisms is the active efflux of the toxic substance by dedicated transporters located in the cell membrane. Even though membrane transporters involved in efflux of various drugs, antibiotics and heavy metals have been identified, our understanding of their mechanism of function is still limited. The goal of the present chapter is to present the current state of knowledge on the structure and function of different arsenic transporters found in microorganisms.

Initial studies on arsenic resistance were focused on the plasmid-encoded *ars* operon found in *Escherichia coli* [2, 3]. More recently, *ars* homologs have been identified in other bacteria and in archaea [4]. Since the plasmid-encoded anion transporter of *E. coli* is the best characterized arsenic transporter, it would be the major focus of this chapter. Several reviews have already been written on this transporter [2, 3, 5], however, recently, there has been an influx of new information from biochemical studies and from analysis of the crystal structure of a subunit of this transporter. Furthermore, genome sequencing projects have revealed the existence of homologs of some of the *ars* genes in other organisms, including eukaryotes such as *Caenorhabditis elegans*, *Mus musculatus* (mouse), and *Homo sapiens* (human) [5]. It is not known if all the *ars* homologs confer arsenic resistance in the host organism in which they are found, however, their identification has broadened our view, and availability of this information allows us to speculate on the origin and sequential evolution of multisubunit/multidomain transport complexes and of resistance mechanisms. Significance of some of these findings will be discussed in this chapter. Finally, arsenic resistance mechanisms unrelated to the genes of the *ars* operon have been identified in eukaryotic cells [4] and will also be briefly described here.

17.2
Overall Architecture of the Plasmid-encoded Pump in *Escherichia coli*

The plasmid-encoded *ars* operon was originally isolated from a conjugative R-factor R773 found in a clinical isolate of *E. coli* [2]. The genes of this operon confer resistance in *E. coli* to arsenite and antimonite, both anions of the +3 oxidation state. Early on it was shown that the *ars* genes code for an arsenite efflux system and that the basic efflux pump consists of two subunits – a membrane component that acts as the carrier and a catalytic component that transduces the energy of ATP hydrolysis to transport of the anions [2]. Ars operon consists of five genes in the order *arsRDABC* (Fig. 1). ArsR and arsD are the regulatory genes [6, 7],

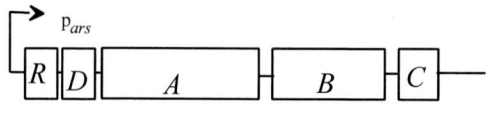

Protein	ArsR	ArsD	ArsA	ArsB	ArsC
Total residues	117	120	583	429	141
Molecular weight	13,198	13,218	63,188	45,598	15,830

Fig. 1. Arrangement of the genes in the *E. coli ars* operon. Five genes of the *E. coli ars* operon are shown. The direction of transcription is shown with the arrow at the promoter, P_{ars}. Five gene products are listed below with the number of amino acid residues and molecular masses in Daltons (Da).

Fig. 2. Model of the plasmid-encoded arsenite/antimonite transporter in *E. coli*. The complex of ArsA and ArsB proteins is shown. ArsB is an integral membrane protein which serves as an anchor for ArsA. Together, ArsA and ArsB form an ATP-dependent efflux pump for oxyanions of the +3 oxidation state. The ArsA protein is shown as a dimer, each monomer consisting of two homologous halves. ArsC associates with the pump and couples the reduction of arsenate to extrusion through the ArsAB pump.

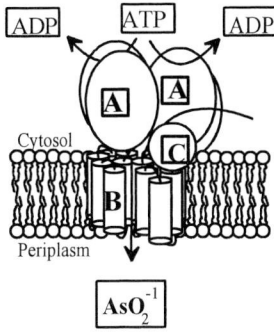

whereas the *arsA* and *arsB* encode the structural components of the pump [2]. ArsC protein is an arsenate reductase that reduces arsenate(+5) to arsenite(+3) which then becomes a substrate for the ArsAB pump [8].

The ability of the gene products of the *ars* operon to form an effective membrane transporter for efflux of arsenite and antimonite depends upon the ability of Ars proteins to sense small amounts of these oxyanions. This latter ability manifests itself at both the transcriptional level and at the biochemical level. At the transcriptional level, the metalloregulatory protein ArsR acts as a very sensitive sensor (described below) and the binding of the anion to ArsR eventually turns on the expression of the downstream *ars* genes [6]. At the biochemical level, ATPase activity of the catalytic subunit, ArsA, is stimulated by binding of the anion [9]. Both ArsR and ArsA contain crucial cysteines that are proposed to form tricoordinate AsS_3 complexes, thus recognizing the soft metal form of arsenic [4].

ArsR is a *trans*-acting repressor protein that negatively regulates the expression of the genes of the *ars* operon [10]. It controls the basal level of expression, whereas ArsD, another *trans*-acting repressor controls the maximal expression of the *ars* genes [11]. Thus, expression of *ars* genes is maintained optimally within a narrow range. Both arsenite and antimonite can act as inducers for expression of the operon. ArsR has been shown to contain a helix-turn-helix DNA-binding domain and a conserved consensus sequence $ELC_{32}VC_{34}DL$ at the N-terminus of the DNA-binding domain [12]. It has been proposed that ArsR responds to arsenite or antimonite by formation of a tricoordinate AsS_3 complex via cysteine thiolates of Cys32, Cys34 and Cys37 [4]. Binding of arsenite to the three cysteines in ArsR is proposed to change the conformation of the DNA-binding site, thus releasing ArsR from the operator. ArsR and ArsD are each 13 kDa homodimeric proteins, however, they share no sequence similarity. Both bind to the same site in the *ars* promoter, however, ArsD binds with twofold lower affinity, thus it is proposed that ArsD controls the maximal level of *ars* operon expression [7]. This may be necessary in order to prevent toxicity resulting from overexpression of the membrane protein ArsB, hence the dual control of gene expression maintains a balance between efflux and expression of transporter genes.

An important question concerning the function of the transporter is the mechanism of energy transduction between the catalytic component ArsA and the integral membrane protein ArsB that serves as an anchor for ArsA and as a carrier for efflux (Fig. 2). In spite of numerous attempts and many different strategies it has not been possible to overexpress ArsB protein, nor have attempts to produce antibodies against this protein been met with any success. Hence, knowledge of the biochemistry of the ArsB protein is still quite rudimentary. The peripheral membrane catalytic component, ArsA, on the other hand, is a model enzyme and very amenable to biochemical manipulation. Upon overexpression, it is present in the soluble fraction of the cell, thus facilitating purification and *in vitro* characterization of the protein. A summary of the present state of understanding of the structure and function of the three structural components of the Ars pump is given below.

17.2.1
ArsA

ArsA, a 63 kDa hydrophilic protein, forms the catalytic component of the arsenite/antimonite transporter. It is an anion-stimulated ATPase, however, it does not bear resemblance to members of the known classes of ATPases (including F-type, V-type, P-type, and ABC-type), nor is it inhibited by the known inhibitors of these four classes [9].

ArsA consists of two homologous halves, which have been designated A1 and A2. Each half contains a nucleotide binding domain (NBD, [13]) consisting of the consensus P-loop or the Walker A motif (15/334GKGGVGKTS/T 23/342). It has been proposed that the *arsA* gene arose by duplication and fusion of an ancestral gene half its size. Since half *arsA* genes have now been discovered in a variety of organisms, its implication is discussed later in Sect. 17.2.1.5. A comparison of the amino acid sequence of the A1 and A2 domains suggests that roughly a 40 amino acid long stretch of residues at the end of A1 might serve as a linker connecting the two domains of ArsA [14]. Hence, ArsA consists of two nucleotide binding domains, a linker connecting the two domains and a site for binding of the anion. An attempt would be made in the following sections to describe the structure of various domains of ArsA and the interactions between them. The complexity of these interactions makes understanding the function of this enzyme a challenging task.

17.2.1.1   The Ligand (Arsenite/Antimonite) Binding Site

The ATPase activity of ArsA is stimulated by the substrates, arsenite or antimonite [9]. Genetic and biochemical experiments suggest that activation of ArsA by arsenite or antimonite occurs via binding to cysteine thiolates. ArsA contains four cysteines, Cys26, Cys113, Cys172, Cys422. To determine if cysteines play a role in the binding of arsenite or antimonite, these residues were changed to serines. It was found that while Cys26S mutation had no effect, mutation in any of the other three cysteines rendered the cells sensitive to arsenite [15]. Purified mutant proteins were found to require a much higher concentration of the anion for activation of the ATPase activity, suggesting that arsenite or antimonite forms a tricoordinate complex with the cysteine thiolates of residues 113, 172 and 422. Cys113 and Cys172 belong to the A1 half of ArsA, while Cys422 belongs to the A2 domain, thus in the three-dimensional structure, these three cysteines must lie close together. Dibromobimane, a bifunctional alkylating agent that reacts with thiol pairs within 3-6 Å to form a fluorescent adduct, has been used to determine proximity of these three cysteines in ArsA. It was found that if two of the three cysteines were changed to serines, reaction with dibromobimane did not result in a fluorescent adduct, suggesting that Cys113, 172 and 422 are in close proximity of each other [16]. It has been suggested that binding of arsenite or antimonite pulls together two halves of ArsA, thus forming an A1/A2 interface [16].

17.2.1.2 The Nucleotide Binding Sites

17.2.1.2.1 Are Both NBDs in ArsA Required?

First question that needs to be answered, of course, is if both nucleotide binding sites in ArsA are required. If so, are both sites catalytic? That both sites are required became amply evident from mutagenesis studies; independent point mutations in the Walker A sequence of either the A1 or the A2 domain knocked out ATPase activity of the protein, and the cells carrying mutant *arsA* alleles were found to be arsenite sensitive [17, 18]. Furthermore, complementation studies showed that the A1 and the A2 mutant alleles (each carrying one functional site) could complement each other, indicating that the two sites are not only required but that they interact [19]. These studies also indicate that ArsA might function as a homodimer so that an active interface between a functional A1 and a functional A2 is formed *in trans* from two subunits. Alternatively, it is also possible that the A1 and A2 domains normally interact *in cis* within a monomer of ArsA; if the two functional sites are not available *in cis*, then they can form an interface *in trans* as is seen in the complementation analysis. This issue still awaits resolution. That both A1 and A2 are required for function of the protein has also been demonstrated by *in vitro* reconstitution of the active ATPase from half A1 and A2 polypeptides while neither the A1 half nor the A2 half by itself shows anion-stimulated activity [20]. One such combination of polypeptides (N18 and C46) that results in successful reconstitution is shown in Fig. 3. The reconstituted complex of N18 and C46 shows ATP-binding and hydrolysis [20]. The question whether both sites in ArsA are catalytic is somewhat more difficult to answer, especially if function of one site depends on the other site, which seems to be the situation with ArsA, as will become clear from the following discussion.

17.2.1.2.2 Asymmetry between the two NBDs

Recent studies from the author's laboratory suggest that the two nucleotide binding sites in ArsA have a built-in asymmetry with respect to their affinity for ATP and their conformation [21]. A1 appears to be the high-affinity ATP-binding site which is filled in the presence of ATP alone. The A2 site, on the other hand,

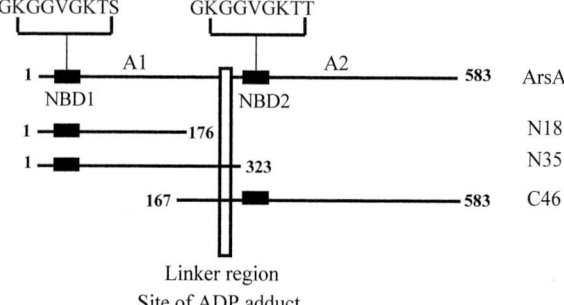

Fig. 3. Subclones of ArsA and domain organization of wild-type ArsA. A1 and A2 domains of the wild-type ArsA with the Walker A sequences (NBD1 and NBD2) and the linker region are shown. Different polypeptides derived from the A1 (N18 and N35) or the A2 (C46) half of ArsA are also shown (the number of the first and the last residue is indicated). N-terminal peptide, N35, forms a UV-activated adduct with ADP, which is proposed to be the result of unisite catalysis from A1. A mixture of N18 and C46 can be reconstituted *in vitro* to yield an active ATPase and the reconstituted protein forms a UV-activated adduct in the linker region contained in C46.

seems to be inaccessible to nucleotide binding, and only in the presence of the ligand, antimonite, does it open up to bind ATP. A model for nucleotide occupancy of A1 and A2 domains in ArsA is shown in Fig. 4. This proposal arises from close examination of the behavior of an ATP analog, FSBA (5'-parafluorosulfonylbenzoyladenosine). Since FSBA binds covalently to ArsA, it allowed identification of certain conformations of ArsA which would otherwise be too fast to be easily accessible for analysis. Curiously, FSBA binds preferentially to the A2 half of ArsA suggesting that the A1 and A2 NBDs have different conformations [21]. This difference in conformation of two sites may be the key to understanding the catalytic cycle of this enzyme.

The observations that proved valuable in proposing the model in Fig. 4 are derived form tryptic analysis of different conformations of ArsA obtained in the presence of ATP, FSBA and antimonite [21]. It was found that at a particular low concentration of trypsin, an initial cleavage in ArsA occurs at Arg290 located in the linker region, thus producing a 32 kDa A1 and a 27 kDa A2 fragment (Fig. 4). The 27 kDa A2 fragment is cleaved further into smaller fragments, whereas the 32 kDa A1 is more compact and resistant to further cleavage. Interestingly, it was observed that binding of FSBA to the A2 half of ArsA (conformation F) results in protection of the 27 kDa fragment. By the same token, it was found that in the presence of ATP (conformation A), the 27 kDa A2 fragment is not protected, indicating that A2 is empty in this situation. Simultaneous presence of ATP and FSBA (conformation AF or FA) or ATP and antimonite (AS) results in complete protection of ArsA so that even the initial attack at Arg290 does not occur. Since ATP and FSBA result in complete protection irrespective of the sequence in which they are added to ArsA, it indicates that in this conformation (AF/FA) ArsA contains ATP in A1 and FSBA in A2 and, hence, complete protection results

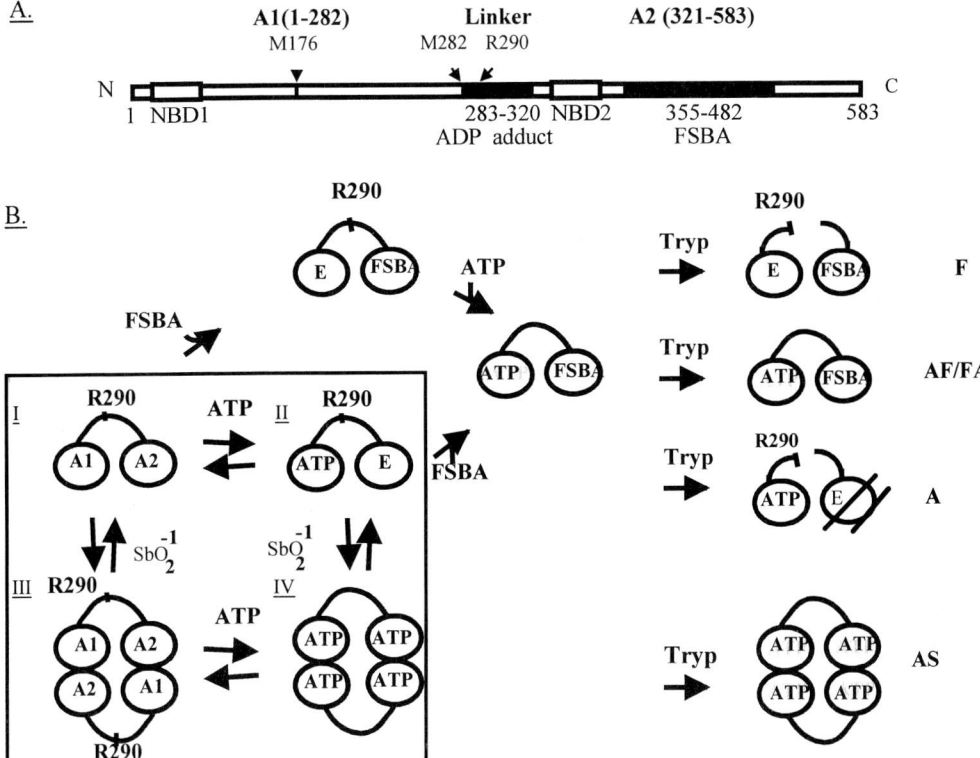

Fig. 4. A model for occupancy of the two nucleotide binding sites in ArsA. **A** Linear depiction of the A1 and A2 domains and the linker region in ArsA. The P-loop regions of A1 and A2, the adenine binding site in the linker (as determined by UV cross-linking) and the FSBA binding site for the A2 domain are shown. Location of amino acid residues M_{176}, M_{282}, and R_{290} is indicated. **B** Conformations of ArsA. The ArsA protein is shown as consisting of two domains A1 and A2 connected by a flexible linker in the middle. This model shows four central conformations (I-IV, boxed) of ArsA based on its trypsin accessibility. Conformation IV is the active conformation and can be achieved via II or III with antimonite binding acting as a switch that regulates ATP binding to A2. A conformational change induced by the oxyanion precedes binding of ATP to A2 and it is shown as the dimerization of ArsA. Additional conformations, such as F, FA/AF are attained by the addition of FSBA and ATP to conformations I or II in the sequence shown. Trypsin sensitivity of different conformations is indicated by availability of residue R_{290} in the linker. An initial cleavage of conformation I at R_{290} cleaves ArsA into a 32 kDa A1 and a 27 kDa A2 fragment. The 27 kDa fragment is cleaved further, whereas the 32 kDa fragment is more compact and less susceptible to cleavage by trypsin. The occupancy of the A2 site by FSBA is seen by protection of the 27 kDa C-terminal fragment on binding FSBA (in conformation F). Complete cleavage of this domain is seen on addition of ATP (conformation A) implying that A2 is unoccupied. Hatched lines within the A2 domain indicate cleavage of that domain. ATP can not access A2 in the absence of the oxyanion, while the ATP analog FSBA can (as in conformation F); hence ATP and FSBA together result in the trypsin resistant conformation AF or FA due to occupancy of A1 by ATP and A2 by FSBA. This conformation is similar to the conformation achieved in the presence of ATP and oxyanion (AS or IV) in being trypsin protected. It is proposed that the completely trypsin protected conformation, AS, results when both sites are occupied by ATP in the presence of antimonite. Tryp, trypsin; F, FSBA; A, ATP; AF, ATP, followed by FSBA; FA, FSBA, followed by ATP; AS, ATP + antimonite.

from simultaneous occupancy of both A1 and A2. Since AF and AS conformations are similar in being completely protected from trypsin, the AS conformation is believed to result from occupancy of both A1 and A2 by ATP in the presence of antimonite. Hence, it appears that binding of the ligand acts as a switch that opens up the A2 site for nucleotide binding. Binding of ATP to A2 then results in catalytic cooperativity and a much higher rate of hydrolysis. Thus, we have proposed the existence of a novel allosteric switch in ArsA which, instead of increasing or decreasing the affinity for the substrate (which is the commonly accepted mechanism for allosteric activation), acts as an "on/off" switch for ATP binding to A2 [21]. The effect of an "on/off" allosteric switch can not be bypassed by increasing the concentration of ATP. It is possible that a similar mechanism of allosteric switching operates in other ligand-stimulated ATPases which contain multiple nucleotide binding sites.

17.2.1.2.3 Unisite and Multisite Catalysis in ArsA

The model in Fig. 4 proposes that the A1 site of ArsA is a high-affinity site and it binds ATP in the absence of antimonite, whereas the A2 site opens up only in the presence of antimonite. This model is supported by the observation that the high-affinity A1 site of ArsA carries out unisite catalysis at a low rate in the absence of antimonite [22]. On addition of antimonite, catalytic cooperativity between A1 and A2 results in accelerated multisite catalysis [22]. Hence, the function of antimonite seems to be to recruit the A2 site into the catalytic reaction. Following experiments proved useful in deducing unisite/multisite catalysis in ArsA:

UV-activated cross-linking between ArsA and ATP is one of the approaches that has been used to study ATP binding to ArsA [23]. UV activation results in immobilization of the nucleotide at its binding site in a protein, hence, it is commonly used to identify nucleotide binding site(s) in proteins [24–26]. ArsA has been shown to form a UV-activated adduct in the presence of ATP and magnesium [23], which is localized in the linker region [14]. Point mutations in the A1 Walker A sequence result in inability of ArsA to form an adduct, whereas point mutations in A2 do not affect adduct formation [17, 18]. These results led us, in the past, to believe that only A1 domain in ArsA is involved in ATP binding. More recently, however, it has been shown that the nature of the adduct with ArsA is actually ADP [22]. This result indicates that ATP hydrolysis occurs in ArsA in the absence of antimonite [22]. Since the N-terminal peptide, N35, (Fig. 3) that contains only the A1 domain forms a UV-activated ADP adduct [19], it has now been proposed that this adduct is the result of unisite catalysis from A1 [22]. Similarly, A2 mutant proteins form the ADP adduct just like the wild-type protein [18]. A2 mutants carry an intact A1 site, thus they are able to carry out unisite catalysis from A1 resulting in the formation of an adduct with ADP. A major difference, however, is observed between N35 or A2 mutants and the wild-type protein on addition of antimonite – wild type protein carries out multisite catalysis resulting in much more ADP cross-linked to the linker region, whereas in A2 mutants or N35 peptide, addition of the ligand has no effect [22]. Based on the available data, it is proposed that in the absence of antimonite, the adduct is the result of unisite catalysis from A1, and in its presence, much more adduct results from multisite catalysis involving both A1 and A2 (Fig. 5). Multisite catalysis is not seen in A2 mutants or the N35 polypeptide that lacks A2. Since mutants in A1 do not form any adduct at all, it indicates that there is a sequence to the catalytic events so that the A1 site must be first in the sequence of events and A2 comes into play later and only if A1 is functional. Precisely for this reason, it has been difficult to determine the catalytic nature of A2.

The unisite and multisite catalytic mechanism of ArsA is similar to the mechanism of the F_1 ATPase [27]. Even though the three catalytic sites in F_1 are equivalent and lie on identical β-subunits, an asymmetry between the sites is introduced by rotation of the γ-subunit with respect to the $\alpha3\beta3$ head and each site in turn becomes the high-affinity site [27]. At substoichiometric concentrations of ATP, the high-affinity site is filled with ATP and is able to carry out unisite catalysis. At higher concentrations, ATP binds to the second and the third sites, and it causes a much higher rate of product release from the first high-affinity site [28]. The A1 site of ArsA is able to carry out unisite catalysis just like each of the three sites in F_1, however, an increase in the ATP concentration is not enough to switch ArsA from unisite to multisite catalysis. Instead, it is the binding of the ligand that appears to recruit A2. Hence, it can be said with a fair amount of certainty that A1 in ArsA is catalytic, however, the contribution of A2 is not clear. It is possible that the A2 site may only play a regulatory role so that ATP binding to A2 allows much faster product release from A1 without A2 itself being catalytic in the process.

Another interesting protein that is relevant to our discussion here is the P-glycoprotein (Pgp) that also contains two nucleotide binding domains and is involved in ATP-dependent export of a variety of drugs [29]. In this respect, Pgp is a structural and functional analog of ArsA. Vanadate (Vi) trapping experiments have been very useful in analyzing the catalytic nature of two NBDs in Pgp. Vi is an analog of phosphate and it inhibits the activity of some ATPases. Upon catalytic turnover,

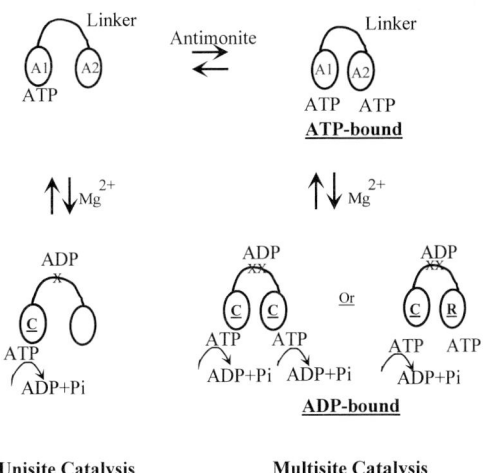

Fig. 5. Unisite and multisite catalysis by ArsA. ArsA carries out unisite catalysis from A1 in the presence of ATP and magnesium. In the presence of antimonite, ATP binds to A2, resulting in catalytic cooperativity and multisite catalysis from A1 and A2. The product of hydrolysis, ADP, is seen (by UV cross-linking) to associate with the linker region in ArsA. Hence, ArsA goes through an ATP-bound and an ADP-bound conformation, which may have significance in the transduction of energy. The A1 ATP-binding site of ArsA has been shown to be catalytic (C), however, it is not clear if the A2 is catalytic (C) or regulatory (R) in nature.

Vi is trapped in the catalytic site of a protein as ADP.Vi complex [30]. Both sites in Pgp have been seen to trap nucleotide equally, suggesting that the two NBDs in Pgp are equivalent and that both sites are also catalytically active [31, 32]. Since the ATPase activity of ArsA is not inhibited by vanadate, trapping experiments with vanadate have not been successful in elucidating the catalytic nature of the two sites in ArsA.

17.2.1.2.4 ATP- and ADP-bound Conformation of ArsA

ArsA seems to interconvert between two different conformations: ATP-bound and ADP-bound. ADP, the product of the catalytic reaction, associates with the linker region connecting the A1 and A2 halves of ArsA. This association is seen in UV cross-linking experiments (described above) under both unisite and multisite catalytic conditions [22]. Since the ATP and ADP-stabilized conformations of ArsA are different (Kaur, unpublished data), it implies that the conformation of ArsA changes upon hydrolysis of ATP. It is possible that the energy of ATP hydrolysis might be used to produce cyclic changes in the conformation of ArsA which may be critical for producing conformational changes in ArsB, thus facilitating transport of the anion.

Switching between ATP-bound and the ADP-bound conformations has been implicated in the energy transduction processes in other nucleotide hydrolyzing proteins, such as RecA, p21, and the motor proteins, myosin and dynein [33, 34]. In RecA, which is a DNA-dependent ATPase, the ATP-stabilized conformation has been shown to have a higher affinity for DNA, whereas the ADP-bound form has a lower affinity. Hence, it has been proposed that the nucleotide cofactors play a role in the cycling of RecA protein on and off ssDNA, with ADP serving as a release factor [34]. A similar switching mechanism between GTP- and GDP-bound forms of Ras has been proposed and domains undergoing conformational changes upon GTP hydrolysis have been identified [33]. These domains may form a likely recognition site for other effector or receptor molecules. Hence, understanding the biochemical basis of catalysis in ArsA and the conformational changes that result from catalysis might eventually lead to an understanding of the mechanism of energy transduction between the catalytic component and the membrane component of the pump.

17.2.1.3 The DTAP Domain in ArsA

In addition to containing a consensus Walker A sequence, each half of ArsA also contains a 12-residue conserved sequence, 142/447DTAPTGHTIRLL153/458. This sequence contains an aspartate residue at its N-terminus which may correspond to the aspartate residue present in the Walker B motif found in most other nucleotide binding proteins and believed to be crucial for magnesium binding [33, 34]. The active-site magnesium ion bridges the β- and γ-phosphate of ATP and at the same time provides a link between the γ-phosphate and the aspartate in the Walker B motif [34]. It has been proposed that the DTAP domain in ArsA is involved in transduction of signal between the metalloid binding site and the nucleotide binding site [35]. Intrinsic fluorescence of single tryptophan residues placed at either end of the DTAP domain was used to report on the environment of this domain. It was found that, during hydrolysis, the C-terminal end of this conserved domain moves into a less polar environment, whereas the N-terminal end moves into a more polar environment, suggesting that the environment of this domain changes on hydrolysis, hence the proposal that this domain may function as a transduction domain [35].

17.2.1.4 The Linker Region in ArsA

Linker region in ArsA is the stretch of residues (roughly extending from residues 280-320) connecting the A1 and A2 half of the protein [14]. This region has become important in studies on ArsA because of two main reasons. One, linker region is the site where ADP, the product of the reaction, associates in ArsA (Sect. 17.2.1.2.3). Hence, ATP is hydrolyzed by ArsA and adduct with ADP is formed in the linker region. Two, residue Arg290, which is the site of initial trypsin attack in ArsA, lies in the linker region, suggesting that this region of the protein is easily accessible. Cleavage at Arg290 results in a 32 kDa A1 and a 27 kDa A2 fragment [21]. However, upon binding ATP and antimonite, ArsA goes through a conformational change, involving the linker, which makes Arg290 inaccessible to trypsin, hence most of the ArsA protein is present in the 63 kDa species (see also Sect. 17.2.1.2.2). Thus, based on these observations, it is likely that the linker may play an important role in the function of the enzyme. Deletion analysis of the linker region carried out by Barry Rosen's group seems to suggest that only the length of the linker is of importance [36]. Hence, five or ten codon deletions, encompassing residues between 295-299, 300-304, 305-309, 297-306, have only a slight effect on the function of the enzyme. A 23-codon deletion (291-313) has the most significant effect on arsenite resistance, thus it has been suggested that linker is not required for catalysis, but it facilitates the ability of the enzyme to form the A1/A2 interface [36]. Recent studies from the author's laboratory, however, show that point mutations in certain residues in the linker region drastically affect function of the enzyme (unpublished data), suggesting that linker region interacts actively with the nucleotide binding domains in the protein and hence, it may play a crucial role in the catalytic mechanism. Further analysis of the role of the linker is warranted.

17.2.1.5 Variations on the ArsA Theme

E. coli ArsA is the first characterized member of a new class of ATPases now known as the A-type ATPases [37]. New members of this class have since been identified among plasmid-encoded proteins from other gram-negative bacteria [38, 39]. These homologs contain two homologous halves, each half containing one NBD and show 82-89% identity with the *E. coli* ArsA. Genome sequencing has also revealed the presence of ArsA homologs in archaea and in eukaryotes, including *C. elegans*, mouse, and human cells [5]. Interestingly, however, the eukaryotic homologs of ArsA are half the size of *E. coli* ArsA and contain only one NBD. These half homologs show about 25-28 % identity with either the A1 or the A2 half of *E. coli* ArsA. The function of the eukaryotic ArsA homologs is not known, although the human homolog has been shown to be an ATPase, albeit a very slow ATPase (18 nmoles min^{-1} mg^{-1}), which shows a stimulation of activity of only about 1.6-fold in the presence of arsenite and none in the presence of antimonite [40]. By contrast, *E. coli* ArsA exhibits about 40-fold stimulation in the presence of antimonite with the stimulated activity of about 1,000 nmoles min^{-1} mg^{-1}. It is also stimulated by arsenite but to a lower extent [9].

Why do eukaryotes, including human cells, contain a half molecule of ArsA? The most likely explanation is that both the bacterial and the eukaryotic *arsA* genes arose from a common ancestor that was only half in size and contained one NBD. Since bacteria are much more vulnerable to the environmental toxins and, hence, are under a greater selective pressure, the bacterial *arsA* gene evolved further by gene duplication and fusion to produce the full-length ArsA (Fig. 6).

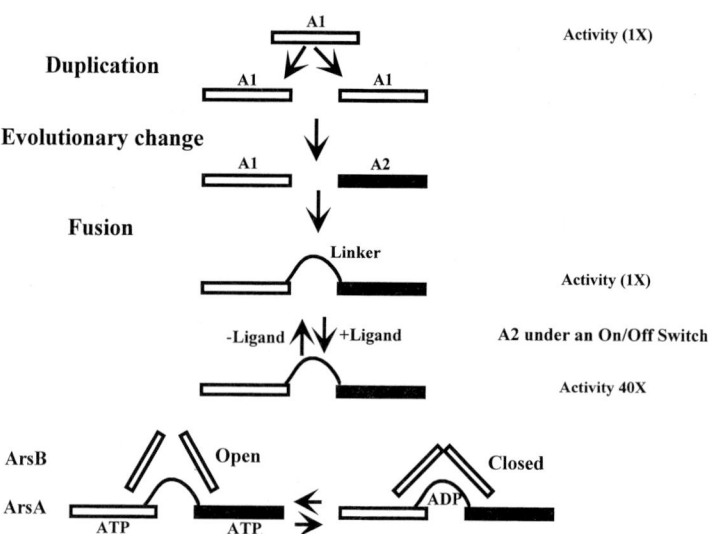

Fig. 6. A proposed scheme for the evolution of the *E. coli* plasmid-encoded ArsA. The ancestral *arsA* gene was half the size of the present day gene and contained one consensus nucleotide binding site. Duplication of the ancestral gene, followed by evolutionary change and fusion of the two copies of the gene, resulted in the present day *arsA* gene with two homologous halves which are connected by a linker. Two nucleotide binding sites in ArsA are asymmetric. During the evolutionary process, A2 was placed under the control of an allosteric "on/off" switch. In the presence of the ligand, A1 and A2 domains interact to give an activity about 40-fold higher (40X, multisite catalysis) than the activity of either half of the protein (unisite catalysis,1X). Since both ATP- and ADP-bound conformations of ArsA can be isolated, it is possible that these two conformations may play a role in the ability of ArsA to interact with the membrane component ArsB. In this schematic, the ATP-bound conformation is shown as the active or the open conformation, while the ADP conformation is in the closed state.

There is little doubt that the full-length version of the enzyme is a far more efficient enzyme. Studies described above (Sect. 17.2.1.2.3) have shown that the E. coli ArsA with two NBDs exhibits catalytic cooperativity, while the A1 half by itself shows low-level catalysis, which has been termed "unisite catalysis" [22]. In combination with the A2 domain and in the presence of the anion, the catalytic activity of ArsA is about 40-fold higher. Since ArsB homologs have not been found in the eukaryotic cells, whether these half ArsA homologs are involved in arsenite resistance in eukaryotic cells is an open question. Arsenic resistance mechanisms unrelated to ArsB have also been found to exist in the eukaryotic cells (see Sect. 17.4), hence, it is possible that prokaryotic and eukaryotic cells brought different players to the forefront in dealing with the problem of environmental arsenic.

What is intriguing is the fact that the archaeal *arsA* homologs, so far identified, are also half molecules with the exception of the *arsA* gene from *Halobacterium*. This might support the early divergence of the bacterial line from the archaeal/eukaryal line as has recently been proposed [41]. The *Halobacterium* ArsA protein, that shows about 32% identity to the E. coli protein, is plasmid-encoded [42], hence it could be the result of horizontal gene transfer at a later point in evolution or result of an independent duplication and fusion event.

17.2.1.6 Insights from the Crystal Structure of ArsA

Crystal structure of ArsA has recently become available at 2.2 Å [43]. These crystals were obtained in the presence of ADP, magnesium, and antimonite. Both the nucleotide binding sites were seen to contain ADP. When the crystals were soaked in ATP, ADP in the A2 site could be exchanged with ATP, however, ADP in the A1 site could not be exchanged, thus suggesting that in this conformation of ArsA, A1 is closed and A2 is in the open state. These results provide support for earlier biochemical studies which suggested that there is an asymmetry between the two nucleotide binding sites [21].

Analysis of the crystal structure also shows that the two NBDs and the metal binding site in ArsA are located at the interface of the A1 and the A2 domains. As expected, the phosphate groups of ADP bound to the A1 or A2 domain interact with the residues in the Walker A sequence in their respective domain. The magnesium ion is coordinated to the β-phosphate of ADP and to two aspartate residues – Asp45 and Asp144 in A1 and to Asp354 and Asp447 in A2. Asp144 and Asp447 are present at the N-terminus of the DTAP sequence (discussed in Sect. 17.2.1.3) in A1 and A2, respectively, and, as suggested earlier, these aspartates may correspond to aspartates in the Walker B motifs of other nucleotide binding proteins. For example, the crystal structure of the RecA protein shows coordination of magnesium ion to Asp142 in the Walker B motif as well as to the β- and γ-phosphate of ATP. Curiously, the metal binding site in ArsA shows three antimonite atoms in the allosteric site; each antimonite is three-coordinate, with one ligand from A1, the second from A2, and the third a non-protein ligand. Of the three antimonite atoms, one is coordinated to His148 and Ser420, one to Cys113 and Cys422, and one to Cys172 and His 453. His148 and His453 are present towards the C-terminus of the DTAP sequence. Since Asp144 and Asp447, present at the N-terminus of DTAP sequence, are coordinated to Mg via water molecules, and the histidines at the C-terminus of DTAP sequence are involved in metal cluster interactions, it provides support for the hypothesis that the DTAP sequence may function as a signal transduction domain between the nucleotide binding site and the metalloid binding site [35].

Overall, the crystal structure shows that the A1 and A2 domains of ArsA form an interface and both the NBDs and the metal binding cluster are located on the interface. It provides support for the earlier hypothesis that the two NBDs in ArsA are asymmetric and are in different conformations. The reason for the presence of three antimonites in the metal cluster is not clear. Biochemical experiments suggest that the antimonite atom is tri-coordinated to the three cysteines in ArsA [16]. In the crystal structure, these three cysteines are seen to participate in the metal cluster interactions, however, tri-coordination of any one antimonite to three cysteines is not seen.

The high-resolution structure is a big step forward for future studies, however, the structure is a single snapshot and it does not provide a mechanism for the catalytic cycle and does not show the conformational changes that occur as a result of hydrolysis. It does not indicate if the A2 site is catalytic, which is a major question at the moment. Furthermore, the linker is not visible in the structure, perhaps due to its flexibility. Hence, further biochemical analysis is essential for understanding the mechanism of catalysis and for interpretation of the structural data. Notably, these crystals were obtained in the presence of MgADP and antimonite, hence it is crucial to be able to obtain crystals under other conditions, including in the absence of antimonite, so that conformational changes resulting from binding of antimonite can be understood.

17.2.2
ArsB

ArsB is a 45 kDa hydrophobic membrane protein that is proposed to consist of 12 transmembrane a-helices [44] and forms the carrier for the transport of the anions. Specific interaction between ArsA and membranes containing ArsB has been shown [45]. This interaction results in the formation of an ATP-driven efflux pump. Since it has not been possible to overexpress ArsB protein, its biochemical characterization is still limited. It has been shown that some regulatory mechanism in the cell results in selective degradation of the message for *arsB*, thus preventing its overexpression [46]. This selective degradation of *arsB* message may have a direct relation to ArsB being a membrane protein, hence its overexpression might be toxic to the cell. ArsB contains a single cysteine which is not required for its function, suggesting that it transports arsenic or antimonite as an oxyanion and not as a soft metal [47].

17.2.3
ArsC

ArsC is a 16 kDa cytoplasmic protein that is shown to be an arsenate reductase. It reduces arsenate(+5) to arsenite(+3), thus increasing the substrate specificity of the Ars pump to include arsenicals of the +5 oxidation state [8]. Arsenate is not an inducer of the expression of the genes of the *ars* operon, nor does it stimulate the catalytic activity of ArsA [9, 10]. Only upon its reduction to arsenite by ArsC does it act as an inducer [10]. For its function, ArsC requires glutathione and glutaredoxin, suggesting that cysteines are involved in the catalytic function of the reductase [8]. Mutagenesis of Cys12 and Cys106, two cysteines present in ArsC, showed that cells containing Cys12S are arsenate sensitive, while the Cys106S mutation has no effect. Hence, Cys12 is an active-site residue which is involved in catalysis. Thus ArsC is proposed to be a single thiol reductase where the arsenylated thiol of Cys12 interacts with glutaredoxin resulting in reduction of arsenate [48].

17.3
Variations on the *Escherichia coli* Arsenic Transporter among Prokaryotes

The plasmid-encoded *ars* operon of *E. coli* is a five-gene operon consisting of two regulatory genes and three structural genes, as described above. Five-gene operons have also been found in plasmids of other gram-negative bacteria, e. g., the plasmid pKW301 [39] and the plasmid R46 [38] contain five-gene *ars* operons with high degree of homology to the *E. coli ars* genes. The first variation to this theme was found in the *ars* operon present in the plasmid, pI258, that confers arsenite resistance in *Staphylococcus aureus* cells [49, 50]. Gene sequencing of pI258 showed that this operon consists of only three genes, *arsRBC* (Fig. 7). The *arsR* and the *arsB* genes of *S. aureus* are 42% and 59% identical to the *E. coli* genes, whereas the *arsC* gene is only 22% identical. Hence, the *arsR* and *arsB* genes of *S. aureus ars* operon seem to share a common ancestor with the *arsR* and *arsB* genes of *E. coli*, whereas ArsC belongs to a distinct family of arsenate reductases. Biochemically also, the two enzymes seem to be different. The *S. aureus* ArsC protein has been shown to require thioredoxin as a reductant [51] in contrast to *E. coli* ArsC that requires glutathione and glutaredoxin to carry out reduction of arsenate [8, 52].

The presence of a three-gene operon lacking the gene for the catalytic subunit, ArsA, poses a bioenergetics problem. How does ArsB carry out efflux of the toxic anion without the energy of ATP hydrolysis? It has been shown that ArsB demonstrates dual mode of energy coupling; by itself, it can function as a pmf-driven secondary carrier, while in the presence of ArsA, it becomes an ATP-driven obligatory primary pump [53]. This has significance in the evolution of multisubunit complexes and it provides an explanation for the ability of the *S. aureus* system to confer arsenite resistance in the absence of the catalytic subunit. Three-gene operons (*arsRBC*) have since also been discovered in the chromosome of all *E. coli* strains [54]. The genes of these operons show a very high degree of homology to

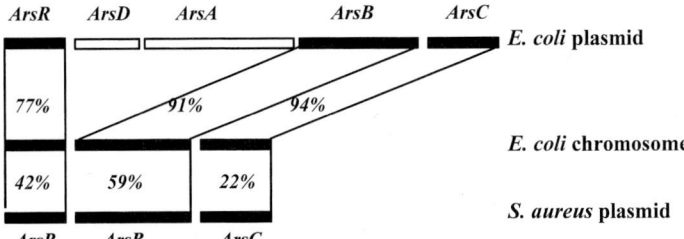

Fig. 7. Different *ars* operons in bacteria. The five-gene operon (from *E. coli* plasmid) and the three-gene operons (from *E. coli* chromosome and the *S. aureus* plasmid) are shown. The percentage homology between the genes is indicated.

the plasmid-encoded *ars* operon (Fig. 7). The presence of such an operon in *E. coli* presumably protects the cell against low levels of environmental arsenic. Knocking out the chromosomal *ars* operon in *E. coli* renders the cells hypersensitive to arsenic and antimony [54].

What is the evolutionary advantage of carrying the five-gene operon vs. the three-gene operon? Arsenite efflux pumps that function with the help of the catalytic subunit – the ArsAB pumps – are much more efficient and confer higher levels of arsenite resistance as compared to the ArsB-only pumps. It is likely that the ArsB-only pumps evolved earlier and ArsA was recruited later with the express purpose of making it a more effective pump. A gene for a soluble ATPase with a function unrelated to arsenic transport could be recruited by the *arsB* system and over time the recruited gene product would evolve to interact specifically with the ArsB protein, resulting in the two-subunit ArsAB pump. A two-gene *ars* operon, consisting of *arsRB*, has also been found in *Bacillus subtilis* [55]. Such an operon would most likely be a starting point for evolution of the *ars* operons that later resulted in the three-gene operon (*arsRBC*) by acquisition of *arsC* and in the five-gene operons (*arsRDABC*) by acquisition of *arsD* and *arsA*. It is not clear if *arsD* and *arsA* were acquired sequentially or in a single step. Interestingly, the five-gene operons have so far only been found on plasmids, whereas the three-gene operons are present in the chromosome of all *E. coli* strains tested and in the *S. aureus* plasmids. It can be speculated, at least in *E. coli*, that the plasmid-encoded *ars* operons have originated from the chromosomal operons. Hence, the acquisition of the *arsA* gene as well as the transfer of the operon from chromosome to a mobile element, such as a plasmid, are both factors that provide a greater selective advantage to the population harboring such an element.

17.4

Other Arsenic Transporters

Efflux-related mechanisms for arsenic resistance are also present in eukaryotic cells, including yeast, protozoa, and human cells, although ArsB homologs have not been found in these organisms. In yeast, *Saccharomyces cerevisiae*, arsenic resistance is coded by genes of the *ACR* (arsenic compounds resistance) family [56]. *ACR3* gene encodes an arsenite efflux protein of 45.8 kDa containing 10 transmembrane domains [56]. It carries out efflux of arsenite but not of antimonite. *ACR3* lies in a locus with *ACR1*, encoding a transcription factor unrelated to ArsR or ArsD, and *ACR2* that codes for an arsenate reductase [57]. Incidentally, Acr3p is homologous to a protein YqcL in *B. subtilis* [58]. YqcL is a membrane protein which is unrelated to ArsB, hence it forms an atypical arsenite transporter in *B. subtilis* in addition to the typical ArsB encoded by the *arsRB* operon also present in the *B. subtilis* chromosome [55]. Gene *yqcL* in *B. subtilis* occurs in an operon with three other genes, *yqcJ* (*E. coli* ArsR homolog), *yqcM* (*S. aureus* ArsC homolog), and *yqcK* (no known homologs). In addition to the *ACR* genes in yeast, arsenite resistance in eukaryotic cells is also known to be conferred by ABC proteins, such as MRP1 (multidrug resistance-associated protein) in human cells [59] and Ycf1p (yeast cadmium factor protein) in *S. cerevisiae* [60]. MRP1, normally involved in export of leukotriene–GSH conjugates [61], has also been shown to catalyze export of arsenite-triglutathione conjugates in human cells [62]. Hence, overexpression of MRP1 may provide a pathway for protecting human cells from arsenic and other heavy metals. Ycf1p, with 63% identity to MRP1, carries out export of glutathione conjugates of cadmium and other compounds from yeast cytosol into the yeast vacuole, thus conferring arsenite resistance. *S. cerevisiae* thus contains two parallel pathways for detoxification of arsenic, Acr3p, in the plasma membrane, couples export to membrane potential and Ycf1p, in the vacuolar membrane, couples ATP to accumulation of arsenite–triglutathione conjugates in the vacuole [63].

17.5
Conclusion

Arsenic resistance mechanisms have, no doubt, evolved more than once. No one solution seems to suffice for all organisms, even though efflux as a mode of resistance seems to be a common theme. Prokaryotes mostly seem to employ ArsB homologs to carry out the task. Eukaryotes, true to their nature, contain a variety of parallel pathways, thus exhibiting some degree of complexity. Homologs of the *E. coli* ArsB protein have not been found in eukaryotic cells. Instead, proteins of the ACR family and the MRP family are seen to be involved. Eukaryotic homologs have only been found for the bacterial ArsA protein. Interestingly, the eukaryotes carry an *arsA* gene which is half the size of the *E. coli* gene and encodes for only one NBD. This suggests that the bacterial and the eukaryotic *arsA* genes evolved from a common ancestor which was half the size of the present day bacterial gene. Hence, the *E. coli* gene appears to have arisen by a gene duplication of an ancestral gene half its size, followed by a fusion event (Fig. 6). Whether the eukaryotic ArsA homologs are involved in conferring arsenite resistance is not known, however, their identification is of paramount significance in understanding the origin and evolution of the *E. coli arsA* gene.

E. coli ArsA contains two active sites. Its activity is allosterically regulated by arsenite or antimonite binding. It has been proposed that the binding of the ligand acts as a novel allosteric switch that controls the accessibility of ATP to the A2 domain, thus effectively maintaining the enzyme at a unisite level of activity in the absence of the ligand and allowing cooperative action of the two sites in its presence. If this hypothesis holds true, this may become a paradigm for the mode of action of ligands in other proteins with multiple nucleotide binding sites.

Another interesting feature that emerges from studying microbial arsenic resistance is the presence in nature of *ars* operons of differing complexity. A basic operon codes for a simple pmf-driven transporter consisting of ArsB, which confers a low-level arsenic resistance. A more complex operon also carries a gene for the arsenate reductase, ArsC. Yet more complexity is seen in operons containing the *arsA* gene, thus forming a two-subunit ATP-driven ArsAB transporter. Hence, the evolutionary process leaves tell-tale signs of its progression, and if we analyze a broad enough sample of organisms, we can begin to understand how transport systems become more efficient and sophisticated.

References

1. Cullen, W. R., Reimer, K. J., *Chem. Rev.* **1989**, *89*, 713-764.
2. Kaur, P., Rosen, B. P., *Plasmid* **1992**, *27*, 29-40.
3. Kaur, P., *Anion Transport Systems* Vol. 23B, Jai Press, **1998**, Greenwich, CT.
4. Rosen, B. P., *Trends Microbiol.* **1999**, *7*, 207-212.
5. Bhattacharjee, H., Zhou, T., Li, J., Gatti, D. L., Walmsley, A. R., Rosen, B. P., *Biochem. Soc. Trans.* **2000**, *28*, 520-526.
6. Rosen, B. P., Bhattacharjee, H., Shi, W., *J. Bioenerg. Biomembr.* **1995**, *27*, 85-91.
7. Chen, Y., Rosen, B. P., *J. Biol. Chem.* **1997**, *272*, 14257-14262.
8. Oden, K. L., Gladysheva, T. B., Rosen, B. P., *Mol. Microbiol.* **1994**, *12*, 301-306.
9. Hsu, C. M., Rosen, B. P., *J. Biol. Chem.* **1989**, *264*, 17349-17354.
10. Wu, J., Rosen, B. P., *J. Biol. Chem.* **1993**, *268*, 52-58.
11. Wu, J., Rosen, B. P., *Mol. Microbiol.* **1993**, *8*, 615-623.
12. Shi, W., Wu, J., Rosen, B. P., *J. Biol. Chem.* **1994**, *269*, 19826-19829.
13. Walker, J. E., Saraste, M., Runswick, M. J., Gay, N. J., *EMBO J.* **1982**, *1*, 945-951.
14. Kaur, P., Rosen, B. P., *Biochemistry* **1994**, *33*, 6456-6461.
15. Bhattacharjee, H., Li, J., Ksenzenko, M. Y., Rosen, B. P., *J. Biol. Chem.* **1995**, *270*, 11245-11250.
16. Bhattacharjee, H., Rosen, B. P., *J. Biol. Chem.* **1996**, *271*, 24465-24470.
17. Karkaria, C. E., Chen, C. M., Rosen, B. P., *J. Biol. Chem.* **1990**, *265*, 7832-7836.
18. Kaur, P., Rosen, B. P., *J. Biol. Chem.* **1992**, *267*, 19272-19277.
19. Kaur, P., Rosen, B. P., *J. Bacteriol.* **1993**, *175*, 351-357.
20. Kaur, P., Rosen, B. P., *J. Biol. Chem.* **1994**, *269*, 9698-9704.
21. Ramaswamy, S., Kaur, P., *J. Biol. Chem.* **1998**, *273*, 9243-9248.
22. Kaur, P., *J. Biol. Chem.* **1999**, *274*, 25849-25854.
23. Rosen, B. P., Weigel, U., Karkaria, C., Gangola P., *J. Biol. Chem.* **1988**, *263*, 3067-3070.
24. Kierdaszuk, B., Eriksson, S., *Biochemistry* **1988**, *27*, 4952-4956.
25. Banks, G. R., Sedgwick, S. G., *Biochemistry* **1986**, *25*, 5882-5889.
26. Yue, V. T., Schimmel, P. R., *Biochemistry* **1977**, *16*, 4678-4684.

27. Abrahams, J. P., Leslie, A. G., Lutter, R., Walker, J. E., *Nature* **1994**, *370*, 621-628.
28. Boyer, P. D., *Annu. Rev. Biochem.* **1997**, *66*, 717-749.
29. Gottesman, M. M., Pastan, I., *Annu. Rev. Biochem.* **1993**, *62*, 385-427.
30. Urbatsch, I. L., Sankaran, B., Weber, J., Senior, A. E., *J. Biol. Chem.* **1995**, *270*, 19383-19390.
31. Hrycyna, C. A., Ramachandra, M., Ambudkar, S. V., Ko, Y. H., Pedersen, P. L., Pastan, I., Gottesman, M. M., *J. Biol. Chem.* **1998**, *273*, 16631-16634.
32. Urbatsch, I. L., Sankaran, B., Bhagat, S., Senior, A. E., *J. Biol. Chem.* **1995**, *270*, 26956-26961.
33. Milburn, M. V., Tong, L., deVos, A. M., Brunger, A., Yamaizumi, Z., Nishimura, S., Kim, S. H., *Science* **1990**, *247*, 939-945.
34. Story, R. M., Steitz, T. A., *Nature* **1992**, *355*, 374-376.
35. Zhou, T., Rosen, B. P., *J. Biol. Chem.* **1997**, *272*, 19731-19737.
36. Li, J., Rosen, B. P., *Mol. Microbiol.* **2000**, *35*, 361-367.
37. Saier, M. H., Jr., *Bioessays* **1994**, *16*, 23-29.
38. Bruhn, D. F., Li, J., Silver, S., Roberto, F., Rosen, B. P., *FEMS Microbiol. Lett.* **1996**, *139*, 149-153.
39. Suzuki, K., Wakao, N., Kimura, T., Sakka, K., Ohmiya, K., *Appl. Environ. Microbiol.* **1998**, *64*, 411-418.
40. Kurdi-Haidar, B., Heath, D., Aebi, S., Howell, S. B., *J. Biol. Chem.* **1998**, *273*, 22173-22176.
41. Olsen, G. J., Woese, C. R., *FASEB J.* **1993**, *7*, 113-123.
42. Ng, W. V., Ciufo, S. A., Smith, T. M., Bumgarner, R. E., Baskin, J., Faust, J., Hall, B., Loretz, C., Seto, J., Slagel, J., Hood, L., DasSarma, S., *Genome Res.* **1998**, *8*, 1131-1141.
43. Zhou, T., Radaev, S., Rosen, B. P., Gatti, D. L., *EMBO J.* **2000**, *19*, 4838-4845.
44. Wu, J., Tisa, L. S., Rosen, B. P., *J. Biol. Chem.* **1992**, *267*, 12570-12576.
45. Tisa, L. S., Rosen, B. P., *J. Biol. Chem.* **1990**, *265*, 190-194.
46. Owolabi, J. B., Rosen, B. P., *J. Bacteriol.* **1990**, *172*, 2367-2371.
47. Chen, Y., Dey, S., Rosen, B. P., *J. Bacteriol.* **1996**, *178*, 911-913.
48. Liu, J., Gladysheva, T. B., Lee, L., Rosen, B. P., *Biochemistry* **1995**, *34*, 13472-13476.
49. Ji, G., Silver, S., *J. Bacteriol.* **1992**, *174*, 3684-3694.
50. Rosenstein, R., Peschel, A., Wieland, B., Gotz, F., *J. Bacteriol.* **1992**, *174*, 3676-3683.
51. Ji, G., Garber, E. A., Armes, L. G., Chen, C. M., Fuchs, J. A., Silver, S., *Biochemistry* **1994**, *33*, 7294-7299.
52. Gladysheva, T. B., Oden, K. L., Rosen, B. P., *Biochemistry* **1994**, *33*, 7288-7293.
53. Dey, S., Rosen, B. P., *J. Bacteriol.* **1995**, *177*, 385-389.
54. Diorio, C., Cai, J., Marmor, J., Shinder, R., DuBow, M. S., *J. Bacteriol.* **1995**, *177*, 2050-2056.
55. Beloin, C., Ayora, S., Exley, R., Hirschbein, L., Ogasawara, N., Kasahara, Y., Alonso, J. C., Hegarat, F. L., *Mol. Gen. Genet.* **1997**, *256*, 63-71.
56. Wysocki, R., Bobrowicz, P., Ulaszewski, S., *J. Biol. Chem.* **1997**, *272*, 30061-30066.
57. Bobrowicz, P., Wysocki, R., Owsianik, G., Goffeau, A., Ulaszewski, S., *Yeast* **1997**, *13*, 819-828.
58. Sato, T., Kobayashi, Y., *J. Bacteriol.* **1998**, *180*, 1655-1661.
59. Borst, P., Schinkel, A. H., *Trends Genet.* **1997**, *13*, 217-222.
60. Li, Z. S., Lu, Y. P., Zhen, R. G., Szczypka, M., Thiele, D. J., Rea, P. A., *Proc. Natl. Acad. Sci. USA* **1997**, *94*, 42-47.
61. Deeley, R. G., Cole, S. P., *Semin. Cancer Biol.* **1997**, *8*, 193-204.
62. Cole, S. P., Sparks, K. E., Fraser, K., Loe, D. W., Grant, C. E., Wilson, G. M., Deeley, R. G., *Cancer Res.* **1994**, *54*, 5902-5910.
63. Ghosh, M., Shen, J., Rosen, B. P., *Proc. Natl. Acad. Sci. USA* **1999**, *96*, 5001-5006.

18
Microbial Nickel Transport

Thomas Eitinger

Abstract

The essential trace element nickel is incorporated into prokaryotic and eukaryotic metalloenzymes that are involved in energy- and nitrogen metabolism, and in the detoxification of harmful metabolites. On the other hand, nickel itself is toxic to various cellular targets. This ambivalent role requires precise control of its cellular concentration. Nickel homeostasis is affected by a variety of non-specific transporters involved in metal uptake or metal extrusion. Biosynthesis of nickel-containing enzymes in the natural environment depends on high-affinity uptake of nickel ion. A subclass of the ATP-binding cassette transport systems and a novel class of transition metal permeases fulfill this task in bacteria. Very limited data are available on the molecular mechanism of high-affinity nickel uptake in archaea and eukaryotes.

18.1
Introduction

Transport of transition metal ions across biomembranes is an important process in all organisms, since these ions are incorporated into apoproteins to form metalloenzymes. The latter are major players in redox and non-redox biochemical reactions. At elevated concentrations, however, transition elements are toxic and the cellular concentration of these metal ions must be adjusted within a narrow range in response to specific requirements. Membrane transporters mediating metal ion uptake into cells, sequestration in subcellular compartments, or expulsion from cells play a central role in metal homeostasis. These transporters fall into several classes of the "functional-phylogenetic classification system of transmembrane solute transporters" (TC, [1], Chap. 1, Fig. 1) Some types of metal transport systems seem to be restricted to certain groups of organisms, while others are widespread and have been identified in all kingdoms of life.

The present chapter focuses on the transition metal nickel, that is required for specific redox-, hydrolysis-, and isomerization reactions in both prokaryotes and eukaryotes. After an overview on the beneficial and hazardous roles of nickel in metabolism, I shall mention various types of metal transporters that directly or indirectly affect the incorporation of nickel ion into metalloenzymes. The central part of this review is dedicated to the structure, distribution, specificity, and regulation of microbial high-affinity nickel uptake systems that allow their hosts to produce nickel-dependent enzymes even in environments where the metal is present only in the nanomolar range.

18.2
Metabolic Roles of Nickel

The various aspects of nickel biochemistry have been the subject of a monography written by R. P. Hausinger in 1993 [2], and the reader is directed to this book and the references therein for a detailed overview of this huge field. The intention of this section is to give a short summary of nickel-dependent reactions including recent findings and to point out known or potential mechanisms of nickel toxicity and nickel resistance.

18.2.1
Nickel as a Cofactor of Metalloenzymes

Nickel is an essential trace element in both prokaryotes and eukaryotes because it is involved in urea hydrolysis, hydrogen oxidation and -production, anaerobic carbon monoxide metabolism, methanogenesis, superoxide detoxification, and most probably also in the conversion of toxic α-keto aldehydes (such as methylglyoxal) into non-toxic α-hydroxycarboxylic acids (Tab. 1). Ureases are the most widespread nickel metalloenzymes and have been identified in prokaryotes and eukaryotes [3, 4]. They allow their hosts to utilize urea, a substance being constantly released in large quantities into the environment by biological processes, as a source of nitrogen. In the model plant *Arabidopsis thaliana*, urease plays an essential role in seedling development [5]. Also, urease seems to be important for crop plants. Mutations in the *Eu2* and *Eu3* genes that prevent proper insertion of nickel ion into the embryo-specific and ubiquitous urease isoenzymes of soybean (*Glycine max*) result in a urease null-phenotype leading to urea accumulation and leaf necrosis [6]. Ureases are important virulence factors in bacterial and fungal pathogens. Bacterial ureases are involved in infections of the oral cavity, the stomach and duodenum, and the urinary tract of humans [7]. Likewise, many fungi pathogenic to humans have urease activity. The importance of urease in pathogenesis was shown for *Cryptococcus neoformans*, a heterothallic yeast that causes severe systemic infections including meningoencephalitis in immunodeficient patients. Comparison of a urease-negative mutant of *C. neoformans* and its parental strain in murine intravenous and inhalational infection models revealed significant differences in survival [8]. The best-investigated urease is the enzyme of the bacterium *Klebsiella aerogenes*. It contains two nickel ions per active site, which are coordinated by a carbamylated lysine, and by histidine and aspartate residues [9]. Very similar metal coordination has been reported for the urease of *Bacillus pasteurii* [10].

Hydrogenases are metalloproteins catalyzing the reversible cleavage of molecular hydrogen into protons and electrons. They oxidize hydrogen to generate an electrochemical potential and/or reducing power in many prokaryotes such as aerobic chemolithoautotrophs, methanogens, acetogens, nitrogen fixers, sulfate reducers, and photosynthetic bacteria, or evolve hydrogen during fermentation or as a redox valve in photosynthesis. Most of these enzymes are structurally related and contain nickel and iron at their active site [11]. An unrelated group (Fe-only hydrogenases) lacks the nickel ion and contains a novel [Fe–S] cluster at the catalytic center. CN^- and CO as unusual iron ligands have been identified by crystal structure analyses and infrared spectroscopy in both types of hydrogenase [12–14].

Nickel-containing carbon monoxide dehydrogenases (CODH) reversibly convert carbon monoxide and water into carbon dioxide and reducing equivalents [15]. CODH of the purple bacterium *Rhodospirillum rubrum* is the best-investigated member of this class of enzymes. It enables *R. rubrum* to grow on carbon monoxide as the energy source in the dark. Recent biochemical, spectroscopic and crystallographic analyses have revealed that *R. rubrum* CODH is a dimer which contains five metal clusters. The active site (cluster C) consists of a [Ni–Fe–S] cluster and contains – like in hydrogenases – at least one CO molecule as a ligand to a Fe atom [16, 17]. More complex forms of nickel-containing CODH are the acetyl-

Tab. 1. Metabolic roles of nickel

A. Cofactor of metalloenzymes

Ureases (archaea, bacteria, algae, fungi, plants, invertebrates)
[NiFe] **Hydrogenases** (archaea, bacteria, algae)
[Ni] **Carbon monoxide dehydrogenases** (archaea, bacteria)
Coenzyme F_{430} (archaea)
[Ni] **Superoxide dismutases** (*Streptomyces* sp., ...?)
Glyoxalase I (*E. coli*, ...?)

B. Hazardous effects at elevated concentrations

Production of **reactive oxygen species** leading to **lipid peroxidation** and **damage** of **nucleic acids** and **proteins**
Interference with **chromatin condensation and methylation** leading to **gene silencing** and **carcinogenesis**

C. Resistance mechanisms

Export from the cell (bacteria)
Complexation with organic acids and **vacuolar sequestration** (yeasts, plants)
Extracellular sequestration (yeasts)

CoA synthetases. In addition to the reversible interconversion of CO and CO_2, these enzymes build or cleave a carbon–carbon bond between the methyl- and the carbonyl group of the acetyl moiety in acetyl-coenzyme A. They play a central role in anaerobic acetotrophic archaea and bacteria, in acetogens, and in methanogens [15].

Methanogens require nickel for incorporation into hydrogenases and acetyl-CoA synthetase. The final step of methanogenesis, the reduction of the methyl group (bound to coenzyme M) to methane is catalyzed by methyl-CoM reductase. This enzyme contains the essential prosthetic group F_{430}, a nickel-containing, highly saturated, cyclic tetrapyrrole [18–20].

Nickel-dependent superoxide dismutases (SOD) that convert toxic superoxide anion radicals into peroxide and molecular oxygen were more recently discovered in various gram-positive, mycelium-forming soil bacteria of the genus *Streptomyces* [21–23]. These organisms are capable of producing a NiSOD and an iron- and zinc-containing SOD in response to metal availability. In *Streptomyces coelicolor* production of the NiSOD is strongly stimulated in the presence of nickel ion both on the transcriptional and posttranslational level [24]. In contrast, transcription of *sodF* encoding FeZnSOD is specifically inhibited by Ni^{2+} [25, 26]. Similar regulation was uncovered in a study on *S. griseus* that identified a dyad-symmetry sequence from −2 to +15 as an operator site involved in nickel-mediated transcriptional repression of *sodF* [27]. The postulated nickel-responsive repressor has not yet been identified. Data on the crystal structure are not available for any NiSOD. Examination of NiSOD of *S. seoulensis* in the "as isolated" state by X-ray absorption- and electron paramagnetic resonance spectroscopy identified a novel class of five-coordinate nickel center with S-, N- and/or O-donor ligands [28].

Recent studies on the detoxification of the natural metabolite methylglyoxal suggest that nickel can be involved in this process. Several routes of methylglyoxal synthesis are known including a side reaction of triose phosphate isomerase and in particular the production from dihydroxyacetone phosphate by methylglyoxal synthase [29]. Methylglyoxal synthase is a widespread enzyme and methylglyoxal production seems to play a significant role in the balance of the glycolytic carbon flux. On the other hand, methylglyoxal is a toxic electrophile that can react with macromolecules and lead to cell death. A major detoxification mechanism comprises the spontaneous reaction of methylglyoxal with glutathione, the glyoxalase I-catalyzed isomerization of the resulting hemithiolacetal to give S-lactoylglutathione, and finally, the glyoxalase II-mediated formation of glutathione and D-lactate. A common feature of glyoxalase I of *Pseudomonas putida*, baker's yeast and humans is the requirement of zinc as a metal cofactor. In contrast, glyoxalase I of *Escherichia coli* is almost completely inactive in the presence of zinc, but is maximally activated by nickel ion [30]. Analysis of the crystal structure of *E. coli* glyoxalase I revealed that two nickel ions are present per homodimer. Each of the nickel ions is coordinated octahedrally by a histidine- and a glutamate residue in the amino-terminal domain of monomer 1, a histidine- and a glutamate residue in the carboxyl-terminal domain of monomer 2, and two water molecules [31].

18.2.2
Nickel Toxicity

Although it is well known and can easily be demonstrated in the laboratory that — in the absence of specific resistance mechanisms (see Sect. 18.2.3) — nickel salts at high concentrations inhibit the growth of microorganisms and kill the cells, surprisingly little is known on the exact biochemical reactions behind nickel toxicity in prokaryotes. Various aspects of nickel toxicity including carcinogenesis, however, have been addressed in eukaryotic systems. Nickel is a cause of allergic contact dermatitis. All insoluble nickel compounds except for metallic nickel are considered as human carcinogens. Both epigenetic and genotoxic mechanisms have been discussed as the basis of nickel-caused carcinogenesis [32, 33]. Epigenetic mechanisms include modification of the expression pattern of transcription factors, interference with cellular Ca^{2+} metabolism, and alteration of DNA methylation and chromatin condensation in heterochromatic regions that can silence neighboring genes including tumor suppressor genes [32, 34]. Nickel was shown to interact with histone proteins. Ni^{2+} strongly stimulates specific hydrolysis of the carboxyl-terminal domain of histone H2A. An eight-residue-long peptide modeling the cutoff product of H2A hydrolysis was shown to form a redox-active complex with Ni^{2+} that promoted oxidative damage of plasmid pUC19 DNA by H_2O_2 [35]. This result is in agreement with the finding that Ni(III) oxides and certain nickel complexes can induce Fenton-like reactions resulting in the production of reactive hydroxyl radicals from hydrogen peroxide. Hydroxyl radical is one of the most popular candidates for the initiation of lipid peroxidation, that is the incorporation of molecular oxygen into polyunsaturated fatty acids to form lipid hydroperoxides [36]. Likewise, hydroxyl radical is considered as a major player in the formation of 8-oxo-7,8-dihydroguanosin, the most common oxidative reaction product of DNA bases, that causes G→T transversions [37]. In addition, Ni^{2+} is an inhibitor of the *E. coli* MutT and human MTH1 8-Oxo-7,8-dihydro-2′-deoxyguanosin-5′-triphosphatases that eliminate 8-oxo-dGTP from the nucleotide pool and thus, prevent its incorporation into DNA [38].

18.2.3
Nickel Resistance

Resistance mechanisms, that prevent the nickel-induced damage to cellular targets outlined in the previous section, have been described for bacteria, yeast and plants (Tab. 1). In bacteria, resistance is mainly mediated by export systems that expel Ni^{2+} ion from the cells. Among these transporters are members of the resistance-nodulation-cell division (RND) permease superfamily and the major facilitator superfamily (MFS) [39–41] (see Sect. 18.3). Eukaryotic nickel resistance mechanisms include complexation of the metal with organic acids and vacuolar sequestration in yeast and plants, and extracellular sequestration in yeast. Growth of baker's yeast *Saccharomyces cerevisiae* on complex medium is inhibited by Ni^{2+} at a concentration of 2.5 mM at pH below 5 [42]. The same study demonstrated that toxicity of nickel and other metals is exacerbated in strains with defects in histidine biosynthesis. *his3* mutants lacking imidazole-glycerolphosphate dehydratase were unable to grow on complex medium in the presence of 2.5 mM NiCl$_2$ unless the *His3$^+$* gene was introduced on a plasmid and the pH was raised to 6.3. Metal toxicity was enhanced by the lack of the vacuolar H$^+$-ATPase suggesting that a functional vacuole is required for resistance. Based on these and previous results which showed that histidine accumulates in the yeast vacuole far more than other amino acids [43], Pearce and Sherman [42] concluded that vacuolar accumulation of histidine is a normal process for detoxification of nickel and other transition metals. In accordance with this assumption, Nishimura et al. [44] observed proton gradient-driven nickel transport into right-side-out vacuolar membrane vesicles of *S. cerevisiae*.

Overproduction of histidine in response to the addition of nickel was reported as a resistance mechanism in the nickel hyperaccumulating plants *Alyssum lesbiacum*, *A. bertolonii*, and *A. murale* that are members of the Brassicaceae [45]. Addition of nickel to *A. lesbiacum* – in contrast to the non-accumulator *A. montanum* – led to a 35-fold increase of the histidine concentration in xylem sap. As a consequence, *A. lesbiacum* tolerated up to 1 mM Ni^{2+} in the hydroponic solution, about two orders of magnitude more than *A. montanum*. Thus, transport of Ni–histidine complex through the xylem into the aerial parts and sequestration in vacuoles of leaf cells is considered as a major nickel resistance mechanism [45]. Nevertheless, additional resistance mechanisms must exist in nickel hyperaccumulators among the Brassicaceae. A study on the effect of nickel on histidine biosynthesis in the hyperaccumulator *Thlaspi goesingense* failed to demonstrate any regulation of the histidine biosynthetic enzymes ATP phosphoribosyltransferase, imidazole-glycerolphosphate dehydratase, and histidinol dehydrogenase. Likewise, no increase of the concentration of histidine in root, shoot, or xylem sap was observed in response to Ni exposure suggesting that an alternative mechanism determines nickel resistance in this organism [46]. Comparison of the nickel content of vacuoles of *T. goesingense* and of the non-tolerant non-accumulator *T. arvense* revealed twice as much nickel in the accumulator when both plants were grown under moderate nickel exposure. The majority of nickel in the leafs of *T. goesingense* was associated with the cell wall, the remaining nickel was complexed with citrate and histidine [47]. The mechanistic details of nickel tolerance in this species need to be elucidated.

An alternative mechanism, that confers resistance to various toxic compounds including transition metal salts, was uncovered in a study on methylglyoxal resistance in *S. cerevisiae* [48]. Homologous expression of a plasmid-borne copy of the methylglyoxal reductase gene, encoding an NADPH-dependent enzyme that reduces methylglyoxal to lactaldehyde, led – for unknown reasons – to the overproduction and excretion of glutathione. As a consequence, these yeast cells were able to grow on supplemented minimal agar plates in the presence of 10 mM NiCl$_2$, whereas the parental strain was not tolerant to concentrations above 300 µM. The data suggest that extracellular sequestration prevents Ni^{2+} uptake into the recombinant yeast cells.

18.3
Transport Systems Involved in Nickel Homeostasis

The previous sections provided an overview on the ambivalent function of nickel in metabolism summarizing its significance as an essential trace element and as a toxic heavy metal. The present section will briefly describe various membrane transport systems from different organisms and cell types that affect the cellular nickel availability directly by uptake or export of Ni^{2+}, or indirectly by import or extrusion of related transition metal cations that can interfer with cellular nickel metabolism (Fig. 1). High-affinity Ni^{2+} uptake systems that are essential for the biosynthesis of nickel-dependent enzymes under natural conditions are discussed more extensively in Sect. 18.4.

Uptake of Ni^{2+} ion by non-specific magnesium transport systems has been described for bacteria, archaea, and yeasts. A detailed description of magnesium transport systems is presented in Chap. 15). The major Mg^{2+} transporter in bacteria and archaea is CorA, a constitutively synthesized uptake system with high capacity and low specificity, that transports a variety of divalent cations including

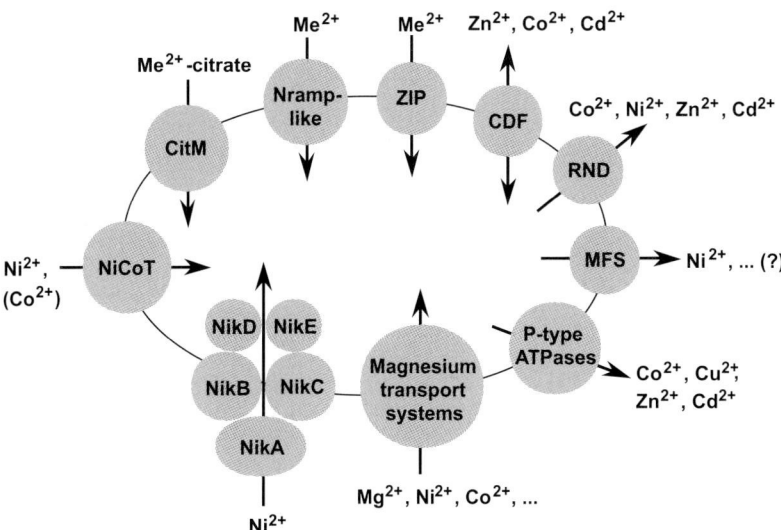

Fig. 1. Transport systems in bacterial and eukaryotic plasma membranes or organellar membranes affecting nickel homeostasis (redrawn and modified from reference [80] with permission of the publisher). See text for details. NiCoT, nickel/cobalt transporter family (TC 2.A.52); CitM, metal–citrate transporter (TC 2.A.11); Nramp, natural resistance-associated macrophage protein family (TC 2.A.55); ZIP, Zrt- and Irt-related proteins (TC 2.A.5); CDF, cation diffusion facilitator family (TC 2.A.4); RND, resistance-nodulation-cell division superfamily (TC 2.A.6); MFS, major facilitator superfamily (TC 2.A.1); P-type ATPase superfamily (TC 3.A.3); magnesium transport systems (CorA, TC 9.A.17; MgtA, MgtB, TC 3.A.3; MgtE, TC 9.A.19); Nik, Ni^{2+}-specific ABC-type transporter (TC 3.A.1.5.3). TC, transporter classification system according to reference [1].

Ni^{2+}. Less widespread Mg^{2+} transporters are the P-type ATPases MgtA and MgtB, and the non-related MgtE system [49]. The physiological role of these transporters is uptake of Mg^{2+} ion. Transport of transition metal ions by these systems is likely to play a minor role in the natural environment, since the affinity of the Mg^{2+} transporters for these ions is far below physiological requirements. At unphysiologically high concentrations of metal ions, however, these systems are major contributors to transition metal toxicity. corA-like sequences have been identified in yeasts during the *Saccharomyces cerevisiae* and *Schizosaccharomyces pombe* genome sequencing projects; they have only recently been discovered in higher eukaryotes including plants and humans. The eukaryotic CorA-like proteins are believed to mediate Mg^{2+} transport across plasma and organeller membranes [50]. Transport of transition metals by Mg^{2+} transporters was reported for both *S. cerevisiae* and *S. pombe*. The absence of the two CorA-like proteins Alr1p and Alr2p in *S. cerevisiae* mutants produced a magnesium-deficient phenotype while conferring resistance to several metals including copper, manganese, nickel, and zinc [51]. A study on nickel uptake in *S. pombe* clearly demonstrated that excess amounts of Mg^{2+} strongly reduce Ni^{2+} uptake to a residual level. The remaining activity is due to Nic1p, a high-affinity nickel permease that is not inhibited by Mg^{2+} [52] (see Sect. 18.4.2).

Transport of nickel ions in complex with citrate anions was reported for the CitM protein of *Bacillus subtilis* [53]. CitM transports citrate in complex with Mg^{2+}, Ni^{2+}, Mn^{2+}, Co^{2+}, and Zn^{2+}, it does not transport free citrate molecules. Thus, magnesium and transition metal uptake by CitM could be significant for the cells [53]. On the other hand, expression of *citM* is under control of carbon catabolite repression and citrate induction suggesting a primary role of CitM in citrate utilization [54]. This view is corroborated by the kinetic parameters of CitM for uptake of the metal–citrate complexes [53]. While the affinity constant for the magnesium–citrate complex (63 µM) is within the range of Mg^{2+} found in the natural environment, that for the respective nickel complex (43 µM) and other transition metal complexes (35–40 µM) exceeds by far environmental concentrations.

Nramp (for "natural resistance-associated macrophage protein") proteins comprise a family of metal transporters whose members have been identified in phylogenetically non-related organisms like mammals (including humans), nematodes, insects, yeasts, plants, and bacteria [55–58]. The physiological function of Nramp proteins is not entirely clear. Nramp1 is located in the endosomal/lysosomal compartment of macrophages. Its absence in mice leads to susceptibility to unrelated intracellular pathogens including the parasite *Leishmania donovani* and the bacteria *Mycobacterium bovis* and *Salmonella typhimurium*. It has been suggested that Nramp1 removes divalent transition metal ions from the phagosome and thus, prevents the pathogens from formation of metalloenzymes (like superoxide dismutase) that protect against the attack of the macrophage by superoxide and other radicals [59]. Nramp2 is the major non-transferrin-dependent iron transporter of mammalian intestine and was shown to transport Zn^{2+}, Mn^{2+}, Co^{2+}, Cd^{2+}, Cu^{2+}, Pb^{2+}, and Ni^{2+}, in addition to Fe^{2+} [60]. Similar selectivity was observed for Smf1p, one of the three Nramp isologs in *S. cerevisiae*. In a more recent study [61] on the function of the Smf transporters in yeast, disruptants with lesions in *SMF1*, *SMF2*, *SMF3*, and all combinations were constructed and analyzed. The investigation revealed that Smf1p and Smf2p are located in the plasma membrane and transport primarily Mn^{2+}, Cu^{2+}, and Fe^{2+} (Smf1p) and Mn^{2+} and Cu^{2+} (Smf2p). Smf3p may function at the Golgi or at post-Golgi vesicles. Its substrate spectrum needs to be established. The function of the bacterial Nramp isologs of *Mycobacterium tuberculosis* [62], *Escherichia coli* [56, 57], and *Salmonella typhimurium* [56] has been investigated experimentally. *M. tuberculosis* Nramp is considered to be a non-specific metal transporter that interacts with Zn^{2+}, Fe^{2+}, Mn^{2+}, and Cu^{2+}. Investigation of the Nramp isolog of *E. coli* (MntH) showed that it acts as a selective divalent metal transporter with highest preference for Mn^{2+} and Fe^{2+}. It is capable of transporting Ni^{2+} and other transition metals [57]. A more detailed analysis of the kinetic parameters of MntH of *E. coli* and *S. typhimurium* suggested, however, that MntH is a selective Mn^{2+} transporter. This conclusion is based on the observation that the affinity constant for Mn^{2+} is approximately 100 nM, while the inhibitory constants for inhibition of $^{54}Mn^{2+}$ uptake by other metal ions including Ni^{2+} were in the range of 20–100 µM or even higher, and are thus too high to be physiologically relevant [56].

The ZIP family comprises zinc- and iron-regulated transporter proteins of protists, fungi, plants, and animals (including humans) [63, 64]. These transporters are predicted to have eight transmembrane segments (TMS) and a histidine-rich stretch in a variable region between TMS 3 and TMS 4. ZIP systems are involved in the transport of a variety of cations including zinc, iron, manganese, and cadmium, and can have extremely low affinity constants (e.g., 10 nM Zn^{2+} for *S. cerevisiae* Zrt1p, [65]) for their substrates. ZIP transporters may be involved in the uptake of nickel ion. In a study on iron uptake in tomato (*Lycopersicon esculentum*), the cDNAs for two iron-regulated ZIP transporters (LeIRT1 and LeIRT2) have been isolated from a library obtained with iron-deficient tomato roots [66]. The cDNAs were individually expressed in *S. cerevisiae* mutants and shown to complement iron-, manganese-, zinc-, and copper deficiency in strains with lesions in the respective transporter genes. LeIRT1- and LeIRT2-mediated Fe^{2+} transport in a yeast strain lacking its high-affinity and low-affinity iron transporters was strongly inhibited by a tenfold excess of Ni^{2+} suggesting that both tomato transporters may be capable of transporting nickel ion [66].

Various types of export systems confer resistance to heavy metals in bacterial cells [67]. The list includes members of the resistance-nodulation-cell-division superfamily (RND), major facilitator superfamily (MFS), cation diffusion facilitator family (CDF), and a subgroup of the P-type ATPases. Intracellular nickel availability is not only affected by those systems that extrude Ni^{2+} ion from the cells, but is also influenced by transporters that export other metals resulting in a relative increase in nickel. This is important, since nickel incorporation into nickel-dependent enzymes is impaired by unbalanced amounts of related transition metals (see Sect. 18.4.2.2). Although nickel metallocenter formation [68, 69] is assisted by auxiliary proteins that are responsible for the incorporation of non-protein ligands and especially for metal specificity, the process depends on balanced intracellular metal concentrations. The RND-type Ni^{2+} export systems Cnr (cobalt–nickel resistance, [70]) and Ncc (nickel–cobalt–cadmium resistance, [71]) have been identified in *Ralstonia metallidurans* CH34 and *Achromobacter xylosoxidans* 31A, respectively. These systems are closely related to each other and to Czc (cobalt–zinc–cadmium resistance) of *R. metallidurans* CH34. CnrA, CzcA and NccA (the RND proteins) are the central parts of these multicomponent exporters. CzcA was shown to operate as a cation/proton antiporter [72]. The function of CzcB and CzcC (like that of CnrB, CnrC and NccB, NccC) is not entirely clear. RND-type exporters confer resistance to mM levels of heavy metal cations. Expression of the *cnr* and *czc* genes is regulated on the transcriptional level by metal induction [73, 74]. Genome sequence analyses have identified related RND systems with a putative function in cation export in various gram-negative bacteria [39, 67].

A putative MFS-type nickel exporter (NrsD) has been analyzed in the cyanobacterium *Synechocystis* sp. strain PCC 6803 [41]. NrsD is predicted to have 12 transmembrane domains and contains a histidine-rich metal binding motif at its carboxyl terminus. It shows significant amino acid sequence identity to NreB, a product of the low-level nickel resistance operon in *Achromobacter xylosoxidans* 31A [71] that is different from the aforementioned high-level nickel resistance (*ncc*) operon.

Cation diffusion facilitators mediating export or import of zinc, cobalt, and cadmium have been identified in archaea, bacteria, yeasts, plants, and animals [58, 75]. Certain members of the family, like Cot1p of *S. cerevisiae*, are located to intracellular membranes and may confer resistance by transport of metal ions into vacuoles. CzcD of *R. metallidurans* CH34 is thought to function as a low-level metal exporter that stimulates transcription of the *czc* genes via a sensory system including an extracellular-function sigma factor [76].

P-type ATPases of the CPx subgroup are involved in uptake and export of monovalent and divalent metal cations in prokaryotes and eukaryotes and play an important role in metal homeostasis [58, 77, 78]. The human hereditary disorders Menkes syndrome and Wilson disease, e.g., correlate with defects in the Cu(I)-translocating P-type ATPases ATP7A and ATP7B that are essential for copper metabolism. The CPx-type ATPases contain metal binding motifs in the hydrophilic amino-terminal domain, a conserved aspartate residue that is phosphorylated during the transport process, and a transmembrane CPC (or CPH, CPS, SPC) motif that could be involved in metal translocation. The significance of metal-exporting CPx-type ATPases for the biosynthesis of a nickel-containing enzyme is discussed in Sect. 18.4.2.2 choosing urease in *Helicobacter pylori* as an example.

18.4
High-affinity Nickel Uptake Systems

Although various types of transporters can be involved in nickel uptake under certain experimental conditions, the biosynthesis of nickel-dependent enzymes in the natural environment depends on highly specific transport systems with an affinity for Ni^{2+} in the nM range. The present section focuses on those high-affinity nickel uptake systems for which molecular biological data are available. Two major types can be distinguished from each other [79, 80]: An ABC-type nickel transporter belonging to the peptide/opine transporter family (TC 3.A.1.5) has been identified in *E. coli*, and related systems seem to be present in other bacteria. The vast majority of nickel transporters identified to date are independent of ATP hydrolysis as an energy source. They form a novel class of permeases, the nickel/cobalt transporter family (TC 2.A.52), whose members are found in gram-negative and gram-positive bacteria. Recently, this type of transporter has been identified in fission yeast [52] and in the euryarchaeon *Thermoplasma acidophilum* [81], and thus, it is present in all kingdoms of life.

18.4.1
ABC-type Nickel Transporters

18.4.1.1 The Nik System of *Escherichia coli*
The *nik* operon was discovered during complementation analyses of a certain class of nickel-deficient hydrogenase mutants of *E. coli*. It encodes a periplasmic binding protein (NikA), two integral membrane proteins (NikB, NikC) and two peripheral membrane proteins containing the motifs typical of ATP-binding cassette transporters [82]. The Nik proteins are most closely related to dipeptide and oligopeptide transporters and it is difficult or even impossible to predict the substrate spectrum of these systems based on amino sequence comparisons. NikA is considered to play a dual role: It is the primary nickel binder in the transport process and it is involved in nickel repellent signaling by the Tar chemotaxis receptor. NikA has been purified to homogeneity and, as expected for the binding protein of a high-affinity nickel transporter, binds Ni^{2+} ion with a very low dissociation constant of less than 100 nM [83]. Expression of the *nik* genes is under dual control. Transcription is stimulated by Fnr, a global regulator of anaerobic metabolism [84]. In addition, the *nik* genes are negatively controlled by the nickel-responsive repressor NikR. The *nikR* gene has been identified adjacent to the *nikABCDE* operon and contains its own promoter [85]. NikR is a novel member of the ribbon-helix-helix family of transcription factors [86] that includes DNA-binding proteins affecting bacteriophage development, methionine biosynthesis, plasmid copy number, conjugative DNA transfer, and alginate production. Purified NikR binds as a dimer to a dyad-symmetric sequence that overlaps with the putative transcriptional start point upstream of the *nikA* reading frame. While the DNA-binding domain is located in the amino-terminal segment, the carboxyl-terminal region is considered to act as a nickel sensor [87]. NikR homologs are widespread among bacteria and archaea (including methanogens). The role of these proteins remains speculative since experimental data on their function are not available.

18.4.1.2 Nik-related Transporters in Prokaryotes

As already mentioned in the previous section, predictions on the physiological role of Nik-related ABC transporters are critical and depend on experimental analyses. Nevertheless, it is likely that additional nickel-specific ABC transporters exist in prokaryotes. NikA homologs displaying from 40 to more than 70% identity on the amino acid level are present in the enteric bacteria *Klebsiella pneumoniae* and *Salmonella* strains, in *Pseudomonas putida*, and in the alkaliphilic spore former *Bacillus halodurans*. Based on preliminary genome sequence data, the structure of the *K. pneumoniae nikABCDE* operon with an adjacent *nikR* gene seems to be very similar to its counterpart in *E. coli*. Moreover, a putative operator site for NikR binding is present upstream of the *K. pneumoniae nikA* [87]. This organism contains urease and has been shown to produce a nickel-dependent, pyridine nucleotide-reducing hydrogenase during citrate fermentation [88]. The *nik*-like operon in the urease producer *B. halodurans* has been considered to encode a nickel transporter [89]. A close relative of *E. coli* NikR is not present in this bacterium.

Investigation of the DNA region containing a hemolysin gene (*trh*) in various clinical isolates of the human pathogen *Vibrio parahaemolyticus* revealed that a *nikABCDE* operon is located between *trh* and the urease operon [90]. Interestingly, the *nik* and the *ure* genes are oriented divergently. A *ureR* gene encoding a putative urea-responsive inducer of the *ure* cluster is located between *nik* and *trh*. *nik*, *ureR*, and *trh* are organized in the same orientation. The structure of this gene region suggests that UreR could be involved in urea-dependent induction of both the *ure* and the *nik* operons. Mutants with individual insertions in *ureC*, *nikD*, *ureR*, and *trh* have been constructed. Disruption of *ureC*, *nikD*, or *ureR* abolishes urease activity but does not affect hemolytic activity. Vice versa, the *trh* mutation prevents hemolysin formation but does not impair urease activity [90]. It remains to be established whether or not nickel transport and urease activity play a role in enterotoxicity.

18.4.2
The Nickel/Cobalt Transporter Family

18.4.2.1 Signature Motifs

The first member (HoxN) of the "nickel/cobalt transporter (NiCoT)" family was identified in the gram-negative aquatic and soil bacterium *Ralstonia eutropha* H16 [91]. Later, isologous transporters have been identified in a number of gram-negative and gram-positive bacteria [80] and recently also in a yeast [52] and in an archaeon [81]. In spite of the family name, HoxN does not transport Co^{2+} (or other divalent metal ions) but is highly specific and has a very low affinity constant (approximately 20 nM) for Ni^{2+} ion ([92]; see Fig. 4). NiCoTs consist of 331 to 405 amino acid residues and display a very similar hydropathy profile ([80], Fig. 2). Since the hydropathy profile represents a link between the amino acid sequence and the three-dimensional structure of a membrane protein [93], this result suggests that the tertiary structures of the NiCoTs are closely related. Topological analyses of HoxN and NixA, the NiCoT of *Helicobacter pylori*, correspondingly

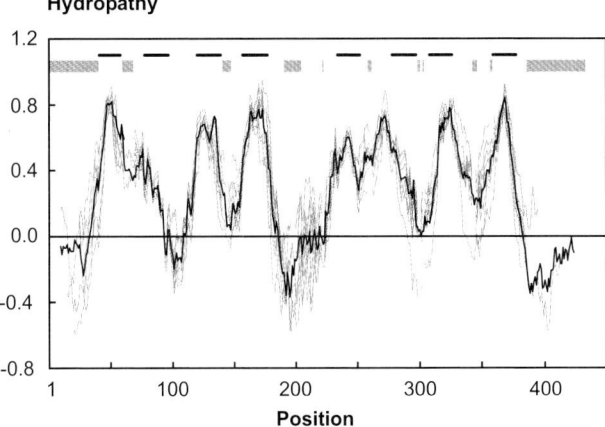

Fig. 2. Hydropathy profile alignment of 10 members of the nickel/cobalt transporter family (reprinted from reference [80], with permission of the publisher). The individual hydropathy profiles are shown as thin lines. The bold line represents the family profile. Horizontal bars show putative transmembrane helices and vertical bars represent gaps that occur at any place in the alignment.

identified eight transmembrane domains [94–96]. Four characteristic amino acid signatures located in transmembrane segments (TMS) are conserved in all members of the NiCoT family, as follows [80]: The motifs (R/K)HAXDADH(I/L) (TMS 2), FXXGHS(T/S)(V/I)V (TMS 3), LGX(D/E)T(A/S)(T/S)E (TMS 5), and GMXXXD(T/S)XD (TMS 6) have been shown to be critical for transport activity [80, 95, 97].

18.4.2.2 Significance in Microorganisms

Experimental analyses have revealed that members of the NiCoT family play a crucial role in microbial biosyntheses of nickel- and cobalt-dependent metalloenzymes (Fig. 3). In the absence of the respective transporter, *Bradyrhizobium japonicum* [98] and *Ralstonia eutropha* [91] are unable to produce enzymatically active [NiFe] hydrogenases under physiological conditions. Likewise, in *R. eutropha* [91], *Helicobacter pylori* [99], and the fission yeast *Schizosaccharomyces pombe* [52], urease activity depends on a functional NiCoT. Cobalt transport activity has been proved for the NhlF protein of the actinomycete *Rhodococcus rhodochrous* J1 [100]. This organism produces two types of non-corrin cobalt-containing nitrile hydratases that are used as industrial catalysts for the conversion of aliphatic and aromatic nitriles into their respective amides [101]. NhlF was originally considered as a selective Co^{2+} transporter [100]. A subsequent study showed, however, that NhlF transports both Co^{2+} and Ni^{2+} ion with very high affinity ([92]; Fig. 4).

Metal transport and urease production in *H. pylori* deserve a special comment. *H. pylori* produces various virulence factors including urease and causes severe gastroduodenal diseases in humans [102]. Moreover, a recent (preliminary and controversial) investigation associated *H. pylori* infection of stomach, trachea, and lung

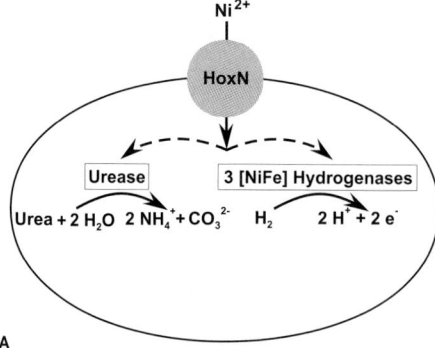
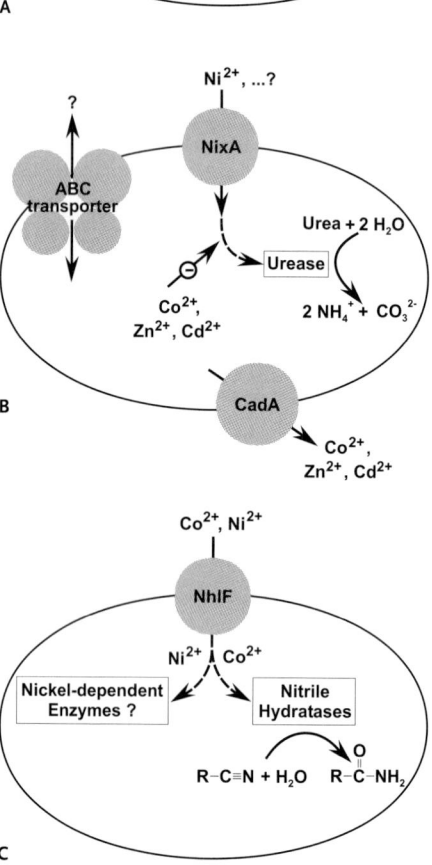

Fig. 3. Metabolic functions of the nickel/cobalt transporters HoxN, NixA, and NhlF in the gram-negative bacteria *Ralstonia eutropha* (**A**) and *Helicobacter pylori* (**B**) and in the gram-positive actinomycete *Rhodococcus rhodochrous* J1 (**C**), respectively.

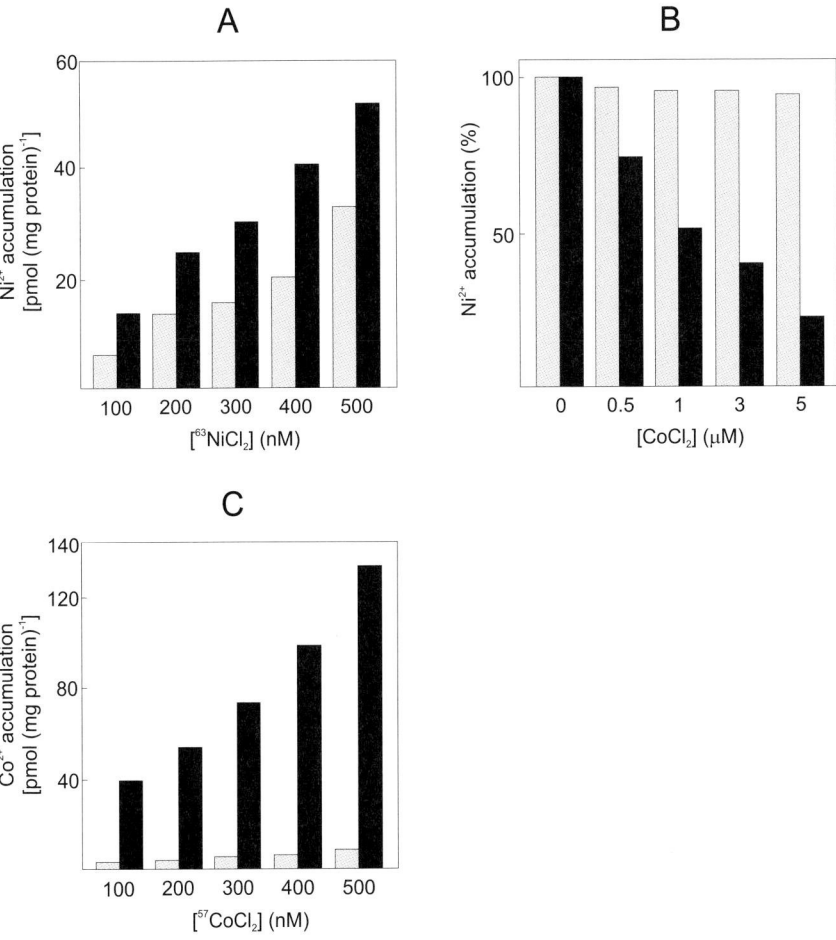

Fig. 4. Selective $^{63}Ni^{2+}$ (**A**, **B**) and $^{57}Co^{2+}$ (**C**) accumulation of *Escherichia coli* cells expressing *hoxN* (shaded boxes) and *nhlF* (solid boxes) during growth in complex medium. The concentration of $^{63}NiCl_2$ in (**B**) was 500 nM.

tissues with "sudden infant death syndrome" [103]. Urease is made in large quantities (about 6% of the cell protein) in this organism. Urease-mediated neutralizing of the acidic environment in the stomach lumen is essential for survival and a prerequisite for colonization of the gastric mucosa. High-affinity nickel uptake by NixA, a member of the NiCoT family ([104], Fig. 3), provides Ni^{2+} ion for intracellular incorporation into urease. Urease metallocenter formation is affected by additional (metal) transporters. Hendricks and Mobley [105] reported that an ABC transporter with limited similarity to the *E. coli* Nik system is important for urease activity in *H. pylori*. The substrate(s) of this transporter, that potentially lacks a periplasmic binding protein, are unknown. It could be involved in exporting divalent metal ions from the cytoplasm thereby preventing them from competing with Ni^{2+} for incorporation into urease. Such a function has been assigned to CadA, a Cd^{2+}–Zn^{2+}–Co^{2+}-exporting CPx-type ATPase in *H. pylori* that mediates heavy metal resistance and is important for high-level activity of urease [106].

A putative nickel transporter (UreH, UreI) of thermophilic *Bacillus* sp. TB-90 has features in common with the NiCoT family but lacks the eight-helix structure and two motifs that serve as family signatures [107]. *ureH* and *ureI* are linked to the urease gene cluster in this *Bacillus* species. When this cluster is expressed in *E. coli*, urease activity depends on *ureH* and *ureI*, unless the cells are grown in the presence of Ni^{2+} at unphysiologically high concentration. UreH consists of 228 amino acid residues and is predicted to contain six transmembrane domains (TMS). The conserved motifs found in TMS 2 and TMS 3 of NiCoTs are present with little deviation in putative TMS 1 and TMS 2 of UreH. UreI is a small protein consisting of only 65 amino acid residues and has no counterparts in the databases.

Data on the regulation of NiCoTs on the transcriptional or posttranslational level have not been reported. Interestingly, a putative operator structure that resembles the NikR-binding site upstream of the *E. coli nikA* reading frame (see Sect. 18.4.1.1) is located upstream of *hupN* encoding a NiCoT in *Bradyrhizobium japonicum* [87]. Due to the lack of experimental results, transcriptional control of NiCoT genes by metal-responsive repressor molecules remains speculative.

18.4.2.3 Substrate Specificity

Members of the NiCoT family display different selectivity among divalent transition metal cations. This has been demonstrated in a comparative analysis of the *R. eutropha* HoxN and the *R. rhodochrous* NhlF upon expression of the transporter genes in *E. coli* [92]. This study took advantage of a previous finding that heterologous expression of *hoxN* allows reproducible measurement of HoxN activity [108]. The different substrate specificity is illustrated in Fig. 4. It is obvious that both permeases are capable of transporting $^{63}Ni^{2+}$ ion. NhlF-mediated nickel uptake is significantly inhibited by Co^{2+}, while HoxN-dependent nickel uptake is not affected. The differences are demonstrated directly by supplying $^{57}Co^{2+}$ ion as the transport substrate: NhlF is able to transport this ion, while HoxN is not. Inhibition studies with NixA of *H. pylori* showed a moderate inhibitory effect of Co^{2+} on $^{63}Ni^{2+}$ uptake when the inhibitor was present at equimolar concentration or at tenfold excess [97]. Preliminary $^{57}Co^{2+}$ uptake assays suggest, however, that NixA does not mediate high-affinity cobalt transport (O. Degen and T. Eitinger, unpublished data). Significant inhibition of $^{63}Ni^{2+}$ uptake by Co^{2+} has been observed for Nic1p of *Schizosaccharomyces pombe* [52]. The implication of Nic1p in Co^{2+} uptake remains to be established.

Interaction of transmembrane segments 1 and 2 in NiCoTs seems to be important for ion selectivity (O. Degen and T. Eitinger, unpublished data). Replacement of TMS 1 of HoxN by the respective segment of NhlF resulted in a chimeric permease that showed low activity but transported both nickel and cobalt ion. TMS 1 of NhlF resembles a transmembrane segment of Cot1p, a CDF-type cobalt transporter of yeast intracellular membranes [109]. NhlF-type transporters of the

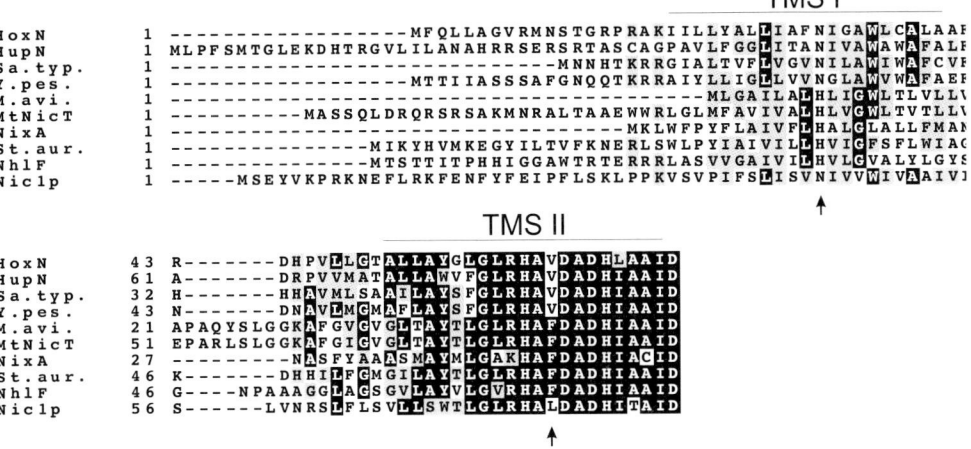

Fig. 5. Alignment of the amino-terminal sequences of 10 members of the nickel/cobalt transporter family from *Ralstonia eutropha* (HoxN), *Bradyrhizobium japonicum* (HupN), *Salmonella enterica* serovar *typhimurium*, *Yersinia pestis*, *Mycobacterium avium*, *Mycobacterium tuberculosis* (MtNicT), *Staphylococcus aureus*, *Rhodococcus rhodochrous* J1 (NhlF) and the fission yeast *Schizosaccharomyces pombe* (Nic1p). The alignment was performed with program CLUSTAL W and the output was processed with program BOXSHADE. TMS, putative transmembrane segment. The effects of replacements in HoxN and NhlF at the positions marked by arrows are discussed in the text.

NiCoT family contain a histidine residue in the center of TMS 1, while HoxN-type permeases have an asparagine residue at the corresponding position (Fig. 5). With the exception of the eukaryotic Nic1p, the histidine correlates with a phenylalanine residue in the central part of TMS 2, while the asparagine correlates with a valine. Site-directed mutagenesis has been applied to construct a number of NhlF and HoxN variants with alterations at these positions. Most interestingly, replacement of Val-64 within TMS 2 of HoxN by Phe led to a threefold increase of nickel transport activity and a significant cobalt transport activity. This result suggests that a few replacements may extend the substrate spectrum of these permeases to additional transition metal cations.

18.5
Perspective

Two major families of high-affinity nickel transporters with representatives in both gram-negative and gram-positive bacteria have been described. It cannot, of course, be excluded that additional systems are involved in the specific supply of nickel ion to prokaryotic cells under physiological conditions. Our knowledge on archaeal nickel transporters is marginal although in one group, the methanogens, nickel plays a central role in energy- and carbon metabolism. Even less information is available on the structure and regulation of nickel transport systems in eukaryotes. The finding that a relative of bacterial nickel permeases operates in fission yeast may stimulate further work. Cellular and systemic nickel transport in higher eukaryotes offers many open questions. The mechanism which enables nickel-specific transporters to distinguish between metal ions that have a very similar size and charge density is still a mystery. Solving this mystery would no doubt contribute in an important way to our understanding of the function of solute transporters in general.

Acknowledgement

I thank Olaf Degen, Marion Müller, and Ute Böhnke for their contributions, and I am grateful to Edward Schwartz for comments on the manuscript and to Bärbel Friedrich for long-term support. Work of my group was funded by grants from the Deutsche Forschungsgemeinschaft.

References

1. M. H. Saier, Jr., *Microbiol. Mol. Biol. Rev.* **2000**, *64*, 354-411.
2. R. P. Hausinger, *Biochemistry of Nickel*, Plenum Publishing Corporation, New York, **1993**.
3. H. L. T. Mobley, R. P. Hausinger, *Microbiol. Rev.* **1989**, *53*, 85-108.
4. H. L. T. Mobley, M. D. Island, R. P. Hausinger, *Microbiol. Rev.* **1995**, *59*, 451-480.
5. L. E. Zonia, N. E. Stebbins, J. C. Polacco, *Plant Physiol.* **1995**, *107*, 1097-1103.
6. S. K. Freyermuth, M. Bacanamwo, J. C. Polacco, *Plant J.* **2000**, *21*, 53-60.
7. R. A. Burne, Y.-Y. M. Chen, *Microbes Infect.* **2000**, *2*, 533-542.
8. G. M. Cox, J. Mukherjee, G. T. Cole, A. Casadevall, J. R. Perfect, *Infect. Immun.* **2000**, *68*, 443-448.
9. E. Jabri, M. B. Carr, R. P. Hausinger, P. A. Karplus, *Science* **1995**, *268*, 998-1004.
10. S. Benini, W. R. Rypniewski, K. S. Wilson, S. Miletti, S. Ciurli, S. Mangani, *Structure* **1999**, *7*, 205-216.
11. S. P. J. Albracht, *Biochim. Biophys. Acta* **1994**, *1188*, 167-204.
12. A. Volbeda, E. Garcin, C. Piras, A. L. de Lacey, V. M. Fernandez, E. C. Hatchikian, M. Frey, J. C. Fontecilla-Camps, *J. Am. Chem. Soc.* **1996**, *118*, 12989-12996.
13. R. P. Happe, W. Roseboom, A. J. Pierik, S. P. J. Albracht, K. A. Bagley, *Nature* **1997**, *385*, 126.
14. Y. Nicolet, B. J. Lemon, J. C. Fontecilla-Camps, J. W. Peters, *Trends Biochem. Sci.* **2000**, *25*, 138-143.
15. J. G. Ferry, *Annu. Rev. Microbiol.* **1995**, *49*, 305-333.
16. J. Heo, C. R. Staples, C. M. Halbleib, P. W. Ludden, *Biochemistry* **2000**, *39*, 7956-7963.
17. P. W. Ludden, J. Heo, C. R. Staples, W. B. Leon, C. L. Drennan, D. C. Rees, *Abstract LA18, 6th International Conference on the Molecular Biology of Hydrogenases*, Potsdam, Germany, **2000**.
18. U. Ermler, W. Grabarse, S. Shima, M. Goubeaud, R. K. Thauer, *Science* **1997**, *278*, 1457-1462.
19. R. K. Thauer, *Microbiology* **1998**, *144*, 2377-2406.
20. R. K. Thauer, L. G. Bonacker, *Ciba Foundation Symposium* **1994**, *180*, 210-227.

21. H.-D. Youn, H. Youn, J.-W. Lee, Y.-I. Yim, J. K. Lee, Y. C. Hah, S.-O. Kang, *Arch. Biochem. Biophys.* **1996**, *334*, 341-348.
22. E.-J. Kim, H.-P. Kim, Y. C. Hah, J.-H. Roe, *Eur. J. Biochem.* **1996**, *214*, 178-185.
23. V. Leclerc, P. Boiron, R. Blondeau, *Curr. Microbiol.* **1999**, *39*, 365-368.
24. E.-J. Kim, H.-J. Chung, B. Suh, Y. C. Hah, J.-H. Roe, *Mol. Microbiol.* **1998**, *27*, 187-195.
25. E.-J. Kim, H.-J. Chung, B. Suh, Y. C. Hah, J.-H. Roe, *J. Bacteriol.* **1998**, *180*, 2014-2020.
26. H.-J. Chung, J.-H. Choi, E.-J. Kim, Y.-H. Cho, J.-H. Roe, *J. Bacteriol.* **1999**, *181*, 7381-7384.
27. J.-S. Kim, J.-H. Jang, J.-W. Lee, S.-O. Kang, K.-S. Kim, J. K. Lee, *Biochim. Biophys. Acta* **2000**, *1493*, 200-207.
28. S. B. Choudhury, J. W. Lee, G. Davidson, Y. I. Yim, K. Bose, M. L. Sharma, S. O. Kang, D. E. Cabelli, M. J. Maroney, *Biochemistry* **1999**, *38*, 3744-3752.
29. G. P. Ferguson, S. Tötemeyer, M. J. MacLean, I. R. Booth, *Arch. Microbiol.* **1998**, *170*, 209-219.
30. S. L. Clugston, J. F. J. Barnard, R. Kinach, D. Miedema, R. Ruman, E. Daub, J. F. Honek, *Biochemistry* **1998**, *37*, 8754-8763 (glyoxalase).
31. M. M. He, S. L. Clugston, J. F. Honek, B. W. Matthews, *Biochemistry* **2000**, *39*, 8719-8727.
32. K. Salnikow, M. Costa, *J. Environ. Pathol. Toxicol. Oncol.* **2000**, *19*, 307-318.
33. J. R. Landolph, in: *Metal Ions in Biological Systems* Vol. 36 (Sigel, A., Sigel, H., Eds.), Marcel Dekker, Switzerland, **1999**, pp. 445-483.
34. Y.-W. Lee, C. B. Klein, B. Kargacin, K. Salnikow, J. Kitahara, K. Dowjat, A. Zhitkovic, N. T. Christie, M. Costa, *Mol. Cell. Biol.* **1995**, *15*, 2547-2557.
35. W. Bal, R. Liang, J. Lukszo, S.-H. Lee, M. Dizdaroglu, K. S. Kasprzak, *Chem. Res. Toxicol.* **2000**, *13*, 616-624.
36. O. Sergent, I. Morel, J. Cillard, in: *Metal Ions in Biological Systems* Vol. 36 (Sigel, A., Sigel, H., Eds.), Marcel Dekker, Switzerland, **1999**, pp. 251-287.
37. B. P. Branchaud, in: *Metal Ions in Biological Systems* Vol. 36 (Sigel, A., Sigel, H., Eds.), Marcel Dekker, Switzerland, **1999**, pp. 79-102.
38. D. W. Porter, H. Yakushiji, Y. Nakabeppu, M. Sekiguchi, M. J. Fivash Jr., K. S. Kasprzak, *Carcinogenesis* **1997**, *18*, 1785-1791.
39. T.-T. Tseng, K. S. Gratwick, J. Kollman, D. Park, D. H. Nies, A. Goffeau, M. H. Saier, Jr., *J. Mol. Microbiol. Biotechnol.* **1999**, *1*, 107-125.
40. M. R. Bruins, S. Kapil, F. Oehme, *Ecotoxicol. Environ. Safety* **2000**, *45*, 198-207.
41. M. Garcia-Dominguez, L. Lopez-Maury, F. J. Florencio, J. C. Reyes, *J. Bacteriol.* **2000**, *182*, 1507-1514.
42. D. A. Pearce, F. Sherman, *J. Bacteriol.* **1999**, *181*, 4774-4779.
43. K. Kitamoto, K. Yoshizawa, Y. Ohsumi, Y. Anraku, *J. Bacteriol.* **1988**, *170*, 2683-2686.
44. K. Nishimura, K. Igarashi, Y. Kakinuma, *J. Bacteriol.* **1998**, *180*, 1962-1964.
45. U. Krämer, J. D. Cotter-Howells, J. M. Charnock, A. J. M. Baker, J. A. C. Smith, *Nature* **1996**, *379*, 635-638.
46. M. W. Persans, X. Yan, J. M. Patnoe, U. Krämer, D. E. Salt, *Plant Physiol.* **1999**, *121*, 1117-1126.
47. U. Krämer, I. J. Pickering, R. C. Prince, I. Raskin, D. E. Salt, *Plant. Physiol.* **2000**, *122*, 1343-1353.
48. K. Murata, Y. Fukuda, M. Shimosaka, K. Watanabe, T. Saikusa, A. Kimura, *Appl. Environ. Microbiol.* **1985**, *50*, 1200-1207.
49. R. L. Smith, M. E. Maguire, *Mol. Microbiol.* **1998**, *28*, 217-226.
50. G. Zsurka, J. Gregáň, R. J. Schweyen, *Genomics* **2001**, *72*, 158-168.
51. C. W. MacDiarmid, R. C. Gardner, *J. Biol. Chem.* **1998**, *273*, 1727-1732.
52. T. Eitinger, O. Degen, U. Böhnke, M. Müller, *J. Biol. Chem.* **2000**, *275*, 18029-18033 (Correction: **2000**, *275*, 33184).
53. B. P. Krom, J. B. Warner, W. N. Konings, J. S. Lolkema, *J. Bacteriol.* **2000**, *182*, 6374-6381.
54. J. B. Warner, B. P. Krom, C. Magni, W. N. Konings, J. S. Lolkema, *J. Bacteriol.* **2000**, *182*, 6099-6105.
55. R. Bellamy, *Microbes Infect.* **1999**, *1*, 23-27.
56. D. G. Kehres, M. L. Zaharik, B. B. Finlay, M. E. Maguire, *Mol. Microbiol.* **2000**, *36*, 1085-1100.
57. H. Makui, E. Roig, S. T. Cole, J. D. Helman, P. Gros, M. F. M. Cellier, *Mol. Microbiol.* **2000**, 1065-1078.
58. L. E. Williams, J. K. Pittman, J. L. Hall, *Biochim. Biophys. Acta* **2000**, *1465*, 104-126.
59. N. Nelson, *EMBO J.* **1999**, *18*, 4361-4371.

60. H. Gunshin, B. Mackenzie, U. V. Berger, Y. Gunshin, M. F. Romero, W. F. Boron, S. Nussberger, J. L. Gollan, M. A. Hediger, *Nature* **1997**, *388*, 482-488.
61. A. Cohen, H. Nelson, N. Nelson, *J. Biol. Chem.* **2000**, *275*, 33388-33394.
62. D. Agranoff, I. M. Monahan, J. A. Mangan, P. D. Butcher, S. Krishna, *J. Exp. Med.* **2000**, *190*, 717-727.
63. B. H. Eng, M. L. Guerinot, D. Eide, M. H. Saier, Jr., *J. Membr. Biol.* **1998**, *166*, 1-7.
64. M. L. Guerinot, *Biochim. Biophys. Acta* **2000**, *1465*, 190-198.
65. H. Zhao, D. J. Eide, *Mol. Cell. Biol.* **1997**, *17*, 5044-5052.
66. U. Eckhardt, A. Mas Marques, T. J. Buckhout, *Plant Mol. Biol.* **2001**, *45*, 437-448.
67. D. H. Nies, *Appl. Microbiol. Biotechnol.*, **1999**, *51*, 730-750.
68. R. P. Hausinger, G. L. Eichhorn, L. G. Marzilli (Eds.), *Mechanisms of Metallocenter Assembly*, VCH, New York, **1996**.
69. R. P. Hausinger, *J. Biol. Inorg. Chem.* **1997**, *2*, 279-286.
70. H. Liesegang, K. Lemke, R. A. Siddiqui, H. G. Schlegel, *J. Bacteriol.* **1993**, *175*, 767-778.
71. T. Schmidt, H. G. Schlegel, *J. Bacteriol.* **1994**, *176*, 7045-7054.
72. M. Goldberg, T. Pribyl, S. Juhnke, D. H. Nies, *J. Biol. Chem.* **1999**, *274*, 26056-26070.
73. G. Grass, C. Große, D. H. Nies, *J. Bacteriol.* **2000**, *182*, 1390-1398.
74. C. Große, G. Grass, A. Anton, S. Franke, A. N. Santos, B. Lawley, N. L. Brown, D. H. Nies, *J. Bacteriol.* **1999**, *181*, 2385-2393.
75. I. T. Paulsen, M. H. Saier, Jr., *J. Membr. Biol.* **1997**, *156*, 99-103.
76. A. Anton, C. Große, J. Reißman, T. Pribyl, D. H. Nies, *J. Bacteriol.* **1999**, *181*, 6876-6881.
77. C. Rensing, M. Ghosh, B. P. Rosen, *J. Bacteriol.* **1999**, *181*, 5891-5897.
78. D. Gatti, B. Mitra, B. P. Rosen, *J. Biol. Chem.*, **2000**, *275*, 34009-34012.
79. T. Eitinger, B. Friedrich, in: *Transition Metals in Microbial Metabolism* (Winkelmann, G., Carrano, C. J., Eds.), Harwood Academic Publishers, The Netherlands, **1997**, pp. 235-256.
80. T. Eitinger, M.-A. Mandrand-Berthelot, *Arch. Microbiol.* **2000**, *173*, 1-9.
81. A. Ruepp, W. Graml, M.-L. Santos-Martinez, K. K. Koretke, C. Volker, H. W. Mewes, D. Frishman, S. Stocker, A. N. Lupas, W. Baumeister, *Nature* **2000**, *407*, 508-513.
82. C. Navarro, L.-F. Wu, M.-A. Mandrand-Berthelot, *Mol. Microbiol.* **1993**, *9*, 1181-1191.
83. K. de Pina, C. Navarro, L. McWalter, D. H. Boxer, N. C. Price, S. M. Kelly, M.-A. Mandrand-Berthelot, L.-F. Wu, *Eur. J. Biochem.* **1995**, *227*, 857-865.
84. L.-F. Wu, M.-A. Mandrand-Berthelot, R. Waugh, C. J. Edmonds, S. E. Holt, D. H. Boxer, *Mol. Microbiol.* **1989**, *3*, 1709-1718.
85. K. de Pina, V. Desjardin, M.-A. Mandrand-Berthelot, G. Giordano, L.-F. Wu, *J. Bacteriol.* **1999**, *181*, 670-674.
86. P. T. Chivers, R. T. Sauer, *Protein Sci.* **1999**, *8*, 2494-2500.
87. P. T. Chivers, R. T. Sauer, *J. Biol. Chem.* **2000**, *275*, 19735-19741.
88. J. Steuber, W. Krebs, M. Bott, P. Dimroth, *J. Bacteriol.* **1999**, *181*, 241-245.
89. H. Takami, K. Nakasone, Y. Takaki, G. Maeno, R. Sasaki, N. Masui, F. Fuji, C. Hirama, Y. Nakamura, N. Ogasawara, S. Kuhara, K. Horikoshi, *Nucleic Acids Res.* **2000**, *28*, 4317-4331.
90. K.-S. Park, T. Iida, Y. Yamaichi, T. Oyagi, K. Yamamoto, T. Honda, *Infect. Immun.* **2000**, *68*, 5742-5748.
91. T. Eitinger, B. Friedrich, *J. Biol. Chem.* **1991**, *266*, 3222-3227.
92. O. Degen, M. Kobayashi, S. Shimizu, T. Eitinger, *Arch. Microbiol.* **1999**, *171*, 139-145.
93. J. S. Lolkema, D.-J. Slotboom, *Mol. Membr. Biol.* **1998**, *15*, 33-42.
94. T. Eitinger, B. Friedrich, *Mol. Microbiol.* **1994**, *12*, 1025-1032.
95. T. Eitinger, L. Wolfram, O. Degen, C. Anthon, *J. Biol. Chem.* **1997**, *272*, 17139-17144.
96. J. F. Fulkerson, Jr., H. L. T. Mobley, *J. Bacteriol.* **2000**, *182*, 1722-1730.
97. J. F. Fulkerson, Jr., R. M. Garner, H. L. T. Mobley, *J. Biol. Chem.* **1998**, *273*, 235-241.
98. C. Fu, S. Javedan, F. Moshiri, R. J. Maier, *Proc. Natl. Acad. Sci. USA* **1994**, *91*, 5099-5103.
99. P. Bauerfeind, R. M. Garner, H. L. T. Mobley, *Infect. Immun.* **1996**, *64*, 2877-2880.
100. H. Komeda, M. Kobayashi, S. Shimizu, *Proc. Natl. Acad. Sci. USA* **1997**, *94*, 36-41.

101. M. Kobayashi, S. Shimizu, *Nature Biotechnol.* **1998**, *16*, 733-736.
102. C. Montecucco, E. Papini, M. de Bernard, M. Zoratti, *FEBS Lett.* **1999**, *452*, 16-21.
103. J. R. Kerr, A. Al-Khattaf, A. J. Barson, J. P. Burnie, *Arch. Dis. Child.* **2000**, *83*, 429-434.
104. H. L. T. Mobley, R. M. Garner, P. Bauerfeind, *Mol. Microbiol.* **1995**, *16*, 97-109.
105. J. K. Hendricks, H. L. T. Mobley, *J. Bacteriol.* **1997**, *179*, 5892-5902.
106. L. Herrmann, D. Schwan, R. Garner, H. L. T. Mobley, R. Haas, K. P. Schäfer, K. Melchers, *Mol. Microbiol.* **1999**, *33*, 524-536.
107. M. Maeda, M. Hidaka, A. Nakamura, H. Masaki, T. Uozumi, *J. Bacteriol.* **1994**, *176*, 432-442.
108. L. Wolfram, B. Friedrich, T. Eitinger, *J. Bacteriol.* **1995**, *177*, 1840-1843.
109. D. S. Conklin, J. A. McMaster, M. R. Culbertson, C. Kung, *Mol. Cell. Biol.* **1992**, *12*, 3678-3688.

19
Mitochondrial Copper Ion Transport

Thalia Nittis, Keith McCall and Dennis R. Winge

19.1
Introduction

Copper ions are required in the mitochondrion for assembly of an active cytochrome *c* oxidase. This review will focus on the mechanism of copper ion delivery to the mitochondrion and insertion into the oxidase, which requires Cu ion transport across a membrane bilayer and insertion into two binding sites, one of which is buried deep in the bilayer. A discussion of these processes necessitates background information on the mitochondrion, protein import into mitochondria, assembly of the cytochrome *c* oxidase complex and copper ion transport in cells. Much is known of the assembly of cytochrome *c* oxidases in bacterial systems. These processes are relevant to the assembly pathway and copper metallation of the eukaryotic oxidase, so bacterial oxidases will be reviewed. Finally, a summary will be presented with a model that encompasses information from many organisms on how copper metallation of cytochrome *c* oxidase may occur.

19.2
Mitochondrial Structure

The mitochondrion is a cellular organelle consisting of a continuous reticulum that makes up nearly 10% of the cell volume in respiring yeast cells [1]. The mitochondrial tubular network is highly dynamic and changes size and shape through fission and fusion events [2]. The organelle is enclosed by a double membrane creating two internal spaces. The space between the two membranes is called the intermembrane space and is interrupted by junction points. The volume enclosed within the inner membrane is the matrix. Proteins are distributed in both membranes as well as in the two soluble spaces. The inner membrane of rat liver mitochondria contains 21% of the total protein, while the outer membrane contains about 5%. The bulk of the remaining protein exists as soluble components within the matrix. The proportion of proteins in the inner membrane increases significantly in heart mitochondria. This makes sense, since one major function of the mitochondrion is the generation of ATP through respiratory reduction of oxygen to water and oxidation of NADH and FADH$_2$. Consistent with its major role in ATP synthesis, the mitochondrial network often localizes near structures that consume large quantities of ATP. The energy of the respiratory chain is converted into a proton gradient across the inner membrane that is used to drive ATP synthesis [3].

The mitochondrion contains its own genome. Normal yeast cells contain between 10–50 mitochondrial genomes, the number varying depending on growth conditions [4]. Mitochondrial DNA is not uniformly distributed in the tubular network of the organelle, rather the genomes are localized in discrete zones within the network [1]. The number of gene products encoded by the mitochondrial genome varies between species. The human mitochondrion encodes 13 known polypeptides of the human electron transfer/oxidative phosphorylation pathway and 22 tRNAs. Of the thirteen known mitochondrially encoded proteins, three are subunits of cytochrome c oxidase.

19.3
Mitochondrial Transport

Copper is an abundant metal ion found in the mitochondrion. The copper content of mitochondria in oxen (*Bos taurus*) is 1.46 nmol Cu mg^{-1} protein compared to 9 nmol iron mg^{-1} protein which is needed for heme formation and iron sulfur cluster assembly [5]. The only known copper metalloenzyme within the organelle is cytochrome c oxidase. Cytochrome c oxidase contains a variety of metal cofactors including two iron-containing hemes, copper, zinc and magnesium ions. Thus, assembly of active cytochrome c oxidase requires specific transport pathways for delivery of multiple metal ions.

Significant advances have been made on the transport of copper ions to the mitochondrion. This process involves the specific transport of cations across bilayers, and may occur through cation permeases, facilitated uptake or diffusion. Mitochondrial copper transport appears to involve facilitated uptake. Thus, understanding the process of mitochondrial protein transport is relevant and will be briefly reviewed.

Mitochondrial function is dependent on transport processes that translocate polypeptides and a range of cofactors from the cytoplasm, and this occurs through at least two pathways. The majority of mitochondrial proteins are nuclear-encoded and are synthesized as preproteins within the cytoplasm prior to translocation into the mitochondrion [6]. Protein import can also occur cotranslationally from ribosomes attached to the mitochondrial surface [7]. The translocation of preproteins occurs through import channels formed in the outer and inner membranes. The outer membrane channel consists of a protein complex designated the TOM complex [8, 9]. An N-terminal precursor sequence or internal signal sequence targets many proteins to the TOM import channel which consists of a receptor complex

(4 proteins) and an insertion complex (4 proteins). The receptor complex, consisting of Tom20 and Tom70, functions in specific recognition of preproteins with targeting sequences [6, 10]. The channel formed by the Tom40 protein creates a hydrophilic 22 Å wide pore [11]. The pore diameter is too small to accommodate folded protein import. Thus, the preproteins are extruded through the channel as unfolded polypeptides [12]. Hsp70 class molecular chaperones facilitate docking of unfolded preproteins to the receptor complex [9].

Intermembrane space chaperones escort import proteins to one of two inner membrane complexes designated TIM complexes [8, 9]. The Tim22/Tim54 complex mediates insertion of import proteins in the inner membrane. Tim10 and Tim12 are also required for this pathway [13]. The second TIM complex Tim23/Tim17 mediates translocation of proteins across the inner membrane and into the matrix. This step is ATP-dependent and requires the membrane potential. Translocation is followed by proteolytic cleavage of the precursor signal sequence and ATP-dependent protein folding mediated by Hsp70 and Tim44 [14, 15].

Several proteins localized within the intermembrane space require only translocation across the outer membrane. Two such enzymes are involved in the covalent attachment of heme to cytochrome *c*. These proteins lack an N-terminal targeting sequence, yet use the TOM complex for entrance into the intermembrane space [9, 16]. This transport is ATP-independent and does not require the membrane potential. Some inner membrane proteins that transit into the matrix project N-terminal domains within the intermembrane space, and, therefore, require an additional transport event to move the N-terminus back across the inner membrane. The yeast protein Oxa1 is required for translocation of N-terminal tails of several proteins, including two mitochondrially encoded subunits of cytochrome *c* oxidase (Cox2 and Cox3), from the matrix to the intermembrane space [17].

The second pathway of import involves cotranslational transport. A number of intrinsic membrane proteins of the mitochondrion appear to be synthesized on ribosomes attached to the outer mitochondrial membrane [18]. The ribosomes attach to regions of the membrane in which the inner and outer membranes form a contact [19]. Two mechanisms of targeting ribosomes to the mitochondrial surface may exist. Nascent chains may initiate interaction with a mitochondrial receptor analogous to the docking of polysomes on the endoplasmic reticulum. Alternatively, ribosomes may attach to the mitochondrial surface through motifs within the mRNA. Transcripts coding for a series of hydrophobic proteins, including Atm1 and Cox10, are exclusively localized to mitochondrial-bound polysomes [18]. The mitochondrial presequence of Atm1 and the 3' untranslated region of the *ATM1* mRNA are sufficient to localize the transcript to the mitochondrial surface [18]. Thus, mRNA-directed localization of polysomes to the mitochondrion followed by cotranslational import may be a key mechanism for uptake of hydrophobic polypeptides.

Non-proteinaceous components are transported across the mitochondrial membranes either by channels or translocases. A second transmembrane channel in the outer membrane is formed by the porin protein which permits diffusion of ions and molecules less than 1 kDa [20]. Yeast contain two porin molecules, Por1 and

Por2. Por1 forms a voltage-dependent channel that transports NADH. Transport of metabolites such as pyruvate, citrate, fatty acids in the form of acylcarnitine derivatives, amino acids and ADP/P into the matrix is facilitated by translocases. The inner membrane is impermeable to most ions, and thus must contain a series of cation permeases. Two potential iron transporters, Mmt1 and Mmt2, are localized within the mitochondrion. The actual membrane of the mitochondrion is unresolved [21].

19.4
Assembly of Mitochondrial Cytochrome c Oxidase

Copper transport to the mitochondrion for metallation of cytochrome c oxidase is intimately related to the assembly of the oxidase complex. The bovine cytochrome c oxidase consists of 13 subunits, most of which are embedded within the inner membrane [22] (Fig. 1A). The 13-subunit complex exists as a 360 kDa dimeric unit sensitive to detergents. Most detergent solubilized isolates of the oxidase complex consist primarily of 200 kDa monomers [23]. Three core polypeptides of the complex, Cox1-3, are encoded by the mitochondrial genome, whereas the remaining 10 subunits are nuclear-encoded. The yeast oxidase complex is oligomeric with 11 subunits [24]. The subunit composition varies somewhat depending on extraction conditions [25]. The mitochondrially encoded core subunits are the largest polypeptides of the complex being 55, 26 and 30 kDa for yeast subunits 1, 2 and 3, respectively [24]. The nuclear-encoded subunits are small peripheral polypeptides with masses of the mature molecules ranging from 6 to 15 kDa. Several of these small subunits consist merely of a single helix spanning the inner membrane (Fig. 1B). These single-helix subunits pack on the outside of the core mitochondrially encoded subunits. Although not all of the peripheral small subunits show sequence conservation between animals and yeast, these small polypeptides are functionally important. Gene disruption of any one of the small 55-59-residue polypeptides in yeast (*COX7*, *COX9* or *COX14*) results in a non-functional enzyme. Thus, assembly of the polypeptide components of cytochrome c oxidase requires coordinated mitochondrial uptake of up to nine polypeptides encoded by the nuclear genome and integration with the three mitochondrially encoded subunits. Many of the nuclear-encoded subunits are hydrophobic prepeptides, so the possibility exists that they are routed to the mitochondrion by targeting of the transcripts as described for *ATM1*. The stability of the oxidase complex is dependent on the presence of the various subunits, and disruption of one of several nuclear-encoded oxidase subunit genes results in accumulation of a low molecular weight intermediate complex. Some of the low molecular weight subunits may have regulatory roles. Yeast subunit Cox5a is a 133-residue polypeptide that contains one transmembrane helix. Yeast cultured in hypoxic conditions use an alternative Cox5 subunit designated Cox5b, which is a related 134-residue polypeptide [26]. Curiously, the efficiency of electron transfer from heme *a* to oxygen is enhanced 3-fold in the Cox5b–enzyme complex [27].

19 Mitochondrial Copper Ion Transport | 423

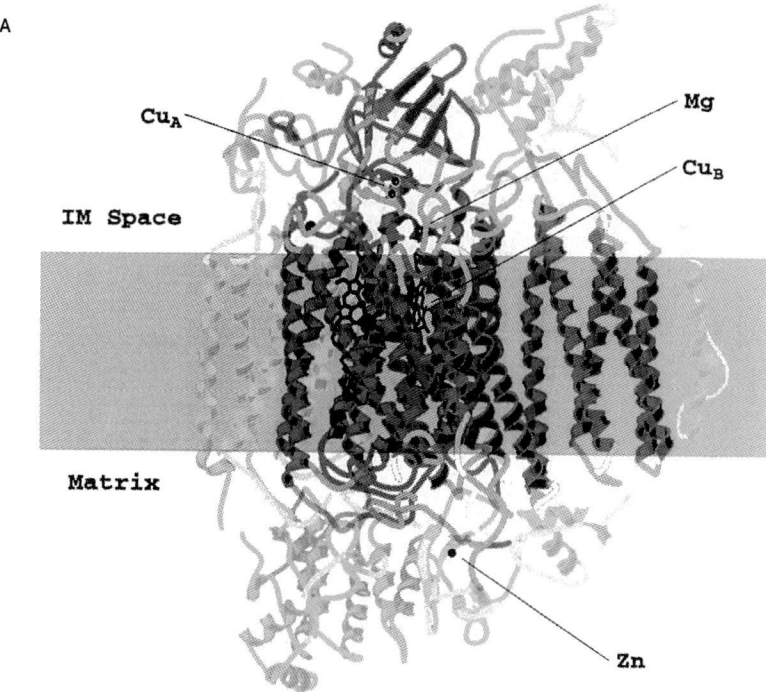

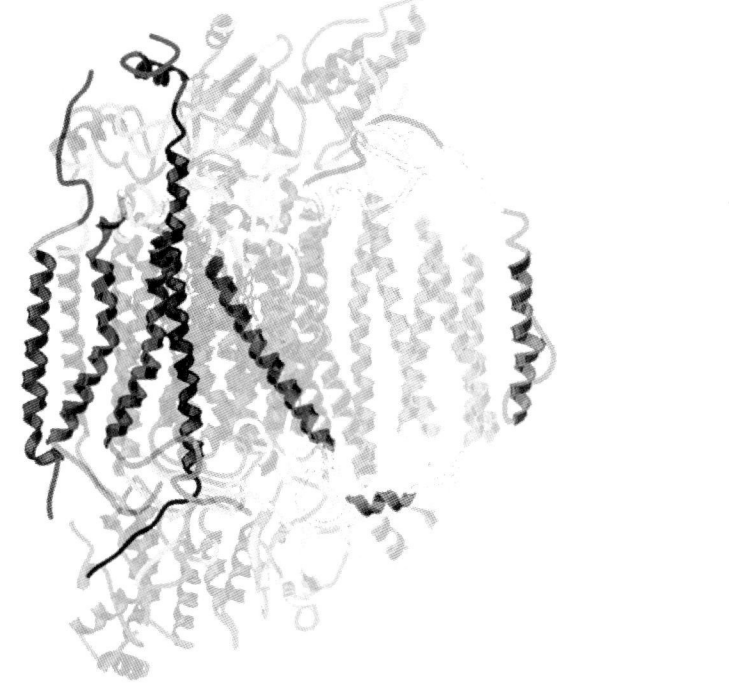

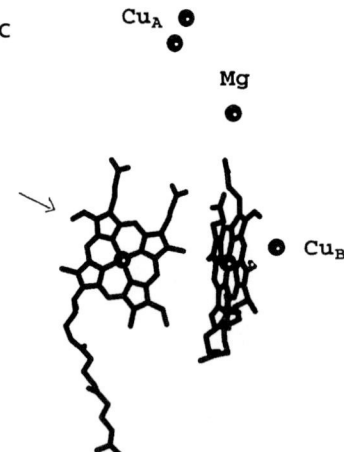

Fig. 1. Structure of cytochrome c oxidase [22]. In **A** the 13 subunits are shown. The phospholipid bilayer is shaded. The binuclear Cu_A site projects into the intermitochondrial membrane space. The binuclear Cu_B-heme a_3 is buried within the bilayer. The Mg(II) and Zn(II) sites are also shown. In **B** the various single helix subunits are shown in the dark shade. In **C** the spatial relationship of the Cu, heme a and Mg ion cofactors is shown. The long chain coming off the heme is the hydroxyethylfarnesyl moiety. The arrow shows the site of formylation.

Cytochrome c oxidase is the terminal electron acceptor of the respiratory chain receiving electrons from cytochrome c and transferring them to molecular oxygen. The oxidase drives proton translocation during electron transfer. Protein pumping is the driving force for ATP synthesis. Recent evidence suggests that cytochrome c oxidase forms a supramolecular complex within the inner membrane with cytochrome c reductase (complex III), and this may also stabilize the oxidase complex [28]. Electrons enter the binuclear Cu_A site and pass through heme a and the heterobinuclear heme a_3-Cu_B site that constitutes the site of oxygen reduction (Fig. 1C). The Cox1 polypeptide binds both hemes and forms the Cu_B site. The Cox2 subunit binds the two mixed valence Cu ions in the Cu_A site. The two hemes in cytochrome c oxidase are modified by two reactions resulting in the addition of a farnesyl isoprene unit to a vinyl group creating a hydroxyethylfarnesyl moiety and oxidation of a pyrrole methyl group to a formyl substituent (Fig. 1C).

The beef heart enzyme is embedded within the inner membrane with a portion of the molecule protruding into the intermembrane space (37 Å) and a separate portion extending into the matrix (32 Å) [30]. The Cu_A site protrudes 8 Å above the inner membrane into the intermembrane space, whereas the heme a_3-Cu_B binuclear site is buried 13 Å below the membrane surface [22] (Fig. 1A). Cytochrome c docks with the oxidase within the intermembrane space, interacting with the exposed subunit 2 close to the Cu_A site [31]. Cu_A site formation likely occurs through metallation of subunit 2 within the intermembrane space. In contrast, Cu_B site formation deep within the membrane bilayer may be more complex. It is unclear whether heme a insertion into the two buried sites of Cox subunit 1 occurs in a preassembled complex in the matrix or after insertion within the inner membrane. Recently, heme a insertion into subunit 1 was found to occur independently of association with subunits 2 or 3 in *Rhodobacter sphaeroides* [32]. Episomal expression of subunit 1 in *R. sphaeroides* resulted in the accumulation of Cox1 with one heme a bound [32]. Co-expression of subunit 2 was necessary for the formation of the Cu$_B$ site and oxidase activity. The single Zn(II) ion in cytochrome c oxidase is bound to bovine subunit Vb (yeast Cox4) on the matrix side of the inner membrane and its function is unknown [22]. A Mg(II) ion is bound near the interface of subunits 1 and 2.

Thus, assembly of eukaryotic cytochrome c oxidase is dependent on the following steps:

(1) import of 8–10 nuclear-encoded Cox polypeptides;
(2) interaction of nuclear- and mitochondrial-encoded subunits;
(3) transport of Cu ions to the intermembrane space and/or the matrix;
(4) transport of Fe ions to the matrix for heme formation by ferrochelatase;
(5) transport of Zn(II) and Mg(II) ions to the mitochondrion;
(6) modification of protoheme to heme a by farnesylation and formylation;
(7) posttranslational modification of Tyr288 to form a covalent link with a Cu$_B$ histidyl residue (His284) [33];
(8) insertion of the metal ions and hemes into oxidase subunits.

The steps outlined above are all posttranslational processes. Additionally, a host of gene products are important for stability and/or translation of *COX1-COX3* mitochondrial genes. The complexity of assembly of cytochrome c oxidase is highlighted by the observation that 34 genes, whose products are not part of the mature oxidase complex, are known that affect oxidase function [34]. The molecules responsible for transport of zinc, magnesium and iron ions into the mitochondrion are not known. Heme a formation requires at least two modifying enzymes, but only the farnesyl transferase (Cox10) has been identified. No information exists on whether the covalent link between Tyr288 and the Cu$_B$ His ligand is enzyme-mediated. A series of other accessory proteins are known to be important for oxidase maturation, but their exact roles are unresolved. These include Shy1 (Surf1 in humans), Cox14, Cox15, Cox18, Cox20, Pet100, Pet117, Pet191 [35–40]. Several of these accessory proteins, including Cox18 and Pet100, appear to function at a late stage of assembly, and four known accessory factors (Cox14, Pet100, Pet117, Pet191) are very small polypeptides (< 13 kDa).

Deletion of oxidase subunit genes or accessory factor genes typically lead to unassembled and dysfunctional oxidase. One common phenotype observed is diminished Cox1 and/or Cox2 polypeptides and decreased heme a levels [41]. To date, no human mutations have been identified in nuclear oxidase subunit genes, although mutations in human genes for the farnesyl transferase (Cox10), Sco1 and Sco2, and Surf1 result in oxidase deficiency and a resulting respiratory defect [42]. The *cox10* mutation resulted in diminution of Cox2 levels. This is curious since heme a is bound only to Cox1. *cox10* mutant patients present with leukodystrophy. Patients with *surf1* mutations present with Leigh subacute necrotizing encephalomyopathy [34]. Cytochrome c oxidase activity is diminished 80-90 % in these patients. Progressive deterioration of the basal ganglion and brain stem occurs, and death usually occurs between 6 months and 12 years of age [43]. The yeast ortholog Shy1 is localized within the inner mitochondrial membrane and is important for efficient electron transfer between complex III and cytochrome c oxidase [44].

19.5
Copper Ion Delivery to Targets other than the Mitochondrion

Studies in yeast revealed that copper metallation of copper binding proteins such as Sod1 and Fet3 rely on metallochaperones (Lys7 and Atx1, respectively) for delivery of the Cu ions. These paradigms appear to be similar for metallation of cytochrome c oxidase and are discussed next.

Copper ions enter *Saccharomyces cerevisiae* through both high-affinity and low-affinity permeases. The high-affinity uptake system consists of two permeases, Ctr1 and Ctr3, as well as metalloreductases [45–47] (Fig. 2). The K_m for high-affinity copper ion uptake is about 2 μM [48]. Genes encoding the permeases and the *FRE1* metalloreductase are induced under conditions of copper deficiency [45, 46, 49–51]. Ctr1 and Ctr3 are functionally redundant; a Cu-deficient state arises only when both permeases are non-functional [52]. Ctr1 and Ctr3 are oligomeric complexes that presumably form import channels [45, 47]. *CTR3* is not expressed in most laboratory yeast strains due to the insertion of a transposable element within the gene [52]. The putative ectodomain of Ctr1 has multiple m-x-x-m sequence motifs that may function as low-affinity Cu(II) or Cu(I) scavenging modules gathering extracellular Cu ions for subsequent transport. Uptake across the lipid bilayer may be facilitated, in part, by binding of Cu ions to two Cys–x–Cys sequence motifs within the candidate cytoplasmic domain. Cu uptake in yeast is energy dependent, although the mechanism of energy-coupled transport is unresolved [48].

Cu ion uptake by the high-affinity Ctr1/Ctr3 system is facilitated by the metalloreductase Fre1. Fre1 is a flavocytochrome-containing NADPH oxidase that pumps a diffusible reductant into the growth medium to mobilize oxidized Cu(II) complexes [53–55]. A second metalloreductase, Fre2, functions in both copper and iron ion uptake, but *FRE2* is only actively expressed under iron limiting conditions [51, 53, 56, 57].

Cells lacking the two permeases gain greater resistance to copper toxicity [45], suggesting that the high-affinity permeases remain partially functional in Cu-

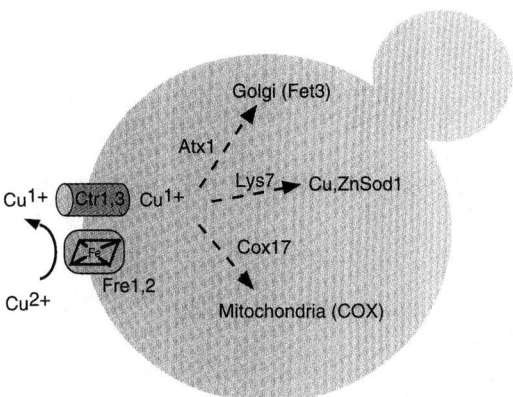

Fig. 2. The copper transport system in *S. cerevisiae*. Plasma membrane Cu ion transport occurs through the high-affinity Ctr1 and Ctr3 permeases along with the metalloreductases (Fre1 and Fre2). Intracellular Cu(I) ions are routed to sites of copper metalloprotein assembly by metallochaperones, Atx1, Lys7 and Cox17.

replete cells. Cu ion uptake can also occur through low-affinity permeases including Fet4, Ctr2, and Smf1 [58–60]. Fet4 is likely to be the predominant low-affinity copper permease with a K_m for Cu ion uptake of 35 µM [60]. Curiously, uptake by Fet4 correlates with the reduced cuprous valence state suggesting the importance of Fre1,Fre2 metalloreductases for Fet4 function [60]. Thus, Cu ions enter the cytoplasm as reduced cuprous ions.

The Cu(I) ions are subsequently transported to sites of utilization by metallochaperones (Fig. 2). The free pool of Cu(I) in the cell is kept at very low levels to minimize cytotoxicity, such as Cu-catalyzed reactions that generate oxygen radicals [61]. The prediction is that other redox active metal ions will also be transported within the cells by metallochaperones.

Two well-characterized metallochaperones exist in yeast, Atx1 and Lys7 (CCS). The Atx1 metallochaperone routes Cu ions to the Ccc2 P-type ATPase transporter localized in post-Golgi vesicles [62–64]. The human Atx1 ortholog, Hah1, transports Cu ions to the Menkes and Wilson P-type ATPases [65]. The P-type ATPases pump Cu(I) ions into the vesicles for incorporation into trinuclear Cu oxidases, ceruloplasmin in animal cells and Fet3 in yeast [66]. Metallated Fet3 is subsequently translocated to the plasma membrane for participation in high-affinity Fe ion uptake [67].

Atx1 binds a single Cu(I) ion within a conserved metal binding motif found also in Cu transporting ATPases [68]. The structures of yeast Atx1 and the human ortholog Hah1 resemble that of one of the repeating metal motifs of the Menkes P-type ATPase, which consists of a four-stranded antiparallel β-sheet covered by two α-helices [68, 69] (Fig. 3). A single diagonally bent Cu(I) site forms with two thiolates from the conserved CxxC sequence motif [68]. The structure of yeast Atx1 shown in Fig. 3 contains a Hg(II) ion coordinated in the Cu(I) site [69]. A physical interaction occurs between Atx1 and the Ccc2 target protein during the Cu transfer step from Atx1 to Ccc2 [62, 65].

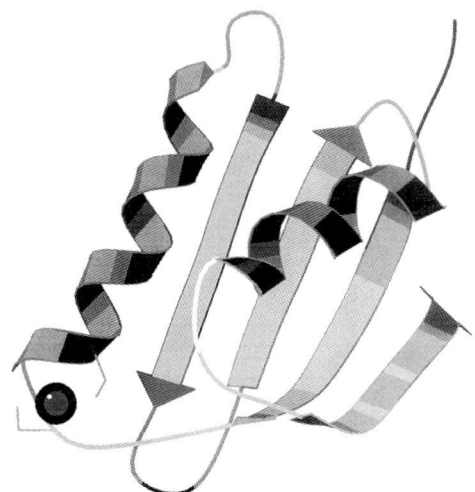

Fig. 3. Structure of the HgAtx1 complex [69]. The single Hg(I) is shown at the bottom left as a sphere. The two cysteinyl ligands are shown in wire frame.

Cu-specific ATPase pumps represent one way of translocating Cu(I) ions across a bilayer. These pumps resemble well-known cation effluxers that translocate cations against their electrochemical potential gradient by using the energy from hydrolysis of ATP [70]. A P-type ATPase prototype is the sarcoplasmic reticulum Ca(II) ATPase that translocates Ca(II) ions from the cytoplasm into the sarcoplasmic reticulum [71]. Clues to the mechanism of Cu(I) translocation by copper-ATPases come from inspection of the Ca(II) ATPase structure recently solved at 2.6 Å [72] (Fig. 4). The 994-residue protein has a cytoplasmic headpiece consisting of three separate domains and a transmembrane segment consisting of ten helices [72]. Two Ca(II) ions diffuse into sites created by six oxygen atoms within the transmembrane region. ATP binding in a cytoplasmic domain results in domain closure and subsequent ATP hydrolysis and phosphorylation of a conserved Asp residue in a separate domain that appears to control diffusion of Ca(II) ions into the transmembrane sites. The phosphorylated Asp is > 25 Å away from the site of ATP binding, so domain closure must occur during ATP hydrolysis. Ca(II) binding to the two sites within the transmembrane region appears to initiate the domain closure that is completed upon phosphorylation, and domain closure is coupled to Ca(II) translocation across the bilayer [71]. Two of the helices in the transmembrane region are unwound in a region where Ca(II) binds. The coordination geometry of the Ca(II) site requires unwinding of the helix [72]. The residues within the unwound segment of transmembrane helix 4 form part of the Ca(II) site 2. The Cu P-type ATPases have a conserved CPC sequence in the corresponding segment. Thus, one clear prediction is that Cu(I) binding by the CPC motif initiates the reaction cycle of ATP hydrolysis and cytoplasmic channel closure. Other features of the catalytic cycle are likely to resemble those of the Ca(II) ATPase. One distinction between the two types of ATPases is the presence of Cu(I)-binding modules at the N-terminal region of Cu ATPases. The Cu ATPase contains multiple N-terminal Cu(I)-binding modules with a

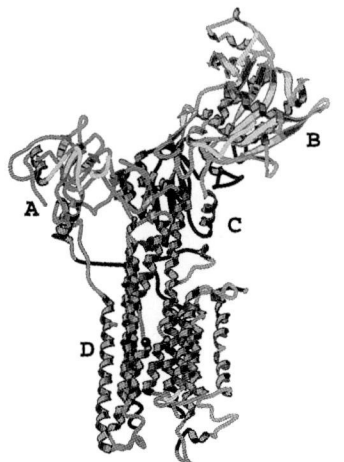

Fig. 4. Structure of Ca-ATPase calcium pump of the sarcoplasmic reticulum [72]. Domain A moves during Ca(II) transport. Domain B is the ATP-binding domain. Domain C contains the Asp that is phosphorylated in a Ca-dependent manner. Segment D is the transmembrane segment of the ATPase. Two Ca(II) ions bind within this segment. The Ca(II) ions are shown in spacefill. Cu-ATPases are predicted to bind Cu(I) in one of the corresponding positions.

conserved CxxC sequence. These modules structurally resemble the Atx1 metallochaperone [69]. The Atx1 motifs in the P-type ATPases appear to be docking sites for a specific metallochaperone prior to transfer of Cu(I) ions to the ATPase [73].

One potential route of copper transport to the mitochondrion, therefore, would be through a specific P-type ATPase. However, no mitochondrial copper P-type ATPases are known. An alternate spliced form of the Wilson's P-type ATPase was reported to be localized to the mitochondrion [74], but no functional studies have appeared. Alternative splicing of the Wilson's disease gene does occur, and a Wilson protein variant is known to be specifically expressed with diurnal rhythm in the pineal gland [75].

The second metallochaperone, Lys7 (CCS), is a key molecule for assembly of an active Cu,Zn-superoxide dismutase (Sod1) in the yeast cytoplasm [76–78]. Yeast lacking Lys7 fail to metallate Sod1 with Cu(I) [61]. Lys7 resembles Atx1, having a similar Cu(I)-binding motif. This N-terminal Cu-binding motif and a C-terminal CxC motif both participate in Cu transfer to superoxide dismutase. The transfer of Cu ions from Lys7 to Sod1 requires a physical interaction [78–80]. Lys7 contains a central domain that adopts a β-barrel conformation, analogous to Sod1, and this domain is likely the interface for binding to Sod1.

19.6
Copper Ion Transport to the Mitochondrion by Cox17

Transport of copper ions to the mitochondrion is proposed to occur through the Cox17 metallochaperone. Cu metallation of cytochrome *c* oxidase likely involves protein–protein interactions as is the case with Atx1 and Lys7 metallochaperones. *COX17* was first cloned and characterized by Glerum and Tzagoloff [81]. *S. cerevisiae* lacking Cox17 are respiratory-deficient and unable to grow on non-fermentable carbon sources. The mutant cells lacked heme *a* but not heme *b* or *c* [81]. Cytochrome *c* oxidase activity was absent in the mutant cells. Mitochondria isolated from the mutant cells contained both mitochondrial and nuclear oxidase subunits, precluding a role for Cox17 in the expression or import of the subunit polypeptides. The mutant phenotypes were suppressed by the addition of 0.4% copper salts to the growth medium. The effect of exogenous Cu(II) was specific to *cox17* mutant cells [81], and was consistent with Cox17 functioning in copper delivery to cytochrome *c* oxidase. Since the mutant cells contained a functional Sod1 superoxide dismutase, the role of Cox17 is specific to cytochrome *c* oxidase and/or mitochondria.

Cox17 is a small, hydrophilic protein of 69 residues. Three of the seven cysteinyl residues are present in a CCxC putative metal binding motif (Fig. 5). The protein, purified as a recombinant molecule in *Escherichia coli*, binds three Cu(I) ions per monomer in a poly-copper cluster, as shown by X-ray absorption spectroscopy [82]. Cox17 exists in a dimer/tetramer equilibrium with a K_d of 20 µM [82]. Thus, Cu(I) may bind in a single hexanuclear cluster or two separate trinuclear clusters.

The Cu(I) cluster(s) exhibit predominantly trigonal Cu(I) coordination by cysteinyl thiolates [82, 83]. Since the apo-Cox17 polypeptide is predominantly monomeric, a reasonable prediction is that Cu(I) binds within a cluster formed at the dimer interface. The tetrameric species is then predicted to be a dimer of dimers. The formation of the Cu(I) thiolate cluster may stabilize a conformation that enables the protein to be imported into the mitochondrion.

Cox17 is localized to both the cytoplasm (40%) and the mitochondrial intermembrane space (60%) [84]. Cox17 in the intermembrane space appears to be predominantly tetrameric, while Cox17 in the cytosol is predicted to be mostly dimeric [82]. Mutation of any of the three Cys residues in the CCxC sequence motif results in loss of function of the protein, even though each mutant is still able to bind Cu(I) and localize within the mitochondrion [82]. However, these mutants fail to form tetramers, suggesting that oligomerization is functionally important. Mutant Cox17 proteins containing a double Cys→Ser mutation in this CCxC motif fail to bind Cu(I) [85]. Three additional Cys residues are conserved in Cox17 besides the CCxC residues (Fig. 5). Cys→Ser substitutions at other positions do not affect function or Cu(I) ion binding [85].

The dual localization of Cox17 in the cytosol and intermitochondrial membrane space is consistent with Cox17 functioning as a shuttle to ferry Cu(I) ions across the outer mitochondrial membrane. The mechanism of mitochondrial targeting of Cox17 is unknown. Cox17 does not contain a classical mitochondrial target sequence. Import through the outer membrane TOM complex is unlikely as most proteins entering through the TOM channel are unfolded. The candidate function of Cox17 is to ferry a cargo (Cu ions) across the bilayer, so unfolding would seem a non-ideal pathway. Preliminary results from mutational analysis of *COX17* suggests that Cox17 import is conferred by C-terminal sequences (Maxfield and Winge, unpublished data).

Alignment of Cox17 Sequences

```
                       *.** * **.*    ** *..  ..    *  .** * **
S.cerevisiae  MTETDKKQEQENHAECEDKPKPCCVCKPE-KEERDTCILFNGQDSEKCKEFIEKYKECMKG
S. pombe      MSSSTEPSTATKVSEPAPIASEEKPKPCCAC-PETKQARDACMLQSSNGPIECAKLIEAHKKCMAQ
human            MPGLVDSNPAPPESQEKKPLKPCCAC-PETKKARDACIIEKGEEH--CGHLIEAHKECMRA
porcine          MPGLAAASPAPAESQEKKPLKPCCAC-PETKKARDACIIEKGEEH--CGHLIEAHKECMRA
rat              MPGLAAASPAPPEAQEKKPLKPCCAC-PETKKARDACIIEKGEEH--CGHLIEAHKECMRA
C.elegans        MPAEPQKSTEAGSVAPEKKLKACCAC-PETKRVRDACIIENGEEK--CGKLIEAHKACMRA

                        *
S. cerevisiae  YGFEVPSANZ
S. pombe       YGYEV
human          LGFKI
porcine        LGFKI
rat            LGRKI
C.elegans      AGFNI
```

Fig. 5. Alignment of Cox17 sequences from various species. Cox17 is not found in prokaryotes.

Cox17 is conserved in eukaryotes. The human ortholog was first identified by functional complementation of the yeast *cox17* null mutant [86], and was later mapped to the long arm of chromosome 3 [87, 88]. Human Cox17 contains six of the seven cysteine residues present in the yeast protein, and is presumably also important in mitochondrial copper transport. Studies to determine whether Cox17 is involved in inherited disorders associated with cytochrome *c* oxidase deficiencies are being conducted, but no disease links have been reported yet [87].

19.7
Co-metallochaperones in Cu Metallation of Cytochrome *c* Oxidase

Factors besides Cox17 are important for Cu metallation of cytochrome *c* oxidase. The products of *SCO1*, *SCO2* and *COX11*, all appear to assist Cox17. *SCO1* and *SCO2* were first implicated in Cu delivery to cytochrome *c* oxidase by the observation that the respiratory-deficient phenotype of a *cox17* point mutant could be suppressed by overexpression of either gene [89]. Despite showing a high degree of homology (overall amino acid identity of 53.8%), and a similar localization to the inner mitochondrial membrane [89, 90], the phenotypes associated with *SCO1* are very different from those of *SCO2*. First, Δ*sco1* cells have a respiratory-deficient phenotype with diminished cytochrome *c* oxidase activity and undetectable hemes *a* and a_3 [91]. On the other hand, Δ*sco2* cells do not have any obvious phenotype. In cells lacking Sco1, oxidase subunits 1 and 2 are unstable and rapidly degraded [91, 92]. Furthermore, overexpression of *SCO1* not only suppresses the *cox17* point mutant, but it also suppresses the null strain. In contrast, Sco2 can only partially restore respiratory growth in Δ*cox17*, and requires addition of Cu salts to the growth medium to do so. Neither excess $CuSO_4$, overexpression of *COX17*, nor a combination of the two, can suppress the Δ*sco1* phenotype. Overexpression of *SCO2* also does not rescue Δ*sco1* cells, though it does rescue one *sco1* point mutant [89]. Based on this evidence it was proposed that Sco1 transfers Cu from Cox17 to Cox1 and/or Cox2, while the role of Sco2 remains less clear.

Further support of the proposed model comes from the fact that Sco1 and Sco2 contain a potential metal binding motif, CxDxC, which is present in the C-terminal half of the proteins (Fig. 6). The C-terminal ends of the proteins are thought to protrude into the mitochondrial intermembrane space [84], which would bring them in close proximity to the Cu_A site of Cox2. Interestingly, Cox2 contains a similar CxExC motif that binds Cu. Cox2 consists of three segments: an N-terminal loop, two transmembrane helices, and a C-terminal globular, hydrophilic domain containing the Cu_A site. The N-terminal loop and the C-terminal domain interact tightly in the intermembrane space. The C-terminal domain contains a 10-stranded β-barrel, which resembles that of type I blue copper proteins, such as plastocyanin and azurin. The two Cu ions are bridged by the two cysteine residues in the CxExC motif. The other ligands involved are a histidine and methionine for one Cu ion, and a second histidine and glutamate for the second Cu ion [22, 93].

The motif in Sco1 may also bind Cu, since both cysteine residues are essential for function [94]. In addition, purified Sco1p binds one Cu(I) ion per molecule, and X-ray absorption spectroscopy suggests the Cu(I) is coordinated by three thiolate ligands (Nittis and Winge, unpublished data).

The Cu$_B$ site in Cox1 is buried within the inner mitochondrial membrane, and would, therefore, be far away from Cu$_A$ and possibly Sco1. Thus, it is possible that Sco1 may not be required for Cu incorporation into Cox1. Support for this theory comes from the work of Mattatall et al. [95]. *Bacillus subtilis* has a Sco1 homolog, YpmQ, but does not have a Cox17 homolog. It also contains two different cyto-

Alignment of Sco Sequences

```
S.pombe       MFRRGLVFSRHCHYSLIRPRFPLNRTCLARFADGRKNLATDNRTQTYQSWRGMISIRR--
Human         MAMLVLVPGRVMRPLGGQLWRFLPRGLEFWGPAEGTARVLLRQFCARQAEAWRASGRPGY
Arabidopsis   MLKMDQRCLLSTSASDTTSKHDSGKPETKSSEKNEKSGGSESSDGGSDHKNERASGKCVR
S.cerevisiae  MLKLSRSANLRLVQLPAARLSGNGAKLLTQRGFFTVTRLWQSNGKKPLSRVPVGGTPIKD
C.elegans     MLRTVSLACSTANLCKNTKPVWTLASAARFSDKNKGDDLETDLQKDLKKLNEILKTGITET


S.pombe       ---------------------------------ALLLAAATSVGLYAYFQHEK
Human         CLGTRPLSTARPPPPWSQKGPGDSTRPSKPGPVSWKSLAITFAIGGALLAGMKHVK
Arabidopsis   -----------------------GGPVSWMSFFLLFATGAGLVYYYDTQKKRHIE
S.cerevisiae  --------------------NGKVREGSIEFSTGKAIALFLAVGGALSYFFNREK
C.elegans     ETTSSPKEPPVDKNFMNFRKQAEQEAFQRSSIFNWKTVLGTFAVGGTCLAALFYIK

                     *.  .** * *         *   *.       *. ...****. ***.**
S.pombe       KKVLERQNDKVLATIGRPQLGGAFSLIDHHGNRVTDNDFKGKFSLIYFGFTRCPDICP
Human         KEKAEKLEKERQRHIGKPLLGGPFSLTTHTGERKTDKDYLGQWLLIYFGFTHCPDVCP
Arabidopsis   DINKNSIAVKEGPSAGKAAIGGPFSLIRDDGKRVTEKNLMGKWTILYFGFTHCPDICP
S.cerevisiae  RRLETQKEAEANRGYGKPSLGGPFHLEDMYGNEFTEKNLLGKFSIIYFGFSNCPDICP
C.elegans     KIRLDEREKHRKQTAGKARIGGEWELMNTDGKMEGSQELRGNWLLMYFGFTNCPDICP

                  .* . *.          . *.**. ** **       . * .     *.**. ...
S.pombe       DELDKMSAAIDIVNNVVGD-VVYPIFITCDPARDPPQEMAEYLEDFNPKIVGLTGSYEEI
Human         EELEKMIQVVDEIDSITTLPDLTPLFISIDPERDTKEAIANYVKEFSPKLVGLTGTREEV
Arabidopsis   DELIKLAAAIDKIKENSGV-DVVPVFISVDPERDTVQQVHEYVKEFHPKLIGLTGSPEEI
S.cerevisiae  DELDKLGLWLNTLSSKYGI-TLQPLFITCDPARDSPAVLKEYLSDFHPSILGLTGTFDEV
C.elegans     DEIEKMVKVVEIIEAKKDATPIVPVFISVDPERDSVARVKEYCSEFSNKLRGFTGTTEQV

                     .***         **.***↓.**.*.    *    *
S.pombe       KDICKKFRVYFSTPKNIDPKKDDYLVDHSVFFYLMDPEGKFIEVFGRN----STSEDLAR
Human         DQVARAYRVYYS-PGPKDED-EDYIVDHTIIMYLIGPDGEFLDYFGQN----KRKGEIAA
Arabidopsis   KSVARSYRVYYMKTE--EEDS-DYLVDHSIVMYLMSPEMNFVKFYGKN----HDVDSLTD
S.cerevisiae  KNACKKYRVYFSTPPNVKPG-QDYLVDHSIFFYLMDPEGQFVDALGRNYDEKTGVDKIVE
C.elegans     NKVAKTFRVYHSQGPRTNKQEDDYIVDHTVIMYLIDPSGQFHDYYGQNRKAEEIANVIEM


S.pombe       AIGSYYLS--RKKQK---------
Human         SIATHMRP-YRKKS----------
Arabidopsis   GVVKEIRQ-YRK------------
S.cerevisiae  HVKSYVPAEQRAKQKEAWYSFLFK
C.elegans     KVLKYQAQNRKSLLNLF
```

Fig. 6. Alignment of Sco protein sequences from various species. Five eukaryotic Sco sequences are shown. The candidate single transmembrane helix is shown in bold. The conserved CxxxC motif is highlighted in bold and underlined in the yeast sequence. The Ser residue mutated in human Sco2 (only the human Sco1 sequence is shown) that results in cardioencephalomyopathy is shown with an arrow.

chrome oxidases: a cytochrome *c* oxidase that has both Cu$_A$ and Cu$_B$ centers, and a menaquinol oxidase, that only has Cu$_B$ [95]. Cells lacking YpmQ show a reduction in cytochrome *c* oxidase, but not menaquinol oxidase. Interestingly, addition of copper to the growth media restores cytochrome *c* oxidase activity [95]. These results suggest that Cox17 delivers Cu to mitochondria, but does not insert Cu directly into Cox2. That step may be carried out by Sco1. So it makes sense that Cox17 would be absent from bacteria. If Sco1 delivers Cu to Cox2, one would predict the proteins to interact physically with one another. Lode et al. [96] have demonstrated such a protein–protein interaction through both affinity chromatography and co-immunoprecipitation experiments.

SCO1 is conserved in both prokaryotes and higher eukaryotes (Fig. 6). There are two homologous genes in humans, found on chromosomes 17 and 22 [97]. Neither gene product complements the lack of yeast Sco1. However, one chimera, consisting of the N-terminal half of yeast Sco1 and the C-terminal half of the chromosome 17 homolog, does substitute for the yeast protein. Thus, this gene was designated *HsSCO1*, even though its sequence homology to *ScSCO1* is no better than that of the chromosome 22 homolog, which has been designated *HsSCO2* [98]. The human homologs show similar divergence from both yeast genes, suggesting they are not orthologous genes that diverged from a common ancestral gene pair; rather they are paralogous genes, resulting from gene duplication events occurring independently in the two species [98].

The importance of *SCO1* in yeast has been proven, while the importance of *SCO2* is still unclear. However, both gene products have recently been shown to be essential in humans. Mutation of either *HsSCO1* or *HsSCO2* leads to decreased cytochrome *c* oxidase activity and death in infants within a few months of birth [98–100]. *sco2* mutations caused cardioencephalomyopathy, while *sco1* mutations caused neonatal hepatic failure and ketoacidotic comas. Interestingly, all patients so far were compound heterozygotes with one mutation in a residue adjacent to the CxDxC motif. However, when the parallel mutation was made in yeast Sco1, the protein was functional. A second pathogenic mutation in HsSco2, located near the C-terminus, did result in loss of function in yeast [101].

Although cells harboring this mutant Sco1 (hS225F) were respiratory-deficient, heme *a* was bound by the cytochrome *c* oxidase, suggesting that Sco1 acts late in the step of cytochrome *c* oxidase assembly [101].

If the Sco proteins are only important in Cu insertion into Cox2, then how is Cu provided to Cox1? This pathway seems to require another inner mitochondrial membrane protein [102], the product of *COX11*. *S. cerevisiae* lacking Cox11 are respiratory-deficient, have impaired cytochrome *c* oxidase activity, have lower levels of Cox1, and lack hemes *a* and *a3* [102]. These data suggest a role in oxidase activity, but it was the work done by Hiser et al. in *R. sphaeroides* that finally defined the precise role of Cox11 and linked it to Cu delivery [103].

Cytochrome oxidase was overexpressed and purified from *R. sphaeroides* cells lacking Cox11, and its properties examined. The mutant enzyme, which is present in smaller amounts in the membrane compared to wild type, does not have Cu$_B$, shows a decrease in magnesium/manganese content, and has altered heme environments. Formation of Cu_A, subunit assembly, and heme synthesis and insertion into cytochrome oxidase are not affected by Cox11. Thus, the absence of Cox11 appears to preclude Cu_B site formation without affecting binding of other cofactors. Cox11 is a conserved protein that, like Sco1, contains a single candidate transmembrane motif downstream of the N-terminal mitochondrial targeting sequence (Fig. 7). Cox11 contains a conserved CxC sequence motif that may be important in Cu(I) binding and Cu(I) ion delivery to the Cu_B site in cytochrome c oxidase. *Rhodobacter* Cox11 resembles the eukaryotic Cox11 molecules (Fig. 8A).

Alignment of Cox11 Sequences

```
                                                                                  *
S. cerevisiae     -------MIRICPIVRSKVPLLG----TFLRSDSWLAPHALALRRAICKNVALRSYSVNS
Neurospora        -----------MTTAAQQQASG-----AKKQQSDRAWRWFSTEGASRRQQQQQTRSQSS
Arabidopsis       -----------------------------------------------MLDSAHRQYSTHS
Human             MGGLWRPGWRCVPFCGWRWIHPGSPTRAAERVEPFLRPEWSGTGGAERGLRWLGTWKRCS

                                                  . .   *  .*     .  .* .**.*.
yeast_Cox11       EQPKHTFDISKLTRNEIQQLRELKRARERKFKDRTVAFYFSSVAVLFLGLAYAAVPLYRA
Neurospora        ------------RTGVSPEMERVRAEYKKRNQS--TMYYVISVILGTVALSYGSVPMYKM
Arabidopsis       -------------PSE------------TKSQK--MLYYLTAVVFGMVGLTYAAVPLYRT
Human             LRARHPALQPPRRPKSSNPFTRAQEE-ERRRQNKTTLTYVAAVAVGMLGASYAAVPLYRL

                   *   ** **                            .   . *  . *   *  * * * *
yeast_Cox11       ICARTGFGGIPITD-------RRKFTDDKLIPVDTEKRIRISFTSEVSQILPWKFVPQQR
Neurospora        ICQTTGWGGQPVRAHGAGGSDSDVDLAAKLEPVRDAKRMRVTFSASVSDVLPWKFVPQQR
Arabidopsis       FCQATGYGGTVQRKETV----EEKIARHSESGTVTEREIVVQFNADVADGMQWKFTPTQR
Human             YCQTTGLGGSAVAG-------HASDKIENMVPVK-DRIIKISFNADVHASLQWNFRPQQT

                   *.  . ***.***** * *     . *..** .  *    ** ***.******.*   *   .
yeast_Cox11       EVYVLPGETALAFYKAKNYSDKDIIGMATYSIAPGEAAQYFNKIQCFCFEEQKLAAGEEI
Neurospora        EVRILPGETALAFYTATNMSDKDIIGVATYSVTPGQVAPYFSKIQCFCFEEQRLNAGETV
Arabidopsis       EVRVKPGESALAFYTAENKSSAPITGVSTYNVTPMKAGVYFNKIQCFCFEEQRLLPGEQI
Human             EIYVVPGETALAFYRAKNPTDKPVIGISTYNIVPFEAGQYFNKIQCFCFEEQRLNPQEEV

                   ****** ..**..    *  *    .    * ****
yeast_Cox11       DMPVFFFIDPDFASDPAMRNIDDIILHYTFFRAHYGDGTAVSDS-----KKEPEMNADEK
Neurospora        DMPVFFYLDPDYLNDLNMKGIETVTLSYTFFSKFCPVLPRVWSVGFWTVETMKGTKADDG
Arabidopsis       DMPVFFYIDPEFETDPRMDGINNLILSYTFFKVSEENTTETVN----------NNNSVP
Human             DMPVFFYIDPEFAEDPRMIKVDLITLSYTFFEAKEGHKLPV----------PGYN----

yeast_Cox11       AASLANAAILSPEVIDTRKDNSN
Neurospora        -SFLTEAKYDDNGVLKGVPGAP-
Arabidopsis       -VQETN-----------------
Human             -----------------------
```

Fig. 7. Alignment of Cox11 sequences from various species. Four eukaryotic Cox11 sequences are shown. Cox11 has a single candidate transmembrane segment shown underlined for the yeast sequence. Cox11 is also found in many prokaryotes. The two cysteines in the highly conserved QCFCF motif are in bold.

A.

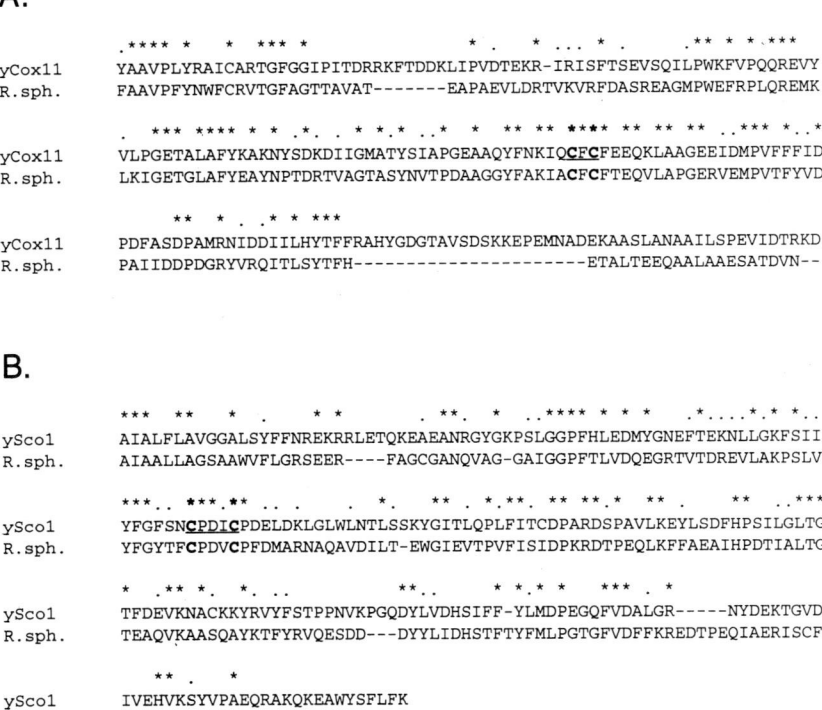

```
              .****  *   *  *** *                 *  .   *  ...  *  .        .** * * .***
yCox11    YAAVPLYRAICARTGFGGIPITDRRKFTDDKLIPVDTEKR-IRISFTSEVSQILPWKFVPQQREVY
R.sph.    FAAVPFYNWFCRVTGFAGTTAVAT-------EAPAEVLDRTVKVRFDASREAGMPWEFRPLQREMK

              .  *** **** *  *  .*. .  * *.*  ..*  *  ** ** **** ** ** **   .*** *..*
yCox11    VLPGETALAFYKAKNYSDKDIIGMATYSIAPGEAAQYFNKIQCFCFEEQKLAAGEEIDMPVFFFID
R.sph.    LKIGETGLAFYEAYNPTDRTVAGTASYNVTPDAAGGYFAKIACFCFTEQVLAPGERVEMPVTFYVD

                  ** *  .* * ***
yCox11    PDFASDPAMRNIDDIILHYTFFRAHYGDGTAVSDSKKEPEMNADEKAASLANAAILSPEVIDTRKD
R.sph.    PAIIDDPDGRYVRQITLSYTFH--------------------ETALTEEQAALAAESATDVN--
```

B.

```
              ***  **   *  .   *  *          .**. *   ..**** * * *    .*....*.* *..
ySco1     AIALFLAVGGALSYFFNREKRRLETQKEAEANRGYGKPSLGGPFHLEDMYGNEFTEKNLLGKFSII
R.sph.    AIAALLAGSAAWVFLGRSEER----FAGCGANQVAG-GAIGGPFTLVDQEGRTVTDREVLAKPSLV

              ***..***.**  ..          *.   **  *.** ** **.*  **        **    ..***
ySco1     YFGFSNCPDICPDELDKLGLWLNTLSSKYGITLQPLFITCDPARDSPAVLKEYLSDFHPSILGLTG
R.sph.    YFGYTFCPDVCPFDMARNAQAVDILT-EWGIEVTPVFISIDPKRDTPEQLKFFAEAIHPDTIALTG

              * .** *.  *. ..          **..    * * *    *** . *
ySco1     TFDEVKNACKKYRVYFSTPPNVKPGQDYLVDHSIFF-YLMDPEGQFVDALGR-----NYDEKTGVDK
R.sph.    TEAQVKAASQAYKTFYRVQESDD---DYYLIDHSTFTYFMLPGTGFVDFFKREDTPEQIAERISCFA

              ** . *
ySco1     IVEHVKSYVPAEQRAKQKEAWYSFLFK
R.sph.    NDSHVSTSFDARAQKSYQASRGKQMGD
```

Fig. 8. Sequence alignment of yeast Cox11 and *Rhodobacter* Cox11 (**A**) and Sco1 and *Rhodobacter* PrrC (**B**). Only the conserved middle segments of each are shown. The sequences for *Rhodobacter* PrrC and Cox11 start at residues 12 and 25, respectively. The sequences shown for yeast Sco1 and Cox11 start at residues 76 and 101, respectively.

19.8
Terminal Oxidases in Prokaryotes

Aerobic respiration in many bacteria involves cytochrome *c* oxidases as the terminal oxidase [104]. Bacterial cytochrome *c* oxidases are usually homologous to the mitochondrial complex in containing heme *a* and copper as essential electron transfer cofactors. The structure of the oxidase from *Paracoccus denitrificans* is known [93]. The enzyme is simpler than the mitochondrial oxidase in containing only four subunits. Three subunits resemble the core three subunits synthesized within the mitochondrion of eukaryotes. As with eukaryotic oxidases, subunit 1 contains the two heme *a* moieties and the Cu_B site. Subunit 2 contains the binuclear Cu_A site. The fourth *Paracoccus* subunit resembles one of the small eukaryotic subunits and forms a single transmembrane helix. The C-terminal globular domain of subunit 2 protrudes into the periplasm where it docks with cytochrome *c*. Cu_A site formation is expected to occur within the periplasm.

Thermus thermophilus contains two distinct cytochrome *c* oxidases. Expression of the two varies depending on growth conditions [105]. The *ba3* enzyme is simpler than the *aa3* enzyme, which resembles the paracoccal enzyme. Three intriguing differences are the presence of a *b*-type cytochrome in place of heme *a* (heme *a* is present in the heterometallic binuclear site with Cu_B), the presence of hydroxyethylgeranylgeranyl moiety instead of the hydroxyethylfarnesyl group, and the absence of a subunit equivalent to the mitochondrial subunit 3 [29]. The third subunit in the *Thermus* enzyme is again a small single membrane spanning helix [29]. A novel aspect of the *aa3* oxidase is that subunit 2 is a fusion polypeptide with its substrate, cytochrome *c* [104].

The genes encoding subunits of the bacterial cytochrome *c* oxidases are organized in one or two operons [104, 106]. One *P. denitrificans* oxidase operon encodes subunits 2 and 3 and, in addition, the two enzymes involved in the conversion of protoheme to heme *a* [104]. Subunits 1 and 4 are encoded elsewhere in the genome. The *B. subtilis* oxidase operon encodes all four subunits and the two heme modifying enzymes.

E. coli lack a cytochrome *c* oxidase. One terminal oxidase is a ubiquinol oxidase. Reduced ubiquinol is oxidized directly by oxygen without involving the cytochrome *bc* complex [107]. Although this enzyme has subunits homologous to the *aa3* enzymes, the quinol oxidase lacks the Cu_A center [108]. Subunit 1 of the enzyme contains a cytochrome *b* and a cytochrome *o*-Cu_B binuclear site. Cytochrome *o* is a modified heme with the hydroxyethylfarnesyl modification but lacking the formylation modification. Subunit 2 is homologous to Cu_A subunit 2 polypeptides, but lacks the ligand residues for the Cu_A site. A series of six mutations made in the *E. coli* quinol oxidase subunit 2 to create the Cu_A ligands resulted in the formation of a Cu_A site after reconstitution [109]. The Cu_A was not metallated *in vivo* unless the cells were cultured in medium containing high levels of Cu(II) [109]. However, the metallated mutant quinol oxidase failed to oxidize reduced cytochrome *c*.

Another type of ubiquinol oxidase is expressed in some fungi and plants under certain growth conditions. *Podospora anserina* cultured in Cu-limiting conditions expresses an alternative oxidase that is homologous to ubiquinol oxidases from plants and *Neurospora crassa* [110]. The terminal oxidase is believed to have a binuclear Fe center and no copper. Proton pumping from complex I provides the driving force for ATP synthesis in this alternative respiration pathway. Cu-limiting conditions result in diminished cytochrome *c* oxidase activity. The mechanism of induction of the alternative oxidase is unclear, but seems to correlate with conditions that result in enhanced life-span of this filamentous fungus [110]. An enhanced life-span is also seen in cells containing a *cox5Δ* mutation [111]. The alternative oxidase does not generate the same level of reactive oxygen species as the usual respiratory chain involving cytochrome *c* oxidase does, so the life-span of this fungus correlates with oxidative stress.

19.9
Metallation of Prokaryotic Terminal Oxidases

Cytochrome c oxidases exist in prokaryotes, though they are simpler oligomers. Many steps in the assembly of these species are known. An understanding of metallation of bacterial oxidases provides insights into the more complex eukaryotic oxidase metallation pathway. In addition to copper metallation of Cu_A and Cu_B sites, more information is available in prokaryotes on heme a biosynthesis. Some data suggest that heme a modification is regulated by the copper status of cells, so the reactions in heme a formation will be discussed first.

Assembly of the bacterial oxidases is less complex than the mitochondrial enzymes in that fewer subunits are involved. Heme a formation is required in most species other than E. coli, which utilizes heme o. The oxidase operons contain only oxidase subunit genes and genes encoding the heme a modifying enzymes (farnesyl transferase and the formylating enzyme) [106]. Mutations in the gene encoding the heme farnesyl transferase (cyoE) in E. coli resulted in the accumulation of a non-functional oxidase containing only heme b. Furthermore, membranes isolated from E. coli overproducing CyoE catalyzed production of heme o from protoheme and farnesyl pyrophosphate [112]. Heme o appears to be a precursor of heme a [112]. The farnesyl transferase (CyoE) is conserved in many species, including S. cerevisiae, where it is designated Cox10 [113]. Cox10 sequences from various species show strong homology in a region containing the candidate protoheme His ligand.

In contrast, the formylating enzyme is poorly characterized. Bacterial mutants defective in the formylation reaction lead to accumulation of the heme o intermediate [114, 115]. Expression of the putative formylating enzyme CtaA from Bacillus stearothermophilus or B. subtilis in E. coli resulted in increased formation of heme a [114, 116]. As mentioned previously, E. coli does not normally contain heme a. Membranes from the transfected E. coli cells catalyzed the conversion of heme o to heme a in vitro [116]. CtaA is a 35 kDa intrinsic membrane protein with seven hydrophobic segments and is believed to be a heme-containing monooxygenase [117]. The results with CtaA strongly suggest that the molecule is the functional heme a synthase. It is unclear whether the heme modifying enzymes have an additional function, inserting heme a into oxidase subunit I sites.

No robust eukaryotic homologs of CtaA exist. The only known heme a modifying enzyme in yeast is Cox10, a farnesyl transferase. Although Cox11 shows no significant sequence homology to CtaA, Cox11 has been suggested to be the CtaA equivalent responsible for formylation. In support of this prediction, analyses of the heme constituents in a cox11 mutant revealed the absence of heme a and presence of heme o [118]. The addition of heme a to cox11 mutants restored growth on non-fermentable carbon sources [37]. The evidence against this prediction is the presence of heme a in a deletion of the COX11 homolog in Rhodobacter [103].

Cu_A and Cu_B site formation in bacteria may also involve accessory proteins. These gene products are distinct from heme a modifying enzyme genes that are frequently part of bacterial oxidase operons. The only candidate Cu metallochaperone that appears in a bacterial oxidase operon is Cox11, which is implicated in Cu_B site formation [103]. *COX11* homologs from *P. denitrificans* (*ctaG*) and *Rhodobacter* are part of the oxidase operons. Cox11 is found in other bacteria such as *Bradyrhizobium* and *Pseudomonas stutzeri*, that contain well characterized cytochrome *c* oxidases. Curiously, Cox11 homologs appear in neither the genomes of *B. subtilis* nor *E. coli*, both of which contain Cu_B site oxidases.

Cu_A site formation in yeast is believed to involve both Cox17 and Sco1/Sco2. No bacterial homologs to Cox17 have been identified. This is not surprising considering its proposed function as a Cu(I) transporter through the cytoplasm and into the mitochondrion. However, bacterial cytochrome *c* oxidases have the Cu_A domain protruding into the periplasm. Thus, Cu transport into the periplasm is expected to be important in Cu_A site formation. Cu ion transport to the periplasm may occur through an outer membrane permease or export from the cytoplasm through a transporter. Details of this step are not known. Organisms such as *P. denitrificans*, *B. subtilis*, and *P. stutzeri* have Cu_A-type oxidases and also contain Sco1 homologs. Sco1 is the putative Cu_A site metallochaperone, thus it is predicted that deletion of the Sco1 genes in these species will abrogate Cu_A formation. Corroborating this, disruption of the Sco1 gene (*ypmQ*) in *B. subtilis* abolishes formation of a functional cytochrome *c* oxidase, and exogenous Cu salts can partially suppress this mutation [95].

The marine *Bacillus* sp. strain SG-1 contains a robust Sco1 homolog designated MnxC. This gene is present in a Mnx operon that also contains a multi-copper oxidase (MnxG) in the laccase family rather than the cytochrome oxidase family. The other members of the operon, MnxB, MnxD, MnxE and MnxF are small, questionable, open reading frames. The operon is functionally important in manganese oxidation, so the copper oxidase is likely to be the functionally important enzyme. Thus, it is possible that the Sco1 homolog is important for the assembly of this oxidase.

A distinct function is suggested for the Sco1 homolog in *R. sphaeroides* [119, 120]. *Rhodobacter* is a purple bacterium capable of both aerobic growth and anoxic photosynthesis [120]. Cells growing aerobically use both an aa_3-type cytochrome *c* oxidase and a cbb_3 oxidase. The cbb_3 enzyme oxidizes reduced cytochrome *c* directly using cytochrome *b* and a second cytochrome *b*-Cu_B binuclear site. The switch from aerobic growth to anoxic growth involves induction of genes whose products form the photosynthetic reaction center within membrane invaginations of the plasma membrane [120]. *R. sphaeroides* Sco1 is part of a gene cluster of a two-component regulatory system (PrrB and PrrA) that controls induction of the photosystem [119]. Electron transport through the respiratory chain and a functional cbb_3 oxidase are important in signaling the PrrBA system to repress the photosystem under aerobic conditions [121]. The cbb_3 oxidase appears to be an oxygen sensor in addition to a terminal oxidase. PrrBA repression also requires functional Sco1 (PrrC) [120] (Fig. 8B). It is not clear how Sco1 mediates signaling to PrrBA. The redox state of the bound Cu ion in Sco1 may be important in signaling. Alternatively, a distinct function in Sco1 which does not involve Cu ion binding may be regulatory. It is also unclear whether *Rhodobacter* Sco1 is functionally important in assembly of the aa_3-type cytochrome c oxidase. A candidate Cu(I) P-type ATPase exists in an operon adjacent to the cbb_3 oxidase operon in R. sphaeroides [122]. Disruption of this putative ATPase gene diminishes cbb_3 oxidase activity, but no information is available on its effect on cytochrome c oxidase activity. It is conceivable that the ATPase provides Cu(I) for formation of the Cu_B site in the cbb_3 oxidase. The R. sphaeroides Cox11 is important for Cu_B site formation in the cytochrome c oxidase [103], yet no information exists on whether it is also important in Cu_B site formation in the cbb_3 oxidase.

Cu_A also exists in the nitrous oxide reductase (N_2OR) in P. stutzeri [123, 124]. This enzyme catalyzes the final step in bacterial denitrification: the reduction of nitrous oxide to dinitrogen. The Cu_A site transfers electrons from an external electron donor to a tetracopper catalytic center (the Cu_Z site). The Cu_A site resembles that of cytochrome c oxidase in ligand arrangement and in the short Cu–Cu distance of 2.5 Å. Despite those similarities, the reductase exhibits only limited sequence homology to subunit 2 of cytochrome c oxidase [125]. Nitrous oxidase reductase is a periplasmic enzyme, like cytochrome c oxidase [123]. Given that Sco1 is likely to be a key factor in Cu_A site formation in cytochrome c oxidase, the possibility exists that Sco1 is also important in Cu_A site formation in N_2OR. P. stutzeri contain a robust Sco1 homolog. Metallation of the Cu_A site in N_2OR occurs after export of the reductase to the periplasm [126, 127]. Three gene products, nosD, nosF, and nosY, have been identified in P. stutzeri that are proposed to function in transport of Cu ions across the plasma membrane [123]. An outer membrane transporter, NosA, is also required for copper metallation of N_2OR [128]. A unifying postulate is that Sco1 functions in Cu_A site formation and NosD, NosF and NosY are functionally important for Cu_Z site formation. NosA may be important in transporting Cu ions from the medium for periplasmic Cu_A site formation, but it is not clear whether NosA is important for cytochrome c oxidase Cu_A site formation. The importance of accessory factors in metallation of N_2OR was highlighted by the observation that no metallation of the reductase occurred in E. coli, although the apo-protein could be reconstituted in vitro [123, 127].

Other binuclear Cu centers are known. Hemocyanin, tyrosinase and catechol oxidases contain binuclear, spin-coupled Cu centers. These centers are distinct from a Cu_A site in each having a Cu(II) ion in the binuclear center coordinated by three histidyl residues and a bridging solvent ion. The Cu ions are separated by 2.9 Å in the oxidized structures. In contrast, Cu_A sites have two bridging cysteinyl thiolates and the two Cu ions are only 2.5 Å apart. It is unlikely that Sco1 is involved in formation of these non-Cu_A binuclear sites, although a formal test of this prediction has not been conducted.

19.10
Postulated Model

Copper ion transport to the mitochondrion is required for assembly of a functional cytochrome c oxidase. Distinct pathways probably exist for the metallation of the Cu_A and Cu_B sites of cytochrome c oxidase. Cox17 and Sco1 appear critical for Cu_A site formation. In contrast, Cox11 is a candidate co-metallochaperone for Cu_B site formation and perhaps functions with Cox17.

The evidence linking Cox17 as a mitochondrial Cu ion shuttle is based on three observations. First, Cox17 exhibits dual localization in the cytoplasm and intermitochondrial membrane space. Second, Cox17 binds Cu(I) ions. Third, exogenous Cu salts suppress a *cox17* null strain. Cox17 presumably ferries Cu(I) ions across the outer membrane as a Cu–protein complex followed by docking with co-metallochaperones Sco1 and/or Cox11. The CuCox17 complex is oligomeric unlike the apo-Cox17 polypeptide. A poly-copper cluster formed at the dimer interface may protect the cluster during import across the mitochondrial outer membrane.

Sco1 was implicated in copper ion delivery by the observation that the respiratory-deficient phenotype of *cox17-1* cells was suppressed by high-copy *SCO1* or the homologous *SCO2* gene [89]. Cox17 is proposed to shuttle Cu ions to the mitochondria and transfer Cu to Sco1 as an intermediate step in Cu metallation of Cu_A. The role of Sco2 is unclear. The identification of respiratory defects in humans with mutations in both *SCO1* and *SCO2* is consistent with both genes being functionally important. Since Sco1 in *R. sphaeroides* is functionally important in oxygen sensing and regulation of photosystem induction, it is conceivable that yeast Sco2 normally functions in a pathway distinct from Cu_A site formation but can replace Sco1 when overexpressed.

As mentioned, Cox11 is critical for Cu_B site formation in *R. sphaeroides* [103]. If yeast Cox11 is a metallochaperone for the Cu_B site of cytochrome oxidase, the possibility exists that CuCox17 delivers Cu(I) to both Sco1 and Cox11 for subsequent donation to the Cu_A and Cu_B sites, respectively.

Assembly of cytochrome c oxidase requires not only Cu(I) ions but heme modified to heme a. An intriguing question is whether heme a modification is coupled to the copper status of cells. Cu-deficient *Candida utilis* cells express oxidase subunits but fail to accumulate heme a [129]. Likewise, Cu-deficient swine fail to accumulate heme a and have diminished cytochrome oxidase activity [130]. Heme a formation was more impaired than heme synthesis, suggesting a potential link between copper status and heme a formation [130].

The assembly of cytochrome oxidase can be modeled as follows. Heme a formation is catalyzed by Cox10 and at least one additional enzyme. Subunit 1 binds heme a independently of other subunits, but assembly of the heme a_3-Cu_B site requires a partially assembled complex [32]. The heme a modification appears to be coupled to the availability of Cu(I). The Cu(I) signal may come from metallation of either Sco1 or Cox11. The small nuclear-encoded subunits stabilize the core complex and Zn(II) binds to stabilize Cox4.

The Cu(I) ions in the intermitochondrial membrane (IM) space enter via the CuCox17 complex (Fig. 9). CuCox17 uptake is predicted to occur without the unfolding which usually accompanies import through the TOM complex. It is unclear whether CuCox17 is threaded through the outer membrane via the TOM channel as a native complex or whether import occurs through a distinct channel. CuCox17 must be actively transported as its concentration in the intermembrane space is significantly elevated over that in the cytoplasm. The enhanced Cox17 concentration is predicted to result in tetramerization of the protein which is expected to be the active form [82].

Another question is how Cox17 transfers Cu(I) to the oxidase complex. Based on the Atx1 and Lys7 metallochaperone systems, Cu(I) transfer is likely to be mediated by protein–protein interactions. Cu(I) transfer to cytochrome c oxidase is predicted to be mediated by Sco1 which interacts directly with Cox2 [96]. Cu_A site formation is predicted to occur in the assembled oligomeric oxidase complex and may involve the dimeric cytochrome c oxidase. The two Cu_A centers in the dimeric bovine enzyme are separated by 74 Å [22]. The dimer interface of the bovine enzyme within the intermitochondrial membrane space is formed by subunit VIb which also interacts tightly with subunit 2 [22]. The N-terminal residues of subunit VIb molecules are only separated by 26 Å (Fig. 10). The corresponding gene in yeast is Cox12. Sco1 may interact as a dimer with the dimeric oxidase, and may interact with both Cox2 and Cox12. The dimer interface formed by Cox12 is suitably positioned to serve as a docking site. The possibility exists that tetrameric Cox17 forms

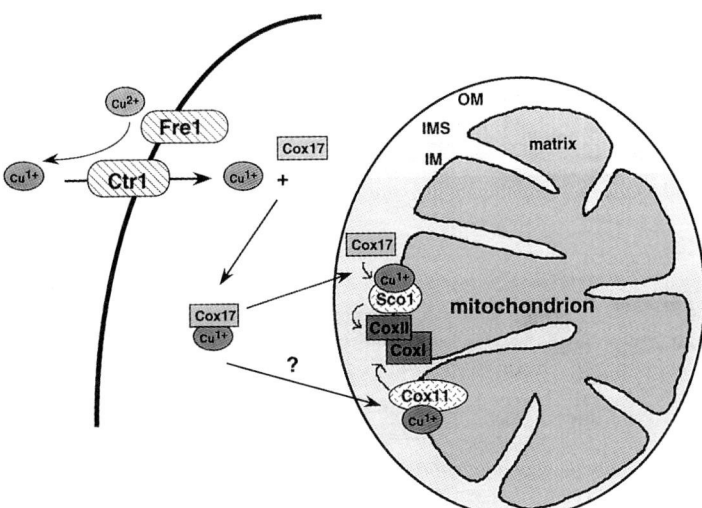

Fig. 9. Proposed model for Cu metallation of mitochondrial cytochrome c oxidase. Cox17 is proposed to be the shuttle transporting Cu(I) across the outer membrane. CuCox17 is proposed to dock with Sco1 prior to Cu(I) transfer to Sco1 and subsequently to Cox2. Sco1 is known to interact with Cox2 [96]. Metallation of Cu_B is known to involve Cox11 [103]. CuCox17 may also transfer Cu(I) to Cox11 prior to donation to the Cu_B in oxidase subunit 1.

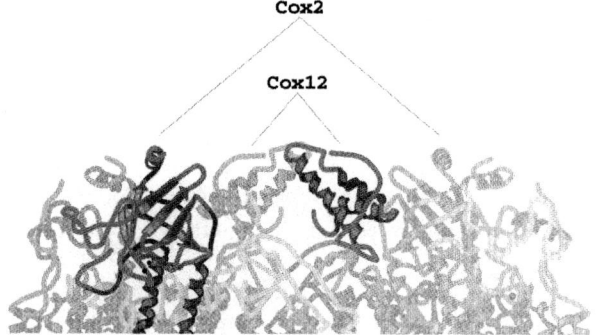

Fig. 10. The domains of cytochrome c oxidase that project into the intermitochondrial membrane space are shown as the oxidase dimer. Two main components of this interface are Cox2 and Cox12. The Cox2 domain in the IM space is shown in dark shade on the monomer at the left. The Cox12 domain in the IM space is shown in dark shade on the monomer at the right. Cox2 and perhaps Cox12 may be important interfaces for assembly of Cu_A through Sco1 and Cox17.

a transient complex with a dimeric Sco1 docked onto Cox2/Cox12 within the IM space. Cu(I) transfer may occur through ligand exchange reactions as predicted for Cu(I) transfer from Atx1 to Ccc2 [62]. It is likely that many of the steps in the assembly of cytochrome c oxidase will be resolved by studies in both yeast and bacterial systems.

References

1. Stevens, B., in: *Mitochondrial Structure. The Molecular Biology of the Yeast Saccharomyces: Life Cycle and Inheritance* (Strathen, J. N., Jones, E. W., Broach, J., Eds.), Cold Spring Harbor Laboratory Press, **1981**.
2. Hermann, G. J., Shaw, J. M., *Annu. Rev. Cell Dev. Biol.* **1998**, *14*, 265-303.
3. Mitchell, P., *Science* **1979**, *206*, 1148.
4. Rickwood, D., Dujon, B., Darley-Usmar, V. M., in: *Yeast Mitochondria. Yeast: A Practical Approach* (Campbell, I., Duffus, J. H., Eds.), IRL Press, Oxford, **1988**.
5. Munn, E. A. *The Structure of Mitochondria*, Academic Press, London, **1974**.
6. Schleiff, E., Turnbull, J. L., *Biochemistry* **1998**, *37*, 13043-13051.
7. Suissa, M., Schatz, G., *J. Biol. Chem.* **1982**, *257*, 13048-13055.
8. Schatz, G., *J. Biol. Chem.* **1996**, *271*, 31763-31766.
9. Neupert, W., *Ann. Rev. Biochem.* **1997**, *66*, 863-917.
10. Stan, T., Ahtitng, U., Dembowski, M., Kunkele, K.-P., Nussberger, S., Neupert, W., Rapaport, D., *EMBO J.* **2000**, *19*, 4895-4902.
11. Hill, K., Model, K., Ryan, M. T., Dietmeier, K., Martin, F., Wagner, R., Pfanner, N., *Nature* **1998**, *395*, 516-521.
12. Schwartz, M. P., Hung, S., Matouschek, A., *J. Biol. Chem.* **1999**, *274*, 12759-12764.
12. Koehler, C. M., Jarosch, E., Tokatlidis, K., Schmid, K., Schweyen, R. J., Schatz, G., *Science* **1998**, *279*, 369-373.
13. Gaume, B., Klaus, C., Ungermann, C., Guiard, B., Neupert, W., Brunner, M., *EMBO J.* **1999**, *17*, 6497-6507.
15. Leuenberger, D., Bally, N. A., Schatz, G., Koehler, C. M., *EMBO J.* **1999**, *18*, 4816-4822.

16. Diekert, K., Kispal, G., Guiard, B., Lill, R., *Proc. Natl. Acad. Sci. USA* **1999**, *96*, 11752-11757.
17. Hell, K., Herrmann, J. M., Pratje, E., Neupert, W., Stuart, R. A., *Proc. Natl. Acad. Sci. USA* **1998**, *95*, 2250-2255.
18. Corral-Debrinski, M., Blugeon, C., Jacq, C., *Mol. Cell. Biol.* **2000**, *20*, 7881-7892.
19. Kellems, R. E., Allison, V. F., Butow, R. A., *J. Cell Biol.* **1975**, *65*, 1-14.
20. Tzagoloff, A. *Mitochondria*, Plenum Press, New York, **1982**.
21. Li, L., Kaplan, J., *J. Biol. Chem.* **1997**, *272*, 28485-28493.
22. Tsukihara, T., Aoyama, H., Yamashita, E., Tomizaki, T., Yamaguchi, H., Shinzawa-Itoh, K., Hakashima, R., Yaono, R., Yoshikawa, S., *Science* **1995**, *269*, 1069-1074.
23. Musatov, A., Ortega-Lopez, J., Robinson, N. C., *Biochemistry* **2000**, *39*, 12996-13004.
24. Geier, B. M., Schagger, H., Ortwein, C., Link, T. A., Hagen, W. R., Brandtt, U., Von Jagow, G., *Eur. J. Biochem.* **1995**, *227*, 296-302.
25. Taanman, J. W., Capaldi, R. A., *J. Biol. Chem.*, **1992**, *267*, 22481-2485.
26. Burke, P. V., Raitt, D. C., Allen, L. A., Kellogg, E. A., Poyton, P. O., *J. Biol. Chem.* **1997**, *272*, 14705-14712.
27. Burke, P. V., Poyton, R. O., *J. Exp. Biol.* **1998**, *201*, 1163-1175.
28. Cruciat, C. M., Brunner, S., Baumann, F., Neupert, W., Stuart, R. A., *J. Biol. Chem.* **2000**, *275*, 18093-18098.
29. Soulimane, T., Buse, G., Bourenkov, G. P., Bartunik, H. D., Huber, R., Than, M. E., *EMBO J.* **2000**, *19*, 1766-1776.
30. Beinert, H., *Chem. Biol.* **1995**, *2*, 781-785.
31. Zhen, Y., Hoganson, C. W., Babcock, G. T., Ferguson-Miller, S., *J. Biol. Chem.* **1999**, *274*, 38032-38041.
32. Bratton, M. R., Hiser, L., Anthroline, W. E., Hoganson, C., Hosler, J. P., *Biochemistry* **2000**, *39*, 12989-12995.
33. Das, T. K., Pecoraro, C., Tomson, F. L., Gennis, R. B., Rousseau, D. L., *Biochemistry* **1998**, *37*, 14471-14476.
34. Poyau, A., Buchet, K., Bouzidi, M. F., Zabot, M.-T., Echenne, B., Yao, J., Shoubridge, E. A., Godinot, C., *Hum. Genet.* **2000**, *106*, 194-205.
35. McEwen, J. E., Hong, K. H., Park, S., Preciado, G. T., *Curr. Genet.* **1993**, *23*, 9-14.
36. Glerum, D. M., Koerner, T. J., Tzagoloff, A., *J. Biol. Chem.* **1995**, *270*, 15585-15590.
37. Church, C., Chapon, C., Poyton, R. O., *J. Biol. Chem.* **1996**, *271*, 18499-18507.
38. Glerum, D. M., Muroff, I., Jin, C., Tzagoloff, A., *J. Biol. Chem.* **1997**, *272*, 19088-19094.
39. Souza, R. L., Green-Willms, N. S., Fox, T. D., Tzagoloff, A., Nobrega, F. G., *J. Biol. Chem.* **2000**, *275*, 148998-14902.
40. Hell, K., Tzagoloff, A., Neupert, W., Stuart, R. A., *J. Biol. Chem.* **2000**, *275*, 4571-4578.
41. Glerum, D. M., and Tzagoloff, A., *FEBS Lett.* **1997**, *412*, 410-414.
42. Valnot, I., Von Kleist-Retzow, J.-C., Barrientos, A., Gorbatyuk, M., Taanman, J.-W., Mehaye, B., Rustin, P., Tzagoloff, A., Munnich, A., Rotig, A., *Hum. Mol. Genet.* **2000**, *9*, 1245-1249.
43. Robinson, B. H., *Pediatric Res.* **2000**, *48*, 581-585.
44. Mashkevich, G., Repetto, B., Glerum, D. M., Jin, C., Tzagoloff, A., *J. Biol. Chem.* **1997**, *272*, 14356-14364.
45. Dancis, A., Haile, D., Yuan, D. S., Klausner, R. D., *J. Biol. Chem.* **1994**, *269*, 25660-25667.
46. Labbe, S., Zhu, Z., Thiele, D. J., *J. Biol. Chem.* **1997**, *272*, 15951-15958.
47. Pena, M. M. O., Puig, S., Thiele, D. J., *J. Biol. Chem.* **2000**, *275*, 33244-33251.
48. Lin, C. M., Kosman, D. J., *J. Biol. Chem.* **1990**, *265*, 9194-9200.
49. Hassett, R., Kosman, D. J., *J. Biol. Chem.* **1995**, *270*, 128-134.
50. Yamaguchi Iwai, Y., Serpe, M., Haile, D., Yang, W., Kosman, D. J., Klausner, R. D., Dancis, A., *J. Biol. Chem.* **1997**, *272*, 17711-17718.
51. Georgatsou, E., Mavrogiananis, L. A., Fragiadakis, G. S., Alexandraki, D., *J. Biol. Chem.* **1997**, *272*, 13786-13792.
52. Knight, S. A. B., Labbe, S., Kwon, L. F., Kosman, D. J., Thiele, D. J., *Genes Devel.* **1996**, *10*, 1917-1929.
53. Shatwell, K. P., Dancis, A., Cross, A. R., Klausner, R. D., Segal, A. W., *J. Biol. Chem.* **1996**, *271*, 14240-14244.
54. Lesuisse, E., Casteras-Simon, M., Labbe, P., *J. Biol. Chem.* **1996**, *271*, 13578-13583.
55. Finegold, A. A., Shatwell, K. P., Segal, A. W., Klausner, R. D., Dancis, A., *J. Biol. Chem.* **1996**, *271*, 31021-31024.
56. Dancis, A., Roman, D. G., Anderson, G. J., Hinnebush, A. G., Klausner, R. D., *Proc. Natl. Acad. Sci. USA* **1990**, *89*, 3869-3873.

57. Georgatsou, E., Alexandraki, D., *Mol. Cell. Biol.* **1994**, *14*, 3065-3073.
58. Kampfenkel, K., Kushnir, S., Babiychuk, E., Inze, D., Van Montagu, M., *J. Biol. Chem.* **1995**, *270*, 28479-28486
59. Liu, X. F., Supek, F., Nelson, N., Culotta, V. C., *J. Biol. Chem.* **1997**, *272*, 11763-11769.
60. Hassett, R., Dix, D. R., Eide, D. J., and Kosman, D. J., *Biochem. J.* **2000**, *351*, 477-484.
61. Rae, R. D., Schmidt, P. J., Pufahl, R. A., Culotta, V. C., O"Halloran, T. V., *Science* **1999**, *284*, 805-807.
62. Pufahl, R. A., Singer, C. P., Peariso, K. L., Lin, S.-J., Schmidt, P., Fahrni, C., Culotta, V. C., Penner-Hahn, J. E., O"Halloran, T. V. O., *Science* **1997**, *278*, 853-856.
63. Lin, S.-J., Pufahl, R. A., Dancis, A., O'Halloran, T. V. O., Culotta, V. C., *J. Biol. Chem.* **1997**, *272*, 9215-9220.
64. Huffman, D. L., O'Halloran, T. V., *J. Biol. Chem.* **2000**, *275*, 18611-18614.
65. Hamza, I., Schaefer, M., Klomp, L. W. J., Gitlin, J. D., *Proc. Natl. Acad. Sci. USA* **1999**, *96*, 13363-13368.
66. Yuan, D. S., Stearman, R., Dancis, A., Dunn, T., Beeler, T., Klausner, R. D., *Proc. Natl. Acad. Sci. USA* **1995**, *92*, 2632-2636.
67. Askwith, C. C., de Silva, D., Kaplan, J., *Mol. Microbiol.* **1996**, *20*, 27-34.
68. Rosenzweig, A. C., and O'Halloran, T. V., *Curr. Opin. Chem. Biol.* **2000**, *4*, 140-147.
69. Rosenzweig, A. C., Huffman, D. L., Hou, M. Y., Wernimont, A. K., Pufahl, R. A., O'Halloran, T. V., *Structure* **1999**, *7*, 605-617.
70. Lutsenko, S., Kaplan, J. H., *Biochemistry* **1995**, *34*, 15607-15613.
71. MacLennan, D. H., Rice, W. J., Green, N. M., *J. Biol. Chem.* **1997**, *272*, 28815-28818.
72. Toyoshima, C., Nakasako, M., Nomura, H., Ogawa, H., *Nature* **2000**, *405*, 647-655.
73. O"Halloran, R. V., Culotta, V. C., *J. Biol. Chem.* **2000**, *275*, 25057-25060.
74. Lutsenko, S., Cooper, M. J., *Proc. Natl. Acad. Sci. USA* **1998**, *95*, 6004-6009.
75. Borjigin, J., Payne, A. S., Deng, J., Li, X., Wang, M. M., Ovodenko, B., Gitlin, J. D., Synder, S. H., *J. Neurosci.* **1999**, *19*, 1018-1026.
76. Culotta, V. C., Klomp, L. W. J., Strain, J., Casareno, R. L. B., Krems, B., Gitlin, J. D., *J. Biol. Chem.* **1997**, *272*, 23469-23472.
77. Gamonet, F., Lauquin, G. J. M., *Eur. J. Biochem.* **1998**, *251*, 716-723.
78. Casareno, R. L. B., Waggoner, D., Gitlin, J. D., *J. Biol. Chem.* **1998**, *273*, 23625-23628.
79. Hall, L. T., Sanchez, R. J., Holloway, S. P., Zhu, H., Stine, J. E., Lyons, T. J., Demeler, B., Schirf, V., Hansen, J. C., Nersissian, A. M., Valentine, J. S., Hart, P. J., *Biochemistry* **2000**, *39*, 3611-3623.
80. Schmidt, P. J., Kunst, C., Culotta, V. C., *J. Biol. Chem.* **2000**, *275*, 33771-33776.
81. Glerum, D. M., Shtanko, A., Tzagoloff, A., *J. Biol. Chem.* **1996**, *271*, 14504-14509.
82. Heaton, D. N., George, G. N., Garrison, G., Winge, D. R. *Biochemistry*, **2001**, *40*, 743-751.
83. Srinivasan, C., Posewitz, M. C., George, G. N., Winge, D. R., *Biochemistry* **1998**, *37*, 7572-7577.
84. Beers, J., Glerum, D. M., Tzagoloff, A., *J. Biol. Chem.* **1997**, *272*, 33191-33196.
85. Heaton, D., Nittis, T., Srinivasan, C., Winge, D. R., *J. Biol. Chem.* **2000**, *275*, 37582-37587.
86. Amaravadi, R., Glerum, D. M., Tzagoloff, A., *Hum. Genet.* **1997**, *99*, 329-333.
87. Horvath, R., Lochmuller, H., Stucka, R., Yao, J., Shoubridge, E. A., Gerbitz, K. D., Jaksch, M., *Biochem. Biophys. Res. Commun.* **2000**, *276*, 530-533.
88. Punter, F. A., Adams, D. L., Glerum, D. M., *Hum. Genet.* **2000**, *107*, 69-74.
89. Glerum, D. M., Shtanko, A., Tzagoloff, A., *J. Biol. Chem.* **1996**, *271*, 20531-20535.
90. Buchwald, P., Krummeck, G., Rodel, G., *Mol. Gen. Genet.* **1991**, *229*, 413-420.
91. Schulze, M., Rodel, G., *Mol. Gen. Genet.* **1988**, *211*, 492-498.
92. Krummeck, G., Rödel, G., *Curr. Genet.* **1990**, *18*, 13-15.
93. Iwata, S., Ostermeier, C., Ludwig, B., Michel, H., *Nature* **1995**, *376*, 660-669.
94. Rentzsch, N., Krummeck-Weiß, G., Hofer, A., Bartuschka, A., Ostermann, K., Rodel, G., *Curr. Genet.* **1999**, *35*, 103-108.
95. Mattatall, N. R., Jazairi, J., Hill, B. C., *J. Biol. Chem.* **2000**, *275*, 28802-28809.
96. Lode, A., Kuschel, M., Paret, C., Rodel, G., *FEBS Lett.* **2000**, *448*, 1-6.
97. Paret, C., Ostermann, K., Krause-Buchholz, U., Rentzsch, A., Rodel, G., *FEBS Lett.* **1999**, *447*, 65-70.
98. Papadopolou, L. C., Sue, C. M., Davidson, M. M., Tanji, K., Nishion, I., Sadlock, J. E. et. al., *Nature Genet.* **1999**, *23*, 333-337.
99. Jaksch, M., Ogilvie, I., Yao, J., Kortenhaus, G., Bresser, H.-G., Gerbitz, K.-D., Shou-

bridge, E. A., *Hum. Mol. Genet.* **2000**, 795-801.
100. Valnot, I., Osmond, S., Gigarel, N., Mehaye, B., Amiel, J., Cormier-Daire, V., Munnich, A., Bonnefont, J.-P., Rustin, P., Rotig, A., *Am. J. Hum. Genet.* **2000**, *67*, 1104-1109.
101. Dickinson, E. K., Adams, D. L., Schon, E. A., Glerum, D. M. *J. Biol. Chem.* **2000**, *275*, 26780-26785.
102. Tzagoloff, A., Capitanio, N., Nobrega, M. P., Gatti, D., *EMBO J.* **1990**, *9*, 2759-2764.
103. Hiser, L., Di Valentin, M., Hamer, A. G., Hosler, J. P., *J. Biol. Chem.* **2000**, *275*, 619-623.
104. Thony-Meyer, L., *Microbio. Mol. Biol. Rev.* **1997**, *61*, 337-376.
105. Slutter, C. E., Sanders, D., Wittungg, P., Malmstrom, B. G., Aasa, R., Richards, J. H., Gray, H. B., Fee, J. A., *Biochemistry* **1996**, *35*, 3387-3395.
106. Mogi, T., Saiki, K., Anraku, Y., *Mol. Microbiol.* **1994**, *14*, 391-398.
107. Anraku, Y., Gennis, R. B., *Trends Biochem. Sci.* **1987**, *12*, 262-266.
108. Puustinen, A., Finel, M., Haltia, T., Gennis, R. B., Wikstrom, M., *Biochemistry* **1991**, *30*, 3939-3942.
109. van der Oost, J., Lappalainen, P., Musacchio, A., Warne, A., Lemieux, L., Rumbley, J., Gennis, R. B., Aasa, R., Pascher, T., Malmstrom, B. G., Saraste, M., *EMBO J.* **1992**, *11*, 3209-3217.
110. Borghouts, C., Werner, A., Elthon, T., Osiewacz, H. D., *Mol. Cell Biol.* **2001**, *21*, 390-399.
111. Dufour, E., Boulay, J., Rincheval, V., Sainsard-Chanet, A., *Proc. Natl. Acad. Sci. USA* **2000**, *97*, 4138-4143.
112. Saiki, K., Mogi, T., Ogura, K., Anraka, Y., *J. Biol. Chem.* **1993**, *268*, 26041-26044.
113. Nobrega, M. P., Nobrega, F. G., Tzagoloff, A., *J. Biol. Chem.* **1990**, *265*, 14220-14226.
114. Svensson, B., Lubben, M., Hederstedt, L., *Mol. Microbiol.* **1993**, *10*, 193-201.
115. Del Arenal, I. P., Contreras, M. L., Svlateorova, B. B., Rangel, P., Lledias, F., Davila, J. R., Escamilla, J. E., *Arch. Microbiol.* **1997**, *167*, 24-31.
116. Sakamoto, J., Hayakawa, A., Uehara, T., Noguchi, S., Sone, N., *Biosci. Biotechnol. Biochem.* **1999**, *63*, 96-103.
117. Svensson, B., Hederstedt, L., *J. Bacteriol.* **1994**, *176*, 6663-6671.
118. Tzagoloff, A., Nobrega, M., Gorman, N., Sinclair, P., *Biochem. Mol. Biol. Int.* **1993**, *31*, 593-598.
119. Eraso, J. M., Kaplan, S., *J. Bacteriol.* **1995**, *177*, 2695-2706.
120. Eraso, J. M., Kaplan, S., *Biochemistry* **2000**, *39*, 2052-2062.
121. Oh, J.-I., Kaplan, S., *EMBO J.* **2000**, *19*, 4237-4247.
122. Roh, J. H., Kaplan, S., *J. Bacteriol.* **2000**, *182*, 3475-3481.
123. Zumft, W. G., Viebrock-Sambale, A., Braun, C., *Eur. J. Biochem.* **1990**, *192*, 591-599.
124. Brown, K., Tegoni, M., Prudencio, M., Pereira, A. S., Besson, S., Moura, J. J., Moura, I., Cambillau, C., *Nature Struct. Biol.* **2000**, *7*, 191-195.
125. Zumft, W. G., Dreusch, A., Lochelt, S., Cuypers, H., Friedrich, B., Schneider, B., *Eur. J. Biochem.* **1992**, *208*, 31-40.
126. Korner, H., Mayer, F., *Arch. Microbiol.* **1992**, *157*, 218-222.
127. Dreusch, A., Burgisser, D. M., Heizmann, C. W., Zumft, W. G., *Biochim. Biophys. Acta* **1997**, *1319*, 311-318.
128. Mokhele, K., Tang, Y. J., Clark, M. A., Ingraham, J. L., *J. Bacteriol.* **1987**, *169*, 5721-5726.
129. Keyhani, E., Keyhani, J., *Arch. Biochem. Biophys.* **1975**, *167*, 596-602.
130. Williams, D. M., Loukopoulos, D., Lee, G. R., Cartwright, G. E., *Blood* **1976**, *48*, 77-84.

20
Iron and Manganese Transporters in Yeast

Matthew E. Portnoy and Valeria C. Culotta

20.1
Iron Transport in *Saccharomyces cerevisiae*

Iron is a vital nutrient required by virtually all organisms. Several cellular processes require iron for enzymatic function as well as for structural purposes. In nature, however, iron is largely present as the oxidized insoluble Fe^{3+} hydroxide, Fe(OH)$_3$. Organisms must, therefore, develop systems to solubilize iron before uptake and utilization. The bakers' yeast *Saccharomyces cerevisiae* has developed two major pathways to utilize insoluble extracellular iron. Chap. 21 describes one such solubilization and uptake mechanism through the use of small metal binding molecules called siderophores. This section of this chapter will focus on non-siderophore iron transport both at the cell surface and via intracellular transporters.

20.1.1
Reduction of Iron at the Cell Surface

The uptake of non-siderophore iron in bakers' yeast *S. cerevisiae* first requires reduction of the insoluble ferric iron at the cell surface to the ferrous form of the metal. This is largely accomplished through the action of the *FRE1* gene product, a plasma membrane ferric reductase [1, 2]. Fre1p contains binding sites for flavin adenine dinucleotide (FAD) and nicotinamide adenine dinucleotide phosphate (NADPH), which presumably carry out the reduction of Fe^{3+}. In fact, partially purified Fre1p requires FAD and NADPH for reductase activity [3]. Yeast null mutants lacking *FRE1* exhibit a greatly reduced Fe^{3+} reductase activity and are unable to grow on iron limited media [4, 5]. However, the *fre1*Δ mutants still express a residual level of iron-regulated Fe^{3+} reductase activity [1, 6] and this is attributable to a second reductase, the *FRE2* gene product, also localized at the cell surface [7]. Both Fre1p and Fre2p have also been shown to reduce Cu^{2+} to Cu^{1+} [8, 9] and, therefore, may also play a role in high-affinity copper uptake. *FRE2* is related to *FRE1* by approximately 25 % amino acid identity and a genomic search has revealed five additional genes with identity to *FRE1* and

FRE2, designated as *FRE3-7*. The *FRE3* and *FRE4* gene products have recently been proposed to function in siderophore mediated iron uptake and Fre3p has been localized to the cell surface [10]. To date, the function of remainder of the *FRE* genes (*FRE5-7*) in iron metabolism is not completely clear [11, 12].

An additional gene which may play a role in *FRE1*-dependent iron reduction is *UTR1* [13]. The *UTR1* gene product is predicted to have a cytosolic localization and strains deleted for *UTR1* retain only 5% of their Fe^{3+} reductase activity [14]. While the function of Utr1p remains unknown, it has been suggested that phosphorylation of Utr1p by a cAMP/protein kinase A pathway may be involved in modulating ferric reductase activity [14].

The majority of the *S. cerevisiae FRE* genes are regulated at the transcriptional level by iron. When iron is abundant, *FRE1-6* are repressed and the genes are induced under iron starvation conditions [11]. A number of other *S. cerevisiae* iron homeostasis genes fall under this iron regulon and will be discussed below (Sect. 20.1.4).

Once the insoluble environmental ferric iron is reduced to ferrous iron by Fre1p and Fre2p, the soluble Fe^{2+} ion is thought to be recognized by systems for the high- and low-affinity uptake of iron into the cell.

20.1.2
Iron Translocation across the Plasma Membrane

20.1.2.1 High-affinity Iron Uptake: The Requirement for a Multi-copper Oxidase

High-affinity iron uptake at the cell surface in yeast is accomplished through the use of a bipartite transporter system utilizing the *FET3* and *FTR1* gene products [15, 16]. These genes are induced under iron limiting conditions and their role is to facilitate iron entry into the cell. The *FET3* gene product contains a single transmembrane domain and is localized to the plasma membrane [17]. Fet3p is a member of a large family of multi-copper oxidases including ascorbate oxidase, laccase, and the mammalian ceruloplasmin [15, 17]. Multi-copper oxidases carry out the sequential 4-electron oxidation of a substrate, followed by the 4-electron reduction of molecular oxygen to generate oxidized substrate and water. The *FET3* gene product has the capacity to oxidize Fe^{2+} to Fe^{3+} with a K_m of 0.15 μM Fe [18]. Presumably, the Fe^{2+} product of the Fre1/2p reaction serves as the substrate for Fet3p, and the re-oxidized form of the metal is in turn utilized by the iron permease, Ftr1p. It is curious that the cell reduces Fe^{3+} to Fe^{2+}, just to oxidize the ion back to Fe^{3+} prior to transmembrane transport. Although the rationale is not completely understood, it has been proposed that the ferroxidase activity of Fet3p may impart specificity to the high-affinity iron transporter and that only the oxidized form of the metal can be translocated [19, 20].

20.1.2.2 The Iron–Copper Connection for High-affinity Iron Uptake

The requirement for a multi-copper oxidase in iron uptake in *S. cerevisiae* has effected an intriguing connection between iron and copper uptake in this organism. In a search for iron uptake genes in *S. cerevisiae*, Andy Dancis and coworkers unraveled a series of genes involved in copper transport. Included in this list were *CTR1*, encoding the cell surface copper permease and *CCC2*, encoding an intracellular transporter of copper [21, 22]. Because of the stringent requirement of copper ions for Fet3p activity, any major perturbation in copper trafficking results in a defect in iron uptake [15, 22].

Fet3p receives its copper specifically in a late/post-Golgi compartment of the cell [23]. This compartment receives its copper through the action of Ccc2p, a Golgi copper transporting P-type ATPase that represents the homolog to the Menkes/Wilson disease gene products in humans [22–25]. The Ccc2p-dependent copper loading of Fet3p is also dependent on chloride ions in the Golgi provided by the Gef1p voltage-gated chloride channel [26].

The copper ions taken up by the cell via cell surface permeases (Ctr1p and Ctr3p) are delivered to Ccc2p in the Golgi through the action of a soluble protein termed Atx1p [27, 28]. Atx1p represents the founding member of a family of soluble copper carriers termed metallochaperones (for a review see [29]). As is the case with other proteins involved in copper uptake and trafficking, a deletion in yeast Atx1p results in a marked reduction in high-affinity iron transport, due to a defect in copper activation of Fet3p [27, 30, 31].

The intriguing connection between copper and iron transport exemplified in yeast may also be expanded to other eukaryotes. For example, mammalian ATOX1 (homolog to Atx1p) is responsible for delivering copper to the human Wilson gene product (homolog to Ccc2p), which in turn delivers copper in the Golgi to ceruloplasmin, a multi-copper oxidase involved in cellular iron transport [30, 32–34].

20.1.2.3 Iron Transport by the Cell Surface Permease, FTR1

The translocation of iron across the plasma membrane in yeast is largely accomplished by the *FTR1* gene product, an iron permease localized at the cell surface [16]. Ftr1p harbors an iron binding consensus motif REGLE which is conserved in the mammalian L-chain ferritin. Mutation of the glutamate residues in this motif in Ftr1p was shown to abrogate high-affinity iron uptake activity [16]. Ftr1p is believed to transport the oxidized Fe^{3+} received from Fet3p with a K_m of transport of 0.15 µM [18, 19].

Ftr1p appears to form a complex with Fet3p that is not only needed for iron uptake at the cell surface, but also for proper trafficking of the two proteins through the secretory pathway. For example, Ftr1p is necessary for the plasma membrane localization of Fet3p. In the absence of Ftr1p, Fet3p remains in the secretory pathway and exhibits low copper-dependent ferroxidase activity [16]. Conversely, in cells lacking Fet3p, Ftr1p remains in the endoplasmic reticulum [16]. This suggests that Fet3p and Ftr1p must move together as a complex through the secretory pathway to be correctly localized to the cell surface. Even though Fet3p mediates the proper targeting of Ftr1p to the cell surface, this does not represent the sole function of Fet3p in iron uptake. Specifically, mutations which abrogate ferroxidase activity of Fet3p eliminate high-affinity iron uptake, even though Ftr1p is still properly trafficked to the cell surface [19].

The driving force for iron translocation by Ftr1p is not understood. As of yet, there are no reports of requirements for ATP hydrolysis or coupling of transport with other ions. Perhaps Ftr1p operates through a thermodynamic gradient to drive the translocation of ferric atoms in the direction of the strongly reducing cell cytosol, where the oxidized form of the metal may be of very low abundance.

20.1.2.4 Low-affinity Iron Uptake at the Cell Surface

The high-affinity iron uptake system of Fet3p/Ftr1p is required only under iron limiting conditions. Under iron replete conditions, the high-affinity system is down-regulated (see Sect. 20.1.4) and a low-affinity iron uptake system serves to transport iron at the plasma membrane [5, 35, 36]. The *FET4* gene product encodes an iron transporter for the low-affinity iron uptake system [35]. Fet4p is localized at the cell surface and transports ferrous iron (Fe^{2+}) with a K_m of approximately 30 µM [37]. Transport is inhibitable by other metals such as cobalt, cadmium, and to some extent nickel, suggesting that Fet4p may be a relatively non-specific divalent metal ion transporter [35, 37]. Most recently, Fet4p was shown to also directly transport copper at a low affinity [38], implicating Fet4p in copper as well as iron homeostasis.

20.1.3
Intracellular Iron Transport

Once iron enters the cell, it must be distributed to various intracellular sites for the incorporation into iron-containing enzymes and for storage. Iron is a highly toxic as well as essential metal ion and the surplus metal that is not utilized for biological processes must be sequestered. Yeast cells do not express a ferritin-type of protein for the storage of iron atoms. Therefore, there must be alternative means for storing and sequestering the metal. One possible reservoir for iron includes the yeast vacuole [39–41]. Several genes have been recently localized to the vacuolar membrane and have been proposed to function as iron transporters involved in the export of iron out of the vacuole [42, 43].

FTH1 and *FET5* encode homologs of the cell surface high-affinity iron transport proteins Ftr1p and Fet3p, respectively [16, 42, 44]. *FTH1* is proposed to encode an iron permease and *FET5*, a multi-copper oxidase; both gene products are localized to the vacuolar membrane [42]. As has been shown with the cell surface Ftr1p/Fet3p pair, the trafficking of Fth1p and Fet5p to the vacuolar membrane appears to be interdependent on one another; i.e., deletion of *FET5* causes the accumulation of Fth1p in the endoplasmic reticulum [42]. Predictions based on protein topology have indicated that the Fth1/Fet5p complex is oriented in the vacuolar membrane such that the direction of iron transport is from the lumen of the vacuole to the cytosol [42]. Therefore, it is conceivable that this protein pair functions in the export of vacuolar stores of iron.

Another proposed iron transporter on the vacuole membrane is the *SMF3* gene product. *SMF3* encodes one of three *S. cerevisiae* members of the evolutionarily conserved Nramp family of metal ion transporters [45, 46]. Like all members of the Nramp family of metal transporters, the yeast Smf proteins all contain numerous transmembrane (TM) spanning domains and a consensus transporter signature sequence found in a predicted cytosolic loop between TM8 and TM9 [47]. Yeast Smf1p and Smf2p appear to function primarily in manganese transport and are discussed below in Sect. 20.2.1; Smf3p by comparison evidently acts in iron transport and utilization and the gene product was recently localized to the vacuolar membrane [43]. The predicted membrane topology of Smf3p is consistent with metal transport from the lumen of the vacuole to the cytosol, therefore, we have hypothesized that Smf3p may serve to transport iron out of the vacuole [43]. Consistent with this notion, we noted that yeast cells deleted for *SMF3* have an intracellular iron starvation signal [43]. It is possible that both Smf3p and the Fth1p/Fet5p system serve to transport iron out of the vacuole, perhaps under differing metabolic conditions.

Iron is required in the mitochondria for the biosynthesis of heme and Fe–S clusters. How iron gets into the mitochondria is not clear but some recent work in yeast has shed some light on mitochondrial iron transport. Two genes were recently identified (*MMT1* and *MMT2*) that affect the distribution of iron between mitochondrial and cytosolic pools [48]. Additionally, yeast Yfh1p, a homolog to human frataxin responsible for Friedreich's Ataxia [49], has been proposed to participate directly in mitochondrial iron efflux [50, 51].

20.1.4
Regulation of Iron Transport

Iron transport in *S. cerevisiae* is regulated primarily through iron availability in the growth media. Under low iron conditions, several iron transport genes are transcriptionally induced including *FRE1-6*, *FET3*, *FTR1*, *FET5*, *FTH1*, and a new class of proteins involved in siderophore mediated iron uptake, *ARN1-4* [1, 7, 15, 44, 52–54]. Low-iron conditions also induce the copper transport genes *ATX1* and *CCC2*, further clarifying their role in iron uptake [27, 52]. This transcriptional iron regulation is accomplished primarily through the iron sensitive transcription factor Aft1p [52, 55]. Under iron limiting conditions, Aft1p binds to a consensus sequence in the promoter of iron-regulated genes and activates their transcription. Under iron replete conditions, iron is thought to bind to Aft1p and prevent its binding to promoter DNA [52, 55].

In addition to Aft1p-dependent regulation, an Aft1p-independent mechanism also appears to control iron metabolism in yeast. For example, the polypeptide levels of the Smf3p metal transporter are strongly induced under iron starvation conditions and this induction is not affected by an *aft1Δ* mutation [43]. Additionally, the low-affinity iron transporter Fet4p is up-regulated under iron starvation [36, 37], however, there appears to be no control at the transcriptional level by Aft1p (C. Philpott, personal communication). The mechanism of this *AFT1*-independent iron regulation currently remains unknown.

20.2
Manganese Transport in *Saccharomyces cerevisiae*

As is the case with iron, manganese is a trace element required for life by virtually all organisms. Essential roles for the metal include antioxidant defense, protein processing in the mitochondria, protein glycosylation in the secretory pathway, and cell division. Because manganese performs so many vital cellular functions, all organisms have evolved with specialized systems devoted to the uptake and intracellular trafficking of the metal. But as with iron, manganese is also potentially toxic and accumulation of the metal must fall under tight control. Currently, very little is understood about how eukaryotic organisms regulate the homeostasis, transport and trafficking of manganese. However, some recent clues have been obtained using the baker's yeast *S. cerevisiae* as a model. A number of membrane-bound proteins have been identified that participate in the uptake of manganese and partitioning of the metal between organelle and cytosolic pools.

20.2.1
The Smf1p and Smf2p Members of the Nramp Family of Ion Transporters

The bulk of what is currently understood regarding the fungal transport of manganese has emerged through studies of Smf1p and Smf2p, two members of the Nramp family of metal transporters. As described above in Sect. 20.1.3, *S. cerevisiae* expresses three members of the Nramp family: Smf1p, Smf2p, and Smf3p. Smf3p was recently shown to function in iron metabolism in yeast [43], while Smf1p and Smf2p largely participate in manganese homeostasis. Yeast lacking Smf1p and Smf2p exhibit sensitivity towards the metal chelator EGTA, and this defect is complemented by expression of the mammalian Nramp2 protein, demonstrating the evolutionarily conserved role of mammalian and fungal Nramp transporters [56].

20.2.1.1 Transport of Heavy Metals by Smf1p and Smf2p

Smf1p was originally identified as one of two genes that when overexpressed, would suppress a manganese-dependent protein processing defect in the mitochondria [57]. Subsequently, Smf1p was demonstrated by Nelson and coworkers to be a high-affinity transporter for manganese [58, 59]. Curiously, we identified Smf1p through a study of genes that affect the oxygen resistance of yeast strains lacking superoxide dismutase 1 (SOD1) [60]. These *sod1Δ* mutants are highly sensitive to oxygen toxicity, and in a screen for genes that suppress this oxidative damage, we identified *S. cerevisiae BSD2* (bypass SOD1 defect) [61]. Mutations in *BSD2* reverse oxygen toxicity through activation of Smf1p, which in turn causes high cellular accumulation of manganese ions [60]. It is the manganese ions that suppress oxidative damage, perhaps through neutralization of the toxic superoxide anion [62–64]. The means by which Bsd2p represses Smf1p will be the subject of discussion in Sect. 20.2.1.2 below.

As with mammalian Nramp2, Smf1p is thought to have a broad substrate range specificity. In yeast, Smf1p showed evidence of transporting manganese, copper, cadmium, and iron [60, 65]. Through expression studies in oocytes, Smf1p was found to act on manganese, iron, zinc, cadmium, and copper through proton coupled uptake [66, 67]. Interestingly, sodium and potassium ions were noted to "leak" through Smf1p with no coupling to proton transport. Sodium and potassium ions can compete with the heavy metals for transport by Smf1p and as such, it has been proposed that Smf1p binding to these cations *in vivo* may guard against overload from toxic metals such as iron [66, 67].

The second yeast Nramp transporter, Smf2p, was also originally identified as a suppressor of a manganese-dependent processing defect in mitochondria [57]. Overall, Smf1p and Smf2p share nearly 50% identity at the amino acid level. As with Smf1p, Smf2p can contribute to yeast cell accumulation of manganese, cadmium, copper, and iron, and Smf2p also falls under negative regulation by Bsd2p [43, 60, 65]. In spite of the commonalties between Smf1p and Smf2p, these two are not redundant metal transporters. First, Smf2p contributes to cobalt accumulation, whereas Smf1p does not [60]. Secondly, mutations in *SMF2* have a drastic effect on manganese accumulation and prohibit manganese activation of superoxide dismutase 2 in the mitochondria, whereas mutations in *SMF1* do not [68]. Thirdly, these transporters localize to distinct cellular compartments. Smf1p can be found at the cell surface [58, 69, 70], whereas Smf2p is confined to intracellular vesicles [43]. Therefore, Smf1p appears to act in uptake of manganese from the growth medium whereas Smf2p may act in the vesicular storage and trafficking of the metal.

20.2.1.2 Regulation of Smf1p and Smf2p by Bsd2p and Manganese Ions

As described above, the yeast *BSD2* gene product is a negative regulator of Smf1p and Smf2p. This regulation does not occur at the mRNA level, but rather at the level of Smf1p and Smf2p protein stability and sorting of these proteins through the secretory pathway [69]. Under normal growth conditions when manganese ions are ample, Bsd2p mediates the trafficking of Smf1p and Smf2p directly from the Golgi to the vacuole where these metal transporters are degraded by vacuolar proteases [43, 69]. Bsd2p is localized in the endoplasmic reticulum [60] and the mechanism by which it targets Smf1p and Smf2p to the vacuole is still unknown. Yeast *bsd2Δ* mutants show no vacuolar localization for Smf1p and Smf2p; instead the bulk of these transport proteins can be found within compartments of the secretory pathway [69]. In the case of Smf1p, a small fraction of the transporter can also be found at the cell surface which is presumably responsible for the hyperaccumulation of metals in strains lacking Bsd2p [69].

Smf1p and Smf2p are also regulated by the availability of manganese in the growth medium [69]. When manganese ions are abundant, Smf1p and Smf2p are targeted to the vacuole for degradation as described above. However, under manganese starvation conditions, Smf1p and Smf2p fail to arrive at the vacuole, even when Bsd2p is present. Instead, Smf1p moves to the cell surface, while Smf2p is redirected towards intracellular vesicles [43, 69]. This control of Smf1p and Smf2p by metal conditions is relatively specific for manganese ions. Iron treatment does effect a low level of Smf1p and Smf2p repression through protein degradation, but supplementation with copper, zinc and cobalt has no effect on these proteins [43, 69].

Precisely how metals and Bsd2p regulate Smf1p and Smf2p is unknown, but our recent evidence supports a model in which the fate of these transport proteins is determined by conformational changes in the corresponding polypeptides. Our site-directed mutagenesis studies strongly indicate that transporter activity is necessary for transporter regulation. Mutations that abolished transport activity also abrogated the targeting of Smf1p to the vacuole by Bsd2p and manganese [70]. We additionally identified through mutagenesis, a set of Smf1p mutants that exhibited wild-type manganese transport and regulation by Bsd2p; however, these mutants showed no regulation by manganese ions. These "non-responsive" Smf1p mutants appear frozen in a conformational state that is always recognized by Bsd2p for degradation in the vacuole, irrespective of the manganese status of the cell [70]. Together, these observations have led us to develop a model for regulation of Smf1p and Smf2p by Bsd2p and manganese. When manganese is abundant, the transporters maintain an active conformation that is recognized by Bsd2p for targeting to the vacuole. Manganese binding directly to Smf1p and Smf2p may produce this conformation. Conversely, when manganese ions are not available, a unique conformation is adopted (perhaps that of apo Smf1p and Smf2p) that escapes recognition by Bsd2p, and also mediates trafficking of Smf1p to the cell surface and Smf2p to intracellular vesicles.

Why are Smf1p and Smf2p regulated in this manner by manganese ions? Like other Nramp transporters, [71], Smf1p and Smf2p are somewhat non-specific for metal ion substrate and have the capacity to transport toxic metals such as cadmium [60, 61, 65]. Therefore, the cell must express these transporters only when necessary (e. g., under manganese starvation conditions) to prevent accumulation of toxic metal ions. When cells are starved for manganese, the proteins are rapidly diverted from the vacuole and redirected to cellular locations that facilitate the uptake and distribution of the much-needed manganese ions. In this manner, the cell can rapidly respond to changes in manganese status without the need for new protein synthesis.

20.2.2
Manganese Transport in the Golgi Apparatus

One essential role of manganese involves activation of manganese-dependent enzymes involved in protein glycosylation. A variety of such enzymes are present in the Golgi complex and are responsible for decorating proteins in the secretory pathway with complex sugars [72–78]. Therefore, it is critical that manganese ions are delivered into the lumen of the Golgi. Yeast Pmr1p represents one method by which this can be accomplished.

20.2.2.1 Pmr1p: A Manganese Transporting ATPase

Pmr1p is a member of the large family of P-type ATPase ion transporters. These ion pumps have been identified in all organisms and can drive transmembrane ion gradients up to 10,000-fold [79]. Yeast Pmr1p was first identified as a calcium transporting P-type ATPase localized to the Golgi [80–82]. We subsequently isolated Pmr1p as a manganese homeostasis protein that dramatically affects resistance to oxidative stress. As is the case with *BSD2* (see above), mutations in *PMR1* were found to overcome all the oxidative damage associated with loss of SOD1 [83]. Inactivation of Pmr1p also caused cells to accumulate high levels of manganese [83, 84], which in turn suppressed the oxidative damage. Consistent with the role of Pmr1p as a manganese transporter, manganese ions are an effective competitive inhibitor of calcium uptake by Pmr1p [82]. Pmr1p is now described as a Golgi ATPase that drives the transport of both calcium and manganese ions, and as such, Pmr1p is needed for protein processing and sorting in the Golgi and for proper protein degradation in the endoplasmic reticulum [85].

An interesting series of studies recently conducted by Rao and coworkers have dissected at the molecular level, the disparate roles of Pmr1p in calcium and manganese uptake. Pmr1p contains an N-terminal EF hand-like motif that is characteristic of mammalian SERCA Ca^{2+} ATPases and is believed to bind calcium. A specific mutation in the Pmr1p EF hand (D93A) had a dramatic effect on calcium transport, but did not alter the manganese pumping activity of this transporter [86]. Another region of the protein thought to participate in ion translocation includes the transmembrane spanning domains TM4 to TM8. Rao and coworkers found that specific mutations in this region abolished interaction with both manganese and calcium, whereas others (e.g., Q783A in M6) uniquely affected manganese transport by Pmr1p, but not calcium [87, 88]. Therefore, distinct, but overlapping cation binding sites permit Pmr1p to act on both manganese and calcium. It is noteworthy that mammalian homologs to Pmr1p have been identified [89] and are presumed to play parallel roles in the pumping of both manganese and calcium ions into the Golgi for the quality control, processing and trafficking of proteins in the secretory pathway.

20.2.2.2 Ccc1p: A Manganese Homeostasis Protein Localized in the Golgi

In 1994, Dunn and Beeler described the yeast *CCC1* gene product as a calcium trafficking protein [90]. We subsequently noted that as with Pmr1p, Ccc2p controls the trafficking of manganese as well, and appears to sequester these ions in Golgi-like vesicles [91]. A similar notion has been proposed by Cyert and coworkers [92]. Ccc2p evidently acts synergistically with Pmr1p to deplete the cytosol of manganese ions. Cytosolic manganese ions help to combat oxidative damage and as such, overexpression of either Ccc2p or Pmr1p is lethal to cells that lack cytosolic superoxide dismutase [91]. Yet contrary to results obtained with Pmr1p, defects in Ccc2p do not impair protein processing and secretion [91]. It is possible that Pmr1p and Ccc2p operate at distinct locations of the secretory pathway to partition manganese ions.

Very recently, Ccc1p was found to participate in iron homeostasis in yeast and it has been proposed that Ccc1p may actually be an iron transporter that plays a secondary role in manganese transport [93]. The exact function of Ccc1p remains to be resolved.

20.2.2.3 Atx2p: An Antagonizer of Pmr1p?

We isolated *ATX2* in a genetic screen for genes that when overexpressed would suppress the oxidative damage of yeast lacking SOD1 [94]. The *ATX2* gene product was found to encode a Golgi-localized membrane protein that seems to play an important role in maintaining cytosolic pools of manganese [94]. Our studies indicate that Atx2p functions opposite to Pmr1p: while Pmr1p depletes cytosolic manganese, Atx2p helps ensure that cytosolic manganese levels are maintained at an adequate level.

It should be noted that the precise role of Atx2p and Ccc2p in partitioning manganese ions in the secretory pathway is unknown. These two proteins exhibit no obvious homology to known metal transporters and it is unclear whether they directly participate in the membrane translocation of manganese ions. It is possible that Atx2p and Ccc2p are not metal transporters, but rather modulators of manganese ion transport and trafficking, as is the case with yeast Bsd2p.

20.2.3
Homeostasis of Cytosolic Manganese: A Possible Role for the *CDC1* Gene Product

CDC1 is one of many so-called <u>c</u>ell <u>d</u>ivision <u>c</u>ontrol genes assumed to play a vital role in the yeast cell cycle or cell division process [95]. The *CDC1* gene product is in fact essential for viability in yeast [95]. More recent work has demonstrated that *CDC1* actually participates in manganese homeostasis. First, the conditional growth defect of a *cdc1* temperature sensitive (ts) mutant can be rescued by supplementing manganese to the growth medium [96]. Furthermore, Supek and Nelson noted that this *cdc1* ts mutant is sensitive to metal chelators, and this defect could be suppressed by overexpression of the Smf1p manganese transporter [58]. Most recently, Paidungat and Garrett have provided strong evidence that Cdc1p does not directly participate in cell cycle control, but instead functions to maintain cytosolic pools of manganese that are somehow critical in the cell division process [97]. Cdc1p does not harbor any potential membrane spanning domains and as such, does not appear to be a membrane manganese transporter [97]. It is possible that Cdc1p is a cytosolic carrier for manganese or a regulator of other manganese homeostasis factors that control the critical cytosolic pools of manganese.

20.2.4
The Yeast Vacuole and Manganese

As described above in Sect. 20.1.3, the yeast vacuole is thought to play an important role in the homeostasis of metal ions such as iron. There is evidence that the vacuole also participates in manganese storage and distribution [98]. Hence, disruption of vacuolar function should affect the homeostasis of manganese ions. In 1998, Paidhungat and Garrett isolated a number of genes involved in vacuolar function through a screen for mutational suppressors of the *cdc1* growth defect [99]. One such gene identified was *COS16*, encoding a vacuolar membrane protein [99]. A deletion of *COS16* together with a disruption of *PMR1* led to severe manganese toxicity [99]. Therefore, it has been proposed that Cos16p helps to remove cytosolic manganese by sequestering these ions in the vacuole [99]. The precise role of Cos16p in manganese homeostasis has not been elucidated.

20.3
Conclusions and Directions for the Future

Iron and manganese ions clearly play vital roles in a number of cellular processes, but since these metals are also potentially toxic, elaborate mechanisms must exist to ensure the precise trafficking of the metal to preclude reactivity with non-productive cellular sites. A number of cell surface and organelle transport systems have evolved to control the proper intracellular partitioning of the metal and in many cases, these transporters are regulated in accordance with the bioavailability of the respective metal ion. In the case of iron, yeast cells express an iron-sensing transcription factor (Aft1p) that allows for expression of iron transport genes only under iron starvation conditions. In the case of manganese, no equivalent transcription factor has been identified, but methods to control manganese transport at the level of transporter protein stability are evident.

A large number of iron and manganese homeostasis factors have been identified in yeast (summarized in Tab. 1). However, there is still much to be uncovered. For example, how does the metal enter the mitochondria, an organelle where both metal ions play critical roles in forming cofactor and enzyme prosthetic groups. Also, how do iron and manganese ions translocate across the cytosol to the various cellular locations? In the case of copper, a family of so-called copper chaperones have been identified that function in the cytosolic transport of the metal [29]. It is presumed that an analogous family of molecules exist to escort manganese and iron atoms, but as yet, none have been identified. In any case, the baker's yeast *S. cerevisiae* has always been at the forefront for the identification of eukaryotic metal transport and trafficking proteins, primarily due to the ease by which genes in yeast can be identified through classical genetics and bioinformatic approaches. It is, therefore, likely that new systems for the transport and trafficking of manganese and iron atoms will continue to emerge through future studies with this small, but very important eukaryotic organism.

Tab. 1. *Saccharomyces cerevisiae* genes involved in iron and manganese transport

	Gene	Proposed Function	Localization
Iron	AFT1	iron-responsive transcription factor	nucleus
	ATX1	copper metallochaperone for Ccc2p	cytoplasm
	CCC2	copper donor for Fet3p	late/post Golgi
	FET3	high-affinity multi-copper oxidase	plasma membrane
	FET4	low-affinity iron/copper transporter	plasma membrane
	FET5	proposed multi-copper oxidase	vacuolar membrane
	FRE1	Fe^{3+} and Cu^{2+} reductase	plasma membrane
	FRE2	Fe^{3+} and Cu^{2+} reductase	plasma membrane
	FRE3	implicated in siderophore iron uptake	plasma membrane
	FRE4	implicated in siderophore iron uptake	unknown
	FRE5-FRE7	unknown	unknown
	FTH1	proposed iron transporter	vacuolar membrane
	FTR1	high-affinity iron transporter	plasma membrane
	MFT1	mitochondrial iron transporter	mitochondria membrane
	MFT2	mitochondrial iron transporter	mitochondria membrane
	SMF3	proposed iron transporter	vacuolar membrane
	UTR1	involved in Fe^{3+}/Cu^{2+} reduction	cytoplasm
	YFH1	frataxin homolog, mitochondrial iron	mitochondria

Tab. 1. continued

Gene	Proposed Function	Localization
Manganese		
ATX2	manganese transport from Golgi to cytosol	Golgi
BSD2	negative regulator of SMF1 and SMF2	endoplasmic reticulum
CCC1	manganese transport from cytosol to Golgi	Golgi
CDC1	cytosolic manganese trafficking	unknown
COS16	vacuolar transport of manganese	vacuolar membrane
PMR1	manganese transport from cytosol to Golgi	Golgi
SMF1	cell surface uptake of manganese	plasma membrane
SMF2	vesicular transport of manganese	intracellular vesicles

Acknowledgement

Many of the findings presented here on yeast iron and manganese homeostasis genes was supported by the Johns Hopkins University Center for Environmental Health Sciences and by NIH grant ES 08996 (to V. C. C.). M. E.P. was supported by a EPA star fellowship U915646. We also wish to thank Caroline Philpott for personal communications of unpublished findings.

References

1. Dancis, A., Roman, D. G., Anderson, G. J., Hinnebusch, A. G., Klausner, R. D., *Proc. Natl. Acad. Sci. USA* **1992**, *89*, 3869-3873.
2. Eide, D. J., *Annu. Rev. Nutr.* **1998**, *18*, 441-469.
3. Lesuisse, E., Crichton, R. R., Labbe, P., *Biochim Biophys Acta* **1990**, *1038*, 253-259.
4. Dancis, A., Klausner, R. D., Hinnebusch, A. G., Barriocanal, J. G., *Mol. Cell. Biol.* **1990**, *10*, 2294-2301.
5. Eide, D., Davis-Kaplan, S., Jordan, I., Sipe, D., Kaplan, J., *J. Biol. Chem.* **1992**, *267*, 20774-20781.
6. Anderson, G. J., Lesuisse, E., Dancis, A., Roman, D. G., Labbe, P., Klausner, R. D., *J. Inorg. Biochem.* **1992**, *47*, 249-255.
7. Georgatsou, E., Alexandraki, D., *Mol. Cell. Biol.* **1994**, *14*, 3065-3073.
8. Georgatsou, E., Mavrogiannis, L. A., Fragiadakis, G. S., Alexandraki, D., *J. Biol. Chem.* **1997**, *272*, 13786-13792.
9. Hassett, R., Kosman, D. J., *J. Biol. Chem.* **1995**, *270*, 128-134.
10. Yun, C. W., Bauler, M., Moore, R. E., Klebba, P. E., Philpott, C. C., *J. Biol. Chem.* **2001**, *276*, 10218-10223.
11. Martins, L. J., Jensen, L. T., Simon, J. R., Keller, G. L., Winge, D. R., Simons, J. R., *J. Biol. Chem.* **1998**, *273*, 23716-2371.
12. Georgatsou, E., Alexandraki, D., *Yeast* **1999**, *15*, 573-584.
13. Anderson, G. J., Dancis, A., Roman, D. G., Klausner, R. D., *Adv. Exp. Med. Biol.* **1994**, *356*, 81-89.
14. Lesuisse, E., Casteras-Simon, M., Labbe, P., *J. Biol. Chem.* **1996**, *271*, 13578-13583.
15. Askwith, C., Eide, D., V-Ho, A., Bernard, P. S., Li, L., Davis-Kaplan, S., Sipe, D. M., Kaplan, J., *Cell* **1994**, *76*, 403-410.
16. Stearman, R., Yuan, D., Yamaguchi-Iwan, Y., Klausner, R. D., Dancis, A., *Science* **1996**, *271*, 1552-1557.
17. De-Silva, D. M., Askwith, C. C., Eide, D., Kaplan, J., *J. Biol. Chem.* **1995**, *270*, 1098-1101.
18. Radisky, D., Kaplan, J., *J. Biol. Chem.* **1999**, *274*, 4481-4484.
19. Askwith, C. C., Kaplan, J., *J. Biol. Chem.* **1998**, *273*, 22415-22419.
20. di Patti, M. C., Pascarella, S., Catalucci, D., Calabrese, L., *Protein Eng.* **1999**, *12*, 895-897.
21. Dancis, A., Yuan, S., Haile, D., Askwith, C., Eide, D., Moehle, C., Kaplan, J., Klausner, R., *Cell* **1994**, *76*, 393-402.
22. Yuan, D. S., Stearman, R., Dancis, A., Dunn, T., Beeler, T., Klausner, R. D., *Proc. Natl. Acad. Sci. USA* **1995**, *92*, 2632-2636.
23. Yuan, D. S., Dancis, A., Klausner, R. D., *J. Biol. Chem.* **1997**, *272*, 25787-25793.
24. Fu, D., Beeler, T. J., Dunn, T. M., *Yeast* **1995**, *11*, 283-293.
25. Bull, P. C., Cox, D. W., *Trends Genet.* **1994**, *10*, 246-252.
26. Davis-Kaplan, S. R., Askwith, C. C., Bengtzen, A. C., Radisky, D., Kaplan, J., *Proc. Natl. Acad. Sci. USA* **1998**, *95*, 13641-13645.
27. Lin, S. J., Pufahl, R., Dancis, A., O'Halloran, T. V., Culotta, V. C., *J. Biol. Chem.* **1997**, *272*, 9215-9220.
28. Pufahl, R., Singer, C., Peariso, K. L., Lin, S. J., Schmidt, P., Fahrni, C., Culotta, V. C., Penner-Hahn, J. E., O'Halloran, T. V., *Science* **1997**, *278*, 853-856.
29. O'Halloran, T. V., Culotta, V. C., *J. Biol. Chem.* **2000**, *275*, 25057-25060.
30. Klomp, L. W. J., Lin, S. J., Yuan, D., Klausner, R. D., Culotta, V. C., Gitlin, J. D., *J. Biol. Chem.* **1997**, *272*, 9221-9226.
31. Portnoy, M. E., Rosenzweig, A. C., Rae, T., Huffman, D. L., O'Halloran, T. V., Culotta, V. C., *J. Biol. Chem.* **1999**, *274*, 15041-15045.
32. Hamza, I., Schaefer, M., Klomp, L. W. J., Gitlin, J. D., *Proc. Natl. Acad. Sci USA* **1999**, *96*, 13363-13368.
33. Harris, Z. L., Takahashi, Y., Miyajima, H., Serizawa, M., MacGillivray, R. T. A., Gitlin, J. D., *Proc. Natl. Acad. Sci. USA* **1995**, *92*, 2539-2543.
34. Harris, Z. L., Durley, A. P., Man, T. K., Gitlin, J. D., *Proc. Natl. Acad. Sci. USA* **1999**, *96*, 10812-10817.
35. Dix, D., Bridgham, J. T., Broderius, M. A., Byersdorfer, C. A., Eide, D. J., *J. Biol. Chem.* **1994**, *269*, 26092-26099.
36. Li, L., Kaplan, J., *J Biol Chem* **1998**, *273*, 22181-22187.
37. Dix, D., Bridgham, J., Broderius, M., Eide, D., *J. Biol. Chem.* **1997**, *272*, 11770-11777.

38. Hassett, R., Dix, D. R., Eide, D. J., Kosman, D. J., *Biochem. J.* **2000**, *351*, 477-484.
39. Kitamoto, K., Yoshizawa, K., Ohsumi, Y., Anraku, Y., *J. Bacteriol.* **1988**, *170*, 2687-2691.
40. Raguzzi, F., Lesuisse, E., Crichton, R. R., *FEBS Lett.* **1988**, *231*, 253-258.
41. Bode, H. P., Dumschat, M., Garotti, S., Fuhrmann, G. F., *Eur. J. Biochem.* **1995**, *228*, 337-342.
42. Urbanowski, J. L., Piper, R. C., *J. Biol. Chem.* **1999**, *274*, 38061-38070.
43. Portnoy, M. E., Liu, X. F., Culotta, V. C., *Mol. Cell. Biol.* **2000**, *20*, 7893-7902.
44. Spizzo, T., Byersdorfer, C., Duesterhoeft, S., Eide, D., *Mol. Gen. Genet.* **1997**, *256*, 547-556.
45. Fleming, M. D., Andrews, N. C., *J. Lab. Clin. Med.* **1998**, *132*, 464-468.
46. Andrews, N. C., *Int. J. Biochem. Cell Biol.* **1999**, *31*, 991-994.
47. Cellier, M., Prive, G., Belouchi, A., Kwan, T., Rodrigues, V., Chia, W., Gros, P., *Proc. Natl. Acad. Sci. USA* **1995**, *92*, 10089-10093.
48. Li, L., Kaplan, J., *J. Biol. Chem.* **1997**, *272*, 28485-28493.
49. Campuzano, V., Montermini, L., Molto, M. D., Pianese, L., Cossee, M., Cavalcanti, F., Monros, E., Rodius, F., Duclos, F., Monticelli, A. et al., *Science* **1996**, *271*, 1423-1427.
50. Radisky, D. C., Babcock, M. C., Kaplan, J., *J. Biol. Chem.* **1999**, *274*, 4497-4499.
51. Babcock, M., Silva, D. d., R. Oaks, Davis-Kaplan, S., Jiralerspong, S., Montermini, L., Pandolfo, M., Kaplan, J., *Science* **1997**, *276*, 1709-1712.
52. Yamaguchi-Iwai, Y., Stearman, R., Dancis, A., Klausner, R. D., *EMBO J.* **1996**, *15*, 3377-3384.
53. Yun, C., Tiedeman, J., Moore, R., Philpott, C., *J. Biol. Chem.* **2000**, *275*, 16354-16359.
54. Yun, C. W., Ferea, T., Rashford, J., Ardon, O., Brown, P. O., Botstein, D., Kaplan, J., Philpott, C. C., *J. Biol. Chem.* **2000**, *275*, 10709-10715.
55. Yamaguchi-Iwai, Y., Dancis, A., Klausner, R., *EMBO J.* **1995**, *14*, 1231-1239.
56. Pinner, E., Gruenheid, S., Raymond, M., Gros, P., *J. Biol. Chem.* **1997**, *272*, 28933-28938.
57. West, A. H., Clark, D. J., Martin, J., Neupert, W., Hart, F. U., Horwich, A. L., *J. Biol. Chem.* **1992**, *267*, 24625-24633.
58. Supek, F., Supekova, L., Nelson, H., Nelson, N., *Proc. Natl. Acad. Sci. USA* **1996**, *93*, 5105-5110.
59. Supek, F., Supekova, L., Nelson, H., Nelson, N., *J. Exp. Biol.* **1997**, *200*, 321-330.
60. Liu, X. F., Supek, F., Nelson, N., Culotta, V. C., *J. Biol. Chem.* **1997**, *272*, 11763-11769.
61. Liu, X. F., Culotta, V. C., *Mol. Cell. Biol.* **1994**, *14*, 7037-7045.
62. Faulker, K. M., Stefan, I., Liochev, I., Fridovich, I., *J. Biol. Chem.* **1994**, *269*, 23471-23476.
63. Archibald, F. S., Fridovich., I., *Arch. Biochem. Biophys.* **1982**, *214*, 452-463.
64. Archibald, F. S., Fridovich, I., *J. Bacteriol* **1981**, *146*, 928-936.
65. Cohen, A., Nelson, H., Nelson, N., *J Biol Chem* **2000**, *275*, 33388-33394.
66. Chen, X., Peng, J., Cohen, A., Nelson, H., Nelson, N., Hediger, M. A., *J. Biol. Chem.* **1999**, *274*, 350898-35094.
67. Nelson, N., *EMBO J.* **1999**, *19*, 4361-4371.
68. Luk, E., Tsui, C., Culotta, V. C., in preparation.
69. Liu, X. F., Culotta, V. C., *J. Biol. Chem.* **1999**, *274*, 4863-4868.
70. Liu, X. F., Culotta, V. C., *J. Mol. Biol.* **1999**, *289*, 885-891.
71. Gunshin, H., Mackenzie, B., Berger, U. V., Gushin, Y., Romero, M. F., Boron, W. F., Nussberger, S., Gollan, J. L., Hediger, M. A., *Nature* **1997**, *388*, 482-488.
72. Coste, H., Martel, M. B., Azzar, G., Got, R., *Biochim Biophys Acta* **1985**, *814*, 1-7.
73. Haselbeck, A., Schekman, R., *Proc. Natl. Acad. Sci. USA* **1986**, *83*, 2017-2021.
74. Kaufman, R., Swaroop, M., Murtha-Riel, P., *Biochemistry* **1994**, *33*, 9813-9819.
75. Kuhn, N. J., Ward, S., Leong, W. S., *Eur. J. Biochem.* **1991**, *195*, 243-250.
76. Nakajima, T., Ballou, C. E., *Proc. Natl. Acad. Sci. USA* **1975**, *72*, 3912-3916.
77. Parodi, A. J., *J. Biol. Chem.* **1979**, *254*, 8343-8352.
78. Ram, B. P., Munjal, D. D., *CRC Crit. Rev. Biochem.* **1985**, *17*, 257-311.
79. Maeda, M., Hamano, K., Hirano, Y., Suzuki, M., Takahashi, E., Terada, T., Futai, M., Sato, R., *Cell Struct. Funct.* **1988**, *23*, 315-323.
80. Antebi, A., Fink, G. R., *Mol. Biol. Cell* **1992**, *3*, 633-654.
81. Rudolph, H. K., Antebi, A., Fink, G. R., Buckley, C. M., Dorman, T. E., LeVitre, J., Davidow, L. S., Mao, J. I., Moir, D. T., *Cell* **1989**, *58*, 133-145.
82. Sorin, A., Ross, G., Rao, R., *J. Biol. Chem.* **1997**, *272*, 9895-9901.

83. Lapinskas, P. J., Cunningham, K. W., Liu, X. F., Fink, G. R., Culotta, V. C., *Mol. Cell. Biol.* **1995**, *15*, 1382-1388.
84. Lapinskas, P. J. *Characterization of Genes Involved in the Homeostasis of Oxygen Free Radicals and Metal Ions in Saccharomyces cerevisiae*; Johns Hopkins University Press, **1995**.
85. Durr, G., Strayle, J., Plemper, R., Elbs, S., Klee, S. K., Catty, P., Wolf, D. H., Rudolph, H. K., *Mol. Biol. Cell* **1998**, *9*, 1149-1162.
86. Wei, Y., Marchi, V., Wang, R., Rao, R., *Biochemistry* **1999**, *38*, 14534-14541.
87. Mandal, D., Woolf, T. B., Rao, R., *J. Biol. Chem.* **2000**, *31*, 23933-23938.
88. Wei, Y., Chen, J., Rosas, G., Tompkins, D. A., Holt, P. A., Rao, R., *J. Biol. Chem.* **2000**, *275*, 23927-23832.
89. Gunteski-Hamblin, A. M., Clarke, D. M., Shull, G. E., *Biochemistry* **1992**, *31*, 7600-7608.
90. Fu, D., Beeler, T. J., Dunn, T. M., *Yeast* **1994**, *10*, 515-521.
91. Lapinskas, P. J., Lin, S. J., Culotta, V. C., *Mol. Microbiol.* **1996**, *21*, 519-528.
92. Pozos, T., Sekler, I., Cyert, M. S., *Mol. Cell. Biol.* **1996**, *16*, 3730-3741.
93. Chen, O. S., Kaplan, J., *J. Biol. Chem.* **2000**, *275*, 7626-7632.
94. Lin, S. J., Culotta, V. C., *Mol. Cell. Biol.* **1996**, *16*, 6303-6312.
95. Hartwell, L., Culotti, J., Reid, B. J., *Genetics* **1970**, *74*, 267-286.
96. Loukin, S., Kung, C., *J. Cell Biol.* **1995**, *131*, 1025-1037.
97. Paidhungat, M., Garrett, S., *Genetics* **1998**, *148*, 1777-1786.
98. Okorokov, L. A., Lichko, L. P., Kadomtseva, V. M., Kholodenko, V. P., Titovsky, V. T. et al., *Eur. J. Biochem.* **1977**, *75*, 373-377.
99. Paidhungat, M., Garrett, S., *Genetics* **1998**, *148*, 1787-1798.

21
Siderophore Transport in Fungi

Günther Winkelmann

21.1
Introduction

While most metal ions occur as ionic species in a hydroxo-aquo form, iron complexed to microbial siderophores represents a special case. Due to the organic shell wrapped around the iron center during complexation, the ferric ion becomes an organic molecule which can be treated by microorganisms like other organic nutrients in an aqueous environment. Thus, the biological advantage of excreting a chelator for collecting iron from the environment is twofold: besides the obviously entropically driven solubilization of iron hydroxides, the molecule adopts a new conformation with altered surface properties allowing new modes of interaction with membrane located transport systems. In-depth discussions of the structures, coordination chemistry, thermodynamics and kinetics of iron chelation and release may be found in two recent monographs [1, 2]. Another important aspect, which constitutes the subject of the present review, concerns the specificity of uptake. Previous reports have shown that specificity of siderophore uptake depends on the different fungal genera or species, resulting in recognition only of endogenous vs. exogenous siderophores [3, 4] or their enantiomeric structures [5, 6]. The present chapter will highlight some recent results on the specificity of single siderophore transporters in yeast. Based on the complete genome sequence of yeast, functional genomics allows to determine the function of single genes via PCR-mediated gene replacement (disruption) techniques [7]. Although the complete genome sequence of *Saccharomyces cerevisiae* revealed that some 5,800 genes have been estimated to be encoded [8], only 28% of the known open reading frames (ORFs) could be assigned to proteins of known function [9]. Of the presently known 2,300 membrane proteins nearly 200 [10] represent proteins of the Major Facilitator Superfamily (MFS). Four of the six genes of the previously designated Unknown Major Facilitator superfamily (UMF), which are now collectively termed the Siderophore Iron Transport (SIT) family, turned out to be siderophore transporters. Therefore, much of the work discussed in this review is concerned with siderophore-iron transport in *S. cerevisiae*, that is the identification of novel

SIT transporters by gene disruption studies and the physiological analysis of their specificity using growth promotion and transport assays.

21.2
Siderophore Classes and Properties

The most common siderophore classes (Fig. 1) in microorganisms are catecholates, hydroxamates, and polycarboxylates [1]. However, so far only the hydroxamates and polycarboxylates have been found in fungi, while catecholate siderophores seem to prevail in the bacterial genera. Although a variety of different catecholate-type compounds have been described in fungi, such as telephoric acid, xerocomic acid, variegatic acid [11] or pigments from *Cortinarius violaceus* [12], these have never been shown to be excreted in order to function as siderophores. The reason for this might be that iron(III) complexes with catecholates are most stable under neutral to alkaline conditions while hydroxamates and polycarboxylates are most stable under more acidic conditions (pH 2–5). Acid conditions prevail in most fungal cultures after prolonged cultivation and are even more pronounced under low-iron conditions [42]. In the case of the hexadentate tris-hydroxamates the 3+ charge of ferric iron is neutralized by the three deprotonated N–OH groups resulting in an uncharged octahedrally coordinated 1:1 Fe–tris-hydroxamate complex. Catecholates and polycarboxylates preferentially give negatively charged 1:1 ferric siderophore complexes. Thus enterobactin, possessing six negative charges of the three catecholate groups, gives a 3− negatively charged molecule after complexing with ferric iron. Polycarboxylates like citrate or rhizoferrin [13, 14] may give rise to negatively charged or neutral complexes depending on the pH. Tetradentate dihydroxamic acids are unable to satisfy the preferred octahedral coordination geometry of Fe(III) by forming simple 1:1 complexes and must, therefore, form bimetallic complexes with a stoichiometry of Fe_2L_3 [15]. Thus for the fungal siderophore, rhodotorulic acid, multiple tris- and bis-chelated mono- and diferric species are present in solution depending on the pH and the Fe/ligand ratio. While structural differences in the produced ligands are easily determined by modern spectroscopic techniques, the coordination chemistry and the aqueous solution speciation of the metal siderophore complexes require pH-dependent analysis of mono-, bis- and tris-chelated complexes, which are more difficult to analyze because of the variety of species possible with the dihydroxamate and carboxylate siderophores [16, 17]. With regard to their function as iron-scavenging compounds, the stability or formation constants are of the most important properties. The stability constants for the fungal hexadentate trihydroxamates fall within the range of $10^{29-32.5}$ [1], while the stability of the fungal rhizoferrin is only 10^{25} [14]. The much lower stability of the Fe(II) hydroxamate complexes compared to Fe(III) hydroxamate complexes allows iron removal by ligand exchange after reduction, which is the preferred mechanism of reductive iron release in the cytoplasm or on the fungal cell membrane. Reduction potentials of fungal hydroxamate siderophores range from −468 mV vs. NHE (triacetylfusarinine C) to −400 mV vs. NHE (ferrichrome), which is within the range of physiological reductants like NADH$_2$, while rhizoferrin has a reduction potential of -82 mV [1]. As pointed out by Albrecht-Gary and Crumbliss [1], pH and hydrophobicity may change the redox potential, so that even enterobactin may be amenable to reduction by cellular reductases [47].

Although stability constants and redox potentials represent important properties for iron binding and release, the main feature of siderophore transport, and the basis for successful competition for iron, is specific recognition of siderophore molecules by membrane located transport systems. Thus, competition for iron

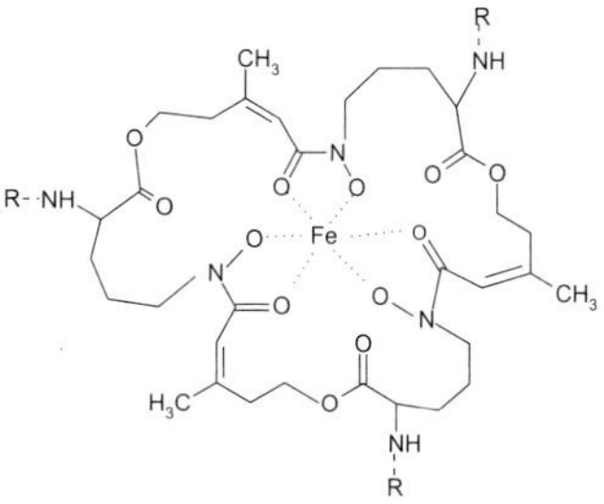

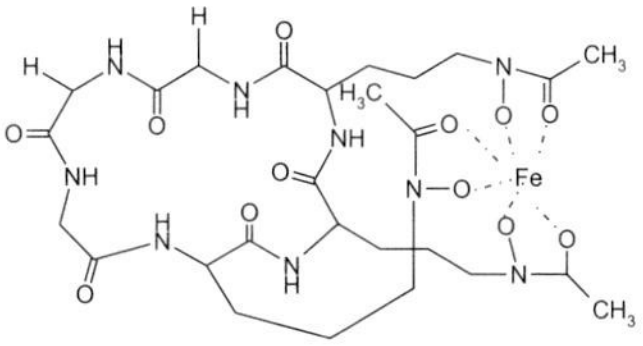

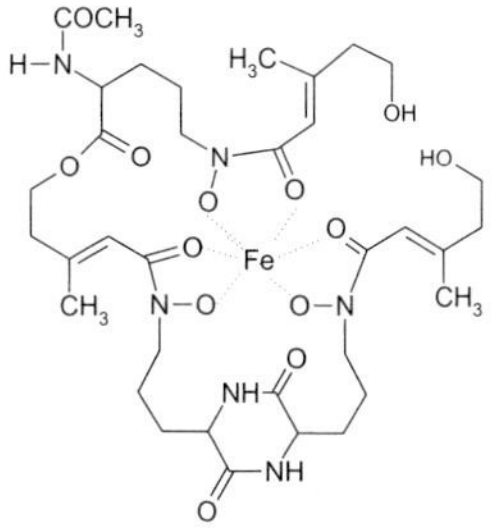

Fig. 1. Representative microbial siderophores. Triacetylfusarinine C (R = acetyl) (I), ferrichrome (II) and coprogen (III) are the most common fungal hydroxamate-type siderophores, synthesized and excreted by strains of *Aspergillus*, *Penicillium*, *Neurospora*, *Ustilago* and other ascomycetous and basidiomycetous fungi. The iron-surrounding N-acetyl residues of ferrichromes (ferrichrome, ferricrocin and ferrichrysin) may be replaced by *trans*- and *cis*-anhydromevalonyl groups (ferrirubin and ferrirhodin) or by methylglutaconyl residues (ferrichrome A). Ferrioxamines (IV) are found in streptomyces as well as in enterobacteria (*Erwinia*, *Hafnia*, *Enterobacter*), while enterobactin (V) is a catecholate-type siderophore produced by enterobacteria (*E. coli* and *Salmonella*). Rhizoferrin (VI) possessing an R, R-configuration, as shown in the structure, is the prototype of fungal carboxylate siderophores produced by members of the Mucorales (Zygomycetes). The corresponding S,S-enantiomer (*enantio*-rhizoferrin) has been recently found in *Ralstonia pickettii* [43].

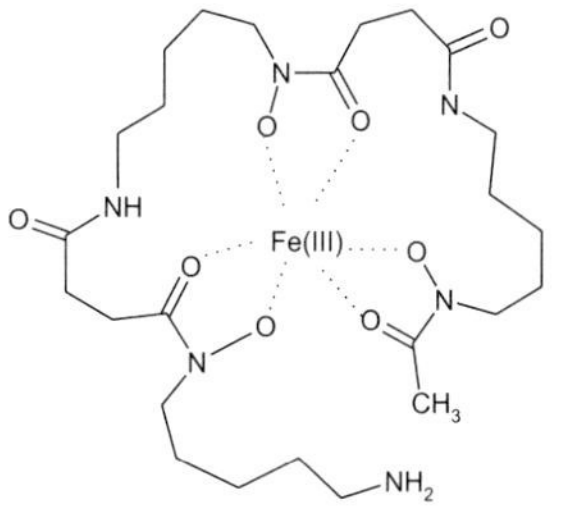

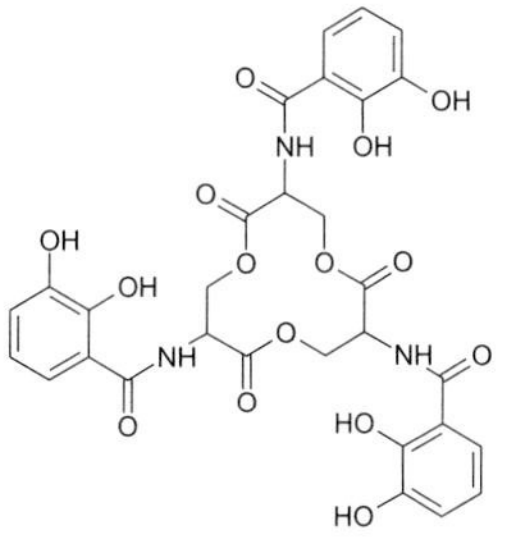

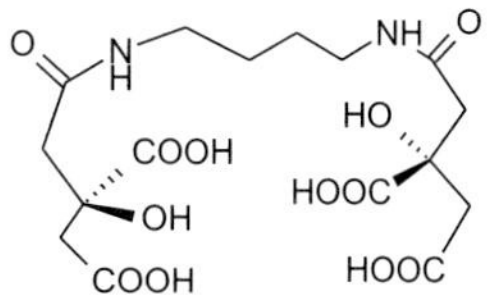

has resulted in a great structural diversity among siderophores and an increasing number of distinct siderophore transport systems. Recognition of siderophores becomes even more important when iron becomes a limiting factor as in an aerobic environment or when a pathogen invades a host where various iron binding proteins like lactoferrin, ovotransferrin or transferrin are present [18].

21.3
Siderophore Production and Biosynthesis

Transport systems for hydroxamate siderophores in fungi may have evolved concomitantly with siderophore biosynthetic pathways. Any mutational change in siderophore biosynthesis may have been followed by an adaption of the transport system, allowing recognition and uptake of the produced siderophore. However, recognition by membrane transport systems is not restricted to siderophores of the producing strains. There are plenty of examples among enteric bacteria, showing that siderophore receptors have evolved for recognition of exogenous hydroxamate siderophores (ferrichromes, coprogen) not synthesized by the organism [19]. Biosynthesis of hydroxamates and their cognate transport systems in fungi seem to be relicts from a prokaryotic world, being absent in higher eukaryotes like plants and animals. It may be assumed that most fungi have retained hydroxamate biosynthesis and transport systems during evolution, although hydroxamate production is most obvious in higher fungi and is absent in the lower fungi (Zygomycota, Chytridiomycota).

Biosynthesis of siderophores has been studied extensively in the basidiomycete *Ustilago maydis* resulting in the detection of two genes, *sid1*, a fungal ornithine-N^5-oxygenase [20] and *sid2*, a possible peptide synthetase gene, as well as a siderophore biosynthesis regulator, *urbs1*. The starting point for hydroxamate biosynthesis is the oxygenation and acylation of the ε- or δ-amino groups of lysine or ornithine [21]. Lysine- and ornithine oxygenase genes have been identified in bacteria and fungi, respectively [22–24]. Iron regulation has been shown to function via iron sensing repressors or activators at the transcriptional level; Fur in enterobacteria [25], AFT1 in yeast [26] and Urbs1 in *Ustilago* [27]. Disruption of the *urbs1* gene in *U. maydis* led to constitutive expression of siderophores, and DNA sequence analysis revealed a single open reading frame of 950 amino acids containing two putative zinc-finger motifs belonging to the GATA family of transcription factors with an unusual tract of 21 histidines within 28 amino acids at the C-terminus. Recently the two putative zinc-finger motifs of Urbs1 were studied by analyzing mutants containing altered finger domains [28], indicating that only the second finger of Urbs1 is required for iron-mediated repression of *sid1* in *U. maydis*. Further GATA factors that control siderophore biosynthesis and siderophore iron transport have been reported in *Neurospora crassa* (SRE) [29, 30], *Penicillium chrysogenum* (SREP) [31] and in *Aspergillus nidulans* (SREA) [32]. Deletion of *sreA* resulted in derepression of the L-ornithine-N^5-oxygenase and constitutive triacetylfusarinine C production in *Aspergillus* even under sufficient iron supply.

As siderophores seem to have invaluable properties for fast growing fungi, the production generally starts from a non-proteinogenic amino acid such as ornithine or its decarboxylation product, putrescine. Regulated biosynthesis and excretion of siderophores under iron limiting conditions allow fungi to overcome many environmentally unfavorable conditions without wasting too much energy on the production of organic molecules. Nearly all fungal hydroxamates are composed of N-hydroxyornithine moieties linked to different acid residues, often acetic acid or anhydromevalonic acid. Both constituents seem to be present in the cell during early growth, leading to a rapid onset of siderophore biosynthesis under iron limiting conditions. The iron requirement of spore forming fungi may vary depending on the developmental stage. While siderophore production by mycelia is strongly dependent on intracellular transcriptional iron regulating factors, such as AFT1 in yeast [26], or SRE, SREP and SREA in filamentous Ascomycetes, germination from spores is not dependent on extracellular iron. This is probably because fungal spores often contain significant iron reserves. In the hydroxamate-producing fungi these consist of stored siderophores as has been shown in *Neurospora* and *Aspergillus* [33, 34], while in the rhizoferrin-producing fungi the iron stores represent ferritins [35, 36]. Therefore, the need to synthesize siderophores and to express the corresponding transport systems during germination from spores is not obvious. Later stages of germinating fungi, however, may profit from the presence of exogenous siderophores, as has been shown by the growth promotion effect of ferricrocin as a germination factor in *N. crassa* [37].

21.4
Evolutionary Aspects of Siderophores

Siderophores have been found in all the three higher fungal groups, the Zygomycota, Ascomycota, and Basidiomycota [38], while they seem to be absent in the Chytridiomycota. The zoospore forming Oomycetes, like *Pythium* and *Saprolegnia*, solubilize iron by simply excreting citric acid [39], confirming their different phylogenetic lineage and relationship to plants and algae [40, 41]. Iron acquisition of the Oomycetes resembles the strategy I plants in their property of acidifying the media and solubilizing environmental iron. The tight phylogenetic relationship of Oomycetes and plants underlines this behavior.

Although citrate is an intracellular primary metabolite, it is also an iron binding compound which, like malate and succinate, is excreted under iron deficiency by fungi in large amounts [42]. From an evolutionary view citrate may be regarded as the simplest siderophore from which several other siderophores such as rhizoferrin, staphyloferrin A, schizokinen, arthrobactin, aerobactin, etc., have evolved [2]. There are many composite ligands produced that have citrate as a building block and that profit from the iron binding property of the a-hydroxycarboxylic function. Irrespective of the question of whether or not citrate itself is a siderophore, it almost certainly represents the beginning of siderophore evolution. The next step in siderophore evolution is the condensation to an amine or amino acid, as seen in rhizoferrin and staphyloferrin, respectively. While in the fungal siderophore, rhizoferrin, citrate is amidically linked to 1,4-butanediamine (putrescine), citrate in bacterial staphyloferrin A is linked to D-ornithine. The configuration in the fungal rhizoferrin is *R,R* as determined by CD measurements, while the configuration of the citrate residues in staphyloferrin has not been determined thus far. However, we have recently reported on the production of an *enantio*-rhizoferrin in *Ralstonia* (*Pseudomonas*) *pickettii* possessing the opposite *S,S*-configuration of the quaternary carbon atoms of the citryl residues [43]. This shows that biosynthesis of rhizoferrin has occurred earlier in the bacterial evolution and is not a fungal invention. However, the opposite configuration of the two rhizoferrins is evidence that citrate activation for rhizoferrin biosynthesis might be well different. This suggests that there is no phylogenetic relationship between the two rhizoferrins and that the basic molecule has been independently invented twice in evolution.

21.5
Siderophore Transporters in *Saccharomyces cerevisiae*

The yeast *S. cerevisiae*, is a fungus that has obviously lost the ability to produce any hydroxamate siderophores and thus represents an exception of most other ascomycetous and basidiomycetous fungi [44]. As *S. cerevisiae* is the only fungus from which the whole genome is known, it is not surprising that the first siderophore transporters have been identified and characterized from it. While the non-siderophore iron transport of *S. cerevisiae* has been described in Chap. 20, iron-siderophore transport will be dealt with here.

21.5.1
SIT1 Transporter

Lesuisse and coworkers were the first to show that *S. cerevisiae* was able to utilize the bacterial hydroxamate siderophore, ferrioxamine B, as an iron source [45]. Although the route of entry was not clear when the first transport studies were done, two mechanisms of siderophore-iron uptake were already proposed, one via transport of the intact molecule and one via a reduction step. The existence of two different uptake mechanisms was subsequently proved by genetic studies, showing that: (1) a *SIT1* gene was required for ferrioxamine B transport [46] and (2) that Fre3p, a member of the plasma membrane reductases, was required for reduction [47] allowing subsequent uptake of ferrioxamine-iron via *FET3*- and *FTR1*-encoded transport into the cells. The original identification of the *SIT1* gene as a member of the major facilitator superfamily (MFS) has subsequently been confirmed using cDNA microarrays of *AFT1*-regulated genes [48]. These authors also showed that the bulk of Sit1p was not on the plasma membrane but predominantly on intracellular vesicles co-sedimenting with endosomal protein Pep12. Although it seems reasonable that endosomal vesicles contain plasma membrane transporters as intermediate carriers, it seems unlikely that a transporter can fulfill its function without integration into the plasma membrane. Therefore, the main function of Sit1p is to integrate into the plasma membrane for translocating external siderophores to the inside. Maybe that the same transporter can translocate the siderophores into the vesicles where desferration may take place as was shown in *Ustilago* [70, 71]. Another question has been raised with respect to the specificity of the Sit1p which besides ferrioxamines also recognizes and transports other hydroxamate siderophores, such as ferrichromes and coprogen. Double disruption of *SIT1* and *ARN1* had shown earlier that Sit1p has a very broad specificity recognizing both ferrioxamines and ferrichromes [48]. We also observed that coprogen is recognized by Sit1p as well.

21.5.2
TAF1 Transporter

Taf1p is a specific membrane transporter for triacetylfusarinine C which has recently been identified in *S. cerevisiae* by disruption of the gene *YHL047c* [49]. Triacetylfusarinine C (TAFC), also named triacetylfusigen, is a typical fungal siderophore isolated form *Aspergillus* and *Penicillium* species [50, 51]. Although linear and cyclic, as well as acetylated and non-acetylated, fusarinines are produced by many fungal genera, the cyclic triacetylated fusarinine C has the highest chemical stability of all fusarinine siderophores. Due to the importance of aspergilli in invasive aspergillosis, siderophore-mediated iron transport, its biosynthesis and regulation is a crucial factor for the establishment and growth in tissues [52]. Although *Aspergillus fumigatus* is the main causative species of invasive aspergillosis, *A. flavus*, *A. nidulans*, *A. niger* and *A. terreus* have also been implicated. Patients with leukemia and lymphoma, and transplant recipients are particularly prone to infection. Ferrichromes and fusarinines are the main siderophores of aspergilli and the specificity of uptake has been documented in early kinetic studies [53]. Production and regulation of triacetylfusarinine C in *A. nidulans* has been studied in a recent article [54]. It is interesting to note that transport of fusarinines has so far not been reported in bacteria, while transport of ferrichromes and coprogen via the outer membrane receptors FhuA and FhuE in enterobacteria is well known [19] (see Chap. 11 and 12).

21.5.3
ARN1 Transporter

Arn1p is a yeast siderophore transporter specific for ferrichromes possessing anhydro-mevalonic acid residues linked to N^δ-ornithine as revealed by loss of function after disruption of the *YHL040c* gene [55].Compared to the ferrichromes (ferrichrome, ferricrocin, and ferrichrysin), the ARN ferrichromes, like ferrirubin and ferrirhodin, possess *trans*- and *cis*-anhydromevalonyl residues (Fig. 2). The ARN ferrichromes occur widely in *Aspergillus*, *Penicillium*, and *Botrytis* [38, 56]. Like other siderophore transporters in *S. cerevisiae*, Arn1p is expressed under conditions of iron deprivation and is regulated by Aft1p [57], the major iron-dependent transcription factor detected earlier in yeast [26]. Besides the anhydromevalonyl residues other acyl residues are recognized by Arn1p, such as the structurally related methylglutaconyl residues, found in ferrichrome A. Moreover, recognition remained active after removing the *cis/trans* configuration of anhydromevalonyl residues by catalytic hydrogenation of the double bond yielding hexahydroferrirhodin. Semi-synthetic butyrylferrichrome and butyrylferrichrysin were also recognized, while the corresponding propionyl derivatives failed, suggesting that a minimum of four C-atoms of the hydroxamic acid chain length is sufficient to allow recogni-

Siderophore	Peptide residues	Acyl residues	Transporter
Ferrichrome	$R_1 = R_2 = H$	$R_3 = R_4 = R_5 = CH_3$	Sit1p, Arn1p
Ferricrocin	$R_1 = H, R_2 = CH_2OH$	$R_3 = R_4 = R_5 = CH_3$	Sit1p, Arn1p
Ferrichrysin	$R_1 = R_2 = CH_2OH$	$R_3 = R_4 = R_5 = CH_3$	Sit1p, Arn1p
Asperchrome D1	$R_1 = R_2 = CH_2OH$	$R_3, R_4, R_5 = $ (2 CH_3, 1A)	Sit1p, Arn1p
Enantio-ferrichrome: containing D – N^δ - acetyl – N^δ - hydroxyornithine			not recognized
Ferrirubin	$R_1 = R_2 = CH_2OH$	$R_3 = R_4 = R_5 = A$	Arn1p
Ferrirhodin	$R_1 = R_2 = CH_2OH$	$R_3 = R_4 = R_5 = B$	Arn1p
Hexahydroferrirhodin	$R_1 = R_2 = CH_2OH$	$R_3 = R_4 = R_5 = D$	Arn1p
Asperchrome B1	$R_1 = R_2 = CH_2OH$	$R_3, R_4, R_5 = $ (1 CH_3, 2 A)	Arn1p
Ferrichrome A	$R_1 = R_2 = CH_2OH$	$R_3 = R_4 = R_5 = C$	Arn1p
Des(diserylglycyl)ferrirhodin: lacking the amino acid sequence ser-ser-gly			Arn1p

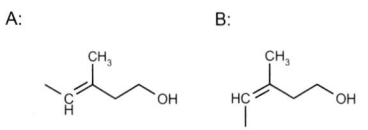

Fig. 2. Ferrichromes containing N-acetyl residues are substrates for both Sit1p and Arn1p transporters in *S. cerevisiae*, while ferrichromes containing Anhydromevalonyl or methylglutaconyl Residues linked to N^δ-ornithine (= ARN ferrichromes, in bold) are transported preferentially via Arn1p.

tion by Arn1p [55]. Disruption of the *ARN1* (*YHL040c*) gene was made in a Δ*fet3* background lacking the high-affinity ferrous transport, although there was no difference in the behavior of the strain whether or not the reduction pathway was active.

21.5.4
Transporter for Ferrichromes

Single disruption of siderophore transporter genes (*SIT1*, *TAFC1*, *ARN1*) failed to detect transporters for ferrichromes possessing acetyl residues linked to N^5-ornithine such as ferrichrome, ferricrocin, ferrichrysin (Fig. 2). This was surprising as of the six unknown major facilitator (UMFs) genes only three could be assigned to specific hydroxamate siderophore classes: *SIT1* encodes a ferrioxamine transporter [46], *TAF1* encodes a fusarinine transporter [49], and *ARN1* encodes a transporter for ARN ferrichromes like ferrirubin, ferrirhodin, ferrichrome A [55]. Evidence for the transport of ferrichrome came from loss of function results after double disruption studies [57]. These studies showed that ferrichrome transport is abolished only after disruption of both *SIT1* and *ARN1*, suggesting that either of these transporters, in addition to the above described specific binding sites for ferrioxamines and ARN ferrichromes, respectively, possess further recognition sites for ferrichrome (and related ferrichromes, like ferricrocin, ferrichrysin, etc.). Thus Sit1p and Arn1p seem to have multiple recognition sites (Fig. 3). Structurally, ferrioxamines and ferrichromes have little in common, but are of similar molecular dimension and hydrophobicity. Competition studies in the authors laboratory have shown that ferrichrysin transport via Arn1p was not inhibited by ferrirubin, confirming the view that different siderophore binding sites may exist on Arn1p. On the other hand Sit1p seem to have a common recognition site for different hydroxamate-type siderophores (ferrioxamines, ferrichromes, and coprogen) as competitive inhibition of all three siderophores occurred when transport of ^{55}Fe-labeled siderophores was followed (unpublished data). A similar observation had been made earlier by Lesuisse showing that ferrioxamine transport was competitively inhibited by ferricrocin [46]. Another interesting observation is the fact that besides ferrirhodin a natural derivative of ferrirhodin, des(diserylglycyl)ferrirhodin, lacking the peptide backbone, is also transported by Arn1p, suggesting that the iron-surrounding anhydromevalonyl residues are involved in binding to the transporter. This has been confirmed by further studies with semi-synthetic propionyl and butyryl ferrichromes, indicating that a chain length of four C-atoms as iron surrounding acyl residues is required for recognition by Arnp1 [55]. As shown in earlier transport studies with *N. crassa*, recognition of siderophores by fungal membranes is highly stereospecific [5, 6], indicating that the binding to the transporter is not dependent on the overall size or hydrophobicity but on specific binding sites of the transporters. As shown in Fig. 2, this also holds true for the siderophore iron transporters in *S. cerevisiae*, where *enantio*-ferrichrome was not recognized. Thus changing the configuration of the iron center from Λ-*cis* to Δ-*cis* by replacing L-ornithine by D-ornithine, prevents from recognition by MFS transpor-

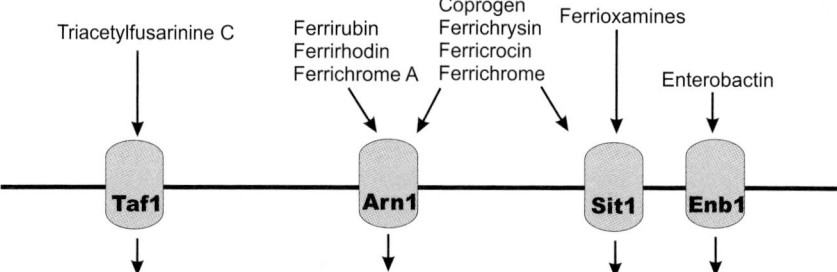

Fig. 3. Specificity of transporters of siderophore-iron transporters in *S. cerevisiae*.

ters. This again shows that siderophore transporters in fungi are generally stereoselective.

21.5.5
Transporter for Coprogens

Since anhydromevalonyl residues are characteristic recognition motifs of both Arn1p and Taf1p transporters, we assumed that coprogen possessing anhydromevalonyl residues might be likewise recognized by these transporters. However, single disruption of either of these transporters failed to inhibit coprogen uptake by bioassays. Double disruption experiments performed in the authors laboratory using the Cre-recombinase to remove the kanMX-marker of single disruption cassettes [58] revealed that coprogen transport could only be abolished after disruption of both, Arn1p and Sit1p. This again shows that recognition sites for siderophores on different transporters exist, of which only one (Arn1p) might recognize anhydromevalonyl residues while the other recognition motif site on Sit1p remains unknown. Coprogen has three anhydromevalonyl residues, one inserted centrally in the backbone and two as outer residues. It is tempting to speculate that the two outer ones are equivalent to the anhydromevalonyl residues of ferrirubin. Irrespective of the question whether or not the binding sites for coprogen are identical on both transporters, the results showed that some of the MFS transporters seem to be rather unspecific with regard to certain siderophores. A similar observation has been obtained for the transport of ferricrocin (ferrichrome, ferrichrysin), which could be inhibited only after disruption of both Arn1p and Sit1p, indicating binding sites on both transporters (Fig. 3).

21.5.6
ENB1 transporter

Disruption of the *YOL158c Sce* gene of the major facilitator superfamily in a Δ*fet3* background enabled the identification of an Enb1p transporter for ferric enterobactin in *S. cerevisiae* [59]. Due to the fact that Fe-enterobactin is a specific siderophore

Fig. 4. Evolutionary tree illustrating the relationship between the four identified siderophore-iron transporters (Sit1p, Taf1p, Arn1p, and Enb1p), including the two functionally unidentified MFS proteins.

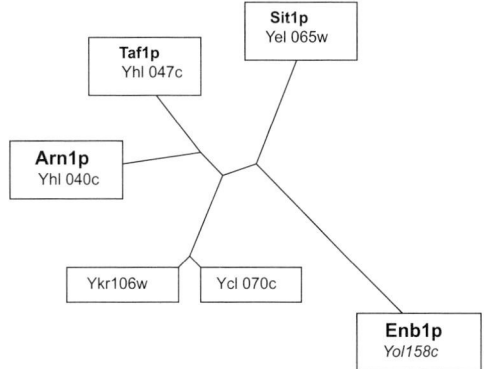

of enterobacteria its active transport in yeast was surprising and was described as an non-specific diffusion in an earlier report [46]. However, bioassays and short time frame transport experiments using radioactively labeled ^{55}Fe-enterobactin clearly confirmed the active transport of enterobactin in yeast. The YOL158c (ENB1) gene encoding the Fe-enterobactin transporter and the YEL065w (SIT1) gene encoding the ferrioxamine transporter are the most divergent members of the SIT family [60] reflecting the structural diversity of catecholate and hydroxamate siderophores.

The phylogenetic relationships of yeast genes of the Siderophore-Iron Transporter (SIT) family, previously designated UMFs [60], are summarized in Fig. 4 of which four genes encode siderophore transporters, while two further genes, YKR106w and YCL073c, could not be assigned a function in siderophore transport. Although we have screened a variety of known siderophores, it remains to be determined whether the collection of existing siderophores is greater than believed so far or whether these genes are required for other nutrients.

21.6
Energetics and Mechanisms

Although the mechanisms by which MFS transporters in yeast translocate the various siderophores has not been analyzed in detail so far, there is evidence that respiratory poisons like azide [59] or lack of glucose inhibit siderophore transport [46]. If we assume that siderophore transport in yeast is analogous to siderophore transport in filamentous fungi, then previous results on the energy requirements of the ascomycetous fungus, N. crassa, may be valid as well. Thus, earlier results have shown that siderophore transport in N. crassa [61] was inhibited by 2,4-dinitrophenol, sodium azide, or potassium cyanide, which interfere with mitochondrial oxidative phosphorylation or respiration. Although all these inhibitors have a different target within fungal cells, the final result is a decrease of the internal ATP level which is required to maintain a membrane potential of the plasma membrane of

200 mV (inside negative) as determined by microelectrodes [62]. The fungal membrane ATPase is a proton translocating ATPase, generating a transmembrane H^+ gradient, which provides the driving force for proton symport-driven uptake of various nutrients. Uptake of siderophores in alkaline solutions (pH 7–8) decreases to very low levels compared to uptake in slightly acidic solution, which makes sense as fungi tend to acidify their culture medium down to pH 3–5 and are known to excrete a variety of organic acids [42].

Evidence that the membrane potential may be the driving force of siderophore transport [61] came from experiments with dicyclohexylcarbodiimide (DCCD) and diethylstilbestrol (DES), which both inhibit fungal membrane ATPase [63]. The actual mechanism of transport of siderophores via MFS transporters in yeast is still unresolved, but by analogy to filamentous fungi a proton symport is the most probable mechanism [64]. MFS transporters are a large (several hundreds) and diverse family of about 600 amino acids in length possessing 12 or 14 putative transmembrane a-helical spans [65]. Because of the involvement of MFS in multidrug resistance (MDR), this group of transporters was also named MFS–MDR family. The role of the MDR1 in fluconazole resistance of clinical *Candida albicans* isolates has been shown recently by disruption studies, resulting in enhanced susceptibility of the mutants against fluconazole [66, 67].

MFS transporters catalyze in principle three types of transport mechanisms: uniport, symport or antiport, carrying sugars, oligosaccharides, polyols, amino acids, peptides, nucleosides, organic acids, anions, cations, and a variety of other small molecules. However, none has yet been found to be capable of transporting macromolecules as found in ABC transport systems [60]. The predicted topology of membrane spans of the four MFS transporters Taf1p, Arn1p, Sit1p and Enb1p (Fig. 5) correspond well with those of cluster III previously analyzed by Goffeau and coworkers [65]. However, there are variations of the predicted 14 spans which may increase to 15 spans (Taf1p) or which decrease to 13 spans as seen for Enb1p (Fig. 5) leading to odd numbers. Furthermore, as some of the spans predicted by hydropathy plots may overlap with predicted N-glycosylation sites, prediction of membrane spans remains still highly speculative.

21.7
FRE Reductases in Siderophore Transport

Iron uptake by fungi involves reduction either before or after entry into the cells. *Neurospora crassa* has been shown to possess soluble NADH-linked siderophore reductases which seem to be rather unspecific with regard to the different siderophores [68]. The mechanism of reduction after uptake seems to prevail with most siderophore-producing fungi, although reduction before uptake has been observed for certain siderophores like ferrichrome A [69] and ferrioxamine B [70] in *Ustilago sphaerogena* and *U. maydis*, respectively. By using fluorescently labeled ferrichrome analogs (B9-LRB, B9-ANT) Shanzer and coworkers were able to show that the siderophore molecules were concentrated in vesicles in a deferrated

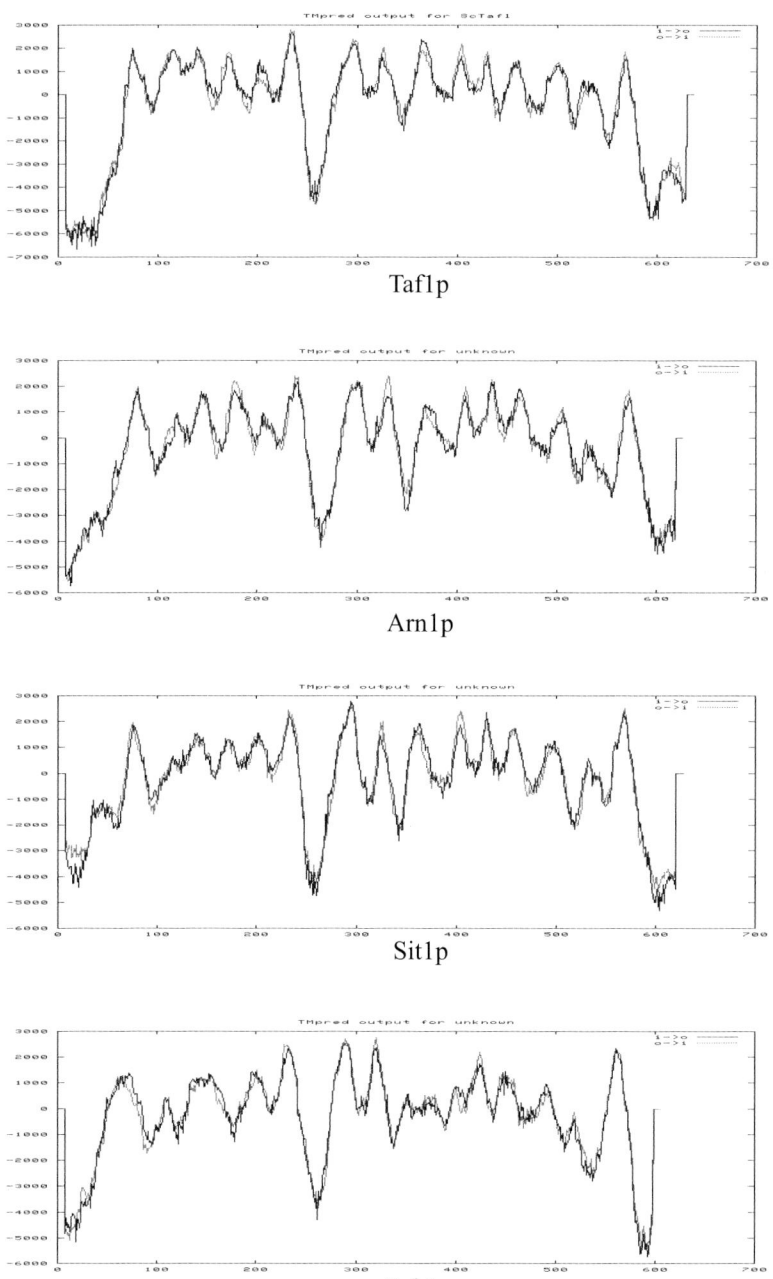

Fig. 5. Hydropathy plots of siderophore-iron transporters (Taf1p, Arn1p, Sit1p, and Enb1p).

state [70, 71]. While the B9-LRB analog was not quenched upon iron binding, the B9-ANT analog was, making the latter a suitable agent for monitoring deferration inside the cells. A further result of these studies was the fact that a fluorescently labeled ferrioxamine B analog (CAT18) behaved differently, delivering only the iron to the cells and leaving the fluorescent ligand outside on the membrane. Thus, ferrioxamine B-mediated iron transport involved an extracellular reduction mechanism. While growth promotion tests cannot distinguish between transport of the intact siderophore–iron complex and reductive iron uptake, the use of fluorescently labeled siderophores clearly shows the different pathways. The existence of non-specific reductive mechanisms of obtaining iron from exogenous siderophores has been suggested earlier [72].

The role of reductases in siderophore uptake by *S. cerevisiae* had already been emphasized by Lesuisse who showed that two mechanisms were active during ferrioxamine B-mediated iron transport [73]. Although the involvement of intracellular hydroxamate reductases had been shown earlier in the filamentous fungus *N. crassa*, there is a difference between *S. cerevisae* and *N. crassa* siderophore reductases with respect to location. While siderophores of *N. crassa* and probably most other filamentous fungi have to be first internalized before reduction can take place, yeasts possess cell surface reductases (FRE) that are able to reduce ionic and complexed ferric iron [74–76]. The role of the FRE family of plasma membrane reductases in the uptake of siderophore-iron in *S. cerevisiae* has been analyzed recently [47]. Since soluble ferric iron is virtually absent in extracellular fluids, ferric complexes of organic acids, such as citrate, malate and lactate prevail in natural culture fluids from fruits and plant materials. The flavocytochrome containing plasma membrane reductases are not only designed to reduce ferric citrate but also reduce a variety of siderophores originating from bacteria and filamentous fungi, such as enterobactin, ferrioxamines, ferrichromes, fusarinines, and Fe-rhodotorulic acid. Reduction and utilization of a broad range of ferric complexes enable yeasts to compete successfully with other siderophore producing microorganisms. Strains deleted for *FRE1* exhibit only 10 % of the ferric citrate reductase activity but still can grow with Fe-hydroxamates and Fe-enterobactin [47]. However, deletion of *FRE1* and *FRE2* abolished growth on Fe-enterobactin as an iron source, while growth on hydroxamates, such as ferrioxamines, ferrichromes, Fe-triacetyl-fusarinine C and Fe-rhodotorulic acid was still possible. After deletion of *FRE1*, *FRE2* and *FRE3* all of the Fe-hydroxamates except Fe-rhodotorulic acid failed to support growth, indicating that *FRE3* was sufficient to reduce most of the hydroxamate siderophores. *FRE4* could facilitate utilization of ferric rhodotorulate when the siderophore was present in higher concentrations. These results showed that *S. cerevisiae* has reductases of different reduction capacities, which may correlate with the different reduction potentials of the various siderophores.

21.8
Conclusions

It is expected that much of the results obtained from *S. cerevisiae* will be widely applicable to other classes of fungi, some of which are medically significant pathogens refractory to present treatment regimens. A thorough understanding of the underlying mechanisms of transport may help identify new strategies to deal with drug resistance or provide new targets for antifungal compounds. One such approach would be to prepare siderophore–antibiotic conjugates, "Trojan horses", which piggyback drugs into the fungi via these transporters [77]. Much more detail needs to be elucidated before such compounds could be developed into effective drugs.

Acknowledgement

I gratefully acknowledge the cooperation of many friends and colleagues during the recent years of siderophore research at the University of Tübingen. I thank G. Jung and H. Drechsel for their extensive help in structure elucidation and J. F. Ernst and P. Heymann for the development of the ideas and results on yeast genetics. I am indebted to C. J. Carrano and D. van der Helm for helpful comments.

References

1. A. Albrecht-Gary, A. L. Crumbliss, in: *Metal Ions in Biological Systems* Vol. 35 (Sigel, A., Sigel, H., Eds.), Marcel Dekker, New York, **1998**, pp. 239-327.
2. H. Drechsel, G. Winkelmannn, in: *Transition Metals in Microbial Metabolism* (Winkelmann, G., Carrano, C. J., Eds.), Harwood Academic Publishers, Amsterdam **1997**, pp. 1-49.
3. H. Huschka, M. A. F. Jalal, D. van der Helm, G. Winkelmann, *J. Bacteriol.* **1986**, *167*, 1020-1024.
4. H. Huschka, H. U. Naegeli, H. Leuenberger-Ryf, W. Keller-Schierlein, G. Winkelmann, *J. Bacteriol.* **1985**, *162*, 715-721.
5. G. Winkelmann, *FEBS Lett.* **1979**, *97*, 43-46.
6. G. Winkelmann, V. Braun, *FEMS Microbiol. Lett.* **1981**, *11*, 237-241.
7. A. Goffeau, *Nature* **1994**, *369*, 101-102.
8. A. Goffeau, *Science* **1996**, *274*, 546-567.
9. H. Feldmann, in: *Molecular Fungal Biology* (Oliver, R. P., Schweizer, M., Eds.), Cambridge University Press, Cambridge, **1999**, pp.78-134.
10. B. Nelissen, R. De Wachter, A. Goffeau, *FEMS Microbiol. Rev.* **1997**, *21*, 113-134.
11. G. Winkelmann, H. Drechsel, in: *Biotechnology* Vol. 7, *Products of Secondary Metabolism* (Rehm, H.-J., Reed, G., Pühler, A., Stadler, P.,Eds.) **1997**, pp. 199-246.
12. F. Nussbaum, P. Spiteller, M. Rüth, W. Steglich, G. Wanner, B. Gamblin, L. Stievano, F. E. Wagner, *Angew. Chem.* **1998**, *110*, 3483-3485.
13. H. Drechsel, G. Jung, G. Winkelmann, *BioMetals* **1992**, *5*, 141-148.
14. C. J. Carrano, H. Drechsel, D. Kaiser, G. Jung, B. Matzanke, G. Winkelmann, N. Rochel, A. Albrecht-Gary, *Inorg. Chem.* **1996**, *35*, 6429-6436.

15. C. J. Carrano, K. N. Raymond, *J. Am. Chem. Soc.* **1978**, *100*, 5371-5374.
16. I. Spasojević, H. Boukhalfa, R. D. Stevens, A. L. Crumbliss, *Inorg. Chem.* **2001**, *40*, 49-58.
17. C. J. Carrano, H. Drechsel, D. Kaiser, G. Jung, B. Matzanke, G. Winkelmann, N. Rochel, A. Albrecht-Gary, *Inorg. Chem.* **1996**, *35*, 6429-6436.
18. E. Griffith, J. J. Bullen, in: *Iron and Infection* (Bullen, J. J., Griffith, E., Eds.), John Wiley & Sons, Chichester, **1999**, pp. 87-212.
19. V. Braun, K. Hantke, in: *Transition Metals in Microbial Metabolism* (Winkelmann, G., Carrano, C. J., Eds.),Harwood Academic Publishers, Amsterdam, **1997**, pp. 81-116.
20. B. Mei, A. D. Budde, S. A. Leong, *Proc. Natl. Acad. Sci. USA* **1993**, *90*, 903-907.
21. H. J. Plattner, H. Diekmann, in: *Metal Ions in Fungi* (Winkelmann, G., Winge, D. R., Eds.), Marcel Dekker, New York, **1994**, pp. 99-116.
22. R. Gross, F. Engelbrecht, V. Braun, *Mol. Gen. Genet.* **1985**, *201*, 204-212.
23. H. J. Plattner, P. Pfefferle, A. Romaguera, S. Waschütza, H. Diekmann, *BioMetals* **1989**, *2*, 1-5.
24. Z. An, B. Mei, W. M. Yuan S. A. Leong, *EMBO J.* **1997**, *16*, 1742-1750.
25. K. Hantke, *Mol. Gen. Genet.* **1987**, *210*, 135-139.
26. Y. Yamaguichi-Iwai, A. Dancis, R. D. Klausner, *EMBO J.* **1995**, *14*, 1231-1239.
27. B. Mei, S. A. Leong, in *Metal Ions in Fungi* (Winkelmann, G., Winge, D. R., Eds.), Marcel Dekker, New York, **1994**, pp.117-147.
28. Z. An, Q. Zhao, J. McEvoy, W. M. Yuan, J. L. Markley, S. A. Leong, *Proc. Natl. Acad. Sci. USA* **1997**, *94*, 5882-5887.
29. L. Zhou, H. Haas, G. A. Marzluf, *Mol. Gen. Genet.* **1998**, *259*, 532-540.
30. L. Zhou, G. A. Marzluf, *Biochemistry* **1999**, *38*, 4335-4341.
31. H. Haas, K. Angermayr, G. Stöffler, *Gene* **1997**, *184*, 33-37.
32. H. Haas, I. Zadra, G. Stöffler, K. Angermayr, *J. Biol. Chem.* **1999**, *274*, 4613-4619.
33. B. F. Matzanke, E. Bill, A. X. Trautwein, G. Winkelmann, *J. Bacteriol.* **1987**, *169*, 5873-5876.
34. B. F. Matzanke, E. Bill, A. X. Trautwein, G. Winkelmann, *BioMetals* **1988**, *1*, 18-25.
35. C. J. Carrano, R. Böhnke, B. F. Matzanke, *FEBS Lett.* **1996**, *390*, 251-264.
36. B. F. Matzanke, in: *Transition Metals in Microbial Metabolism* (Winkelmann, G., Carrano, C. J., Eds.), Harwood Academic Publishers, Amsterdam, **1997**, pp. 117-157.
37. N. H. Horowitz, G. Charlang, G. Horn, N. P. Williams, *J. Bacteriol.* **1976**, *127*, 135-140.
38. D. van der Helm, G. Winkelmann, in: *Metal Ions in Fungi* (Winkelmann, G., Winge, D. R., Eds.), Marcel Dekker, New York, **1994**, pp. 39-98.
39. J. R. Boelaert, P. Pootrakul, A. Chaiprasert, H. W. Van Landuyt, A. Lambert, G. Winkelmann, *Abstract* of BioIron99, May 23-28, **1999**, Sorrento, Italy.
40. L. Margulis, K. V. Schwartz, *Five Kingdoms*, W. H. Freeman, New York, **1988**.
41. M. L. Brebee, J. W. Taylor, in: *Molecular Fungal Biology* (Oliver, R. P., Schweizer, M., Eds.), Cambridge University Press, Cambridge, **1999**, pp. 21-77.
42. G. Winkelmann, *Arch. Microbiol.* **1974**, *121*, 43-51.
43. M. Münzinger, K. Taraz, H. Budzikiewicz, H. Drechsel, P. Heymann, G. Winkelmann, J.-M. Meyer, *BioMetals* **1999**, *12*, 189-193.
44. J. B. Neilands, K. Konopka, B. Schwyn, M. Coy, R. T. Francis, B. H. Paw, A. Bagg, in: *Iron Transport in Microbes, Plants and Animals* (Winkelmann, G., van der Helm, D., Neilands, J. B., Eds.), VCH, Weinheim, **1987**, pp. 3-33.
45. E. Lesuisse, P. Labbe, *J. Gen. Microbiol.* **1989**, *135*, 257-263.
46. E. Lesuisse, M. Simon-Casteras, P. Labbe, *Microbiology* **1998**, *144*, 3455-3462.
47. C-W. Yun, M. Bauler, R. E. Moore, P. E. Klebba, C. C. Philpott, *J. Biol. Chem.* **2001**, *276*, 10218-10223.
48. C. W. Yun, T. Ferea, J. Rashford, O. Ardon, P. O. Brown, D. Botstein, J. Kaplan, C. C. Philpott. *J. Biol. Chem.* **2000**, *275*, 10709-10715.
49. P. Heymann, J. F. Ernst, G. Winkelmann, *BioMetals* **1999**, *12*, 301-306.
50. H. Diekmann, H. Zähner, *Eur. J. Biochem.* **1967**, *3*, 213-218.
51. M. B. Hossain, D. L. Eng-Wilmot, R. A. Loghry, D. van der Helm, *J. Am. Chem. Soc.* **1980**, *102*, 5766-5773.
52. D. H. Howard, *Clin. Microbiol. Rev.* **1999**, *12*, 394-404.
53. C. Wiebe, G. Winkelmann, *J. Bacteriol.* **1975**, *123*, 837-842.

54. H. Haas, I. Zadra, G. Stoffler, K. Angermayr, *J. Biol. Chem.* **1999**, *19*, 4613-4619.
55. P. Heymann, J. F. Ernst, G. Winkelmann, *FEMS Microbiol. Lett.* **2000**, *186*, 221-227.
56. S. Konetschny-Rapp, G. Jung, H. Huschka, G. Winkelmann, *BioMetals* **1988**, *1*, 90-98.
57. C.-W. Yun, J. S. Tiedeman, R. E. Moore, C. C. Philpott, *J. Biol. Chem.* **2000**, *275*, 16354-16359
58. U. Güldner, S. Heck, T. Fiedler, J. Beinhauer, J. H. Hegemann, *Nucleic Acids Res.* **1996**, *24*, 2519-2524.
59. P. Heymann, J. F. Ernst, G. Winkelmann, *BioMetals* **2000**, *13*, 65-72.
60. S. S. Pao, I. T. Paulsen, M. H. Saier Jr., *Microbiol. Mol. Biol. Rev.* **1998**, 1-34.
61. H. Huschka, G. Müller, G. Winkelmann, *FEMS Microbiol. Lett.* **1983**, *20*, 125-129.
62. C. L. Slayman, C. W. Slayman, *Proc. Natl. Acad. Sci. USA* **1974**, *71*, 1935-1939.
63. B. J. Bowmann, S. E. Mainzer, K. E. Allen, C. W. Slayman, *Biochim. Biophys. Acta* **1978**, *512*, 13-28.
64. G. Winkelmann, H. Huschka, *J. Plant Nutr.* **1984**, *7*, 479-487.
65. A. Goffeau, J. Park, I. T. Paulsen, J. Jonniaux, T. Dinh, P. Mordant, M. Saier, *Yeast* **1997**, *13*, 43-54.
66. S. Wirsching, S. Michel, J. Morschhäuser, *Mol. Microbiol.* **2000**, *36*, 856-865.
67. S. Wirsching, S. Michel, G. Köhler, J. Morschhäuser, *J. Bacteriol.* **2000**, *182*, 400-404.
68. J. F. Ernst, G. Winkelmann, *Biochim. Biophys. Acta* **1977**, *500*, 27-41.
69. D. J. Ecker, T. Emery, *J. Bacteriol.* **1983**, *155*, 616-622.
70. O. Ardon, R. Nudelman, C. Caris, J. Libman, A. Shanzer, Y. Chen, Y Hadar, *J. Bacteriol.* **1998**, *180*, 2021-2026.
71. O. Ardon, H. Weizman, J. Libman, A. Shanzer, Y. Shen, Y. Hadar, *Microbiology* **1997**, *143*, 3625-3631.
72. T. Emery, in: *Iron Transport in Microbes, Plants and Animals* (Winkelmann, G., van der Helm, D., Neilands, J. B., Eds.), VCH, Weinheim, **1987**, pp. 235-250.
73. E. Lesuisse, P. Labbe, in: *Metal Ions in Fungi* (Winkelmann, G., Winge, D. R., Eds.), Marcel Dekker, New York, **1994**, pp. 149-178.
74. E. Lesuisse, F. Raguzzi, R. R. Crichton, *J. Gen. Microbiol.* **1987**, *133*, 3229-3236.
75. E. Lesuisse, P. Labbe, *J. Gen. Microbiol.* **1989**, *135*, 257-263.
76. A. Dancis, R. D. Klausner, A. G. Hinnebusch, J. G. Barricocanal, *Mol. Cell. Biol.* **1990**, *10*, 2294-2301.
77. A. Gosh, M. Gosh, C. Niu, F. Malouin, U. Möllmann, M. J. Miller, *Chem. Biol.* **1996**, *3*, 1011-1019.

Note added in proof: see page 161.

Glycylsarcosine (GSar) and glycylproline (GP), the prototypic substrates for peptide transporters, were found by use of conformational analyisis to be able to adopt only A7 conformers and not A4, A10, leading to the prediction that they would act as substrates for Dpp-type but not Tpp-type transporters, which was confirmed using transport assays with *E. coli* tansport mutants [1]. This finding may have significance for assays of substrate transport in, e. g., human peptide transporters, in which competition with GSar and GP is used to evaluate absorption of various peptide-based therapeutic agents. The importance of electrostatic charge and dielectric constant in conformational analysis of biologically active dipeptides that are substrates for peptide transporters has been described [2].

Two of the three OppA proteins of the *opp* operon of *Borrelia burgdorferi*, and its two plasmid-encoded OppA paralogs, have been shown to complement *E. coli* OppBCDF in an *opp* mutant [3]. The substrate specificities of the OppAs are not identical and it will be intriguing to determine whether their specificities match those of any of the well-characterized transporters of enteric bacteria, e.g., Dpp, Tpp and Opp, or those of other bacteria. In *Lactococcus lactis*, characterization of genes encoding a binding-protein dependent ABC transporter for dipeptides (Dpp) has appeared together with information on the complex interrelated regulation of Dpp and the proteolytic system in this organism [4], and further studies have described the coregulation of casein hydrolysis an activity of the DtpT peptide transporter [5, 6]. In the pathogen *Listeria monocytogenes*, an ABC-type *opp* operon has been identified, which, in addition to a main role in peptide transport, was also implicated in bacterial growth at low temperature and pleiotropic effects on the behavior of the pathogen in the environment and in its host [7]. The synthesis of the dipeptide antibiotic bacilysin by *Bacillus subtilis*, was found to be dependent upon the oligopeptide permease *oppA*, raising the speculation that its synthesis may be linked to the quorum sensing pathway involved in sporulation, competence development and sufactin biosynthesis [8].

Inspection of genome sequences has led to the suggestion that gram-negative bacteria as well as gram-positive ones may use post-translational processing and transport of peptide pheromones in mechanisms of quorum sensing, in addition to the N-acyl homoserine lactones identified previously [9]. In the multiple auxotrophic bacterium *Streptococcus pyogenes* growing under conditions of amino acid deprivation, a *relA*-independent stringent response was found comprising transcriptional up-regulation of a specific subset of genes involved in pathogenesis, which included the oligopeptide and dipeptide permeases and the *pepB* gene, putatively involved in intracellular hydrolysis of transported peptides; overall, this response enables the pathogen to respond to the protein/peptide status of its environment [10]. Interestingly, the up-regulation of *opp* expression involved stimulated read-through transcription of an operon-internal *oppA* terminator, allowing increased synthesis of Opp membrane proteins relative to OppA. A further study on the multiplicity and regulation of genes encoding peptide transporters in *Saccharomyces cerevisiae* has appeared [11]. Finally, the roles and properties of ABC transporters specific for peptides have been surveyed [12].

[1] J. W. Payne, G. M. Payne, S. Gupta, N. J. Marshall, B. M. Grail, *Biochim. Biophys. Acta*, **2001**, in press.

[2] N. J. Marshall, J. W. Payne, *J. Mol. Model.* **2001**, *7*, 112-119.

[3] B. Lin, S. A. Short, M. Eskildsen, M. S. Klempner, L. T. Hu, *Biochim. Biophys. Acta*, **2001**, *1499*, 222-231.

[4] Y. Sanz, F. C. Lanfermeijer, P. Renault, A. Bolotin, W. N. Konings, B. Poolman, *Arch. Microbiol.* **2001**, *175*, 334-343.

[5] E. Guedon, P. Renault, S. D. Ehrlich, C. Delorme, *J. Bacteriol.* **2001**, *183*, 3614-3622.

[6] E. Guedon, C. Martin, F. X. Gobert, S. D. Ehrlich, P. Renault, C. Delorme, *Lait*, **2001**, *81*, 65-74.

[7] E. Borezee, E. Pellegrini, P. Berche, *Infect. Immun.* **2000**, *68*, 7069-7077.

[8] A. Yazgan, G. Ozcengiz, M. A. Marahiel, *Biochim. Biophys. Acta*, **2001**, *1518*, 87-94.

[9] J. Michiels, G. Dirix, J. Vanderleyden, C. W. Xi, *Trends Microbiol.* **2001**, *9*, 164-168.

[10] K. Steiner, H. Malke, *Mol. Microbiol.* **2000**, *38*, 1004-1016.

[11] M. Hauser, V. Narita, A. M. Donhardt, F. Naider, J. M. Becker, *Mol. Memb. Biol.* **2001**, *18*, 105-112.

[12] F. J. M. Detmers, F. C. Lanfermeijer, B. Poolman, *Res. Microbiol.* **2001**, *152*, 245-258.

Index

a

A-type ATPases 387
ABC (ATP Binding Cassette) 77
ABC secretion 184 ff
　diversity of proteins secreted 186
ABC secretion pathway, schematic
　representation of 191
ABC subunit, sequence 97 ff
ABC systems, definition 79
ABC transporters 77 ff
　associated proteins 84
　binding proteins 85 ff
　components of 85 ff
　composition of 80 ff
　crystal structure 88 ff
　integral transmembrane domains
　　(TMDs) 91 ff
　listing of 89
　metal- 316
ABC-type efflux permeases 13
ABC-type nickel transporter 407
ABC-type uptake permeases 12
Acidithiobacillus ferrooxidans 332
ACR (arsenic compounds resistance)
　family 393
ACR genes, in yeast 393
aerolysin 215 ff
Aeromonas hydrophila, channel formation by
　aerolysin 216
AFT1 in yeast 466
alignment studies, on TonB-dependent
　receptors 282
alkaliphilic *Bacillus* 50
ALLBP 89
α-toxin 217
α-type channels 5
δ-aminolevulinic acid, transport of 154

anion-stimulated ATPase 380
anthrax protective antigen 218
anthrax toxin 218
antimonite transporters 377 ff
antiporters 8 ff
aquaglyceroporins 251
aquaporin AQP1 248
aquaporin Z, of *E. coli* 250, 255
aquaporin 1 254
aquaporins 247 ff
　MIP-related- 248
　plant- 249
AraF 89
Archaeoglobus fulgidus 49
ARN1 transporter 469
ars operon
　in bacteria 392
　of *E. coli* 378, 391
arsA 379
　ATPase activity of 380
　crystal structure of 389
　DTAP domain 386
　nucleotide binding sites 381 ff
　of *E. coli* 388
　unisite and multisite catalysis in 384
ArsAB pumps 392
arsB 379, 390
ArsC 379, 391
arsenite efflux 392
arsenite resistance 392
arsenite transporters 377 ff
ArsR 379
asperchrome B1 470
asperchrome D1 470
ATP hydrolysis 98 ff
　by FhuC 298
ATP synthases 23 ff

Index

ATPase activity of ArsA 380
ATPases, phylogram 363
Atx1 metallochaperone 427
Atx2p 456

b

β-barrel porins 6
Bacillus stearothermophilus 49
Bacillus subtilis, CitM protein 404
   saturable Mn permease 330
Bacillus thuringiensis Cry toxins 223
bacterial copper transport 361 ff
bacterial cytochrome *c* oxidases 436
bacterial iron transport 289 ff
bacterial zinc transport 313 ff
binding protein-dependent Zn^{2+} uptake
   in gram-negative bacteria 320
   in gram-positive bacteria 316 ff
Bradyrhizobium japonicum (HupN) 413
Bsd2p and manganese ions 453

c

Ca^{2+}-ATPase, of sarcoplasmic reticulum 364
Ca(II) ATPase 428
Ca-ATPase calcium pump, structure of 428
Caloramator fervidus 49
Candida albicans peptide transport gene
   (*Ca Ptr2*) 155
carbon monoxide dehydrogenases [Ni] 399
catalysis within the F_1 complex 27 ff
cation diffusion facilitator 315
Ccc1p: a manganese homeostasis protein 456
Ccc2 P-type ATPase transporter 427
CCC2, intracellular transporter of copper 449
Ccc2p 456
CDC1 gene product 456
cell surface permease FTR1 449
channel formation steps 212 ff
channel forming colicins 222
channel forming protein toxins 209 ff
   α-toxin 217
   aerolysin 215 ff
channel forming proteins 209 ff
   classification of 211
cholesterol-dependent toxins 219
CitM protein of *Bacillus subtilis* 404
citrate fermentation 51
classes of transporters 4 ff
Clostridium fervidus 49
Cnr (cobalt–nickel resistance) 406
coenzyme F_{430} 399
colicins 222

copA 368
copB 368
cop operon of *Enterococcus hirae* 368
copY 368
copZ 368
CopA, in copper uptake 369
CopB, in copper excretion 369
CopB copper ATPase, of *Enterococcus hirae* 363
copper ATPases
   of *Helicobacter pylori* 373
   of *Listeria monocytogenes* 373
copper chaperones 366
copper excretion, CopB 369
copper ion transport, mitochondrial- 419 ff
copper resistance
   in *Escherichia coli* 371
   plasmid-encoded- 374
copper transport 361 ff
copper transport system, in *S. cerevisiae* 426
copper uptake, CopA 369
coprogen transporter 472
CopY repressor protein 370
CorA
   of *Methanococcus jannaschii* 353
   of *Salmonella typhimurium* 352 ff
CorA magnesium transporter 351 ff
Corynebacterium glutamicum 53
COS16 457
Cox polypeptides 425
Cox1 polypeptide 424
Cox2 subunit 424
Cox11 sequences 434
Cox17 metallochaperone 429
Cox17 sequences 430
CPx motif 364
CPx-type ATPases 362
   sequence motifs 365
crystal structures of DtxR 305
CTR1, cell surface copper permease 449
Ctr1/Ctr3 system 426
Cu metallation, of cytochrome *c* oxidase 431 ff
Cu metallation, of mitochondrial cytochrome *c* oxidase 441
Cu-specific ATPase pumps 428
CxxC motifs 366
cytochrome *c* oxidase, structure of 424
cytochrome *c* oxidase
   Cu metallation of 431 ff
   Cu metallation of mitochondrial- 441
   mitochondrial- 422 ff
cytochrome *c* oxidases, bacterial- 436
Czc (cobalt–zinc–cadmium resistance) 406

d

des(diserylglycyl)ferrirhodin 470
dipeptide permease (Dpp) 144
diphtheria toxin 304
DMT1 329
Dpp-type 159
DppA 89
DTAP domain, in ArsA 386
DtxR, crystal structures 305
DtxR/IdeR family 338

e

EI 118
enantio-ferrichrome 470
enantio-rhizoferrin 468
ENB1 transporter 472
energy transduction, ATP synthase 31 ff
enteric bacteria, peptide transport in 143 ff
Enterococcus hirae cop operon 368
Enterococcus hirae CopB copper ATPase 363
Escherichia coli HlyA 221
    aquaporin Z 250, 255
    ars operon 378, 391
    ABC transporters 80 ff
    ArsA 388
    ATP synthase 24 ff, 36 ff
    ATP transporter binding proteins 90
    copper transport in 371
    FhuABCD activities 292
    FhuD protein 297
    LamB 240
    MalK 105
    melibiose transporter 54
    Na^+/melibiose transporter 58
    Na^+/proline transporter 55, 59 ff, 65 ff
    Na^+/substrate transport in 52
    Nik system 407
    OmpF 234
    peptide transport 143 ff
    Tat pathway in 174
    TolC trimer 242
extracellular loops
    in FepA 270 ff
    in FhuA 270 ff

f

F_1 subunits 25 ff
F_1/F_O interface 31 ff
F_1F_O ATP synthase 23 ff
    model of energy transduction 39 ff
F_O subunits 33 ff
    proton translocation pathway 42
F-ATPases 23 ff

F-type ATPases 23 ff
families of transporters 1 ff, 5 ff
Fe^{3+}-citrate 306
Fe^{3+}-dicitrate 300
Fe^{3+} regulatory mechanism 306
Fe^{3+}-siderophore transport 291 ff
Fe-enterobactin transporter 473
FecI and FecR proteins 306
feo genes 304
FepA 264 ff
    β-barrel 265 ff
    conserved residues 284
    extracellular loops 270 ff
    N-terminal domain 267 ff
    ribbon diagram 264
    structure determination of 275
FepA mutants 285
Ferric Uptake Repressor (Fur) 337
ferric–citrate transport system 306
ferrichrome 470
ferrichrome A 470
ferrichrome transporter 471
ferrichrysin 470
ferricrocin 470
ferrirhodin 470
ferrirubin 470
ferrous iron transport system 304
FET3 448
FET4 450
FET5 450
FhuA 264 ff
    barrel structure 266
    conserved residues 284
    extracellular loops 270 ff
    N-terminal domain 267 ff
    structures with ligand 272 ff
FhuA protein 265, 292
FhuABCD activities, of *E. coli* 292
FhuB transport protein 298
FhuC, ATP hydrolysis by 298
FhuD 89, 295 ff
    crystal structure 297
Fibrobacter succinogenes 50
FRE genes, of *Saccharomyces cerevisiae* 448
FRE reductases, in siderophore transport 474 ff
FRE1 447
FRE2 447
FRE3 448
FRE4 448
FTR1 448
fungal ornithine-N^5-oxygenase 466
fungi, siderophore transport in 463 ff

g

GATA family of transcription factors 466
GBP 89
General Diffusion Porins 234
GlnBP 89
GlpFs 251
glucose transport 115 ff
glutathione transport 154
glycerol conducting channels 247, 256
glyoxalase I 399
group translocation 115

h

H^+ cycle 52
Haemophilus influenzae
  ATP transporter binding proteins 90
  heme transport systems 301
  hFBP 297
Halobacteriales 50
Halobacterium 50
HbpA 89
heavy metal ATPases 362
heavy metals, transport of 452
Helicobacter pylori, nickel/cobalt transporters 409
Helicobacter pylori copper ATPases 373
heme, bacterial use of 300
heme transport systems 301 ff
hexahydroferrirhodin 470
HgAtx1 complex, structure of 427
high-affinity nickel uptake systems 406
HisJ 89
HoxN 410
hpCopA 373
HPr 118
hydrogenases [NiFe] 399
hydroxamate reductases 476

i

IIA 118
IIA^{Glc} subunit 120
  regulatory role 131
IIB 118
IIC 118
$IICB^{Glc}$ mutants 124 ff
  listing of 125 ff
$IICB^{Glc}$ subunit 121 ff
  regulatory role 132
  topology of 122
insecticidal crystal (Cry) proteins 223
integral transmembrane domains (TMDs)
  ABC transporters 91 ff
  sequence 96
ion selectivity of porins 233

iron metabolism, in eukaryotic cells 305
iron transport 261 ff, 289 ff
  Saccharomyces cerevisiae genes involved in 458
  regulation of 451
iron transporters, in yeast 447 ff
iron uptake, low-affinity- 450
iron-dependent regulatory protein (IRP) 305

k

Klebsiella pneumoniae
  citrate fermentation in 51
  melibiose transporter 54
  Na^+/citrate transporter 56, 59, 61, 69

l

lactic acid bacteria, peptide transport in 148 ff
Lactobacillus plantarum 328
Lactobacillus plantarum MntA, gene 331
Lactococcus lactis
  ABC transporters 81 ff
  peptide transporters 148 ff
lactoferrin 299
LamB, of *E. coli* 240
LamB channel 238
LAO 89
LbpA 299
LbpB 299
lipid bilayer membranes 231
Listeria monocytogenes, copper ATPase 373
LivJ 89
LivK 89
low-affinity iron uptake 450
low-affinity Zn^{2+} uptake systems 321
Lys7 metallochaperones 429

m

magnesium transporters 347 ff
  CorA 351 ff
  MgtE 350
Major Intrinsic Proteins (MIPs) superfamily 247
malK dimer, asymmetry within 105 ff
MalK from *Thermococcus litoralis* 101 ff
maltoporin 237
maltose binding protein from *E. coli* 90
maltose binding protein MBP or MalE 239
manganese, in bacteria 326
manganese accumulation 325 ff
manganese homeostasis 452
manganese homeostasis protein, Ccc1p 456

manganese transport
  Saccharomyces cerevisiae genes involved in 458
  in *Saccharomyces cerevisiae* 452 ff
  in bacteria 330 ff
manganese transport ATPase, Pmr1p 455
manganese transporters, in yeast 447 ff
manganese uptake, in bacteria 325
MBP 89
MBP (MalE) 89
melibiose transporter of *E. coli* (EcMelB) 54, 62 ff
melibiose transporter of *K. pneumoniae* (KnMelB) 54, 62
membrane topology of CPx-type ATPases 363
metal ABC transporters 316
metalloenzymes, nickel as cofactor 398 ff
Methanobacterium thermoautotrophicum 49
Methanococcus jannaschii 49, 353
MFS-type nickel exporter (NrsD) 406
Mg^{2+} transporter families: MgtE, CorA, and MgtA/B 347
MgtA/MgtB Mg^{2+} transporters 355 ff
MgtC protein 357
MgtE magnesium transporters 350
microbial nickel transport 397 ff
microbial siderophores 465
MIP-like channel genes 252
MIP-related aquaporins 248
MIP-related sequences, phylogenetic tree 252
mitochondrial copper ion transport 419 ff
mitochondrial cytochrome *c* oxidase 422 ff
MMT1 451
MMT2 451
Mn(II) ABC transporters 328
Mn(II) as a Fur co-repressor 338
Mn(II) primary transporter 328
Mn(II) permeases 328
Mn(II) transport in *S. aureus* 330
MntABC uptake system 333
MntH 329
MntH proteins 336
ModA 89
molecular recognition templates (MRT) 156 ff
molecular recognition templates (MRTs), optimal features of 159
MRP1 (multidrug resistance-associated protein) 393
multi-copper oxidase 448
Mycobacterium tuberculosis (MtNicT) 413

n

N-terminal domain
  of FepA 267 ff
  of FhuA 267 ff
Na^+ cycle 52
Na^+/citrate transporter (CitS) of *K. pneumoniae* 56, 59, 61, 69
Na^+/H^+ antiporter 52
Na^+/K^+ ATPase 23
Na^+/melibiose transporter of *E. coli* 58
Na^+/proline transporter 53
Na^+/proline transporter (PutP) of *E. coli* 59 ff, 55, 65 ff
Na^+/substrate transport in *Escherichia coli* 52
Na^+/substrate transport systems 48 ff
Natural Resistance-Associated Macrophage Protein (Nramp1) 329
Ncc (nickel–cobalt–cadmium resistance) 406
NhlF 410
Nic1p, of *Schizosaccharomyces pombe* 412
nickel, as a cofactor of metalloenzymes 398 ff
nickel/cobalt transporter family 408 ff
  hydropathy profile alignment 409
  in *Bradyrhizobium japonicum* 412
nickel homeostasis 403 ff
nickel resistance 401
nickel toxicity 401
nickel transport 397 ff
nickel transporter (UreH, UreI) of thermophilic *Bacillus* sp. 412
nickel transporters, ABC-type- 407
nickel uptake, high-affinity systems 406
nickel/cobalt transporters, metabolic functions 410
NiCoT in *Bradyrhizobium japonicum*, nickel/cobalt transporter in 412
Nik system, *Escherichia coli* 407
Nik-related transporters 408
nikABCDE operon 407
nikR gene 407
nitrous oxide reductase (N_2OR), in *P. stutzeri* 439
NixA 410
(Nramp)/Divalent Metal Transporter (DMT) 329
Nramp family 452
Nramp1 protein 339
Nramp2 329
nucleotide binding sites, in ArsA 381 ff

o

oligopeptide permease (Opp)   145 ff
oligopeptide transport in sporulation   151
oligopeptide transporter family (OPT)   142
OmpF of *E. coli*   234
Opp-type   159
OppA   89
ornithine-N^5-oxygenase, fungal-   466
osmoregulator   53
osmosensor   53
osmotic stress   53

p

P-type ATPase, superfamily   331
P-P-bond-hydrolysis-driven transporters   12
P-type ATPase superfamily   355
P-type ATPases   315, 362
   sequence motifs   365
P. aeruginosa exotoxin A   304
Pelobacter venetianus porins   235
PepT family   141
peptide-acetyl-CoA transporter (PAT) family   142
peptide transport   139 ff
   in eukaryotic microorganisms   155
peptide transport systems, classification of   140 ff
peptide-uptake permease (PUP) family   142
perfringolysin O   219
periplasmic $Zn^{2+}/Mn^{2+}/Fe^{?}$ binding proteins   319
peroxide stress regulator PerR   338
pertussis CyaA   221
phage adsorption sites   293
Phosphotransferase System (PTS)   115 ff
   components   117 ff
   PTS proteins   118
   PTS transporters   119 ff
   regulation of   129 ff
plant aquaporins   249
plasma membrane ferric reductase   447
plasma membranes, transport systems in   403
plasmid-encoded *ars* operon of *E. coli*   391
plasmid-encoded copper resistance   374
Pmr1p: a manganese transporting ATPase   455
pore-forming toxins   6
porin channels, function of   230 ff
porin pores, reconstitution of   232
porins   227 ff
   Rhodobacter capsulatus   234 ff
   ion selectivity of   233
   isolation of   229
   of *Pelobacter venetianus*   235
   solute selectivity of   230
   specific-   237 ff
Porphyromonas gingivalis, heme transport system   302
porters   8 ff
PotD   89
PotF   89
Propionigenium modestum,
   ATP synthase   36 ff, 40
protein export   165 ff
   Sec pathway   168 ff
   Tat pathway   173 ff
protein secretion   165 ff
   pathways found in gram-negative bacteria   167
   Sec-dependent pathway   178 ff
   Sec-independent pathways   184 ff
   type III secretion pathway   192 ff
   type IV secretion systems   198 ff
proton motive force (pmf)   48
proton-dependent manganese transporter   329
proton-dependent oligopeptide transporter family (POT)   141
PsaA and TroA proteins   332
Pseudomonas stutzeri, nitrous oxide reductase (N_2OR)   439
PstS   89
PTR (Peptide Transport)   155
PTR2 transporter   155
PUP (peptide-uptake permease)   152
Pyrococcus furiosus   86
   ATP transporter binding proteins   90
Pyrococcus horikoshii   49

r

Ralstonia eutropha, nickel/cobalt transporters   409
Ralstonia eutropha (HoxN)   413
RbsB   89
reconstitution, of porin pores   232
reduction of arsenate   391
regulation
   by Fe^{3+} 306
   by Fe^{3+}-siderophores   306
repressor protein *CopY*   370
rhizoferrin   468
Rhodobacter capsulatus porin   234 ff
Rhodococcus rhodochrous, nickel/cobalt transporters   409
RND family of exporters   313 ff
RND-type Ni^{2+} export systems   406
RTX toxins   220

s

S. gordonii ScaABC system 333
S. marcescens cytotoxin 304
Saccharomyces cerevisiae
  FRE genes 448
  copper transport system 426
  genes involved in iron and manganese transport 458
  iron transport in 447 ff
  manganese transport in 452 ff
Salmonella SitA periplasmic protein 334
Salmonella typhimurium 52, 88, 352 ff
  peptide transport 143 ff
sap (sensitivity to antimicrobial peptides) 152
sarcoplasmic reticulum, Ca^{2+}-ATPase 364
SBP 89
ScaABC system, of S. gordonii 333
Schizosaccharomyces pombe, Nic1p 412
Sco protein sequences 432
Sec-dependent pathway:type II secretion pathway 178 ff
Sec pathway 168 ff
  schematic representation of 171
Sec proteins 166
Sec translocase 169 ff
secretion proteins 166
Selenomonas ruminantium 50
SfuA 300
Shigella toxin 304
shuttle mechanism 295
siderophore classes 464 ff
Siderophore Iron Transport (SIT) family 463
siderophore receptors 261 ff
  genetic and biochemical studies on 276 ff
  mechanism 279 ff
siderophore transport
  FRE reductases 474 ff
  in fungi 463 ff
siderophores, representative microbial- 465
SIT1 transporter 468
SitA periplasmic protein, of Salmonella 334
Smf1p 452
Smf2p 452
SMF3 451
Smf3p 452
sodium motive force, (smf) 48
sodium/substrate transport 47 ff
solute selectivity, of porins 230
specific porins 237 ff
SRE 467
SREA 467
SREP 467

Staphylococcus aureus, Mn(II) transport 330
staphyloferrin 468
Streptococcus bovi 50
Streptococcus pyogenes 328
subunit γ rotation 29 ff
subunits IIA^{Glc} and $IICB^{Glc}$ 119 ff
subunits $IIAB^{Man}$, IIC^{Man} and IID^{Man} 119
Sulfolobus solfataricus, ABC transporters 81 ff
superoxide dismutases [Ni] 399
symporters 8 ff
synechococcal copper ATPases 372

t

TAF1 transporter 469
Tat (twin arginine transfer) pathway 173 ff
Tat pathway, in Escherichia coli 174
Tat proteins 175
Tat signal peptide 176
TbpA 299
TbpB 299
Thermococcus litoralis 86
  ATP transporter binding proteins 90
  MalK 101 ff
Thermoplasma acidophilum 332
TIM complex 421
TMBP 89
TolC, of E. coli 241
TOM complex 420
TonB, functions of 281
TonB box 294
TonB-dependent phages 294
TonB-dependent transport 281
Tpp-type 159
transferrin 299
transport, of Fe^{3+}-siderophores 291 ff
transport proteins, for heme 301 ff
transport systems, in bacterial and eukaryotic plasma membranes 403
transporter, for ferrichromes 471
Transporter Classification (TC) 1 ff
transporter for coprogens 472
Treponema, TroA protein 332
tripeptide permease (Tpp) 147
TroA protein from Treponema 332
type I secretion pathway 184 ff
  schematic representation of 191
type II secretion pathways 178 ff
  schematic representation of 183
type III secretion pathway 192 ff
  schematic representation of 197
type IV secretion pathways, schematic representation of 200
type IV secretion system 198 ff

u

ubiquinol oxidase 436
uniporters 8 ff
Urbs1 in *Ustilago* 466
ureases 399
Ustilago, Urbs1 466

v

Vibrio alginolyticus 50
Vibrio parahaemolyticus 50
virulence factors 304
vitamin B_{12}, binding of 294

w

Walker A and B sequences 98
Walker A sequence 386
water channels 247
Wilson copper ATPase (ATP7B) 367

y

yeast
  ACR genes 393
  AFT1 in 466
  iron transporters 447 ff
  manganese transporters 447 ff
yeast oxidase complex 422
yeast vacuole and manganese 457
Yersinia pestis YfeABCD system 334
YfeABCD system, of *Yersinia pestis* 334

z

Zinc transport 313 ff
Zinc uptake, low-affinity systems 321
ZIP transporters 405
Zn^{2+} export 314
Zn^{2+} transport systems, regulators of 323
Zn^{2+} uptake, binding protein-dependent- 316 ff
znuA (zinc uptake) gene 321